SOLUTIONS TO EXERCISES

ROXY WILSON
UNIVERSITY OF ILLINOIS

SEVENTH  EDITION

CHEMISTRY
THE CENTRAL SCIENCE

BROWN LEMAY BURSTEN

PRENTICE HALL Upper Saddle River, NJ 07458

Acquisitions Editor: **Ben Roberts**
Associate Editor: **Mary Hornby**
Production Editor: **Kimberly Dellas**
Cover Designer: **Patricia Gutierrez**
Special Projects Manager: **Barbara A. Murray**
Supplements Manager: **Paul Gourhan**
Production Coordinator/Buyer: **Benjamin D. Smith**

Printed in the United States of America

10 9 8 7 6 5 4 3 2 1

ISBN 0-13-578311-9

Prentice-Hall International (UK) Limited, *London*
Prentice-Hall of Australia Pty. Limited, *Sydney*
Prentice-Hall Canada, Inc., *Toronto*
Prentice-Hall Hispanoamericana, S.A., *Mexico*
Prentice-Hall of India Private Limited, *New Delhi*
Prentice-Hall of Japan, Inc., *Tokyo*
Simon & Schuster Asia Pte. Ltd., *Singapore*
Editora Prentice-Hall do Brasil, Ltda., *Rio de Janeiro*

Contents

Introduction

Chemistry: The Central Science, 7th edition, contains nearly 2200 end-of-chapter exercises. Considerable attention has been given to these exercises because one of the best ways for students to master chemistry is by solving problems. Grouping the exercises according to subject matter is intended to aid the student in selecting and recognizing particular types of problems. Within each subject matter group, similar problems are arranged in pairs. This provides the student with an opportunity to reinforce a particular kind of problem. There are also a substantial number of general exercises in each chapter to supplement those grouped by topic. Integrative exercises, which require students to integrate concepts from several chapters, are a new feature of the 7th edition. Answers to the odd numbered topical exercises plus selected general and integrative exercises, about 900 in all, are provided in the text. These appendix answers help to make the text a useful self-contained vehicle for learning.

This manual, **Solutions to Exercises in Chemistry: The Central Science, 7th edition**, was written to enhance the end-of-chapter exercises by provided documented solutions. The manual assists the instructor by saving time spent generating solutions for assigned problem sets and aids the student by offering a convenient independent source to check their understanding of the material. Most solutions have been worked in the same detail as the in-chapter sample exercises to help guide students in their studies.

When using this manual, keep in mind that the numerical result of any calculation is influenced by the precision of the numbers used in the calculation. In this manual, for example, atomic masses and physical constants are typically expressed to four significant figures, or at least as precisely as the data given in the problem. If students use slightly different values to solve problems, their answers will differ slightly from those listed in the appendix of the text or this manual. This is a normal and a common occurrence when comparing results from different calculations or experiments.

Rounding methods are another source of differences between calculated values. In this manual, when a solution is given in steps, intermediate results will be rounded to the correct number of significant figures; however, unrounded numbers will be used in subsequent calculations. By following this scheme, calculators need not be cleared to re-enter rounded intermediate results in the middle of a calculation sequence. The final answer will appear with the correct number of significant figures. This may result in a small discrepancy in the last significant digit between student-calculated answers and those given in this manual. Variations due to rounding can occur in any analysis of numerical data.

The first step in checking your solution and resolving differences between your answer and the listed value is to look for similarities and differences in problem-solving methods. Ultimately, resolving the small numerical differences described above is less important than understanding the general method for solving a problem. The goal of this manual is to provide a reference for sound and consistent problem-solving methods in addition to accurate answers to text exercises.

Extraordinary efforts have been made to keep this manual as error-free as possible. All exercises were worked by at least two chemists and proof-read by two others to ensure clarity in methods and accuracy in mathematics. The work of Julie Grundman, Leslie Kinsland, Jennifer Ridlen and Scott Wilson has been invaluable to this project. However, in a written work as technically challenging as this manual, typos and errors inevitably creep in. Please help us find and eliminate them. We hope that both instructors and students will find this manual accurate, helpful and instructive.

Roxy B. Wilson
University of Illinois
School of Chemical Sciences
601 S. Mathews Ave.
Urbana, IL 61801

1 Introduction: Matter and Measurement

Classification and Properties of Matter

1.1   (a) gas   (b) solid   (c) liquid   (d) gas

1.2   (a) gas   (b) solid   (c) liquid   (d) solid

1.3   (a) heterogeneous mixture   (b) homogeneous mixture   (If there are undissolved particles, such as sand or decaying plants, the mixture is heterogeneous.)   (c) pure substance   (d) homogeneous mixture

1.4   (a) homogeneous mixture (clean air is homogeneous)
(b) pure substance   (c) pure substance   (d) heterogeneous mixture
(There are usually distinguishable regions, such as solid garlic in Italian dressing; French dressing appears homogeneous, but is really a colloid.)

1.5   Pure water is a pure substance, while a solution of salt in water is a mixture. We should be able to separate the components of the mixture by a physical process such as evaporation. Take a small quantity of the liquid and allow it to evaporate. If the liquid is salt water, there will be a solid white residue (salt). If the liquid is water, there will be no residue.

1.6   The two solids in this mixture have different physical properties, including different solubilities in water. Add water to the mixture. Sugar will dissolve and sand will not. Filter the mixture. Solid sand will remain on the filter paper and sugar water will be in the flask (Figure 1.8). Evaporate the water from the flask to recover solid sugar.

1.7   (a) Al   (b) Na   (c) Fe   (d) K   (e) P   (f) Br   (g) N   (h) Hg

1.8   (a) C   (b) Cd   (c) Cr   (d) Zn   (e) I   (f) S   (g) O   (h) Ne

1.9   (a) hydrogen   (b) magnesium   (c) lead   (d) silicon   (e) fluorine   (f) tin
(g) copper   (h) calcium

1.10   (a) sodium   (b) cobalt   (c) manganese   (d) sulfur   (e) phosphorus   (f) nickel
(g) silver   (h) titanium

1.11   A(s) $\xrightarrow{\text{heat}}$ B(s) + C(g)

When carbon(s) is burned in excess oxygen the two elements combine to form a gaseous compound, carbon dioxide. Clearly substance C is this compound.

Since C is produced when A is heated in the absence of oxygen (from air), both the carbon and oxygen in C must have been present in A originally. A is, therefore, a compound composed of two or more elements chemically combined. Without more information on the chemical or physical properties of B, we cannot determine absolutely whether it is an element or a compound. However, few if any elements exist as white solids, so B is probably also a compound.

1.12   Before modern instrumentation, the classification of a pure substance as an element was determined by whether it could be broken down into component elements. Scientists subjected the substance to all known chemical means of decomposition, and if the results were negative, the substance was an element. Classification by negative results was somewhat ambiguous, since an effective decomposition technique might exist, but not yet have been discovered.

1.13   <u>Physical properties</u>: silvery white (color); lustrous; melting point = 649°C; boiling point = 1105°C; density at 20°C = 1.738 g/mL; pounded into sheets (malleable); drawn into wires (ductile); good conductor. <u>Chemical properties</u>: burns in air to give intense white light; reacts with Cl_2 to produce brittle white solid.

1.14   <u>Physical properties</u>: reddish-brown color; liquid; boils at 58.9°C; freezes at -7.2°C; density at 20°C is 3.12 g/mL. <u>Chemical properties</u>: corrodes (reacts with) metals; rapid reaction with Al to form aluminum bromide.

1.15   (a) chemical   (b) physical   (c) physical   (d) chemical   (e) chemical

1.16   (a) chemical   (b) physical   (c) physical (The production of H_2O is a chemical change, but its **condensation** is a physical change.)   (d) physical (The production of soot is a chemical change, but its **deposition** is a physical change.)

Units and Measurement

1.17   (a) 1×10^{-1} (b) 1×10^{-2} (c) 1×10^{-15} (d) 1×10^{-6} (e) 1×10^{6} (f) 1×10^{3}
      (g) 1×10^{-9} (h) 1×10^{-3} (i) 1×10^{-12}

1.18   (a) 3.4 <u>p</u>m   (b) 4.8 <u>n</u>L   (c) 7.23 <u>k</u>g

      (d) $2.35 \times 10^{-6} \, m^3 \times \dfrac{1 \, cm^3}{(1 \times 10^{-2})^3 \, m^3} \times \dfrac{1 \, mL}{1 \, cm^3} = 2.35$ mL

      (e) 5.8 <u>n</u>s   (f) 3.45 <u>m</u>mol

1.19 (a) $454 \text{ mg} \times \dfrac{1 \times 10^{-3} \text{ g}}{1 \text{ mg}} = 0.454 \text{ g}$

(b) $5.0 \times 10^{-9} \text{ m} \times \dfrac{1 \text{ pm}}{1 \times 10^{-12} \text{ m}} = 5.0 \times 10^{3} \text{ pm}$

(c) $3.5 \times 10^{-2} \text{ mm} \times \dfrac{1 \times 10^{-3} \text{ m}}{1 \text{ mm}} \times \dfrac{1 \, \mu\text{m}}{1 \times 10^{-6} \text{ m}} = 35 \, \mu\text{m}$

1.20 (a) $3.05 \times 10^{5} \text{ g} \times \dfrac{1 \text{ kg}}{1 \times 10^{3} \text{ g}} = 3.05 \times 10^{2} \text{ kg} \ (305 \text{ kg})$

(b) $0.00035 \text{ mm} \times \dfrac{1 \times 10^{-3} \text{ m}}{1 \text{ mm}} \times \dfrac{1 \text{ nm}}{1 \times 10^{-9} \text{ m}} = 3.5 \times 10^{2} \text{ nm}$

(c) $3.45 \times 10^{-1} \text{ s} \times \dfrac{1 \text{ ms}}{1 \times 10^{-3} \text{ s}} = 3.45 \times 10^{2} \text{ ms} \quad (345 \text{ ms})$

1.21 (a) time   (b) density   (c) length   (d) area   (e) temperature
      (f) volume   (g) temperature

1.22 (a) volume   (b) area   (c) volume   (d) density   (e) time
      (f) length   (g) temperature

1.23 (a) density $= \dfrac{\text{mass}}{\text{volume}} = \dfrac{39.75 \text{ g}}{25.0 \text{ mL}} \times \dfrac{1 \text{ mL}}{1 \text{ cm}^{3}} = 1.59 \text{ g/cm}^{3}$

The units cm^{3} and mL will be used interchangeably in this manual.

(b) $75.0 \text{ cm}^{3} \times 23.4 \dfrac{\text{g}}{\text{cm}^{3}} = 1.76 \times 10^{3} \text{ g} \ (1.76 \text{ kg})$

(c) $275 \text{ g} \times \dfrac{1 \text{ cm}^{3}}{1.74 \text{ g}} = 158 \text{ cm}^{3} \ (158 \text{ mL})$

1.24 (a) volume = length^3 (cm^3); density = mass/volume (g/cm^3)

volume = $(1.5)^{3} \text{ cm}^{3} = 3.375 = 3.4 \text{ cm}^{3}$

density = $\dfrac{1.9 \text{ g}}{3.375 \text{ cm}^{3}} = 0.56 \text{ g/cm}^{3}$ plastic

The plastic is less dense than water (1.0 g/cm^3), so the object will float.

Note: This is the first exercise where "intermediate rounding" occurs. In this manual, when a solution is given in steps, the intermediate result will be rounded to the correct number of significant figures. However, the **unrounded** number will be used in subsequent calculations. The final answer will appear with the correct number of significant figures. That is, calculators need not be cleared and new numbers entered in the middle of a calculation sequence. This may result in a small discrepancy in the last significant digit between student-calculated answers and those given in the solution manual. These variations occur in any analysis of numerical data.

For example, in this exercise $(1.5)^3$ cm^3 = 3.375 cm^3 is rounded to 3.4 cm^3 in the first step, but 3.375 is retained in the subsequent calculation of the density, 0.56 g/cm^3. In this case, 1.9 g/3.4 cm^3 also yields 0.56 g/cm^3. In other exercises, the correctly rounded results of the two methods may not be identical.

(b) $0.250 \text{ L} \times \dfrac{1000 \text{ mL}}{1 \text{ L}} \times \dfrac{3.12 \text{ g}}{1 \text{ mL}} = 7.80 \times 10^2$ g bromine

(c) $5.74 \text{ kg} \times \dfrac{1000 \text{ g}}{1 \text{ kg}} \times \dfrac{1 \text{ cm}^3}{1.20 \text{ g}} = 4.78 \times 10^3$ cm^3 (4.78 L) ebony

1.25 (a)   density = $\dfrac{38.5 \text{ g}}{45 \text{ mL}}$ = 0.86 g/mL

The substance is probably toluene, density = 0.866 g/mL.

(b) $45.0 \text{ g} \times \dfrac{1 \text{ mL}}{1.114 \text{ g}} = 40.4$ mL ethylene glycol

(c) $(5.00)^3 \text{ cm}^3 \times \dfrac{8.90 \text{ gm}}{1 \text{ cm}^3} = 1.11 \times 10^3$ g (1.11 kg) nickel

1.26 (a) $\dfrac{21.95 \text{ g}}{25.0 \text{ mL}}$ = 0.878 g/mL

The tabulated value has 4 significant figures, while the experimental value has 3. The tabulated value rounded to 3 figures is 0.879. The values agree within 1 in the last significant figure of the experimental value; the two results agree. The liquid could be benzene.

(b) $15.0 \text{ g} \times \dfrac{1 \text{ mL}}{0.7781 \text{ g}} = 19.3$ mL cyclohexane

(c)   r = d/2 = 5.0 cm/2 = 2.5 cm

V = 4/3 π r^3 = 4/3 $\times$ π $\times$ $(2.5)^3$ cm^3 = 65 cm^3

$65.4498 \text{ cm}^3 \times \dfrac{11.34 \text{ g}}{\text{cm}^3} = 7.4 \times 10^2$ g

(The answer has 2 significant figures because the diameter had only 2 figures.)

1.27   Calculate the volume of the aluminum foil:

$5.175 \text{ g} \times \dfrac{1 \text{ cm}^3}{2.70 \text{ g}} = 1.92$ cm^3

Divide volume by area to get thickness

$1.92 \text{ cm}^3 \times \dfrac{1}{12.0 \text{ in}} \times \dfrac{1}{15.5 \text{ in}} \times \dfrac{1 \text{ in}^2}{(2.54)^2 \text{ cm}^2} \times \dfrac{10 \text{ mm}}{1 \text{ cm}} = 1.60 \times 10^{-2}$ mm

1.28   thickness = volume/area

$$\text{volume} = 1.00 \text{ g} \times \frac{1 \text{ cm}^3}{19.32} = 0.0518 \text{ cm}^3$$

$$\text{area} = 8.0 \text{ ft} \times 5.0 \text{ ft} \times \frac{12^2 \text{ in}^2}{1 \text{ ft}^2} \times \frac{2.54^2 \text{ cm}^2}{\text{in}^2} = 3.7 \times 10^4 \text{ cm}^2$$

$$\text{thickness} = \frac{0.05176 \text{ cm}^3}{37{,}161 \text{ cm}^2} = 1.4 \times 10^{-6} \text{ cm}$$

$$1.4 \times 10^{-6} \text{ cm} \times \frac{1 \times 10^{-2} \text{ m}}{1 \text{ cm}} \times \frac{1 \text{ nm}}{1 \times 10^{-9} \text{ m}} = 14 \text{ nm thick}$$

1.29   (a) $°C = 5/9 (°F - 32°)$; $5/9 (62 - 32) = 17°C$

        (b) $°F = 9/5 (°C) + 32°$; $9/5 (-16.7) + 32 = 1.9°F$

        (c) $K = °C + 273.15$; $-33°C + 273.15 = 240 \text{ K}$

        (d) $315 \text{ K} - 273 = 42°C$; $9/5 (42°C) + 32 = 108°F$

        (e) $°C = 5/9 (°F - 32°)$; $5/9 (2500 - 32) = 1371°C$; $1371°C + 273 = 1644 \text{ K}$

            (assuming 2500 °C has 4 sig figs)

1.30   (a) $°C = 5/9 (82°F - 32°) = 28°C$

        (b) $K = 804°C + 273 = 1077 \text{ K}$

        (c) $234.28 \text{ K} - 273.15 = -38.87°C$; $°F = 9/5 (-38.87°C) + 32 = -37.97°F$

        (d) $K = 25°C + 273 = 298 \text{ K}$; $°F = 9/5 (25°C) + 32 = 77°F$

        (e)   melting point = $-248.6°C + 273.15 = 24.6 \text{ K}$
            boiling point = $-246.1°C + 273.15 = 27.1 \text{ K}$

Uncertainty In Measurement

1.31   Exact: (c), (d), (e), (f)

1.32   Exact: (b), (e)   (The number of students is exact on any given day.)

1.33   (a) 4   (b) 3   (c) 4   (d) 3   (e) 5

1.34   (a) 5   (b) 2   (c) ambiguous; 5, 6 or 7   (d) 4   (e) 6

1.35   (a) 3.002×10^2 (b) 4.565×10^5 (c) 6.543×10^{-3}
        (d) 9.578×10^{-4} (e) 5.078×10^4 (f) -3.500×10^{-2}

1.36   (a) 1.00×10^2 (b) 5.00×10^{-3} (c) 7.30×10^4
        (d) 1.56×10^5 (e) 8.85×10^3 (f) -1.24×10^4

1.37   (a) 77.04   (b) -51   (c) 9.995×10^4 (d) 3.13×10^4

1.38   (a) -2.3×10^3 (The intermediate result has 2 significant figures, so only the thousand and hundred places in the answer are significant.)

   (b) $[285.3 \times 10^5 - 0.01200 \times 10^5] \times 2.8954 = 8.260 \times 10^7$ (Since subtraction depends on decimal places, both numbers must have the same exponent to determine decimal places/sig figs. The intermediate result has 1 decimal place and 4 sig figs, so the answer has 4 sig figs)

   (c) $(0.0045 \times 20,000.0)$ + (2813×12) $= 3.4 \times 10^4$
       2 sig figs /0 dec pl   2 sig figs /first 2 digits

   (d)   863   ×   [1255   - (3.45×108)] $= 7.62 \times 10^5$
                                3 sig figs /0 dec pl
       3 sig figs   ×   0 dec pl/3 sig figs   = 3 sig figs

Dimensional Analysis

1.39   Arrange conversion factors so that the starting units cancel and the new units remain in the appropriate place, either numerator or denominator.

1.40 $\dfrac{1.609 \text{ km}}{1 \text{ mi}}$;   when converting miles to kilometers, miles goes in the denominator so that it cancels the original unit, leaving km in the numerator.

1.41   (a) $0.076 \text{ L} \times \dfrac{1000 \text{ mL}}{1 \text{ L}} = 76 \text{ mL}$

   (b) $5.0 \times 10^{-8} \text{ m} \times \dfrac{1 \text{ nm}}{1 \times 10^{-9} \text{ m}} = 50. \text{ nm}$

   (c) $6.88 \times 10^5 \text{ ns} \times \dfrac{1 \times 10^{-9} \text{ s}}{1 \text{ ns}} = 6.88 \times 10^{-4} \text{ s}$

   (d) $1.55 \dfrac{\text{kg}}{\text{m}^3} \times \dfrac{1000 \text{ g}}{1 \text{ kg}} \times \dfrac{1 \text{ m}^3}{(10)^3 \text{ dm}^3} \times \dfrac{1 \text{ dm}^3}{1 \text{ L}} = 1.55 \text{ g/L}$

1.42   (a) $2.3 \times 10^{-8} \text{ cm} \times \dfrac{1 \times 10^{-2} \text{ m}}{1 \text{ cm}} \times \dfrac{1 \text{ pm}}{1 \times 10^{-12} \text{ m}} = 2.3 \times 10^2 \text{ pm}$

   (b) $1.35 \times 10^9 \text{ km}^3 \times \dfrac{(1000)^3 \text{ m}^3}{1 \text{ km}^3} \times \dfrac{(10)^3 \text{ dm}^3}{1 \text{ m}^3} \times \dfrac{1 \text{ L}}{1 \text{ dm}^3} = 1.35 \times 10^{21} \text{ L}$

   (c) $\dfrac{200 \text{ mg}}{100 \text{ mL}} \times \dfrac{1000 \text{ mL}}{\text{L}} \times 5.0 \text{ L} \times \dfrac{1 \text{ g}}{1000 \text{ mg}} = 10. \text{ g cholesterol}$

1.43 (a) $8.60 \text{ mi} \times \dfrac{1.609 \text{ km}}{1 \text{ mi}} \times \dfrac{1000 \text{ km}}{1 \text{ m}} = 1.38 \times 10^4 \text{ m}$

(b) $3.00 \text{ days} \times \dfrac{24 \text{ hr}}{1 \text{ day}} \times \dfrac{60 \text{ min}}{1 \text{ hr}} \times \dfrac{60 \text{ s}}{1 \text{ min}} = 2.59 \times 10^5 \text{ s}$

(c) $\dfrac{\$1.55}{\text{gal}} \times \dfrac{1 \text{ gal}}{4 \text{ qt}} \times \dfrac{1.057 \text{ qt}}{1 \text{ L}} = \dfrac{\$0.410}{\text{L}}$

(d) $\dfrac{5.0 \text{ pm}}{\text{ms}} \times \dfrac{1 \times 10^{-12} \text{ m}}{1 \text{ pm}} \times \dfrac{1 \text{ ms}}{1 \times 10^{-3} \text{ s}} = 5.0 \times 10^{-9} \text{ m/s}$

(e) $\dfrac{75.00 \text{ mi}}{\text{hr}} \times \dfrac{1.609 \text{ km}}{1 \text{ mi}} \times \dfrac{1000 \text{ m}}{1 \text{ km}} \times \dfrac{1 \text{ hr}}{60 \text{ min}} \times \dfrac{1 \text{ min}}{60 \text{ s}} = 33.52 \text{ m/s}$

(f) $55.35 \text{ ft}^3 \times \dfrac{(12)^3 \text{ in}^3}{1 \text{ ft}^3} \times \dfrac{(2.54)^3 \text{ cm}^3}{1 \text{ in}^3} = 1.567 \times 10^6 \text{ cm}^3$

1.44 (a) $9.5 \text{ ft} \times \dfrac{12 \text{ in}}{1 \text{ ft}} \times \dfrac{2.54 \text{ cm}}{1 \text{ in}} = 2.9 \times 10^2 \text{ cm}$

(b) $4.95 \text{ qt} \times \dfrac{1 \text{ L}}{1.057 \text{ qt}} \times \dfrac{1000 \text{ mL}}{1 \text{ L}} = 4.68 \times 10^3 \text{ mL}$

(c) $\dfrac{45.7 \text{ in}}{\text{hr}} \times \dfrac{2.54 \text{ cm}}{1 \text{ in}} \times \dfrac{10 \text{ mm}}{1 \text{ cm}} \times \dfrac{1 \text{ hr}}{60 \text{ min}} \times \dfrac{1 \text{ min}}{60 \text{ s}} = 0.322 \text{ mm/s}$

(d) $7.00 \text{ yd}^3 \times \dfrac{1 \text{ m}^3}{(1.094)^3 \text{ yd}^3} = 5.35 \text{ m}^3$

(e) $\dfrac{\$2.99}{\text{lb}} \times \dfrac{100\cent}{\$1} \times \dfrac{1 \text{ lb}}{453.6 \text{ g}} = \dfrac{0.659 \cent}{\text{g}}$

(f) $\dfrac{2.57 \text{ g}}{\text{mL}} \times \dfrac{1 \text{ kg}}{1000 \text{ g}} \times \dfrac{1000 \text{ mL}}{1 \text{ L}} \times \dfrac{1 \text{ L}}{\text{dm}^3} \times \dfrac{10^3 \text{ dm}^3}{1 \text{ m}^3} = \dfrac{2.57 \times 10^3 \text{ kg}}{\text{m}^3}$

1.45 (a) $31 \text{ gal} \times \dfrac{4 \text{ qt}}{1 \text{ gal}} \times \dfrac{1 \text{ L}}{1.057 \text{ qt}} = 1.2 \times 10^2 \text{ L}$

(b) $\dfrac{6 \text{ mg}}{\text{kg (body)}} \times \dfrac{1 \text{ kg}}{2.205 \text{ lb}} \times 150 \text{ lb} = 4 \times 10^2 \text{ mg}$

(c) $\dfrac{254 \text{ mi}}{11.2 \text{ gal}} \times \dfrac{1.609 \text{ km}}{1 \text{ mi}} \times \dfrac{1 \text{ gal}}{4 \text{ qt}} \times \dfrac{1.057 \text{ qt}}{1 \text{ L}} = \dfrac{9.64 \text{ km}}{\text{L}}$

(d) $\dfrac{50 \text{ cups}}{1 \text{ lb}} \times \dfrac{1 \text{ qt}}{4 \text{ cups}} \times \dfrac{1 \text{ L}}{1.057 \text{ qt}} \times \dfrac{1000 \text{ mL}}{1 \text{ L}} \times \dfrac{1 \text{ lb}}{453.6 \text{ g}} = \dfrac{26 \text{ mL}}{\text{g}}$

1.46 (a) $12 \text{ gal} \times \dfrac{4 \text{ qt}}{1 \text{ gal}} \times \dfrac{0.946 \text{ L}}{1 \text{ qt}} = 45 \text{ L}$

(b) $\dfrac{3.4 \text{ m}}{\text{s}} \times \dfrac{1.094 \text{ yd}}{1 \text{ m}} \times \dfrac{1 \text{ mi}}{1760 \text{ yd}} \times \dfrac{60 \text{ s}}{1 \text{ min}} \times \dfrac{60 \text{ min}}{1 \text{ hr}} = 7.6 \text{ mi/hr}$

(c) $320 \text{ in}^3 \times \dfrac{(2.54)^3 \text{ cm}^3}{1 \text{ in}^3} \times \dfrac{1 \text{ dm}^3}{(10)^3 \text{ cm}^3} \times \dfrac{1 \text{ L}}{1 \text{ dm}^3} = 5.24 \text{ L}$

(d) $2.4 \times 10^5 \text{ barrels} \times \dfrac{42 \text{ gal}}{1 \text{ barrel}} \times \dfrac{4 \text{ qt}}{1 \text{ gal}} \times \dfrac{1 \text{ L}}{1.057 \text{ qt}} = 3.8 \times 10^7 \text{ L}$

1.47 $12.5 \text{ ft} \times 15.5 \text{ ft} \times 8.0 \text{ ft} = 1.6 \times 10^3 \text{ ft}^3$ (2 sig figs)

$1550 \text{ ft}^3 \times \dfrac{(1 \text{ yd})^3}{(3 \text{ ft})^3} \times \dfrac{(1 \text{ m})^3}{(1.0936)^3 \text{ yd}^3} \times \dfrac{10^3 \text{ dm}^3}{1 \text{ m}^3} \times \dfrac{1 \text{ L}}{1 \text{ dm}^3} \times \dfrac{1.19 \text{ g}}{\text{L}} \times \dfrac{1 \text{ kg}}{1000 \text{ g}} = 52 \text{ kg air}$

1.48 $8 \text{ ft} \times 12 \text{ ft} \times 20 \text{ ft} = 1920 \text{ ft}^3$

$1920 \text{ ft}^3 \times \dfrac{(1 \text{ yd})^3}{(3 \text{ ft})^3} \times \dfrac{(1 \text{ m})^3}{(1.094 \text{ yd})^3} \times \dfrac{10 \text{ mg CO}}{1 \text{ m}^3} \times \dfrac{1 \text{ g}}{1000 \text{ mg}} = 0.54 \text{ g CO}$

(rounding to one significant figure gives 0.5 g CO)

1.49   A wire is a very long, thin cylinder of volume, $V = \pi r^2 h$, where h is the length of the wire and πr^2 is the cross-sectional area of the wire.

Strategy: 1) Calculate total volume of copper in cm^3 from mass and density

2) h (length in cm) = $\dfrac{V}{\pi r^2}$

3) Change cm → ft

$150 \text{ lb Cu} \times \dfrac{453.6 \text{ g}}{1 \text{ lb Cu}} \times \dfrac{1 \text{ cm}^3}{8.94 \text{ g}} = 7.61 \times 10^3 \text{ cm}^3$

$r = d/2 = 8.25 \text{ mm} \times \dfrac{1 \text{ cm}}{10 \text{ mm}} \times \dfrac{1}{2} = 0.413 \text{ cm}$

$h = \dfrac{V}{\pi r^2} = \dfrac{7610.7 \text{ cm}^3}{\pi (0.4125)^2 \text{ cm}^2} = 1.42 \times 10^4 \text{ cm}$

$1.4237 \times 10^4 \text{ cm} \times \dfrac{1 \text{ in}}{2.54 \text{ cm}} \times \dfrac{1 \text{ ft}}{12 \text{ in}} = 467 \text{ ft}$

1.50   (a) $26.73 \text{ g total} \times \dfrac{0.90 \text{ g Ag}}{1 \text{ g total}} \times \dfrac{1 \text{ tr oz}}{31.1 \text{ g Ag}} \times \dfrac{\$1.18}{1 \text{ tr oz}} = \$0.91$

(b) $\$25.00 \times \dfrac{1 \text{ tr oz}}{\$3.70} \times \dfrac{31.1 \text{ g}}{1 \text{ tr oz}} \times \dfrac{1 \text{ g total}}{0.90 \text{ g Ag}} \times \dfrac{1 \text{ coin}}{26.73 \text{ g}} = 8.7 \text{ coins}$

Since coins come in integer numbers, 9 coins are required.

Additional Exercises

1.51   Composition is the contents of a substance, the kinds of elements that are present and their relative amounts. Structure is the arrangement of these contents.

1.52   A gold coin is probably a **solid solution**. Pure gold (element 79) is too soft and too valuable to be used for coinage, so other metals are added. However, the simple term "gold coin" does not give a specific indication of the other metals in the mixture.

A cup of coffee is a **solution** if there are no suspended solids (coffee grounds). It is a heterogeneous mixture if there are grounds. If cream or sugar are added, the homogeneity of the mixture depends on how thoroughly the components are mixed.

A wood plank is a **heterogeneous mixture** of various cellulose components. The different domains in the mixture are visible as wood grain or knots.

The ambiguity in each of these examples is that the name of the substance does not provide a complete description of the material. We must rely on mental images and these vary from person to person.

1.53   (a)   A **hypothesis** is a possible explanation for certain phenomena based on preliminary experimental data. A **theory** may be more general, and has a significant body of experimental evidence to support it; a theory has withstood the test of experimentation.

           (b)   A scientific **law** is a summary or statement of natural behavior; it tells how matter behaves. A **theory** is an explanation of natural behavior, it attempts to explain why matter behaves the way it does.

1.54   Any sample of vitamin C has the same relative amount of carbon and oxygen; the ratio of oxygen to carbon in the isolated sample is the same as the ratio in synthesized vitamin C.

$$\frac{2.00\,g\,O}{1.50\,g\,C} = \frac{x\,g\,O}{6.35\,g\,C}; \quad x = \frac{(2.00\,g\,O)(6.35\,g\,C)}{1.50\,g\,C} = 8.47\,g\,O$$

This illustrates the *law of constant composition*.

1.55   Intensive (does not depend on amount): (b) density; (c) temperature; (e) color

1.56   (a) $\dfrac{m}{s^2}$ (b) $\dfrac{kg \cdot m}{s^2}$ (c) $\dfrac{kg \cdot m}{s^2} \times m = \dfrac{kg \cdot m^2}{s^2}$

        (d) $\dfrac{kg \cdot m}{s^2} \times \dfrac{1}{m^2} = \dfrac{kg}{m \cdot s^2}$ (e) $\dfrac{kg \cdot m^2}{s^2} \times \dfrac{1}{s} = \dfrac{kg \cdot m^2}{s^3}$

1.57   Magnesium is *less dense* than steel. That is, a unit volume of magnesium weighs less than a unit volume of steel.

1.58   In the solid state, molecules are in a regular orderly array. In the liquid state, molecules are moving relative to each other. The fixed molecular orientation of the solid state tends to minimize empty space between molecules. For a specified volume of substance, there is more mass (less empty space) in the solid state than the liquid state. Thus, the density of the solid state is greater

1.59 $K = {}^\circ C + 273.15;$ $K = -268.9{}^\circ C + 273.15 = 4.3\ K$

$^\circ F = 9/5\ (^\circ C) + 32;$ $^\circ F = 9/5\ (-268.9) + 32 = -452.0^\circ F$

(We consider 32 to be exact, so the result has 4 significant figures, as does the data.)

1.60   (a)   I. $(3.20 + 3.15 + 3.22)/3 = 3.19\%$

II. $(3.65 + 3.58 + 3.45)/3 = 3.56\%$

Based on the average, set II is more accurate. That is, it is closer to the true value of 3.55%.

(b)   Average derivation = Σ | value-average |/3

I. $|3.20\text{-}3.19| + |3.15\text{-}3.19| + |3.22\text{-}3.19|/3 = 0.03$

II. $|3.65\text{-}3.56| + |3.58\text{-}3.56| + |3.45\text{-}3.56|/3 = 0.07$

Set I is more precise than set II. That is, the values in set I are closer to each other than the values in set II.

1.61   (a) Inappropriate - The circulation of a widely read publication would vary over a year's time, and could simply not be counted to the nearest single subscriber. Probably about four significant figures would be appropriate. (b) Appropriate - It might be possible to do better than two significant figures, but an estimate to two significant figures should easily be possible. (c) Rainfall can be measured to within 0.02 in., but it is probably not possible to record an entire year's rainfall to the nearest 0.01 in. Further, the variation from year to year is sufficiently large that it does not make much sense to report the annual average to this number of significant figures. Probably two significant figures would be appropriate. (d) Inappropriate - The population of a city is not constant during a year, and probably cannot be counted at any one time to an accuracy of one person. An appropriate estimate would be 51,000.

1.62   (a) $25.83 \times 10^9\ lb \times \dfrac{453.6\ g}{1\ lb} = 1.172 \times 10^{13}\ g\ NaOH$

(b) $1.17165 \times 10^{13}\ g \times \dfrac{1\ cm^3}{2.130\ g} \times \dfrac{1\ m^3}{(100)^3\ cm^3} \times \dfrac{1\ km^3}{(1000)^3\ m^3} = 5.501 \times 10^{-3}\ km^3$

1.63   (a)   density = $(5.26\ g - 3.01\ g)/2.36\ mL = 0.953\ g/mL$

(b) $34.5\ kg \times \dfrac{1000\ g}{1\ kg} \times \dfrac{1\ mL}{13.6\ g} \times \dfrac{1\ L}{1000\ mL} = 2.54\ L$

(c) $V = 4/3\ \pi\ r^3 = 4/3\ \pi\ (2.37)^3\ cm^3 = 55.8\ cm^3$

$55.76135\ cm^3 \times \dfrac{8.47\ g}{cm^3} = 472\ g$

1.64   mass of benzene = 24.54 g - 8.47 g = 16.07 g

volume of benzene = $16.07 \text{ g} \times \dfrac{1 \text{ mL}}{0.879 \text{ g}} = 18.3 \text{ mL}$

volume of solid = 25.00 mL - 18.3 mL = 6.7 mL

density of solid = $\dfrac{8.47 \text{ g}}{6.7 \text{ mL}} = 1.3 \text{ g/mL}$

1.65   There are 209.1 degrees between the freezing and boiling points on the Celsius (C) scale and 100 degrees on the glycol (G) scale. Also, $-11.5°C = 0°G$. By analogy with °F and °C,

$$°G = \dfrac{100}{209.1}(°C + 11.5) \text{ or } °C = \dfrac{209.1}{100}(°G) - 11.5$$

These equations correctly relate the freezing point and boiling point of ethylene glycol on the two scales.

f.p. of H_2O: $°G = \dfrac{100}{209.1}(0°C + 11.5) = 5.50°G$

b.p. of H_2O: $°G = \dfrac{100}{209.1}(100°C + 11.5) = 53.3°G$

1.66 $13 \text{ s} \times \dfrac{1 \text{ min}}{60 \text{ s}} = 0.22 \text{ min};$ the total time is 20.22 min

$$\dfrac{10,000 \text{ m}}{20.22 \text{ min}} \times \dfrac{1 \text{ km}}{1000 \text{ m}} \times \dfrac{1 \text{ mi}}{1.6093 \text{ km}} \times \dfrac{60 \text{ min}}{1 \text{ hr}} = 18.44 \text{ mi/hr}$$

(Since the time is known to the single second, there are 4 sig figs in the time and 4 in the final answer.)

1.67   (a) $41 \text{ L} \times \dfrac{1.057 \text{ qt}}{1 \text{ L}} \times \dfrac{1 \text{ gal}}{4 \text{ qt}} \times \dfrac{29 \text{ mi}}{1 \text{ gal}} = 3.1 \times 10^2 \text{ mi} \ (310 \text{ mi})$

(b) $\dfrac{\$1.39}{1 \text{ gal}} \times \dfrac{1 \text{ gal}}{29 \text{ mi}} \times 650 \text{ mi} = \$31$

1.68   (a) $2.4 \times 10^5 \text{ mi} \times \dfrac{1.609 \text{ km}}{1 \text{ mi}} \times \dfrac{1000 \text{ m}}{1 \text{ km}} = 3.9 \times 10^8 \text{ m}$

(b) $2.4 \times 10^5 \text{ mi} \times \dfrac{1.609 \text{ km}}{1 \text{ mi}} \times \dfrac{1 \text{ hr}}{2.4 \times 10^3 \text{ km}} \times \dfrac{60 \text{ min}}{1 \text{ hr}} \times \dfrac{60 \text{ s}}{1 \text{ min}} = 5.8 \times 10^5 \text{ s}$

1.69   (a) $575 \text{ ft} \times \dfrac{12 \text{ in}}{1 \text{ ft}} \times \dfrac{2.54 \text{ cm}}{1 \text{ in}} \times \dfrac{10 \text{ mm}}{1 \text{ cm}} \times \dfrac{1 \text{ quarter}}{1.55 \text{ mm}} = 1.13 \times 10^5 \text{ quarters}$

(b) $1.13 \times 10^5 \text{ quarters} \times \dfrac{5.67 \text{ g}}{1 \text{ quarter}} = 6.41 \times 10^5 \text{ g} \ (641 \text{ kg})$

(c) $1.13 \times 10^5 \text{ quarters} \times \dfrac{1 \text{ dollar}}{4 \text{ quarters}} = \$2.83 \times 10^4$

($28,250 but the result has only 3 significant figures)

(d) $\$4.6 \times 10^{12} \times \dfrac{1 \text{ stack}}{\$2.83 \times 10^4} = 1.6 \times 10^8$ stacks   (approximately 160 million stacks)

1.70   (a) $\dfrac{\$2000}{\text{acre} \cdot \text{ft}} \times \dfrac{1 \text{ acre}}{4840 \text{ yd}^2} \times \dfrac{3 \text{ ft}}{1 \text{ yd}} \times \dfrac{(1.094 \text{ yd})^3}{(1 \text{ m})^3} \times \dfrac{(1 \text{ m})^3}{(10 \text{ dm})^3} \times \dfrac{(1 \text{ dm})^3}{1 \text{ L}} =$

            $\$1.6 \times 10^{-3}$/L or 0.16¢ / L

       (b) $\dfrac{\$2000}{\text{acre} \cdot \text{ft}} \times \dfrac{1 \text{ acre} \cdot \text{ft}}{2 \text{ households} \cdot \text{year}} \times \dfrac{1 \text{ year}}{365 \text{ days}} \times 1 \text{ household} = \dfrac{\$2.74}{\text{day}}$

1.71   (a)   volume = $\pi r^2 h = \pi \times (16.5 \text{ cm})^2 \times 22.3 \text{ cm} = 1.91 \times 10^4 \text{ cm}^3$

       (b) $r = d/2 = 2.0 \text{ ft}/2 = 1.0 \text{ ft} = 12 \text{ in}$

            $V = \pi (12 \text{ in})^2 \times \dfrac{(2.54 \text{ cm})^2}{1 \text{ in}^2} \times \dfrac{(1 \text{ m})^2}{(100 \text{ cm})^2} \times 6.3 \text{ ft} \times \dfrac{1 \text{ yd}}{3 \text{ ft}} \times \dfrac{1 \text{ m}}{1.094 \text{ yd}} = 0.56 \text{ m}^3$

       (c) $0.560 \text{ m}^3 \times \dfrac{(100 \text{ cm})^3}{(1 \text{ m})^3} \times \dfrac{13.6 \text{ g Hg}}{1 \text{ cm}^3} \times \dfrac{1 \text{ kg}}{1000 \text{ g}} = 7.6 \times 10^3 \text{ kg Hg}$

1.72 $9.64 \text{ g ethanol} \times \dfrac{1 \text{ cm}^3}{0.789 \text{ g ethanol}} = 12.2 \text{ cm}^3$, volume of cylinder

      $V = \pi r^2 h; \ r = (V/\pi h)^{1/2} = \left[\dfrac{12.218 \text{ cm}^3}{\pi \times 15.0 \text{ cm}} \right]^{1/2} = 0.509 \text{ cm}$

      $d = 2r = 1.02 \text{ cm}$

1.73   (a)   Let x = mass of Au in jewelry

            9.85 - x = mass of Ag in jewelry

            The total volume of jewelry = volume of Au + volume of Ag

            $0.675 \text{ cm}^3 = x \text{ g} \times \dfrac{1 \text{ cm}^3}{19.3 \text{ g}} + (9.85 - x) \text{ g} \times \dfrac{1 \text{ cm}^3}{10.5 \text{ g}}$

            $0.675 = \dfrac{x}{19.3} + \dfrac{9.85 - x}{10.5}$ (To solve, multiply both sides by (19.3)(10.5))

            $0.675 \, (19.3)(10.5) = 10.5 \, x + (9.85 - x)(19.3)$

                  $136.79 = 10.5 \, x + 190.105 - 19.3 \, x$

                    $-53.315 = -8.8 \, x$

          x = 6.06 g Au;   9.85 g total - 6.06 g Au = 3.79 g Ag

          mass % Au = $\dfrac{6.06 \text{ g Au}}{9.85 \text{ g jewelry}} \times 100 = 61.5\%$ Au

       (b)   24 karats × 0.615 = 15 karat gold

1.74   A solution can be separated into components by physical means, so separation would be attempted. If the liquid is a solution, the solute could be a solid or a liquid; these two kinds of solutions would be separated differently. Therefore, divide the liquid into several samples and do different tests on each. Try evaporating the solvent from one sample. If a solid remains, the liquid is a solution and the solute is a solid. If the result is negative, try distilling a sample to see if two or more liquids with different boiling points are present. If this result is negative, the liquid is probably a pure substance, but negative results are never entirely conclusive. We might not have tried the appropriate separation technique.

1.75   The separation is successful if two distinct spots are seen on the paper. To quantify the characteristics of the separation, calculate a reference value for each spot that is

$$\frac{\text{distance travelled by spot}}{\text{distance travelled by solvent}}$$

If the values for the two spots are fairly different, the separation is successful. (One could measure the distance between the spots, but this would depend on the length of paper used and be different for each experiment. The values suggested above are independent of the length of paper.)

1.76   The densities are:

carbon tetrachloride (methane, tetrachloro) - 1.5940 g/cm^3

hexane - 0.6603 g/cm^3

benzene - 0.87654 g/cm^3

methylene iodide (methane, diiodo) - 3.3254 g/cm^3

The volume of a sphere is $4/3 \, \pi \, r^3$.

$V = 4/3 \, \pi \, (0.56)^3 \text{ cm}^3 = 0.74 \text{ cm}^3$; density $= \dfrac{2.00 \text{ g}}{0.7356 \text{ cm}^3} = 2.7 \text{ g/cm}^3$

The marble will float on the methylene iodide only.

1.77   (a) osmium, density $= 22.6 \text{ g/cm}^3$

         (b) tungsten, m.p. $= 3410°C$

         (c) helium, b.p. $= -268.9°C$

         (d) mercury and bromine - both have freezing points below room temperature and boiling points above it

2 Atoms, Molecules, and Ions

Atomic Theory and Atomic Structure

2.1   Postulate 4 of the atomic theory is the *law of constant composition*. It states that the relative number and kinds of atoms in a compound are constant, regardless of the source. Therefore, 1.0 g of pure water should always contain the same relative amounts of hydrogen and oxygen, no matter where or how the sample is obtained.

2.2   (a)   1.250 g compound - 0.074 g hydrogen = 1.176 g oxygen

(b)   *Conservation of mass*

(c)   According to postulate 3 of the atomic theory, atoms are neither created nor destroyed during a chemical reaction. If 0.074 g of H are recovered from a compound that contains only H and O, the remaining mass must be oxygen.

2.3   (a) $\dfrac{17.37 \text{ g oxygen}}{15.20 \text{ g nitrogen}} = \dfrac{1.143 \text{ g O}}{1 \text{ g N}}$; $1.143/1.143 = 1.0$; $1.0 \times 2 = 2$

$\dfrac{34.74 \text{ g oxygen}}{15.20 \text{ g nitrogen}} = \dfrac{2.286 \text{ g O}}{1 \text{ g N}}$; $2.286/1.143 = 2.0$; $2.0 \times 2 = 4$

$\dfrac{43.43 \text{ g oxygen}}{15.20 \text{ g nitrogen}} = \dfrac{2.857 \text{ g O}}{1 \text{ g N}}$; $2.857/1.143 = 2.5$; $2.5 \times 2 = 5$

(b)   These masses of oxygen per one gram nitrogen are in the ratio of 2:4:5 and thus obey the *law of multiple proportions*.

2.4   (a)   A: $\dfrac{55.0 \text{ g fluorine}}{23.2 \text{ g sulfur}} = 2.37$ g fluorine/1 g sulfur

B: $\dfrac{9.8 \text{ g fluorine}}{16.6 \text{ g sulfur}} = 0.59$ g fluorine/1 g sulfur

C: $\dfrac{68.6 \text{ g fluorine}}{19.3 \text{ g sulfur}} = 3.55$ g fluorine/1 g sulfur

(b)   Dividing through by 0.59 to look for integer relationships between these values, the ratios are A:B:C as 4:1:6. Yes, the different masses of fluorine that will combine with one gram of sulfur as calculated in (a) are in the ratio of small whole numbers and, therefore, obey the *law of multiple proportions*.

2.5   Evidence that cathode rays were negatively charged particles was (1) that electric and magnetic fields deflected the rays in the same way they would deflect negatively charged particles and (2) that a metal plate exposed to cathode rays acquired a negative charge.

2.6   (a)   The electron itself has a negative charge so it is repelled by the negatively charged plate and attracted to the positively charged plate.

         (b)   As the charge on the plates is increased, the respective repulsion and attraction of the electron increases (Coloumb's Law) and the amount of bend should increase.

         (c)   As the mass of the particle increases, a greater force is required to deflect it. If the strength of the magnetic field is constant, the bend should decrease as the mass increases.

2.7   The droplets contain different charges because there may be 1, 2, 3 or more excess electrons on the droplet. The electronic charge is likely to be the lowest common factor in all the observed charges. Assuming this is so, we calculate the apparent electronic charge from each drop as follows:

A: $1.60 \times 10^{-19} / 1 = 1.60 \times 10^{-19}$ C

B: $3.15 \times 10^{-19} / 2 = 1.58 \times 10^{-19}$ C

C: $4.81 \times 10^{-19} / 3 = 1.60 \times 10^{-19}$ C

D: $6.31 \times 10^{-19} / 4 = 1.58 \times 10^{-19}$ C

The reported value is the average of these four values. Since each calculated charge has three significant figures, the average will also have three significant figures.

$(1.60 \times 10^{-19}$ C $+ 1.58 \times 10^{-19}$ C $+ 1.60 \times 10^{-19}$ C $+ 1.58 \times 10^{-19}$ C $) / 4 = 1.59 \times 10^{-19}$ C

2.8   Each drop must carry an integer number of electrons. The integer relationship among the charges of the 3 drops is not obvious.

$$\frac{3.62 \times 10^{-15} \text{ groks}}{3.62 \times 10^{-15}} = 1; \ 1 \times 2 = 2; \ 2 \times 3 = 6$$

$$\frac{5.93 \times 10^{-15} \text{ groks}}{3.62 \times 10^{-15}} = 1.64; \ 1.64 \times 2 = 3.3; \ 3.3 \times 3 = 10$$

$$\frac{9.05 \times 10^{-15} \text{ groks}}{3.62 \times 10^{-15}} = 2.5; \ 2.5 \times 2 = 5; \ 5 \times 3 = 15$$

It appears that the drops carried 6, 10 and 15 electrons, respectively.

$$\frac{3.62 \times 10^{-15} \text{ groks}}{6 \text{ electrons}} = 6.03 \times 10^{-16} \text{ groks}$$

$$\frac{5.93 \times 10^{-15} \text{ groks}}{10 \text{ electrons}} = 5.93 \times 10^{-16} \text{ groks}$$

$$\frac{9.05 \times 10^{-15} \text{ groks}}{15 \text{ electrons}} = 6.03 \times 10^{-16} \text{ groks}$$

The average charge of an electron is 6.00×10^{-16} groks.

The conversion factor is $\dfrac{6.00 \times 10^{-16} \text{ groks}}{1.59 \times 10^{-19} \text{ C}} = 3.77 \times 10^3$ groks/C

2.9   The Be nuclei have a much smaller volume and positive charge than the Au nuclei; the charge repulsion between the alpha particles and the Be nuclei will be less, and there will be fewer direct hits because the Be nuclei have an even smaller volume than the Au nuclei. Fewer alpha particles will be scattered in general and fewer will be strongly back scattered.

2.10   In the scattering experiment, most alpha particles, massive positively charged helium nuclei, passed directly through the foil, but a few were deflected at large angles. The diffuse positive charge in the "plum pudding" model would not have produced a repulsion strong enough to deflect the alpha particles at large angles. In Rutherford's model, the positive charge in the atom is concentrated in the small nucleus. If an alpha particle strikes a gold nucleus directly, it is deflected at a large angle.

2.11   (a) $2.4 \text{ Å} \times \dfrac{1 \times 10^{-8} \text{ cm}}{1 \text{ Å}} \times \dfrac{1 \text{ m}}{100 \text{ cm}} \times \dfrac{1 \text{ nm}}{1 \times 10^{-9} \text{ m}} = 0.24 \text{ nm}$

                 $2.4 \text{ Å} \times \dfrac{1 \times 10^{-10} \text{ m}}{1 \text{ Å}} \times \dfrac{1 \text{ pm}}{1 \times 10^{-12} \text{ m}} = 2.4 \times 10^2$ or 240 pm (1 Å = 100 pm)

     (b) $1.0 \text{ cm} \times \dfrac{1 \text{ Å}}{1 \times 10^{-8} \text{ cm}} \times \dfrac{1 \text{ Cr atom}}{2.4 \text{ Å}} = 4.2 \times 10^7$ Cr atoms

2.12 $\dfrac{44 \text{ m}}{1.0 \times 10^{11} \text{ Ba atoms}} \times \dfrac{1 \text{ pm}}{1 \times 10^{-12} \text{ m}} = 4.4 \times 10^2$ pm

           $\dfrac{44 \text{ m}}{1.0 \times 10^{11} \text{ Ba atoms}} \times \dfrac{1 \text{ Å}}{1 \times 10^{-10} \text{ m}} = 4.4 \text{ Å}$

2.13   p = protons, n = neutrons, e = electrons

     (a) ^{40}Ar has 18 p, 22 n, 18 e   (b) ^{55}Mn has 25 p, 30 n, 25 e

     (c) ^{65}Zn has 30 p, 35 n, 30 e   (d) ^{79}Se has 34 p, 45 n, 34 e

     (e) ^{184}W has 74 p, 110 n, 74 e   (f) ^{235}U has 92 p, 143 n, 92 e

2.14   (a) ^{32}P has 15 p, 17 n   (b) ^{51}Cr has 24 p, 27 n   (c) ^{60}Co has 27 p, 33 n

     (d) ^{99}Tc has 43 p, 56 n   (e) ^{131}I has 53 p, 78 n   (f) ^{201}Tl has 81 p, 120 n

2.15

Symbol	^{39}K	^{55}Mn	^{112}Cd	^{137}Ba	^{207}Pb
Protons	19	25	48	56	82
Neutrons	20	30	64	81	125
Electrons	19	25	48	56	82
Mass no.	39	55	112	137	207

2.16

Symbol	^{70}Ga	^{51}V	^{79}Se	^{197}Au	^{222}Rn
Protons	31	23	34	79	86
Neutrons	39	28	45	118	136
Electrons	31	23	34	79	86
Atomic no.	31	23	34	79	86
Mass no.	70	51	79	197	222

2.17 (a) $^{23}_{11}$Na (b) $^{51}_{23}$V (c) $^{4}_{2}$He (d) $^{37}_{17}$Cl (e) $^{24}_{12}$Mg

2.18 (a) $^{235}_{92}$U, $^{238}_{92}$U

The Periodic Table; Molecules and Ions

2.19 (a) Ag (metal) (b) He (nonmetal) (c) P (nonmetal) (d) Cd (metal)
(e) Ca (metal) (f) Br (nonmetal) (g) As (metalloid)

2.20 (a) strontium (metal) (b) silicon (metalloid) (c) sulfur (nonmetal) (d) samarium (metal)
(e) antimony (metalloid) (f) scandium (metal) (g) selenium (nonmetal)

2.21 (a) K, alkali metals (metal) (b) I, halogens (nonmetal) (c) Mg, alkaline earth metals
(metal) (d) Ar, noble gases (nonmetal) (e) S, chalcogens (nonmetal)

2.22 O, oxygen, nonmetal; S, sulfur, nonmetal; Se, selenium, nonmetal; Te, tellurium, metalloid;
Po, polonium, metal; although Po is in the position of a metalloid, its properties are those of
a metal

2.23 A structural formula contains the most information. It shows the total number and kinds of
atoms (the molecular formula, from which the empirical formula can be deduced) and how
these atoms are connected.

2.24 No. Two molecules with the same empirical formula can have different molecular formulas
if the integer number of empirical formula units in the two molecules is different. For example,
CH_2O is the empirical formula for both formaldehyde, CH_2O, and glucose, $C_6H_{12}O_6$.

2.25 (a) 6 (b) 6 (c) 12

2.26 (a) 4 (b) 6 (c) 9

2.27 (a) C_2H_6O

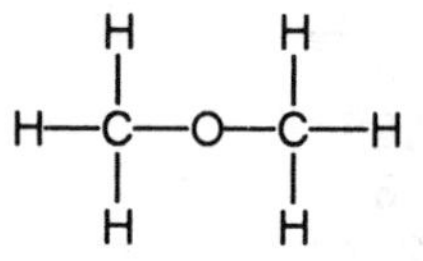

```
      H       H
      |       |
  H—C—O—C—H
      |       |
      H       H
```

(b) C_2H_6O

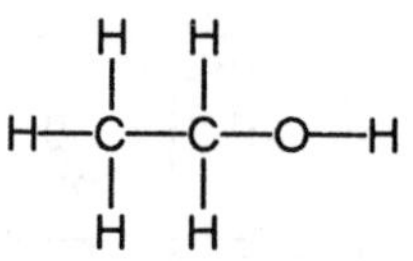

```
      H   H
      |   |
  H—C—C—O—H
      |   |
      H   H
```

(c) CH_4O

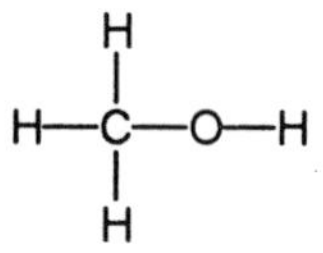

```
      H
      |
  H—C—O—H
      |
      H
```

(d) PCl_3

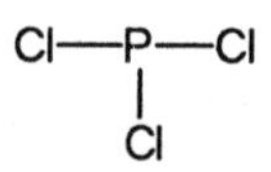

```
  Cl—P—Cl
      |
      Cl
```

2.28 (a) C_2H_5Br

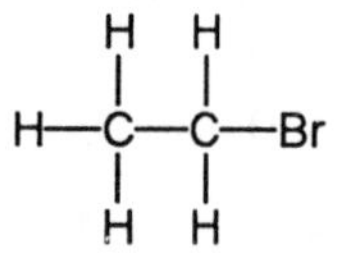

```
      H   H
      |   |
  H—C—C—Br
      |   |
      H   H
```

(b) C_2H_7N

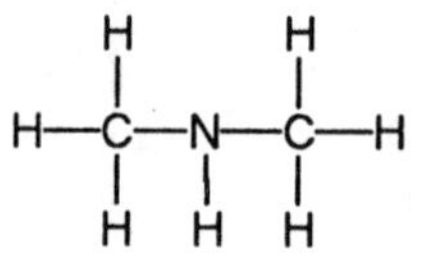

```
      H       H
      |       |
  H—C—N—C—H
      |   |   |
      H   H   H
```

(c) CH_2Cl_2

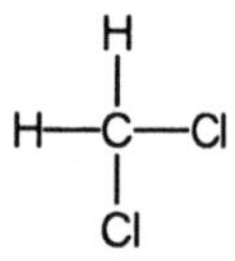

```
      H
      |
  H—C—Cl
      |
      Cl
```

(d) NH_2Cl

```
  H—N—Cl
      |
      H
```

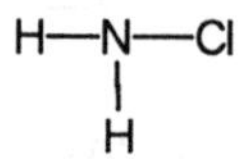

2.29 CH: C_2H_2, C_6H_6

CH$_2$: C_2H_4, C_3H_6, C_4H_8

NO$_2$: N_2O_4, NO_2

2.30 (a) NO_2 (b) CH_2 (c) C_2HO_2 (d) P_2O_5 (e) CH_2O (f) SO_3

2.31

Symbol	$^{31}P^{3-}$	$^{40}Ca^{2+}$	$^{51}V^{2+}$	$^{79}Se^{2-}$	$^{59}Ni^{2+}$
Protons	15	20	23	34	28
Neutrons	16	20	28	45	31
Electrons	18	18	21	36	26
Net Charge	3-	2+	2+	2-	2+

2.32

Symbol	$^{52}Cr^{3+}$	$^{130}I^-$	$^{107}Ag^+$	$^{119}Sn^{2+}$	$^{75}As^{3-}$
Protons	24	53	47	50	33
Neutrons	28	77	60	69	42
Electrons	21	54	46	48	36
Net Charge	3+	1-	1+	2+	3-

2.33 (a) Al^{3+} (b) Ca^{2+} (c) S^{2-} (d) I^- (e) Cs^+

2.34 (a) Rb^+ (b) Sr^{2+} (c) Sc^{3+} (d) Se^{2-} (e) At^-

2.35   (a) GaF_3, gallium(III) fluoride   (b) LiH, lithium hydride
       (c) AlI_3, aluminum iodide   (d) K_2S, potassium sulfide

2.36   (a) CaS   (b) NaF   (c) Mg_3N_2 (d) Al_2O_3

2.37   (a) $CaBr_2$ (b) NH_4Cl (c) $Al(C_2H_3O_2)_3$ (d) K_2SO_4 (e) $Mg_3(PO_4)_2$

2.38   (a) $Mg(NO_3)_2$ (b) Na_2CO_3 (c) $Ba(OH)_2$ (d) $(NH_4)_3PO_4$ (e) $Hg_2(ClO_3)_2$

2.39   Molecular (all elements are nonmetals): (a) B_2H_6 (b) CH_3OH (f) NOCl   (g) NF_3

       Ionic (formed by a cation and an anion, usually contains a metal cation): (c) $LiNO_3$,
       (d) Sc_2O_3, (e) CsBr, (h) Ag_2SO_4

2.40   Molecular (all elements are nonmetals): (a) CO, (d) CS_2, (e) PCl_3 (h) $(NH_2)_2CO$

       Ionic (formed from ions, usually contains a metal cation): (b) Na_2O, (c) $Ba(NO_3)_2$, (f) LaP,
       (g) $CoCO_3$

Naming lorganic Compounds

2.41   (a) ClO_2^- (b) Cl^- (c) ClO_3^- (d) ClO_4^- (e) ClO^-

2.42   sulfate - SO_4^{2-}; sulfite - SO_3^{2-} sulfite has one less oxygen atom.
       sulfide - S^{2-}; sulfide has no oxygen atoms;
       hydrogen sulfate - HSO_4^-, H^+ added to SO_4^{2-} increases the net charge by +1.

2.43   (a) aluminum fluoride   (b) iron(II) hydroxide (ferrous hydroxide)
       (c) copper(II) nitrate (cupric nitrate)   (d) barium perchlorate   (e) lithium phosphate
       (f) mercury(I) sulfide (mercurous sulfide)   (g) calcium acetate   (h) chromium(III) carbonate
       (chromic carbonate)   (i) potassium chromate   (j) ammonium sulfate

2.44   (a) lead(II) sulfide   (b) aluminum sulfate   (c) calcium chlorite   (d) nickel(II) carbonate
       (e) cobalt(II) cyanide (cobaltous cyanide)   (f) tin(II) bromide (stannous bromide
       (g) potassium permanganate   (h) zinc dihydrogen phosphate   (i) silver sulfite
       (j) ammonium dichromate

2.45   (a) Cu_2O (b) K_2O_2 (c) $Al(OH)_3$ (d) $Zn(NO_3)_2$ (e) Hg_2Br_2 (f) $Fe_2(CO_3)_3$ (g) NaBrO

2.46   (a) Mg_3N_2 (b) $FeSO_3$ (c) $Cr_2(CO_3)_3$ (d) CaH_2 (e) $Mg(HCO_3)_2$
       (f) KClO   (g) $Cu(C_2H_3O_2)_2$

2.47   (a) bromic acid   (b) hydrobromic acid   (c) phosphoric acid   (d) HClO   (e) HIO_3
       (f) H_2SO_3

2.48   (a) H_2SO_4 (b) HNO_2 (c) HI   (d) carbonic acid   (e) perchloric acid   (f) acetic acid

2.49   (a) sulfur hexafluoride   (b) iodine pentafluoride   (c) xenon trioxide   (d) N_2O_4 (e) HCN
       (f) P_4S_6

2.50   (a) iodine trichloride (b) silicon tetrabromide (c) dinitrogen pentoxide (d) CS_2 (e) H_2Te
        (f) Cl_2O_7

2.51   (a) $ZnCO_3$, ZnO, CO_2 (b) HF, SiO_2, SiF_4, H_2O (c) SO_2, H_2O, H_2SO_3
        (d) H_3P (or PH_3)   (e) $HClO_4$, Cd, $Cd(ClO_4)_2$ (f) VBr_3

2.52   (a) $KClO_3(s)$, $O_2(g)$ (b) $NaClO(aq)$ (c) $NH_3(g)$, $NH_4NO_3(s)$ (d) $HF(aq)$
        (e) $H_2S(g)$ (f) $HCl(aq)$, $NaHCO_3(s)$, $CO_2(g)$

Additional Exercises

2.53   (a)   Based on data accumulated in the late eighteenth century on how substances react with one another, **Dalton** postulated the atomic theory. Dalton's theory is based on the indivisible atom as the smallest unit of an element that can combine with other elements.

          (b)   By determining the effects of electric and magnetic fields on cathode rays, **Thomson** measured the mass-to-charge ratio of the electron. He also proposed the "plum pudding" model of the atom in which most of the space in an atom is occupied by a diffuse positive charge in which the tiny negatively charged electrons are imbedded.

          (c)   By observing the rate of fall of oil drops in and out of an electric field, **Millikin** measured the charge of an electron.

          (d)   After observing the scattering of alpha particles at large angles when the particles struck gold foil, **Rutherford** postulated the nuclear atom. In Rutherford's atom, most of the mass of the atom is concentrated in a small dense region called the nucleus and the tiny negatively charged electrons are moving through empty space around the nucleus.

2.54   (a)   Most of the volume of an atom is empty space in which electrons move. Most alpha particles passed through this space. The path of the massive alpha particle would not be significantly altered by interaction with a "puny" electron.

          (b)   Most of the mass of an atom is contained in a very small, dense area called the nucleus. The few alpha particles that hit the massive, positively charged gold nuclei were strongly repelled and essentially deflected back in the direction they came from.

2.55   The charge of an alpha particle is 2+ (Section 2.2).

$$\frac{1\ g}{4.822 \times 10^4\ C} \times \frac{1.602 \times 10^{-19}\ C}{1+ \text{ charge}} \times \frac{2+ \text{ charge}}{\alpha \text{ particle}} = 6.645 \times 10^{-24}\ g$$

2.56 mass of proton = 1.0073 amu

1.0073 amu $\times$ 1.66054 $\times$ 10^{-24} g/amu = 1.6727 $\times$ 10^{-24} g

diameter = 1.0 $\times$ 10^{-15} m, radius = 0.50 $\times$ 10^{-15} m $\times$ $\dfrac{100 \text{ cm}}{1 \text{ m}}$ = 5.0 $\times$ 10^{-14} cm

Assuming a proton is a sphere, V = 4/3 π r^3.

density = $\dfrac{\text{g}}{\text{cm}^3}$ = $\dfrac{1.6727 \times 10^{-24} \text{ g}}{4/3 \, \pi \, (5.0 \times 10^{-14})^3 \text{ cm}^3}$ = 3.2 $\times$ 10^{15} g/cm^3

2.57 (a) 5 significant figures. $^1H^+$ is a bare proton with mass 1.0073 amu. ^1H is a hydrogen atom, with 1 proton and 1 electron. The mass of the electron is 5.486 $\times$ 10^{-4} or 0.0005486 amu. Thus the mass of the electron is significant in the fourth decimal place or fifth significant figure in the mass of ^1H.

(b) Mass of ^1H = 1.0073 amu   (proton)

           <u>0.0005486 amu</u>   (electron)

           1.0078 amu   (We have not rounded up to 1.0079 since

                              49 < 50 in the final sum.)

Mass % of electron = $\dfrac{\text{mass of } e^-}{\text{mass of } {}^1H}$ $\times$ 100 = $\dfrac{5.486 \times 10^{-4} \text{ amu}}{1.0078 \text{ amu}}$ $\times$ 100 = 0.05444%

2.58 (a) deuterium - ^2_1H,   tritium - ^3_1H

(b) Both deuterium and tritium have 1 proton and 1 electron; their atomic number is 1. They differ in their number of neutrons and, therefore, have different mass numbers and nuclear composition.

2.59 copper - Cu, 1B (coinage metals, transition metal)
tin - Sn, 4A
zinc - Zn, 2B
phosphorus - P, 5A
lead - Pb, 4A

2.60 (a) an alkali metal - K   (b) an alkaline earth metal - Ca   (c) a noble gas - Ar
(d) a halogen - Br   (e) a metalloid - Ge   (f) a nonmetal in 1A - H
(g) a metal that forms a 3+ ion - Al   (h) a nonmetal that forms a 2– ion - O
(i) an element that resembles Al - Ga

2.61 (a) $\dfrac{0.077 \text{ g H}}{0.923 \text{ g C}}$ = $\dfrac{x \text{ g H}}{1.000 \text{ g C}}$; 0.083 g H/1 g C

(b) methane: $\dfrac{0.250 \text{ g H}}{0.750 \text{ g C}}$ = $\dfrac{x \text{ g H}}{1.000 \text{ g C}}$; 0.333 g H/1 g C

ethylene: $\dfrac{0.143 \text{ g H}}{0.857 \text{ g C}}$ = $\dfrac{x \text{ g H}}{1.000 \text{ g C}}$; 0.167 g H/1 g C ethane: $\dfrac{0.200 \text{ g H}}{0.800 \text{ g C}} = \dfrac{x \text{ g H}}{1.000 \text{ g C}}$; 0.250 g H/1 g C

propane: $\dfrac{0.182 \text{ g H}}{0.818 \text{ g C}} = \dfrac{x \text{ g H}}{1.000 \text{ g C}}$; 0.222 g H/1 g C

butane: $\dfrac{0.172 \text{ g H}}{0.828 \text{ g C}} = \dfrac{x \text{ g H}}{1.000 \text{ g C}}$; 0.208 g H/1 g C

(c)   If 0.083 g H/1 g C is a 1:1 combining ratio, dividing the g H/1 g C obtained in (b) by 0.083 should indicate the ratio of H:1C in the other compounds. (d) Empirical formulas follow:

methane: $\dfrac{0.333 \text{ g H/1 g C}}{0.083 \text{ g H/1 g C}} = 4\text{H} : 1\text{C}; \ CH_4$

ethylene: $\dfrac{0.167 \text{ g H/1 g C}}{0.083 \text{ g H/1 g C}} = 2\text{H} : 1\text{C}; \ CH_2$

ethane: $\dfrac{0.250 \text{ g H/1 g C}}{0.083 \text{ g H/1 g C}} = 3\text{H} : 1\text{C}; \ CH_3$

propane: $\dfrac{0.222 \text{ g H/1 g C}}{0.083 \text{ g H/1 g C}} = 2.67\text{H}: 1\text{C} = 8/3 \text{ H} : 1\text{C} = 8\text{H} : 3\text{C}; \ C_3H_8$

butane: $\dfrac{0.208 \text{ g H/1 g C}}{0.083 \text{ g H/1 g C}} = 2.5 \text{ H} : 1\text{C} = 5/2 \text{ H} : 1\text{C} = 5\text{H} : 2\text{C}; \ C_2H_5$

2.62

Symbol	^{106}Pd	$^{122}Sb^{3-}$	$^{184}W^{3+}$	$^{207}Pb^{2+}$	$^{232}Th^{4+}$
Protons	46	51	74	82	90
Neutrons	60	71	110	125	142
Electrons	46	54	71	80	86
Net Charge	0	3-	2+	2+	4+

2.63   perchlorate - ClO_4^-, sulfate - SO_4^{2-}, phosphate - PO_4^{3-}, silicate - SiO_4^{4-}

2.64   (a) IO_3^- (b) IO_4^- (c) IO^- (d) HIO (e) HIO_4 or (H_5IO_6)

2.65   (a) perbromate   (b) selenite   (c) AsO_4^{3-} (d) $HTeO_4^-$

2.66   Cations of calcium can only adopt a 2+ charge, so the name calcium chloride is unambiguous. Cations of cobalt can have either 2+ or 3+ charges. The name cobalt chloride does not specify the charge on the cobalt cation or the number of chloride ions in the molecule. The unambiguous name of $CoCl_2$ is cobalt(II) chloride.

2.67   (a) sodium chloride   (b) sodium bicarbonate (or sodium hydrogen carbonate)
             (c) sodium hypochlorite   (d) sodium hydroxide   (e) ammonium carbonate
             (f) calcium sulfate

2.68   (a) potassium nitrate   (b) sodium carbonate   (c) calcium oxide
       (d) hydrochloric acid   (e) magnesium sulfate   (f) magnesium hydroxide

2.69   (a) CaS, $Ca(HS)_2$ (b) HBr, $HBrO_3$ (c) AlN, $Al(NO_2)_3$ (d) FeO, Fe_2O_3
       (e) NH_3, NH_4^+ (f) K_2SO_3, $KHSO_3$ (g) Hg_2Cl_2, $HgCl_2$ (h) $HClO_3$, $HClO_4$

2.70   (a) There are 10 known isotopes of sulfur, ranging from ^{29}S to ^{38}S.

       (b) The four most abundant are: $^{32}_{16}S$ - 95.0%, $^{34}_{16}S$ - 4.22%, $^{33}_{16}S$ - 0.76%, $^{36}_{16}S$ - 0.14%

2.71 PF_3 - phosphorus trifluoride; density = 3.907 g/mL; m.p. = 151.5°C; b.p. = -101.5°C

3 Stoichiometry: Calculations with Chemical Formulas and Equations

Balancing Chemical Equations

3.1   (a)   In balancing chemical equations, the *law of conservation of mass*, that atoms are neither created nor destroyed during the course of a reaction, is observed. This means that the **number** and **kinds** of atoms on both sides of the chemical equation must be the same.

         (b)   gases - (g); liquids - (l); solids - (s); aqueous solutions - (aq)

         (c) P_4 indicates that there are four phosphorus atoms bound together by chemical bonds into a single molecule; 4P denotes four separate phosphorus atoms.

3.2   (a)   Reactants are the substances we start with and products are what we end up with. Reactants appear to the left of the arrow in a chemical equation and products are written on the right.

         (b)   Subscripts in chemical formulas should not be changed when balancing equations because changing the subscript changes the identity of the compound (*law of constant composition*).

         (c)   This equation is **inconsistent** with the *law of conservation of mass* because the reactants side has two more H atoms and one more O atom than the products side; $\underline{2}H_2O(l)$ are needed.

3.3   (a) $1\ NH_4NO_3(s) \rightarrow 1\ N_2O(g) + 2\ H_2O(l)$

         (b) $1\ La_2O_3(s) + 3\ H_2O(l) \rightarrow 2\ La(OH)_3(aq)$

         (c) $1\ Mg_3N_2(s) + 6\ H_2O(l) \rightarrow 3\ Mg(OH)_2(s) + 2\ NH_3(aq)$

         (d) $1\ NCl_3(aq) + 3\ H_2O(l) \rightarrow 1\ NH_3(aq) + 3\ HOCl(aq)$

         (e) $1\ Al(OH)_3(s) + 3\ HNO_3(aq) \rightarrow 1\ Al(NO_3)_3(aq) + 3\ H_2O(l)$

         (f) $2\ C_6H_6(l) + 15\ O_2(g) \rightarrow 12\ CO_2(g) + 6\ H_2O(l)$

         (g) $4\ CH_3NH_2(g) + 9\ O_2(g) \rightarrow 4\ CO_2(g) + 10\ H_2O(g) + 2\ N_2(g)$

3.4   (a) $N_2O_5(g) + H_2O(l) \rightarrow$ **2**$HNO_3(aq)$

          (b)   **4**$FeO(s) + O_2(g) \rightarrow$ **2**$Fe_2O_3(s)$

          (c)   **1**$PCl_5(l) +$ **4**$H_2O(l) \rightarrow$ **1**$H_3PO_4(aq) +$ **5**$HCl(aq)$

          (d)   **1**$Ni(NO_3)_2(aq) +$ **2**$NaOH(aq) \rightarrow$ **1**$Ni(OH)_2(s) +$ **2**$NaNO_3(aq)$

          (e)   **2**$C_5H_{10}O_2(l) +$ **13**$O_2(g) \rightarrow$ **10**$CO_2(g) +$ **10**$H_2O(l)$

          (f)   **1**$(NH_4)_2Cr_2O_7(s) \rightarrow$ **1**$N_2(g) +$ **1**$Cr_2O_3(s) +$ **4**$H_2O(l)$

          (g)   **2**$Fe(OH)_3(s) +$ **3**$H_2SO_4(aq) \rightarrow Fe_2(SO_4)_3(aq) +$ **6**$H_2O(l)$

3.5   (a)   **2**$NH_3(g) +$ **2**$Na(l) \rightarrow$ **2**$NaNH_2(s) + H_2(g)$

          (b) $Zn(s) + H_2SO_4(aq) \rightarrow ZnSO_4(aq) + H_2(g)$

          (c)   **2**$KNO_3(s) \overset{\Delta}{\rightarrow}$ **2**$KNO_2(s) + O_2(g)$

          (d) $PCl_3(l) +$ **3**$H_2O(l) \rightarrow H_3PO_3(aq) +$ **3**$HCl(aq)$

          (e) $Cu(s) +$ **2**$H_2SO_4(aq) \rightarrow CuSO_4(aq) + SO_2(g) +$ **2**$H_2O(l)$

3.6   (a) $SO_3(g) + H_2O(l) \rightarrow H_2SO_4(aq)$

          (b) $B_2S_3(s) +$ **6**$H_2O(l) \rightarrow$ **2**$H_3BO_3(aq) +$ **3**$H_2S(g)$

          (c)   **4**$PH_3(g) +$ **8**$O_2(g) \rightarrow$ **6**$H_2O(g) + P_4O_{10}(s)$

          (d)   **2**$Hg(NO_3)_2(s) \rightarrow$ **2**$HgO(s) +$ **4**$NO_2(g) + O_2(g)$

          (e)   **3**$H_2S(g) +$ **2**$Fe(OH)_3(s) \rightarrow Fe_2S_3(s) +$ **6**$H_2O(g)$

Patterns of Chemical Reactivity

3.7   (a) $C_6H_{12}(l) +$ **9**$O_2(g) \rightarrow$ **6**$CO_2(g) +$ **6**$H_2O(l)$

          (b) $C_2H_5OC_2H_5(l) +$ **6**$O_2(g) \rightarrow$ **4**$CO_2(g) +$ **5**$H_2O(l)$

          (c)   **2**$Cs(s) +$ **2**$H_2O(l) \rightarrow$ **2**$CsOH(aq) + H_2(g)$

          (d) $Mg(s) + Br_2(l) \rightarrow MgBr_2(s)$

3.8   (a) $C_8H_8(l) +$ **10**$O_2(g) \rightarrow$ **8**$CO_2(g) +$ **4**$H_2O(l)$

          (b)   **2**$C_5H_{12}O(l) +$ **15**$O_2(g) \rightarrow$ **10**$CO_2(g) +$ **12**$H_2O(l)$

          (c)   **2**$Li(s) +$ **2**$H_2O(l) \rightarrow$ **2**$LiOH(aq) + H_2(g)$

          (d) $PbCO_3(s) \overset{\Delta}{\rightarrow} PbO(s) + CO_2(g)$

3.9   (a) $C_5H_6O(l) +$ **6**$O_2(g) \rightarrow$ **5**$CO_2(g) +$ **3**$H_2O(l)$ combustion

          (b)   **2**$H_2O_2(l) \rightarrow$ **2**$H_2O(l) + O_2(g)$ decomposition

          (c) $CaO(s) + H_2O(l) \rightarrow Ca(OH)_2(aq)$ combination

          (d)   **6**$Li(s) + N_2(g) \rightarrow$ **2**$Li_3N(s)$ combination

3.10   (a) $2KClO_3(s) \rightarrow 2KCl(s) + 3O_2(g)$ decomposition

(b) $C_7H_8O_2(l) + 8O_2(g) \rightarrow 7CO_2(g) + 4H_2O(l)$ combustion

(c) $2Cr(s) + 3Cl_2(g) \rightarrow 2CrCl_3(s)$ combination

(d) $2SO_3(g) \rightarrow 2SO_2(g) + O_2(g)$ decomposition

3.11   (a) $2K(s) + 2NH_3(l) \rightarrow 2KNH_2(solv) + H_2(g)$

(b) $4CH_3NO_2(g) + 7O_2(g) \rightarrow 4NO_2(g) + 4CO_2(g) + 6H_2O(l)$

(c) $CH_4(g) + 4F_2(g) \rightarrow CF_4(g) + 4HF(g)$

3.12   (a) $2Si_2H_6(g) + 7O_2(g) \rightarrow 4SiO_2(s) + 6H_2O(l)$

(b) $CH_3SH(g) + 3O_2(g) \rightarrow SO_2(g) + CO_2(g) + 2H_2O(l)$

(c) $2Al(s) + 3Br_2(l) \rightarrow 2AlBr_3(s)$

Atomic and Molecular Weights

3.13   (a) $^{12}_{6}C$ (b) An atomic mass unit is exactly 1/12 of the mass of one atom of $^{12}_{6}C$, or 1.66054×10^{-24} g.

3.14   (a) 12 amu   (b) The mass of an average H atom is 1.0079 amu.

If the mass of a carbon-12 atom were redefined to 20 nmu, the mass of an average H atom would increase proportionally.

$$\frac{20 \text{ nmu}}{12 \text{ amu}} = \frac{x \text{ nmu}}{1.0079 \text{ amu}}; \; x = 1.6798 \text{ nmu}$$

3.15   The **average** mass of a boron atom is given by the sum of the mass of each isotope present in the sample times its **fractional** abundance.

average atomic mass = 0.1978(10.013 amu) + 0.8022(11.009 amu)
(atomic weight, AW) = 1.981 amu + 8.831 amu = 10.812 amu

3.16   Average atomic mass = 0.7870(23.9924) + 0.1013(24.9938) + 0.1117(25.9898) = 24.32 amu

3.17   FW in amu to 1 decimal place (see Sample Exercise 3.5)

(a) P_2O_3: 2(31.0) + 3(16.0) = 110.0 amu

(b) $BaSO_4$: 1(137.3) + 1(32.1) + 4(16.0) = 233.4 amu

(c) $Mg(C_2H_3O_2)_2$: 1(24.3) + 4(12.0) + 6(1.0) + 4(16.0) = 142.3 amu

(d) $(NH_4)_2Cr_2O_7$: 2(14.0) + 8(1.0) + 2(52.0) + 7(16.0) = 252.0 amu

(e) HNO_3: 1(1.0) + 1(14.0) + 3(16.0) = 63.0 amu

(f) Li_2CO_3: 2(6.9) + 1(12.0) + 3(16.0) = 73.8 amu

(g) Si_2Cl_6: 2(28.1) + 6(35.5) = 269.2 amu

3.18   FW in amu to 1 decimal place

      (a) C_2H_2: $2(12.0) + 2(1.0) = 26.0$ amu

      (b) $(NH_4)_2SO_4$: $2(14.0) + 8(1.0) + 1(32.1) + 4(16.0) = 132.1$ amu

      (c) $C_6H_8O_6$: $6(12.0) + 8(1.0) + 6(16.0) = 176.0$ amu

      (d) $PtCl_2(NH_3)_2$: $1(195.1) + 2(35.5) + 2(14.0) + 6(1.0) = 300.1$ amu

      (e) $C_{19}H_{28}O_2$: $19(12.0) + 28(1.0) + 2(16.0) = 288.0$ amu

3.19   Calculate the formula weight, then the mass % of oxygen in the compound.

      (a) SO_3: FW = $1(32.1) + 3(16.0) = 80.1$ amu

$$\% \text{ O} = \frac{3(16.0) \text{ amu}}{80.1 \text{ amu}} \times 100 = 59.9\%$$

      (b) CH_3COOH: FW = $2(12.0) + 4(1.0) + 2(16.0) = 60.0$ amu

$$\% \text{ O} = \frac{2(16.0) \text{ amu}}{60.0 \text{ amu}} \times 100 = 53.3\%$$

      (c) $Ca(NO_3)_2$: FW = $1(40.1) + 2(14.0) + 6(16.0) = 164.1$ amu

$$\% \text{ O} = \frac{6(16.0) \text{ amu}}{164.1 \text{ amu}} \times 100 = 58.5\%$$

      (d) $(NH_4)_2SO_4$: FW = 132.1 amu (Exercise 3.18(b))

$$\% \text{ O} = \frac{4(16.0) \text{ amu}}{132.1 \text{ amu}} \times 100 = 48.4\%$$

3.20   (a) N_2O: FW = $2(14.0) + 1(16.0) = 44.0$ amu

$$\% \text{ N} = \frac{28.0 \text{ amu}}{44.0 \text{ amu}} \times 100 = 63.6\%$$

      (b) $C_7H_6O_2$: $7(12.0) + 6(1.0) + 2(16.0) = 122.0$ amu

$$\% \text{ C} = \frac{7(12.0) \text{ amu}}{122.0 \text{ amu}} \times 100 = 68.9\%$$

      (c) $Mg(OH)_2$: $1(24.3) + 2(16.0) + 2(1.0) = 58.3$ amu

$$\% \text{ Mg} = \frac{24.3 \text{ amu}}{58.3 \text{ amu}} \times 100 = 41.7\%$$

      (d) $(NH_2)_2CO$: $2(14.0) + 4(1.0) + 1(12.0) + 1(16.0) = 60.0$ amu

$$\% \text{ N} = \frac{2(14.0) \text{ amu}}{60.0 \text{ amu}} \times 100 = 46.7\%$$

      (e) $CH_3CO_2C_5H_{11}$: $7(12.0) + 14(1.0) + 2(16.0) = 130.0$ amu

$$\% \text{ H} = \frac{14(1.0) \text{ amu}}{130.0 \text{ amu}} \times 100 = 10.8\%$$

3.21   (a) CO_2: FW = $1(12.0) + 2(16.0) = 44.0$ amu

$$\% \text{ C} = \frac{12.0 \text{ amu}}{44.0 \text{ amu}} \times 100 = 27.3\%$$

(b) C_2H_6: FW = 2(12.0) + 6(1.0) = 30.0

$\%\ C = \dfrac{2(12.0)\ amu}{30.0\ amu} \times 100 = 80.0\%$

(c) CH_3OH: FW = 1(12.0) + 4(1.0) + 1(16.0) = 32.0 amu

$\%\ C = \dfrac{12.0\ amu}{32.0\ amu} \times 100 = 37.5\%$

(d) $CS(NH_2)_2$: FW = 1(12.0) + 1(32.1) + 2(14.0) + 4(1.0) = 76.1 amu

$\%\ C = \dfrac{12.0\ amu}{76.1\ amu} \times 100 = 15.8\%$

3.22   (a) C_7H_6O: FW = 7(12.0) + 6(1.0) + 1(16.0) = 106.0 amu

$\%C = \dfrac{7(12.0)\ amu}{106.0\ amu} \times 100 = 79.2\%$

(b) $C_8H_8O_4$: FW = 8(12.0) + 8(1.0) + 4(16.0) = 168.0 amu

$\%\ C = \dfrac{8(12.0)\ amu}{168.0\ amu} \times 100 = 57.1\%$

(c) $C_7H_{14}O_2$: FW = 7(12.0) + 14(1.0) + 2(16.0) = 130.0 amu

$\%\ C = \dfrac{7(12.0)\ amu}{130.0\ amu} \times 100 = 64.6\%$

3.23   (a)   Three peaks: ^1H - ^1H, ^1H - ^2H, ^2H - ^2H

(b) ^1H - ^1H = 2(1.00783) = 2.01566 amu

^1H - ^2H = 1.00783 + 2.01411 = 3.02194 amu

^2H - ^2H = 2(2.01411) = 4.02822 amu

[The mass ratios are 1 : 1.49923 : 1.99846 or 1 : 1.5 : 2]

(c) ^1H - ^1H is largest, because there is the greatest chance that two atoms of the more abundant isotope will combine.

^2H - ^2H is the smallest, because there is the least chance that two atoms of the less abundant isotope will combine.

3.24   (a)   A Br_2 molecule could consist of two atoms of the same isotope or one atom of each of the two different isotopes. This second possibility is twice as likely as the first. Therefore, the second peak (twice as large as peaks 1 and 3) represents a Br_2 molecule containing different isotopes. The mass numbers of the two isotopes are determined from the masses of the two smaller peaks. Since 157.84 ≈ 158, the first peak represents a ^{79}Br - ^{79}Br molecule. Peak 3, 161.81 ≈ 162, represents a ^{81}Br - ^{81}Br molecule. Peak 2 then contains one atom of each isotope, ^{79}Br - ^{81}Br, with an approximate mass of 160 amu.

(b)   The mass of the lighter isotope is 157.84 amu/2 atoms, or 78.92 amu/atom. For the heavier one, 161.84 amu/2 atoms = 80.92 amu/atom.

(c)   The relative size of the three peaks in the mass spectrum of Br_2 indicates their relative abundance. The average mass of a Br_2 molecule is
    $0.2534(157.84) + 0.5000(159.84) + 0.2466(161.84) = 159.83$
(Each product has four significant figures and two decimal places, so the answer has two decimal places.)

(d) $\dfrac{159.83 \text{ amu}}{\text{avg. } Br_2 \text{ molecule}} \times \dfrac{1 \; Br_2 \text{ molecule}}{2 \text{ Br atoms}} = 79.915 \text{ amu}$

(e)   Let x = the abundance of ^{79}Br, 1-x = abundance of ^{81}Br. From (b), the masses of the two isotopes are 78.92 amu and 80.92 amu, respectively. From (d), the mass of an average Br atom is 79.915 amu.

   $x(78.92) + (1 - x)(80.92) = 79.915, \; x = 0.5025 = 0.503$
   $^{79}Br = 50.3\%, \;\; ^{81}Br = 49.7\%$

The Mole

3.25   (a)   A *mole* is the amount of matter that contains as many objects as the number of atoms in exactly 12 g of ^{12}C.

(b) 6.022×10^{23}. This is the number of objects in a mole of anything.

(c)   The formula weight of a substance in amu has the same numerical value as the molar mass expressed in grams.

3.26   (a)   <u>exactly</u> 12 g   (b) 6.0221367×10^{23}, Avogadro's number

(c) $\dfrac{12 \text{ g } ^{12}C}{1 \text{ mol } ^{12}C} \times \dfrac{1 \text{ mol } ^{12}C}{6.0221367 \times 10^{23} \text{ atoms}} = 1.9926482 \times 10^{-23} \text{ g/}^{12}C \text{ atom}$

3.27   (a) $\dfrac{12 \text{ H atoms}}{6 \text{ C atoms}} = \dfrac{2 \text{ H}}{1 \text{ C}} \times 4.0 \times 10^{22} C \text{ atoms} = 8.0 \times 10^{22} \text{ H atoms}$

(b) $\dfrac{1 \; C_6H_{12}O_6 \text{ molecule}}{6 \text{ C atoms}} \times 4.0 \times 10^{22} \text{ C atoms} = 6.67 \times 10^{21}$

   $= 6.7 \times 10^{21} \; C_6H_{12}O_6 \text{ molecules}$

(c) $6.67 \times 10^{21} \; C_6H_{12}O_6 \text{ molecules} \times \; = 1.1 \times 10^{-2} \text{ mol } C_6H_{12}O_6$

(d)   1 mole of $C_6H_{12}O_6$ weighs 180.0 g (Sample Exercise 3.8)

   $1.1 \times 10^{-2} \text{ mol } C_6H_{12}O_6 \times \dfrac{180.0 \text{ g } C_6H_{12}O_6}{1 \text{ mol}} = 2.0 \text{ g } C_6H_{12}O_6$

3.28   (a) $3.0 \times 10^{20} \text{ H atoms} \times \dfrac{18 \text{ C atoms}}{24 \text{ H atoms}} = 2.3 \times 10^{20} \text{ C atoms}$

(b) 3.0×10^{20} H atoms $\times \dfrac{1\ C_{18}H_{24}O_2\ \text{molecule}}{24\ \text{H atoms}} = 1.25 \times 10^{19}$

$= 1.3 \times 10^{19}\ C_{18}H_{24}O_2$ molecules

(c) $1.25 \times 10^{19}\ C_{18}H_{24}O_2$ molecules $\times \dfrac{1\ \text{mol}}{6.022 \times 10^{23}\ \text{molecules}} = 2.08 \times 10^{-5}$

$= 2.1 \times 10^{-5}$ mol

(d) molar mass $= 18(12.0) + 24(1.01) + 2(16.0) = 272$ g

2.08×10^{-5} mol $\times \dfrac{272\ \text{g}}{\text{mol}} = 5.7 \times 10^{-3}$ g estradiol

3.29 (a) molar mass: $63.55 + 2(14.01) + 6(16.00) = 187.57$ g

(b) 0.120 mol $Cu(NO_3)_2 \times \dfrac{187.57\ \text{g}}{1\ \text{mol}} = 22.5$ g $Cu(NO_3)_2$

(c) 3.15 g $Cu(NO_3)_2 \times \dfrac{1\ \text{mol}}{187.57\ \text{g}} = 1.68 \times 10^{-2}$ mol

(d) 1.25 mg $\times \dfrac{1 \times 10^{-3}\text{g}}{1\ \text{mg}} \times \dfrac{1\ \text{mol}}{187.57\ \text{g}} \times \dfrac{6.022 \times 10^{23}\ \text{molecules}}{\text{mol}} \times \dfrac{2\ \text{N atoms}}{1\ \text{molecule}}$

$= 8.03 \times 10^{18}$ N atoms

3.30 (a) molar mass $= 14(12.01) + 18(1.008) + 2(14.01) + 5(16.00) = 294.3$ g

(b) 5.67 g aspartame $\times \dfrac{1\ \text{mol}}{294.3\ \text{g}} = 1.93 \times 10^{-2}$ mol aspartame

(c) 0.122 mol aspartame $\times \dfrac{294.3\ \text{g}}{\text{mol}} = 35.9$ g aspartame

(d) 2.13 mg aspartame $\times \dfrac{1 \times 10^{-3}\ \text{g}}{1\ \text{mg}} \times \dfrac{1\ \text{mol}}{294.3\ \text{g}} \times \dfrac{18\ \text{mol H}}{1\ \text{mol aspartame}}$

$\times \dfrac{6.022 \times 10^{23}\ \text{H atoms}}{1\ \text{mol H}} = 7.85 \times 10^{19}$ H atoms

3.31 (a) 0.00150 mol $SO_2 \times \dfrac{64.06\ \text{g SO}_2}{1\ \text{mol SO}_2} = 0.0961$ g SO_2

(b) 2.98×10^{21} Ar atom $\times \dfrac{1\ \text{mol}}{6.022 \times 10^{23}\ \text{atoms}} \times \dfrac{39.95\ \text{g Ar}}{\text{mol Ar}} = 0.198$ g Ar

(c) 1.05×10^{20} molecules $\times \dfrac{1\ \text{mol}}{6.022 \times 10^{23}\ \text{molecules}} \times \dfrac{194.2\ \text{g } C_8H_{10}N_4O_2}{1\ \text{mol caffeine}}$

$= 3.39 \times 10^{-2}$ g $C_8H_{10}N_4O_2$

3.32 (a) 0.0287 mol aspirin $\times \dfrac{180.2\ \text{g}}{1\ \text{mol}} = 5.17$ g $C_9H_8O_4$

(b) 1.75×10^{21} molecules $O_3 \times \dfrac{1\ \text{mol } O_3}{6.022 \times 10^{23}\ \text{molecules}} \times \dfrac{48.00\ \text{g } O_3}{1\ \text{mol } O_3} = 0.139$ g O_3

(c) 3.69×10^{24} molecules $C_{27}H_{46}O \times \dfrac{1\ mol}{6.022 \times 10^{23}\ molecules} \times \dfrac{386.7\ g\ C_{27}H_{46}O}{1\ mol}$

$= 2.37 \times 10^{3}\ g = 237\ kg\ C_{27}H_{46}O$

3.33   (a) $0.0666\ mol\ C_{3}H_{8} \times \dfrac{6.022 \times 10^{23}\ molecules}{1\ mol} = 4.01 \times 10^{22}\ C_{3}H_{8}$ molecules

(b) $50.0\ mg\ C_{8}H_{9}O_{2}N \times \dfrac{1 \times 10^{-3}\ g}{1\ mg} \times \dfrac{1\ mol\ C_{8}H_{9}O_{2}N}{151.2\ g\ C_{8}H_{9}O_{2}N} \times \dfrac{6.022 \times 10^{23}\ molecules}{1\ mol}$

$= 1.99 \times 10^{20}\ C_{8}H_{9}O_{2}N$ molecules

(c) $10.5\ g\ C_{12}H_{22}O_{11} \times \dfrac{1\ mol\ C_{12}H_{22}O_{11}}{342.3\ g\ C_{12}H_{22}O_{11}} \times \dfrac{6.022 \times 10^{23}\ molecules}{1\ mol}$

$= 1.85 \times 10^{22}\ C_{12}H_{22}O_{11}$ molecules

3.34   (a) $0.135\ mol\ C_{2}H_{2} \times \dfrac{6.022 \times 10^{23}\ molecules}{1\ mol} = 8.13 \times 10^{22}\ C_{2}H_{2}$ molecules

(b) $250\ mg\ C_{6}H_{8}O_{6} \times \dfrac{1 \times 10^{-3}\ g}{1\ mg} \times \dfrac{1\ mol\ C_{6}H_{8}O_{6}}{176.1\ g\ C_{6}H_{8}O_{6}} \times \dfrac{6.022 \times 10^{23}\ molecules}{1\ mol}$

$= 8.55 \times 10^{20}\ C_{6}H_{8}O_{6}$ molecules

(c) $40 \times 10^{-5}\ g\ H_{2}O \times \dfrac{1\ mol\ H_{2}O}{18.02\ g\ H_{2}O} \times \dfrac{6.022 \times 10^{23}\ molecules}{1\ mol}$

$= 1.3 \times 10^{19}\ H_{2}O$ molecules

3.35   Strategy: volume $H_{2}O$ → mass $H_{2}O$ → moles $H_{2}O$ → molecules $H_{2}O$

$350\ ft \times 99\ ft \times 6.0\ ft \times \dfrac{12^{3}\ in^{3}}{1\ ft^{3}} \times \dfrac{2.54^{3}\ cm^{3}}{in^{3}} \times \dfrac{1\ mL}{1\ cm^{3}} = 5.887 \times 10^{9} = 5.9 \times 10^{9}\ mL$

$5.887 \times 10^{9}\ mL \times \dfrac{1.0\ g}{mL} \times \dfrac{1\ mol\ H_{2}O}{18.02\ g\ H_{2}O} \times 6.022 \times 10^{23} = 2.0 \times 10^{32}\ H_{2}O$ molecules

3.36   Strategy: volume soln → mass soln → mass Sevin (201.2 g/mol) → moles Sevin

→ molecules Sevin

0.1% Sevin = 0.1 g Sevin / 100 g soln

$250\ mL\ soln \times \dfrac{1.0\ g\ soln}{mL\ soln} \times \dfrac{0.1\ g\ Sevin}{100\ g\ soln} \times \dfrac{1\ mol\ Sevin}{201.2\ g\ Sevin} \times \dfrac{6.022 \times 10^{23}\ molecules}{mol}$

$= 7.48 \times 10^{20} = 7 \times 10^{20}$ molecules

3.37 $\dfrac{2.05 \times 10^{-6}\ g\ C_{2}H_{3}Cl}{1\ L} \times \dfrac{1\ mol\ C_{2}H_{3}Cl}{62.50\ g\ C_{2}H_{3}Cl} = 3.280 \times 10^{-8} = 3.28 \times 10^{-8}\ mol\ C_{2}H_{3}Cl/L$

$\dfrac{3.280 \times 10^{-8}\ mol\ C_{2}H_{3}Cl}{1\ L} \times \dfrac{6.022 \times 10^{23}\ molecules}{1\ mol} = 1.97 \times 10^{16}$ molecules/L

3.38 25×10^{-6} g $C_{21}H_{30}O_2 \times \dfrac{1 \text{ mol } C_{21}H_{30}O_2}{314.5 \text{ g } C_{21}H_{30}O_2} = 7.95 \times 10^{-8} = 8.0 \times 10^{-8}$ mol $C_{21}H_{30}O_2$

7.95×10^{-8} mol $C_{21}H_{30}O_2 \times \dfrac{6.022 \times 10^{23} \text{ molecules}}{1 \text{ mol}} = 4.8 \times 10^{16}$ $C_{21}H_{30}O_2$ molecules

Empirical Formulas

3.39   An *empirical formula* gives the relative number and kind of each atom in a compound, but a *molecular formula* gives the actual number of each kind of atom, and thus the molecular weight.

3.40   Percent composition gives the **relative** mass of each element in a compound. These percents yield **relative** numbers of each atom in the molecule, the empirical formula.

3.41   (a)   Find the **simplest ratio of moles** by dividing by the smallest number of moles present.

   0.0130 mol C / 0.0065 = 2
   0.039 mol H / 0.0065 = 6
   0.0065 mol O / 0.0065 = 1

   The empirical formula is C_2H_6O.

   (b)   Calculate the moles of each element present, then the simplest ratio of moles.

   11.66 g Fe $\times \dfrac{1 \text{ mol Fe}}{55.85 \text{ g Fe}} = 0.2088$ mol Fe; 0.2088 / 0.2088 = 1

   5.01 g O $\times \dfrac{1 \text{ mol O}}{16.00 \text{ g O}} = 0.3131$ mol O; 0.3131 / 0.2088 $\approx$ 1.5

   Multiplying by two, the integer ratio is 2 Fe : 3 O; the empirical formula is Fe_2O_3.

   (c)   Assume 100 g sample, calculate moles of each element, find the simplest ratio of moles.

   40.0 g C $\times \dfrac{1 \text{ mol C}}{12.01 \text{ g C}} = 3.33$ mol C; 3.33 / 3.33 = 1

   6.7 g H $\times \dfrac{1 \text{ mol H}}{1.008 \text{ mol H}} = 6.65$ mol H; 6.65 / 3.33 $\approx$ 2

   53.3 g O $\times \dfrac{1 \text{ mol O}}{16.00 \text{ mol O}} = 3.33$ mol O; 3.33 / 3.33 = 1

   The empirical formula is CH_2O.

3.42   (a)   Calculate the simplest ratio of moles.

0.104 mol K / 0.052 = 2
0.052 mol C / 0.052 = 1
0.156 mol O / 0.052 = 3

The empirical formula is K_2CO_3.

(b)   Calculate moles of each element present, then the simplest ratio of moles.

$5.28 \text{ g Sn} \times \dfrac{1 \text{ mol Sn}}{118.7 \text{ g Sn}} = 0.04448 \text{ mol Sn}; \; 0.04448 / 0.04448 = 1$

$3.37 \text{ g F} \times \dfrac{1 \text{ mol F}}{19.00 \text{ g F}} = 0.1774 \text{ mol F}; \; 0.1774 / 0.04448 \approx 4$

The integer ratio is 1 Sn : 4 F; the empirical formula is SnF_4.

(c)   Assume 100 g sample, calculate moles of each element, find the simplest ratio of moles.

$87.5\% \text{ N} = 87.5 \text{ g N} \times \dfrac{1 \text{ mol N}}{14.01 \text{ g}} = 6.25 \text{ mol N}; \; 6.25 / 6.25 = 1$

$12.5\% \text{ H} = 12.5 \text{ g H} \times \dfrac{1 \text{ mol}}{1.008 \text{ g}} = 12.4 \text{ mol H}; \; 12.4 / 6.25 \approx 2$

The empirical formula is NH_2.

3.43   The procedure in all these cases is to assume 100 g of sample, calculate the number of moles of each element present in that 100 g, then obtain the ratio of moles as smallest whole numbers.

(a) $32.79 \text{ g Na} \times \dfrac{1 \text{ mol Na}}{22.99 \text{ g Na}} = 1.426 \text{ mol Na}; \; 1.426 / 0.4826 \approx 3$

$13.02 \text{ g Al} \times \dfrac{1 \text{ mol Al}}{26.98 \text{ g Al}} = 0.4826 \text{ mol Al}; \; 0.4826 / 0.4826 = 1$

$54.19 \text{ g F} \times \dfrac{1 \text{ mol F}}{19.00 \text{ g F}} = 2.852 \text{ mol F}; \; 2.852 / 0.4826 \approx 6$

The empirical formula is Na_3AlF_6.

(b) $62.1 \text{ g C} \times \dfrac{1 \text{ mol C}}{12.01 \text{ g C}} = 5.17 \text{ mol C}; \; 5.17 / 0.864 \approx 6$

$5.21 \text{ g H} \times \dfrac{1 \text{ mol H}}{1.008 \text{ g H}} = 5.17 \text{ mol O}; \; 5.17 / 0.864 \approx 6$

$12.1 \text{ g N} \times \dfrac{1 \text{ mol N}}{14.01 \text{ g N}} = 0.864 \text{ mol N}; \; 0.864 / 0.864 = 1$

$20.7 \text{ g O} \times \dfrac{12 \text{ mol O}}{16.00 \text{ g O}} = 1.29 \text{ mol O}; \; 1.29 \times 0.864 \approx 1.5$

Multiplying by two, the formula is $C_{12}H_{12}N_2O_3$.

3.44   Assume 100 g sample in the following problems.

(a) $10.4 \text{ g C} \times \dfrac{1 \text{ mol C}}{12.01 \text{ g C}} = 0.866 \text{ mol C}; \quad 0.866 / 0.866 = 1$

$27.8 \text{ g S} \times \dfrac{1 \text{ mol S}}{32.07 \text{ g S}} = 0.867 \text{ mol S}; \quad 0.867 / 0.866 \approx 1$

$61.7 \text{ g Cl} \times \dfrac{1 \text{ mol Cl}}{35.45 \text{ g Cl}} = 1.74 \text{ mol Cl}; \quad 1.74 / 0.866 \approx 2$

The empirical formula is $CSCl_2$.

(b) $21.7 \text{ g C} \times \dfrac{1 \text{ mol C}}{12.01 \text{ g C}} = 1.81 \text{ mol C}; \quad 1.81 / 0.600 \approx 3$

$9.6 \text{ g O} \times \dfrac{1 \text{ mol O}}{16.00 \text{ g O}} = 0.600 \text{ mol O}; \quad 0.600 / 0.600 = 1$

$68.7 \text{ g F} \times \dfrac{1 \text{ mol F}}{19.00 \text{ g F}} = 3.62 \text{ mol F}; \quad 3.62 / 0.600 \approx 6$

The empirical formula is C_3OF_6.

3.45   (a)   Calculate the empirical formula weight (FW) and its ratio with the molar mass.

$\text{FW} = 12.01 + 1.01 = 13.02; \quad \dfrac{\text{MM}}{\text{FW}} = \dfrac{78}{13} = 6$

The subscripts of the empirical formula are multiplied by 6. The molecular formula is C_6H_6.

(b) $\text{FW} = 14.01 + 2(16.00) = 46.01; \quad \dfrac{\text{MM}}{\text{FW}} = \dfrac{92.02}{46.01} = 2$

The molecular formula is N_2O_4.

3.46   (a) $\text{FW} = 12.01 + 2(1.008) = 14.03 \quad \dfrac{\text{MM}}{\text{FW}} = \dfrac{70.14}{14.03} = 5$

The molecular formula is C_5H_{10}.

(b) $\text{FW} = 14.01 + 2(1.008) + 35.45 = 51.48 \quad \dfrac{\text{MM}}{\text{FW}} = \dfrac{51.5}{51.5} = 1$

The empirical and molecular formulas are NH_2Cl.

3.47   Assume 100 g in the following problems.

(a) $38.7 \text{ g C} \times \dfrac{1 \text{ mol C}}{12.01 \text{ g C}} = 3.22 \text{ mol C}; \quad 3.22 / 3.22 = 1$

$9.7 \text{ g H} \times \dfrac{1 \text{ mol C}}{1.008 \text{ g H}} = 9.62 \text{ mol H}; \quad 9.62 / 3.22 \approx 3$

$51.6 \text{ g O} \times \dfrac{1 \text{ mol O}}{16.00 \text{ g O}} = 3.23 \text{ mol O}; \quad 3.23 / 3.23 \approx 1$

Thus, CH_3O, formula weight = 31. If the molar mass is 62.1 amu, a factor of 2 gives the molecular formula $C_2H_6O_2$.

(b) $49.5\text{ g C} \times \dfrac{1\text{ mol C}}{12.01\text{ g C}} = 4.12\text{ mol C};\quad 4.12 / 1.03 \approx 4$

$5.15\text{ g H} \times \dfrac{1\text{ mol H}}{1.008\text{ g H}} = 5.11\text{ mol H};\quad 5.11 / 1.03 \approx 5$

$28.9\text{ g N} \times \dfrac{1\text{ mol N}}{14.01\text{ g N}} = 2.06\text{ mol N};\quad 2.06 / 1.03 \approx 2$

$16.5\text{ g O} \times \dfrac{1\text{ mol O}}{16.00\text{ g O}} = 1.03\text{ mol O};\quad 1.03 / 1.03 = 1$

Thus, $C_4H_5N_2O$, FW = 97. If the molar mass is about 195, a factor of 2 gives the molecular formula $C_8H_{10}N_4O_2$.

3.48   Assume 100 g in the following problems.

(a) $59.0\text{ g C} \times \dfrac{1\text{ mol C}}{12.01\text{ g C}} = 4.91\text{ mol C};\quad 4.91 / 0.550 \approx 9$

$7.1\text{ g H} \times \dfrac{1\text{ mol H}}{1.008\text{ g H}} = 7.04\text{ mol H};\quad 7.04 / 0.550 \approx 13$

$26.2\text{ g O} \times \dfrac{1\text{ mol O}}{16.00\text{ g O}} = 1.64\text{ mol O};\quad 6.64 / 0.550 \approx 3$

$7.7\text{ g N} \times \dfrac{1\text{ mol N}}{14.01\text{ g N}} = 0.550\text{ mol N};\quad 0.550 / 0.550 = 1$

Thus, the empirical formula is $C_9H_{13}O_3N$. This corresponds to a formula weight of 183, so it is also the molecular formula.

(b) $74.1\text{ g C} \times \dfrac{1\text{ mol C}}{12.01\text{ g C}} = 6.17\text{ mol C};\quad 6.17 / 1.23 \approx 5$

$8.6\text{ g H} \times \dfrac{1\text{ mol H}}{1.008\text{ g H}} = 8.53\text{ mol H};\quad 8.53 / 1.23 \approx 7$

$17.3\text{ g N} \times \dfrac{1\text{ mol N}}{14.01\text{ g N}} = 1.23\text{ mol N};\quad 1.23 / 1.23 \approx 1$

Thus, C_5H_7N. This corresponds to a formula weight of 81. If the molar mass is 160 ± 5 g, a factor of 2 gives the molecular formula $C_{10}H_{14}N_2$.

3.49   We can calculate the mass of O by subtraction, but first the masses of C and H in the sample must be found.

$6.32 \times 10^{-3}\text{ g CO}_2 \times \dfrac{12.01\text{ g C}}{44.01\text{ g CO}_2} = 1.725 \times 10^{-3}\text{ g C} = 1.73\text{ mg C}$

$2.58 \times 10^{-3}\text{ g H}_2O \times \dfrac{2.016\text{ g H}}{18.02\text{ g H}_2O} = 2.886 \times 10^{-4}\text{ g H} = 0.289\text{ mg H}$

mass of O = 2.78 mg sample - (1.725 mg C + 0.289 mg H) = 0.77 mg O

$$1.725 \times 10^{-3} \text{ g C} \times \frac{1 \text{ mol C}}{12.01 \text{ g C}} = 1.44 \times 10^{-4} \text{ mol C}; \quad 1.44 \times 10^{-4} / 4.81 \times 10^{-5} \approx 3$$

$$2.886 \times 10^{-4} \text{ g C} \times \frac{1 \text{ mol H}}{1.008 \text{ g H}} = 2.86 \times 10^{-4} \text{ mol H}; \quad 2.86 \times 10^{-4} / 4.81 \times 10^{-5} \approx 6$$

$$7.7 \times 10^{-4} \text{ g O} \times \frac{1 \text{ mol O}}{16.00 \text{ g H}} = 4.81 \times 10^{-5} \text{ mol O}; \quad 4.81 \times 10^{-5} / 4.81 \times 10^{-5} = 1$$

The empirical formula is C_3H_6O.

3.50   Calculate the masses of C and H in the sample and get the mass of O by subtraction.

$$0.2829 \text{ g CO}_2 \times \frac{12.01 \text{ g C}}{44.01 \text{ g CO}_2} = 0.07720 \text{ g C}$$

$$0.1159 \text{ g H}_2\text{O} \times \frac{2.016 \text{ g H}}{18.02 \text{ g H}_2\text{O}} = 0.01297 \text{ g H}$$

mass O = 0.1005 g sample - (0.07720 g C + 0.01297 g H) = 0.01033 g O

$$0.07720 \text{ g C} \times \frac{1 \text{ mol C}}{12.01 \text{ g C}} = 6.428 \times 10^{-3} \text{ mol C}; \quad \frac{6.428 \times 10^{-3}}{6.456 \times 10^{-4}} \approx 10$$

$$0.01297 \text{ g H} \times \frac{1 \text{ mol H}}{1.008 \text{ g H}} = 1.287 \times 10^{-2} \text{ mol H}; \quad \frac{1.287 \times 10^{-2}}{6.456 \times 10^{-4}} \approx 20$$

$$0.01033 \text{ g O} \times \frac{1 \text{ mol O}}{16.00 \text{ g O}} = 6.456 \times 10^{-4} \text{ mol O}; \quad \frac{6.456 \times 10^{-4}}{6.456 \times 10^{-4}} = 1$$

The empirical formula is $C_{10}H_{20}O$.

$$\text{FW} = 10(12.01) + 20(1.008) + 16.00 = 156; \quad \frac{\text{MM}}{\text{FW}} = \frac{156}{156} = 1$$

The molecular formula is the same as the empirical formula, $C_{10}H_{20}O$.

3.51   The reaction involved is $Na_2CO_3 \cdot xH_2O(s) \rightarrow Na_2CO_3(s) + xH_2O(g)$.

Calculate the mass of H_2O lost and then the mole ratio of Na_2CO_3 and H_2O.

g H_2O lost = 2.558 g sample - 0.948 g Na_2CO_3 = 1.610 g H_2O

$$0.948 \text{ g Na}_2\text{CO}_3 \times \frac{1 \text{ mol Na}_2\text{CO}_3}{106.0 \text{ g Na}_2\text{CO}_3} = 0.00894 \text{ mol Na}_2\text{CO}_3$$

$$1.610 \text{ g H}_2\text{O} \times \frac{1 \text{ mol H}_2\text{O}}{18.02 \text{ g H}_2\text{O}} = 0.08935 \text{ mol H}_2\text{O}$$

$$\frac{\text{mol H}_2\text{O}}{\text{mol Na}_2\text{CO}_3} = \frac{0.08935}{0.00894} = 9.99; \quad x = 10.$$

The formula is $Na_2CO_3 \cdot \underline{\textbf{10}} \; H_2O$.

3.52   The reaction involved is $MgSO_4 \cdot xH_2O(s) \rightarrow MgSO_4(s) + xH_2O(g)$. First, calculate the number of moles of product $MgSO_4$; this is the same as the number of moles of starting hydrate.

$$2.472 \text{ g } MgSO_4 \times \frac{1 \text{ mol } MgSO_4}{120.4 \text{ g } MgSO_4} \times \frac{1 \text{ mol } MgSO_4 \cdot xH_2O}{1 \text{ mol } MgSO_4} = 0.02053 \text{ mol } MgSO_4 \cdot xH_2O$$

Thus, $\dfrac{5.061 \text{ g } MgSO_4 \cdot xH_2O}{0.02053} = 246.5 \text{ g/mol} = \text{FW of } MgSO_4 \cdot xH_2O$.

FW of $MgSO_4 \cdot xH_2O$ = FW of $MgSO_4$ + x(FW of H_2O).

$246.5 = 120.4 + x(18.02)$. $x = 6.998$. The hydrate formula is $MgSO_4 \cdot \underline{7}H_2O$.

Alternatively, we could calculate the number of moles of water represented by weight loss: $(5.061 - 2.472) = 2.589 \text{ g } H_2O$ lost.

$$2.589 \text{ g } H_2O \times \frac{1 \text{ mol } H_2O}{18.02 \text{ g } H_2O} = 0.1437 \text{ mol } H_2O; \quad \frac{\text{mol } H_2O}{\text{mol } MgSO_4} = \frac{0.1437}{0.02053} = 7.000$$

Again the correct formula is $MgSO_4 \cdot \underline{7}H_2O$.

Calculations Based on Chemical Equations

3.53   The mole ratios implicit in the coefficients of a balanced chemical equation are essential for solving stoichiometry problems. If the equation is not balanced, the mole ratios will be incorrect and lead to erroneous calculated amounts of reactants and/or products.

3.54   The **integer coefficients** immediately preceeding each molecular formula in a chemical equation give information about relative numbers of moles of reactants and products involved in a reaction.

3.55   **Apply the mole ratio** of moles CO_2/moles C_2H_5OH to calculate moles CO_2 produced. This is the heart of every stoichiometry problem.

$$C_2H_5OH(l) + 3O_2(g) \rightarrow 2CO_2(g) + 3H_2O(l)$$

(a) $3.00 \text{ mol } C_2H_5OH \times \dfrac{2 \text{ mol } CO_2}{1 \text{ mol } C_2H_5OH} = 6.00 \text{ mol } CO_2$

(b) $\text{g } C_2H_5OH \longrightarrow \text{mol } C_2H_5OH \xrightarrow{\text{mole ratio}} \text{mol } CO_2 \longrightarrow \text{g } CO_2$

$$3.00 \text{ g } C_2H_5OH \times \frac{1 \text{ mol } C_2H_5OH}{46.07 \text{ } C_2H_5OH} \times \frac{2 \text{ mol } CO_2}{1 \text{ mol } C_2H_5OH} \times \frac{44.01 \text{ g } CO_2}{1 \text{ mol } CO_2} = 5.73 \text{ g } CO_2$$

3.56 $2C_8H_{18}(l) + 25O_2(g) \rightarrow 16CO_2(g) + 18H_2O(l)$

(a) $2.00 \text{ mol } C_8H_{18} \times \dfrac{25 \text{ mol } O_2}{2 \text{ mol } C_8H_{18}} = 25.0 \text{ mol } O_2$

(b)   The molar mass of C_8H_{18} = 8(12.01) + 18(1.008) = 114.2 g

$2.00 \text{ g } C_8H_{18} \times \dfrac{1 \text{ mol } C_8H_{18}}{114.2 \text{ g } C_8H_{18}} \times \dfrac{25 \text{ mol } O_2}{2 \text{ mol } C_8H_{18}} \times \dfrac{32.00 \text{ g } O_2}{1 \text{ mol } O_2} = 7.01 \text{ g } O_2$

(c) $1.00 \times 10^3 \text{ mL } C_8H_{18} \times \dfrac{0.692 \text{ g } C_8H_{18}}{1 \text{ mL } C_8H_{18}} \times \dfrac{1 \text{ mol } C_8H_{18}}{114.2 \text{ g } C_8H_{18}} \times \dfrac{25 \text{ mol } O_2}{2 \text{ mol } C_8H_{18}}$

$\times \dfrac{32.00 \text{ g } O_2}{1 \text{ mol } O_2} = 2.42 \times 10^3 \text{ g } O_2$

3.57   (a) $CaH_2(s) + 2H_2O(l) \rightarrow Ca(OH)_2(aq) + 2H_2(g)$

(b) $10.0 \text{ g } H_2 \times \dfrac{1 \text{ mol } H_2}{2.016 \text{ g } H_2} \times \dfrac{1 \text{ mol } CaH_2}{2 \text{ mol } H_2} \times \dfrac{42.10 \text{ g } CaH_2}{1 \text{ mol } CaH_2} = 104 \text{ g } CaH_2$

3.58   (a) $Al_2S_3(s) + 6H_2O(l) \rightarrow 2Al(OH)_3(s) + 3H_2S(g)$

(b) $155 \text{ g } Al_2S_3 \times \dfrac{1 \text{ mol } Al_2S_3}{150.2 \text{ g } Al_2S_3} \times \dfrac{2 \text{ mol } Al(OH)_3}{1 \text{ mol } Al_2S_3} \times \dfrac{78.00 \text{ g } Al(OH)_3}{1 \text{ mol } Al(OH)_3} = 161 \text{ g } Al(OH)_3$

3.59 $C_6H_{12}O_6 \rightarrow 2C_2H_5OH(aq) + 2CO_2(g)$

(a) $0.330 \text{ mol } C_6H_{12}O_6 \times \dfrac{2 \text{ mol } CO_2}{1 \text{ mol } C_6H_{12}O_6} = 0.660 \text{ mol } CO_2$

(b) $2.00 \text{ mol } C_2H_5OH \times \dfrac{1 \text{ mol } C_6H_{12}O_6}{2 \text{ mol } C_2H_5OH} \times \dfrac{180.2 \text{ g } C_6H_{12}O_6}{1 \text{ mol } C_6H_{12}O_6} = 180 \text{ g } C_6H_{12}O_6$

(c) $2.00 \text{ g } C_2H_5OH \times \dfrac{1 \text{ mol } C_2H_5OH}{46.07 \text{ g } C_2H_5OH} \times \dfrac{2 \text{ mol } CO_2}{2 \text{ mol } C_2H_5OH} \times \dfrac{44.01 \text{ g } CO_2}{1 \text{ mol } CO_2} = 1.91 \text{ g } CO_2$

3.60 $Na_2SiO_3(s) + 8HF(aq) \rightarrow H_2SiF_6(aq) + 2NaF(aq) + 3H_2O(l)$

(a) $0.50 \text{ mol } Na_2SiO_3 \times \dfrac{8 \text{ mol } HF}{1 \text{ mol } Na_2SiO_3} = 4.0 \text{ mol } HF$

(b) $0.300 \text{ mol } HF \times \dfrac{2 \text{ mol } NaF}{8 \text{ mol } HF} \times \dfrac{41.99 \text{ g } NaF}{1 \text{ mol } NaF} = 3.15 \text{ g } NaF$

(c) $0.300 \text{ g } HF \times \dfrac{1 \text{ mol } HF}{20.01 \text{ g } HF} \times \dfrac{1 \text{ mol } Na_2SiO_3}{8 \text{ mol } HF} \times \dfrac{122.1 \text{ g } Na_2SiO_3}{1 \text{ mol } Na_2SiO_3} = 0.229 \text{ g } Na_2SiO_3$

3.61   (a) $2NaN_3(s) \rightarrow 2Na(s) + 3N_2(g)$

(b) $5.00 \text{ g } N_2 \times \dfrac{1 \text{ mol } N_2}{28.01 \text{ g } N_2} \times \dfrac{2 \text{ mol } NaN_3}{3 \text{ mol } N_2} \times \dfrac{65.01 \text{ g } NaN_3}{1 \text{ mol } NaN_3} = 7.74 \text{ g } NaN_3$

(c)   First determine how many g N_2 are in 10.0 ft^3, using the density of N_2.

$$\frac{1.25 \text{ g}}{1 \text{ L}} \times \frac{1 \text{ L}}{1000 \text{ cm}^3} \times \frac{(2.54)^3 \text{ cm}^3}{1 \text{ in}^3} \times \frac{(12)^3 \text{ in}^3}{1 \text{ ft}^3} \times 10.0 \text{ ft}^3 = 354.0 = 354 \text{ g } N_2$$

$$354.0 \text{ g } N_2 \times \frac{1 \text{ mol } N_2}{28.01 \text{ g } N_2} \times \frac{2 \text{ mol NaN}_3}{3 \text{ mol } N_2} \times \frac{65.01 \text{ g NaN}_3}{1 \text{ mol NaN}_3} = 548 \text{ g NaN}_3$$

3.62   (a) $Al(OH)_3(s) + 3HCl(aq) \rightarrow AlCl_3(aq) + 3H_2O(l)$

       (b) $2.50 \text{ g HCl} \times \dfrac{1 \text{ mol HCl}}{36.46 \text{ g HCl}} \times \dfrac{1 \text{ mol } Al(OH)_3}{3 \text{ mol HCl}} \times \dfrac{78.00 \text{ g } Al(OH)_3}{1 \text{ mol } Al(OH)_3} = 1.78 \text{ g } Al(OH)_3$

Limiting Reactants; Theoretical Yields

3.63   (a)   The *limiting reactant* determines the maximum number of product moles resulting from a chemical reaction; any other reactant is an *excess reactant*.

       (b)   The limiting reactant regulates the amount of products because it is completely used up during the reaction; no more product can be made when one of the reactants is unavailable.

3.64   (a)   *Theoretical yield* is the maximum amount of product possible, as predicted by stoichiometry, assuming that the limiting reactant is converted entirely to product.

              Actual yield is the amount of product actually obtained, less than or equal to the theoretical yield. *Percent yield* is the ratio of (actual yield to theoretical yield) × 100.

       (b)   No reaction is perfect. Not all reactant molecules come together effectively to form products; alternative reaction pathways may produce secondary products and reduce the amount of desired product actually obtained, or it might not be possible to completely isolate the desired product from the reaction mixture. In any case, these factors reduce the actual yield of a reaction.

3.65   (a)   Each bicycle needs 2 wheels, 1 frame and 1 set of handlebars. A total of 5350 wheels corresponds to 2675 pairs of wheels. This is fewer than the number of frames but more than the number of handlebars. The 2655 handlebars determine that 2655 bicycles can be produced.

       (b)   3023 frames - 2655 bicyles = 368 frames left over

              (2675 pairs of wheels - 2655 bicycles) $\times \dfrac{2 \text{ wheels}}{\text{pair}}$ = 40 wheels left over

       (c)   The handlebars are the "limiting reactant" in that they determine the number of bicycles that can be produced.

3.66   (a)   51,575 L beverage $\times \dfrac{1 \text{ bottle}}{0.355 \text{ L}}$ = 1.45 × 10^5 (145,000) portions of beverage

              120,550 bottles; 123,000 caps

              120,550 bottles can be filled and capped.

(b)   123,000 caps - 120,550 bottles = 2,450 caps remain

$$120{,}550 \text{ bottles} \times \frac{0.355 \text{ L}}{1 \text{ bottle}} = 42795 = 4.28 \times 10^4 \text{ L used}$$

51,575 L total - 42,795 L used = 8,780 = 8.8×10^3 L beverage left over

(c)   The empty bottles limit production.

3.67 $SiO_2(s) + 3C(s) \rightarrow SiC(s) + 2CO(g)$

(a)   Follow the approach in Sample Exercise 3.17.

$$3.00 \text{ g } SiO_2 \times \frac{1 \text{ mol } SiO_2}{60.09 \text{ g } SiO_2} \times \frac{1 \text{ mol SiC}}{1 \text{ mol } SiO_2} \times \frac{40.10 \text{ g SiC}}{1 \text{ mol SiC}} = 2.00 \text{ g SiC}$$

$$4.50 \text{ g C} \times \frac{1 \text{ mol C}}{12.01 \text{ g C}} \times \frac{1 \text{ mol SiC}}{3 \text{ mol C}} \times \frac{40.10 \text{ g SiC}}{1 \text{ mol SiC}} = 5.01 \text{ g SiC}$$

The lesser amount, 2.00 g SiC, can be produced.

(b) SiO_2 is the limiting reactant (it leads to the smaller amount of product) and C is the excess reactant.

(c) $$3.00 \text{ g } SiO_2 \times \frac{1 \text{ mol } SiO_2}{60.09 \text{ g } SiO_2} \times \frac{3 \text{ mol C}}{1 \text{ mol } SiO_2} \times \frac{12.01 \text{ g C}}{1 \text{ mol C}} = 1.80 \text{ g C}$$

4.50 g C initially present - 1.80 g C consumed = 2.70 g C remain

3.68 $4NH_3(g) + 5O_2(g) \rightarrow 4NO(g) + 6H_2O(g)$

(a)   Follow the approach in Sample Exercise 3.17.

$$1.50 \text{ g } NH_3 \times \frac{1 \text{ mol } NH_3}{17.03 \text{ g } NH_3} \times \frac{4 \text{ mol NO}}{4 \text{ mol } NH_3} \times \frac{30.01 \text{ g NO}}{1 \text{ mol NO}} = 2.64 \text{ g NO}$$

$$1.85 \text{ g } O_2 \times \frac{1 \text{ mol } O_2}{32.00 \text{ g } O_2} \times \frac{4 \text{ mol NO}}{5 \text{ mol } O_2} \times \frac{30.01 \text{ g NO}}{1 \text{ mol NO}} = 1.39 \text{ g NO}$$

The lesser amount, 1.39 g NO, is produced.

(b) O_2 is the limiting reactant and NH_3 is present in excess.

(c) $$1.85 \text{ g } O_2 \times \frac{1 \text{ mol } O_2}{32.0 \text{ g } O_2} \times \frac{4 \text{ mol } NH_3}{5 \text{ mol } O_2} \times \frac{17.03 \text{ g } NH_3}{1 \text{ mol } NH_3} = 0.788 \text{ g } NH_3 \text{ consumed}$$

1.50 g NH_3 initial - 0.788 g NH_3 consumed = 0.71 g NH_3 remain.

3.69 $H_2S(g) + 2NaOH(aq) \rightarrow Na_2S(s) + 2H_2O(l)$

$$1.50 \text{ g } H_2S \times \frac{1 \text{ mol } H_2S}{34.09 \text{ g } H_2S} \times \frac{1 \text{ mol } Na_2S}{1 \text{ mol } H_2S} \times \frac{78.05 \text{ g } Na_2S}{1 \text{ mol } Na_2S} = 3.43 \text{ g } Na_2S$$

$$1.65 \text{ g NaOH} \times \frac{1 \text{ mol NaOH}}{40.00 \text{ g NaOH}} \times \frac{1 \text{ mol } Na_2S}{2 \text{ mol NaOH}} \times \frac{78.05 \text{ g } Na_2S}{1 \text{ mol } Na_2S} = 1.61 \text{ g } Na_2S$$

The lesser amount, 1.61 g Na_2S is formed.

3.70 $C_2H_4(g) + 3O_2(g) \rightarrow 2CO_2(g) + 2H_2O(g)$

$$1.93 \text{ g } C_2H_4 \times \frac{1 \text{ mol } C_2H_4}{28.05 \text{ g } C_2H_4} \times \frac{2 \text{ mol } CO_2}{1 \text{ mol } C_2H_4} \times \frac{44.01 \text{ g } CO_2}{1 \text{ mol } CO_2} = 6.06 \text{ g } CO_2$$

$$3.75 \text{ g } O_2 \times \frac{1 \text{ mol } O_2}{32.00 \text{ g } O_2} \times \frac{2 \text{ mol } CO_2}{3 \text{ mol } O_2} \times \frac{44.01 \text{ g } CO_2}{1 \text{ mol } CO_2} = 3.44 \text{ g } CO_2$$

The lesser amount, 3.44 g CO_2 is formed.

3.71 Strategy: Write balanced equation; determine limiting reactant; calculate amounts of excess reactant remaining and products, based on limiting reactant.

$$2AgNO_3(aq) + Na_2CO_3(aq) \rightarrow Ag_2CO_3(s) + 2NaNO_3(aq)$$

Follow the method in Sample Exercise 3.16.

$$5.00 \text{ g } AgNO_3 \times \frac{1 \text{ mol } AgNO_3}{169.9 \text{ g } AgNO_3} \times \frac{1 \text{ mol } Na_2CO_3}{2 \text{ mol } AgNO_3} \times \frac{106.0 \text{ g } Na_2CO_3}{1 \text{ mol } Na_2CO_3} = 1.56 \text{ g } Na_2CO_3$$

5.00 g $AgNO_3$ will react completely with 1.56 g Na_2CO_3.

$AgNO_3$ is the limiting reactant and Na_2CO_3 is present in excess.

mass Na_2CO_3 remaining = 5.00 g initial - 1.56 g reacted = 3.44 g

$$5.00 \text{ g } AgNO_3 \ \frac{1 \text{ mol } AgNO_3}{169.9 \text{ g } AgNO_3} \times \frac{1 \text{ mol } Ag_2CO_3}{2 \text{ mol } AgNO_3} \times \frac{275.8 \text{ g } Ag_2CO_3}{1 \text{ mol } Ag_2NO_3} = 4.06 \text{ g } Ag_2CO_3$$

$$5.00 \text{ g } AgNO_3 \times \frac{1 \text{ mol } AgNO_3}{169.9 \text{ g } AgNO_3} \times \frac{2 \text{ mol } NaNO_3}{2 \text{ mol } AgNO_3} \times \frac{85.00 \text{ g } NaNO_3}{1 \text{ mol } NaNO_3} = 2.50 \text{ g } NaNO_3$$

After evaporation there will be no $AgNO_3$ remaining (limiting reactant), 3.44 g Na_2CO_3 (excess reactant), 4.06 g Ag_2CO_3 and 2.50 g $NaNO_3$. The sum of the amounts is 10.00 g; mass is conserved.

3.72 Strategy: See Exercise 3.71.

$$H_2SO_4(aq) + Pb(C_2H_3O_2)_2(aq) \rightarrow PbSO_4(s) + 2HC_2H_3O_2(aq)$$

Following the method in Sample Exercise 3.16,

$$10.0 \text{ g } H_2SO_4 \times \frac{1 \text{ mol } H_2SO_4}{98.09 \text{ g } H_2SO_4} \times \frac{1 \text{ mol } Pb(C_2H_3O_2)_2}{1 \text{ mol } H_2SO_4} \times \frac{325.3 \text{ g } Pb(C_2H_3O_2)_2}{1 \text{ mol } Pb(C_2H_3O_2)_2}$$
$$= 33.2 \text{ g } Pb(C_2H_3O_2)_2$$

10.0 g H_2SO_4 could consume 33.2 g $Pb(C_2H_3O_2)_2$, but we have only 10.0 g $Pb(C_2H_3O_2)_2$, so it is the limiting reactant.

$$10.0 \text{ g Pb(C}_2\text{H}_3\text{O}_2)_2 \times \frac{1 \text{ mol Pb(C}_2\text{H}_3\text{O}_2)_2}{325.3 \text{ g Pb(C}_2\text{H}_3\text{O}_2)_2} \times \frac{1 \text{ mol H}_2\text{SO}_4}{1 \text{ mol Pb(C}_2\text{H}_3\text{O}_2)_2} \times \frac{98.09 \text{ g H}_2\text{SO}_4}{1 \text{ mol H}_2\text{SO}_4}$$

$$= 3.02 \text{ g H}_2\text{SO}_4$$

mass H_2SO_4 remaining = 10.0 g initial - 3.02 g reacted = 6.98 g = 7.0 g

$$10.0 \text{ g Pb(C}_2\text{H}_3\text{O}_2)_2 \times \frac{1 \text{ mol Pb(C}_2\text{H}_3\text{O}_2)_2}{325.3 \text{ g Pb(C}_2\text{H}_3\text{O}_2)_2} \times \frac{1 \text{ mol PbSO}_4}{1 \text{ mol Pb(C}_2\text{H}_3\text{O}_2)_2} \times \frac{303.3 \text{ g PbSO}_4}{1 \text{ mol PbSO}_4}$$

$$= 9.32 \text{ g PbSO}_4$$

$$10.0 \text{g Pb(C}_2\text{H}_3\text{O}_2)_2 \times \frac{1 \text{ mol Pb(C}_2\text{H}_3\text{O}_2)_2}{325.3 \text{ g Pb(C}_2\text{H}_3\text{O}_2)_2} \times \frac{2 \text{ mol HC}_2\text{H}_3\text{O}_2}{1 \text{ mol Pb(C}_2\text{H}_3\text{O}_2)_2} \times \frac{60.05 \text{ g HC}_2\text{H}_3\text{O}_2}{1 \text{ mol HC}_2\text{H}_3\text{O}_2}$$

$$= 3.69 \text{ g HC}_2\text{H}_3\text{O}_2$$

After evaporation there will be no $Pb(C_2H_3O_2)_2$ remaining (limiting reactant), 7.0 g H_2SO_4 (excess reactant), 9.32 g $PbSO_4$ and 3.69 g $HC_2H_3O_2$. The sum of these amounts is 20.0 g; mass is conserved.

3.73 $\quad$ Strategy: Calculate theoretical (stoichiometric) yield and then the % yield.

$$2.00 \text{ g KClO}_3 \times \frac{1 \text{ mol KClO}_3}{122.6 \text{ g KClO}_3} \times \frac{3 \text{ mol O}_2}{2 \text{ mol KClO}_3} \times \frac{32.00 \text{ g O}_2}{1 \text{ mol O}_2} = 0.7830 = 0.783 \text{ g O}_2$$

$$\% \text{ yield} = \frac{\text{actual yield}}{\text{theoretical yield}} \times 100 = \frac{0.720 \text{ g O}_2}{0.783 \text{ g O}_2} \times 100 = 92.0\%$$

Two reasons might be uneven heating of the $KClO_3$ and thus incomplete reaction, or loss of O_2 before weighing.

3.74 $\quad$ (a) $\quad$ Since we want amount of product, use the method in Sample Exercise 3.17.

$$30.0 \text{ g C}_6\text{H}_6 \times \frac{1 \text{ mol C}_6\text{H}_6}{78.11 \text{ g C}_6\text{H}_6} \times \frac{1 \text{ mol C}_6\text{H}_5\text{Br}}{1 \text{ mol C}_6\text{H}_6} \times \frac{157.0 \text{ g C}_6\text{H}_5\text{Br}}{1 \text{ mol C}_6\text{H}_5\text{Br}} = 60.30$$

$$= 60.3 \text{ g C}_6\text{H}_5\text{Br}$$

$$65.0 \text{ g Br}_2 \times \frac{1 \text{ mol Br}_2}{159.8 \text{ g Br}_2} \times \frac{1 \text{ mol C}_6\text{H}_5\text{Br}}{1 \text{ mol Br}_2} \times \frac{157.0 \text{ g C}_6\text{H}_5\text{Br}}{1 \text{ mol C}_6\text{H}_5\text{Br}} = 63.9 \text{ g C}_6\text{H}_5\text{Br}$$

C_6H_6 is the limiting reactant and 60.3 g C_6H_5Br is the theoretical yield.

(b) $\quad$ % yield = $\dfrac{56.7 \text{ g C}_6\text{H}_5\text{Br actual}}{60.3 \text{ g C}_6\text{H}_5\text{Br theoretical}} \times 100 = 94.0\%$

Additional Exercises

3.75 $\quad$ (a) $\qquad$ $C_4H_8O_2(l) + 5O_2(g) \rightarrow 4CO_2(g) + 4H_2O(l)$

(b) $\qquad$ $Cu(OH)_2(s) \rightarrow CuO(s) + H_2O(g)$

(c) $\qquad$ $Zn(s) + 2HCl(aq) \rightarrow ZnCl_2(aq) + H_2(g)$

3.76   In the current system, the mass of ^{9}Be = 9.01218 amu (its atomic weight, since ^{9}Be is the only naturally occuring isotope).

If we call the newly defined mass unit an nmu, the amu/nmu ratio is:

$$\frac{9.01218 \text{ amu}}{9 \text{ nmu}} = \frac{1.00135 \text{ amu}}{1 \text{ nmu}} \text{ or 1 nmu = 1.00135 amu.}$$

The nmu is a larger unit of mass than the amu.

3.77   (a)   The 68.926 amu isotope has a mass number of 69, with 31 protons, 38 neutrons and the symbol $^{69}_{31}$Ga. The 70.926 amu isotope has a mass number of 71, 31 protons, 40 neutrons, and symbol $^{71}_{31}$Ga. (All Ga atoms have 31 protons.)

      (b)   The average mass of a Ga atom (given on the inside cover of the text) is 69.72 amu. Let x = abundance of the lighter isotope, 1-x = abundance of the heavier isotope. Then x(68.926) + (1-x)(70.926) = 69.72; x = 0.603, ^{69}Ga = 60.3%, ^{71}Ga = 39.7%

3.78   (a)   4 peaks; there are 4 possible combinations of isotopes.

      (b) ^{1}H^{35}Cl   1.0078 amu + 34.969 amu = 35.977 amu

                ^{2}H^{35}Cl   2.0140 amu + 34.969 amu = 36.983 amu

                ^{1}H^{37}Cl   1.0078 amu + 36.966 amu = 37.974 amu

                ^{2}H^{37}Cl   2.0140 amu + 36.966 amu = 38.980 amu

      (c)   The intensities will be the products of the natural abundances of the two isotopes.

                ^{1}H^{35}Cl   (99.985)(0.7553) = 75.52

                ^{2}H^{35}Cl   (0.015)(0.7553) = 0.011

                ^{1}H^{37}Cl   (99.985)(0.2447) = 24.47

                ^{2}H^{37}Cl   (0.015)(0.2447) = 0.0037

                Dividing through by 0.0037, the relative intensities in order of increasing mass are 20,000 : 3.0 : 6600 : 1. The two ^{2}HCl peaks are very small relative to the other peaks.

      (d)   Assuming the very small peaks due to ^{2}HX can be observed, element **X** has 4 naturally occurring isotopes.

3.79   Strategy: mass N → mol N → mol ratio → mass compound

$$\text{NH}_3: 1.00 \times 10^3 \text{ g N} \times \frac{1 \text{ mol N}}{14.01 \text{ g N}} \times \frac{1 \text{ mol NH}_3}{1 \text{ mol N}} \times \frac{17.03 \text{ g NH}_3}{1 \text{ mol NH}_3} \times \frac{1 \text{ kg}}{1000 \text{ g}} = 1.22 \text{ kg NH}_3$$

$$\text{NH}_4\text{NO}_3: 1.00 \times 10^3 \text{ g N} \times \frac{1 \text{ mol N}}{14.01 \text{ g N}} \times \frac{1 \text{ mol NH}_4\text{NO}_3}{2 \text{ mol N}} \times \frac{80.05 \text{ g NH}_4\text{NO}_3}{1 \text{ mol NH}_4\text{NO}_3} \times \frac{1 \text{ kg}}{1000 \text{ g}}$$

$$= 2.86 \text{ kg NH}_4\text{NO}_3$$

$(NH_4)_2SO_4:$ 1.00×10^3 g N $\times \dfrac{1 \text{ mol N}}{14.01 \text{ g N}} \times \dfrac{1 \text{ mol} (NH_4)_2SO_4}{2 \text{ mol N}} \times \dfrac{132.2 \text{ g} (NH_4)_2SO_4}{1 \text{ mol} (NH_4)_2SO_4}$

$\times \dfrac{1 \text{ kg}}{1000 \text{ g}} = 4.72 \text{ kg} (NH_4)_2SO_4$

$(NH_2)_2CO:$ 1.00×10^3 g N $\times \dfrac{1 \text{ mol N}}{14.01 \text{ g N}} \times \dfrac{1 \text{ mol} (NH_2)_2CO}{2 \text{ mol N}} \times \dfrac{60.06 \text{ g} (NH_2)_2CO}{1 \text{ mol} (NH_2)_2CO} \times \dfrac{1 \text{ kg}}{1000 \text{ g}}$

$= 2.14 \text{ kg} (NH_2)_2CO$

3.80   (a)   1.00 mg $\times \dfrac{1 \text{ g}}{1000 \text{ mg}} \times \dfrac{1 \text{ mol } C_{12}H_4Cl_4O_2}{322.0 \text{ g } C_{12}H_4Cl_4O_2} \times \dfrac{6.022 \times 10^{23} \text{ molecules}}{1 \text{ mol}}$

$= 1.87 \times 10^{18} \; C_{12}H_4Cl_4O_2 \text{ molecules}$

(b) $10.0 \text{ cm}^3 \times \dfrac{3.97 \text{ g } Al_2O_3}{1 \text{ cm}^3} \times \dfrac{1 \text{ mol } Al_2O_3}{102.0 \text{ g } Al_2O_3} \times \dfrac{2 \text{ mol Al}}{1 \text{ mol } Al_2O_3} \times \dfrac{6.022 \times 10^{23} \text{ Al atoms}}{1 \text{ mol}}$

$= 4.69 \times 10^{23} \text{ Al atoms}$

3.81   (a) $\dfrac{5.342 \times 10^{-21} \text{ g}}{1 \text{ molecule penicillin G}} \times \dfrac{6.0221 \times 10^{23} \text{ molecules}}{1 \text{ mol}} = 3217 \text{ g/mol penicillin G}$

(b)   1.00 g hemoglobin (hem) contains 3.40×10^{-3} g Fe.

$\dfrac{1.00 \text{ g hem}}{3.40 \times 10^{-3} \text{ g Fe}} \times \dfrac{55.85 \text{ g Fe}}{1 \text{ mol Fe}} \times \dfrac{4 \text{ mol Fe}}{1 \text{ mol hem}} = 6.57 \times 10^4 \text{ g/mol hemoglobin}$

3.82   0.50 mol $H_2O \times \dfrac{6.022 \times 10^{23} \text{ molecules}}{1 \text{ mol}} \times \dfrac{1 \text{ O atom}}{1 \; H_2O \text{ molecule}} = 3.0 \times 10^{23} \text{ O atoms}$

15.0 g $NO_2 \times \dfrac{1 \text{ mol } NO_2}{46.01 \text{ g } NO_2} \times \dfrac{6.022 \times 10^{23} \text{ molecules}}{1 \text{ mol}} \times \dfrac{2 \text{ O atoms}}{1 \; NO_2 \text{ molecule}}$

$= 3.9 \times 10^{23} \text{ O atoms}$

$3.0 \times 10^{23} \; SO_3 \text{ molecules} \times \dfrac{3 \text{ O atoms}}{1 \; SO_3 \text{ molecule}} = 9.0 \times 10^{23} \text{ O atoms}$

$3.0 \times 10^{23} \; SO_3$ molecules contains more O atoms

3.83   (a)   0.068 g $C_5H_5N \times \dfrac{1 \text{ mol } C_5H_5N}{79.1 \text{ g } C_5H_5N} \times \dfrac{6.022 \times 10^{23} \text{ molecules}}{1 \text{ mol}} = 5.18 \times 10^{20}$

$= 5.2 \times 10^{20} \; C_5H_5N \text{ molecules}$

(b)   5.0 g ZnO $\times \dfrac{1 \text{ mol Zn}}{81.4 \text{ g ZnO}} \times \dfrac{6.022 \times 10^{23} \text{ molecules}}{1 \text{ mol}} = 3.70 \times 10^{22}$

$= 3.7 \times 10^{22} \text{ ZnO formula units}$

There is one C_5H_5N molecule for each $3.70 \times 10^{22}/5.18 \times 10^{20} = 71$ ZnO units

(c) $5.0 \text{ g ZnO} \times \dfrac{48 \text{ m}^2}{1 \text{ g ZnO}} \times \dfrac{1}{5.18 \times 10^{20} \text{ C}_5\text{H}_5\text{N molecules}} = 4.63 \times 10^{-19}$

$$= 4.6 \times 10^{-19} \text{ m}^2/\text{molecule}$$

$$\dfrac{4.63 \times 10^{-19} \text{ m}^2}{\text{C}_5\text{H}_5\text{N molecule}} \times \dfrac{(1 \times 10^{10} \text{ Å})^2}{1 \text{ m}^2} = 46 \text{ Å}^2 / \text{C}_5\text{H}_5\text{N molecule}$$

3.84   (a)   Let AW = the atomic weight of X.

According to the chemical reaction, moles XI_3 reacted = moles XCl_3 produced

$0.5000 \text{ g } XI_3 \times 1 \text{ mol } XI_3 / (AW + 389.70) \text{ g } XI_3$

$$= 0.2360 \text{ g } XCl_3 \times \dfrac{1 \text{ mol } XCl_3}{(AW + 106.35) \text{ g } XCl_3}$$

$0.5000 (AW + 106.35) = 0.2360 (AW + 380.70)$
$0.5000 \text{ AW} + 53.175 = 0.2360 \text{ AW} + 89.845$
$0.2640 \text{ AW} = 36.67; \quad AW = 138.9 \text{ g}$

(b)   X is lanthanum, La, atomic number 57.

3.85   (a) $0.7787 \text{ g C} \times \dfrac{1 \text{ mol C}}{12.01 \text{ g C}} = 0.06484 \text{ mol C}$

$0.1176 \text{ g H} \times \dfrac{1 \text{ mol H}}{1.008 \text{ g H}} = 0.1167 \text{ mol H}$

$0.1037 \text{ g O} \times \dfrac{1 \text{ mol C}}{16.00 \text{ g O}} = 0.006481 \text{ mol O}$

Dividing through by the smallest of these values we obtain $C_{10}H_{18}O$.

(b)   The formula weight of $C_{10}H_{18}O$ is 154. Thus, the empirical formula is also the molecular formula.

3.86   Since all the C in the vanillin must be present in the CO_2 produced, get g C from g CO_2.

$2.43 \text{ g CO}_2 \times \dfrac{1 \text{ mol CO}_2}{44.01 \text{ g CO}_2} \times \dfrac{12.01 \text{ g C}}{1 \text{ mol C}} = 0.6631 = 0.663 \text{ g C}$

Since all the H in vanillin must be present in the H_2O produced, get g H from g H_2O.

$0.50 \text{ g H}_2\text{O} \times \dfrac{1 \text{ mol H}_2\text{O}}{18.02 \text{ g H}_2\text{O}} \times \dfrac{2 \text{ mol H}}{1 \text{ mol H}_2\text{O}} \times \dfrac{1.008 \text{ g H}}{1 \text{ mol H}} = 0.0559 = 0.056 \text{ g H}$

Get g O by subtraction. (Since the analysis was performed by combustion, an unspecified amount of O_2 was a reactant, and thus not all the O in the CO_2 and H_2O produced came from vanillin.)

1.05 g vanillin - 0.663 g C - 0.056 g H = 0.331 g O

$0.6631 \text{ g C} \times \dfrac{1 \text{ mol C}}{12.01 \text{ g C}} = 0.0552 \text{ mol C}; \quad 0.0552 / 0.0207 = 2.67$

$$0.0559 \text{ g H} \times \frac{1 \text{ mol H}}{1.008 \text{ g H}} = 0.0555 \text{ mol C}; \quad 0.0556 / 0.0207 = 2.68$$

$$0.331 \text{ g O} \times \frac{1 \text{ mol O}}{16.00 \text{ g O}} = 0.0207 \text{ mol O}; \quad 0.0207 / 0.0207 = 1.00$$

Multiplying the numbers above by **3** to obtain an integer ratio of moles, the empirical formula of vanillin is $C_8H_8O_3$.

3.87   The mass percentage is determined by the relative number of atoms of the element times the atomic weight, divided by the total formula mass. Thus, the mass percent of bromine in $KBrO_x$ is given by $0.5292 = \dfrac{79.91}{39.10 + 79.91 + x(16.00)}$. Solving for x, we obtain x = 2.00. Thus, the formula is $KBrO_2$.

3.88   Let x = mol Sn, y = mol Zn, g Sn = 118.71x, g Zn = 65.39y;   118.71x + 65.39y = 1.540 g

mol SnF_4 = mol Sn = x,   mol ZnF_2 = mol Zn = y

g SnF_4 = 194.70x, g ZnF_2 = 103.39 y;   194.70x + 103.39y = 2.489 g

103.39 (118.71x + 65.39y) = 103.39 (1.540)
<u>- 65.39 (194.70x + 103.39y) = 65.39 (2.489)</u>

  12273.4 x = 159.22
<u>-12731.4 x = -162.76</u>
   -458 x   =   -3.54;   x = 0.00773 = 7.7×10^{-3} mol Sn

(159.2 - 162.8 = 3.6 leads to 2 sig figs in this result and the final % Sn)

$$7.73 \times 10^{-3} \text{ mol Sn} \times \frac{118.71 \text{ g Sn}}{\text{mol Sn}} \times \frac{1}{1.540 \text{ g alloy}} \times 100 = 59.6 = 60\% \text{ Sn, } 40\% \text{ Zn}$$

3.89   Strategy: Because different sample sizes were used to analyze the different elements, calculate mass % of each element in the sample.

    i.   Calculate mass % C from g CO_2.
    ii.   Calculate mass % Cl from AgCl.
    iii.   Get mass % H by subtraction.
    iv.   Calculate mole ratios and the empirical formulas.

    i. $3.52 \text{ g } CO_2 \times \dfrac{1 \text{ mol } CO_2}{44.01 \text{ g } CO_2} \times \dfrac{1 \text{ mol C}}{1 \text{ mol } CO_2} \times \dfrac{12.01 \text{ g C}}{1 \text{ mol C}} = 0.9606 = 0.961 \text{ g C}$

            $\dfrac{0.9606 \text{ g C}}{1.50 \text{ g sample}} \times 100 = 64.04 = 64.0\% \text{ C}$

    ii. $1.27 \text{ g AgCl} \times \dfrac{1 \text{ mol AgCl}}{143.3 \text{ g AgCl}} \times \dfrac{1 \text{ mol Cl}}{1 \text{ mol AgCl}} \times \dfrac{35.45 \text{ g Cl}}{1 \text{ mol Cl}} = 0.3142 = 0.314 \text{ g Cl}$

            $\dfrac{0.3142 \text{ g Cl}}{1.00 \text{ g sample}} \times 100 = 31.42 = 31.4\% \text{ Cl}$

    iii.   % H = 100.0 - (64.04% C + 31.42% Cl) = 4.54 = 4.5% H

iv. Assume 100 g sample.

$$64.04 \text{ g C} \times \frac{1 \text{ mol C}}{12.01 \text{ g C}} = 5.33 \text{ mol C}; \quad 5.33 / 0.886 = 6.02$$

$$31.42 \text{ g Cl} \times \frac{1 \text{ mol Cl}}{35.45 \text{ g Cl}} = 0.886 \text{ mol Cl}; \quad 0.886 / 0.886 = 1.00$$

$$4.54 \text{ g H} \times \frac{1 \text{ mol H}}{1.008 \text{ g H}} = 4.50 \text{ mol H}; \quad 4.50 / 0.886 = 5.08$$

The empirical formula is probably C_6H_5Cl.

The subscript for H, 5.08, is relatively far from 5.00, but C_6H_5Cl makes chemical sense. More significant figures in the mass data are required for a more accurate mole ratio.

3.90 $1.00 \text{ g NaHCO}_3 \times \dfrac{1 \text{ mol NaHCO}_3}{84.01 \text{ g/mol}} = 1.190 \times 10^{-2} = 1.19 \times 10^{-2} \text{ mol NaHCO}_3$

Abbreviate citric acid as H_3Cit

$1.190 \times 10^{-2} \text{ mol NaHCO}_3 \times \dfrac{1 \text{ mol NaHCO}_3}{3 \text{ mol NaHCO}_3} \times \dfrac{192.1 \text{ g H}_3\text{Cit}}{1 \text{ mol H}_3\text{Cit}} = 0.762 \text{ g citric acid}$

3.91 $2NaCl(aq) + 2H_2O \rightarrow 2NaOH(aq) + H_2(g) + Cl_2(g)$

Calculate mol Cl_2 and relate to mol H_2, mol NaOH.

$1.4 \times 10^6 \text{ kg} \times \dfrac{1000 \text{ g}}{1 \text{ kg}} \times \dfrac{1 \text{ mol Cl}_2}{70.91 \text{ g Cl}_2} = 1.974 \times 10^7 = 2.0 \times 10^7 \text{ mol Cl}_2$

$1.974 \times 10^7 \text{ mol Cl}_2 \times \dfrac{1 \text{ mol H}_2}{1 \text{ mol Cl}_2} \times \dfrac{2.016 \text{ g H}_2}{1 \text{ mol H}_2} = 3.98 \times 10^7 \text{ g H}_2 = 4.0 \times 10^4 \text{ kg H}_2$

$4.0 \times 10^7 \text{ g} \times \dfrac{1 \text{ metric ton}}{1 \times 10^6 \text{ g (1 Mg)}} = 40 \text{ metric tons H}_2$

$1.974 \times 10^7 \text{ mol Cl}_2 \times \dfrac{2 \text{ mol NaOH}}{1 \text{ mol Cl}_2} \times \dfrac{40.0 \text{ g NaOH}}{1 \text{ mol NaOH}} = 1.58 \times 10^9 = 1.6 \times 10^9 \text{ g NaOH}$

$1.6 \times 10^9 \text{ g NaOH} = 1.6 \times 10^6 \text{ kg NaOH} = 1.6 \times 10^3 \text{ metric tons NaOH}$

3.92 $2C_{57}H_{110}O_6 + 163O_2 \rightarrow 114CO_2 + 110H_2O$

molar mass of fat $= 57(12.01) + 110(1.008) + 6(16.00) = 891.5$

$1.0 \text{ kg fat} \times \dfrac{1000 \text{ g}}{1 \text{ kg}} \times \dfrac{1 \text{ mol fat}}{891.5 \text{ g fat}} \times \dfrac{110 \text{ mol H}_2\text{O}}{2 \text{ mol fat}} \times \dfrac{18.02 \text{ g H}_2\text{O}}{1 \text{ mol H}_2\text{O}} \times \dfrac{1 \text{ kg}}{1000 \text{ g}} = 1.1 \text{ kg H}_2\text{O}$

3.93 (a) Strategy: Calculate the total mass of C from g CO and g CO_2. Calculate the mass of H from g H_2O. Calculate mole ratios and the empirical formula.

$0.467 \text{ g CO} \times \dfrac{1 \text{ mol CO}}{28.01 \text{ g CO}} \times \dfrac{1 \text{ mol C}}{1 \text{ mol CO}} \times 12.01 \text{ g C} = 0.200 \text{ g C}$

$$0.733 \text{ g CO}_2 \times \frac{1 \text{ mol CO}_2}{44.01 \text{ g CO}_2} \times \frac{1 \text{ mol C}}{1 \text{ mol CO}_2} \times 12.01 \text{ g C} = 0.200 \text{ g C}$$

Total mass C is 0.200 g + 0.200 g = 0.400 g C.

$$0.450 \text{ g H}_2\text{O} \times \frac{1 \text{ mol H}_2\text{O}}{18.02 \text{ g H}_2\text{O}} \times \frac{2 \text{ mol H}}{1 \text{ mol H}_2\text{O}} \times \frac{1.008 \text{ g H}}{1 \text{ mol H}} = 0.0503 \text{ g H}$$

(Since hydrocarbons contain only the elements C and H, g H can also be obtained by subtraction: 0.450 g sample - 0.400 g C = 0.050 g H.)

$$0.400 \text{ g C} \times \frac{1 \text{ mol C}}{12.01 \text{ g C}} = 0.0333 \text{ mol C}; \ 0.0333 / 0.0333 = 1.0$$

$$0.0503 \text{ g H} \times \frac{1 \text{ mol H}}{1.008 \text{ g H}} = 0.0499 \text{ mol H}; \ 0.0499 / 0.0333 = 1.5$$

Multiplying by a factor of 2, the emprical formula is C_2H_3.

(b)   Mass is conserved. Total mass products - mass sample = mass O_2 consumed.
0.467 g CO + 0.733 g CO_2 + 0.450 g H_2O - 0.450 g sample = 1.200 g O_2 consumed

(c)   For complete combustion, 0.467 g CO must be converted to CO_2.
$2CO(g) + O_2(g) \rightarrow 2CO_2(g)$

$$0.467 \text{ g CO} \times \frac{1 \text{ mol CO}}{28.01 \text{ g C}} \times \frac{1 \text{ mol O}_2}{2 \text{ mol CO}} \times \frac{32.00 \text{ g O}_2}{1 \text{ mol O}_2} = 0.267 \text{ g O}_2$$

The total mass of O_2 required for complete combustion is
1.200 g + 0.267 g = 1.467 g O_2.

3.94 $N_2(g) + 3H_2(g) \rightarrow 2NH_3(g)$

Determine the moles of N_2 and H_2 required to form the 2.0 moles of NH_3 present after the reaction has stopped.

$$2.0 \text{ mol NH}_3 \times \frac{3 \text{ mol H}_2}{2 \text{ mol NH}_3} = 3.0 \text{ mol H}_2 \text{ reacted}$$

$$2.0 \text{ mol NH}_3 \times \frac{1 \text{ mol N}_2}{2 \text{ mol NH}_3} = 1 \text{ mol N}_2 \text{ reacted}$$

mol H_2 initial = 2.0 mol H_2 remain + 3.0 mol H_2 reacted = 5.0 mol H_2
mol N_2 initial = 2.0 mol N_2 remain + 1.0 mol N_2 reacted = 3.0 mol N_2

In tabular form:	$N_2(g)$ +	$3H_2(g)$ →	$2NH_3(g)$
initial	3.0 mol	5.0 mol	0 mol
reaction	-1.0 mol	-3.0 mol	+2.0 mol
final	2.0 mol	2.0 mol	2.0 mol

(Tables like this will be extremely useful for solving chemical equilibrium problems in Chapter 15.)

3.95   All of the O_2 is produced from $KClO_3$; get g $KClO_3$ from g O_2. All of the H_2O is produced from $KHCO_3$; get g $KHCO_3$ from g H_2O. The g H_2O produced also reveals the g CO_2 from the decomposition of $NaHCO_3$. The remaining CO_2 (13.2 g CO_2 - g CO_2 from $NaHCO_3$) is due to K_2CO_3 and g K_2CO_3 can be derived from it.

$$4.00 \text{ g } O_2 \times \frac{1 \text{ mol } O_2}{32.00 \text{ g } O_2} \times \frac{2 \text{ mol } KClO_3}{3 \text{ mol } O_2} \times \frac{122.6 \text{ g } KClO_3}{1 \text{ mol } KClO_3} = 10.22 = 10.2 \text{ g } KClO_3$$

$$1.80 \; H_2O \times \frac{1 \text{ mol } H_2O}{18.02 \text{ g } H_2O} \times \frac{2 \text{ mol } KHCO_3}{1 \text{ mol } H_2O} \times \frac{100.1 \text{ g } KHCO_3}{1 \text{ mol } KHCO_3} = 20.00 = 20.0 \text{ g } KHCO_3$$

$$1.80 \text{ g } H_2O \times \frac{1 \text{ mol } H_2O}{18.02 \text{ g } H_2O} \times \frac{2 \text{ mol } CO_2}{1 \text{ mol } H_2O} \times \frac{44.01 \text{ g } CO_2}{1 \text{ mol } CO_2} = 8.792 = 8.79 \text{ g } CO_2 \text{ from } KHCO_3$$

13.20 g CO_2 total - 8.792 CO_2 from $KHCO_3$ = 4.408 = 4.41 g CO_2 from K_2CO_3

$$4.408 \text{ g } CO_2 \times \frac{1 \text{ mol } CO_2}{44.01 \text{ g } CO_2} \times \frac{1 \text{ mol } K_2CO_3}{1 \text{ mol } CO_2} \times \frac{138.2 \text{ g } K_2CO_3}{1 \text{ mol } K_2CO_3} = 13.84 = 13.8 \text{ g } K_2CO_3$$

100.0 g mixture - 10.22 g $KClO_3$ - 20.00 g $KHCO_3$ - 13.84 g K_2CO_3 = 56.0 g KCl

3.96   (a) $2C_2H_2(g) + 5O_2(g) \rightarrow 4CO_2(g) + 2H_2O(g)$

       (b)   Following the approach in Sample Exercise 3.16,

$$10.0 \text{ g } C_2H_2 \times \frac{1 \text{ mol } C_2H_2}{26.04 \text{ g } C_2H_2} \times \frac{5 \text{ mol } O_2}{2 \text{ mol } C_2H_2} \times \frac{32.00 \text{ g } O_2}{1 \text{ mol } O_2} = 30.7 \text{ g } O_2 \text{ required}$$

             Only 10.0 g O_2 are available, so O_2 limits.

       (c)   Since O_2 limits, 0.0 g O_2 remain.

             Next, calculate the g C_2H_2 consumed and the amounts of CO_2 and H_2O produced by reaction of 10.0 g O_2.

$$10.0 \text{ g } O_2 \times \frac{1 \text{ mol } O_2}{32.00 \text{ g } O_2} \times \frac{2 \text{ mol } C_2H_2}{5 \text{ mol } O_2} \times \frac{26.04 \text{ g } C_2H_2}{1 \text{ mol } C_2H_2} = 3.26 \text{ g } C_2H_2 \text{ consumed}$$

             10.0 g C_2H_2 initial - 3.26 g consumed = 6.74 = 6.7 g C_2H_2 remain

$$10.0 \text{ g } O_2 \times \frac{1 \text{ mol } O_2}{32.00 \text{ g } O_2} \times \frac{4 \text{ mol } CO_2}{5 \text{ mol } O_2} \times \frac{44.01 \text{ g } CO_2}{1 \text{ mol } CO_2} = 11.0 \text{ g } CO_2 \text{ produced}$$

$$10.0 \text{ g } O_2 \times \frac{1 \text{ mol } O_2}{32.00 \text{ g } O_2} \times \frac{2 \text{ mol } H_2O}{5 \text{ mol } O_2} \times \frac{18.02 \text{ g } H_2O}{1 \text{ mol } H_2O} = 2.25 \text{ g } H_2O \text{ produced}$$

3.97   (a) $1.5 \times 10^5 \text{ g } C_9H_8O_4 \times \dfrac{1 \text{ mol } C_9H_8O_4}{180.2 \text{ g } C_9H_8O_4} \times \dfrac{1 \text{ mol } C_7H_6O_3}{1 \text{ mol } C_9H_8O_4} \times \dfrac{138.1 \text{ g } C_7H_6O_3}{1 \text{ mol } C_7H_6O_3}$

$$= 1.1496 \times 10^2 = 1.1 \times 10^2 \text{ kg } C_7H_6O_3$$

(b)   If only 80 percent of the acid reacts, then we need $1/0.80 = 1.25$ times as much to obtain the same mass of product: $1.25 \times 1.15 \times 10^2$ kg $= 1.4 \times 10^2$ kg.

(c)   Calculate the number of moles of each reactant:

$$1.85 \times 10^5 \text{ g } C_7H_6O_3 \times \frac{1 \text{ mol } C_7H_6O_3}{138.1 \text{ g } C_7H_6O_3} = 1.340 \times 10^3 = 1.34 \times 10^3 \text{ mol } C_7H_6O_3$$

$$1.25 \times 10^5 \text{ g } C_4H_6O_3 \times \frac{1 \text{ mol } C_4H_6O_3}{102.1 \text{ g } C_4H_6O_3} = 1.224 \times 10^3 = 1.22 \times 10^3 \text{ mol } C_4H_6O_3$$

We see that $C_4H_6O_3$ limits, because equal numbers of moles of the two reactants are consumed in the reaction.

$$1.224 \times 10^3 \text{ mol } C_4H_6O_3 \times \frac{1 \text{ mol } C_9H_8O_4}{1 \text{ mol } C_7H_6O_3} \times \frac{180.2 \text{ g } C_9H_8O_4}{1 \text{ mol } C_9H_8O_4} = 2.206 \times 10^5$$

$$= 2.21 \times 10^5 \text{ g } C_9H_8O_4$$

(d)   percent yield $= \dfrac{1.82 \times 10^5 \text{ g}}{2.206 \times 10^5 \text{ g}} \times 100 = 82.5\%$

Integrative Exercises

3.98   Strategy: volume of alloy sphere $\xrightarrow{\text{density}}$ mass of alloy $\xrightarrow{\text{mass \% Cu}}$ mass of Cu

$$V = 4/3 \, \pi \, r^3 = 4/3 \, \pi \, (2.0)^3 \text{ in}^3 \times \frac{(2.54)^3 \text{ cm}^3}{1 \text{ in}^3} = 549 = 5.5 \times 10^2 \text{ cm}^3$$

$$549 \text{ cm}^3 \times \frac{10.3 \text{ g sterling}}{1 \text{ cm}^3} \times \frac{7.5 \text{ g Cu}}{100 \text{ g sterling}} \times \frac{1 \text{ mol Cu}}{63.55 \text{ g Cu}} \times \frac{6.022 \times 10^{23} \text{ atoms}}{\text{mol}}$$

$$= 4.0 \times 10^{24} \text{ g Cu atoms}$$

3.99   Strategy: volume cube $\xrightarrow{\text{density}}$ mass $CaCO_3$ → moles $CaCO_3$ → moles O → O atoms

$$(1.25)^3 \text{ in}^3 \times \frac{(2.54)^3 \text{ cm}^3}{1 \text{ in}^3} \times \frac{2.71 \text{ g } CaCO_3}{1 \text{ cm}^3} \times \frac{1 \text{ mol } CaCO_3}{100.1 \text{ g } CaCO_3} \times \frac{3 \text{ mol O}}{1 \text{ mol } CaCO_3}$$

$$\times \frac{6.022 \times 10^{23} \text{ O atoms}}{1 \text{ mol O}} = 1.57 \times 10^{24} \text{ O atoms}$$

3.100   Strategy: $m^3 \longrightarrow L \xrightarrow{\text{density}} \text{g } C_8H_{18}$ g $\xrightarrow[\text{ratio}]{\text{mole}}$ g $CO_2 \longrightarrow$ kg CO_2

$$0.15 \text{ m}^3 \, C_8H_{18} \times \frac{10^3 \text{ dm}^3}{1 \text{ m}^3} \times \frac{1 \text{ L}}{1 \text{ dm}^3} \times \frac{1000 \text{ mL}}{1 \text{ L}} \times \frac{0.69 \text{ g } C_8H_{18}}{1 \text{ mL } C_8H_{18}} = 1.035 \times 10^5$$

$$= 1.0 \times 10^5 \text{ g } C_8H_{18}$$

$$2C_8H_{18}(l) + 25O_2(g) \rightarrow 16CO_2(g) + 18H_2O(l)$$

$$1.035 \times 10^5 \text{ g } C_8H_{18} \times \frac{1 \text{ mol } C_8H_{18}}{114.2 \text{ g } C_8H_{18}} \times \frac{16 \text{ mol } CO_2}{2 \text{ mol } C_8H_{18}} \times \frac{44.01 \text{ g } CO_2}{1 \text{ mol } CO_2} \times \frac{1 \text{ kg}}{1000 \text{ g}}$$

$$= 3.2 \times 10^2 \text{ kg } CO_2$$

3.101   We can proceed by writing the ratio of masses of Ag to $AgNO_3$, where y is the atomic mass of nitrogen.

$$\frac{Ag}{AgNO_3} = 0.634985 = \frac{107.8682}{107.8682 + 3(15.9994) + y}$$

Solve for y to obtain y = 14.0088. This is to be compared with the currently accepted value of 14.0067.

3.102   (a) $S(s) + O_2(g) \rightarrow SO_2(g)$; $SO_2(g) + CaO(s) \rightarrow CaSO_3(s)$

       (b) $\dfrac{2000 \text{ tons coal}}{\text{day}} \times \dfrac{2000 \text{ lb}}{1 \text{ ton}} \times \dfrac{1 \text{ kg}}{2.20 \text{ lb}} \times \dfrac{1000 \text{ g}}{1 \text{ kg}} \times \dfrac{0.028 \text{ g S}}{1 \text{ g coal}} \times \dfrac{1 \text{ mol S}}{32.1 \text{ g S}}$

$$\times \frac{1 \text{ mol } SO_2}{1 \text{ mol S}} \times \frac{1 \text{ mol } CaSO_3}{1 \text{ mol } SO_2} \times \frac{120 \text{ g } CaSO_3}{1 \text{ mol } CaSO_3} \times \frac{1 \text{ kg } CaSO_3}{1000 \text{ g } CaSO_3}$$

$$= 1.9 \times 10^5 \text{ kg } CaSO_3/\text{day}$$

This corresponds to over 200 tons of $CaSO_3$ per day as a waste product.

3.103   (a)   Strategy: calculate the kg of air in the room and then the mass of HCN required to produce a dose of 300 mg HCN/kg air.

12 ft × 15 ft × 8.0 ft = 1440 $= 1.4 \times 10^3 \text{ ft}^3$ of air in the room

$$1440 \text{ ft}^3 \text{ air} \times \frac{(12 \text{ in})^3}{1 \text{ ft}^3} \times \frac{(2.54 \text{ cm})^3}{1 \text{ in}^3} \times \frac{0.00118 \text{ g air}}{1 \text{ cm}^3 \text{ air}} \times \frac{1 \text{ kg}}{1000 \text{ g}} = 48.12$$

$$= 48 \text{ kg air}$$

$$48.12 \text{ kg air} \times \frac{300 \text{ mg HCN}}{1 \text{ kg air}} \times \frac{1 \text{ g}}{1000 \text{ mg}} = 14.43 = 14 \text{ g HCN}$$

       (b) $2NaCN(s) + H_2SO_4(aq) \rightarrow Na_2SO_4(aq) + 2HCN(g)$

The question can be restated as: What mass of NaCN is required to produce 14 g of HCN according to the above reaction?

$$14.43 \text{ g HCN} \times \frac{1 \text{ mol HCN}}{27.03 \text{ g HCN}} \times \frac{2 \text{ mol NaCN}}{2 \text{ mol HCN}} \times \frac{49.01 \text{ g NaCN}}{1 \text{ mol NaCN}} = 26.2 = 26 \text{ g NaCN}$$

(c) $12 \text{ ft} \times 15 \text{ ft} \times \dfrac{1 \text{ yd}^2}{9 \text{ ft}^2} \times \dfrac{30 \text{ oz}}{1 \text{ yd}^2} \times \dfrac{1 \text{ lb}}{16 \text{ oz}} \times \dfrac{454 \text{ g}}{1 \text{ lb}} = 17{,}025$

$$= 1.7 \times 10^4 \text{ g acrilan in the room}$$

50% of the carpet burns, so the starting amount of CH_2CHCN is
$0.50(17{,}025) = 8{,}513 = 8.5 \times 10^3 \text{ g}$

$8{,}513 \text{ g } CH_2CHCN \times \dfrac{50.9 \text{ g HCN}}{100 \text{ g } CH_2CHCH} = 4333 = 4.3 \times 10^2 \text{ g HCN possible}$

If the actual yield of combustion is 20%,

Actual g HCN = $4{,}333(0.20) = 866.6 = 8.7 \times 10^2$ g HCN produced

From part (a), 14 g of HCN is a lethal dose. The fire produces much more than a lethal dose of HCN.

4 Aqueous Reactions and Solution Stoichiometry

Solution Composition: Molarity

4.1   Concentration is the **ratio** of the amount of solute present in a certain quantity of solvent or solution. This ratio remains constant regardless of how much solution is present. Thus, concentration is an intensive property. The absolute concentration does depend on the amount of solute present, but once this ratio is established, it doesn't vary with the volume of solution present.

4.2   The concentration of the remaining solution is unchanged, assuming the original solution was thoroughly mixed. Molar concentration is a **ratio** of moles solute to liters solution. Although there are fewer moles solute remaining in the flask, there is also less solution volume, so the ratio of moles solute/solution volume remains the same.

4.3   The second solution is 5 times as concentrated as the first. An equal volume of the more concentrated solution will contain 5 times as much solute (5 times the number of moles and also 5 times the mass) as the 0.50 M solution. Thus, the mass of solute in the 2.50 M solution is 5 × 4.5 g = 22.5 g.

Mathematically:

$$\frac{\dfrac{2.50 \text{ mol solute}}{1 \text{ L solution}}}{\dfrac{0.50 \text{ mol solute}}{1 \text{ L solution}}} = \frac{x \text{ grams solute}}{4.5 \text{ g solute}}$$

$$\frac{2.50 \text{ mol solute}}{0.50 \text{ mol solute}} = \frac{x \text{ g solute}}{4.5 \text{ g solute}}; \quad 5.0(4.5 \text{ g solute}) = 23 \text{ g solute}$$

The result has 2 sig figs; 22.5 rounds to 23 g solute

4.4   The term *0.50 mol HCl* defines an amount (~18 g) of the pure substance HCl. The term 0.50 M HCl is a ratio; it indicates that there are 0.50 mol of HCl solute in 1.0 liter of solution. This same ratio of moles solute to solution volume is present regardless of the volume of solution under consideration.

4.5   (a) $M = \dfrac{\text{mol solute}}{\text{L solution}}$; $\dfrac{0.0345 \text{ mol NH}_4\text{Cl}}{400 \text{ mL}} \times \dfrac{1000 \text{ mL}}{1 \text{ L}} = 0.0863 \ M \ \text{NH}_4\text{Cl}$

(b) $mol = M \times L$; $\dfrac{2.20 \text{ mol HNO}_3}{1 \text{ L}} \times 0.0350 \text{ L} = 0.0770 \text{ mol HNO}_3$

(c) $L = \dfrac{mol}{M}$; $\dfrac{0.125 \text{ mol KOH}}{1.50 \text{ mol KOH/L}} = 0.0833 \text{ L or } 83.3 \text{ mL of } 1.50 \text{ } M \text{ KOH}$

4.6   (a) $M = \dfrac{\text{mol solute}}{\text{L solution}}$; $\dfrac{0.0715 \text{ mol Na}_2\text{SO}_4}{0.650 \text{ L}} = 0.110 \text{ } M \text{ Na}_2\text{SO}_4$

(b) $mol = M \times L$; $\dfrac{0.0850 \text{ mol KMnO}_4}{1 \text{ L}} \times 0.0750 \text{ L} = 6.38 \times 10^{-3} \text{ mol KMnO}_4$

(c) $L = \dfrac{mol}{M}$; $\dfrac{0.105 \text{ mol HCl}}{11.6 \text{ mol HCl/L}} = 9.05 \times 10^{-3} \text{ L or } 9.05 \text{ mL}$

4.7 $M = \dfrac{mol}{L}$; $mol = \dfrac{g}{MM}$ (MM is the symbol for molar mass in this manual.)

(a) $\dfrac{0.150 \text{ } M \text{ KBr}}{1 \text{ L}} \times 0.250 \text{ L} \times \dfrac{119.0 \text{ g KBr}}{1 \text{ mol KBr}} = 4.46 \text{ g KBr}$

(b) $4.75 \text{ g Ca(NO}_3)_2 \times \dfrac{1 \text{ mol Ca(NO}_3)_2}{164.1 \text{ g Ca(NO}_3)_2} \times \dfrac{1}{0.200 \text{ L}} = 0.145 \text{ } M \text{ Ca(NO}_3)_2$

(c) $5.00 \text{ g Na}_3\text{PO}_4 \times \dfrac{1 \text{ mol Na}_3\text{PO}_4}{163.9 \text{ g Na}_3\text{PO}_4} \times \dfrac{1 \text{ L}}{1.50 \text{ mol Na}_3\text{PO}_4} \times \dfrac{1000 \text{ mL}}{1 \text{ L}}$

$= 20.3 \text{ mL solution}$

4.8 $M = \dfrac{mol}{L}$; $mol = \dfrac{g}{MM}$ (MM is the symbol for molar mass in this manual.)

(a) $\dfrac{1.33 \text{ mol CoSO}_4}{1 \text{ L}} \times 50.0 \text{ mL} \times \dfrac{1 \text{ L}}{1000 \text{ mL}} \times \dfrac{155.0 \text{ g CoSO}_4}{1 \text{ mol CoSO}_4} = 10.3 \text{ g CoSO}_4$

(b) $1.50 \text{ g (NH}_4)_2\text{SO}_4 \times \dfrac{1 \text{ mol (NH}_4)_2\text{SO}_4}{132.2 \text{ g (NH}_4)_2\text{SO}_4} \times \dfrac{1}{250. \text{ mL}} \times \dfrac{1000 \text{ mL}}{1 \text{ L}}$

$= 0.0454 \text{ } M \text{ (NH}_4)_2\text{SO}_4$

(c) $1.00 \text{ g NiCl}_2 \times \dfrac{1 \text{ mol NiCl}_2}{129.6 \text{ g NiCl}_2} \times \dfrac{1 \text{ L}}{0.510 \text{ } M \text{ NiCl}_2} \times \dfrac{1000 \text{ mL}}{1 \text{ L}} = 15.1 \text{ mL solution}$

4.9   The number of moles of sucrose needed is

$\dfrac{0.150 \text{ mol}}{1 \text{ L}} \times 0.125 \text{ L} = 0.01875 = 0.0188 \text{ mol}$

Weigh out $0.01875 \text{ mol C}_{12}\text{H}_{22}\text{O}_{11} \times \dfrac{342.3 \text{ g C}_{12}\text{H}_{22}\text{O}_{11}}{1 \text{ mol C}_{12}\text{H}_{22}\text{O}_{11}} = 6.42 \text{ g C}_{12}\text{H}_{22}\text{O}_{11}$

Add this amount of solid to a 125 mL volumetric flask, dissolve in a small volume of water, and add water to the mark on the neck of the flask. Agitate thoroughly to ensure total mixing.

4.10   The amount of $AgNO_3$ needed is: $0.08000\ M \times 0.1000\ L = 8.000 \times 10^{-3}\ mol\ AgNO_3$

$$8.000 \times 10^{-3}\ mol\ AgNO_3 \times \frac{169.88\ g\ AgNO_3}{1\ mol\ AgNO_3} = 1.359\ g\ AgNO_3$$

Add this amount of solid to a 100 mL volumetric flask, dissolve in a small amount of water, bring the total volume to exactly 100 mL and agitate well.

4.11   Calculate the moles of solute present in the final 400 mL of 0.100 $M\ C_{12}H_{22}O_{11}$ solution:

$$moles\ C_{12}H_{22}O_{11} = M \times L = \frac{0.100\ mol\ C_{12}H_{22}O_{11}}{1\ L} \times 0.400\ L = 0.0400\ mol\ C_{12}H_{22}O_{11}$$

Calculate the volume of 1.50 M glucose solution that would contain 0.0400 mol $C_{12}H_{22}O_{11}$:

$$L = moles/M;\ 0.0400\ mol\ C_{12}H_{22}O_{11} \times \frac{1\ L}{1.50\ mol\ C_{12}H_{22}O_{11}} = 0.02667 = 0.0267\ L$$

$$0.02667\ L \times \frac{1000\ mL}{1\ L} = 26.7\ mL$$

Thoroughly rinse, clean and fill a 50 mL buret with the 1.50 $M\ C_{12}H_{22}O_{11}$. Dispense 26.7 mL of this solution into a 400 mL volumetric container, add water to the mark and mix thoroughly. (26.7 mL is a difficult volume to measure with a pipette.)

4.12   Dilute the 6.0 $M\ HNO_3$ to prepare 200 mL of 1.0 $M\ HNO_3$. To determine the volume of 6.0 $M\ HNO_3$ needed, calculate the moles HNO_3 present in 200 mL of 1.0 $M\ HNO_3$ and then the volume of 6.0 M solution that contains this number of moles.

$0.200\ L \times 1.0\ L = 0.200\ mol\ HNO_3$ needed;

$$L = \frac{mol}{M};\ L\ 6.0\ M\ HNO_3 = \frac{0.200\ mol\ needed}{6.0\ M} = 0.033\ L = 33\ mL$$

Thoroughly clean, rinse and fill a 50 mL buret with the 6.0 $M\ HNO_3$, taking precautions appropriate for working with a relatively concentrated acid. Dispense 33 mL of the 6.0 M acid into a 200 mL volumetric flask, add water to the mark and mix thoroughly. (A pipette could be used to measure the 33 mL of concentrated acid, but a buret is more safe and convenient for this volume of acid.)

4.13   Calculate the mass of acetic acid, $HC_2H_3O_2$, present in 10.0 mL of the pure liquid.

$$10.00\ mL\ acetic\ acid \times \frac{1.049\ g\ acetic\ acid}{1\ mL\ acetic\ acid} = 10.49\ g\ acetic\ acid$$

$$10.49\ g\ HC_2H_3O_2 \times \frac{1\ mol\ HC_2H_3O_2}{60.05\ g\ HC_2H_3O_2} = 0.17469 = 0.1747\ mol\ HC_2H_3O_2$$

$$M = mol/L = \frac{0.17469\ mol\ HC_2H_3O_2}{0.1000\ L\ solution} = 1.747\ M\ HC_2H_3O_2$$

4.14 $40.000 \text{ mL glycerol} \times \dfrac{1.2656 \text{ g glycerol}}{1 \text{ mL glycerol}} = 50.624 \text{ g glycerol}$

         $50.624 \text{ g } C_3H_8O_3 \times \dfrac{1 \text{ mol } C_3H_8O_3}{92.094 \text{ g } C_3H_8O_3} = 0.549699 = 0.54970 \text{ mol } C_3H_8O_3$

         $M = \dfrac{0.549699 \text{ mol } C_3H_8O_3}{0.25000 \text{ L solution}} = 2.1988 \; M \; C_3H_8O_3$

Electrolytes

4.15   Tap water contains enough dissolved electrolytes to conduct a significant amount of electricity. Thus, water can complete a circuit between an electrical appliance and our body, producing a shock.

4.16   When an ionic compound dissolves in water, water molecules surround and separate ions from the solid lattice and disperse them into the solution. The negative (O) ends of water molecules point toward cations and the positive (H) ends of water molecules point toward anions.

4.17   (a)   HF -- weak   (b) C_2H_5OH -- non   (c) NH_3 -- weak

        (d) $KClO_3$ -- strong   (e) $Cu(NO_3)_2$ -- strong

4.18   (a)   HBrO -- weak   (b) HNO_3 -- strong   (c) KOH -- strong

        (d) $CoSO_4$ -- strong   (e) $C_{12}H_{22}O_{11}$ -- non   (f) O_2 -- non

4.19   (a) $0.14 \; M \; Na^+$, $0.14 \; M \; OH^-$

        (b) $0.25 \; M \; Ca^{2+}$, $0.50 \; M \; Cl^-$

        (c) $0.25 \; M$ (CH_3OH is a molecular solute)

        (d) $M_2 = M_1V_1/V_2$, where V_2 is the total solution volume.

            K^+: $\dfrac{0.20 \; M \; 0.050 \text{ L}}{0.075 \text{ L}} = 0.133 = 0.13 \; M$

            ClO_3^-: concentration ClO_3^- = concentration $Na^+ = 0.13 \; M$

            SO_4^{2-}: $\dfrac{0.20 \; M \; 0.0250 \text{ L}}{0.075 \text{ L}} = 0.0667 = 0.067 \; M \; SO_4^{2-}$

            Na^+: concentration $Na^+ = 2 \times$ concentration $SO_4^{2-} = 0.13 \; M$

4.20   (a) H^+: $\dfrac{0.100 \; M \times 20.0 \text{ mL} + 0.220 \; M \times 10.0 \text{ mL}}{30.0 \text{ mL}} = 0.140 \; M \; H^+$

            Cl^-: concentration Cl^- = concentration $H^+ = 0.140 \; M \; Cl^-$

(b) Na^+: $\dfrac{2(0.300\ M \times 15.0\ mL)\ +\ (0.100\ M \times 10.0\ mL)}{25.0\ mL} = 0.400\ M$

SO_4^{2-}: $\dfrac{0.300\ M \times 15.0\ mL}{25.0\ mL} = 0.180\ M$; Cl^-: $\dfrac{0.100\ M \times 10.0\ mL}{25.0\ mL} = 0.0400\ M$

(c) K^+: $\dfrac{3.50\ g\ KCl}{0.060\ L} \times \dfrac{1\ mol}{74.55\ g} = 0.782\ M$; Ca^{2+}: $0.500\ M$

Cl^-: $0.782\ M$ (from KCl(s)) + $1.000\ M$ (from $CaCl_2$(aq)) = $1.782\ M$

4.21 $KCl \rightarrow K^+ + Cl^-$; $0.20\ M\ KCl = 0.20\ M\ K^+$

$K_2Cr_2O_7 \rightarrow 2\ K^+ + Cr_2O_7^{2-}$; $0.15\ M\ K_2Cr_2O_7 = 0.30\ M\ K^+$

$K_3PO_4 \rightarrow 3\ K^+ + PO_4^{3-}$; $0.080\ M\ K_3PO_4 = 0.24\ M\ K^+$

$0.15\ M\ K_2Cr_2O_7$ has the highest K^+ concentration.

4.22 $NaCl \rightarrow Na^+ + Cl^-$; $0.35\ M \times 0.040\ L = 0.014\ mol\ Cl^-$

$CaCl_2 \rightarrow Ca^{2+} + 2Cl^-$; $0.25\ M \times 0.025\ L = 0.00625\ mol\ CaCl_2 \times 2 = 0.0125 = 0.013\ mol\ Cl^-$

The NaCl solution has slightly more moles Cl^-.

Acids, Bases and Salts

4.23 Since the solution does conduct some electricity, but less than an equimolar NaCl solution (a strong electrolyte) the unknown solute must be a weak electrolyte. The weak electrolytes in the list of choices are NH_3 and H_3PO_3; since the solution is acidic, the unknown must be **H_3PO_3**.

4.24 (a) HF -- acid, mixture of ions and molecules (weak electrolyte)

(b) CH_3CN -- none of the above, entirely molecules (nonelectrolyte)

(c) $NaClO_4$ -- salt, entirely ions (strong electrolyte)

(d) $Ba(OH)_2$ -- base, entirely ions (strong electrolyte)

4.25 (a) A *monoprotic acid* has one ionizable (acidic) H and a *diprotic acid* has two.

(b) A *strong acid* is completely ionized in aqueous solution whereas only a fraction of *weak acid* molecules are ionized.

(c) An *acid* is an H^+ donor, a substance that increases the concentration of H^+ in aqueous solution. A *base* is an H^+ acceptor and thus increases the concentration of OH^- in aqueous solution.

4.26   (a) NH_3 produces OH^- in aqueous solution by reacting with H_2O (hydrolysis):

                     $NH_3(aq) + H_2O(l) \rightleftharpoons NH_4^+(aq) + OH^-(aq)$. The OH^- causes the solution to be basic.

         (b)   The term "weak" refers to the tendency of HF to dissociate into H^+ and F^- in aqueous solution, not its reactivity toward other compounds.

4.27   In aqueous solution, HNO_3 exists entirely as H^+ and NO_3^- ions; the single arrow denotes complete ionization. HCN is only dissociated to a small extent; the double arrow denotes a mixture of H^+ ions, CN^- ions and neutral undissociated HCN molecules in solution.

4.28 H_2SO_4 is a **diprotic** acid; it has two ionizable hydrogens. The first hydrogen completely dissociates to form H^+ and HSO_4^-, but HSO_4^- only **partially** ionizes into H^+ and SO_4^{2-} (HSO_4^- is a weak electrolyte). Thus, a 0.1 M solution of H_2SO_4 contains a mixture of H^+, HSO_4^- and SO_4^{2-}, with the concentration of H^+ greater than 0.1 M but less than 0.2 M.

4.29   (a) $2HBr(aq) + Ca(OH)_2(aq) \rightarrow CaBr_2(aq) + 2H_2O(l)$

         (b) $Cu(OH)_2(s) + 2HClO_4(aq) \rightarrow Cu(ClO_4)_2(aq) + 2H_2O(l)$

         (c) $2Fe(OH)_3(s) + 3H_2SO_4(aq) \rightarrow Fe_2(SO_4)_3(aq) + 6H_2O(l)$

4.30   (a) $KC_2H_3O_2$ (b) $Ca(NO_3)_2$ (c) $(NH_4)_2SO_4$

Ionic Equations; Metathesis Reactions

4.31   (a) $Pb^{2+}(aq) + SO_4^{2-}(aq) \rightarrow PbSO_4(s)$; spectators: Na^+, NO_3^-

         (b) $2Al(s) + 6H^+(aq) \rightarrow 2Al^{3+}(aq) + 3H_2(g)$; spectator: Cl^-

         (c) $FeO(s) + 2H^+(aq) \rightarrow H_2O(l) + Fe^{2+}(aq)$; spectator: ClO_4^-

4.32   (a) $Cr(OH)_3(s) + 3HNO_3(aq) \rightarrow 3H_2O(l) + Cr(NO_3)_3(aq)$

                   $Cr(OH)_3(s) + 3H^+(aq) \rightarrow 3H_2O(l) + Cr^{3+}(aq)$

         (b) $Na_2CO_3(aq) + 2HCl(aq) \rightarrow 2NaCl(aq) + H_2O(l) + CO_2(g)$

                   $CO_3^{2-}(aq) + 2H^+(aq) \rightarrow H_2O(l) + CO_2(g)$

         (c) $CuBr_2(aq) + 2NaOH(aq) \rightarrow Cu(OH)_2(s) + 2NaBr(aq)$

                   $Cu^{2+}(aq) + 2OH^-(aq) \rightarrow Cu(OH)_2(s)$

4.33   The driving force in a metathesis reaction is the formation of a product that removes ions from solution. This means that at least some reactant species present as ions exist in a different form in the products, as indicated by the net ionic equation.

        Driving forces in Exercise 4.32:

        (a) $H_2O(l)$ -- nonelectrolyte

        (b) $H_2O(l)$, $CO_2(g)$ -- nonelectrolyte and gas, respectively

        (c) $Cu(OH)_2(s)$ -- precipitate

4.34   There are many examples for each of these driving forces.

   (a) $AgNO_3(aq) + NaCl(aq)$ → **$AgCl(s)$** $+ NaNO_3(aq)$

   (b) $NaHCO_3(aq) + HCl(aq)$ → $NaCl(aq) + H_2O(l) +$ **$CO_2(g)$**

   (c) $HNO_2 + KOH(aq)$ → $KNO_2(aq) +$ **$H_2O(l)$**

          (or any neutralization reaction where $H_2O(l)$ is a product)

4.35   Follow the guidelines in Table 4.3.

   (a)   **$NiCl_2$** - soluble   (b) Ag_2S - insoluble

   (c)   **Cs_3PO_4** - soluble (Cs^+ is an alkali metal cation)

   (d)   **$SrCO_3$** - insoluble   (e) $(NH_4)_2$**SO_4** - soluble

4.36   According to Table 4.3:

   (a) K_3PO_4 -- soluble   (b) $Pb(C_2H_3O_2)_2$ -- soluble

   (c) $Ga(OH)_3$ -- insoluble   (d) $NaCN$ -- soluble; cations in group 1A (including Na^+)

   (e) $BaSO_4$ -- insoluble   produce soluble salts of most anions

4.37 Br^- and NO_3^- can be ruled out because the Ba^{2+} salts are soluble. (Actually all NO_3^- salts are soluble.) SO_4^{2-} forms insoluble salts with the three cations given; it must be the anion in question.

4.38 Pb^{2+} is not present or an insoluble hydroxide would have formed. $BaSO_4$ is insoluble and $Ba(OH)_2$ is soluble, so the solution must contain Ba^{2+}. It could also contain K^+, but since we are dealing with a single salt, we will assume that only Ba^{2+} is present.

4.39   (a) $Ba^{2+}(aq) + HSO_4^-(aq)$ → $BaSO_4(s) + H^+(aq)$

   (b)   No reaction. Both components are soluble electrolytes, as are possible metathesis reaction products.

   (c) $2Ag^+(aq) + CO_3^{2-}(aq)$ → $Ag_2CO_3(s)$

   (d) $H^+(aq) + OH^-(aq)$ → $H_2O(l)$

   (e) $HC_2H_3O_2(aq) + OH^-(aq)$ → $H_2O(l) + C_2H_3O_2^-(aq)$

   (f) $Pb^{2+}(aq) + SO_4^{2-}(aq)$ → $PbSO_4(s)$

4.40   (a) $ZnS(s) + 2H^+(aq)$ → $H_2S(g) + Zn^{2+}(aq)$

   (b) $Ba^{2+}(aq) + CO_3^{2-}(aq)$ → $BaCO_3(s)$

   (c) $3H^+(aq) + PO_4^{3-}(aq)$ ⇌ $H_3PO_4(aq)$ (H_3PO_4 is a relatively weak acid.)

   (d) $H^+(aq) + OH^-(aq)$ → $H_2O(l)$

   (e) $Sr^{2+}(aq) + SO_4^{2-}(aq)$ → $SrSO_4(s)$

   (f) $Pb^{2+}(aq) + H_2S(aq)$ → $PbS(s) + 2H^+(aq)$

   (g) $Fe(OH)_3(s) + 3H^+(aq)$ → $Fe^{3+}(aq) + 3H_2O(l)$

Oxidation - Reduction Reactions

4.41   Corrosion was one of the first oxidation-reduction processes to be studied in detail. During corrosion, a metal reacts with some environmental agent, usually oxygen, to form a metal compound. Thus, reaction of a metal with oxygen causes the metal to lose electrons; reaction with oxygen is logically called *oxidation*. Eventually the term oxidation was generalized to mean any process where a substance loses electrons, whether or not the substance is a metal or oxygen is a reactant.

4.42   Oxidation and reduction can only occur together, not separately. When a metal reacts with oxygen, the metal atoms lose electrons and the oxygen atoms gain electrons. Free electrons do not exist under normal conditions. If electrons are lost by one substance they must be gained by another, and vice versa.

4.43   The most easily oxidized metals are near the bottom of groups on the left side of the chart, especially groups 1A and 2A. The least easily oxidized metals are on the lower right of the transition metals, particularly those near the bottom of groups 8B and 1B.

4.44   (a)   Platinum and gold are called the noble metals because they are especially unreactive and difficult to oxidize.

         (b)   The alkali and alkaline earth metals are called active because they are very easily oxidized and chemically reactive.

4.45   (a) $2HCl(aq) + Ni(s) \rightarrow NiCl_2(aq) + H_2(g)$; $Ni(s) + 2H^+(aq) \rightarrow Ni^{2+}(aq) + H_2(g)$

         (b) $H_2SO_4(aq) + Fe(s) \rightarrow FeSO_4(aq) + H_2(g)$; $Fe(s) + 2H^+(aq) \rightarrow Fe^{2+}(aq) + H_2(g)$

         (c) $2HBr(aq) + Zn(s) \rightarrow ZnBr_2(aq) + H_2(g)$; $Zn(s) + 2H^+(aq) \rightarrow Zn^{2+}(aq) + H_2(g)$

         (d) $2HC_2H_3O_2(aq) + Mg(s) \rightarrow Mg(C_2H_3O_2)_2(aq) + H_2(g)$;

                $Mg(s) + 2HC_2H_3O_2(aq) \rightarrow Mg^{2+}(aq) + 2C_2H_3O_2^-(aq) + H_2(g)$

4.46   (a) $Mn(s) + H_2SO_4(aq) \rightarrow MnSO_4(aq) + H_2(g)$; $Mn(s) + 2H^+(aq) \rightarrow Mn^{2+}(aq) + H_2(g)$

         (b) $2Cr(s) + 6HBr(aq) \rightarrow 2CrBr_3(aq) + 3H_2(g)$; $2Cr(s) + 6H^+(aq) \rightarrow 2Cr^{3+}(aq) + 3H_2(g)$

         (c) $Sn(s) + 2HCl(aq) \rightarrow SnCl_2(aq) + H_2(g)$; $Sn(s) + 2H^+(aq) \rightarrow Sn^{2+}(aq) + H_2(g)$

         (d) $2Al(s) + 6HCHO_2(aq) \rightarrow 2Al(CHO_2)_3(aq) + 3H_2(g)$;

                $2Al(s) + 6HCHO_2(aq) \rightarrow 2Al^{3+}(aq) + 6CHO_2^-(aq) + 3H_2(g)$

4.47   (a) $2Al(s) + 3NiCl_2(aq) \rightarrow 2AlCl_3(aq) + 3Ni(s)$

         (b) $Ag(s) + Pb(NO_3)_2(aq) \rightarrow$ no reaction

         (c) $2Cr(s) + 3NiSO_4(aq) \rightarrow Cr_2(SO_4)_3(aq) + 3Ni(s)$

         (d) $Mn(s) + 2HBr(aq) \rightarrow MnBr_2(aq) + H_2(g)$

         (e) $H_2(g) + CuCl_2(aq) \rightarrow Cu(s) + 2HCl(aq)$

         (f) $Ba(s) + 2H_2O(l) \rightarrow Ba(OH)_2(aq) + H_2(g)$

            (The most active metals can displace H^+ from H_2O as well as from acids.)

4.48   (a) $Zn(s) + 2AgNO_3(aq) \rightarrow 2Ag(s) + Zn(NO_3)_2(aq)$

      (b) $Fe(s) + Al_2(SO_4)_3(aq) \rightarrow$ no reaction

      (c) $Co(s) + 2HCl(aq) \rightarrow CoCl_2(aq) + H_2(g)$

      (d) $FeCl_2(aq) + H_2(g) \rightarrow$ no reaction

      (e) $2Li(s) + 2H_2(l) \rightarrow 2LiOH(aq) + H_2(g)$

             (The most active metals can displace H^+ from H_2O as well as from acids.)

4.49   (a)   i. $Zn(s) + Cd^{2+}(aq) \rightarrow Cd(s) + Zn^{2+}(aq)$

             ii. $Cd(s) + Ni^{2+}(aq) \rightarrow Ni(s) + Cd^{2+}(aq)$

      (b)   According to Table 4.5, the most active metals are most easily oxidized, and Zn is more active than Ni. Observation (i) indicates that Cd is less active than Zn; (ii) indicates that Cd is more active than Ni. Cd is between Zn and Ni on the activity series.

      (c)   Place an iron strip in $CdCl_2(aq)$. If Cd(s) is deposited, Cd is less active than Fe; if there is no reaction, Cd is more active than Fe. Do the same test with Co if Cd is less active than Fe or with Cr if Cd is more active than Fe.

4.50   (a) $Br_2 + 2NaI \rightarrow 2NaBr + I_2$ indicates that Br_2 is more easily reduced than I_2.
             $Cl_2 + 2NaBr \rightarrow 2NaCl + Br_2$ shows that Cl_2 is more easily reduced than Br_2.
             The order for ease of reduction is $Cl_2 > Br_2 > I_2$. Conversely, the order for ease of oxidation is $I^- > Br^- > Cl^-$.

      (b)   Since the halogens are nonmetals, they tend to form anions when they react chemically. Nonmetallic character decreases going down a family and so does the tendency to gain electrons during a chemical reaction. Thus, the ease of reduction of the halogen, X_2, decreases going down the family and the ease of oxidation of the halide, X^-, increases going down the family.

      (c) $Cl_2 + 2KI \rightarrow 2KCl + I_2$; $Br_2 + LiCl \rightarrow$ no reaction

Solution Stoichiometry; Titrations

4.51   (a)   Write the balanced equation for the reaction in question:
             $HClO_4(aq) + NaOH(aq) \rightarrow NaClO_4(aq) + H_2O(l)$

             Calculate the moles of the known substance, in this case NaOH.

$$\text{moles NaOH} = M \times L = \frac{0.0875 \text{ mol NaOH}}{1 \text{ L}} \times 0.0500 \text{ L} = 0.004375$$

$$= 0.00438 \text{ mol NaOH}$$

             Apply the mole ratio (mol unknown/mol known) from the chemical equation.

$$0.004375 \text{ mol NaOH} \times \frac{1 \text{ mol } HClO_4}{1 \text{ mol NaOH}} = 0.004375 \text{ mol } HClO_4$$

Calculate the desired quantity of unknown, in this case the volume of 0.115 M $HClO_4$ solution.

$$L = mol/M; \quad L = 0.004375 \text{ mol } HClO_4 \times \frac{1 \text{ L}}{0.115 \text{ mol HCl}} = 0.0380 \text{ L} = 38.0 \text{ mL}$$

(b)   Following the procedure outlined in part (a):

$$2HCl(aq) + Mg(OH)_2(s) \rightarrow MgCl_2(aq) + 2H_2O(l)$$

$$2.87 \text{ g } Mg(OH)_2 \times \frac{1 \text{ mol } Mg(OH)_2}{58.33 \text{ g } Mg(OH)_2} = 0.04920 = 0.04920 \text{ mol } Mg(OH)_2$$

$$0.0492 \text{ mol } Mg(OH)_2 \times \frac{2 \text{ mol HCl}}{1 \text{ mol } Mg(OH)_2} = 0.0984 \text{ mol HCL}$$

$$L = mol/M = 0.09840 \text{ mol HCl} \times \frac{1 \text{ L HCl}}{0.128 \text{ mol HCl}} = 0.769 \text{ L} = 769 \text{ mL}$$

(c) $AgNO_3(aq) + KCl(aq) \rightarrow AgCl(s) + KNO_3(aq)$

$$785 \text{ mg KCl} \times \frac{1 \times 10^{-3} \text{ g}}{1 \text{ mg}} \times \frac{1 \text{ mol KCl}}{74.55 \text{ g KCl}} \times \frac{1 \text{ mol } AgNO_3}{1 \text{ mol KCl}} = 0.01053$$

$$= 0.0105 \text{ mol } AgNO_3$$

$$M = mol/L = \frac{0.01053 \text{ mol } AgNO_3}{0.0258 \text{ L}} = 0.408 \; M \; AgNO_3$$

(d) $HCl(aq) + KOH(aq) \rightarrow KCl(aq) + H_2O(l)$

$$\frac{0.108 \text{ mol HCl}}{1 \text{ L}} \times 0.0453 \text{ L} \times \frac{\text{mol KOH}}{\text{mol HCl}} \times \frac{56.11 \text{ g KOH}}{1 \text{ mol KOH}} = 0.275 \text{ g KOH}$$

4.52   (a) $2HCl(aq) + Ba(OH)_2(aq) \rightarrow BaCl_2(aq) + 2H_2O(l)$

$$\frac{0.101 \text{ mol } Ba(OH)_2}{1 \text{ L } Ba(OH)_2} \times 0.0350 \text{ L } Ba(OH)_2 \times \frac{2 \text{ mol HCl}}{1 \text{ mol } Ba(OH)_2}$$

$$\times \frac{1 \text{ L HCl}}{0.155 \text{ mol HCl}} = 0.0456 \text{ L or } 45.6 \text{ mL HCl soln}$$

(b) $H_2SO_4(aq) + 2NaOH(aq) \rightarrow Na_2SO_4(aq) + 2H_2O(l)$

$$75.0 \text{ g NaOH} \times \frac{1 \text{ mol NaOH}}{40.00 \text{ g NaOH}} \times \frac{1 \text{ mol } H_2SO_4}{2 \text{ mol NaOH}} \times \frac{1 \text{ L } H_2SO_4}{2.50 \text{ mol } H_2SO_4}$$

$$= 0.375 \text{ L or } 375 \text{ mL } H_2SO_4 \text{ soln}$$

(c) $BaCl_2(aq) + Na_2SO_4(aq) \rightarrow BaSO_4(s) + 2NaCl(aq)$

$$544 \text{ mg} = 0.544 \text{ g } Na_2SO_4 \times \frac{1 \text{ mol } Na_2SO_4}{142.1 \text{ g } Na_2SO_4} \times \frac{1 \text{ mol } BaCl_2}{1 \text{ mol } Na_2SO_4} \times \frac{1}{0.0558 \text{ L}}$$

$$= 0.0686 \ M \ BaCl_2$$

(d) $2HCl(aq) + Ca(OH)_2(aq) \rightarrow CaCl_2(aq) + 2H_2O(l)$

$$0.0375 \text{ L HCl} \times \frac{0.250 \text{ mol HCl}}{1 \text{ L HCl}} \times \frac{1 \text{ mol } Ca(OH)_2}{2 \text{ mol HCl}} \times \frac{74.10 \text{ g } Ca(OH)_2}{1 \text{ mol } Ca(OH)_2}$$

$$= 0.347 \text{ g } Ca(OH)_2$$

4.53   See Exercise 4.51(a) for a more detailed approach.

$$\frac{6.0 \text{ mol } H_2SO_4}{1 \text{ L}} \times 0.035 \text{ L} \times \frac{2 \text{ mol } NaHCO_3}{1 \text{ mol } H_2SO_4} \times \frac{84.01 \text{ g } NaHCO_3}{1 \text{ mol } NaHCO_3} = 35 \text{ g } NaHCO_3$$

4.54   See Exercise 4.51 (a) for a more detailed approach.

$$\frac{0.0960 \text{ mol NaOH}}{1 \text{ L}} \times 0.0349 \text{ L} \times \frac{1 \text{ mol } HC_2H_3O_2}{1 \text{ mol NaOH}} \times \frac{60.05 \text{ g } HC_2H_3O_2}{1 \text{ mol } HC_2H_3O_2}$$

$$= 0.2012 = 0.201 \text{ g } HC_2H_3O_2 \text{ in } 2.50 \text{ mL}$$

$$1.00 \text{ qt vinegar} \times \frac{1 \text{ L}}{1.057 \text{ qt}} \times \frac{1000 \text{ mL}}{1 \text{ L}} \times \frac{0.2012 \text{ g } HC_2H_3O_2}{2.50 \text{ mL vinegar}} = 76.1 \text{ g } HC_2H_3O_2/\text{qt}$$

4.55   The neutralization reaction here is:

$2HBr(aq) + Ca(OH)_2(aq) \rightarrow CaBr_2(aq) + 2H_2O(l)$

$$0.0488 \text{ L HBr soln} \times \frac{5.00 \times 10^{-2} \text{ mol HBr}}{1 \text{ L soln}} \times \frac{1 \text{ mol } Ca(OH)_2}{2 \text{ mol HBr}} \times \frac{1}{0.100 \text{ L of } Ca(OH)_2}$$

$$= 1.220 \times 10^{-2} = 1.22 \times 10^{-2} \ M \ Ca(OH)_2$$

From the molarity of the saturated solution, we can calculate the gram solubility of $Ca(OH)_2$ in 100 mL of H_2O.

$$0.100 \text{ L soln} \times \frac{1.220 \times 10^{-2} \text{ mol } Ca(OH)_2}{1 \text{ L soln}} \times \frac{74.10 \text{ g } Ca(OH)_2}{1 \text{ mol } Ca(OH)_2}$$

$$= 0.0904 \text{ g } Ca(OH)_2 \text{ in } 100 \text{ mL soln}$$

4.56   The balanced equation for the titration is:

$Sr(NO_3)_2(aq) + Na_2CrO_4(aq) \rightarrow SrCrO_4(s) + 2NaNO_3(aq)$

Beginning with a 0.100 L sample, we can do the following conversions:

volume soln $\rightarrow$ g $Sr(NO_3)_2$ $\rightarrow$ mol $Sr(NO_3)_2$ $\rightarrow$ mol Na_2CrO_4 $\rightarrow$ vol Na_2CrO_4 soln

$$0.100 \text{ L soln} \times \frac{6.67 \text{ g } Sr(NO_3)_2}{0.750 \text{ L soln}} \times \frac{1 \text{ mol } Sr(NO_3)_2}{211.6 \text{ g } Sr(NO_3)_2} \times \frac{1 \text{ mol } Na_2CrO_4}{1 \text{ mol } Sr(NO_3)_2}$$

$$\times \frac{1 \text{ L soln}}{0.0460 \text{ mol } Na_2CrO_4} = 0.0914 \text{ L } Na_2CrO_4 \text{ soln}$$

Additional Exercises

4.57   (a) $0.0500 \text{ L soln} \times \dfrac{0.200 \text{ mol NaCl}}{1 \text{ L soln}} = 1.00 \times 10^{-2} \text{ mol NaCl}$

$0.1000 \text{ L soln} \times \dfrac{0.100 \text{ mol NaCl}}{1 \text{ L soln}} = 1.00 \times 10^{-2} \text{ mol NaCl}$

Total moles NaCl = 2.00×10^{-2}, total volume = 0.0500 L + 0.1000 L = 0.1500 L

$\text{Molarity} = \dfrac{2.00 \times 10^{-2} \text{ mol}}{0.150 \text{ L}} = 0.133 \; M$

(b) $0.0245 \text{ L soln} \times \dfrac{1.50 \text{ mol NaOH}}{1 \text{ L soln}} = 0.03675 = 0.0368 \text{ mol NaOH}$

$0.0250 \text{ L soln} \times \dfrac{0.850 \text{ mol NaOH}}{1 \text{ L soln}} = 0.017425 = 0.0174 \text{ mol NaOH}$

Total moles NaOH = 0.054175 = 0.0542, total volume = 0.0450 L

$\text{Molarity} = \dfrac{0.054175 \text{ mol NaOH}}{0.0450 \text{ L}} = 1.20 \; M$

4.58 $\dfrac{6.00 \text{ mol HCl}}{1 \text{ L}} \times 0.0500 \text{ L} = 0.300 \text{ mol HCl}$

$\dfrac{1.00 \text{ mol HCl}}{1 \text{ L}} \times 0.1000 \text{ L} = 0.100 \text{ mol HCl}$

$M = \dfrac{\text{total mol HCl}}{0.250 \text{ L solution}} = \dfrac{0.400 \text{ mol HCl}}{0.250 \text{ L solution}} = 1.60 \; M \text{ HCl}$

4.59   (a) $\dfrac{50 \text{ pg}}{1 \text{ mL}} \times \dfrac{1 \times 10^{-12} \text{ g}}{1 \text{ pg}} \times \dfrac{1 \times 10^3 \text{ mL}}{\text{L}} \times \dfrac{1 \text{ mol Na}}{23.0 \text{ g Na}} = 2.17 \times 10^{-9} = 2.2 \times 10^{-9} \; M$

(b) $\dfrac{2.17 \times 10^{-9} \text{ mol Na}}{1 \text{ L soln}} \times \dfrac{1 \text{ L}}{1 \times 10^3 \text{ cm}^3} \times \dfrac{6.02 \times 10^{23} \text{ Na atom}}{1 \text{ mol Na}}$

$= 1.3 \times 10^{12} \text{ atoms or Na}^+ \text{ ions/cm}^3$

4.60 Na^+ must replace the total + charge due to Ca^{2+} and Mg^{2+}. Think of this as moles of charge rather than moles of particles.

$$\frac{0.010 \text{ mol } Ca^{2+}}{1 \text{ L water}} \times 1.0 \times 10^3 \text{ L} \times \frac{2 \text{ mol + charge}}{1 \text{ mol } Ca^{2+}} = 20 \text{ mol of + charge}$$

$$\frac{0.0050 \text{ mol } Mg^{2+}}{1 \text{ L water}} \times 1.0 \times 10^3 \text{ L} \times \frac{2 \text{ mol + charge}}{1 \text{ mol } Mg^{2+}} = 10 \text{ mol of + charge}$$

30 moles of + charge must be replaced; 30 mol Na^+ are needed.

4.61   (a,b)   Expt. 1   No reaction

                     Expt. 2 $2Ag^+(aq) + CrO_4^{2-}(aq) \rightarrow Ag_2CrO_4(s)$ red precipitate

                     Expt. 3   No reaction

                     Expt. 4 $2Ag^+(aq) + C_2O_4^{2-}(aq) \rightarrow AgC_2O_4(s)$ white precipitate

                     Expt. 5 $Ca^{2+}(aq) + C_2O_4^{2-}(aq) \rightarrow CaC_2O_4(s)$ white precipitate

                     Expt. 6 $Ag^+(aq) + Cl^-(aq) \rightarrow AgCl(s)$ white precipitate

          (c)   The silver salts of both ions are insoluble, but many silver salts are insoluble (Expt. 6). The calcium salt of CrO_4^{2-} is soluble (Expt. 3), while the calcium salt of $C_2O_4^{2-}(aq)$ is insoluble (Expt. 5). Thus, chromate salts appear more soluble than oxalate salts.

4.62   (a) $Al(OH)_3(s) + 3H^+(aq) \rightarrow Al^{3+}(aq) + 3H_2O(l)$

          (b) $Mg(OH)_2(s) + 2H^+(aq) \rightarrow Mg^{2+}(aq) + 2H_2O(l)$

          (c) $MgCO_3(s) + 2H^+(aq) \rightarrow Mg^{2+}(aq) + H_2O(l) + CO_2(g)$

          (d) $NaAl(CO)_3(OH)_2(s) + 4H^+(aq) \rightarrow Na^+(aq) + Al^{3+}(aq) + 3H_2O(l) + CO_2(g)$

          (e) $CaCO_3(s) + 2H^+(aq) \rightarrow Ca^{2+}(aq) + H_2O(l) + CO_2(g)$

     [In (c), (d) and (e), one could also write the equation for formation of bicarbonate, e.g., $MgCO_3(s) + H^+(aq) \rightarrow Mg^{2+} + HCO_3^-(aq)$.]

4.63   The two precipitates formed are due to $AgCl(s)$ and $SrSO_4(s)$. Since no precipitate forms on addition of hydroxide ion to the remaining solution, the other two possibilities, Ni^{2+} and Mn^{2+}, are absent.

4.64   (a) $2H^+(aq) + SO_3^{2-}(aq) \rightarrow H_2SO_3(aq)$; sulfurous acid

          (b) $H_2SO_3(aq) \rightarrow H_2O(l) + SO_2(g)$; sulfur dioxide

          (c)   The boiling point of $SO_2(g)$ is $-10°C$. It is a gas at room temperature ($23°C$) and pressure (1 atm).

          (d)   (i) $Na_2SO_3(aq) + 2HCl(aq) \rightarrow 2NaCl(aq) + H_2O(l) + SO_2(g)$

                               $SO_3^{2-}(aq) + 2H^+(aq) \rightarrow H_2O(l) + SO_2(g)$

                 (ii) $Ag_2SO_3(s) + 2HCl(aq) \rightarrow 2AgCl(s) + H_2O(l) + SO_2(g)$

                               $Ag_2SO_3(s) + 2H^+(aq) + 2Cl^-(aq) \rightarrow 2AgCl(s) + H_2O(l) + SO_2(g)$

(iii) $KHSO_3(s) + HCl(aq) \rightarrow KCl(aq) + H_2O(l) + SO_2(g)$

$KHSO_3(s) + H^+(aq) \rightarrow K^+(aq) + H_2O(l) + SO_2(g)$

(iv) $ZnSO_3(aq) + 2HCl(aq) \rightarrow ZnCl_2(aq) + H_2O(l) + SO_2(g)$

$SO_3^{2-}(aq) + 2H^+(aq) \rightarrow H_2O(l) + SO_2(g)$

4.65   (a) $2Ti^{4+}(aq) + Zn(s) \rightarrow 2Ti^{3+}(aq) + Zn^{2+}(aq)$

The coefficients are required because the number of electrons lost by Zn ($2e^-$) must equal the number of electrons gained by Ti^{+4} ($2 \times 1e^-$).

(b)   This reaction does not indicate the position of Ti on the activity series in Table 4.4. All reactions in Table 4.4 involve oxidation of an elemental metal to a metal ion or compound. This reaction does not produce Ti metal and thus cannot be compared to the other reactions.

4.66   A metal on Table 4.4 is able to displace the metal cations below it from their compounds. That is, zinc will reduce the cations below it to their metals.

(a) $Zn(s) + Na^+(aq) \rightarrow$ no reaction

(b) $Zn(s) + Pb^{2+}(aq) \rightarrow Zn^{2+}(aq) + Pb(s)$

(c) $Zn(s) + Mg^{2+}(aq) \rightarrow$ no reaction

(d) $Zn(s) + Fe^{2+}(aq) \rightarrow Zn^{2+}(aq) + Fe(s)$

(e) $Zn(s) + Cu^{2+}(aq) \rightarrow Zn^{2+}(aq) + Cu(s)$

(f) $Zn(s) + Al^{3+}(aq) \rightarrow$ no reaction

4.67   (a)   A - La_2O_3 Metals often react with the oxygen in air to produce metal oxides.

B - $La(OH)_3$ When metals react with water (HOH) to form H_2, OH^- remains.

C - $LaCl_3$ Most chlorides are soluble.

D - $La_2(SO_4)_3$ Sulfuric acid provides SO_4^{2-} ions.

(b) $4La(s) + 3O_2(g) \rightarrow 2La_2O_3(s)$

$2La(s) + 6HOH(l) \rightarrow 2La(OH)_3(s) + 3H_2(g)$

(There are no spectator ions in either of these reactions.)

molecular: $La_2O_3(s) + 6HCl(aq) \rightarrow 2LaCl_3(aq) + 3H_2O(l)$

net ionic: $La_2O_3(s) + 6H^+(aq) \rightarrow 2La^{3+}(aq) + 3H_2O(l)$

molecular: $La(OH)_3(s) + 3HCl(aq) \rightarrow LaCl_3(aq) + 3H_2O(l)$

net ionic: $La(OH)_3(s) + 3H^+(aq) \rightarrow La^{3+}(aq) + 3H_2O(l)$

molecular: $2LaCl_3(aq) + 3H_2SO_4(aq) \rightarrow La_2(SO_4)_3(s) + 6HCl(aq)$

net ionic: $2La^{3+}(aq) + 3SO_4^{2-}(aq) \rightarrow La_2(SO_4)_3(s)$

(c)   La metal is oxidized by water to produce $H_2(g)$, so La is definitely above H on the activity series. [In fact, since an acid is not required to oxidize La, it is probably one of the more active metals. We know Al is not oxidized by water, or it could not be used in cookware. La is somewhere above Al on the activity series.]

4.68 $H_2C_4H_4O_6 + 2OH^-(aq) \rightarrow C_4H_4O_6^{2-}(aq) + 2H_2O(l)$

$$0.02262 \text{ L NaOH soln} \times \frac{0.2000 \text{ mol NaOH}}{1 \text{ L}} \times \frac{1 \text{ mol } H_2C_4H_4O_6}{2 \text{ mol NaOH}} \times \frac{1}{0.04000 \text{ L } H_2C_4H_4O_6}$$

$$= 0.05655 \ M \ H_2C_4H_4O_6 \text{ soln}$$

4.69   mol OH^- from NaOH(aq) + mol OH^- from $Zn(OH)_2$(s) = mol H^+ from HBr

mol H^+ = M HBr × L HBr = 0.550 M HBr × 0.400 L HBr = 0.220 mol H^+

mol OH^- from NaOH = M NaOH × L NaOH = 0.500 M NaOH × 0.165 L NaOH

$$= 0.0825 \text{ mol } OH^-$$

mol OH^- from $Zn(OH)_2$(s) = 0.220 mol H^+ − 0.0825 mol OH^- from NaOH = 0.1375

$$= 0.138 \text{ mol } OH^- \text{ from } Zn(OH)_2$$

$$0.1375 \text{ mol } OH^- \times \frac{1 \text{ mol } Zn(OH)_2}{2 \text{ mol } OH^-} \times \frac{99.41 \text{ g } Zn(OH)_2}{1 \text{ mol } Zn(OH)_2} = 6.83 \text{ g } Zn(OH)_2$$

Integrative Exercises

4.70   Strategy: M × L = mol Na_3PO_4 → mol Na^+ → Na^+ ions

$$\frac{0.0100 \text{ mol } Na_3PO_4}{1 \text{ L solution}} \times 1.00 \text{ mL} \times \frac{1 \text{ L}}{1000 \text{ mL}} \times \frac{3 \text{ mol } Na^+}{1 \text{ mol } Na_3PO_4} \times \frac{6.022 \times 10^{23} \ Na^+ \text{ ions}}{1 \text{ mol } Na^+}$$

$$= 1.81 \times 10^{19} \ Na^+ \text{ ions}$$

4.71   (a)   At the equivalence point of a titration, mol NaOH added = mol H^+ present

$$M_{NaOH} \times L_{NaOH} = \frac{\text{g acid}}{\text{MM acid}} \text{ (for an acid with 1 acidic hydrogen)}$$

$$\text{MM acid} = \frac{\text{g acid}}{M_{NaOH} \times L_{NaOH}} = \frac{0.2053 \text{ g}}{0.1008 \ M \times 0.0150 \text{ L}} = 136 \text{ g/mol}$$

     (b)   Assume 100 g of acid.

$$70.6 \text{ g C} \times \frac{1 \text{ mol C}}{12.01 \text{ g C}} = 5.88 \text{ mol C}; \ 5.88 \ / \ 1.47 \approx 4$$

$$5.89 \text{ g H} \times \frac{1 \text{ mol H}}{1.008 \text{ g H}} = 5.84 \text{ mol H}; \ 5.84 \ / \ 1.47 \approx 4$$

$$23.5 \text{ g O} \times \frac{1 \text{ mol O}}{16.00 \text{ g O}} = 1.47 \text{ mol O}; \ 1.47 \ / \ 1.47 = 1$$

The empirical formula is C_4H_4O.

$$\frac{\text{MM}}{\text{FW}} = \frac{136}{68.1} = 2; \text{ the molecular formula is } 2 \times \text{ the empirical formula.}$$

The molecular formula is $C_8H_8O_2$.

4.72 $Ba^{2+}(aq) + SO_4^{2-}(aq) \rightarrow BaSO_4(s)$

$$0.4123 \text{ g BaSO}_4 \times \frac{137.3 \text{ g Ba}}{233.4 \text{ g BaSO}_4} = 0.2425 \text{ g Ba}$$

$$\text{mass \%} = \frac{\text{g Ba}}{\text{g sample}} \times 100 = \frac{0.24254 \text{ g Ba}}{6.977 \text{ g sample}} \times 100 = 3.476\% \text{ Ba}$$

4.73   Strategy: $M = \dfrac{\text{mol Br}^-}{\text{L sea water}}$;   mg Br$^-$ $\rightarrow$ g Br$^-$ $\rightarrow$ mol Br$^-$;

1 kg sea water $\rightarrow$ g $\xrightarrow{\text{density}}$ water mL water $\rightarrow$ L water

$$65 \text{ mg Br}^- \times \frac{1 \text{ g Br}^-}{1000 \text{ mg Br}^-} \times \frac{1 \text{ mol Br}^-}{79.90 \text{ g Br}^-} = 8.135 \times 10^{-4} = 8.1 \times 10^{-4} \text{ mol Br}^-$$

$$1 \text{ kg sea water} \times \frac{1000 \text{ g}}{1 \text{ kg}} \times \frac{1 \text{ mL water}}{1.025 \text{ g water}} \times \frac{1 \text{ L}}{1000 \text{ mL}} = 0.9756 \text{ L}$$

$$M \text{ Br}^- = \frac{8.135 \times 10^{-4} \text{ mol Br}^-}{0.9756 \text{ L sea water}} = 8.3 \times 10^{-4} \, M \text{ Br}^-$$

4.74 $Ag^+(aq) + Cl^-(aq) \rightarrow AgCl(s)$

$$\frac{0.2997 \text{ mol Ag}^+}{1 \text{ L}} \times 0.04258 \text{ L} \times \frac{1 \text{ mol Cl}^-}{1 \text{ mol Ag}^+} \times \frac{35.453 \text{ g Cl}^-}{1 \text{ mol Cl}^-} = 0.45242 = 0.4524 \text{ g Cl}^-$$

$$25.00 \text{ mL sea water} \times \frac{1.025 \text{ g}}{\text{mL}} = 25.625 = 25.63 \text{ g sea water}$$

$$\text{mass \% Cl}^- = \frac{0.45242 \text{ g Cl}^-}{25.625 \text{ g sea water}} = 1.766\% \text{ Cl}$$

4.75 $$0.0250 \text{ L soln} \times \frac{0.102 \text{ mol Ag}^+}{1 \text{ L soln}} \times \frac{1 \text{ mol Ag}_3\text{AsO}_4}{3 \text{ mol Ag}^+} \times \frac{1 \text{ mol As}}{1 \text{ mol Ag}_3\text{AsO}_4} \times \frac{74.92 \text{ g As}}{1 \text{ mol As}}$$

$$= 0.06368 = 0.0637 \text{ g As}$$

$$\text{mass percent} = \frac{0.06368 \text{ g As}}{1.22 \text{ g sample}} \times 100 = 5.22\% \text{ As}$$

4.76   (a)   mol HCl initial - mol NH$_3$ from air = mol HCl remaining

= mol NaOH required for titration

mol NaOH = 0.0588 M × 0.0131 L = 7.703×10^{-4} = 7.70×10^{-4} mol NaOH

= 7.70×10^{-4} mol HCl remain

mol HCl initial - mol HCl remaining = mol NH$_3$ from air

(0.0105 M HCl × 0.100 L) - 7.703×10^{-4} mol HCl = mol NH$_3$

10.5×10^{-4} mol HCl - 7.703×10^{-4} mol HCl = 2.80×10^{-4} = 2.8×10^{-4} mol NH$_3$

$$2.8 \times 10^{-4} \text{ mol NH}_3 \times \frac{17.03 \text{ g NH}_3}{1 \text{ mol NH}_3} = 4.77 \times 10^{-3} = 4.8 \times 10^{-3} \text{ g NH}_3$$

(b)   ppm is defined as mL $NH_3/1 \times 10^6$ mL air. Calculate "mL" NH_3 from g NH_3 and density.

$$4.77 \times 10^{-3} \text{ g NH}_3 \times \frac{1 \text{ L NH}_3}{0.771 \text{ g NH}_3} \times \frac{1000 \text{ mL}}{1 \text{ L}} = 6.19 = 6.2 \text{ mL NH}_3$$

Calculate total volume of air processed.

$$\frac{10.0 \text{ L}}{1 \text{ min}} \times 10.0 \text{ min} \times \frac{1000 \text{ mL}}{1 \text{ L}} = 1.00 \times 10^5 \text{ mL air}$$

$$\frac{6.19 \text{ mL NH}_3}{1.00 \times 10^5 \text{ mL air}} \times \frac{10 \text{ mL}}{10 \text{ mL}} = \frac{61.9 \text{ mL NH}_3}{1.00 \times 10^6 \text{ mL air}} = 62 \text{ ppm}$$

(c)   62 ppm > 50 ppm. The manufacturer is **not** in compliance.

5 Thermochemistry

Nature of Energy

5.1   (a)   *Heat* is the transfer of energy due to a difference in temperature.

       (b)   Heat is transferred from one system to another until the two systems are at the same temperature.

5.2   (a)   *Work* is a force applied over a distance.

       (b)   The amount of work done is the magnitude of the force times the distance over which it is applied. $w = F \times d$.

5.3   (a)   Gravity; work is done because the force of gravity is opposed and the pencil is lifted.

       (b)   Mechanical force; work is done because the force of the coiled spring is opposed as the spring is compressed over a distance.

5.4   (a)   Electrostatic attraction; no work is done because the particles are held stationary.

       (b)   Magnetic attraction; work is done because the nail is moved a distance.

5.5   (a)   Since $1 \text{ J} = 1 \text{ kg} \cdot \text{m}^2/\text{s}^2$, convert g → kg to obtain E_k in joules.

$$E_k = 1/2\, mv^2 = 1/2 \times 45 \text{ g} \times \frac{1 \text{ kg}}{1000 \text{ g}} \times \left(\frac{61 \text{ m}}{1 \text{ s}}\right)^2 = \frac{84 \text{ kg} \cdot \text{m}^2}{1 \text{ s}^2} = 84 \text{ J}$$

       (b) $83.72 \text{ J} \times \dfrac{1 \text{ cal}}{4.184 \text{ J}} = 20 \text{ cal}$

       (c)   As the ball hits the tree, its speed (and hence its kinetic energy) drops to zero. Most of the kinetic energy is transferred to the potential energy of a slightly deformed golf ball, some is absorbed by the tree and some is released as heat. As the ball bounces off the tree, its potential energy is reconverted to kinetic energy.

5.6   (a)   mi/hr → m/s

$$1050 \, \frac{\text{mi}}{\text{hr}} \times \frac{1.6093 \text{ km}}{1 \text{ m}} \times \frac{1000 \text{ m}}{1 \text{ km}} \times \frac{1 \text{ hr}}{3600 \text{ s}} = 469.38 = 469.4 \text{ m/s}$$

       (b)   Find the mass of one N_2 molecule in kg.

$$\frac{28.0134 \text{ g } N_2}{1 \text{ mol}} \times \frac{1 \text{ mol}}{6.022 \times 10^{23} \text{ molecules}} \times \frac{1 \text{ kg}}{1000 \text{ g}} = 4.6518 \times 10^{-26}$$

$$= 4.652 \times 10^{-26} \text{ kg}$$

$$E_k = 1/2 \, mv^2 = 1/2 \times 4.6518 \times 10^{-26} \text{ kg} \times (469.38 \text{ m/s})^2$$

$$= 5.1244 \times 10^{-21} \frac{\text{kg} \times \text{m}^2}{\text{s}^2} = 5.124 \times 10^{-21} \text{ J}$$

(c) $\dfrac{5.1244 \times 10^{21} \text{ J}}{\text{molecule}} \times \dfrac{6.022 \times 10^{23} \text{ molecules}}{1 \text{ mol}} = 3086 \text{ J/mol} = 3.086 \text{ kJ/mol}$

5.7   Find: J/Btu

Given: heat capacity of water = 1 Btu/lb•°F

Know: heat capacity of water = 4.184 J/g•°C

Strategy: $\dfrac{\text{J}}{\text{g} \cdot {}^\circ\text{C}} \rightarrow \dfrac{\text{J}}{\text{lb} \cdot {}^\circ\text{F}} \rightarrow \dfrac{\text{J}}{\text{Btu}}$

This strategy requires changing °F to °C. Since this involves the magnitude of a degree on each scale, rather than a specific temperature, the 32 in the temperature relationship is not needed.

100 °C = 180 °F;   5 °C = 9 °F

$\dfrac{4.184 \text{ J}}{\text{g} \cdot {}^\circ\text{C}} \times \dfrac{453.6 \text{ g}}{\text{lb}} \times \dfrac{5 \, {}^\circ\text{C}}{9 \, {}^\circ\text{F}} \times \dfrac{1 \text{ lb} \cdot {}^\circ\text{F}}{1 \text{ Btu}} = 1054 \text{ J/Btu}$

5.8   Given: 1 kwh

1 watt = 1 J/s;   1 watt • s = 1 J

Strategy: kwh → wh → ws → J

$1 \text{ kwh} \times \dfrac{1000 \text{ w}}{1 \text{ kw}} \times \dfrac{60 \text{ min}}{\text{h}} \times \dfrac{60 \text{ s}}{\text{min}} \times \dfrac{1 \text{ J}}{1 \text{ w} \cdot \text{s}} = 3.6 \times 10^6 \text{ J}$

1 kwh = 3.6 × 10⁶ J

First Law of Thermodynamics

5.9   (a)   In thermodynamics, the *system* is the well-defined part of the universe whose energy changes are being studied.

         (b)   A closed system can exchange heat but not mass with its surroundings.

5.10   (a)   The solution is the system.

         (b)   The surroundings are the flask, the stopper, the air around the flask and the remainder of the universe.

         (c)   This system is closed, because the stopper prevents mass transfer with the surroundings. Energy can be transferred as heat to the flask and other surroundings.

5.11   (a)   In any chemical or physical change, energy can be neither created nor destroyed, but it can be changed in form.

(b)   The total *internal energy* (E) of a system is the sum of all the kinetic and potential energies of the system components.

(c)   The internal energy of a system increases when work is done on the system by the surroundings and/or when heat is transferred to the system from the surroundings (the system is heated).

5.12   (a) $\Delta E_{sys} = -\Delta E_{surr}$; $\Delta E_{sys} = q + w$

(b)   It is very difficult to measure the absolute internal energy of a system because it has so many components. It encompasses the kinetic energy of all moving particles in the system, including subatomic particles, and the electrostatic potential energies between all these particles. It is possible to measure the changes in internal energy (ΔE) when a system undergoes a chemical or physical change.

(c)   The quantities q and w are negative when the system loses heat to the surroundings (it cools), or does work on the surroundings.

5.13   In each case, evaluate q and w in the expression $\Delta E = q + w$. For an exothermic process, q is negative; for an endothermic process, q is positive.

(a)   q is positive because the system gains heat and w is negative because the system does work. $\Delta E = +327$ kJ $- 430$ kJ $= -103$ kJ. The process is endothermic.

(b) $\Delta E = -1.15$ kJ $- 934$ J $= -1.15$ kJ $- 0.934$ kJ $= -2.08$ kJ   The process is exothermic.

(c)   q is negative because the system loses heat and w is positive because the work is done on the system. $\Delta E = -245$ J $+ 97$ J $= -148$ J. The process is exothermic.

5.14   In each case, evaluate q and w in the expression $\Delta E = q + w$. For an exothermic process, q is negative; for an endothermic process, q is positive.

(a)   q is positive and w is negative. $\Delta E = +240$ J $- 135$ J $= +105$ J. The process is endothermic.

(b)   q is negative and w is essentially zero. $\Delta E = -225$ J. The process is exothermic.

(c)   q is negative and w is zero. $\Delta E = -5.75$ kJ. The process is exothermic.

5.15   (a)   A *state function* is a property of a system that depends only on the physical state (pressure, temperature, etc.) of the system, not on the route used by the system to get to the current state.

(b)   Internal energy and enthalpy <u>are</u> state functions; work <u>is not</u> a state function.

(c)   Temperature is a state function; regardless of how hot or cold the sample has been, the temperature depends only on its present condition.

5.16   (a)   Independent. Potential energy is a state function.

(b)   Dependent. Some of the energy released could be employed in performing work, as is done in the body when sugar is metabolized; heat is not a state function.

(c)   Dependent. The work accomplished depends on whether the gasoline is used in an engine, burned in an open flame, or in some other manner. Work is not a state function.

Enthalpy

5.17   (a)   When a process occurs under constant external pressure, the enthalpy change (ΔH) equals the amount of heat transferred. $\Delta H = q_p$.

(b)   No. Enthalpy is a state function, so it is totally defined by the current conditions (state) of the system, not the history of how the system arrived at its current state.

(c) $\Delta H = q_p$. If the system absorbs heat, q and ΔH are positive and the enthalpy of the system increases.

5.18   (a)   For the many laboratory and real world processes that occur at constant atmospheric pressure, the enthalpy change is a meaningful measure of the energy change associated with the process. At constant pressure, most of the energy change is transferred as heat ($\Delta H = q_p$), even if gases are involved in the process.

(b)   Only under conditions of constant pressure is ΔH for a process equal to the heat transferred during the process.

(c)   If ΔH is negative, the enthalpy of the system decreases and the process is exothermic.

5.19   (a) $CH_3OH(l) + 3/2\ O_2(g) \rightarrow 2\ H_2O(l) + CO_2(g)$

$\Delta H = -726.7$ kJ

(b) $CH_3OH(l) + 3/2\ O_2(g)$

$\Delta H = -726.7$ kJ

$2\ H_2O(l) + CO_2(g)$

5.20   (a) $2NH_3(g) \rightarrow N_2(g) + 3H_2(g)$

$NH_3(g) \rightarrow 1/2\ N_2(g) + 3/2\ H_2(g)$

$\Delta H = 46.19$ kJ

(b) $1/2\ N_2(g) + 3/2\ H_2(g)$

$\Delta H = 46.19$ kJ

$NH_3(g)$

5.21   (a)   Exothermic (ΔH is negative).

(b) $5.6\text{ g Na} \times \dfrac{1\text{ mol}}{22.99\text{ g}} \times \dfrac{-821.8\text{ kJ}}{2\text{ mol Na}} = 1.0 \times 10^2 \text{ kJ}$

(c) $16.5\text{ kJ} \times \dfrac{2\text{ mol NaCl}}{821.8\text{ kJ}} \times \dfrac{58.44\text{ g NaCl}}{1\text{ mol NaCl}} = 2.35\text{ g NaCl produced}$

(d) $2NaCl(s) \rightarrow 2Na(s) + Cl_2(g)$ $\Delta H = +821.8$ kJ

This is the reverse of the reaction given above, so the sign of ΔH is reversed.

$$44.1g\ NaCl \times \frac{1\ mol\ NaCl}{58.44\ g\ NaCl} \times \frac{821.8\ kJ}{2\ mol\ NaCl} = 310\ kJ\ absorbed$$

5.22   (a)   Endothermic (ΔH is positive).

(b) $25.0\ g\ N_2O \times \dfrac{1\ mol}{44.02\ g} \times \dfrac{163.2\ kJ}{2\ mol\ N_2O} = +46.3$ kJ of heat transferred   (ΔH_{sys} increases)

(c) $5.00\ kJ \times \dfrac{2\ mol\ N_2}{163.2\ kJ} \times \dfrac{28.02\ g\ N_2}{1\ mol\ N_2} = 1.72\ g\ N_2$

(d)   Decomposition is the reverse of the given reaction, so the sign of ΔH is reversed.

$$10.0\ g\ N_2O \times \frac{1\ mol}{44.02\ g} \times \frac{-163.2\ kJ}{2\ mol\ N_2O} = -18.5\ kJ\ of\ heat\ produced\ \ (\Delta H_{sys}\ decreases)$$

5.23   (a) $0.715\ mol\ O_2 \times \dfrac{-89.4\ kJ}{3\ mol\ O_2} = -21.3$ kJ

(b) $6.14\ g\ KCl \times \dfrac{1\ mol\ KCl}{74.55\ g\ KCl} \times \dfrac{-89.4\ kJ}{2\ mol\ KCl} = -3.68$ kJ

(c) $12.3\ g\ KClO_3 \times \dfrac{1\ mol\ KClO_3}{122.6g\ KClO_3} \times \dfrac{+89.4\ kJ}{2\ mol\ KClO_3} = +4.48$ kJ   (sign of ΔH reversed)

5.24   (a) $0.500\ mol\ AgCl \times \dfrac{-65.5\ kJ}{mol\ AgCl} = -32.8$ kJ

(b) $7.00\ g\ AgCl \times \dfrac{1\ mol\ AgCl}{143.3\ g\ AgCl} \times \dfrac{-65.5\ kJ}{mol\ AgCl} = -3.20$ kJ

(c) $0.650\ mol\ AgCl \times \dfrac{+65.5\ kJ}{1\ mol\ AgCl} = +42.6$ kJ (sign of ΔH reversed)

5.25   Enthalpy of $H_2O(s) < H_2O(l) < H_2O(g)$. Heat must be added to convert $s \rightarrow l \rightarrow g$.

5.26 O_3 has the higher enthalpy (ΔH is positive).

5.27   At constant pressure, $\Delta E = \Delta H - P\Delta V$. In order to calculate ΔE, more information about the conditions of the reaction must be known. For an ideal gas at constant pressure and temperature, $P\Delta V = RT\Delta n$. The values of either P and ΔV or T and Δn must be known to calculate ΔE from ΔH.

5.28   (a)   At constant volume, q_v is not a direct measure of ΔH.

(b)   At constant volume, $q_v = \Delta E$. For this reaction, the sign of q_v is negative because heat is transferred to the surroundings. $\Delta E = -35$ kJ.

5.29   q = -135 kJ (heat is given off by the system), w = -63 kJ (work is done by the system).

ΔE = q + w = -135 kJ - 63 kJ = -198 kJ. ΔH = q = -135 kJ (at constant pressure).

5.30   The gas is the system. If 600 J of heat is added, q = +600 J. Work done by the system decreases the overall energy of the system, so w = -140 J.

ΔE = q + w = +600 J - 140 J = +460 J. At constant pressure, ΔH = q; ΔH = +600 J.

5.31   (a) $2CO_2(g) + 3H_2O(g) \rightarrow C_2H_6(g) + 7/2\ O_2(g)$ ΔH = +1430 kJ

(b) $2C_2H_6(g) + 7O_2(g) \rightarrow 4CO_2(g) + 6H_2O(g)$ ΔH = 2(-1430) kJ = -2860 kJ

(c)   The exothermic forward reaction is more likely to be thermodynamically favored.

(d)   Vaporization (liquid → gas) is endothermic so the reverse process, condensation (gas → liquid) is exothermic. If the product was $H_2O(l)$, the reaction would be more exothermic and ΔH would have a larger negative value.

5.32   (a) $C_2H_2(g) \rightarrow 1/3\ C_6H_6(l)$ ΔH = -210 kJ

(b) $C_6H_6(l) \rightarrow 3C_2H_2(g)$ ΔH = 3(+210) kJ = +630 kJ

(c)   The exothermic reverse reaction is more likely to be thermodynamically favored.

(d)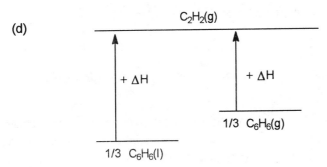

If the reactant is in the higher enthalpy gas phase, the overall ΔH for the reaction has a smaller positive value.

Calorimetry

The specific heat of water to four significant figures, **4.184 J/g • K**, will be used in many of the following exercises; temperature units of K and °C will be used interchangeably.

5.33   (a) $\dfrac{4.184\ J}{1\ g \cdot K}$ or $\dfrac{4.184\ J}{1\ g \cdot °C}$ (b) $\dfrac{265\ g\ H_2O \times 4.184\ J}{1\ g \cdot °C} = \dfrac{1.11 \times 10^3\ J}{°C} = 1.11\ kJ/°C$

(c) $1.00\ kg\ H_2O \times \dfrac{1000\ g}{1\ kg} \times \dfrac{4.184\ J}{1\ g \cdot °C} \times \dfrac{1\ kJ}{1000\ J} \times 25.0\ °C = 105\ kJ$

5.34   (a) $\dfrac{4.184 \text{ J}}{1 \text{ g} \cdot {}^\circ\text{C}} \times \dfrac{18.02 \text{ g H}_2\text{O}}{1 \text{ mol H}_2\text{O}} = \dfrac{75.40 \text{ J}}{\text{mol} \cdot {}^\circ\text{C}}$

        (b) $2.15 \text{ mol H}_2\text{O} \times \dfrac{75.40 \text{ J}}{\text{mol} \cdot {}^\circ\text{C}} = 162 \text{ J/}{}^\circ\text{C}$

        (c) $138 \text{ mol H}_2\text{O} \times \dfrac{75.40 \text{ J}}{\text{mol} \cdot {}^\circ\text{C}} \times (77.1{}^\circ\text{C} - 10.5{}^\circ\text{C}) = 6.93 \times 10^5 \text{ J} = 693 \text{ kJ}$

5.35 $156 \text{ g Si} \times \dfrac{0.702 \text{ J}}{\text{g} \cdot \text{K}} \times (37.5 \ {}^\circ\text{C} - 25.0 \ {}^\circ\text{C}) = 1.37 \times 10^3 \text{ J}$

5.36 $193 \text{ g C}_2\text{H}_5\text{Br} \times \dfrac{0.924 \text{ J}}{\text{g} \cdot \text{K}} \times (35.0 \ {}^\circ\text{C} - 19.0 \ {}^\circ\text{C}) = 2.85 \times 10^3 \text{ J}$

5.37   Since the temperature of the water increases, the dissolving process is exothermic and the sign of ΔH is negative. The heat lost by the NaOH(s) dissolving equals the heat gained by the solution.

Calculate the heat gained by the solution. The temperature change is $37.8 - 21.6 = 16.2{}^\circ\text{C}$. The total mass of solution is (100.0 g H_2O + 6.50 g NaOH) = 106.5 g.

$106.5 \text{ g solution} \times \dfrac{4.184 \text{ J}}{1 \text{ g} \cdot {}^\circ\text{C}} \times 16.2{}^\circ\text{C} \times \dfrac{1 \text{ kJ}}{1000 \text{ J}} = 7.219 = 7.22 \text{ kJ}$

This is the amount of heat lost when 6.50 g of NaOH dissolves.

The heat loss per mole NaOH is

$\dfrac{-7.219 \text{ kJ}}{6.50 \text{ g NaOH}} \times \dfrac{40.00 \text{ g NaOH}}{1 \text{ mol NaOH}} = -44.4 \text{ kJ/mol} \quad \Delta H = q_p = -44.4 \text{ kJ/mol NaOH}$

5.38   Following the logic in Exercise 5.37, the dissolving process is endothermic, ΔH is positive. The total mass of the solution is (60.0 g H_2O + 4.25 g NH_4NO_3) = 64.25 g. The temperature change of the solution is $22.0 - 16.9 = 5.10{}^\circ\text{C}$. The heat lost by the water is

$64.25 \text{ g H}_2\text{O} \times \dfrac{4.184 \text{ J}}{1 \text{ g} \cdot {}^\circ\text{C}} \times 5.10 \ {}^\circ\text{C} \times \dfrac{1 \text{ kJ}}{1000 \text{ J}} = 1.371 = 1.37 \text{ kJ}$

Thus, 1.37 kJ is absorbed when 4.25 g NH_4NO_3(s) dissolves.

$\dfrac{1.371 \text{ kJ}}{4.25 \text{ NH}_4\text{NO}_3} \times \dfrac{80.05 \text{ g NH}_4\text{NO}_3}{1 \text{ mol NH}_4\text{NO}_3} = +25.8 \text{ kJ/mol NH}_4\text{NO}_3$

5.39 $q_{bomb} = -q_{rxn};\ \Delta T = 30.57{}^\circ\text{C} - 23.44{}^\circ\text{C} = 7.13{}^\circ\text{C}$

$q_{bomb} = \dfrac{7.854 \text{ kJ}}{1{}^\circ\text{C}} \times 7.13{}^\circ\text{C} = 56.00 = 56.0 \text{ kJ}$

At constant volume, $q_v = \Delta E$. ΔE and ΔH are very similar.

$$\Delta E_{rxn} = q_{rxn} = -q_{bomb} = \frac{-56.0 \text{ kJ}}{2.20 \text{ g C}_6\text{H}_4\text{O}_2} = -25.454 = -25.5 \text{ kJ/g C}_6\text{H}_4\text{O}_2$$

$$\Delta E_{rxn} = \frac{-25.454 \text{ kJ}}{1 \text{ g C}_6\text{H}_4\text{O}_2} \times \frac{108.1 \text{ g C}_6\text{H}_4\text{O}_2}{1 \text{ mol C}_6\text{H}_4\text{O}_2} = -2.75 \times 10^3 \text{ kJ/mol C}_6\text{H}_4\text{O}_2$$

5.40 $q_{bomb} = -q_{rxn}$; $\Delta T = 28.78°C - 21.36°C = 7.42°C$

$$q_{bomb} = \frac{11.66 \text{ kJ}}{1°C} \times 7.42°C = 86.52 = 86.5 \text{ kJ}$$

At constant volume, $q_v = \Delta E$. ΔE and ΔH are very similar.

$$\Delta E_{rxn} = q_{rxn} = -q_{bomb} = \frac{-86.52 \text{ kJ}}{1.80 \text{ g C}_8\text{H}_{18}} = -48.07 = -48.1 \text{ kJ/g C}_8\text{H}_{18}$$

$$\Delta E_{rxn} = \frac{-48.07 \text{ kJ}}{1 \text{ g C}_8\text{H}_{18}} \times \frac{114.2 \text{ g C}_8\text{H}_{18}}{1 \text{ mol C}_8\text{H}_{18}} = \frac{-5.490 \times 10^3 \text{ kJ}}{\text{mol C}_8\text{H}_{18}} = -5.49 \times 10^3 \text{ kJ/mol C}_8\text{H}_{18}$$

5.41   (a) $C_{total} = 2.500 \text{ g glucose} \times \frac{15.57 \text{ kJ}}{1 \text{ g glucose}} \times \frac{1}{2.70°C} = 14.42 = 14.4 \text{ kJ/°C}$

(b) $C_{H_2O} = 2.700 \text{ kg H}_2\text{O} \times \frac{4.184 \text{ kJ}}{1 \text{ kg} \cdot °C} = 11.30 \text{ kJ/°C}$

$$C_{empty \text{ calorimeter}} = \frac{14.42 \text{ kJ}}{1°C} - \frac{11.30 \text{ kJ}}{1°C} = 3.12 = 3.1 \text{ kJ/°C}$$

(c) $q = 2.500 \text{ g glucose} \times \frac{15.57 \text{ kJ}}{1 \text{ g glucose}} = 38.93 \text{ kJ produced}$

$$C_{H_2O} = 2.000 \text{ kg H}_2\text{O} \times \frac{4.184 \text{ kJ}}{1 \text{ kg} \cdot °C} = 8.368 \text{ kJ/°C}$$

$$C_{total} = \frac{8.368 \text{ kJ}}{1°C} + \frac{3.12 \text{ kJ}}{1°C} = 11.49 = 11.5 \text{ kJ/°C}$$

$$38.98 \text{ kJ} = \frac{11.49 \text{ kJ}}{°C} \times \Delta T; \ \Delta T = 3.39°C$$

5.42   (a) $C = 1.200 \text{ g HC}_7\text{H}_5\text{O}_2 \times \frac{26.38 \text{ kJ}}{1 \text{ g HC}_7\text{H}_5\text{O}_2} \times \frac{1}{3.65°C} = 8.673 = 8.67 \text{ kJ/°C}$

(b) $C_{total} = C_{H_2O} + C_{empty \text{ calorimeter}}$

$$C_{H_2O} = 1.500 \text{ kg H}_2\text{O} \times \frac{4.184 \text{ kJ}}{1 \text{ kg} \cdot °C} = 6.276 \text{ kJ/°C}$$

$$C_{empty \text{ calorimeter}} = \frac{8.673 \text{ kJ}}{1°C} - \frac{6.276 \text{ kJ}}{1°C} = 2.397 = 2.40 \text{ kJ/°C}$$

(c) $q = 1.200 \text{ g } HC_7H_5O_2 \times \dfrac{26.38 \text{ kJ}}{1 \text{ g } HC_7H_5O_2} = 31.66 \text{ kJ produced}$

$$C_{H_2O} = 1000 \text{ g } H_2O \times \frac{4.184 \text{ kJ}}{1 \text{ kg} \cdot {}^\circ C} = \frac{4.184 \text{ kJ}}{{}^\circ C}; \quad C_{empty \ calorimeter} = 2.40 \text{ kJ/}^\circ C$$

$$31.66 \text{ kJ} = \left(\frac{4.184 \text{ kJ}}{1^\circ C} + \frac{2.397 \text{ kJ}}{1^\circ C} \right) \times \Delta T; \quad \Delta T = 4.81\,^\circ C$$

Hess's Law

5.43   If a reaction can be described as a series of steps, ΔH for the reaction is the sum of the enthalpy changes for each step. As long as we can describe a route where ΔH for each step is known, ΔH for any process can be calculated.

5.44   Hess's Law is a consequence of the fact that enthalpy is a state function. Since ΔH is independent of path, we can describe a process by any series of steps that add up to the overall process and ΔH for the process is the sum of the ΔH values for the steps.

5.45   (a) $A \rightarrow B$ $\Delta H = +30 \text{ kJ}$
                  $\underline{B \rightarrow C \quad \Delta H = +60 \text{ kJ}}$
                  $A \rightarrow C$ $\Delta H = +90 \text{ kJ}$

      (b)

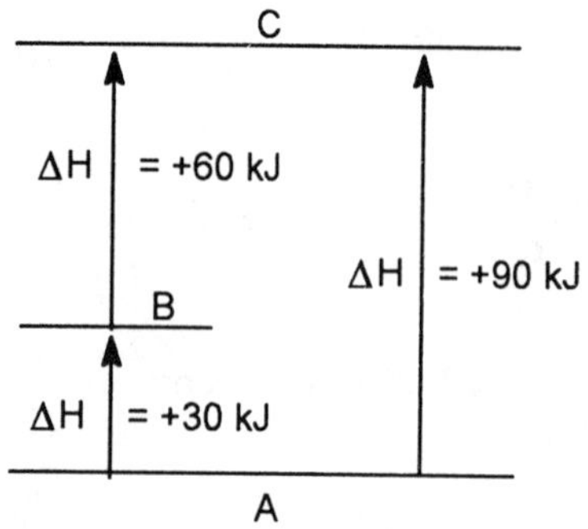

The process of A forming C can be described as A forming B and B forming C.

5.46   (a) $Y \rightarrow X$ $\Delta H = +40 \text{ kJ}$
                  $\underline{X \rightarrow Z \quad \Delta H = -95 \text{ kJ}}$
                  $Y \rightarrow Z$ $\Delta H = -55 \text{ kJ}$

      (b)

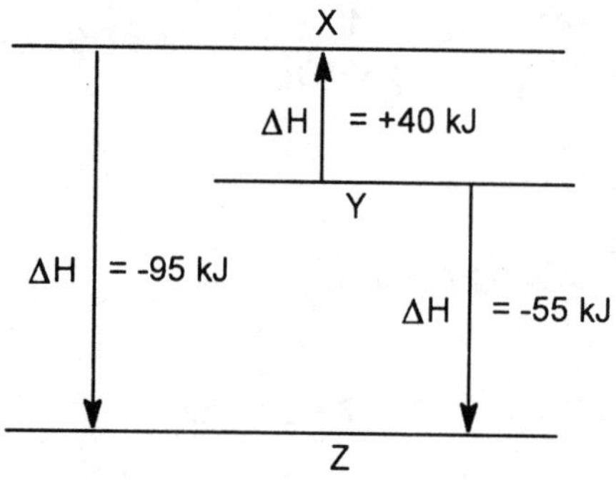

The process of Y forming Z can be described as Y forming X and X forming Z.

5.47 $N_2(g) + 2O_2(g) \rightarrow 2NO_2(g)$ $\Delta H = + 67.6$ kJ

              $2NO_2(g) \rightarrow O_2(g) + 2NO(g)$ $\Delta H = +113.2$ kJ

       $N_2(g) + O_2(g) \rightarrow 2NO(g)$ $\Delta H = +180.8$ kJ

5.48 $2S(s) + 3O_2(g) \rightarrow 2SO_3(g)$ $\Delta H = -790$ kJ

           $2SO_3(g) \rightarrow 2SO_2(g) + O_2(g)$ $\Delta H = +196$ kJ

      $2S(s) + 2O_2(g) \rightarrow 2SO_2(g)$ $\Delta H = -594$ kJ

      $S(s) + O_2(g) \rightarrow SO_2(g)$ $\Delta H = 1/2\,(-594$ kJ$) = -297$ kJ

5.49 $C_2H_4(g) \rightarrow 2\,H_2(g) + 2C(s)$ $\Delta H = -52.3$ kJ

       $2C(s) + 4F_2(g) \rightarrow 2CF_4(g)$ $\Delta H = 2(-680$ kJ$)$

     $2H_2(g) + 2F_2(g) \rightarrow 4HF(g)$ $\Delta H = 2(-537$ kJ$)$

    $C_2H_4(g) + 6F_2(g) \rightarrow 2CF_4(g) + 4HF(g)$ $\Delta H = -2.49 \times 10^3$ kJ

5.50 $2CO_2(g) + 3H_2O(l) \rightarrow C_2H_6(g) + 7/2\,O_2(g)$ $\Delta H = 1/2\,(+3120$ kJ$)$

        $2C(s) + 2O_2(g) \rightarrow 2CO_2(g)$ $\Delta H = 2(-394$ kJ$)$

    $3H_2(g) + 3/2\,O_2(g) \rightarrow 3H_2O(l)$ $\Delta H = 3/2\,(-572$ kJ$)$

     $2C(s) + 3H_2(g) \rightarrow C_2H_6(g)$ $\Delta H = -86$ kJ

Enthalpies of Formation

5.51   (a)   *Standard conditions* for enthalpy changes are usually P = 1 atm and T = 298 K. For the purpose of comparison, standard enthalpy changes, $\Delta H°$, are tabulated for reactions at these conditions.

        (b)   *Enthalpy of formation*, ΔH_f, is the enthalpy change that occurs when a compound is formed from its component elements.

        (c)   *Standard enthalpy of formation*, $\Delta H_f°$, is the enthalpy change that accompanies formation of one mole of a substance from elements in their standard states.

5.52   (a)   A standard enthalpy (or any standard quantity) is indicated with a small superscript zero following the symbol, as in $\Delta H°$.

        (b)   Tables of $\Delta H_f°$ are useful because, according to Hess's law, the standard enthalpy of any reaction can be calculated from the standard enthalpies of formation for the reactants and products.

$$\Delta H_{rxn}° = \sum \Delta H_f° \text{ (products)} - \sum \Delta H_f° \text{ (reactants)}$$

        (c)   The standard enthalpy of formation for any element in its standard state is zero. Elements in their standard states are the reference point for the enthalpy of formation scale.

5.53   (a) $H_2(g) + O_2(g) \rightarrow H_2O_2(g)$ $\Delta H_f^\circ = -136.10$ kJ

        (b) $N_2(g) + 1/2\ O_2(g) \rightarrow N_2O(g)$ $\Delta H_f^\circ = +81.6$ kJ

        (c) $Pb(s) + C(s, gr) + 3/2\ O_2(g) \rightarrow PbCO_3(s)$ $\Delta H_f^\circ = -699.1$ kJ

        (d) $Na(s) + 1/2\ H_2(g) + C(s, gr) + 3/2\ O_2(g) \rightarrow NaHCO_3(s)$ $\Delta H_f^\circ = -947.7$ kJ

5.54   (a) $2Fe(s) + 3/2\ O_2 \rightarrow Fe_2O_3(s)$ $\Delta H_f^\circ = -822.16$ kJ

        (b) $Si(s) + 2Cl_2(g) \rightarrow SiCl_4(l)$ $\Delta H_f^\circ = -640.1$ kJ

        (c) $2C(s, gr) + 3H_2(g) + 1/2\ O_2(g) \rightarrow C_2H_5OH(g)$ $\Delta H_f^\circ = -235.1$ kJ

        (d) $Pb(s) + N_2(g) + 3O_2(g) \rightarrow Pb(NO_3)_2(s)$ $\Delta H_f^\circ = -451.9$ kJ

5.55   Use heats of formation to calculate ΔH° for the combustion of butane.

$$C_4H_{10}(l) + 13/2\ O_2(g) \rightarrow 4CO_2(g) + 5H_2O(l)$$

$$\Delta H_{rxn}^\circ = 4\Delta H_f^\circ\ CO_2(g) + 5\Delta H_f^\circ\ H_2O(l) - \Delta H_f^\circ\ C_4H_{10}(l)$$

$$\Delta H_{rxn}^\circ = 4(-393.5\ kJ) + 5(-285.83\ kJ) - (-147.6\ kJ) = -2855.6 = -2856\ kJ/mol\ C_4H_{10}$$

$$1.0\ g\ C_4H_{10} \times \frac{1\ mol\ C_4H_{10}}{58.123\ g\ C_4H_{10}} \times \frac{-2855.6\ kJ}{1\ mol\ C_4H_{10}} = 49\ kJ$$

5.56 $\Delta H_{rxn}^\circ = \Delta H_f^\circ\ Al_2O_3(s) + 2\Delta H_f^\circ\ Fe(s) - \Delta H_f^\circ\ Fe_2O_3 - \Delta H_f^\circ\ Al(s)$

         $\Delta H_{rxn}^\circ = (-1669.8\ kJ) + 2(0) - (-822.16\ kJ) - 2(0) = -847.6\ kJ$

5.57   (a) $\Delta H_{rxn}^\circ = \Delta H_f^\circ\ N_2O_4(g) - 2\Delta H_f^\circ\ NO_2(g)$

                $= 9.66\ kJ - 2(33.84\ kJ) = -58.02\ kJ$

        (b) $\Delta H_{rxn}^\circ = \Delta H_f^\circ\ CaO(s) + \Delta H_f^\circ\ CO_2(g) - \Delta H_f^\circ\ CaCO_3(s)$

                $= -635.5\ kJ + (-393.5\ kJ) - (-1207.1\ kJ) = +178.1\ kJ$

        (c) $\Delta H_{rxn}^\circ = 2\Delta H_f^\circ\ NH_3(g) - \Delta H_f^\circ\ N_2(g) - 3\Delta H_f^\circ\ H_2(g)$

                $= 2(-46.19\ kJ) - 0 - 3(0) = -92.38\ kJ$

        (d) $\Delta H_{rxn}^\circ = 2\Delta H_f^\circ\ FeCl_3(s) + 3\Delta H_f^\circ\ H_2O(g) - \Delta H_f^\circ\ Fe_2O_3(s) - 6\Delta H_f^\circ\ HCl(g)$

                $= 2(-400\ kJ) + 3(-241.82\ kJ) - (-822.16\ kJ) - 6(-92.30\ kJ) = -150\ kJ$

5.58   (a) $\Delta H_{rxn}^\circ = \Delta H_f^\circ\ NH_4Cl(s) - \Delta H_f^\circ\ NH_3(g) - \Delta H_f^\circ\ HCl(g)$

                $= -314.4\ kJ - (-46.19\ kJ) - (-92.30\ kJ) = -175.9\ kJ$

        (b) $\Delta H_{rxn}^\circ = 2\Delta H_f^\circ\ CH_3OH(l) - \Delta H_f^\circ\ C_2H_4(g) - 2\Delta H_f^\circ\ H_2O(g)$

                $= 2(-238.6\ kJ) - (52.30\ kJ) - 2(-241.82\ kJ) = -45.9\ kJ$

(c) $\Delta H^{\circ}_{rxn} = 3\Delta H^{\circ}_f\ POCl_3(l) + \Delta H^{\circ}_f\ KCl(s) - 3\Delta H^{\circ}_f\ PCl_3(l) - \Delta H^{\circ}_f\ KClO_3(s)$

       $= 3(-597.0\ kJ) + (-435.9\ kJ) - 3(-319.6\ kJ) - (-391.2\ kJ) = -876.9\ kJ$

(d) $\Delta H^{\circ}_{rxn} = 4\Delta H^{\circ}_f\ NO(g) + 6\Delta H^{\circ}_f\ H_2O(g) - 4\Delta H^{\circ}_f\ NH_3(g) - 5\Delta H^{\circ}_f\ O_2(g)$

       $= 4(90.37\ kJ) + 6(-241.82\ kJ) - 4(-46.19\ kJ) - 5(0) = -904.7\ kJ$

5.59 $\Delta H^{\circ}_{rxn} = 3\Delta H^{\circ}_f\ CO_2(g) + 3\Delta H^{\circ}_f\ H_2O(l) - \Delta H^{\circ}_f\ C_3H_6O(l)$

      $-1790\ kJ = 3(-393.5\ kJ) + 3(-285.83\ kJ) - \Delta H^{\circ}_f\ C_3H_6O(l)$

      $\Delta H^{\circ}_f\ C_3H_6O(l) = -248\ kJ$

5.60 $\Delta H^{\circ}_{rxn} = \Delta H^{\circ}_f\ Ca(OH)_2(s) + \Delta H^{\circ}_f\ C_2H_2(g) - 2\Delta H^{\circ}_f\ H_2O(l) - \Delta H^{\circ}_f\ CaC_2(s)$

      $-127.2\ kJ = -986.2\ kJ + 226.7\ kJ - 2(-285.83\ kJ) - \Delta H^{\circ}_f\ CaC_2(s)$

      ΔH°_f for $CaC_2(s) = -60.6\ kJ$

5.61   (a) $C_8H_{18}(l) + 25/2\ O_2(g) \rightarrow 8CO_2(g) + 9H_2O(g)$ $\Delta H^{\circ} = -5069\ kJ$

      (b) $8C(s, gr) + 9H_2(g) \rightarrow C_8H_{18}(l)$ $\Delta H^{\circ}_f = ?$

      (c) $\Delta H^{\circ}_{rxn} = 8\Delta H^{\circ}_f\ CO_2(g) + 9\Delta H^{\circ}_f\ H_2O(g) - \Delta H^{\circ}_f\ C_8H_{18}(l) - 25/2\ \Delta H^{\circ}_f\ O_2(g)$

         $-5069\ kJ = 8(-393.5\ kJ) + 9(-241.82\ kJ) - \Delta H^{\circ}_f\ C_8H_{18}(l) - 0$

         $\Delta H^{\circ}_f\ C_8H_{18}(l) = 8(-393.5\ kJ) + 9(-241.82\ kJ) + 5069\ kJ = -255\ kJ$

5.62   (a) $C_2H_6O_2(l) + 5/2\ O_2(g) \rightarrow 2CO_2(g) + 3H_2O(l)$ $\Delta H^{\circ} = -1190\ kJ$

      (b) $2C(s, gr) + 3H_2(g) + O_2(g) \rightarrow C_2H_6O_2(l)$ $\Delta H^{\circ}_f = ?$

      (c) $\Delta H^{\circ}_{rxn} = 2\Delta H^{\circ}_f\ CO_2(g) + 3\Delta H^{\circ}_f\ H_2O(l) - \Delta H^{\circ}_f\ C_2H_6O_2(l) - 5/2\ \Delta H^{\circ}_f\ O_2(g)$

         $-1190\ kJ = 2(-393.5\ kJ) + 3(-285.83\ kJ) - \Delta H^{\circ}_f\ C_2H_6O_2(l) - 0$

         $\Delta H^{\circ}_f\ C_2H_6O_2(l) = 2(-393.5\ kJ) + 3(-285.83\ kJ) + 1190\ kJ = -454\ kJ$

5.63 $2B(s) + 3/2\ O_2(g) \rightarrow B_2O_3(s)$ $\Delta H^{\circ} = 1/2(-2509.1\ kJ)$

      $3H_2(g) + 3/2\ O_2(g) \rightarrow 3H_2O(l)$ $\Delta H^{\circ} = 3/2(-571.7\ kJ)$

      $B_2O_3(s) + 3H_2O(l) \rightarrow B_2H_6(g) + 3O_2(g)$ $\Delta H^{\circ} = -(-2147.5\ kJ)$

      ————————————————————————

      $2B(s) + 3H_2(g) \rightarrow B_2H_6(g)$ $\Delta H^{\circ}_f = +35.4\ kJ$

5.64 $Mg(s) + 1/2\ O_2(g) \rightarrow MgO(s)$ $\Delta H^{\circ} = 1/2(-1203.6\ kJ)$

      $MgO(s) + H_2O(l) \rightarrow Mg(OH)_2(s)$ $\Delta H^{\circ} = -(37.1\ kJ)$

      $H_2(g) + 1/2\ O_2(g) \rightarrow H_2O(l)$ $\Delta H^{\circ} = 1/2(-571.7\ kJ)$

      ————————————————————————

      $Mg(s) + O_2(g) + H_2(g) \rightarrow Mg(OH)_2(s)$ $\Delta H^{\circ}_f = -924.8\ kJ$

Foods and Fuels

5.65   Calculate the Cal (kcal) due to each nutritional component of the Campbell's® soup, then sum.

$$9 \text{ g carbohydrates} \times \frac{17 \text{ kJ}}{1 \text{ g carbohydrate}} = 153 \text{ or } 2 \times 10^2 \text{ kJ}$$

$$1 \text{ g protein} \times \frac{17 \text{ kJ}}{1 \text{ g protein}} = 17 \text{ or } 0.2 \times 10^2 \text{ kJ}$$

$$7 \text{ g fat} \times \frac{38 \text{ kJ}}{1 \text{ g fat}} = 266 \text{ or } 3 \times 10^2 \text{ kJ}$$

total energy = 153 kJ + 17 kJ + 266 kJ = 436 or 4×10^2 kJ

$$436 \text{ kJ} \times \frac{1 \text{ kcal}}{4.184 \text{ kJ}} \times \frac{1 \text{ Cal}}{1 \text{ kcal}} = 104 \text{ or } 1 \times 10^2 \text{ Cal/serving}$$

5.66   Calculate the fuel value in a pound of M&M® candies.

$$96 \text{ fat} \times \frac{38 \text{ kJ}}{1 \text{ g fat}} = 3648 \text{ kJ} = 3.6 \times 10^3$$

$$320 \text{ g carbohydrate} \times \frac{17 \text{ kJ}}{1 \text{ g carbohydrate}} = 5440 \text{ kJ} = 5.4 \times 10^3$$

$$21 \text{ g carbohydrate} \times \frac{17 \text{ kJ}}{1 \text{ g carbohydrate}} = 357 \text{ kJ} = 3.6 \times 10^2 \text{ kJ}$$

total fuel value = 3648 kJ + 5440 kJ + 357 kJ = 9445 kJ = 9.4×10^3 kJ/lb

$$\frac{9445 \text{ kJ}}{\text{lb}} \times \frac{1 \text{ lb}}{453.6 \text{ g}} \times \frac{42 \text{ g}}{\text{serving}} = 874.5 \text{ kJ} = 8.7 \text{ kJ/serving}$$

$$\frac{874.5 \text{ kJ}}{\text{serving}} \times \frac{1 \text{ kcal}}{4.184 \text{ kJ}} \times \frac{1 \text{ Cal}}{1 \text{ kcal}} = 209.0 \text{ Cal} = 2.1 \times 10^2 \text{ Cal/serving}$$

5.67 $16.0 \text{ g } C_6H_{12}O_6 \times \dfrac{1 \text{ mol } C_6H_{15}O_6}{180.2 \text{ g } C_6H_{12}O_6} \times \dfrac{2812 \text{ kJ}}{\text{mol } C_6H_{12}O_6} \times \dfrac{1 \text{ Cal}}{4.184 \text{ kJ}} = 59.7 \text{ Cal}$

5.68 $355 \text{ mL} \times \dfrac{1.0 \text{ g beer}}{1 \text{ mL}} \times \dfrac{0.037 \text{ g ethanol}}{1 \text{ g beer}} \times \dfrac{1 \text{ mol ethanol}}{46.1 \text{ g ethanol}} \times \dfrac{1371 \text{ kJ}}{1 \text{ mol ethanol}} \times \dfrac{1 \text{ Cal}}{4.184 \text{ kJ}}$
$$= 93 \text{ Cal}$$

Notice that this is more than half the calorie content of ordinary beers (about 150 Cal/12 oz bottle), and nearly all the calorie content of so-called "light" beers.

5.69   Propyne: $C_3H_4(g) + 4O_2(g) \rightarrow 3CO_2(g) + 2H_2O(g)$

(a) $\Delta H = 3(-393.5 \text{ kJ}) + 2(-241.82 \text{ kJ}) - (185.4 \text{ kJ}) = -1849.5 = -1850 \text{ kJ/mol } C_3H_4$

(b) $\dfrac{-1849.5 \text{ kJ}}{1 \text{ mol } C_3H_4} \times \dfrac{1 \text{ mol } C_3H_4}{40.065 \text{ g } C_3H_4} \times \dfrac{1000 \text{ g } C_3H_4}{1 \text{ kg } C_3H_4} = -4.616 \times 10^4 \text{ kJ/kg } C_3H_4$

Propylene: $C_3H_6(g) + 9/2\ O_2(g) \rightarrow 3CO_2(g) + 3H_2O(g)$

(a) $\Delta H = 3(-393.5\ kJ) + 3(-241.82\ kJ) - (20.4\ kJ) = -1926.4 = -1926\ kJ/mol\ C_3H_6$

(b) $\dfrac{-1926.4\ kJ}{1\ mol\ C_3H_6} \times \dfrac{1\ mol\ C_3H_6}{42.080\ g\ C_3H_6} \times \dfrac{1000\ g\ C_3H_6}{1\ kg\ C_3H_6} = -4.578 \times 10^4\ kJ/kg\ C_3H_6$

Propane: $C_3H_8(g) + 5O_2(g) \rightarrow 3CO_2(g) + 4H_2O(g)$

(a) $\Delta H = 3(-393.5\ kJ) + 4(-241.82\ kJ) - (-103.8\ kJ) = -2044.0 = -2044\ kJ/mol\ C_3H_8$

(b) $\dfrac{-2044.0\ kJ}{1\ mol\ C_3H_8} \times \dfrac{1\ mol\ C_3H_8}{44.096\ g\ C_3H_8} \times \dfrac{1000\ g\ C_3H_8}{1\ kg\ C_3H_8} = -4.635 \times 10^4\ kJ/kg\ C_3H_8$

(c)   These three substances yield nearly identical quantities of heat per unit mass, but propane is marginally higher than the other two.

5.70   For nitroethane:

$\dfrac{1348\ kJ}{1\ mol\ C_2H_5NO_2} \times \dfrac{1\ mol\ C_2H_5NO_2}{75.072\ g\ C_2H_5NO_2} \times \dfrac{1.052\ g\ C_2H_5NO_2}{1\ cm^3} = 18.89\ kJ/cm^3$

For ethanol:

$\dfrac{1371\ kJ}{1\ mol\ C_2H_5OH} \times \dfrac{1\ mol\ C_2H_5OH}{46.069\ g\ C_2H_5OH} \times \dfrac{0.789\ g\ C_2H_5OH}{1\ cm^3} = 23.5\ kJ/cm^3$

For diethylether:

$\dfrac{2727\ kJ}{1\ mol\ (C_2H_5)_2O} \times \dfrac{1\ mol\ (C_2H_5)_2O}{74.124\ g\ (C_2H_5)_2O} \times \dfrac{0.714\ g\ (C_2H_5)_2O}{1\ cm^3} = 26.3\ kJ/cm^3$

Thus, **diethyl ether** would provide the most energy per unit volume.

Additional Exercises

5.71   In the process described, one mole of solid CO_2 is converted to one mole of gaseous CO_2. The volume of the gas is much greater than the volume of the solid. Thus the system (that is, the mole of CO_2) must work against atmospheric pressure. To accomplish this work while maintaining a constant temperature requires the absorption of additional heat beyond that required to increase the internal energy of the CO_2. The remaining energy is turned into work.

5.72 $w = \Delta E - q;\ \Delta E = +1.32\ kJ,\ q = +1.55\ kJ;\ w = 1.32\ kJ - 1.55\ kJ = -0.23\ kJ$

The negative sign for w indicates that work is done by the system on the surroundings.

5.73   Freezing is an exothermic process (the opposite of melting, which is clearly endothermic). When the system, the soft drink, freezes, it releases energy to the surroundings, the can. Some of this energy does the work of splitting the can.

5.74   Equation 5.4 describes energy changes to the system. Specifically, it says any change in the internal energy of the system is manifested as heat transfer or work. The first law requires that an equal but opposite change occur to the surroundings.

$$\Delta E_{sys} = -\Delta E_{surr}$$
$$q_{sys} + w_{sys} = -(q_{surr} + w_{surr})$$

5.75 $\Delta E = q + w = +38.95 \text{ kJ} - 2.47 \text{ kJ} = +36.48 \text{ kJ}$

        $\Delta H = q_p = +38.95 \text{ kJ}$

5.76   Find the heat capacity of 1000 gal H_2O.

$$C_{H_2O} = 1000 \text{ gal } H_2O \times \frac{4 \text{ qt}}{1 \text{ gal}} \times \frac{1 \text{ L}}{1.06 \text{ qt}} \times \frac{1 \times 10^3 \text{ cm}^3}{1 \text{ L}} \times \frac{1 \text{ g}}{1 \text{ cm}^3} \times \frac{4.184 \text{ J}}{1 \text{ g} \cdot °C}$$

        $= 1.58 \times 10^7 \text{ J/}°C = 1.58 \times 10^4 \text{ kJ/}°C$; then,

$$\frac{1.58 \times 10^7 \text{ J}}{1°C} \times \frac{1°C \cdot g}{0.85 \text{ J}} \times \frac{1 \text{ kg}}{1 \times 10^3 \text{ g}} \times \frac{1 \text{ brick}}{1.8 \text{ kg}} = 1.0 \times 10^4 \text{ or } 10{,}000 \text{ bricks}$$

5.77   (a) $q_{Cu} = \dfrac{0.386 \text{ J}}{g \cdot K} \times 121.0 \text{ g Cu} \times (30.1°C - 100.4°C) = -3.28 \times 10^3 \text{ J}$

            The negative sign indicates the 3.28×10^3 J are lost by the Cu block.

        (b) $q_{H_2O} = \dfrac{4.184 \text{ J}}{g \cdot K} \times 150.0 \text{ g } H_2O \times (30.1°C - 25.1°C) = +3.1 \times 10^3 \text{ J}$

            The positive sign indicates that 3.14×10^3 J are gained by the H_2O.

        (c)   The difference in the heat lost by the Cu and the heat gained by the water is $3.283 \times 10^3 \text{ J} - 3.138 \times 10^3 \text{ J} = 0.145 \times 10^3 \text{ J} = 1 \times 10^2$ J. The temperature change of the calorimeter is 5.0°C. The heat capacity of the calorimeter in J/K is $0.145 \times 10^3 \text{ J} \times \dfrac{1}{5.0°C} = 3 \times 10$ J/K.

            If the rounded results from (a) and (b) are used, $C_{calorimeter} = \dfrac{0.2 \times 10^3 \text{ J}}{5.0 \text{ °C}} = 4 \times 10$ J/K.

        (d)   If only one cup were used, more heat would be lost through the cups; the difference between the $-q_{Cu}$ and $+q_{H_2O}$ would be greater and the calculated $C_{calorimeter}$ would be greater.

        (e) $q_{H_2O} = 3.283 \times 10^3 \text{ J} = \dfrac{4.184 \text{ J}}{g \cdot K} \times 150.0 \text{ g} \times (\Delta T)$

            $\Delta T = 5.23°C$; $T_f = 25.1°C + 5.23°C = 30.3°C$

5.78 $q = -6.2°C \times 100 \text{ g} \times \dfrac{4.184 \text{ J}}{1 \text{ g} \cdot °C} = -2.6 \times 10^3 \text{ J} = -2.6 \text{ kJ}$

        The reaction as carried out involves only 0.050 mol of $CuSO_4$ and the stoichiometrically equivalent amount of KOH. On a molar basis,

        $\Delta H = \dfrac{-2.6 \text{ kJ}}{0.050 \text{ mol}} = -52 \text{ kJ}$ for the reaction as written

5.79 (a) From the mass of benzoic acid that produces a certain temperature change, we can calculate the heat capacity of the calorimeter.

$$\frac{0.235 \text{ g benzoic acid}}{1.642°C \text{ change observed}} \times \frac{26.38 \text{ kJ}}{1 \text{ g benzoic acid}} = \frac{3.78 \text{ kJ}}{°C}$$

Now we can use this experimentally determined heat capacity with the data for caffeine.

$$\frac{1.525°C \text{ rise}}{0.265 \text{ g caffeine}} \times \frac{3.78 \text{ kJ}}{1°C} \times \frac{194.2 \text{ g caffeine}}{1 \text{ mol caffeine}} = 4.22 \times 10^3 \text{ kJ/mol caffeine}$$

(b) The overall uncertainty is approximately equal to the sum of the uncertainties due to each effect. The uncertainty in the mass measurement is 0.235/0.001 or 0.265/0.001, about 1 part in 235 or 1 part in 265. The uncertainty in the temperature measurements is 1.642/0.002 or 1.525/0.002, about 1 part in 820 or 1 part in 760. Thus the uncertainty in heat of combustion from each measurement is

$$\frac{4220}{235} = 18 \text{ kJ}; \quad \frac{4220}{265} = 16 \text{ kJ}; \quad \frac{4220}{820} = 5 \text{ kJ}; \quad \frac{4220}{760} = 6 \text{ kJ}$$

The sum of these uncertainties is 45 kJ. In fact, the overall uncertainty is less than this because independent errors in measurement do tend to partially cancel.

5.80 $\Delta H_f^° \text{ CH}_3\text{OH(g)} = -201.2 \text{ kJ}$ $\quad\quad \Delta H_f^° \text{ CH}_3\text{OH(l)} = -238.6 \text{ kJ}$

The liquid phase has lower absolute enthalpy, since the reactants for formation of both phases are the same, $\Delta H_f^°$ for the liquid state is more negative. The liquid has a lower enthalpy because condensation is an exothermic process. Heat is given off as intermolecular attractions, in this case hydrogen bonds, are formed when the liquid condenses.

5.81 $\Delta H_{rxn}^° = \Delta H_f^° \text{ TiO}_2\text{(s)} + 4\Delta H_f^° \text{ HCl(g)} - \Delta H_f^° \text{ TiCl}_4\text{(l)} - 2\Delta H_f^° \text{ H}_2\text{O(l)}$

$\Delta H_f^° \text{ TiO}_2\text{(s)} = \Delta H_{rxn}^° + \Delta H_f^° \text{ TiCl}_4\text{(l)} + 2\Delta H_f^° \text{ H}_2\text{O(l)} - 4\Delta H_f^° \text{ HCl(g)}$
$\quad\quad\quad = 67.0 \text{ kJ} + (-804.2 \text{ kJ}) + 2(-285.83 \text{ kJ}) - 4(-92.30 \text{ kJ})$
$\quad\quad\quad = -939.7 \text{ kJ}$

5.82 (a) $\text{C}_6\text{H}_{12}\text{O}_6\text{(s)} + 6\text{O}_2\text{(g)} \rightarrow 6\text{CO}_2\text{(g)} + 6\text{H}_2\text{O(l)}$
$\Delta H_{rxn}^° = 6\Delta H_f^° \text{ CO}_2\text{(g)} + 6\Delta H_f^° \text{ H}_2\text{O(l)} - \Delta H_f^° \text{ C}_6\text{H}_{12}\text{O}_6\text{(s)} - 6\Delta H_f^° \text{ O}_2\text{(g)}$
$\quad\quad\quad = 6(-393.5 \text{ kJ}) + 6(-285.83 \text{ kJ}) - (-1260 \text{ kJ}) - 6(0)$
$\quad\quad\quad = -2816 \text{ kJ/mol C}_6\text{H}_{12}\text{O}_6$
$\text{C}_{12}\text{H}_{22}\text{O}_{11}\text{(s)} + 12\text{O}_2\text{(g)} \rightarrow 12\text{CO}_2\text{(g)} + 11\text{H}_2\text{O(l)}$
$\Delta H_{rxn}^° = 12\Delta H_f^° \text{ CO}_2\text{(g)} + 11\Delta H_f^° \text{ H}_2\text{O(l)} - \Delta H_f^° \text{ C}_{12}\text{H}_{22}\text{O}_{11}\text{(s)} - 12\Delta H_f^° \text{ O}_2\text{(g)}$
$\quad\quad\quad = 12(-393.5 \text{ kJ}) + 11(-285.83 \text{ kJ}) - (-2221 \text{ kJ}) - 12(0)$
$\quad\quad\quad = -5645 \text{ kJ/mol C}_{12}\text{H}_{22}\text{O}_{11}$

(b)
$$\frac{-2816 \text{ kJ}}{1 \text{ mol C}_6\text{H}_{12}\text{O}_6} \times \frac{1 \text{ mol C}_6\text{H}_{12}\text{O}_6}{180.2 \text{ g C}_6\text{H}_{12}\text{O}_6} = \frac{15.63 \text{ kJ}}{1 \text{ g C}_6\text{H}_{12}\text{O}_6} \rightarrow 16 \text{ kJ/g C}_6\text{H}_{12}\text{O}_6$$

$$\frac{-5645 \text{ kJ}}{1 \text{ mol C}_{12}\text{H}_{22}\text{O}_{11}} \times \frac{1 \text{ mol C}_{12}\text{H}_{22}\text{O}_{11}}{342.3 \text{ g C}_{12}\text{H}_{22}\text{O}_{11}} = \frac{16.49 \text{ kJ}}{1 \text{ g C}_{12}\text{H}_{22}\text{O}_{11}} \rightarrow 16 \text{ kJ/g C}_{12}\text{H}_{22}\text{O}_{11}$$

(c)   The average fuel value of carbohydrates (Section 5.8) is 17 kJ/g. These two carbohydrates have fuel values (16 kJ/g) slightly lower but in line with this average. (More complex carbohydrates supply more energy and raise the average value.)

5.83   (a)   For comparison, balance the equations so that 1 mole of CH_4 is burned in each.

$CH_4(g) + O_2(g) \rightarrow C(s) + 2H_2O(l)$ $\Delta H° = -496.8$ kJ

$CH_4(g) + 3/2\, O_2(g) \rightarrow CO(g) + 2H_2O(l)$ $\Delta H° = -607.2$ kJ

$CH_4(g) + 2O_2(g) \rightarrow CO_2(g) + 2H_2O(l)$ $\Delta H° = -890.2$ kJ

(b) $\Delta H°_{rxn} = \Delta H°_f\, C(s) + 2\Delta H°_f\, H_2O(l) - \Delta H°_f\, CH_4(g) - \Delta H°_f\, O_2(g)$

           $= 0 + 2(-285.83 \text{ kJ}) - (-74.8) - 0 = -496.9$ kJ

$\Delta H°_{rxn} = \Delta H°_f\, CO(g) + 2\Delta H°_f\, H_2O(l) - \Delta H°_f\, CH_4(g) - 3/2\, \Delta H°_f\, O_2(g)$

             $= (-110.5 \text{ kJ}) + 2(-285.83 \text{ kJ}) - (-74.8 \text{ kJ}) - 3/2(0) = -607.4$ kJ

$\Delta H°_{rxn} = \Delta H°_f\, CO_2(g) + 2\Delta H°_f\, H_2O(l) - \Delta H°_f\, CH_4(g) - \Delta H°_f\, O_2(g)$

             $= -393.5 \text{ kJ} + 2(-285.83 \text{ kJ}) - (-74.8 \text{ kJ}) - 2(0) = -890.4$ kJ

(c)   Assuming that $O_2(g)$ is present in excess, the reaction that produces $CO_2(g)$ represents the most negative ΔH per mole of CH_4 burned. More of the potential energy of the reactants is released as heat during the reaction to give products of lower potential energy. The reaction that produces $CO_2(g)$ is the most "downhill" in enthalpy.

5.84   (a) $3C_2H_2(g) \rightarrow C_6H_6(l)$

$\Delta H°_{rxn} = \Delta H°_f\, C_6H_6(l) - 3\Delta H°_f\, C_2H_2(g) = 49.0 \text{ kJ} - 3(226.7 \text{ kJ}) = -631.1$ kJ

(b)   Since the reaction is exothermic (ΔH is negative), the product, 1 mole of $C_6H_6(l)$, has less enthalpy than the reactants, 3 moles of $C_2H_2(g)$.

(c)   The fuel value of a substance is the amount of heat (kJ) produced when 1 gram of the substance is burned. Calculate the molar heat of combustion (kJ/mol) and use this to find kJ/g of fuel.

$C_2H_2(g) + 5/2\, O_2(g) \rightarrow 2CO_2(g) + H_2O(l)$

$\Delta H°_{rxn} = 2\Delta H°_f\, CO_2(g) + \Delta H°_f\, H_2O(l) - \Delta H°_f\, C_2H_2(g) - 5/2\, \Delta H°_f\, O_2(g)$

         $= 2(-393.5 \text{ kJ}) + (-285.83 \text{ kJ}) - 226.7 \text{ kJ} - 5/2\, (0) = -1299.5 \text{ kJ/mol C}_2\text{H}_2$

$$\frac{-1299.5 \text{ kJ}}{1 \text{ mol C}_2\text{H}_2} \times \frac{1 \text{ mol C}_2\text{H}_2}{26.036 \text{ g C}_2\text{H}_2} = 49.912 = 50 \text{ kJ/g}$$

$$C_6H_6(g) + 15/2\ O_2(g) \rightarrow 6CO_2(g) + 3H_2O(l)$$

$$\Delta H^{\circ}_{rxn} = 6\Delta H^{\circ}_f\ CO_2(g) + 3\Delta H^{\circ}_f\ H_2O(l) - \Delta H^{\circ}_f\ C_6H_6(l) - 15/2\ \Delta H^{\circ}_f\ O_2(g)$$

$$= 6(-393.5\ kJ) + 3(-285.83\ kJ) - 49.0\ kJ - 15/2\ (0) = -3267.5\ kJ/mol\ C_6H_6$$

$$\frac{-3267.5\ kJ}{1\ mol\ C_6H_6} \times \frac{1\ mol\ C_6H_6}{78.114\ g\ C_6H_6} = 41.830 = 42\ kJ/g$$

5.85   The reaction for which we want ΔH is:

$$4NH_3(l) + 3O_2(g) \rightarrow 2N_2(g) + 6H_2O(g)$$

Before we can calculate ΔH for this reaction, we must calculate ΔH_f for $NH_3(l)$.

We know that ΔH_f for $NH_3(g)$ is -46.2 kJ, and that for $NH_3(l) \rightarrow NH_3(g)$, ΔH = 4.6 kJ

Thus, $\Delta H_{rxn} = \Delta H_f\ NH_3(g) - \Delta H_f\ NH_3(l)$.

    4.6 kJ = -46.2 kJ - $\Delta H_f\ NH_3(l)$; $\Delta H_f\ NH_3(l)$ = -50.8 kJ/mol

Then for the overall reaction, the enthalpy change is:

ΔH_{rxn} = 6 $\Delta H_f\ H_2O(g)$ + 2$\Delta H_f\ N_2(g)$ - 4$\Delta H_f\ NH_3(l)$ - 3$\Delta H_f\ O_2$

    = 6(-241.82 kJ) + 0 - 4(-50.8 kJ) - 0 = -1247.7 kJ

$$\frac{-1247.7\ kJ}{4\ mol\ NH_3} \times \frac{1\ mol\ NH_3}{17.0\ g\ NH_3} \times \frac{0.81\ g\ NH_3}{1\ cm^3} \times \frac{1000\ cm^3}{1\ L} = \frac{1.5 \times 10^4\ kJ}{L\ NH_3}$$

(This result has 2 significant figures because the density is expressed to 2 figures.)

$$2CH_3OH(l) + 3O_2(g) \rightarrow 2CO_2(g) + 4H_2O(g)$$

$$\Delta H = 2(-393.5\ kJ) + 4(-241.82\ kJ) - 2(-239\ kJ) = -1276\ kJ$$

$$\frac{-1276\ kJ}{2\ mol\ CH_3OH} \times \frac{1\ mol\ CH_3OH}{32.04\ g\ CH_3OH} \times \frac{0.792\ g\ CH_3OH}{1\ cm^3} \times \frac{1000\ cm^3}{1\ L} = \frac{1.58 \times 10^4\ kJ}{1\ L\ CH_3OH}$$

In terms of heat obtained per unit volume of fuel, methanol is a slightly better fuel than liquid ammonia.

5.86   **1,3-butadiene**, C_4H_6, MM = 54.092 g/mol

    (a) $C_4H_6(g) + 11/2\ O_2(g) \rightarrow 4CO_2(g) + 3H_2O(l)$

$$\Delta H^{\circ}_{rxn} = 4\Delta H^{\circ}_f\ CO_2(g) + 3\Delta H^{\circ}_f\ H_2O(l) - \Delta H^{\circ}_f\ C_4H_6(g) + 11/2\ \Delta H^{\circ}_f\ O_2(g)$$

$$= 4(-393.5\ kJ) + 3(-285.83\ kJ) - 111.9\ kJ + 11/2\ (0) = -2543.4\ kJ/mol\ C_4H_6$$

    (b) $\dfrac{-2543.4\ kJ}{1\ mol\ C_4H_6} \times \dfrac{1\ mol\ C_4H_6}{54.092\ g} = 47.020 \rightarrow 47\ kJ/g$

    (c)   % H = $\dfrac{6(1.008)}{54.092} \times 100 = 11.18\%\ H$

1-butene, C_4H_8, MM = 56.108 g/mol

(a) $C_4H_8(g) + 6O_2(g) \rightarrow 4CO_2(g) + 4H_2O(l)$

       $\Delta H^\circ_{rxn} = 4\Delta H^\circ_f\, CO_2(g) + 4\Delta H^\circ_f\, H_2O(l) - \Delta H^\circ_f\, C_4H_8(g) - 6\Delta H^\circ_f\, O_2(g)$

          $= 4(-393.5 \text{ kJ}) + 4(-285.83 \text{ kJ}) - 1.2 \text{ kJ} - 6(0) = -2718.5 \text{ kJ/mol } C_4H_8$

(b) $\dfrac{-2718.5 \text{ kJ}}{1 \text{ mol } C_4H_8} \times \dfrac{1 \text{ mol } C_4H_8}{56.108 \text{ g } C_4H_8} = 48.451 \rightarrow 48 \text{ kJ/g}$

(c) $\% \text{ H} = \dfrac{8(1.008)}{56.108} \times 100 = 14.37\% \text{ H}$

n-butane, $C_4H_{10}(g)$, MM = 58.124 g/mol

(a) $C_4H_{10}(g) + 13/2\ O_2(g) \rightarrow 4CO_2(g) + 5H_2O(l)$

       $\Delta H^\circ_{rxn} = 4\Delta H^\circ_f\, CO_2(g) + 5\Delta H^\circ_f\, H_2O(l) - \Delta H^\circ_f\, C_4H_{10}(g) - 13/2\ \Delta H^\circ_f\, O_2(g)$

          $= 4(-393.5 \text{ kJ}) + 5(-285.83 \text{ kJ}) - (-124.7 \text{ kJ}) - 3/2\ (0) = -2878.5 \text{ kJ/mol } C_4H_{10}$

(b) $\dfrac{-2878.5 \text{ kJ}}{1 \text{ mol } C_4H_{10}} \times \dfrac{1 \text{ mol } C_4H_{10}}{58.124 \text{ g } C_4H_{10}} = 49.523 \rightarrow 50 \text{ kJ/g}$

(c) $\% \text{ H} = \dfrac{10(1.008)}{58.124} \times 100 = 17.34\% \text{ H}$

(d)   It is certainly true that as the mass % H increases, the fuel value (kJ/g) of the hydrocarbon increases, given the same number of C atoms. A graph of the data in parts (b) and (c) (see below) suggests that mass % H and fuel value are directly proportional when the number of C atoms is constant.

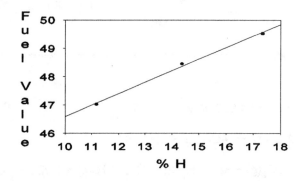

Integrative Exercises

5.87   (a) $CH_4(g) + 2O_2(g) \rightarrow CO_2(g) + 2H_2O(l)$

         $\Delta H^\circ = \Delta H^\circ_f\, CO_2(g) + 2\Delta H^\circ_f\, H_2O(l) - \Delta H^\circ_f\, CH_4(g) - \Delta H^\circ_f\, O_2(g)$

            $= -393.5 \text{ kJ} + 2(-285.83 \text{ kJ}) - (-74.8 \text{ kJ}) - 2(0) = -890.36 = -890.4 \text{ kJ/mol } CH_4$

$$\frac{-890.36 \text{ kJ}}{\text{mol CH}_4} \times \frac{1000 \text{ J}}{1 \text{ kJ}} \times \frac{1 \text{ mol}}{6.022 \times 10^{23} \text{ molecules CH}_4} = 1.4785 \times 10^{-18}$$

$$= 1.479 \times 10^{-18} \text{ J/molecule}$$

(b) 1eV = 96.485 kJ/mol

$$8 \text{ keV} \times \frac{1000 \text{ eV}}{1 \text{ keV}} \times \frac{96.485 \text{ kJ}}{\text{eV} \cdot \text{mol}} \times \frac{1 \text{ mol}}{6.022 \times 10^{23}} \times \frac{1000 \text{ J}}{\text{kJ}} = 1.282 \times 10^{-15}$$

$$= 1 \times 10^{-15} \text{ J/molecule}$$

The X-ray has approximately 1000 times more energy than is produced by the combustion of 1 molecule of $CH_4(g)$.

5.88 (a),(b) $Ag^+(aq) + Li(s) \rightarrow Ag(s) + Li^+(aq)$

$\Delta H° = \Delta H_f° Li^+(aq) - \Delta H_f° Ag^+(aq)$

   = -278.5kJ - 105.90 kJ = -384.4 kJ

$Fe(s) + 2Na^+(aq) \rightarrow Fe^{2+}(aq) + 2Na(s)$

$\Delta H° = \Delta H_f° Fe^{2+}(aq) - 2\Delta H_f° Na^+(aq)$

   = -87.86 kJ - 2(-240.1 kJ) = +392.3 kJ

$2K(s) + 2H_2O(l) \rightarrow 2KOH(aq) + H_2(g)$

$\Delta H° = 2\Delta H_f° KOH(aq) - 2\Delta H_f° H_2O(l)$

   = 2(-482.4 kJ) - 2(-285.83 kJ) = -393.1 kJ

(c) Exothermic reactions are more likely to be favorable, so the first and third reactions should be favorable and the second reaction should be unfavorable.

(d) In the activity series of metals, Table 4.4, any metal can be oxidized by the cation of a metal below it on the table.

   Ag^+ is below Li, so the first reaction will occur.
   Na^+ is above Fe, so the second reaction will not occur.
   H^+ (formally in H_2O) is below K, so the third reaction will occur.

5.89 (a) At constant pressure, $\Delta E = \Delta H - P\Delta V$. The difference between ΔE and ΔH is the expansion work done on or by the system. If all reactants and products are liquids or solids, very little volume change is possible. $P\Delta V$ is essentially zero and $\Delta E \approx \Delta H$.

(b) The reaction that produces the insoluble gas will have a greater difference between ΔE and ΔH, because the gas causes a significant $+\Delta V$, while the aqueous weak electrolyte does not.

5.90 (a) $\Delta H° = \Delta H_f° NaNO_3(aq) + \Delta H_f° H_2O(l) - \Delta H_f° HNO_3(aq) - \Delta H_f° NaOH(aq)$

   $\Delta H° = -446.2$ kJ - 285.83 kJ - (-206.6 kJ) - (-469.6 kJ) = -55.8 kJ

   $\Delta H° = \Delta H_f° NaCl(aq) + \Delta H_f° H_2O(l) - \Delta H_f° HCl(aq) - \Delta H_f° NaOH(aq)$

   $\Delta H° = -407.1$ kJ - 285.83 kJ - (-167.2 kJ) - (-469.6 kJ) = -56.1 kJ

$\Delta H° = \Delta H_f° \; NH_3(aq) + \Delta H_f° \; Na^+(aq) + \Delta H_f° \; H_2O(l) - \Delta H_f° \; NH_4^+(aq) - \Delta H_f° \; NaOH(aq)$

$= -80.29 \; kJ - 240.1 \; kJ - 285.83 \; kJ - (-132.5 \; kJ) - (-469.6 \; kJ) = -4.1 \; kJ$

(b) $H^+(aq) + OH^-(aq) \rightarrow H_2O(l)$ is the net ionic equation for both reactions.

(c) The $\Delta H°$ values for the first two reactions are nearly identical, -55.9 kJ and -56.2 kJ. The spectator ions by definition do not change during the course of a reaction, so $\Delta H°$ is the enthalpy change for the net ionic equation. Since the first two reactions have the same net ionic equation, it is not surprising that they have the same $\Delta H°$.

(d) Strong acids are more likely than weak acids to donate H^+. The neutralization of the two strong acids is energetically favorable, while the third reaction is not. $NH_4^+(aq)$ is probably a weak acid.

5.91 (a) $21.83 \; g \; CO_2 \times \dfrac{1 \; mol \; CO_2}{44.01 \; g \; CO_2} \times \dfrac{1 \; mol \; C}{1 \; mol \; CO_2} \times \dfrac{12.01 \; g \; C}{1 \; mol \; C} = 5.9572 = 5.957 \; g \; C$

$4.47 \; g \; H_2O \times \dfrac{1 \; mol \; H_2O}{18.02 \; g \; H_2O} \times \dfrac{2 \; mol \; H}{1 \; mol \; H_2O} \times \dfrac{1.008 \; g \; H}{mol \; H} = 0.5001 = 0.500 \; g \; H$

The sample mass is (5.9572 + 0.5001) = 6.457 g

(b) $5.957 \; g \; C \times \dfrac{1 \; mol \; C}{12.01 \; g \; C} = 0.4960 \; mol \; C; \; 0.4960/0.496 = 1$

$0.500 \; g \; H \times \dfrac{1 \; mol \; H}{1.008 \; g \; H} = 0.496 \; mol \; H; \; 0.496/0.496 = 1$

The empirical formula of the hydrocarbon is CH.

(c) Calculate the $\Delta H_f°$ for 6.457 g of the sample.

6.457 g sample + $O_2(g) \rightarrow$ 21.83 g $CO_2(g)$ + 4.47 g $H_2O(g)$, $\Delta H° = -311 \; kJ$

$\Delta H_{comb}° = \Delta H_f° \; CO_2(g) + \Delta H_f° \; H_2O(g) - \Delta H_f° \; sample - \Delta H_f° \; O_2(g)$

$\Delta H_f° \; sample = \Delta H_f° \; CO_2(g) + \Delta H_f° \; H_2O(g) - \Delta H_{comb}° \; sample$

$\Delta H_f° \; CO_2(g) = 21.83 \; g \; CO_2 \times \dfrac{1 \; mol \; CO_2}{44.01 \; g \; CO_2} \times \dfrac{-393.5 \; kJ}{mol \; CO_2} = -195.185 = -195.2 \; kJ$

$\Delta H_f° \; H_2O(g) = 4.47 \; g \; H_2O \times \dfrac{1 \; mol \; H_2O}{18.02 \; g \; H_2O} \times \dfrac{-241.82 \; kJ}{mol \; H_2O} = 59.985 = -60.0 \; kJ$

$\Delta H_f° \; sample = -195.185 \; kJ - 59.985 \; kJ - (-311 \; kJ) = 55.83 = 56 \; kJ$

$\dfrac{55.83 \; kJ}{6.457 \; g \; sample} \times \dfrac{13.02 \; g}{CH \; unit} = 112.6 = 1.1 \times 10^2 \; kJ/CH \; unit$

(d)   The hydrocarbons in Appendix C with empirical formula CH are C_2H_2 and C_6H_6.

substance	ΔH_f° / mol	ΔH_f° / CH unit
$C_2H_2(g)$	226.7 kJ	113.4 kJ
$C_6H_6(g)$	82.9 kJ	13.8 kJ
$C_6H_6(l)$	49.0 kJ	8.17 kJ
sample		1.1×10^2 kJ

The calculated value of ΔH_f°/CH unit for the sample is a good match with acetylene, $C_2H_2(g)$.

5.92   (a) $AgNO_3(aq) + NaCl(aq) \rightarrow NaNO_3(aq) + AgCl(s)$

net ionic equation: $Ag^+(aq) + Cl^-(aq) \rightarrow AgCl(s)$

$\Delta H^\circ = \Delta H_f^\circ \, AgCl(s) - \Delta H_f^\circ \, Ag^+(aq) - \Delta H_f^\circ \, Cl^-(aq)$

$\Delta H^\circ = -127.0 \text{ kJ} - (105.90 \text{ kJ}) - (-167.2 \text{ kJ}) = -65.7 \text{ kJ}$

(b) ΔH° for the complete molecular equation will be the same as ΔH° for the net ionic equation. $Na^+(aq)$ and $NO_3^-(aq)$ are spectator ions; they appear on both sides of the chemical equation. Since the overall enthalpy change is the enthalpy of the products minus the enthalpy of the reactants, the contributions of the spectator ions cancel.

(c) $\Delta H^\circ = \Delta H_f^\circ \, NaNO_3(aq) + \Delta H_f^\circ \, AgCl(s) - \Delta H_f^\circ \, AgNO_3(aq) - \Delta H_f^\circ \, NaCl(aq)$

$\Delta H_f^\circ \, AgNO_3(aq) = \Delta H_f^\circ \, NaNO_3(aq) + \Delta H_f^\circ \, AgCl(s) - \Delta H_f^\circ \, NaCl(aq) - \Delta H^\circ$

$\Delta H_f^\circ \, AgNO_3(aq) = -446.2 \text{ kJ} + (-127.0 \text{ kJ}) - (-407.1 \text{ kJ}) - (-65.7 \text{ kJ})$

$\Delta H_f^\circ \, AgNO_3(aq) = -100.4 \text{ kJ/mol}$

6 Electronic Structure of Atoms

Radiant Energy

6.1 (a) meters (m) (b) 1/seconds (s^{-1}) (c) meters/second ($m \cdot s^{-1}$ or m/s)

6.2 Wavelength (λ) and frequency (ν) are inversely proportional; the proportionality constant is the speed of light (c). $\nu = c/\lambda$.

The range of wavelengths in the visible portion of the electromagnetic spectrum is 400-700 nm.

6.3 (a) False. Only a small fraction of electromagnetic radiation (visible light, 400-700 nm) is visible.
   (b) True
   (c) False. Ultraviolet light has shorter wavelengths than does visible light.

6.4 (a) True
   (b) False. All electromagnetic radiation moves at the same speed ($c = 3.0 \times 10^8$ m/s).
   (c) True

6.5 Wavelength of (a) gamma rays < (d) yellow (visible) light < (e) red (visible) light < (b) 93.1 MHz FM (radio) waves < (c) 680 kHz or 0.680 MHz AM (radio) waves

6.6 Wavelength of (b) X-rays < (c) ultraviolet < (e) visible < (a) infrared < (d) microwave

6.7 (a) $\nu = c/\lambda$; $\dfrac{2.998 \times 10^8 \text{ m}}{\text{s}} \times \dfrac{1}{0.589 \text{ pm}} \times \dfrac{1 \text{ pm}}{1 \times 10^{-12} \text{ m}} = 5.09 \times 10^{20} \text{ s}^{-1}$

   (b) $\lambda = c/\nu$; $\dfrac{2.998 \times 10^8 \text{ m}}{\text{s}} \times \dfrac{1 \text{ s}}{5.11 \times 10^{11}} = 5.87 \times 10^{-4} \text{ m}$ (587 μm)

   (c) No. The radiation in (a) is gamma rays and in (b) is infrared. Neither is visible to humans.

   (d) $6.54 \text{ s} \times \dfrac{2.998 \times 10^8 \text{ m}}{\text{s}} = 1.96 \times 10^9 \text{ m}$

6.8 (a) $\nu = c/\lambda$; $\dfrac{2.998 \times 10^8 \text{ m}}{\text{s}} \times \dfrac{1}{1.73 \text{ nm}} \times \dfrac{1 \text{ nm}}{1 \times 10^{-9} \text{ m}} = 1.73 \times 10^{17} \text{ s}^{-1}$

(b) $\lambda = c/v$; $\dfrac{2.998 \times 10^8 \text{ m}}{\text{s}} \times \dfrac{1 \text{ s}}{9.83 \times 10^9} = 3.05 \times 10^{-2} \text{ m}$

(c)   Yes. The radiation in (b) is in the microwave range.

(d) $90.0 \text{ fs} \times \dfrac{1 \times 10^{-15} \text{ s}}{1 \text{ fs}} \times \dfrac{2.998 \times 10^8 \text{ m}}{\text{s}} = 2.70 \times 10^{-5} \text{ m}$ (27.0 μm)

6.9 $v = c/\lambda$; $\dfrac{2.998 \times 10^8 \text{ m}}{1 \text{ s}} \times \dfrac{1}{436 \text{ nm}} \times \dfrac{1 \text{ nm}}{1 \times 10^{-9} \text{ m}} = 6.88 \times 10^{14} \text{ s}^{-1}$;
The color is blue. (See Figure 24.23.)

6.10 $\lambda = c/v$; $\dfrac{2.998 \times 10^8 \text{ m}}{1 \text{ s}} \times \dfrac{1 \text{ s}}{4.47 \times 10^{14}} \times \dfrac{1 \text{ nm}}{1 \times 10^{-9} \text{ m}} = 671 \text{ nm}$
The color is orange-red.

Quantized Energy and Photons

6.11   (a)   *Quantization* means that energy can only be absorbed or emitted in specific amounts or multiples of these amounts. This minimum amount of energy is called a quantum and is equal to a constant times the frequency of the radiation absorbed or emitted. **$E = hv$**.

(b)   In everyday activities, we deal with macroscopic objects such as our bodies or our cars, which gain and lose total amounts of energy much larger than a single quantum, hv. The gain or loss of the relatively minuscule quantum of energy is unnoticed.

6.12   (a)   A xylophone can be considered a quantized instrument. Striking each key produces sound of a certain frequency. Only these frequencies are allowed.

(b)   The digital watch is more quantized. Time change is displayed only in increments of minutes or seconds (depending on the setting and precision of the watch). In an analog watch, time change is displayed continuously as the hands move.

6.13   (a) $E = hv = hc/\lambda = 6.626 \times 10^{-34} \text{ J} \bullet \text{s} \times \dfrac{2.998 \times 10^8 \text{ m}}{1 \text{ s}} \times \dfrac{1}{645 \text{ nm}} \times \dfrac{1 \text{ nm}}{1 \times 10^{-9} \text{ m}}$
$= 3.08 \times 10^{-19} \text{ J}$

(b) $E = hv = 6.626 \times 10^{-34} \text{ J} \bullet \text{s} \times \dfrac{2.85 \times 10^{12}}{1 \text{ s}} = 1.89 \times 10^{-21} \text{ J}$

(c) $\lambda = hc/E = 6.626 \times 10^{-34} \text{ J} \bullet \text{s} \times \dfrac{2.998 \times 10^8 \text{ m}}{1 \text{ s}} \times \dfrac{1}{8.23 \times 10^{-19} \text{ J}} = 2.41 \times 10^{-7} \text{ m}$
$= 241 \text{ nm}$

This radiation is in the ultraviolet region.

6.14   (a) $E = hc/\lambda = 6.626 \times 10^{-34}$ J•s $\times \dfrac{2.998 \times 10^8 \text{ m}}{1 \text{ s}} \times \dfrac{1}{146 \text{ nm}} \times \dfrac{1 \text{ nm}}{1 \times 10^{-9} \text{ m}}$

$$= 1.36 \times 10^{-18} \text{ J}$$

       (b) $E = h\nu = 6.626 \times 10^{-34}$ J•s $\times \dfrac{99.7 \times 10^6}{1 \text{ s}} = 6.61 \times 10^{-26}$ J

       (c) $\nu = E/h = \dfrac{6.10 \times 10^{-21} \text{ J}}{6.626 \times 10^{-34} \text{ J•s}} = 9.21 \times 10^{12} \text{ s}^{-1}$

         $\lambda = hc/E = 3.26 \times 10^{-5}$ m; the radiation is infrared.

6.15   (a) $E = hc/\lambda = 6.626 \times 10^{-34}$ J•s $\times \dfrac{2.998 \times 10^8 \text{ m}}{1 \text{ s}} \times \dfrac{1}{3.3 \text{ μm}} \times \dfrac{1 \text{ μm}}{1 \times 10^{-6} \text{ m}}$

$$= 6.0 \times 10^{-20} \text{ J}$$

         $E = hc/\lambda = 6.626 \times 10^{-34}$ J•s $\times \dfrac{2.998 \times 10^8 \text{ m}}{1 \text{ s}} \times \dfrac{1}{0.154 \text{ nm}} \times \dfrac{1 \text{ nm}}{1 \times 10^{-9} \text{ m}}$

$$= 1.29 \times 10^{-15} \text{ J}$$

       (b)   The 3.3 μm photon is in the infrared and the 0.154 nm (1.54×10^{-10} m) photon is in the X-ray region; the X-ray photon has the greater energy.

6.16 $E = h\nu$

       AM: 6.626×10^{-34} J•s $\times \dfrac{820 \times 10^3}{\text{s}} = 5.43 \times 10^{-28}$ J

       FM: 6.626×10^{-34} J•s $\times \dfrac{89.7 \times 10^6}{\text{s}} = 5.94 \times 10^{-26}$ J

       The FM photon has 100 times more energy than the AM photon.

6.17 $E_{photon} = hc/\lambda = \dfrac{6.626 \times 10^{-34} \text{J•s}}{785 \text{ nm}} \times \dfrac{2.998 \times 10^8 \text{ m}}{1 \text{ s}} \times \dfrac{1 \text{ nm}}{1 \times 10^{-9} \text{ m}} = 2.531 \times 10^{-19}$

$$= 2.53 \times 10^{-19} \text{ J/photon}$$

       $\dfrac{E_{total}}{E_{photon}} = $ # of photons;   31 J $\times \dfrac{1 \text{ photon}}{2.531 \times 10^{-19} \text{ J}} = 1.2 \times 10^{20}$ photons

6.18   The energy of each photon is hc/λ:

       6.626×10^{-34} J•s $\times \dfrac{2.998 \times 10^8 \text{ m}}{1 \text{ s}} \times \dfrac{1}{515 \text{ nm}} \times \dfrac{1 \text{ nm}}{1 \times 10^{-9} \text{ m}} = 3.857 \times 10^{-19}$

$$= 3.86 \times 10^{-19} \text{ J/photon}$$

       $\dfrac{3.65 \times 10^{-17} \text{ J signal}}{3.857 \times 10^{-19} \text{ J/photon}} = 95$ photons (The answer must be an integer; energy is quantized.)

6.19 $\dfrac{495 \times 10^3 \text{ J}}{\text{mol O}_2} \times \dfrac{1 \text{ mol}}{6.022 \times 10^{23} \text{ photons}} = 8.220 \times 10^{-19} = 8.22 \times 10^{-19}$ J/photon

$\lambda = hc/E = \dfrac{6.626 \times 10^{-34} \text{ J} \cdot \text{s}}{8.220 \times 10^{-19} \text{ J}} \times \dfrac{2.998 \times 10^8 \text{ m}}{1 \text{ s}} = 2.42 \times 10^{-7}$ m = 242 nm

According to Figure 6.4, this is ultraviolet radiation.

6.20 $\dfrac{80 \text{ kJ}}{1 \text{ mol}} \times \dfrac{1000 \text{ J}}{1 \text{ kJ}} \times \dfrac{1 \text{ mol}}{6.022 \times 10^{23} \text{ photons}} = 1.328 \times 10^{-19} = 1.33 \times 10^{-19}$ J/photon

$\lambda = hc/E = 6.626 \times 10^{-34} \text{ J} \cdot \text{s} \times \dfrac{2.998 \times 10^8 \text{ m}}{1 \text{ s}} \times \dfrac{1}{1.328 \times 10^{-19} \text{ J}} = 1.50 \times 10^{-6}$ m

According to Figure 6.4, the film could be used for infrared photography.

6.21   (a) $E = h\nu = 6.626 \times 10^{-34} \text{ J} \cdot \text{s} \times 1.09 \times 10^{15} \text{ s}^{-1} = 7.22 \times 10^{-19}$ J

        (b) $\lambda = c/\nu = \dfrac{2.998 \times 10^8 \text{ m}}{1 \text{ s}} \times \dfrac{1 \text{ s}}{1.09 \times 10^{15}} = 2.75 \times 10^{-7}$ m = 275 nm

        (c) $E_{120} = hc/\lambda = 6.626 \times 10^{-34} \text{ J} \cdot \text{s} \times \dfrac{2.998 \times 10^8 \text{ m}}{1 \text{ s}} \times \dfrac{1}{120 \text{ nm}} \times \dfrac{1 \text{ nm}}{1 \times 10^{-9} \text{ m}}$

$$= 1.655 \times 10^{-18} = 1.66 \times 10^{-18} \text{ J}$$

The excess energy of the 120 nm photon is converted into the kinetic energy of the emitted electron.

$E_k = E_{120} - E_{min} = 16.55 \times 10^{-19} \text{ J} - 7.22 \times 10^{-19} \text{ J} = 9.3 \times 10^{-19}$ J/electron

6.22 $\dfrac{222 \text{ kJ}}{\text{mol K}} \times \dfrac{1 \text{ mol}}{6.022 \times 10^{23} \text{ photons}} \times \dfrac{1000 \text{ J}}{1 \text{ kJ}} = 3.686 \times 10^{-19} = 3.69 \times 10^{-19}$ J/photon

        (a) $\nu = E/h = \dfrac{3.686 \times 10^{-19} \text{ J}}{6.626 \times 10^{-34} \text{ J} \cdot \text{s}} = 5.56 \times 10^{14} \text{ s}^{-1}$

        (b) $\lambda = hc/E = \dfrac{6.626 \times 10^{-34} \text{ J} \cdot \text{s} \times 2.998 \times 10^8 \text{ m/s}}{3.686 \times 10^{-19} \text{ J}} = 5.39 \times 10^{-7}$ m = 539 nm

        (c) $E_{350 \text{ nm}} = hc/\lambda = \dfrac{6.626 \times 10^{-34} \text{ J} \cdot \text{s} \times 2.998 \times 10^8 \text{ m/s}}{350 \text{ nm}} \times \dfrac{1 \text{ nm}}{1 \times 10^{-9} \text{ m}} = 5.676 \times 10^{-19}$

$$= 5.68 \times 10^{-19}$$

The excess energy of the 350 nm photon is converted into the kinetic energy of the emitted electron. The maximum possible kinetic energy/electron is the difference between the minimum energy per photon (3.69×10^{-19} J) and the energy of the 350 nm photon (5.68×10^{-19} J).

maximum E_k/electron = $5.676 \times 10^{-19} \text{ J} - 3.686 \times 10^{-19} \text{ J} = 1.99 \times 10^{-19}$ J/electron

Bohr's Model; Matter Waves

6.23   When applied to atoms, the notion of quantized energies means that only certain energies can be gained or lost, only certain values of ΔE are allowed. The allowed values of ΔE are represented by the lines in the emission spectra of excited atoms.

6.24   Balmer observed that the frequencies of the lines in the visible spectrum of hydrogen could be fit to a very simple formula (Equation 6.3) that contained the variable n, which could have any integer value. These integers correspond to the n values or principle quantum numbers of the Bohr orbits.

6.25   An isolated electron is assigned an energy of zero; the closer the electron comes to the nucleus, the more negative its energy. Thus, as an electron moves closer to the nucleus, the energy of the electron decreases and the excess energy is emitted. Conversely, as an electron moves further from the nucleus, the energy of the electron increases and energy must be absorbed.

   (a)   As the principle quantum number increases, the electron moves away from the nucleus and energy is **absorbed**.

   (b)   A decrease in the radius of the orbit means the electron moves closer to the nucleus; energy is **emitted**.

   (c)   Totally removing an electron from the atom requires that energy is **absorbed**.

6.26   (a) emitted   (b) absorbed   (c) Absorbed, but at these large n values, the electron is essentially removed from the atom and ΔE is very small.

6.27   (a) $\Delta E = R_H \left[\dfrac{1}{n_i^2} - \dfrac{1}{n_f^2} \right] = 2.18 \times 10^{-18}\ \text{J}\ (1/25 - 1/1) = -2.093 \times 10^{-18} = -2.09 \times 10^{-18}\ \text{J}$

   $\nu = E/h = \dfrac{2.093 \times 10^{-18}\ \text{J}}{6.626 \times 10^{-34}\ \text{J} \cdot \text{s}} = 3.158 \times 10^{15} = 3.16 \times 10^{15}\ \text{s}^{-1}$

   $\lambda = c/\nu = \dfrac{2.998 \times 10^8\ \text{m}}{1\ \text{s}} \times \dfrac{1\ \text{s}}{3.158 \times 10^{15}} = 9.49 \times 10^{-8}\ \text{m}$

   Since the sign of ΔE is negative, radiation is emitted.

   (b) $\Delta E = 2.18 \times 10^{-18}\ \text{J}\ (1/36 - 1/4) = -4.844 \times 10^{-19} = -4.84 \times 10^{19}\ \text{J}$

   $\nu = \dfrac{4.844 \times 10^{-19}\ \text{J}}{6.626 \times 10^{-34}\ \text{J} \cdot \text{s}} = 7.311 \times 10^{14} = 7.31 \times 10^{14}\ \text{s}^{-1};\ \lambda = \dfrac{2.998 \times 10^8\ \text{m/s}}{7.311 \times 10^{14}/\text{s}}$

   $= 4.10 \times 10^{-7}\ \text{m}$

   Visible radiation is emitted.

(c) $\Delta E = 2.18 \times 10^{-18}$ J(1/16 -1/25) = 4.095 × 10^{-20} = 4.91 × 10^{-20} J

$$\nu = \frac{4.095 \times 10^{-20}\, J}{6.626 \times 10^{-34}\, J \cdot s} = 7.403 \times 10^{13} = 7.40 \times 10^{13}\, s^{-1};\ \lambda = \frac{2.998 \times 10^8\, m/s}{7.403 \times 10^{13}/s}$$

$$= 4.05 \times 10^{-6}\, m$$

Radiation is absorbed.

6.28   (a) $\Delta E = R_H \left[\dfrac{1}{n_i^2} - \dfrac{1}{n_f^2} \right] = 2.18 \times 10^{-18}$ J (1/4 - 1/49) = 5.005 × 10^{-19} = 5.01 × 10^{-19} J

$$\nu = E/h = \frac{5.005 \times 10^{-19}\, J}{6.626 \times 10^{-34}\, J \cdot s} = 7.554 \times 10^{14} = 7.55 \times 10^{14}\, s^{-1}$$

$$\lambda = c/\nu = \frac{2.998 \times 10^8\, m/s}{7.554 \times 10^{14}/s} = 3.97 \times 10^{-7}\, m$$

Visible radiation is absorbed.

(b) $\Delta E = 2.18 \times 10^{-18}$ J (1/25 - 1/36) = 2.664 × 10^{-20} = 2.66 × 10^{-20} J

$$\nu = \frac{2.664 \times 10^{-20}\, J}{6.626 \times 10^{-34}\, J \cdot s} = 4.021 \times 10^{13} = 4.02 \times 10^{13}\, s^{-1};\ \lambda = \frac{2.998 \times 10^8\, m/s}{4.021 \times 10^{13}/s}$$

$$= 7.46 \times 10^{-6}\, m$$

Radiation is absorbed.

(c) $\Delta E = 2.18 \times 10^{-18}$ J (1/36 - 1/9) = - 1.817 × 10^{-19} = -1.82 × 10^{-19} J

$$\nu = \frac{1.817 \times 10^{-19}\, J}{6.626 \times 10^{-34}\, J \cdot s} = 2.742 \times 10^{14} = 2.74 \times 10^{14}\, s^{-1};$$

$$\lambda = \frac{2.998 \times 10^8\, m/s}{2.742 \times 10^{14}/s} = 1.09 \times 10^{-6}\, m$$

Radiation is emitted.

6.29   (a)   Only lines with n_f = 2 represent ΔE values and wavelengths that lie in the visible portion of the spectrum. Lines with n_f = 1 have larger ΔE values and shorter wavelengths that lie in the ultraviolet. Lines with $n_f > 2$ have smaller ΔE values and lie in the lower energy longer wavelength regions of the electromagnetic spectrum.

(b) $n_i = 3,\ n_f = 2;\ \Delta E = R_H \left[\dfrac{1}{n_i^2} - \dfrac{1}{n_f^2} \right] = 2.18 \times 10^{-18}$ J (1/9 - 1/4)

$$\lambda = hc/E = \frac{6.626 \times 10^{-34}\, J \cdot s \times 2.998 \times 10^8\, m/s}{2.18 \times 10^{-18}\, J\ (1/9 - 1/4)} = 6.56 \times 10^{-7}\, m$$

This is the yellow line at 656 nm.

$$n_i = 4,\ n_f = 2;\ \lambda = hc/E = \frac{6.626 \times 10^{-34}\,\text{J}\cdot\text{s} \times 2.998 \times 10^{8}\,\text{m/s}}{2.18 \times 10^{-18}\,\text{J}\,(1/16 - 1/4)} = 4.86 \times 10^{-7}\,\text{m}$$

This is the blue-green (cyan) line at 486 nm.

$$n_i = 5,\ n_f = 2;\ \lambda = hc/E = \frac{6.626 \times 10^{-34}\,\text{J}\cdot\text{s} \times 2.998 \times 10^{8}\,\text{m/s}}{2.18 \times 10^{-18}\,\text{J}\,(1/25 - 1/4)} = 4.34 \times 10^{-7}\,\text{m}$$

This is the blue line at 434 nm.

6.30   (a)   Transitions with $n_f = 1$ have larger ΔE values and shorter wavelengths than those with $n_f = 2$. These transitions will lie in the ultraviolet region.

        (b) $n_i = 2,\ n_f = 1;\ \lambda = hc/E = \dfrac{6.626 \times 10^{-34}\,\text{J}\cdot\text{s} \times 2.998 \times 10^{8}\,\text{m/s}}{2.18 \times 10^{-18}\,\text{J}\,(1/4 - 1/1)} = 1.22 \times 10^{-7}\,\text{m}$

                $n_i = 3,\ n_f = 1;\ \lambda = hc/E = \dfrac{6.626 \times 10^{-34}\,\text{J}\cdot\text{s} \times 2.998 \times 10^{8}\,\text{m/s}}{2.18 \times 10^{-18}\,\text{J}\,(1/9 - 1/1)} = 1.03 \times 10^{-7}\,\text{m}$

                $n_i = 4,\ n_f = 1;\ \lambda = hc/E = \dfrac{6.626 \times 10^{-34}\,\text{J}\cdot\text{s} \times 2.998 \times 10^{8}\,\text{m/s}}{2.18 \times 10^{-18}\,\text{J}\,(1/16 - 1/1)} = 0.972 \times 10^{-7}\,\text{m}$

6.31   (a) $93.8\,\text{nm} \times \dfrac{1 \times 10^{-9}\,\text{m}}{1\,\text{nm}} = 9.38 \times 10^{-8}\,\text{m}$; this line is in the ultraviolet region.

        (b)   Only lines with $n_f = 1$ have a large enough ΔE to lie in the ultraviolet region. Solve Equation 6.6 for n_i, recalling that ΔE is negative for emission.

$$-hc/\lambda = R_H \left[\frac{1}{n_i^2} - \frac{1}{n_f^2} \right]; \quad -\frac{hc}{\lambda \times R_H} = \left[\frac{1}{n_i^2} - 1 \right]; \quad \frac{1}{n_i^2} = 1 - \frac{hc}{\lambda \times R_H}$$

$$n_i^2 = \left(1 - \frac{hc}{\lambda \times R_H} \right)^{-1}; \quad n_i = \left(1 - \frac{hc}{\lambda \times R_H} \right)^{-1/2}$$

$$n_i = \left(1 - \frac{6.626 \times 10^{-34}\,\text{J}\cdot\text{s} \times 2.998 \times 10^{8}\,\text{m/s}}{9.38 \times 10^{-8}\,\text{m} \times 2.18 \times 10^{-18}\,\text{J}} \right)^{-1/2} = 6 \ (n \text{ values must be integers})$$

$n_i = 6,\ n_f = 1$

6.32   (a) $1282\,\text{nm} \times \dfrac{1 \times 10^{-9}\,\text{m}}{1\,\text{nm}} = 1.282 \times 10^{-6}\,\text{m}$; this line is in the infrared.

        (b)   Absorption lines with $n_i = 1$ are in the ultraviolet and with $n_i = 2$ are in the visible. Thus, $n_i \geq 3$, but we do not know the exact value of n_i. Calculate the longest wavelength with $n_i = 3$ ($n_f = 4$). If this is less than 1282 nm, $n_i = 4$.

$$\lambda = hc/E = \frac{6.626 \times 10^{-34} \text{ J} \cdot \text{s} \times 2.998 \times 10^{8} \text{ m/s}}{2.18 \times 10^{-18} \text{ J } (1/16 - 1/9)} = 1.875 \times 10^{-6} \text{ m}$$

This wavelength is longer than 1.282×10^{-6} m, so $n_i = 3$, but $n_f > 4$. Solve for n_f as in Exercise 6.31.

$$n_f = \left(\frac{1}{n_i^2} - \frac{hc}{\lambda \times R_H} \right)^{-1/2} = \left(1/9 - \frac{6.626 \times 10^{-34} \text{ J} \cdot \text{s} \times 2.998 \times 10^{8} \text{ m/s}}{1.282 \times 10^{-6} \text{ m} \times 2.18 \times 10^{-18} \text{ J}} \right)^{-1/2} = 5$$

$n_f = 5$, $n_i = 3$

6.33 $\lambda = \dfrac{h}{mv}$; $1 \text{ J} = \dfrac{1 \text{ kg} \cdot \text{m}^2}{\text{s}^2}$; Change mass to kg and velocity to m/s in each case.

(a) $\dfrac{60 \text{ km}}{1 \text{ hr}} \times \dfrac{1000 \text{ m}}{1 \text{ km}} \times \dfrac{1 \text{ hr}}{60 \text{ min}} \times \dfrac{1 \text{ min}}{60 \text{ s}} = 16.67 = 17$ m/s

$$\lambda = \frac{6.626 \times 10^{-34} \text{ kg} \cdot \text{m}^2 \cdot \text{s}}{1 \text{ s}^2} \times \frac{1}{85 \text{ kg}} \times \frac{1 \text{ s}}{16.67 \text{ m}} = 4.7 \times 10^{-37} \text{ m}$$

(b) $50 \text{ g} \times \dfrac{1 \text{ kg}}{1000 \text{ g}} = 0.050 \text{ kg}$

$$\lambda = \frac{6.626 \times 10^{-34} \text{ kg} \cdot \text{m}^2 \cdot \text{s}}{1 \text{ s}^2} \times \frac{1}{0.050 \text{ kg}} \times \frac{1 \text{ s}}{400 \text{ m}} = 3.3 \times 10^{-35} \text{ m}$$

(c)   We need to calculate the mass of a single Li atom in kg.

$$\frac{6.94 \text{ g Li}}{1 \text{ mol Li}} \times \frac{1 \text{ kg}}{1000 \text{ g}} \times \frac{1 \text{ mol}}{6.022 \times 10^{23} \text{ Li atoms}} = 1.152 \times 10^{-26} = 1.15 \times 10^{-26} \text{ kg}$$

$$\lambda = \frac{6.626 \times 10^{-34} \text{ kg} \cdot \text{m}^2 \cdot \text{s}}{1 \text{ s}^2} \times \frac{1}{1.152 \times 10^{-26} \text{ kg}} \times \frac{1 \text{ s}}{6.5 \times 10^{5} \text{ m}} = 8.8 \times 10^{-14} \text{ m}$$

6.34 $\lambda = h/mv$; Change mass to kg and velocity to m/s.

(a) $3000 \text{ lb} \times \dfrac{1 \text{ kg}}{2.205 \text{ lb}} = 1361 \text{ kg}$

$\dfrac{55 \text{ mi}}{1 \text{ hr}} \times \dfrac{1 \text{ hr}}{3600 \text{ s}} \times \dfrac{1.6093 \text{ km}}{1 \text{ mi}} \times \dfrac{1000 \text{ m}}{1 \text{ km}} = 24.59 = 25$ m/s

$$\lambda = \frac{6.626 \times 10^{-34} \text{ kg} \cdot \text{m}^2 \cdot \text{s}}{1 \text{ s}^2} \times \frac{1}{1361 \text{ kg}} \times \frac{1 \text{ s}}{24.59 \text{ m}} = 2.0 \times 10^{-38} \text{ m}$$

(b) $5.0 \text{ oz} \times \dfrac{28.3 \text{ g}}{1 \text{ oz}} \times \dfrac{1 \text{ kg}}{1000 \text{ g}} = 0.1415 = 0.14 \text{ kg}$

$\dfrac{89 \text{ mi}}{1 \text{ hr}} \times \dfrac{1 \text{ hr}}{3600 \text{ s}} \times \dfrac{1.6093 \text{ km}}{1 \text{ mi}} \times \dfrac{1000 \text{ m}}{1 \text{ km}} = 39.79 = 40$ m/s

$$\frac{6.626 \times 10^{-34}\,\text{kg} \cdot \text{m}^2 \cdot \text{s}}{1\,\text{s}^2} \times \frac{1}{0.1415\,\text{kg}} \times \frac{1\,\text{s}}{39.79\,\text{m}} = 1.2 \times 10^{-34}\,\text{m}$$

(c) $\dfrac{4.002\,\text{g He}}{1\,\text{mol He}} \times \dfrac{1\,\text{kg}}{1000\,\text{g}} \times \dfrac{1\,\text{mol He}}{6.022 \times 10^{23}\,\text{atoms}} = 6.646 \times 10^{-27}\,\text{kg/He atom}$

$$\lambda = \frac{6.626 \times 10^{-34}\,\text{kg} \cdot \text{m}^2 \cdot \text{s}}{1\,\text{s}^2} \times \frac{1}{6.646 \times 10^{-27}\,\text{kg}} \times \frac{1\,\text{s}}{8.5 \times 10^5\,\text{m}} = 1.2 \times 10^{-13}\,\text{m}$$

6.35 $v = h/m\lambda$; $\lambda = 0.88\,\text{Å} \times \dfrac{1 \times 10^{-10}\,\text{m}}{1\,\text{Å}} = 8.8 \times 10^{-11}\,\text{m}$; $m = 1.67 \times 10^{-27}\,\text{kg}$

$$v = \frac{6.626 \times 10^{-34}\,\text{kg} \cdot \text{m}^2 \cdot \text{s}}{1\,\text{s}^2} \times \frac{1}{1.67 \times 10^{-27}\,\text{kg}} \times \frac{1\,\text{s}}{8.8 \times 10^{-11}\,\text{m}} = 4.5 \times 10^3\,\text{m/s}$$

6.36 $m_e = 9.109 \times 10^{-31}\,\text{kg}$ (back cover of text)

$$\lambda = \frac{6.626 \times 10^{-34}\,\text{kg} \cdot \text{m}^2 \cdot \text{s}}{1\,\text{s}^2} \times \frac{1}{9.109 \times 10^{-31}\,\text{kg}} \times \frac{1\,\text{s}}{5.93 \times 10^6\,\text{m}} = 1.23 \times 10^{-10}\,\text{m}$$

$$1.23 \times 10^{-10}\,\text{m} \times \frac{1\,\text{Å}}{1 \times 10^{-10}\,\text{m}} = 1.23\,\text{Å}$$

Since atomic radii and interatomic distances are on the order of 1-5 Å (Section 2.3), the wavelength of this electron is comparable to the size of atoms.

6.37   The uncertainty principle states that there is a limit to how precisely we can know the simultaneous position and momentum of an electron. The more precisely we know the position the greater the uncertainty in the momentum, and vice versa. In the Bohr model, electrons move in exact spherical paths that have known energy, implying that the position and momentum of an electron can be known exactly and simultaneously. This violates the uncertainty principle and is the reason that the Bohr model is an inadequate description of the electronic structure of atoms.

6.38   Whenever light interacts with electrons, the accuracy of the measurement is limited by the wavelength of the light. Short wavelengths alter the momentum of the moving electron. Long wavelengths produce less accurate position measurements. We cannot know the exact position and speed (momentum) of a moving particle.

Quantum Mechanics and Atomic Orbitals

6.39   In the Bohr model, the electron is treated as a small object that moves about the nucleus in circular orbits. A Bohr orbit specifies the exact path and energy of the electron. In the quantum-mechanical model, the wave properties of the electron are considered; any attempt to describe the exact path of an electron is inconsistent with the Heisenberg uncertainty principle. The quantum mechanical model is a statistical model that tells us the probability of finding an electron in certain regions around the nucleus. Thus, quantum mechanics would give the probability of finding the electron at 0.53 Å and this probability would always be less than 100%.

6.40   The square of the wave function has the physical significance of an amplitude, or probability. The quantity ψ^2 at a given point in space is the probability of locating the electron within a small volume element around that point at any given instant. The total probability, that is, the sum of ψ^2 over all the space around the nucleus, must equal 1.

6.41   (a) $n = 4, l = 3, 2, 1, 0$ (b) $l = 2, m_l = -2, -1, 0, 1, 2$

6.42   (a) $n = 5, l = 4, 3, 2, 1, 0$ (b) $l = 3, m_l = -3, -2, -1, 0, 1, 2, 3$

6.43   (a)   2, 1, 1;   2, 1, 0;   2, 1 -1

       (b)   5, 2, 2;   5, 2, 1;   5, 2, 0;   5, 2, -1;   5, 2, -2

6.44   (a)   4, 3, 3;   4, 3, 2;   4, 3, 1;   4, 3, 0;   4, 3, -1;   4, 3, -2;   4, 3, -3

       (b)   3s: 3, 0, 0;   3p: 3, 1, 1;   3, 1, 0;   3, 1, -1

           3d: 3, 2, 2;   3, 2, 1;   3, 2, 0;   3, 2, -1;   3, 2, -2

6.45   (a)   permissible, 2p   (b)   forbidden, for $l = 0$, m_l can only equal 0
       (c)   permissible, 4d   (d)   forbidden, for n = 3, the largest l value is 2

6.46   (a)   forbidden, for $n = 1$, l can only equal 0   (b)   allowed, 3s   (c)   allowed, 4p
       (d)   forbidden, for $l = 1$, the maximum value for m_l is 1

6.47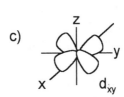

6.48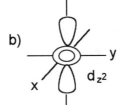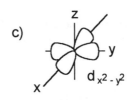

6.49   (a)   The 2s and 3s orbitals of a hydrogen atom have the same overall spherical shape, but the 3s orbital has a larger radial extension and one more node than the 2s orbital. Since the 3s orbital is "larger" than the 2s, there is a greater probability of finding an electron further from the nucleus in the 3s orbital.

      (b)   The shapes of the hydrogen 2s and 2p orbitals are quite different (spherical vs dumbbell), while the average distance from the nucleus of an electron occupying either orbital is similar.

      (c)   In the hydrogen atom, orbitals with the same n value are degenerate and energy increases with increasing n value. Thus, 2s and 2p have the same energy and 3s is at a higher energy.

6.50   (a)   In an s orbital, there are (n-1) nodes.

      (b)   In the quantum-mechanical description of the atom, the average distance of an electron from the nucleus is greater in a 2s orbital than in a 1s orbital.

      (c)   In the hydrogen atom, the $2p_x$ and $3p_x$ orbitals have a similar dumbbell shape, but the average distance of an electron from the nucleus is greater for the $3p_x$ than the $2p_x$ orbital. The $2p_x$ has a smaller radial extension and a lower energy than the $3p_x$.

Many-Electron Atoms; Electron Spin

6.51   (a)   In the hydrogen atom, orbitals with the same principle quantum number, n, are degenerate.

      (b)   In a many-electron atom, orbitals with the same principle and azimuthal quantum numbers, n and l, are degenerate.

6.52   Within a given shell, the energies of the subshells increase in the order s < p < d < f. Within a subshell, the energies of the orbitals are the same.

6.53   The 2p electron in boron is shielded from the full charge of the nucleus by the 2s electrons. (Both the 2s and 2p electrons are shielded by the 1s electrons). Thus, the 2p electron experiences a smaller nuclear charge than the 2s electrons.

6.54   A 3d electron in iron is shielded from the nuclear charge by 3s (and to some extent by 3p) electrons. The 3d electron experiences a smaller nuclear charge and therefore is at a greater average distance from the nucleus than the 3s electron.

6.55   A 2p electron in Ne experiences a greater effective nuclear charge. The shielding experienced by a 2p electron in the two atoms is similar, so the electron in the atom with the larger Z ($Z_{Ne} = 10$, $Z_O = 8$) experiences the larger effective nuclear charge.

6.56   A 3s electron in magnesium experiences a nuclear charge (Z = 12) one higher than that in sodium (Z = 11). This added unit of nuclear charge is partially, but not completely, canceled by the presence of a second 3s electron. The overall effect is an increase in the effective nuclear charge experienced by a 3s electron in Mg.

6.57   (a)   +1/2, - 1/2

        (b)   Electrons with opposite spins are affected differently by a strong inhomogeneous magnetic field. An apparatus similar to that in Figure 6.25 can be used to distinguish electrons with opposite spins.

        (c)   The Pauli exclusion principle states that no two electrons can have the same four quantum numbers. Two electrons in a 1s orbital have the same, n, l and m_l values. They must have different m_s values.

6.58   (a)   Spectral lines that appeared to be single were actually closely spaced pairs of lines. This experimental evidence indicated that there were twice as many electronic energy states as the theory predicted.

        (b)   The **rate** of rotation of a globe spinning on its axis is continuous, not quantized. But the direction of the rotation is quantized and there are only two possibilities, clockwise or counterclockwise. This is a physical model for electron spin, which also is quantized with two possible values.

        (c)   The Pauli exclusion principle dictates that the maximum number of electrons that can occupy a single orbital is two. Thus, electrons will occupy the lowest energy available orbital, but when all degenerate orbitals contain two electrons, a higher energy orbital or orbitals must be filled.

6.59   (a) 10   (b) 2   (c) 6   (d) 14

6.60   (a) 18   (b) 10   (c) 2   (d) 1

6.61   2, 1, 1, 1/2; 2, 1, 1, -1/2; 2, 1, 0, 1/2; 2, 1, 0, -1/2; 2, 1, -1, 1/2; 2, 1, -1, -1/2

6.62   B - $1s^2 2s^2 2p^1$

    1s electrons: 1, 0, 0, 1/2; 1, 0, 0, -1/2; 2s electrons: 2, 0, 0, 1/2; 2, 0, 0, -1/2

    2p electrons: 2, 1, 1, 1/2 or 2, 1, 1, -1/2 or 2, 1, 0, 1/2 or 2, 1, 0, -1/2 or 2, 1, -1, 1/2
               or 2, 1, -1, -1/2

Electron Configurations

6.63   (a)   Each box represents an orbital.

        (b)   Electron spin is represented by the direction of the half-arrows.

        (c)   No. The electron configuration of Be is $1s^2 2s^2$. There are no electrons in subshells that have degenerate orbitals, so Hund's rule is not used.

6.64   (a)   2

      (b)   A paired electron has the opposite spin of the other electron in the pair; the arrows point in opposite directions. Unpaired electrons in degenerate orbitals have the same spin; the arrows point in the same direction.

      (c)   Yes. The electron configuration of Si is $[Ne]3s^23p^2$. Hund's rule requires that the two 3p electrons are unpaired and in two different 3p orbitals.

6.65   (a)   Rb - $[Kr]5s^1$ (b)   Se - $[Ar]4s^23d^{10}4p^4$ (c)   Zn - $[Ar]4s^23d^{10}$

      (d)   V - $[Ar]4s^23d^3$ (e)   Pb - $[Xe]6s^24f^{14}5d^{10}6p^2$ (f)   Yb - $[Xe]6s^24f^{14}$

6.66   (a)   K - $1s^22s^22p^63s^23p^64s^1$ (b)   Al - $1s^22s^22p^63s^23p^1$ (c)   S - $1s^22s^22p^63s^23p^4$

      (d)   Mn - $1s^22s^22p^63s^23p^64s^23d^5$ (e)   Y - $1s^22s^22p^63s^23p^64s^23d^{10}4p^65s^24d^1$

      (f)   Nd - $1s^22s^22p^63s^23p^64s^23d^{10}4p^65s^24d^{10}5p^66s^24f^4$

6.67   (a)   As   [4s: ↿⇂] [3d: ↿⇂ ↿⇂ ↿⇂ ↿⇂ ↿⇂] [4p: ↿ ↿ ↿]   3 unpaired electrons

      (b)   Te   [5s: ↿⇂] [4d: ↿⇂ ↿⇂ ↿⇂ ↿⇂ ↿⇂] [5p: ↿⇂ ↿ ↿]   2 unpaired electrons

      (c)   Sn   [5s: ↿⇂] [4d: ↿⇂ ↿⇂ ↿⇂ ↿⇂ ↿⇂] [5p: ↿ ↿]   2 unpaired electrons

      (d)   Ag   [5s: ↿] [4d: ↿⇂ ↿⇂ ↿⇂ ↿⇂ ↿⇂]   1 unpaired electron

      (e)   Nb   [$5s^2$: ↿⇂] [4d: ↿ ↿ ↿]   3 unpaired electrons

6.68   (a)   Ge   [4s: ↿⇂] [3d: ↿⇂ ↿⇂ ↿⇂ ↿⇂ ↿⇂] [4p: ↿ ↿]   2 unpaired electrons

      (b)   In   [5s: ↿⇂] [4d: ↿⇂ ↿⇂ ↿⇂ ↿⇂ ↿⇂] [5p: ↿]   1 unpaired electron

      (c)   Ni   [4s: ↿⇂] [3d: ↿⇂ ↿⇂ ↿⇂ ↿ ↿]   2 unpaired electrons

      (d)   Kr   [5s: ↿⇂] [4d: ↿⇂ ↿⇂ ↿⇂ ↿⇂ ↿⇂] [5p: ↿⇂ ↿⇂ ↿⇂]   0 unpaired electrons

      (e)   Br   [4s: ↿⇂] [3d: ↿⇂ ↿⇂ ↿⇂ ↿⇂ ↿⇂] [4p: ↿⇂ ↿⇂ ↿]   1 unpaired electron

6.69   (a)   Mg   (b) Al   (c) Cr   (d) Te

6.70   (a)   7A (halogens)   (b) 4B   (c) 3A   (d) the f-block elements Sm and Pu

Additional Exercises

6.71   (a) $\lambda_A = 6.0 \times 10^{-7}$ m, $\lambda_B = 12 \times 10^{-7}$ m

(b) $\nu = c/\lambda$; $\nu_A = \dfrac{2.998 \times 10^8 \text{ m}}{1 \text{ s}} \times \dfrac{1}{6.0 \times 10^{-7} \text{ m}} = 5.0 \times 10^{14} \text{ s}^{-1}$

$\nu_B = \dfrac{2.998 \times 10^8 \text{ m}}{1 \text{ s}} \times \dfrac{1}{12 \times 10^{-7} \text{ m}} = 2.5 \times 10^{14} \text{ s}^{-1}$

(c)   A - visible, B - infrared

6.72   (a)   Ag -- violet (this wavelength is technically in the UV region),

Au -- violet (UV), Ba -- dark blue, Ca -- dark blue,

Cu -- violet (UV), Fe -- violet (UV), K -- dark blue,

Mg -- violet (UV), Na -- yellow/orange, Ni -- violet (UV)

(b)   Au -- shortest wavelength, highest energy;

Na -- longest wavelength, lowest energy

(c) $\lambda = c/\nu = \dfrac{2.998 \times 10^8 \text{ m/s}}{6.59 \times 10^{14}\text{/s}} \times \dfrac{1 \text{ nm}}{1 \times 10^{-9} \text{ m}} = 455$ nm, **Ba**

6.73   The round trip is 4.8×10^5 mi. Light travels at 2.998×10^8 m/s.

$4.8 \times 10^5 \text{ mi} \times \dfrac{1.6093 \text{ km}}{1 \text{ mi}} \times \dfrac{1000 \text{ m}}{1 \text{ km}} \times \dfrac{1 \text{ s}}{2.998 \times 10^8 \text{ m}} = 2.6$ s

6.74   (a) $\nu = c/\lambda = \dfrac{2.998 \times 10^8 \text{ m/s}}{320 \text{ nm}} \times \dfrac{1 \text{ nm}}{1 \times 10^{-9} \text{ m}} = 9.37 \times 10^{14} \text{ s}^{-1}$

(b) $E = hc/\lambda = \dfrac{6.626 \times 10^{-34} \text{ J} \cdot \text{s} \times 2.998 \times 10^8 \text{ m/s}}{3.20 \times 10^{-7} \text{ m}} \times \dfrac{1 \text{ kJ}}{1000 \text{ J}} \times \dfrac{6.022 \times 10^{23} \text{ photons}}{\text{mole}}$

$= 374$ kJ/mol

(c)   UV-B photons have shorter wavelength and higher energy.

(d)   Yes. The higher energy UV-B photons would be more likely to cause sunburn.

6.75   Find the energy of one photon.

$E = hc/\lambda = \dfrac{6.626 \times 10^{-34} \text{ J} \cdot \text{s} \times 2.998 \times 10^8 \text{ m/s}}{6300 \text{ Å}} \times \dfrac{1 \text{ Å}}{1 \times 10^{-10} \text{ m}} = 3.153 \times 10^{-19}$ J

$1 \text{ W} = 1 \text{ J/s}$, $1 \text{ W} \cdot \text{s} = 1$ J

$5.0 \text{ mW} \times \dfrac{1 \times 10^{-3} \text{ W}}{1 \text{ mW}} \times \dfrac{1 \text{ J}}{1 \text{ W} \cdot \text{s}} \times 2.0 \text{ min} \times \dfrac{60 \text{ s}}{1 \text{ min}} \times \dfrac{1 \text{ photon}}{3.153 \times 10^{-19} \text{ J}} = 1.9 \times 10^{18}$ photons

6.76   We know the wavelength of microwave radiation, the volume of water to be heated and the desired temperature change. We need to calculate: (a) the total energy required to heat the water and (b) the energy of a single photon in order to find (c) the number of photons required.

(a)   From Chapter 5, the heat capacity of liquid water is 4.184 J/g°C.

To find the mass of 100 mL of water at 20°C, use the density of water given in Appendix B.

$$100 \text{ mL} \times \frac{0.997 \text{ g}}{1 \text{ mL}} = 99.7 \text{ g } H_2O$$

$$\frac{4.184 \text{ J}}{1 \text{ g °C}} \times 99.7 \text{ g} \times (100°C - 20°C) = 3.337 \times 10^4 \text{ J} = 33 \text{ kJ}$$

(b) $E = hc/\lambda = 6.626 \times 10^{-34} \text{ J} \cdot \text{s} \times \frac{2.998 \times 10^8 \text{ m}}{1 \text{ s}} \times \frac{1}{0.125 \text{ m}} = \frac{1.59 \times 10^{-24} \text{ J}}{1 \text{ photon}}$

(c) $3.337 \times 10^4 \text{ J} \times \frac{1 \text{ photon}}{1.589 \times 10^{-24} \text{ J}} = 2.1 \times 10^{28}$ photons

(The answer has 2 sig figs because the temperature change, 80°C, has 2 sig figs)

6.77 $\dfrac{\Delta E}{1 \text{ molecule}} = E_{absorbed} - E_{emitted} = \dfrac{hc}{460 \times 10^{-9} \text{ m}} - \dfrac{hc}{660 \times 10^{-9} \text{ m}}$

$$\frac{\Delta E}{1 \text{ molecule}} = 6.626 \times 10^{-34} \text{ J} \cdot \text{s} \times \frac{2.998 \times 10^8 \text{ m}}{1 \text{ s}} \left(\frac{1}{460 \times 10^{-9} \text{ m}}\right) - \left(\frac{1}{660 \times 10^{-9} \text{ m}}\right)$$

$$\Delta E = \frac{1.309 \times 10^{-19} \text{ J}}{1 \text{ molecule}} \times \frac{6.022 \times 10^{23} \text{ molecules}}{1 \text{ mol}} = +78.8 \text{ kJ/mol}$$

Net energy is absorbed by the system.

6.78 $\dfrac{8.6 \times 10^{-13} \text{ C}}{1 \text{ s}} \times \dfrac{1 e^-}{1.602 \times 10^{-19} \text{ C}} \times \dfrac{1 \text{ photon}}{1 e^-} = 5.368 \times 10^6 = 5.4 \times 10^6$ photons/s

$$\frac{E}{\text{photon}} = hc/\lambda = \frac{6.626 \times 10^{-34} \text{ J} \cdot \text{s}}{550 \text{ nm}} \times \frac{2.998 \times 10^8 \text{ m}}{1 \text{ s}} \times \frac{1 \text{ nm}}{1 \times 10^{-9} \text{ m}} \times \frac{5.368 \times 10^6 \text{ photon}}{s}$$

$$= 1.9 \times 10^{-12} \text{ J/s}$$

6.79 $\Delta H°_{rxn} = \Delta H°_f \ O_2(g) + \Delta H°_f \ O(g) - \Delta H°_f \ O_3 \ (g)$

$= \quad 0 \quad + 247.5 \text{ kJ} - 142.3 \text{ kJ}$

$\Delta H°_{rxn} = +105.2 \text{ kJ}$

$$\frac{105.2 \text{ kJ}}{\text{mol } O_3} \times \frac{1 \text{ mol } O_3}{6.022 \times 10^{23} \text{ molecules}} \times \frac{1000 \text{ J}}{1 \text{ kJ}} = \frac{1.747 \times 10^{-19} \text{ J}}{O_3 \text{ molecule}}$$

$$\Delta E = hc/\lambda \; ; \quad \lambda = \frac{hc}{\Delta E} = \frac{6.626 \times 10^{-34} \, J \cdot s \times 2.998 \times 10^8 \, m/s}{1.747 \times 10^{-19} \, J} = 1.137 \times 10^{-6} \, m$$

Radiation with this wavelength is in the infrared portion of the spectrum. (Clearly, processes other than simple photodissociation cause O_3 to absorb ultraviolet radiation.)

6.80   (a)   When atoms are exposed to light, as He atoms in the solar atmosphere are exposed to sunlight, interaction may or may not occur. Electrons in atoms exist in discrete allowed energy states. That is, their energy is quantized. If the light has exactly the right amount of energy to cause an electron transition from one allowed energy state to another, that wavelength is absorbed. Other wavelengths are not absorbed. In an absorption spectrum the wavelengths that are absorbed are the black lines.

        (b)   Sunlight is a continuous spectrum of wavelengths. Those wavelengths that correspond to allowed ΔE's for He atoms are absorbed and are the black lines in absorption spectra.

               Excited He atoms relax to the ground state by emitting certain wavelengths of light. These correspond to electron transitions from a higher allowed energy state to a lower one. Since the allowed energy states are the same whether an atom absorbs or emits light, the allowed ΔE's are the same. The wavelengths of the lines in the emission spectrum of excited He atoms are the same as those in the absorption spectrum.

6.81   (a) $\Delta E = R_H \left[\dfrac{1}{n_i^2} - \dfrac{1}{n_f^2} \right] = 2.18 \times 10^{-18} \, J \left(\dfrac{1}{1} - \dfrac{1}{\infty} \right) = 2.18 \times 10^{-18} \, J/atom$

               $2.18 \times 10^{-18} \, J/atom \times \dfrac{6.022 \times 10^{23} \, atoms}{mol} \times \dfrac{1 \, kJ}{1000 \, J} = 1.31 \times 10^3 \, kJ/mol$

        (b) $\lambda = hc/E = \dfrac{6.626 \times 10^{-34} \, J \cdot s \times 2.998 \times 10^8 \, m/s}{2.18 \times 10^{-18} \, J} = 9.11 \times 10^{-8} \, m = 91.1 \, nm$

        (c) ΔE is positive, so light is absorbed. This is reasonable, since energy must be supplied to overcome the electrostatic attraction of the electron for the nucleus.

        (d) $\Delta E = R_H \left[\dfrac{1}{2^2} - \dfrac{1}{\infty} \right] = 2.18 \times 10^{-18} \, J \, (1/4) = 5.45 \times 10^{-19} \, J$

6.82   (a) He^+ is hydrogen-like because it is a one-electron particle. An He atom has two electrons. The Bohr model is based on the interaction of a single electron with the nucleus, but does not accurately account for additional interactions when two or more electrons are present.

(b) Divide each energy by the smallest value to find the integer relationship.

H: $2.18 \times 10^{-18}/2.18 \times 10^{-18} = 1$; $Z = 1$

He$^+$: $8.72 \times 10^{-18}/2.18 \times 10^{-18} = 4$; $Z = 2$

Li^{2+}: $1.96 \times 10^{-17}/2.18 \times 10^{-18} = 9$; $Z = 3$

The ground-state energies are in the ratio of 1:4:9, which is also the ratio Z^2, the square of the nuclear charge for each particle.

The ground state energy for hydrogen-like particles is:

$E = R_HZ^2$. (By definition, n = 1 for the ground state of a one-electron particle.)

(c) $Z = 92$ for U. $E = -2.18 \times 10^{-18}$ J $(92)^2 = -1.85 \times 10^{-14}$ J

6.83 $\lambda = h/mv$; $v = h/m\lambda$. $\lambda = 0.711 \text{ Å} \times \dfrac{1 \times 10^{-10} \text{ m}}{1 \text{ Å}} = 7.11 \times 10^{-11}$ m; $m_e = 9.1094 \times 10^{-31}$ kg

$$v = \frac{6.626 \times 10^{-31} \text{ J} \cdot \text{s}}{9.1094 \times 10^{-31} \text{ kg} \times 7.11 \times 10^{-11} \text{ m}} \times \frac{1 \text{ kg} \cdot \text{m}^2/\text{s}^2}{1 \text{ J}} = 1.02 \times 10^7 \text{ m/s}$$

6.84 Strategy: Change keV to J/electron. Calculate v from kinetic energy. $\lambda = h/mv$.

$$20.0 \text{ keV} \times \frac{1000 \text{ eV}}{\text{keV}} \times \frac{96.485 \text{ kJ}}{1 \text{ eV} \cdot \text{mol}} \times \frac{1000 \text{ J}}{1 \text{ kJ}} \times \frac{1 \text{ mol}}{6.022 \times 10^{23} \text{ electrons}}$$

$$= 3.204 \times 10^{-15} = 3.20 \times 10^{-15} \text{ J/electron}$$

$E_k = mv^2/2$; $v^2 = 2E_k/m$; $v = \sqrt{2E_k/m}$

$$v = \left(\frac{2 \times 3.204 \times 10^{-15} \text{ kg} \cdot \text{m}^2/\text{s}^2}{9.1094 \times 10^{-31} \text{ kg}} \right)^{1/2} = 8.388 \times 10^7 = 8.39 \times 10^7 \text{ m/s}$$

$$\lambda = h/mv = \frac{6.626 \times 10^{-34} \text{ J} \cdot \text{s}}{9.1094 \times 10^{-31} \text{ kg} \times 8.388 \times 10^7 \text{ m/s}} \times \frac{1 \text{ kg} \cdot \text{m}^2/\text{s}^2}{1 \text{ J}} = 8.67 \times 10^{-12} \text{ m} = 8.67 \text{ pm}$$

6.85 (a) $\Delta x \geq \dfrac{h}{2\pi (\Delta p)}$; $\Delta x \geq \dfrac{h}{2\pi (m\Delta v)}$; $\Delta v = 0.010 \, (3.0 \times 10^6 \text{ m/s}) = 3.0 \times 10^4 \text{ m/s}$

$$\Delta x \geq \frac{6.626 \times 10^{-34} \text{ kg} \cdot \text{m}^2}{1 \text{ s}} \times \frac{1}{6.283} \times \frac{1}{9.109 \times 10^{-31} \text{ kg}} \times \frac{1 \text{ s}}{3.0 \times 10^4 \text{ m}} = 3.9 \times 10^{-9} \text{ m}$$

(b) $\Delta v = 0.010 \, (200 \text{ m/s}) = 2.0 \text{ m/s}$

$$\Delta x \geq \frac{6.626 \times 10^{-34} \text{ kg} \cdot \text{m}^2}{1 \text{ s}} \times \frac{1}{6.283} \times \frac{1}{0.012 \text{ kg}} \times \frac{1 \text{ s}}{2.0 \text{ m}} = 4.4 \times 10^{-33} \text{ m}$$

(c)　　Since Δx and Δp are inversely proportional, the electron, with the smaller Δp, has the larger Δx. The diameter of an atom is on the order of Angstroms (10^{-10} m) and Δx for the electron is similar. The Δx of the bullet (4×10^{-33} m) is insignificant compared to the size of a bullet, which is on the order of millimeters.

6.86　(a) l　(b) n and l　(c) m_s　(d) m_l

6.87　(a) 2s　(b) 4d　(c) 5p　(d) 3d　(e) 4f

6.88　(a) 25　(b) 3　(c) 7　(d) 1　(e) 1　(f) 5

6.89　(a)　The p_z orbital has a nodal plane where $z = 0$. This is the xy plane.

　　　(b)　The d_{xy} orbital has 4 lobes and 2 nodal planes, the two planes where $x = 0$ and $y = 0$. These are the yz and xz planes.

　　　(c)　The $d_{x^2-y^2}$ has 4 lobes and 2 nodal planes, the planes where $x^2 - y^2 = 0$. These are the planes that bisect the x and y axes and contain the z axis.

　　　(d)　Following the pattern above, the 3 nodal planes will have $x = 0$, $y = 0$ and $z = 0$, respectively. These are the yz, xz and xy planes.

　　　(e)　Since the planes that contain the x, y and z axes are nodal planes, the 8 lobes of the f_{xyz} orbital will point between the x, y and z axes, one in each octant of the Cartesian coordinate system.

6.90　(d) 3p < (a) 4s < (b) 3d = (c) 3d

6.91　(a) Cd - $[Kr]5s^24d^{10}$　(b) Sb - $[Kr]5s^24d^{10}5p^3$　(c) La - $[Xe]6s^25d^1$

　　　(d) Pd - $[Kr]5s^24d^8$　(e) Ra - $[Rn]7s^2$

6.92　Although Cr and Mo are exceptions to the filling order, W is not. Assume *seaborgium* fills like W.

　　　$[Rn]7s^24f^{14}5d^4$

6.93　(a) Ca, Zn, Kr　(b) K, Sc, Cu (one unpaired 4s electron), Ga, Br　(c) Cr, with the electron configuration $[Ar]4s^13d^5$, has **six** unpaired electrons; Mn, $[Ar]4s^23d^5$, has only five.

6.94　(a) O - excited　(b) Br - ground　(c) P - excited　(d) In - ground

7 Periodic Properties of the Elements

Periodic Table; Electron Shells; Atomic Radii

7.1   Mendeleev insisted that elements with similar chemical and physical properties be placed within a family or column of the table. Since many elements were as yet undiscovered, Mendeleev left blanks. He predicted properties for the "blanks" based on properties of other elements in the family.

7.2   (a)   Mendeleev noted that certain chemical and physical properties recur periodically when the elements are arranged by increasing **atomic weight**. He arranged the known elements by increasing atomic weight so that elements with similar properties were in the same family or vertical column.

       (b)   Shortly after Rutherford postulated the nuclear model of the atom, Mosely proposed that each element be assigned an integer or atomic number which corresponded to the charge on the nucleus of the atom. If elements are arranged by increasing atomic number, a few seeming contradictions in the Mendeleev table (the positions of Ar and K or Te and I) are eliminated.

7.3   The quantum mechanical model describes electron structure in terms of the probability of finding electrons in some volume element of space. On a plot of radial electron density (the electron density on the surface of a sphere whose radius is a certain distance from the nucleus) there are certain radii with high electron densities. The number of these maxima corresponds to the number of "electron shells" for a particular atom as proposed by Lewis. This is also the number of principal quantum levels in the atom. In a multielectron atom, the total of the electron densities of all orbitals or subshells in a principal quantum level is roughly spherical, corresponding to the spherical electron shell.

7.4

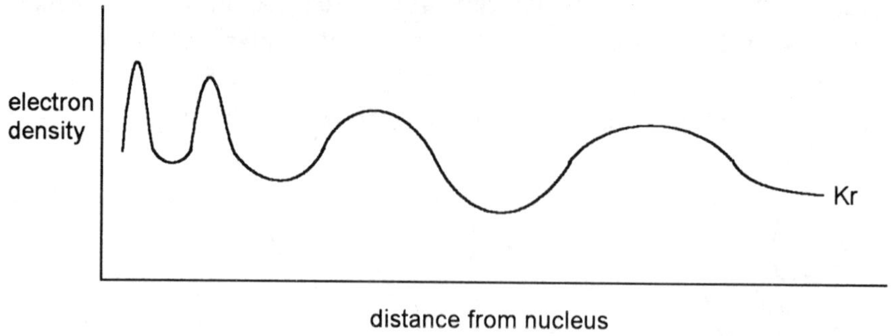

distance from nucleus

Kr has electrons in four shells, so there are four peaks in the radial electron density graph.

7.5   Krypton has a larger nuclear charge (Z = 36) than argon (Z = 18). The shielding of the n = 3 shells by the 1s and 2s electrons in the two atoms is approximately equal, so the n = 3 electrons in Kr experience a greater effective nuclear charge and are thus situated closer to the nucleus.

7.6   Rh < Ti < K < P <Mg

7.7   Since the quantum mechanical description of the atom does not specify the exact location of electrons, there is no specific distance from the nucleus where the last electron can be found. Rather, the electron density decreases gradually as the distance from the nucleus increases. There is no quantum mechanical "edge" of an atom.

7.8   When assigning covalent radii to atoms, we assume that: 1) atoms are spherical objects that touch each other when they bond, 2) atomic radii are constant over time and 3) atomic radii are unchanged by bonding.

7.9   The distance between Cr atoms in Cr metal will be 2 times the atomic radius of Cr, 2 × 1.25 Å = 2.50 Å.

7.10   The atomic radius of Au is the interatomic Au-Au distance divided by 2, 2.88 Å/2 = 1.44 Å.

7.11   From atomic radii, As-I = 1.21 Å + 1.33 Å = 2.54 Å. This is very close to the experimental value of 2.55 Å.

7.12   The atomic radius of Bi is the Bi-Br distance minus the radius of Br. 2.63 Å - 1.14 Å = 1.49 Å.

7.13   (a)   Atomic radii **decrease** moving from left to right across a row and (b) **increase** from top to bottom within a group.

(c)   F < S < P <As. The order is unambiguous according to the trends of increasing atomic radius moving down a column and to the left in a row of the table.

7.14   (a)   The change in atomic radius (Δr) is usually greater moving down one element in a column than moving one element to the right in a row of the table.

(b)   According to the data in Figure 7.5, the representative elements have radii in the range 0.32 Å - 1.63 Å, while transition metal radii are in the range 1.24 Å - 1.78 Å. Clearly the representative elements have a larger variation in radii, owing to the greater range of n values for the p-block elements.

(c)   B < Ga < Ca < Cs. Size increases going down and to the left of the periodic table.

7.15   (a)   The electrons in a He atom experience a nuclear charge of 2 and are drawn closer to the nucleus than the electron in H, which feels a nuclear charge of only 1.

(b)   Even though Z = 10 for Ne and Z = 2 for He, the valence electrons in Ne are in $n = 2$ and those of He are closer to the nucleus in $n = 1$. Also, the 1 s electrons in Ne shield the valence electrons from the full nuclear charge. This shielding, coupled with the larger n value of the valence electrons in the Ne, means that its atomic radius is larger than that of He.

7.16   (a)   Moving from left to right, the value of Z increases and the value of n remains unchanged. Since valence electrons do not effectively shield each other, those of elements on the right side of a row experience a greater nuclear charge than those on the left of a row. As effective nuclear charge increases, size decreases.

(b)   Moving from top to bottom in a family, Z increases but so does n. As n increases, the valence electrons are effectively shielded by the core electrons. This has the effect of increasing the distance of the valence electrons from the nucleus while leaving the effective nuclear charge essentially constant. Thus, size increases going down a family.

Ionization Energies;   Electron Affinities

7.17 $Te(g) \rightarrow Te^+(g) + 1e^-$; $Te^+(g) \rightarrow Te^{2+}(g) + 1e^-$; $Te^{2+}(g) \rightarrow Te^{3+}(g) + 1e^-$

7.18   (a) $Pb^+(g) \rightarrow Pb^{2+}(g) + 1e^-$ (b) $Os^{3+}(g) \rightarrow Os^{4+}(g) + 1e^-$

7.19   The electron configuration of Li^+ is $1s^2$ or [He] and that of Be^+ is $[He]2s^1$. Be^+ has one more valence electron to lose while Li^+ has the stable noble gas configuration of He. It requires much more energy to remove a 1s core electron close to the nucleus of Li^+ than a 2s valence electron further from the nucleus of Be^+.

7.20 $Sr - [Kr]5s^2$. I_1 and I_2 for Sr both involve removing valence electrons from the 5s orbital. I_3 involves removing a core electron from the complete 4p subshell. Because it is closer to the nucleus and less shielded than a 5s electron, removing a 4p electron requires much more energy. The difference between I_2 and I_3 will be much larger than the difference between I_1 and I_2.

7.21   Moving from He to Rn in group 8A, first ionization energies decrease and atomic size increases, because the n value and thus distance of the valence electrons from the nucleus increases going down a group.

7.22   First ionization energies increase slightly going from K to Ar and atomic sizes decrease. As valence electrons are drawn closer to the nucleus (atom size decreases), it requires more energy to completely remove them from the atom (first ionization energy increases). Each trend has a discontinuity at Ga, owing to the increased shielding of the 4p electrons by the filled 3d subshell.

7.23   (a) Ne   (b) Mg   (c) Cr   (d) Br   (e) Ge

7.24   (a)   Cl - As the effective nuclear charge increases in moving from left to right in the third row, the energy required to remove an electron increases.

(b)   Al (slightly) - Al and Ga are rather similar; the buildup of nuclear charge due to the presence of 3d electrons in Ga offsets the larger value of the principle quantum number for its valence electrons.

(c)   La - La and Cs compare as two elements in the same row, with the effects of increasing effective nuclear charge dominating their periodic properties.

(d)   N - Valence electrons in N are closer to the nucleus (n = 2) and are shielded only by the [He] core, so they experience greater attraction for the nucleus and have a higher ionization energy.

7.25   (a)   Na - In an isoelectronic series, all electronic effects (shielding and repulsion) are the same, so the particle with the smallest Z will have the smallest effective nuclear charge.

(b) Si^{3+} - Si has the largest Z and effective nuclear charge.

(c)   The greater the effective nuclear charge experienced by a valence electron, the larger the ionization energy for that electron. According to Table 7.2, I_1 for Na is 496 kJ/mol. I_4 for Si is 3230 kJ/mol.

7.26   To form AlO_2, the aluminum would need to lose four electrons. This would mean removing a 2p core electron from Al. The energy cost of doing this is given by I_4 in Table 7.2. Note that it is much greater than I_4 for Si, because the effective nuclear charge experienced by the 2p electron in the Al^{3+} ion is very high. The high energy cost of removing this electron rules out chemical behavior that corresponds to an Al^{4+} species.

7.27   Ionization energy: $Se(g)$ → $Se^+(g) + 1e^-$
                                    $[Ar]4s^2 3e^{10} 4p^4$ $[Ar]4s^2 3d^{10} 4p^3$

Electron affinity: $Se(g) + 1e^-$ → $Se^-(g)$
                                    $[Ar]4s^2 3d^{10} 4p^4$ $[Ar]4s^2 3d^{10} 4p^5$

7.28   Ionization energy: Kr → Kr^+ $+ 1e^-$
                                    $[Ar]4s^2 3d^{10} 4p^6$ → $[Ar]4s^2 3d^{10} 4p^5$

Ionization energy is always positive. Electrons in atoms are stabilized (lower in energy) because of their electrostatic attraction to the nucleus. Energy is required to overcome this stabilization.

Electron affinity: Kr $+ 1e^-$ → Kr^-
                                $[Ar]4s^2 3d^{10} 4p^6$ $[Ar]4s^2 3d^{10} 4p^6 5s^1$

Energy is required to add an electron to a Kr atom; Kr^- has a higher energy than the isolated Kr atom and free electron. In Kr^- the added electron would have to occupy the higher energy 5s orbital; a 5s electron is effectively shielded by the spherical Kr core and is not stabilized by the nucleus.

7.29   F   + 1e⁻ →   F⁻   ;   Ne   + 1e⁻ →   Ne⁻

$[He]2s^22p^5$ $[He]2s^22p^6$ $[He]2s^22p^6$ $[He]2s^22p^63s^1$

Adding an electron to F completes the $n = 2$ shell; F⁻ has a stable noble gas electron configuration and ΔE for the process is negative. An extra electron in Ne would occupy the higher energy 3s orbital; adding an electron increases the energy of the system and ΔE for the process is positive.

7.30   Li   + 1e⁻ →   Li⁻   ;   Be   + 1e⁻ →   Be⁻

$[He]2s^1$ $[He]2s^2$ $[He]2s^2$ $[He]2s^22p^1$

Adding an electron to Li completes the 2s subshell. The added electron experiences essentially the same effective nuclear charge as the other valence electron, except for the repulsion of pairing electrons in an orbital. There is an overall stabilization; ΔE is negative.

An extra electron in Be would occupy the higher energy 2p subshell. This electron is shielded from the full nuclear charge by the 2s electrons and does not experience a stabilization in energy; ΔE is positive.

Properties of Metals and Nonmetals

7.31 $O_2 < Br_2 < K < Mg$. O_2 and Br_2 are nonmetals; at standard conditions, O_2 is a gas and Br_2 is a liquid, so O_2 has the lower melting point. Nonmetallic character increases going up and to the right on the chart. The greater the nonmetallic character, the lower the boiling point. O_2 is above but to the left of Br_2. As with other trends, the vertical difference is larger and determines the relationship. K and Mg are solid metals. K has the greater metallic character (further down and to the left on the chart) and the lower melting point.

7.32   Metals are good conductors and nonmetals are poor ones. Metallic character increases going from right to left across a row and from top to bottom in a column. The order of increasing metallic character and electrical conductivity is S < Si <Ge <Ca.

7.33   Metallic character increases moving down a family and to the left in a period.

(a)   Li   (b) Na   (c) Sn   (d) Al

7.34   (a)   Metallic character increases going down a group: N < P < As <Sb <Bi

(b)   Nonmetallic character increases going up and to the right in the chart.
Hg < In < Ge < S < F

7.35   Ionic: MgO, Li_2O, Y_2O_3; molecular: SO_2, P_2O_5, N_2O, XeO_3

Ionic compounds are formed by combining a metal and a nonmetal; molecular compounds are formed by two or more nonmetals.

7.36 Solids: CuS, $CrCl_3$, Na_3P, TiO_2; gases: NO_2, PF_3, CO_2

Ionic compounds are high melting solids, while covalent compounds exist in all three states. Covalent compounds with low molecular weight, such as NO_2, PF_3 and CO_2 are often gases.

7.37 When dissolved in water, an "acidic oxide" produces an acidic (pH < 7) solution. Oxides of nonmetals are acidic. Example: $SO_3(g)$. A "basic oxide" dissolved in water produces a basic (pH > 7) solution. Oxides of metals are basic. Example: CaO (quick lime).

7.38 The more **nonmetallic** the central atom, the more acidic the oxide. In order of increasing acidity: $BaO < MgO < Al_2O_3 < SiO_2 < CO_2 < P_2O_5 < SO_3$

7.39 (a) $CaO(s) + H_2O(l) \rightarrow Ca(OH)_2(aq)$

(b) $CuO(s) + 2HNO_3(aq) \rightarrow Cu(NO_3)_2(aq) + H_2O(l)$

(c) $SO_3(g) + H_2O(l) \rightarrow H_2SO_4(aq)$

(d) $CO_2(g) + 2NaOH(aq) \rightarrow Na_2CO_3(aq) + H_2O(l)$

7.40 (a) $Li_2O(s) + H_2O(l) \rightarrow 2LiOH(aq)$

(b) $Cl_2O_7(g) + H_2O(l) \rightarrow 2HClO_4(aq)$

(c) $V_2O_3(s) + 6HNO_3(aq) \rightarrow 2V(NO_3)_3(aq) + 3H_2O(l)$

(d) $SeO_2(s) + 2KOH(aq) \rightarrow K_2SeO_3(aq) + H_2O(l)$

Group Trends in Metals and Nonmetals

7.41

	Na	**Mg**
(a)	$[Ne]\,3s^1$	$[Ne]\,3s^2$
(b)	+1	+2
(c)	+496 kJ/mol	+738 kJ/mol
(d)	very reactive	reacts with steam, but not $H_2O(l)$
(e)	1.86 Å	1.60 Å

(b)   When forming ions, both adopt the stable configuration of Ne, but Na loses one electron and Mg two electrons to achieve this configuration.

(c),(e)   The nuclear charge of Mg (Z = 12) is greater than that of Na, so it requires more energy to remove a valence electron with the same *n* value from Mg than Na. It also means that the 2s electrons of Mg are held closer to the nucleus, so the atomic radius (e) is smaller than that of Na.

(d)   Mg is less reactive because it has a filled subshell and it has a higher ionization energy.

7.42

	Rb	**Ag**
(a)	$[Kr]\,5s^1$	$[Kr]\,5s^1 4d^{10}$
(b)	+1	+1
(c)	extremely reactive	unreactive
(d)	2.2 Å	1.3 Å

(b)   Each has 1 electron in the 5s subshell which it loses to form a +1 ion.

(c),(d) The 5s electron in Ag experiences a much greater Z_{eff} than the 5s electron in Rb because Z is much larger (47 vs 37) and the two electrons are approximately the same distance from the nucleus. (The extra shielding by the 4d electrons in Ag does not compensate for the higher Z value). Thus, Ag is less reactive (less likely to lose an electron) and has a smaller atomic radius.

7.43   (a)   Ca and Mg are both metals; they tend to lose electrons and form cations when they react. Ca is more reactive because it has a lower ionization energy than Mg. The Ca valence electrons in the 4s orbital are less tightly held because they are farther from the nucleus and experience more shielding by core electrons than the 3s valence electrons of Mg.

(b)   K and Ca are both metals; they tend to lose electrons and form cations when they react. K is more reactive because it has a lower ionization energy. The 4s valence electron in K is less tightly held because it experiences a smaller nuclear charge (Z = 19 for K versus Z = 20 for Ca) with similar shielding effects than the 4s valence electrons of Ca.

7.44   (a)   Cs is much more reactive toward H_2O than Li is because its valence electron is less tightly held (greater n value) than that of Li, and it is more easily oxidized.

(b)   K(s) reacts with O_2(g) to form both K_2O_2, potassium peroxide, and KO_2, potassium superoxide, compounds where O is not in its typical -2 state.

$$2K(s) + O_2(g) \rightarrow K_2O_2(s)$$
$$K(s) + O_2(g) \rightarrow KO_2(s)$$

This is surprising, since most metals form metal oxides (O^{2-}) when reacted with O_2.

7.45   (a) $2K(s) + 2H_2O(l) \rightarrow 2KOH(aq) + H_2(g)$

(b) $Ba(s) + 2H_2O(l) \rightarrow Ba(OH)_2(aq) + H_2(g)$

(c) $6Li(s) + N_2(g) \rightarrow 2Li_3N(s)$

(d) $2Mg(s) + O_2(g) \rightarrow 2MgO(s)$

7.46   (a) $2Na(g) + Br_2(g) \rightarrow 2NaBr(s)$

(b) $2Na(l) + H_2(g) \rightarrow 2NaH(s)$

(c) $4Li(s) + O_2(g) \rightarrow 2Li_2O(s)$

(d) $SrO(s) + H_2O(l) \rightarrow Sr(OH)_2(aq)$

7.47 $H - 1s^1$; $Li - [He] 2s^1$; $F - [He] 2s^2 2p^6$. Like Li, H has only one valence electron, and its most common oxidation number is +1, which both H and Li adopt after losing the single valence electron. Like F, H needs only one electron to adopt the stable electron configuration of the nearest noble gas. Both H and F can exist in the -1 oxidation state, when they have gained an electron to complete their valence shells.

7.48   (a) $2Na(s) + Cl_2(g) \rightarrow 2NaCl(s)$

$H_2(g) + Cl_2(g) \rightarrow 2HCl(g)$

Each product is diatomic and contains one atom from each of the reactants. NaCl is an ionic solid, while HCl is a (polar) covalent gas.

(b) $Ca(s) + F_2(g) \rightarrow CaF_2(s)$

$Ca(s) + H_2(g) \rightarrow CaH_2(s)$

Both products are ionic solids containing Ca^{2+} and the corresponding anion in a 1:2 ratio.

7.49

	F	**Cl**
(a)	[He] $2s^2 2p^5$	[Ne] $3s^2 3p^5$
(b)	-1	-1
(c)	1681 kJ/mol	1256 kJ/mol
(d)	reacts exothermically to form HF	reacts slowly to form HCl
(e)	-332 kJ/mol	-349 kJ/mol
(f)	0.72 Å	0.99 Å

(b)   F and Cl are in the same group, have the same valence electron configuration and common ionic charge.

(c),(f)   The n = 2 valence electrons in F are closer to the nucleus and more tightly held than the n = 3 valence electrons in Cl. Therefore, the ionization energy of F is greater, and the atomic radius is smaller.

(d)   In its reaction with H_2O, F is reduced; it gains an electron. Although the electron affinity, a gas phase single atom property, of F is less negative than that of Cl, the tendency of F to hold its own electrons (high ionization energy) coupled with a relatively large exothermic electron affinity makes it extremely susceptible to reduction and chemical bond formation. Cl is unreactive to water because it is less susceptible to reduction.

(e)   Although F has a larger Z_{eff} than Cl, its small atomic radius gives rise to large repulsions when an extra electron is added, so the overall electron affinity of F is smaller (less exothermic) than that of Cl.

(f)   The n = 2 valence electrons in F are closer to the nucleus so the atomic radius is smaller than that of Cl.

7.50

	O	**F**
(a)	[He] $2s^2 2p^4$	[He] $2s^2 2p^5$
(b)	-2	-1
(c)	+1314 kJ/mol	+1681 kJ/mol
(d)	unreactive	reacts exothermically to produce HF
(e)	reacts to form H_2O	reacts to form HF
(f)	0.73 Å	0.72 Å

(b)   When forming ions, both adopt the stable electron configuration of Ne; O atoms gain two electrons and F atoms gain one electron.

(c)   The ionization energy of F is greater because its valence electrons experience a greater Z_{eff} ($Z_F > Z_O$) and the average distance of the electrons from the nucleus is approximately the same.

(d),(e)   The electron affinity of F is greater (more exothermic) than that of O, and F_2 can remove electrons from almost any substance, including H_2O and H_2.

(f)   From trends in atomic size, one would expect F to be somewhat smaller than O. However, electron-electron repulsions are so great in the relatively cramped 2p orbitals of F that the larger Z_{eff} does not lead to a significantly smaller radius.

7.51   In 1962, N. Bartlett discovered that Xe, which has the lowest ionization energy of the nonradioactive Noble gases, would react with substances having a strong tendency to remove electrons, such as F_2. Thus, the term "inert" no longer described all the Group 8A elements. (Kr also reacts with F_2, but reactions of Ar, Ne and He are as yet unknown.)

7.52   Xe has a lower ionization energy than Ne. The valence electrons in Xe are much further from the nucleus than those of Ne ($n = 5$ vs $n = 2$) and much less tightly held by the nucleus; they are more "willing" to be shared than those in Ne. Also, Xe has empty 5d orbitals that can help to accommodate the bonding pairs of electrons, while Ne has all its valence orbitals filled.

7.53   (a) $2Li(s) + Cl_2(g) \rightarrow 2LiCl(s)$

(b) $S_8(s) + 8O_2(g) \rightarrow 8SO_2(g)$

(c) $2KI(l) \rightarrow 2K(l) + I_2(g)$

(d) $Cl_2(g) + H_2O(l) \rightarrow HCl(aq) + HOCl(aq)$

7.54   (a) $2O_3(g) \rightarrow 3O_2(g)$

(b) $Xe(g) + F_2(g) \rightarrow XeF_2(g)$

$Xe(g) + 2F_2(g) \rightarrow XeF_4(s)$

$Xe(g) + 3F_2(g) \rightarrow XeF_6(s)$

(c) $2NaCl(aq) + 2H_2O(l) \rightarrow 2NaOH(aq) + Cl_2(g) + H_2(g)$

(d) $S_8(s) + 16Li(s) \rightarrow 8Li_2S(s)$

7.55   (a)   Valence electrons in F atoms experience a much greater attraction for the nucleus than those in I atoms, because they are much closer to the nucleus. Thus, F atoms have a much greater tendency to gain electrons and are more reactive toward relatively inert substances such as Xe.

(b) O_2 is the allotrope routinely found in the atmosphere (21% of air is O_2). O_3 is produced only under special conditions.

(c) F_2 is extremely reactive because it can remove electrons from almost any substance, so a special apparatus that will not react with F_2 and will not leak F_2 must be used to carry out reactions.

7.56 (a) Te has more metallic character and is a better electrical conductor.

(b) At room temperature, oxygen molecules are diatomic and exist in the gas phase. Sulfur molecules are 8-membered rings and exist in the solid state.

(c) Chlorine is generally more reactive than bromine because Cl atoms have a greater (more exothermic) electron affinity than Br atoms.

Additional Exercises

7.57 (a) *eka - aluminum* is gallium (Ga), Z = 31.

(b) From the **Handbook of Chemistry and Physics**, 74th edition, the physical properties of Ga are:

atomic weight = 69.92 g/mol   m.p. = 29.78°C
density = 5.904 g/mL (s)   b.p. = 2403°C
       = 6.095 g/mL (l)   oxide = Ga_2O_3, Ga_2O

The atomic weight, density and formula of oxide all agree with Mendeleev's predictions. Gallium does have a low melting point and a high boiling point.

7.58 Up to Z = 83, there are four instances where atomic weights are reversed relative to atomic numbers: Ar and K; Co and Ni; Te and I; Ce and Pr.

7.59 The nuclear charge of an atom increases as the atomic number increases. If the accompanying electron is added to the same shell as the element of the proceeding atomic number, there is an increase in the effective nuclear charge. As effective nuclear charge increases, the size of an atom decreases. If the accompanying electron is added to a new shell, it is by definition further from the nucleus and the size of the atom increases.

7.60 According to Figure 7.5, the atomic radius of the most stable form of Fe at room temperature is 1.24 Å. This is derived from an Fe-Fe distance of 2.48 Å, which corresponds to α-iron.

7.61 Zr - [Kr] $5s^24d^2$; Hf - [Xe] $6s^24f^{14}5d^2$. The 6s electrons in Hf are incompletely shielded by the 4f electrons, so the much larger value of Z (72 for Hf, 40 for Zr) is not totally offset by shielding and the greater distance of the 6s electrons from the nucleus. That is, the 6s electrons in Hf experience a slightly greater **attraction** to the nucleus than those of Zr and the atomic radius of Hf is smaller.

7.62 (a) Increasing atomic radius: F < O < C < Si ≈ Se
F, O, C and Si can be ordered according to the trends for size and ionization energy. Se is difficult to place because it is both to the right and below Si; these two directions

have conflicting trends. Also, it is in the fourth row and subject to the effects of the filling of the 3d subshell and accompanying increase in Z and Z_{eff}. Referring to the data in Figures 7.5 and 7.7, the atomic radii of Si and Se are approximately equal. The ionization energy of Se is greater than that of Si and less than that of C.

(b)   Increasing ionization energy: Si < Se < C < O < F

7.63   (a)   The atomic radius of Ca is greater than that of Zn because Zn has a significantly greater Z and Z_{eff}, which draws the 4s valence electrons closer to the nucleus.

(b) Ca^{+2} - [Ar]; Zn^{+2} - [Ar] $3d^{10}$. In both Ca and Zn atoms, the outermost electrons are in the 4s subshell. The much larger Z_{eff} for Zn causes it to have a much smaller radius. In the +2 ions, both have lost their 4s electrons. The outermost electrons in Zn are in the 3d sublevel and are significantly shielded by the [Ar] core electrons. Thus, the radius of Zn^{2+} is closer to the radius of Ca^{2+} than the radii of the neutral ions.

7.64   The neutral atom (on which I_1 is measured) always has more electrons than the 1^+ ion (on which I_2 is measured). With more electrons there is more electron-electron repulsion, making it easier to remove an electron. (This matter can also be viewed from the perspective of the screening effect. With more electrons there is greater shielding giving rise to a reduced effective nuclear charge. When the effective nuclear charge is smaller, it is easier to remove an electron.)

7.65   (a)   Y - [Kr]$5s^2 4d^1$. Each succeeding ionization energy is larger than the previous one, due to increases in effective nuclear charge as electrons are removed (Exercise 7.64). I_1, I_2 and I_3 involve removing valence electrons and the size of the increases is similar. I_4 involves removing an inner-shell electron from the stable Kr core and requires significantly more energy than I_3.

(b)   In Table 7.2, Al has the pattern of ionization energies most similar to yttrium. That is, I_1, I_2 and I_3 increase smoothly and there is a jump from I_3 to I_4. The actual values of I_2, I_3 and I_4 are greater for Al because the electrons being removed are closer to the nucleus and experience a greater effective nuclear charge.

7.66   (a)   Al - [Ne]$3s^2 3p^1$; Mg - [Ne]$3s^2$. I_1 for Al involves removing a single 3p electron which is effectively shielded from the nuclear charge by the 3s electrons. This shielding offsets the increase in Z going from Mg to Al and causes I_1 for Al to be smaller.

(b)   As - [Ar]$4s^2 3d^{10} 4p^3$; Se - [Ar]$4s^2 3d^{10} 4p^4$. In As, each 4p electron occupies a different orbital (Hund's Rule); in Se, two of the 4p electrons are paired. Pairing of electrons increases repulsion, decreases effective nuclear charge and raises the energy of the valence electrons. Less energy is required to remove an electron from Se, even though Z is greater.

7.67   (a)   P - [Ne] $3s^2 3p^3$; S - [Ne] $3s^2 3p^4$. In P, each 3p orbital contains a single electron, while in S one 3p orbital contains a pair of electrons. Removing an electron from S eliminates the need for electron pairing and reduces electrostatic repulsion, so the overall energy required to remove the electron is smaller than in P, even though Z is greater.

        (b)   C - [He] $2s^2 2p^2$; N - [He] $2s^2 2p^3$; O - [He] $2s^2 2p^4$. An electron added to a N atom must be paired in a relatively small 2p orbital, so the additional electron-electron repulsion more than compensates for the increase in Z and the electron affinity is smaller (less exothermic) than that of C. In an O atom, one 2p orbital already contains a pair of electrons, so the additional repulsion from an extra electron is offset by the increase in Z and the electron affinity is greater (more exothermic). Note from Figure 7.8 that the electron affinity of O is only slightly more exothermic than that of C, although the value of Z has increased by 2.

        (c) O^+ - [Ne] $2s^2 2p^3$; O^{2+} - [Ne] $2s^2 2p^2$; F^+ - [Ne] $2s^2 2p^4$; F^{2+} - [Ne] $2s^2 2p^3$. The decrease in electron-electron repulsion going from F^+ to F^{2+} energetically favors ionization and causes it to be less endothermic than the corresponding process in O, where there is no significant decrease in repulsion.

        (d) Mn^{2+} - [Ar]$3d^5$; Mn^{3+} - [Ar] $3d^4$; Cr^{2+} - [Ar] $3d^4$; Cr^{3+} - [Ar] $3d^3$; Fe^{2+} - [Ar] $3d^6$; Fe^{3+} - [Ar] $3d^5$. The third ionization energy of Mn is expected to be larger than that of Cr because of the larger Z value of Mn. The third ionization energy of Fe is **less** than that of Mn because going from $3d^6$ to $3d^5$ reduces electron repulsions, making the process less endothermic than predicted by nuclear charge arguments.

7.68   (a)   The group 2B metals have complete (*n*-1)d subshells. An additional electron would occupy an *n*p subshell and be substantially shielded by both *n*s and (*n*-1) d electrons. Overall this is not a lower energy state than the neutral atom and a free electron.

        (b)   Valence electrons in Group 1B elements experience a relatively large effective nuclear charge due to the build-up in Z with the filling of the (*n*-1)d subshell. Thus, the electron affinities are large and negative. Group 1B elements are exceptions to the usual electron filling order and have the generic electron configuration $ns^1(n-1)d^{10}$. The additional electron would complete the *n*s subshell and experience repulsion with the other *n*s electrons. Going down the group, size of the *n*s subshell increases and repulsion effects decrease. That is, effective nuclear charge is greater going down the group because it is less diminished by repulsion, and electron affinities become more negative.

7.69   (a)   F and O^- have the same electron configuration, [He] $2s^2 2p^5$; they are isoelectronic.

        (b)   In the first process, an electron is added to a neutral F atom, while in the second an electron is added to a negative O^- ion. Clearly, there is electrostatic repulsion to overcome in the second process. Also, since O^- and F are isoelectronic, but F has one more proton, an electron added to F experiences a greater effective nuclear charge. Both factors contribute to the $+\Delta H$ for the second process.

(c)   Following the same trend, $N^{2-}(g)$ will experience greater electrostatic repulsion and smaller Z_{eff} when gaining an electron, so the electron affinity will become smaller (less exothermic).

7.70   (a) $K^+(g) + 1e^- \rightarrow K(g)$

(b) -419 kJ/mol. The electron affinity of K^+ is the reverse of the ionization energy of K, so the two processes are equal in magnitude but opposite in sign.

(c)   The electron affinity of any 1+ cation is exothermic (has a negative sign) and equal in magnitude to the ionization energy of the neutral atom.

7.71   Since Xe reacts with F_2, and O_2 has approximately the same ionization energy as Xe, O_2 will probably react with F_2. Possible products would be O_2F_2, analogous to XeF_2, or OF_2.

$$O_2(g) + F_2(g) \rightarrow O_2F_2(g)$$

$$O_2(g) + 2F_2(g) \rightarrow 2OF_2(g)$$

7.72   Moving one place to the right in a horizontal row of the table, for example, from Li to Be, there is an increase in ionization energy. Moving downward in a given family, for example from Be to Mg, there is usually a decrease in ionization energy. Similarly, atomic size decreases in moving one place to the right and increases in moving downward. Thus, two elements such as Li and Mg that are diagonally related tend to have similar ionization energies and atomic sizes. This in turn gives rise to some similarities in chemical behavior. Note, however, that the valences expected for the elements are not the same. That is, lithium still appears as Li^+, magnesium as Mg^{2+}.

7.73   Fr (1A), Ra (2A), Po (6A), At (7A), Rn (8A)

(a)   most metallic character – **Fr** (Metallic character decreases from left to right in a row.)
(b)   most nonmetallic character -- **Rn**
(c)   largest ionization energy -- **Rn** (Ionization energy increases from left to right in a row.)
(d)   smallest ionization energy -- **Fr**
(e)   greatest electron affinity -- **At** (Electron affinity becomes more exothermic from left to right in a row.)
(f)   largest atomic radius -- **Fr** (Size decreases from left to right in a row.)
(g)   appears least like the element above it – **Fr** (According to the trends in melting point, Fr may be a gas at room temperature, while Cs is a liquid. The others should be in the same state as the element above them.)
(h)   highest melting point -- **Ra** (The melting points of the group 2A metals are much higher than those of the other groups, even though the values decrease going down the group.)
(i)   react most readily with H_2O -- **Fr** (It has the lowest ionization energy.)

7.74 Element	Density (g/cm³)	Melting Point(°C)	Boiling Point(°C)	I_1	I_2	Formulas of Halides	Electron Configuration
Sb	6.684	630.5	1380	833.5	1592	SbX_3, SbX_5	[Kr] $5s^2 4d^{10} 5p^3$
I	4.93	113.5	184.35	1009	1846	IX, IX_3, IX_5, IX_7	[Kr]$5s^2 4d^{10} 5p^5$
Se	4.81	60 - 217	685	940.7	2074	Se_2X_2, SeX_4, SeX_6	[Ar]$4s^2 3d^{10} 4p^4$
Te	6.25	452	1390	869.3	1795	TeX_2, TeX_4, TeX_6	[Kr]$5s^2 4d^{10} 5p^4$

Because Te is a semimetal, its physical properties do not necessarily match the other elements in its group. In fact, with respect to density, boiling point and I_1, it is most similar to its left neighbor, Sb. (There are no obvious similarities in melting point, and the values for I_2 follow the predicted trend, increasing from left to right across a row.) Chemically, one could expect Te to resemble Se, since they have the same valence electron configuration. This is born out by the observed formulas of the halides for the various elements.

Integrative Exercises

7.75   (a) $\nu = c/\lambda$; $1 \text{ Hz} = 1s^{-1}$

Ne: $\nu = \dfrac{2.998 \times 10^8 \text{ m/s}}{14.610 \text{ Å}} \times \dfrac{1 \text{ Å}}{1 \times 10^{-10} \text{ m}} = 2.052 \times 10^{17} \text{ s}^{-1} = 2.052 \times 10^{17} \text{ Hz}$

Ca: $\nu = \dfrac{2.998 \times 10^8 \text{ m/s}}{3.358 \times 10^{-10} \text{ m}} = 8.928 \times 10^{17} \text{ Hz}$

Zn: $\nu = \dfrac{2.998 \times 10^8 \text{ m/s}}{1.435 \times 10^{-10} \text{ m}} = 20.89 \times 10^{17} \text{ Hz}$

Zr: $\nu = \dfrac{2.998 \times 10^8 \text{ m/s}}{0.786 \times 10^{-10} \text{ m}} = 38.14 \times 10^{17} = 38.1 \times 10^{17} \text{ Hz}$

Sn: $\nu = \dfrac{2.998 \times 10^8 \text{ m/s}}{0.491 \times 10^{-10} \text{ m}} = 61.06 \times 10^{17} = 61.1 \times 10^{17} \text{ Hz}$

(b)

Element	Z	ν	$\nu^{1/2}$
Ne	10	2.052×10^{17}	4.530×10^8
Ca	20	8.928×10^{17}	9.449×10^8
Zn	30	20.89×10^{17}	14.45×10^8
Zr	40	38.14×10^{17}	19.5×10^8
Sn	50	61.06×10^{17}	24.7×10^8

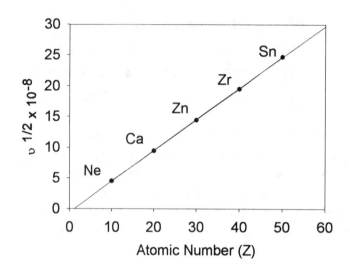

(c)   The plot in part (b) indicates that there is a linear relationship between atomic number and the square root of the frequency of the X-rays emitted by an element. Thus, elements with each integer atomic number should exist. This relationship allowed Moseley to predict the existence of elements that filled "holes" or gaps in the periodic table.

(d)   For Fe, Z = 26. From the graph, $\nu^{1/2} = 12.5 \times 10^8$, $\nu = 1.56 \times 10^{18}$ Hz.

$$\lambda = c/\nu = \frac{2.998 \times 10^8 \text{ m/s}}{1.56 \times 10^{18} \text{ s}^{-1}} \times \frac{1 \text{ Å}}{1 \times 10^{-10} \text{ m}} = 1.92 \text{ Å}$$

(e) $\lambda = 0.980$ Å $= 0.980 \times 10^{-10}$ m

$$\nu = c/\lambda = \frac{2.998 \times 10^8 \text{ m/s}}{0.980 \times 10^{-10} \text{ m}} = 30.6 \times 10^{17} \text{ Hz}; \ \nu^{1/2} = 17.5 \times 10^8$$

From the graph, $\nu^{1/2} = 17.5 \times 10^8$, Z = 36. The element is krypton, Kr.

7.76   (a)   E = hc/λ; 1 nm = 1 × 10^{-9} m; 58.4 nm = 58.4 × 10^{-9} m;

1 eV•mol = 96.485 kJ/mol, 1 eV • mol = 96.485 kJ

$$E = \frac{6.626 \times 10^{-34} \text{ J} \bullet \text{s} \times 2.998 \times 10^8 \text{ m/s}}{58.4 \times 10^{-9} \text{ m}} = 3.4015 \times 10^{-18} = 3.40 \times 10^{-18} \text{ J/photon}$$

$$\frac{3.4015 \times 10^{-18} \text{ J}}{\text{photon}} \times \frac{1 \text{ kJ}}{1000 \text{ J}} \times \frac{6.022 \times 10^{23} \text{ photons}}{\text{mol}} \times \frac{1 \text{ eV} \times \text{mol}}{96.485 \text{ kJ}} 124 = 21.230$$

$$= 21.2 \text{ eV}$$

(b) $Hg(g) \rightarrow Hg^+(g) + 1e^-$

(c) $I_1 = E_{58.4} - E_K = 21.23$ eV -10.75 eV $= 10.48 = 10.5$ eV

$$10.48 \text{ eV} \times \frac{96.485 \text{ kJ}}{1 \text{ eV} \cdot \text{mol}} = 1.01 \times 10^3 \text{ kJ/mol}$$

(d)   From Figures 7.6 and 7.7, iodine (I) appears to have the ionization energy closest to that of Hg, approximately 1000 kJ/mol. (Quantitative data is much easier to read on Figure 7.6.)

7.77   (a)

$$Na(g) \rightarrow Na^+(g) + 1e^- \quad \text{(ionization energy of Na)}$$
$$Cl(g) + 1e^- \rightarrow Cl^-(g) \quad \text{(electron affinity of Cl)}$$
$$\overline{}$$
$$Na(g) + Cl(g) \rightarrow Na^+(g) + Cl^-(g)$$

(b) $\Delta H = I_1$ (Na) $+ E_1$(Cl) $= +496$ kJ $- 349$ kJ $= +147$ kJ, endothermic

(c)   The reaction $2Na(s) + Cl_2(g) \rightarrow 2NaCl(s)$ involves many more steps than the reaction in part (a). One important difference is the production of NaCl(s) versus NaCl(g). The condensation NaCl(g) $\rightarrow$ NaCl(s) is very exothermic and is the step that causes the reaction of the elements in their standard states to be exothermic, while the gas phase reaction is endothermic.

7.78   (a) Mg_3N_2

(b) $Mg_3N_2(s) + 3H_2O(l) \rightarrow 3MgO(s) + 2NH_3(g)$
The driving force is the production of $NH_3(g)$.

(c)   After the second heating, all the Mg is converted to MgO.

Calculate the initial mass Mg.

$$0.486 \text{ g MgO} \times \frac{24.305 \text{ g Mg}}{40.305 \text{ g MgO}} = 0.293 \text{ g Mg}$$

x = g Mg converted to MgO;   y = g Mg converted to Mg_3N_2;   x = 0.293 - y

$$g \text{ MgO} = x \left(\frac{40.305 \text{ g MgO}}{24.305 \text{ g Mg}} \right); \quad g \text{ Mg}_3\text{N}_2 = y \left(\frac{100.929 \text{ g Mg}_3\text{N}_2}{72.915 \text{ g Mg}} \right)$$

g MgO + g Mg_3N_2 = 0.470

$$(0.293 - y) \left(\frac{40.305}{24.305} \right) + y \left(\frac{100.929}{72.915} \right) = 0.470$$

$$(0.293 - y)(1.6583) + y(1.3842) = 0.470$$

$$-1.6583\,y + 1.3842\,y = 0.470 - 0.48588$$

$$-0.2741\,y = -0.016$$

$$y = 0.05794 = 0.058 \text{ g Mg in } Mg_3N_2$$

$$\text{g } Mg_3N_2 = 0.05794 \text{ g Mg} \times \frac{100.929 \text{ g } Mg_3N_2}{72.915 \text{ g Mg}} = 0.0802 = 0.080 \text{ g } Mg_3N_2$$

$$\text{mass \% } Mg_3N_2 = \frac{0.0802 \text{ g } Mg_3N_2}{0.470 \text{ g } (MgO + Mg_3N_2)} \times 100 = 17\%$$

(The final mass % has 2 sig figs because the mass of Mg obtained from solving simultaneous equations has 2 sig figs.)

8 Basic Concepts of Chemical Bonding

Lewis Symbols and Ionic Bonding

8.1   (a)   Valence electrons are those that take part in chemical bonding, those in the outermost electron shell of the atom. This usually means the electrons beyond the core noble gas configuration of the atom, although it is sometimes only the outer shell electrons.

       (b)   C - [He] $2s^2 2p^2$ C has 4 valence electrons.
                   |____|
             valence electrons

       (c) $1s^2 2s^2 2p^6$ $3s^2 3p^1$ The atom (Al) has 3 valence electrons.
             |_____| |_____|
               [Ne]   valence electrons

8.2   (a)   Atoms will gain, lose or share electrons to achieve the nearest noble gas electron configuration. Except for He, this corresponds to eight electrons in the valence shell, thus the term **octet** rule.

       (b)   P - [Ne]$3s^2 3p^3$ A phosphorus atom has 5 valence electrons, so it must gain 3 electrons to achieve an octet.

       (c) $1s^2 2s^2 2p^5$ = [He]$2s^2 2p^5$ The atom (F) has 7 valence electrons and must gain 1 electron to achieve an octet.

8.3   (a) $:\overset{..}{\underset{..}{C}}l\cdot$ (b) $M\overset{\cdot}{g}\cdot$ (c) $:\overset{..}{B}r\cdot$ (d) $:\overset{..}{\underset{..}{A}}r:$ or Ar

8.4   (a) $K\cdot$ (b) $:\overset{..}{S}e\cdot$ (c) $[:\overset{..}{\underset{..}{A}}l:]^{3+}$ or Al^{3+} (d) $[S\overset{\cdot}{n}\cdot]^{2+}$

8.5 $Ca\cdot$ + $\cdot\overset{.}{\underset{..}{O}}:$ ⟶ Ca^{2+} + $\left[\,:\overset{..}{\underset{..}{O}}:\,\right]^{2-}$

127

8.6

$$Al \cdot \; + \; :\ddot{F}: \; + \; \cdot \ddot{F}: \; + \; \cdot \ddot{F}: \; \longrightarrow \; Al^{3+} \; + \; 3\left[\; :\ddot{F}:\; \right]^{-}$$

8.7   (a) CaF_2 (b) Na_2S (c) Y_2O_3 (d) Li_3N

8.8   (a)   BaO   (b) Mg_3N_2 (c) Li_2O (d) $ScBr_3$

8.9   (a) Ba^{2+} : [Xe], noble gas configuration

   (b) Cl^- : $[Ne]3s^23p^6$ = [Ar], noble gas configuration

   (c) Te^{2-}: $[Kr]5s^24d^{10}5p^6$ = [Xe], noble gas configuration

   (d) Cr^{2+}: $[Ar]3d^4$

   (e) Sc^{3+} : [Ar], noble gas configuration

   (f) Co^{3+} : $[Ar]3d^6$

8.10   (a) P^{3-} : $[Ne]3s^23p^6$ = [Ar], noble gas configuration

   (b) Al^{3+} : [Ne], noble gas configuration   (c) Au^+ : $[Xe]4f^{14}5d^{10}$

   (d) Ru^{3+} : $[Kr]4d^5$ (e) Pt^{2+} : $[Xe]4f^{14}5d^8$ (f) Ti^{3+} : $[Ar]3d^1$

8.11   (a)   *Lattice energy* is the energy required to totally separate one mole of solid ionic compound into its gaseous ions.

   (b)   The magnitude of the lattice energy depends on the magnitudes of the charges of the two ions, their radii and the arrangement of ions in the lattice. The main factor is the charges, because the radii of ions do not vary over a wide range.

8.12   According to Equation 8.4, electrostatic attraction increases with increasing charges of the ions and decreases with increasing radius of the ions. Thus, lattice energy (a) **increases** as the charges of the ions increase and (b) **decreases** as the sizes of the ions decrease.

8.13   Equation 8.4 predicts that as the oppositely charged ions approach each other, the energy of interaction will be large and negative. This more than compensates for the energy required to form Ca^{2+} and O^{2-} from the neutral atoms (see Figure 8.4 for the formation of NaCl).

8.14 $Ca(s) \rightarrow Ca(g)$; $Br_2(l) \rightarrow 2Br(g)$; $Ca(g) \rightarrow Ca^+(g) + 1e^-$;

   $Ca^+(g) \rightarrow Ca^{2+}(g) + 1e^-$; $2Br(g) + 2e^- \rightarrow 2Br^-(g)$, exothermic;

   $Ca^{2+}(g) + 2Br^-(g) \rightarrow CaBr_2(s)$, exothermic

8.15   NaF - 910 kJ/mol; MgO - 3795 kJ/mol
   The two factors that affect lattice energies are charge and ionic radii. The Na-F and Mg-O separations are similar (Na^+ is larger than Mg^{2+}, but F^- is smaller than O^{2-}). The charges on Mg^{2+} and O^{2-} are twice those of Na^+ and F^-, so according to Equation 8.4, the lattice energy of MgO is approximately four times that of NaF.

8.16   KF - 808 kJ/mol;   CaO - 3414 kJ/mol;   ScN - 7547 kJ/mol

The sizes of the ions vary as follows: $Sc^{3+} < Ca^{2+} < K^+$ and $F^- < O^{2-} < N^{3-}$. Therefore, the interionic distances are similar. According to Coulomb's law for compounds with similar ionic separations, the lattice energies should be related as the product of the charges of the ions. The lattice energies above are approximately related as (1)(1): (2)(2): (3)(3) or 1 : 4 : 9. Slight variations are due to the small differences in ionic separations.

8.17   (a) O^{2-} is smaller than S^{2-} ; the closer approach of the oppositely charged ions in MgO leads to greater electrostatic attraction.

        (b)   The ions have 1+ and 1- charges in both compounds. However, the fact that the ionic radii are much smaller in LiF means that the ions can approach more closely, with a resultant increase in electrostatic attractive forces.

        (c)   The ions in CaO have 2+ and 2- charges as compared with 1+ and 1- charges in KF.

8.18   Since the ionic charges are the same in the two compounds, the K-Br and Cs-Cl separations must be approximately equal. Since the radii are related as $Cs^+ > K^+$ and $Br^- > Cl^-$, the difference between Cs^+ and K^+ must be approximately equal to the difference between Br^- and Cl^-. This is somewhat surprising, since K^+ and Cs^+ are two rows apart and Cl^- and Br^- are only one row apart.

8.19 $RbCl(s) \rightarrow Rb^+(g) + Cl^-(g)$ ΔH (lattice energy) = ?

By analogy to NaCl, Figure 8.4, the lattice energy is

$\Delta H_{latt} = -\Delta H_f^\circ\, RbCl(s) + \Delta H_f^\circ\, Rb(g) + \Delta H_f^\circ\, Cl(g) + I_1\,(Rb) + E\,(Cl)$

          = -(-430.5 kJ) + 85.8 kJ + 121.7 kJ + 403 kJ + (-349 kJ) = +692 kJ

This value is smaller than that for NaCl (+788 kJ) because Rb^+ has a larger ionic radius than Na^+. This means that the value of d in the denominator of Equation 8.4 is larger for RbCl, and the potential energy of the electrostatic attraction is smaller.

8.20   By analogy to Figure 8.4:

$\Delta H_{latt} = -\Delta H_f^\circ\, CaF_2 + \Delta H_f^\circ\, Ca(g) + 2\Delta H_f^\circ\, F(g) + I_1\,(Ca) + I_2\,(Ca) + 2E\,(F)$

          = -(-1219.6 kJ) + 179.3 kJ + 2(80.0 kJ) + 590 kJ + 1145 kJ + 2(-332 kJ) = +2630 kJ

From Table 8.2, the lattice energy of NaF, +910 kJ/mol, is considerably less than that of CaF_2. The 2+ charge of Ca^{2+} leads to much greater electrostatic attractions and a higher lattice energy.

Sizes of Ions

8.21   (a)   Electrostatic repulsions are reduced by removing an electron from a neutral atom, Z_{eff} increases, and the cation is smaller.

(b)   The additional electrostatic repulsion produced by adding an electron to a neutral atom decreases the Z_{eff} of the valence electrons, causing them to be less tightly bound to the nucleus and the size of the anion to be larger.

(c)   Going down a column, valence electrons are further from the nucleus and they experience greater shielding by core electrons. Inspite of an increase in Z, the size of particles with like charge increases.

8.22   (a)   As Z stays constant and the number of electrons increases, the electron-electron repulsions increase, the electrons spread apart and the ions become larger. $I^- > I > I^+$

(b)   Going down a column, the increased distance of the electrons from the nucleus and increased shielding by inner electrons outweighs the buildup of Z and causes the size of particles with like charge to increase. $Ca^{2+} > Mg^{2+} > Be^{2+}$

(c)   Fe: $[Ar]4s^23d^6$; Fe^{2+}: $[Ar]3d^6$; Fe^{3+}: $[Ar]3d^5$. The 4s valence electrons in Fe are on average further from the nucleus than the 3d electrons, so Fe is larger than Fe^{2+}. Since there are five 3d orbitals, in Fe^{2+} at least one orbital must contain a pair of electrons. Removing one electron to form Fe^{3+} significantly reduces repulsion, increasing the nuclear charge experienced by each of the other d electrons and decreasing the size of the ion. $Fe > Fe^{2+} > Fe^{3+}$

8.23   (a)   In an isoelectronic series, the number of electrons in the particles is the same, but the values of Z are different.

(b) Rb^+: **Kr**; Br^-: **Kr**; S^{2-}: **Ar**; Zn^{2+}: **Ni**

8.24   (a) K^+, Ca^{2+} (b) Ca^{2+}, Sc^{3+} (c) S^{2-}, Ar   (d) Co^{3+}, Fe^{2+}

8.25   (a)   Since the number of electrons in an isoelectronic series is the same, repulsion and shielding effects do not vary for the different particles. As Z increases, Z_{eff} increases, the valence electrons are more strongly attracted to the nucleus and the size of the particle decreases.

(b)   Because F^-, Ne and Na^+ have the same electron configuration, the 2p electron in the particle with the largest Z experiences the largest effective nuclear charge. A 2p electron in Na^+ experiences the greatest effective nuclear charge.

8.26   (a)   Cl < S < K   (b) $K^+ < Cl^- < S^{2-}$

(c)   Even though K has the largest Z value, the effective nuclear charge experienced by the 4s valence electron is smallest because the n value is larger than valence electrons in S and Cl and the 4s electron is shielded by the 3s and 3p core electrons. K atoms are largest. When the 4s electron is removed, K^+ is isoelectronic with Cl^- and S^{2-}. The larger Z value causes the 3p electrons in K^+ to experience the largest effective nuclear charge and K^+ is the smallest ion.

8.27   (a) $Li^+ < K^+ < Rb^+$ (b) $Mg^{2+} < Na^+ < Br^-$ (c) $K^+ < Ar < Cl^- < S^{2-}$ (d) $Ar < Cl < Cl^-$

8.28   (a) $Se < Se^{2-} < Te^{2-}$ (b) $Co^{3+} < Fe^{3+} < Fe^{2+}$ (c) $Ti^{4+} < Sc^{3+} < Ca$ d) $Be^{2+} < Na^+ < Ne$

Covalent Bonding, Electronegativity and Bond Polarity

8.29   (a)   A *covalent bond* is the bond formed when two atoms share one or more pairs of electrons.

      (b)   The ionic bonding in NaCl is due to strong electrostatic attraction between oppositely charged Na^+ and Cl^- ions. The covalent bonding in Cl_2 is due to sharing of a pair of electrons by two neutral chlorine atoms.

8.30   (a)   A free Br atom has 7 valence electrons and thus has less than a complete octet. In the Lewis structure of Br_2, each Br atom has 6 nonbonded electrons and shairs 1 pair of electrons. Both electrons in the shared pair can be counted in the octet of either Br atom. The Br atoms have completed their octets by shairing a pair of electrons.

      (b)   In Figure 8.6, the electron density (density of dots) is greater between the two nuclei than away from the nuclei. While there is a finite probability that an electron will be anywhere within a certain distance (radius) of the nucleus, the electron has the highest probability of being between the two nuclei. This is consistent with the Lewis structure that shows the two H atoms sharing a pair of electrons. The implication is that the shared pair will most likely be found between the two nuclei doing the sharing.

8.31

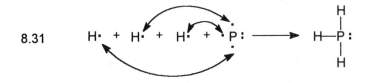

8.32

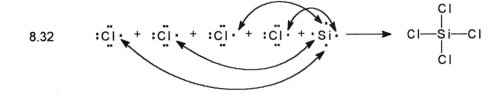

8.33   (a)   :Ö=Ö:

      (b)   A double bond is required because there are not enough electrons to satisfy the octet rule with single bonds and unshared pairs.

      (c)   The greater the number of shared electron pairs between two atoms, the shorter the distance between the atoms. If O_2 has a double bond, the O-O distance will be shorter than the O-O single bond distance.

8.34

$$\overset{..}{\underset{..}{S}}=C=\overset{..}{\underset{..}{S}}$$

The C-S bonds in CS_2 are double bonds, so the C-S distances will be shorter than a C-S single bond distance.

8.35 (a) Electronegativity is the ability of an atom in a molecule (a bonded atom) to attract electrons to itself.

(b) The range of electronegativities on the Pauling scale is 0.7-4.0.

(c) Fluorine, F, is the most electronegative element.

8.36 (a) The electronegativity of the elements increases going from left to right across a row of the periodic chart.

(b) Electronegativity decreases going down a family of the periodic chart.

(c) Generally, the trends in electronegativity are the same as those in ionization energy and opposite those in electron affinity. That is, the more positive the ionization energy and the more negative the electron affinity (omitting a few exceptions), the greater the electronegativity of an element.

8.37 Electronegativity increases going up and to the right of the periodic table.

(a) Cl (b) B (c) As (d) Mg

8.38 Electronegativity increases going up and to the right of the periodic table.

(a) S < O < F (b) Al < Si < C (c) Sn < Te < Br (d) Ca < Be < C

8.39 The bonds in (a), (c) and (e) are polar because the atoms involved differ in electronegativity. The more electronegative element in each polar bond is: (a) Cl (c) F (e) O

8.40 The more different the electronegativity values of the two elements, the more polar the bond.

(a) O-F < H-F < Be-F. Recall that H is a nonmetal and has a much higher electronegativity than its position on the periodic table indicates. Without looking at values, you might predict that Be-F is more polar than H-F because Be is a metal and H is a nonmetal. Electronegativity values confirm this prediction.

(b) C-S < N-O < B-F. You might predict that N-O is least polar since the elements are adjacent on the table. However, the big decrease going from the second row to the third means that the electronegativity of S is not only less than that of O, but essentially the same as that of C. C-S is the least polar.

(c) S-Br < C-P < O-Cl. These pairs of elements have similar electronegativities because the second is one block to the right and one block down from the first. This order is predicted because S and Br are in the third and fourth rows where differences between rows are smaller and O-Cl is in the area where electronegativities are high and differences are large.

Lewis Structure; Resonance Structures

8.41   Counting the **correct number of valence electrons** is the foundation of every Lewis structure.

(a)   Count valence electrons: $4 + (4 \times 1) = 8$ e$^-$, 4 e$^-$ pairs. Follow the procedure in Sample Exercise 8.7.

$$\begin{array}{c} H \\ | \\ H-Si-H \\ | \\ H \end{array}$$

(b)   Valence electrons: $[6 + (2 \times 7)] = 20$ e$^-$, 10 e$^-$ pairs.

$$:\!\overset{\cdot\cdot}{F}\!-\!\overset{\cdot\cdot}{\underset{\cdot\cdot}{S}}\!-\!\overset{\cdot\cdot}{F}\!:$$

i.   Place the S atom in the middle and connect each F atom with a single bond; this requires 2 e$^-$ pairs.

ii.   Complete the octets of the F atoms with lone pairs of electrons; this requires an additional 6 e$^-$ pairs.

iii.   The remaining 2 e$^-$ pairs complete the octet of the central S atom.

(c)   16 valence e$^-$, 8 e$^-$ pairs   (d)   32 valence e$^-$, 16 e$^-$ pairs

$$\overset{\cdot\cdot}{\underset{\cdot\cdot}{N}}\!=\!N\!=\!\overset{\cdot\cdot}{\underset{\cdot\cdot}{O}}$$

There are other resonance structures.

$$\left[\begin{array}{c} :\overset{\cdot\cdot}{O}: \\ | \\ :\overset{\cdot\cdot}{O}\!-\!S\!-\!\overset{\cdot\cdot}{O}\!-\!H \\ | \\ :\overset{\cdot\cdot}{O}: \end{array} \right]^{-}$$

(e)   14 valence e$^-$, 6 e$^-$ pairs   (Choose the Lewis structure that obeys the octet rule, Section 8.8.)

$$\begin{array}{c} H-N-\overset{\cdot\cdot}{\underset{\cdot\cdot}{O}}-H \\ | \\ H \end{array}$$

8.42   (a)   14 e$^-$, 7 e$^-$ pairs   (b) 26 e$^-$, 13 e$^-$ pairs

$$H-\overset{\cdot\cdot}{\underset{\cdot\cdot}{O}}-\overset{\cdot\cdot}{\underset{\cdot\cdot}{Br}}$$

$$:\!\overset{\cdot\cdot}{F}\!-\!\overset{\cdot\cdot}{A}s\!-\!\overset{\cdot\cdot}{F}\!: \\ | \\ :\overset{\cdot\cdot}{F}:$$

(c)   14 e$^-$, 7 e$^-$ pairs   (d) 12 e$^-$, 6 e$^-$ pairs

$$H-\overset{\cdot\cdot}{\underset{\cdot\cdot}{O}}-\overset{\cdot\cdot}{\underset{\cdot\cdot}{O}}-H$$

$$\begin{array}{c} H-C\!=\!\overset{\cdot\cdot}{\underset{\cdot\cdot}{O}} \\ | \\ H \end{array}$$

(e)   10 e$^-$, 5 e$^-$ pairs

$$H-C\!\equiv\!C-H$$

8.43   (a) $16\ e^-$, $8\ e^-$ pairs. The structure on the left minimizes formal charge.

           O=C=O   :O—C≡O:

           0   0   0   -1   0   +1

     (b) $8\ e^-$, $4\ e^-$ pairs 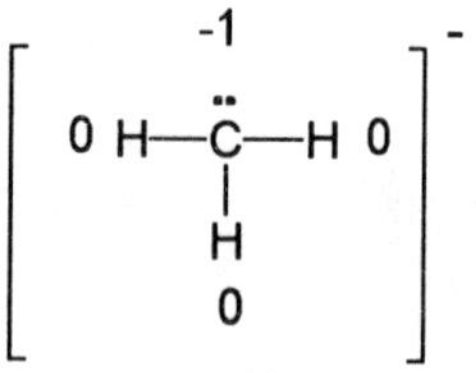   (c) $32\ e^-$, $16\ e^-$ pairs 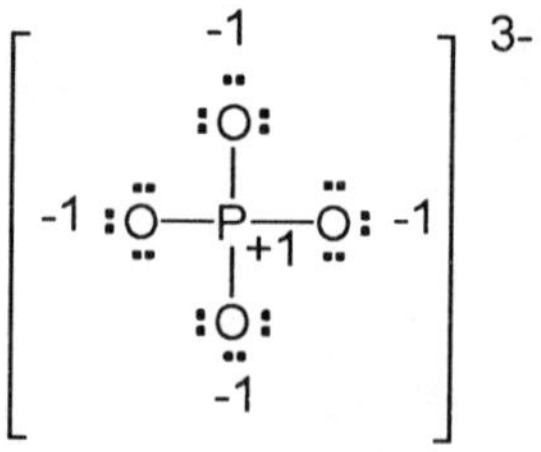   (d) $10\ e^-$, $5\ e^-$ pairs

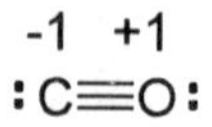

            -1 +1

            :C≡O:

8.44   (a) $18\ e^-$, $9\ e^-$ pairs 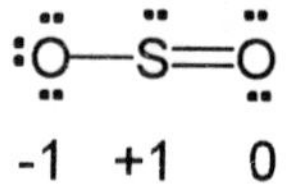   (b) $24\ e^-$, $12\ e^-$ pairs

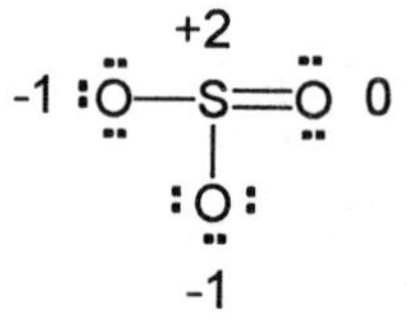

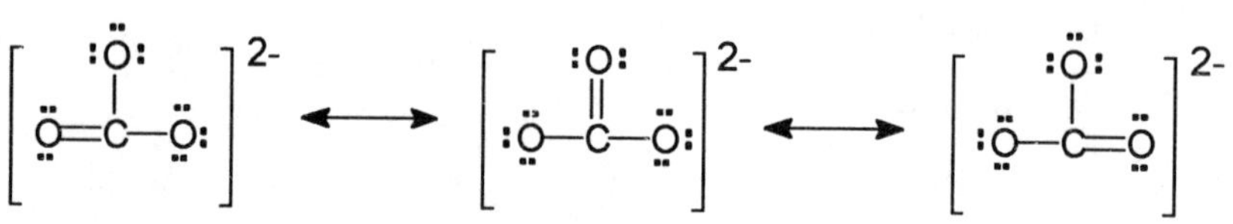

           :O—S=O

           -1   +1   0

     (c) $26\ e^-$, $13\ e^-$ pairs 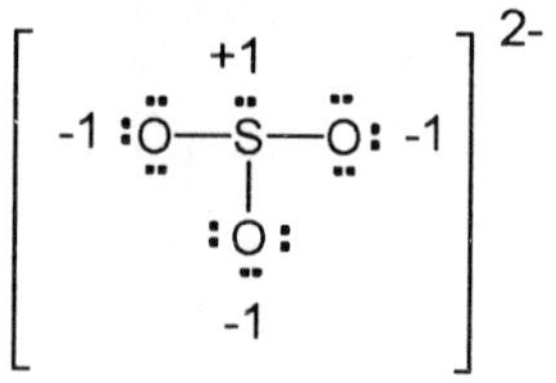   (d) $32\ e^-$, $16\ e^-$ pairs

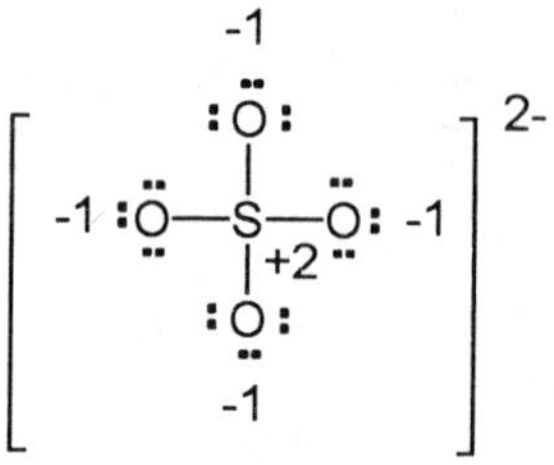

8.45   (a) $18\ e^-$, $9\ e^-$ pairs

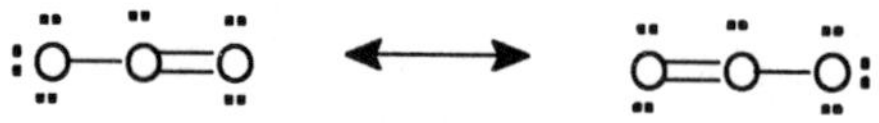

           :O—O=O   ⟷   O=O—O:

     (b) $24\ e^-$, $12\ e^-$ pairs

     (c) $18\ e^-$, $9\ e^-$ pairs

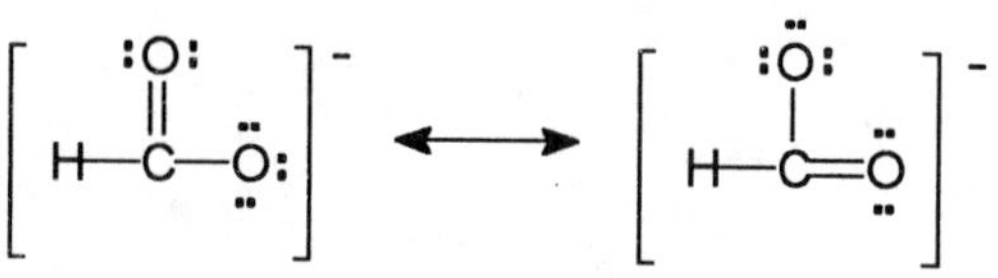

(d)   24 e⁻, 12⁻ pairs

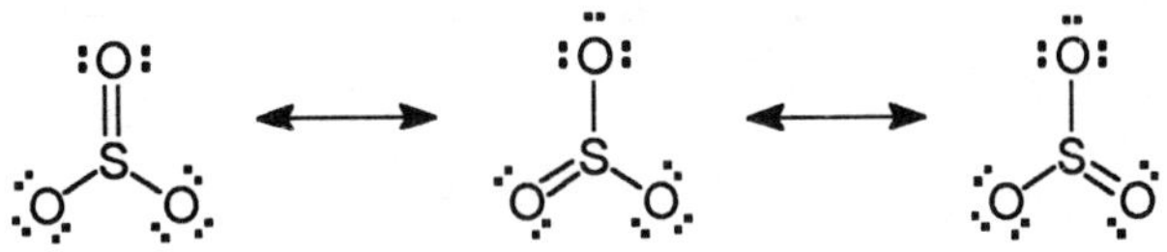

8.46   (a)   24 e⁻, 12 e⁻ pairs

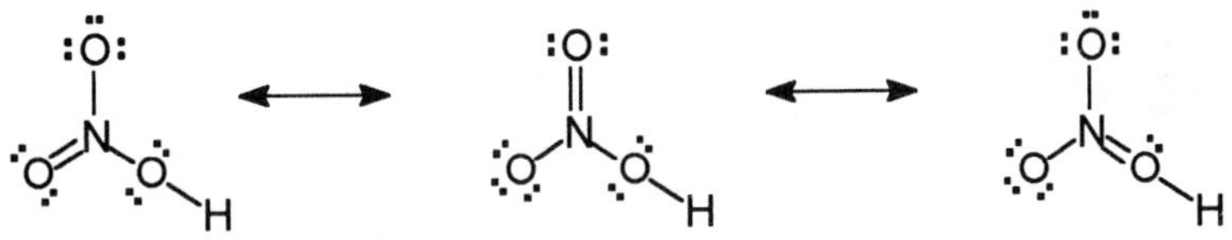

The third form is not a significant contributor because of the nonzero formal charges on all three O atoms.

(b)   18 e⁻, 9 e⁻ pairs

(c)   34 e⁻, 17 e⁻ pairs

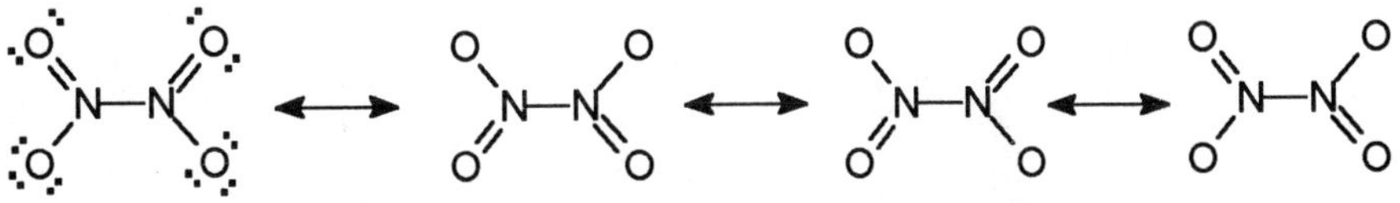

(Nonbonded pairs on terminal O atoms have been omitted on all but the first structure.)

8.47   The Lewis structures are as follows:

5 e⁻ pairs   9 e⁻ pairs

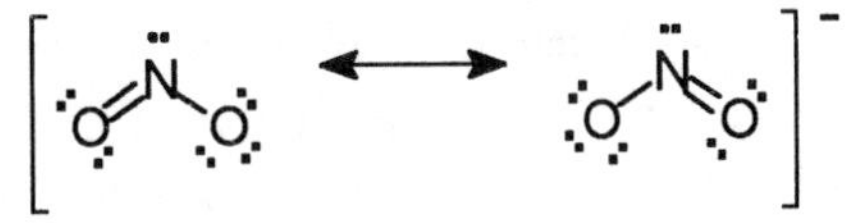

12 e⁻ pairs

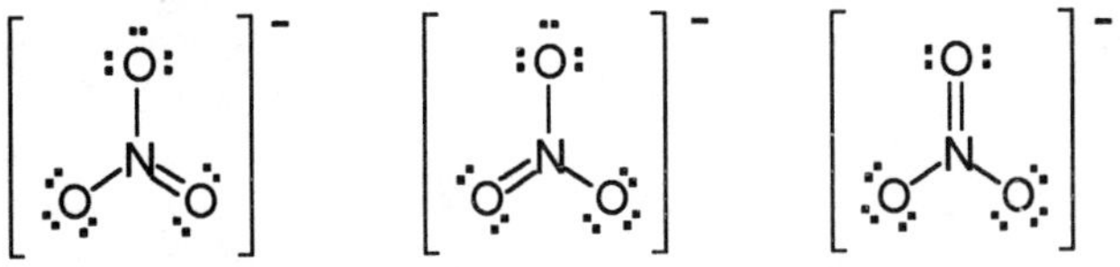

The average number of electron pairs in the N-O bond is 3.0 for NO^+, 1.5 for NO_2^- and 1.33 for NO_3^-. The more electron pairs shared between two atoms, the shorter the bond. Thus the N-O bond lengths vary in the order $NO^+ < NO_2^- < NO_3^-$.

8.48   The Lewis structures are as follows:

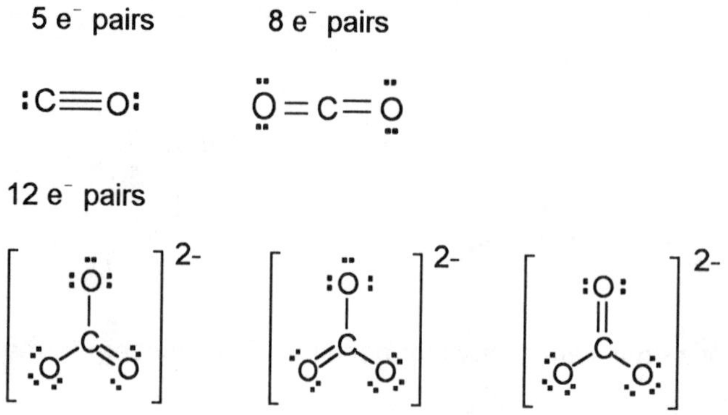

5 e⁻ pairs   8 e⁻ pairs

:C≡O: Ö=C=Ö

12 e⁻ pairs

The more pairs of electrons shared by two atoms, the shorter the bond between the atoms. The average number of electron pairs shared by C and O in the three species is 3 for CO, 2 for CO_2 and 1.33 for CO_3^{2-}. This is also the order of increasing bond length: $CO < CO_2 < CO_3^{2-}$.

8.49   (a)   Two equally valid Lewis structures can be drawn for benzene.

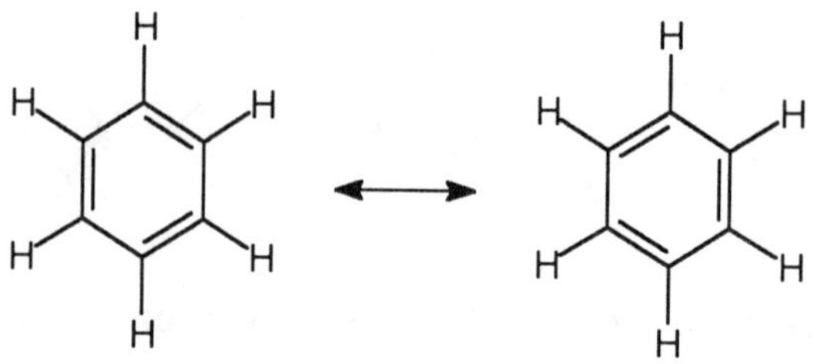

Since the 6 C-C bond lengths are equal, neither of the individual Lewis structures gives an accurate picture of the bonding. The concept of resonance dictates that the true description of bonding is some hybrid or blend of the two Lewis structures.

(b)   The resonance model described in (a) has 6 equivalent C-C bonds, each with some double bond character. That is, more than 1 pair but less than 2 pairs of electrons is involved in each C-C bond. We would expect the C-C bond length in benzene to be shorter than a single bond but longer than a double bond.

8.50   (a)

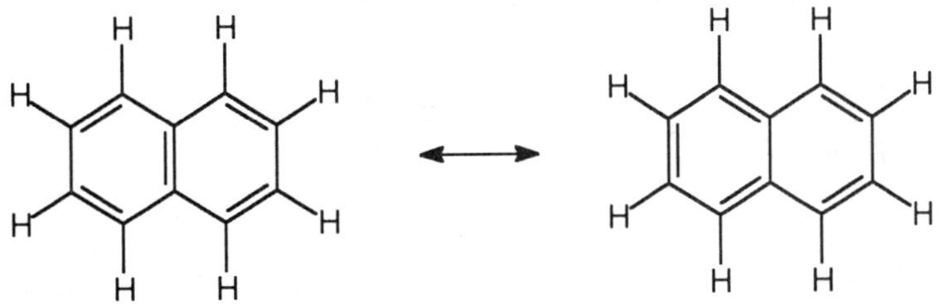

(b)   The resonance model of this molecule has bonds that are neither single nor double, but somewhere in between. This results in bond lengths that are intermediate between C-C single and C=C double bond lengths.

(c)

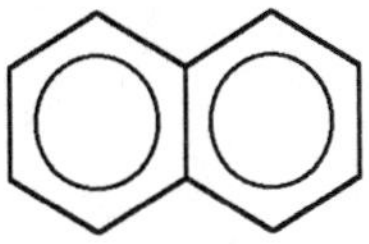

Exceptions to the Octet Rule

8.51   The most common exceptions to the octet rule are molecules with more than eight electrons around one or more atoms, usually the central atom.

8.52   In the third period, atoms have the space and available orbitals to accommodate extra electrons. Since atomic radius increases going down a family, elements in the third period and beyond are less subject to destabilization from additional electron-electron repulsions. Also, the third shell contains d orbitals that are relatively close in energy to 3s and 3p orbitals (the ones that accommodate the octet) and provide an allowed energy state for the extra electrons.

8.53   (a)   26 e⁻, 13 e⁻ pairs   (b)   6 e⁻, 3 e⁻ pairs, impossible to satisfy octet rule with only 6 valence electrons

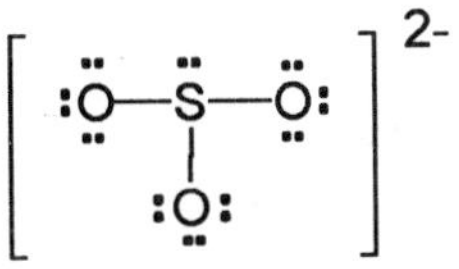

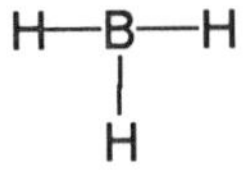

6 electrons around B

(c)   22 e⁻, 11 e⁻ pairs   (d)   48 e⁻ pairs, 24 e⁻ pairs

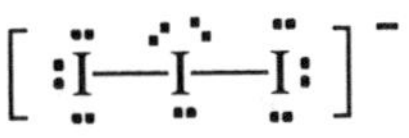

10 e⁻ around central I

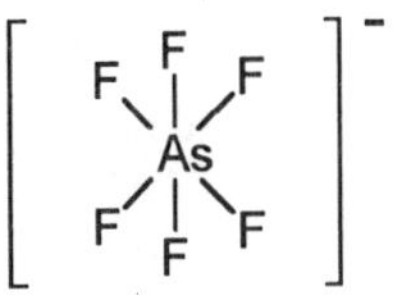

12 e⁻ around As; three nonbonded pairs on each F have been omitted

(e)   17 e⁻, 8.5 e⁻ pairs, odd electron molecule

$$\left[\ddot{:O}-\dot{\ddot{O}}: \right]^{-} \longleftrightarrow \left[:\dot{\ddot{O}}-\ddot{O}: \right]^{-}$$

8.54 (a)   17 e⁻, 8.5 e⁻ pairs;
            odd electron molecule

(b)   32 e⁻, 16 e⁻ pairs

$$O{=}\dot{\ddot{N}}{-}\ddot{\ddot{O}}: \longleftrightarrow :\ddot{\ddot{O}}{-}\dot{N}{=}O$$

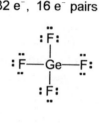

The odd electron is probably on N
because it is less electronegative than O.

(c)   34 e⁻, 17 e⁻ pairs

$$
\begin{array}{c}
:\ddot{F}: \\
| \\
:\ddot{F}{-}\overset{\displaystyle .\,.}{Te}{-}\ddot{F}: \\
| \\
:\ddot{F}:
\end{array}
$$

10 e⁻ around the central Te

(d) 24 e⁻, 12 e⁻ pairs

$$
\begin{array}{c}
:\overset{\displaystyle .\,.}{\ddot{C}l}{-}B{-}\overset{\displaystyle .\,.}{\ddot{C}l}: \\
| \\
:\ddot{C}l:
\end{array}
$$

6 e⁻ around B

(e)   36 e⁻, 18 e⁻ pairs

$$
\begin{array}{c}
:\ddot{F}: \\
| \\
:\ddot{F}{-}\overset{\displaystyle .\,.}{Xe}{-}\ddot{F}: \\
| \\
:\ddot{F}:
\end{array}
$$

12 e⁻ around the central Xe

8.55 (a)   16 e⁻ 8 e⁻ pairs $:\ddot{C}l{-}Be{-}\ddot{C}l:$

This structure violates the octet rule; Be has only 4 e⁻ around it.

(b) $\ddot{C}l{=}Be{=}\ddot{C}l \longleftrightarrow :\ddot{C}l{-}Be{\equiv}Cl: \longleftrightarrow :Cl{\equiv}Be{-}\ddot{C}l:$

(c)   The formal charges on each of the atoms in the four resonance structures are:

$:\overset{\displaystyle .\,.}{\ddot{C}l}{-}Be{-}\overset{\displaystyle .\,.}{\ddot{C}l}:$ $\ddot{C}l{=}Be{=}\ddot{C}l$ $:\ddot{C}l{-}Be{\equiv}Cl:$ $:Cl{\equiv}Be{-}\ddot{C}l:$

0	0	0	+1	-2	+1	0	-2	+2	+2	-2	0

Since formal charges are minimized on the structure that violates the octet rule, this
form is probably most important.

8.56   (a)   19 e⁻, 9.5 e⁻ pairs, odd electron molecule

$$:\ddot{O}-\overset{..}{\underset{..}{Cl}}-\ddot{O}: \longleftrightarrow :\ddot{O}-\overset{..}{\underset{..}{Cl}}-\dot{\ddot{O}} \longleftrightarrow \dot{\ddot{O}}-\overset{..}{\underset{..}{Cl}}-\ddot{O}:$$

(b)   None of the structures satisfies the octet rule. In each structure, one atom has only 7 e⁻ around it. If a molecule has an odd number of electrons in the valence shell, no Lewis structure can satisfy the octet rule.

(c)

$$:\ddot{O}-\overset{..}{Cl}-\ddot{O}: \longleftrightarrow :\ddot{O}-\overset{..}{\underset{..}{Cl}}-\dot{\ddot{O}} \longleftrightarrow \dot{\ddot{O}}-\overset{..}{\underset{..}{Cl}}-\ddot{O}:$$

  -1 +2 -1   -1 +1 0   0 +1 -1

Formal charge arguments predict that the two resonance structures with the odd electron on O are most important. This contradicts electronegativity arguments, which would predict that the less electronegative atom, Cl, would be more likely to have fewer than 8 e⁻ around it.

Bond Enthalpies

8.57   (a) $\Delta H = 4D(C-H) + D(Cl-Cl) - 2D(C-H) - 2D(C-Cl) - D(H-H)$
                = $2D(C-H) + D(Cl-Cl) - 2D(C-Cl) - D(H-H)$
                = $2(413) + 242 - 2(328) - 436 = -24$ kJ

(b) $\Delta H = 6D(C-H) + 2D(C-O) + 2D(O-H) - 6D(C-H) - 2D(C-O) - 2D(O-H)$
                = 0 kJ

(c) $\Delta H = 2D(C-H) + D(C=N) + D(N-H) + 2D(O-H) - 2D(C-H) - D(C=O) - 3D(N-H)$
                = $D(C=N) + 2D(O-H) - D(C=O) - 2D(N-H)$
                = $615 + 2(463) - 799 - 2(391) = -40$ kJ

8.58   (a) $\Delta H = 2D(O-H) + D(O-O) + 4D(C-H) + D(C=C)$
                    $- 2D(O-H) - 2D(O-C) - 4D(C-H) - D(C-C)$
        $\Delta H = D(O-O) + D(C=C) - 2D(O-C) - D(C-C)$
                = $146 + 614 - 2(358) - 348 = -304$ kJ

(b) $\Delta H = 5D(C-H) + D(C\equiv N) + D(C=C) - 5D(C-H) - D(C\equiv N) - 2D(C-C)$
                = $D(C=C) - 2D(C-C) = 614 - 2(348) = -82$ kJ

(c) $\Delta H = 6D(N-Cl) - 3D(Cl-Cl) - D(N\equiv N)$
                = $6(200) - 3(242) - 941 = -467$ kJ

8.59   Draw structural formulas so bonds can be visualized.

(a)   2H-O-O-H → 2H-O-H + O=O
        $\Delta H = D(O-O) - D(O=O) = 2(146) - (495) = -203$ kJ

(b)

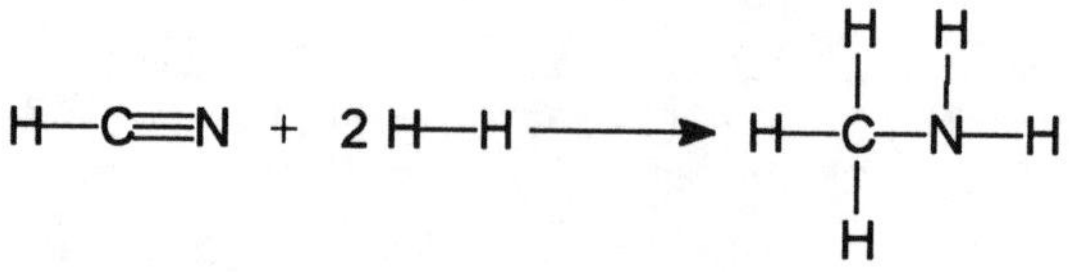

$$H\!-\!C\!\equiv\!N \ + \ 2\,H\!-\!H \longrightarrow H\!-\!\underset{\underset{H}{|}}{\overset{\overset{H}{|}}{C}}\!-\!\underset{}{\overset{\overset{H}{|}}{N}}\!-\!H$$

$$\Delta H = D(C \equiv N) + 2D(H\text{-}H) - 2D(C\text{-}H) - D(C\text{-}N) - 2D(N\text{-}H)$$
$$= 891 + 2(436) - 2(413) - 293 - 2(391) = -138 \text{ kJ}$$

(c)

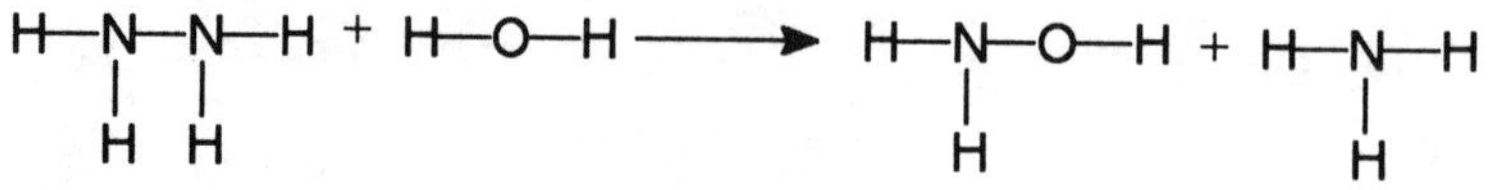

$$H\!-\!\underset{\underset{H}{|}}{N}\!-\!\underset{\underset{H}{|}}{N}\!-\!H \ + \ H\!-\!O\!-\!H \longrightarrow H\!-\!\underset{\underset{H}{|}}{N}\!-\!O\!-\!H \ + \ H\!-\!\underset{\underset{H}{|}}{N}\!-\!H$$

$$\Delta H = D(N\text{-}N) + D(O\text{-}H) - D(N\text{-}O) - D(N\text{-}H)$$
$$= 163 + 463 - 201 - 391 = + 34 \text{ kJ}$$

8.60   Draw structural formulas so bonds can be visualized.

(a)

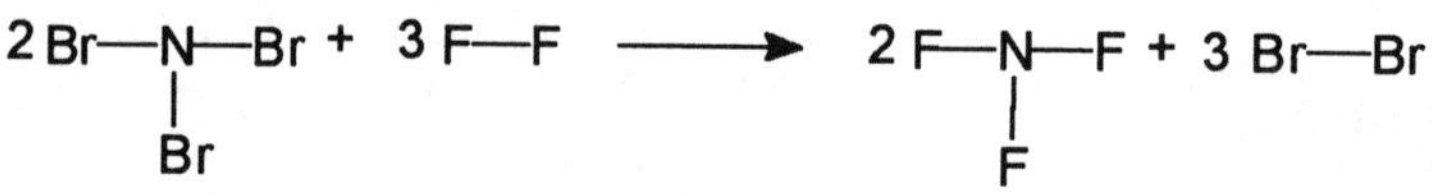

$$2\,Br\!-\!\underset{\underset{Br}{|}}{N}\!-\!Br \ + \ 3\,F\!-\!F \longrightarrow 2\,F\!-\!\underset{\underset{F}{|}}{N}\!-\!F \ + \ 3\,Br\!-\!Br$$

$$\Delta H = 6D(N\text{-}Br) + 3D(F\text{-}F) - 6D(N\text{-}F) - 3D(Br\text{-}Br)$$
$$= 6(243) + 3(155) - 6(272) - 3(193) = -288 \text{ kJ}$$

(b)

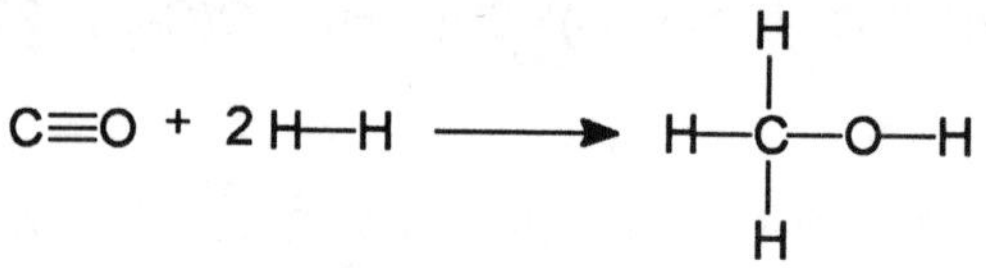

$$C\!\equiv\!O \ + \ 2\,H\!-\!H \longrightarrow H\!-\!\underset{\underset{H}{|}}{\overset{\overset{H}{|}}{C}}\!-\!O\!-\!H$$

$$\Delta H = D(C \equiv O) + 2D(H\text{-}H) - 3D(C\text{-}H) - D(C\text{-}O) - D(O\text{-}H)$$
$$= 1072 + 2(436) - 3(413) - 358 - 463 = -116 \text{ kJ}$$

(c)

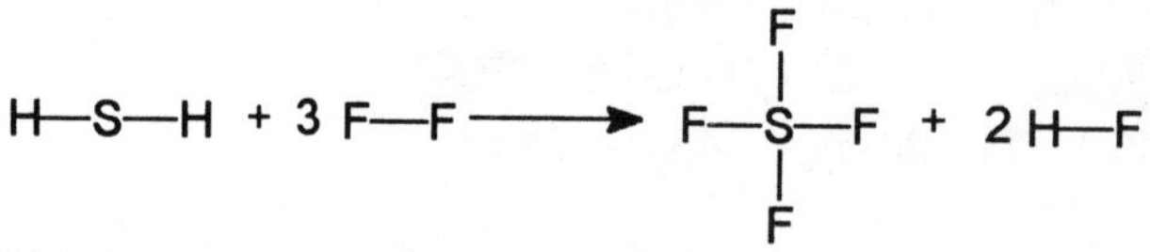

$$H\!-\!S\!-\!H \ + \ 3\,F\!-\!F \longrightarrow F\!-\!\underset{\underset{F}{|}}{\overset{\overset{F}{|}}{S}}\!-\!F \ + \ 2\,H\!-\!F$$

$$\Delta H = 2D(S\text{-}H) + 3D(F\text{-}F) - 4D(S\text{-}F) - 2D(H\text{-}F)$$
$$= 2(339) + 3(155) - 4(327) - 2(567) = -1299 \text{ kJ}$$

8.61   The average Ti-Cl bond dissociation energy is just the average of the four values listed, 430 kJ/mol.

8.62   Cl—P—Cl (g) $\longrightarrow$ P (g) + 3 C l (g)
$\qquad\quad$ |
$\qquad\quad$ Cl

$\Delta H = \Delta H_f^\circ$ P(g) + 3 ΔH_f° Cl(g) - ΔH_f° PCl_3(g)
$\qquad$ = 316.4 kJ + 3(121.7 kJ) - (-288.07 kJ)
$\qquad$ = 969.6 kJ

The reaction above represents the enthalpy of dissociation for 3 P-Cl bonds. The average bond enthalpy of a single P-Cl bond is then $\dfrac{969.6 \text{ kJ}}{3 \text{ P-Cl bonds}}$ = 323.2 kJ/P-Cl bond.

Oxidation Numbers

8.63   (a)   N, 0   (b) C, -4   (c) P, +4   (d) Mn, +7   (e) B, +3
$\qquad$ (f)   Cl, +1   (g) Fe, +3   (h) S, +6

8.64   (a)   O (peroxide), -1   (b) C, +4
$\qquad$ (c)   O, +1 (O is less electronegative than F, so it will have the positive oxidation number.)
$\qquad$ (d)   Co, +3   (e) I, +5   (f) Xe, +6   (g) U, +6   (h) N, +5

8.65   (a)   P +5, -3   (b) C +4, -4   (c) I +7, -1   (d) Cr +6, 0

8.66   (a)   Ba +2, 0   (b) N +5, -3   (c) V +5, 0   (d) Se +6, -2

8.67   (a) MnO_2 (b) Ga_2S_3 (c) SeF_6 (d) copper(I) oxide or cuprous oxide
$\qquad$ (e)   chlorine trifluoride
$\qquad$ (f)   tellurium trioxide or tellurium(VI) oxide

8.68   (a) FeF_3 (b) MoO_3 (c) $AsBr_5$ (d) vanadium(III) oxide
$\qquad$ (e)   cobalt(III) fluoride   (f) manganese(II) sulfide

8.69   (a)   xenon is oxidized from 0 to +4; fluorine is reduced from 0 to -1
$\qquad$ (b)   copper is reduced from +2 to +1; iodine is oxidized from -1 to 0
$\qquad$ (c)   nitrogen is oxidized from -3 to +3; chlorine is reduced from 0 to -1
$\qquad$ (d)   iodine is reduced from 0 to -1; sulfur is oxidized from +4 to +6
$\qquad$ (e)   sulfur is oxidized from -2 to +4; oxygen is reduced from 0 to -2

8.70   (a)   iodine is oxidized from -1 to 0; fluorine is reduced from 0 to -1
$\qquad$ (b)   manganese is reduced from +4 to +2; chlorine is oxidized from -1 to 0
$\qquad$ (c)   oxygen is oxidized from -1 to 0 and reduced from -1 to -2
$\qquad\quad$ $2H_2O_2$(aq) $\rightarrow$ $2H_2O$(l) + O_2(g) (disproportionation)
$\qquad$ (d)   carbon is oxidized from -4 to -2; O is reduced from 0 to -2
$\qquad$ (e)   iron is oxidized from +2 to +3; manganese is reduced from +7 to +2

Additional Exercises

8.71 (a)   Group 4A or 4B   (b) Group 2A   (c) Group 5A or 5B

8.72 (a)   Lattice energy is proportional to Q_1Q_2/d. For each of these compounds, Q_1Q_2 is the same. The anion H^- is present in each compound, but the ionic radius of the cation increases going from Be to Ba. Thus, the value of d (the cation-anion separation) increases and the ratio Q_1Q_2/d decreases. This is reflected in the decrease in lattice energy going from BeH_2 to BaH_2.

(b)   Again, Q_1Q_2 for ZnH_2 is the same as that for the other compounds in the series and the anion is H^-. The lattice energy of ZnH_2, 2870 kJ, is closest to that of MgH_2, 2791 kJ. The ionic radius of Zn^{2+} is similar to that of Mg^{2+}.

8.73 $E = Q_1Q_2/d$; $k = 8.99 \times 10^9$ J $\bullet$ m/coul^2

(a) $E = \dfrac{-8.99 \times 10^9 \text{ J} \bullet \text{m}}{\text{C}^2} \times \dfrac{(1.60 \times 10^{-19} \text{ C})^2}{(0.68 + 1.81) \times 10^{-10} \text{ m}} = -9.24 \times 10^{-19}$ J

The sign of E is negative because one of the interacting ions is an anion; this is an attractive interaction.

On a molar basis: $-9.243 \times 10^{-19} \times 6.022 \times 10^{23} = -5.57 \times 10^5$ J $= -557$ kJ

(b) $E = \dfrac{-8.99 \times 10^9 \text{ J} \bullet \text{m}}{\text{C}^2} \times \dfrac{(1.60 \times 10^{-19} \text{ C})^2}{(1.33 + 2.20) \times 10^{-10} \text{ m}} = -6.52 \times 10^{-19}$ J

On a molar basis: -3.93×10^5 J $= -393$ kJ

(c) $E = \dfrac{-8.99 \times 10^9 \text{ J} \bullet \text{m}}{\text{C}^2} \times \dfrac{(2 \times 1.60 \times 10^{-19} \text{ C})^2}{(0.99 + 1.40) \times 10^{-10} \text{ m}} = -3.85 \times 10^{-18}$ J

On a molar basis: -2.32×10^6 J $= -2.32 \times 10^3$ kJ

8.74 $E = \dfrac{-8.99 \times 10^9 \text{ J} \bullet \text{m}}{\text{C}^2} \times \dfrac{(1.60 \times 10^{-19} \text{ C})^2}{(0.97 + 1.81) \times 10^{-10} \text{ m}} = -8.28 \times 10^{-19}$ J

On a molar basis: $(-8.279 \times 10^{-19} \text{ J})(6.022 \times 10^{23}) = -499$ kJ

Note that its absolute value is less than the lattice energy, 788 kJ/mol. The difference represents the added energy of putting all the Na^+Cl^- ion pairs together in a three-dimensional array, as shown in Figure 8.3.

8.75   (a)

Compound	Lattice Energy (kJ)		Compound	Lattice Energy (kJ)	
NaCl	788		LiCl	834	
NaBr	732	56 kJ	**LiBr**	**779**	55 kJ
Na I	682		Li I	730	

106 kJ (spanning NaCl–Na I); 104 kJ (spanning LiCl–Li I)

The difference in lattice energy between LiCl and LiI is 104 kJ. The difference between NaCl and NaI is 106 kJ; the difference between NaCl and NaBr is 56 kJ, or 53% of the difference between NaCl and NaI. Applying this relationship to the Li salts, 0.53(104 kJ) = 55 kJ difference between LiCl and LiBr. The approximate lattice energy of LiBr is (834 - 55) kJ = 779 kJ.

(b)

Compound	Lattice Energy (kJ)		Compound	Lattice Energy (kJ)	
NaCl	788		CsCl	657	
NaBr	732	56 kJ	**CsBr**	**627**	30 k
Na I	682		Cs I	600	

106 kJ (spanning NaCl–Na I); 57 kJ (spanning CsCl–Cs I)

By analogy to the Na salts, the difference between lattice energies of CsCl and CsBr should be approximately 53% of the difference between CsCl and CsI. The lattice energy of CsBr is approximately 627 kJ.

(c)

Compound	Lattice Energy (kJ)		Compound	Lattice Energy (kJ)	
MgO	3795		$MgCl_2$	2326	
CaO	3414	381 kJ	**$CaCl_2$**	**2195**	131 kJ
SrO	3217		$SrCl_2$	2127	

578 kJ (spanning MgO–SrO); 199 kJ (spanning $MgCl_2$–$SrCl_2$)

By analogy to the oxides, the difference between the lattice energies of $MgCl_2$ and $CaCl_2$ should be approximately 66% of the difference between $MgCl_2$ and $SrCl_2$. That is, 0.66(199 kJ) = 131 kJ. The lattice energy of $CaCl_2$ is approximately (2326 - 131) kJ = 2195 kJ.

8.76   (a) As^{3-} : $[Ar]4s^2 3d^{10}4p^6 = [Kr]$

        (b) Ti^{2+} : $[Ar]3d^2$, not noble gas configuration

        (c) Pt^{4+} : $[Xe]4f^{14}5d^6$, not noble gas configuration

        (d) I^- : $[Kr]5s^2 4d^{10}5p^6 = [Xe]$

        (e) La^{3+} : $[Xe]$

        (f) Rh^{3+} : $[Kr]4d^6$, not noble gas configuration

8.77   P-O > P-N > P-S > P-P. According to periodic trends in electronegativity, O is the most electrongetative element and P the least. Thus, the P-O bond is most polar because its atoms have the greatest difference in electronegativity, ΔEN. The P-P bond is least polar, because ΔEN is necessarily zero. Since electronegativity (and other periodic properties) vary more moving from row 2 to row 3 than moving left to right, P-N is probably more polar than P-S.

8.78 Formal charge (FC) = # valence e^- - (# nonbonding e^- + 1/2 # bonding e^-)

(a)   18 e^-, 9 e^- pairs

FC for the central O = 6 - [2 + 1/2 (6)] = +1

(b)   48 e^-, 24 e^- pairs

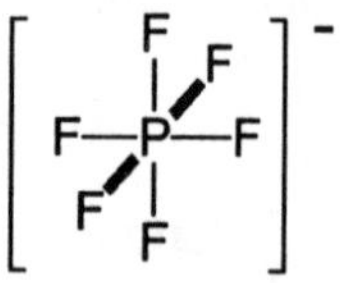

FC for P = 5 - [0 + 1/2 (12)] = -1

The three nonbonded pairs on each F have been omitted.

(c)   17 e^- ; 8 e^- pairs, 1 odd e^-

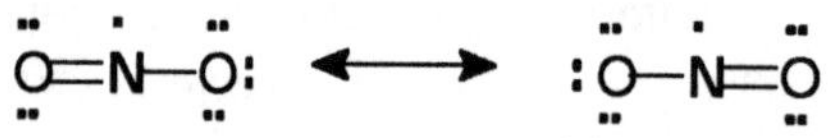

The odd electron is probably on N because it is less electronegative than O.
Assuming the odd electron is on N, FC for N = 5 - [1+ 1/2 (6)] = +1.
If the odd electron is on O, FC for N = 5 - [2 + 1/2 (6)] = 0.

(d)   28 e^-, 14 e^- pairs   (e)   32 e^-, 16 e^- pairs

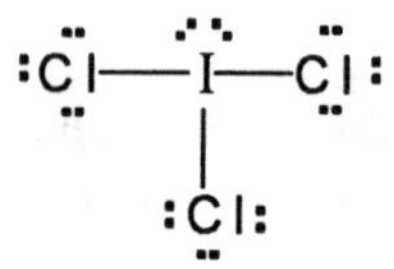

FC for I = 7 - [4 + 1/2 (6)] = 0   FC for Cl = 7 - [0 + 1/2 (8)] = +3

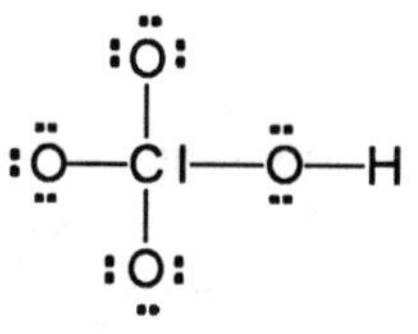

8.79

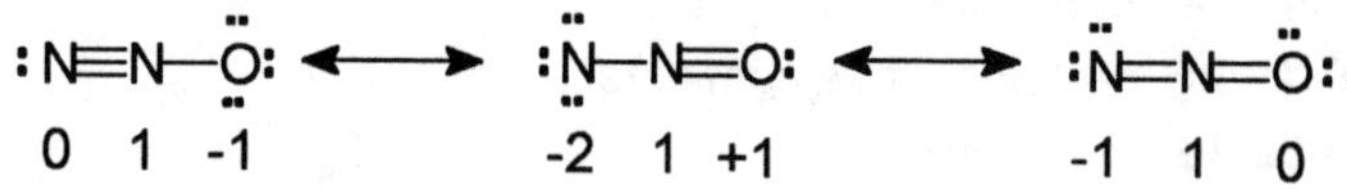

The formal charges are listed below the atoms in the three resonance structures. The structures on the left and right both minimize formal charge to the same extent, so they are likely to be more important to an accurate description of the molecule than the middle structure.

8.80   (a)   42 e⁻, 21 e⁻ pairs

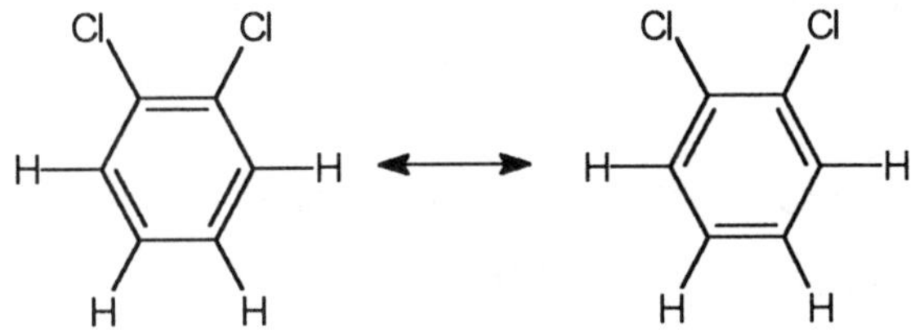

    (b)   Resonance predicts that the true picture of bonding is a hybrid of the two resonance forms. In this description, there are no isolated single and double bonds, but rather 6 equivalent C-C bonds, each with some double bond character (see Exercise 8.49).

    (c)

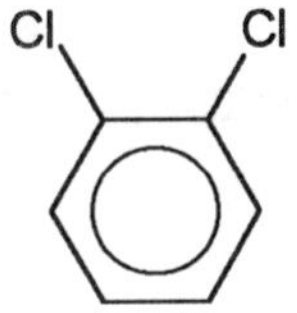

8.81

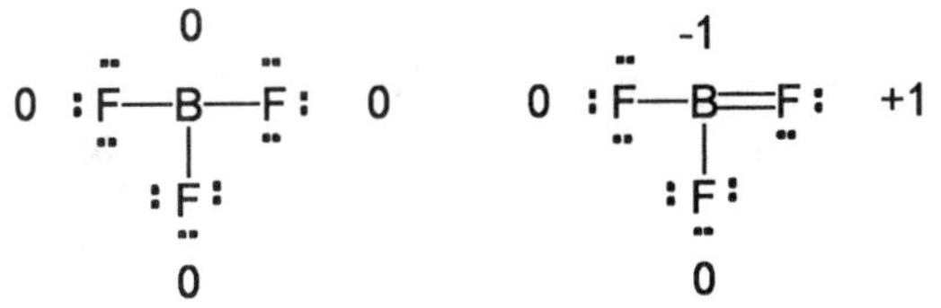

Clearly the structure with less than 8 electrons around B minimizes formal charges. Also, considering the electronegativities, a formal charge of +1 on F seems highly unlikely.

8.82 I_3^- has a Lewis structure with an expanded octet of electrons around the central I.

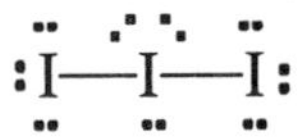

F cannot accommodate an expanded octet because it is too small and has no available d orbitals in its valence shell.

8.83   (a)

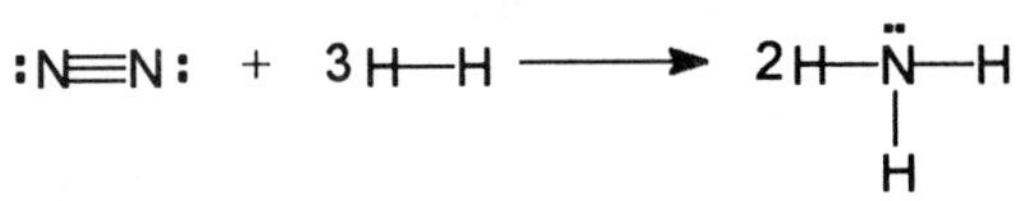

$$\Delta H = D(N \equiv N) + 3D(H\text{-}H) - 6(N\text{-}H) = 941 \text{ kJ} + 3(436 \text{ kJ}) - 6(391 \text{ kJ})$$

$$= -97 \text{ kJ} / 2 \text{ mol } NH_3 ; \textbf{ exothermic}$$

    (b) $\Delta H_f^\circ \ NH_3(g) = -46.19 \text{ kJ}; \ \Delta H = 2(-46.19) = -92.4 \text{ kJ}$

The ΔH calculated from bond energies is slightly more exothermic (more negative) than that obtained using ΔH_f° values.

8.84 $\Delta H = 8D(C-H) - D(C-C) - 6D(C-H) - D(H-H)$

        $= 2D(C-H) - D(C-C) - D(H-H)$

        $= 2(413) - 348 - 436 = +42 \text{ kJ}$

  $\Delta H = 8D(C-H) + 1/2\, D(O=O) - D(C-C) - 6D(C-H) - 2D(O-H)$

        $= 2D(C-H) + 1/2\, D(O=O) - D(C-C) - 2D(O-H)$

        $= 2(413) + 1/2\,(495) - 348 - 2(463) = -200 \text{ kJ}$

The fundamental difference in the two reactions is the formation of 1 mol of H-H bonds versus the formation of 2 mol of O-H bonds. The latter is much more exothermic, so the reaction involving oxygen is more exothermic.

8.85   (a) $\Delta H = 5D(C-H) + D(C-C) + D(C-O) + D(O-H) - 6D(C-H) - 2D(C-O)$

        $= D(C-C) + D(O-H) - D(C-H) - D(C-O)$

        $= 348 \text{ kJ} + 463 \text{ kJ} - 413 \text{ kJ} - 358 \text{ kJ}$

    $\Delta H = +40 \text{ kJ};$ ethanol has the lower enthalpy

    (b) $\Delta H = 4D(C-H) + D(C-C) + 2D(C-O) - 4D(C-H) - D(C-C) - D(C=O)$

        $= 2D(C-O) - D(C=O)$

        $= 2(358 \text{ kJ}) - 799 \text{ kJ}$

    $\Delta H = -83 \text{ kJ};$ acetaldehyde has the lower enthalpy

    (c) $\Delta H = 8D(C-H) + 4D(C-C) + D(C=C) - 8D(C-H) - 2D(C-C) - 2D(C=C)$

        $= 2D(C-C) - D(C=C)$

        $= 2(348 \text{ kJ}) - 614 \text{ kJ}$

    $\Delta H = +82 \text{ kJ};$ cyclopentene has the lower enthalpy

    (d) $\Delta H = 3D(C-H) + D(C-N) + D(C \equiv N) - 3D(C-H) - D(C-C) - D(C \equiv N)$

        $= D(C-N) - D(C-C)$

        $= 293 \text{ kJ} - 348 \text{ kJ}$

    $\Delta H = -55 \text{ kJ};$ acetonitrile has the lower enthalpy

8.86   (a)

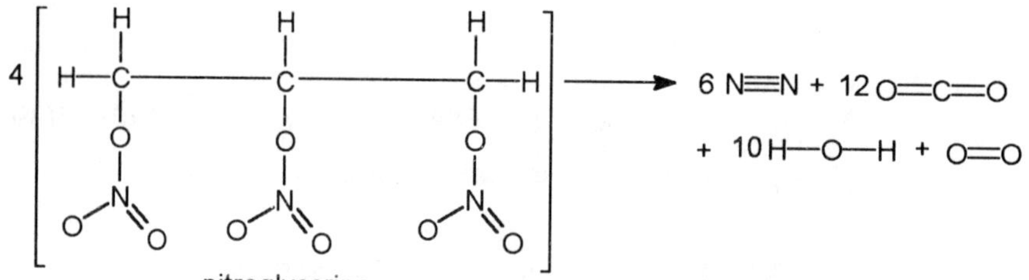

nitroglycerine

$\Delta H = 20D(\text{C-H}) + 8D(\text{C-C}) + 12(\text{C-O}) + 24D(\text{O-N}) + 12D(\text{N=O})$

$= [6D(\text{N}\equiv\text{N}) + 24D(\text{C=O}) + 20D(\text{H-O}) + D(\text{O=O})]$

$= 20(413) + 8(348) + 12(358) + 24(201) + 12(607)$

$- [6(941) + 24(799) + 20(463) + 495]$

$= -7129 \text{ kJ}$

$$1.00 \text{ g } C_3H_5N_3O_9 \times \frac{1 \text{ mol } C_3H_5N_3O_9}{227.1 \text{ g } C_3H_5N_3O_9} \times \frac{-7129 \text{ kJ}}{4 \text{ mol } C_3H_5N_3O_9} = 7.85 \text{ kJ/g } C_3H_5N_3O_9$$

(b) $4C_7H_5N_3O_6(s) \rightarrow 6N_2(g) + 7CO_2(g) + 10H_2O(g) + 21C(s)$

8.87

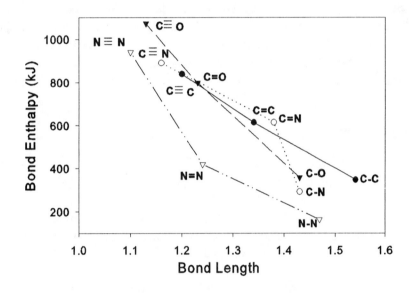

When comparing the same pair of bonded atoms (C-N vs. C=N vs. C $\equiv$ N), the shorter the bond the greater the bond energy, but the two quantities are not necessarily directly proportional. The plot clearly shows that there are no simple length/strength correlations for single bonds alone, double bonds alone, triple bonds alone, or among different pairs of bonded atoms (all C-C bonds vs. all C-N bonds, etc.).

8.88   (a)   14e⁻, 7 e⁻ pairs   32 e⁻, 16 e⁻ pairs

$$\left[\ :\overset{..}{\underset{..}{\text{Cl}}}—\overset{..}{\underset{..}{\text{O}}}: \ \right]^-$$

$$\left[\ \overset{\displaystyle :\overset{..}{\underset{..}{\text{O}}}:}{\underset{\displaystyle :\overset{..}{\underset{..}{\text{O}}}:}{:\overset{..}{\underset{..}{\text{O}}}—\text{Cl}—\overset{..}{\underset{..}{\text{O}}}:}} \ \right]^-$$

FC on Cl = 7 - [6 + 1/2(2)]   FC on Cl = 7 - [0 + 1/2(8)]
         = 0   = +3

(b)   The oxidation number of Cl is +1 in ClO^- and +7 in ClO_4^-.

(c)   The definition of formal charge assumes that all bonding pairs of electrons are equally shared by the two bonded atoms, that all bonds are purely covalent. The definition of oxidation number assumes that the more electronegative element in the bond gets all of the bonding electrons, that the bonds are purely ionic. These two definitions represent the two extremes of how electron density is distributed between bonded atoms.

In ClO^- and ClO_4^-, Cl is the less electronegative element, so the oxidation numbers have a higher positive value than the formal charges. The true description of the electron density distribution is somewhere between the extremes indicated by formal charge and oxidation number.

8.89 SeO_2 and N_2O_3 can be ruled out because both of them are low melting covalent compounds. SrO is a basic oxide and not likely to react with NaOH, so the compound in question is probably GeO_2. Recall that oxides of metals in higher oxidation states are somewhat acidic in character. Thus, it is reasonable that GeO_2 should dissolve slightly in basic solution. Information that may be of interest: SrO, m.p. 2665°C, GeO_2, m.p. 1115°C.

Integrative Exercises

8.90   (a) Ti^{2+} : $[Ar]3d^2$; Ca : $[Ar]4s^2$. Yes. The 2 valence electrons in Ti^{2+} and Ca are in different principle quantum levels and different subshells.

(b)   According to the Aufbau Principle, valence electrons will occupy the lowest energy empty orbital. Thus, in Ca the 4s is lower in energy than the 3d, while in Ti^{2+}, the 3d is lower in energy than the 4s.

(c)   Since there is only one 4s orbital, the 2 valence electrons in Ca are paired. There are 5 degenerate 3d orbitals, so the 2 valence electrons in Ti^{2+} are unpaired. Ca has no unpaired electrons, Ti^{2+} has 2.

8.91   (a)
$Sr(s) \rightarrow Sr(g)$ ΔH_f° Sr(g) [$\Delta H^\circ_{sublimation}$ Sr(s)]
$Sr(g) \rightarrow Sr^+(g) + 1 e^-$ I_1 Sr
$Sr^+(g) \rightarrow Sr^{2+}(g) + 1 e^-$ I_2 Sr
$Cl_2(g) \rightarrow 2Cl(g)$ $2 \Delta H_f^\circ$ Cl(g) [$D(Cl_2)$]
$2Cl(g) + 2 e^- \rightarrow 2Cl^-(g)$ $2E_1$ Cl
$SrCl_2(s) \rightarrow Sr(s) + Cl_2(g)$ $-\Delta H_f^\circ$ $SrCl_2$
___
$SrCl_2(s) \rightarrow Sr^{2+}(g) + 2Cl^-(g)$ ΔH_{latt}

(b) ΔH_f° $SrCl_2$(s) = ΔH_f° Sr(g) + I_1(Sr) + I_2(Sr) + 2ΔH_f° Cl(g) + 2E(Cl) - ΔH_{latt} $SrCl_2$
ΔH_f° $SrCl_2$(s) = 164.4 kJ + 549 kJ + 1064 kJ + 2(121.7) kJ + 2(-349) kJ - 2127 kJ
= -804 kJ

8.92   The pathway to the formation of K_2O can be written:

$2K(s) \rightarrow 2K(g)$	$2\Delta H_f^\circ \ K(g)$
$2K(g) \rightarrow 2K^+(g) + 2\ e^-$	$2\ I_1(K)$
$1/2\ O_2(g) \rightarrow O(g)$	$\Delta H_f^\circ \ O(g)$
$O(g) + 1\ e^- \rightarrow O^-(g)$	$E_1(O)$
$O^-(g) + 1\ e^- \rightarrow O^{2-}(g)$	$E_2(O)$
$2K^+(g) + O^{2-}(g) \rightarrow K_2O(s)$	$-\Delta H_{latt}\ K_2O(s)$
————————————————	———————
$2K(s) + 1/2\ O_2(g) \rightarrow K_2O(s)$	$\Delta H_f^\circ \ K_2O$

$\Delta H_f^\circ \ K_2O(s) = 2\Delta H_f^\circ \ K(g) + 2\ I_1(K) + \Delta H_f^\circ \ O(g) + E_1(O) + E_2(O) - \Delta H_{latt}\ K_2O(s)$

$E_2(O) = \Delta H_f^\circ \ K_2O(s) + \Delta H_{latt}\ K_2O(s) - 2\Delta H_f^\circ \ K(g) - 2\ I_1(K) - \Delta H_f^\circ \ O(g) - E_1(O)$

$E_2(O) = -363.2\ kJ + 2238\ kJ - 2(89.99)\ kJ - 2(419)\ kJ - 247.5\ kJ - (-141)\ kJ$

$\qquad = +750\ kJ$

8.93   (a)

$HF(g) \rightarrow H(g) + F(g)$	D (H-F)	567 kJ
$H(g) \rightarrow H^+(g) + 1\ e^-$	I (H)	1312 kJ
$F(g) + 1\ e^- \rightarrow F^-(g)$	E (F)	-332 kJ
————————————————	———	———
$HF(g) \rightarrow H^+(g) + F^-(g)$	ΔH	1547 kJ

(b) $\Delta H = D(H\text{-}Cl) + I(H) + E(Cl)$
$\Delta H = 431\ kJ + 1312\ kJ + (-349)\ kJ = 1394\ kJ$

(c) $\Delta H = D(H\text{-}Br) + I(H) + E(Br)$
$\Delta H = 366\ kJ + 1312\ kJ + (-325)\ kJ = 1353\ kJ$

8.94   (a) $C_6H_6(g) \rightarrow 6H(g) + 6C(g)$

$\Delta H^\circ = \Delta H_f^\circ \ H(g) + 6\Delta H_f^\circ \ C(g) - \Delta H_f^\circ \ C_6H_6(g)$
$\Delta H^\circ = 6(217.94)\ kJ + 6(718.4)\ kJ - 82.9\ kJ = 5535\ kJ$

(b) $C_6H_6(g) \rightarrow 6CH(g)$

(c)

$C_6H_6(g) \rightarrow 6H(g) + 6C(g)$	ΔH°	5535 kJ
$6H(g) + 6C(g) \rightarrow 6CH(g)$	$-6D(C\text{-}H)$	$-6(413)$ kJ
————————————————		———
$C_6H_6(g) \rightarrow 6CH(g)$		3057 kJ

3057 kJ is the energy required to break the 6 C-C bonds in $C_6H_6(g)$. The average bond dissociation energy for one carbon-carbon bond in $C_6H_6(g)$ is $\dfrac{3057\ kJ}{6\,C\text{-}C\ bonds} =$ 509.5 kJ.

(d)   The value of 509.5 kJ is between the average value for a C-C single bond (348 kJ) and a C=C double bond (614 kJ). It is somewhat greater than the average of these two values, indicating that the carbon-carbon bond in benzene is a bit stronger than we might expect.

8.95   (a) $Br_2(l) \rightarrow 2Br(g)$ $\Delta H° = 2\Delta H°_f \; Br(g) = 2(111.8) \; kJ = 223.6 \; kJ$

       (b) $CCl_4(l) \rightarrow C(g) + 4Cl(g)$

             $\Delta H° = \Delta H°_f \; C(g) + 4\Delta H°_f \; Cl(g) - \Delta H°_f \; CCl_4(l)$
                  $= 718.4 \; kJ + 4(121.7) \; kJ - (-139.3) \; kJ = 1344.5$

             $\dfrac{1344.5 \; kJ}{4 \; C-Cl \; bonds} = 336.1 \; kJ$

       (c) $H_2O_2(l) \rightarrow 2H(g) + 2O(g)$

          $2H(g) + 2O(g) \rightarrow 2OH(g)$

                $\overline{}$

                 $H_2O_2(l) \rightarrow 2OH(g)$

          $D(O-O)(l) = 2\Delta H°_f \; H(g) + 2\Delta H°_f \; O(g) - \Delta H°_f \; H_2O_2(l) - 2D(O-H)(g)$

                    $= 2(217.94) \; kJ + 2(247.5) \; kJ - (-187.8) \; kJ - 2(463) \; kJ$

                    $= 193 \; kJ$

       (d)   The data are listed below.

bond	D gas kJ/mol	D liquid kJ/mol
Br-Br	193	223.6
C-Cl	328	336.1
O-O	146	192.7

Breaking bonds in the liquid requires more energy than breaking bonds in the gas phase. For simple molecules, bond dissociation from the liquid phase can be thought of in two steps:

          molecule (l) $\rightarrow$ molecule (g)
          molecule (g) $\rightarrow$ atoms (g)

The first step is evaporation or vaporization of the liquid and the second is bond dissociation in the gas phase. Average bond dissociation energy from the liquid phase is then the sum of the enthalpy of vaporization for the molecule and the gas phase bond dissociation enthalpies, divided by the number of bonds dissociated. This is greater than the gas phase bond dissociation enthalpy owing to the contribution from the enthalpy of vaporization.

8.96   (a)   Assume 100 g.
          A: 87.7 g In / 114.82 = 0.764 mol In;   0.764 / 0.384 ≈ 2
              12.3 g S /   32.07 = 0.384 mol S;   0.384 / 0.384 = 1
          B: 78.2 g In / 114.82 = 0.681 mol In;   0.681 / 0.680 ≈ 1
              21.8 g S /   32.07 = 0.680 mol S;   0.680 / 0.680 = 1
          C: 70.5 g In / 114.82 = 0.614 mol In;   0.614 / 0.614 = 1
              29.5 g S /   32.07 = 0.920 mol S;   0.920 / 0.614 = 1.5
          A: In_2S;   B: InS;   C: In_2S_3

(b)   A: In(I);   B: In(II);   C: In(III)

(c)   In(I) : $[Kr]5s^2 4d^{10}$

          In(II) : $[Kr]5s^1 4d^{10}$

          In(III) : $[Kr]4d^{10}$

          None of these are noble gas configurations.

(d)   The ionic radius of In^{3+} in compound C will be smallest. Removing successive electrons from an atom reduces electron repulsion, increases the effective nuclear charge experienced by the valence electrons and decreases the ionic radius. The higher the charge on a cation, the smaller the radius.

(e)   Lattice energy is directly related to the charge on the ions and inversely related to the interionic distance. Only the charge and size of the In varies in the three compounds. In(I) in compound A has the smallest charge and the largest ionic radius, so compound A has the smallest lattice energy and the lowest melting point. In(III) in compound C has the greatest charge and the smallest ionic radius, so compound C has the largest lattice energy and highest melting point.

9 Molecular Geometry and Bonding Theories

Molecular Geometry; the VSEPR Model

9.1 (a) No. A set of bonds with particular lengths can be placed in many different relative orientations. Bond lengths alone do not define the size and shape of a molecule.

(b) Yes. This description means that the three terminal atoms point toward the corners of an equilateral triangle and the central atom is in the plane of this triangle. Only 120° bond angles are possible in this arrangement.

9.2 (a) Yes. The bond angle specifies the shape and the bond length tells the size.
(b) Yes. In a perfect tetrahedron, the only possible bond angles are 109.5°.

9.3 (a) trigonal planar (b) tetrahedral (c) trigonal bipyramidal (d) octahedral

9.4 (a) 3 (or 5 if 90° angles are also present)
(b) 2 (or 5 if 120° angles are also present, 6 if more than one 180° angle is present)
(c) 4
(d) 6 (or 5 if 120° angles are also present)

9.5 The electron pair geometry indicated by VSEPR takes into account the total number of bonding (double and triple bonds count as one effective pair) and nonbonding pairs of electrons. The molecular geometry describes just the atomic positions. NH_3 has the Lewis structure given below; there are four pairs of electrons around nitrogen so the electron pair geometry is tetrahedral, but the molecular geometry of the four atoms is trigonal pyramidal.

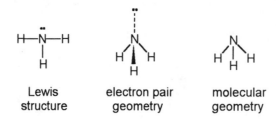

| Lewis structure | electron pair geometry | molecular geometry |

9.6   (a)   Yes. A molecule with tetrahedral electron pair geometry has four bonding and nonbonding electron pairs around the central atom. If two pairs are nonbonding and two pairs are bonding, as in H_2O, the molecular geometry is bent

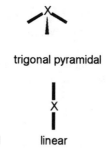

       (b)   If the electron pair geometry is trigonal bipyramidal, there are 5 total regions of electron density around the central atom. An AB_3 molecule has 3 regions of bonding electrons, so there must be 2 nonbonded pairs on A.

9.7

	Lewis structure	electron pair geometry	molecular geometry
(a)		tetrahedral	tetrahedral
(b)		trigonal bipyramidal	T-shaped
(c)		octahedral	square pyramidal

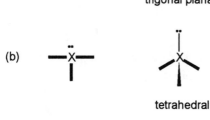

	Lewis structure	electron pair geometry	molecular geometry
9.8 (a)		trigonal planar	trigonal planar
(b)		tetrahedral	trigonal pyramidal
(c)		trigonal bipyramidal	linear

9.9   bent (b), linear (l), octahedral (oh), seesaw(ss), tetrahedral (td), trigonal bipyramidal (tbp), trigonal planar (tr), trigonal pyramidal (tp), T-shaped (T)

	Molecule or ion	Valence electrons	Lewis structure	e⁻ pair geometry	Molecular geometry
(a)	ClO_2^-	20	$[\ddot{\text{O}}-\text{Cl}-\ddot{\text{O}}]^-$	td	b
(b)	SO_3	24	$\ddot{\text{O}}=\text{S}-\ddot{\text{O}}$, $\ddot{\text{O}}$	tr	tr
(c)	PCl_3	26	$\ddot{\text{Cl}}-\text{P}-\ddot{\text{Cl}}$, $\ddot{\text{Cl}}$	td	tp
(d)	BH_4^-	8	$[\text{H}-\text{B(H)(H)}-\text{H}]^-$	td	td
(e)	SO_3^{2-}	26	$[\ddot{\text{O}}-\text{S}-\ddot{\text{O}} , \ddot{\text{O}}]^{2-}$	td	tp
(f)	ICl_3	28	$\ddot{\text{Cl}}-\text{I(Cl)(Cl)}$	tbp	T

9.10   bent (b), linear (l), octahedral (oh), seesaw(ss), tetrahedral (td), trigonal bipyramidal (tbp), trigonal planar (tr), trigonal pyramidal (tp), T-shaped (T)

	Molecule or ion	Valence electrons	Lewis structure	e⁻ pair geometry	Molecular geometry
(a)	CO_3^{2-}	24	$[\ddot{\text{O}}=\text{C}-\ddot{\text{O}} , \ddot{\text{O}}]^{2-}$ (3 res forms)	tr	tr

9.10   (continued)

(b)	SCl_2	20	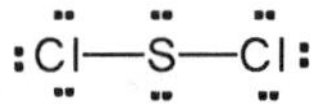 :Cl—S—Cl:	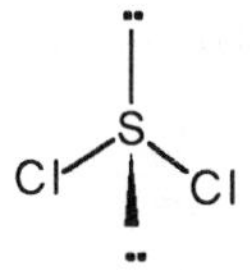 (structure)	td b
(c)	ClO_3^-	26	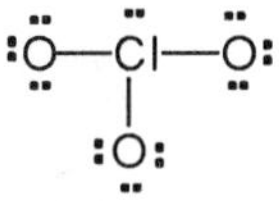 :O—Cl—O: / :O:	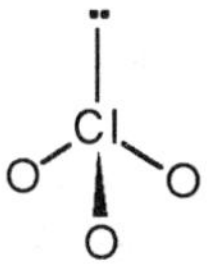 (structure)	td tp
(d)	PF_6^-	48	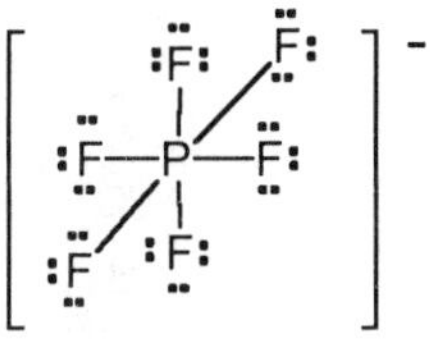 [F—P—F structure]⁻	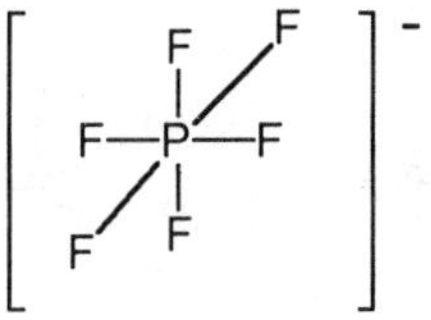 [F—P—F structure]⁻	oh oh
(e)	ICl_2^-	22	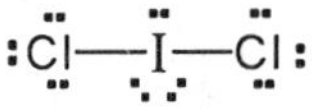 :Cl—I—Cl:	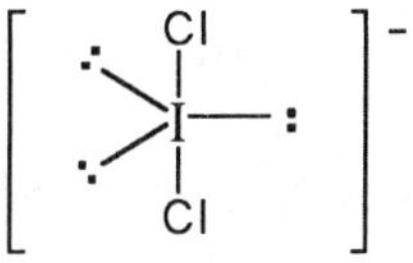 [structure]⁻	tbp l
(f)	TeF_4	34	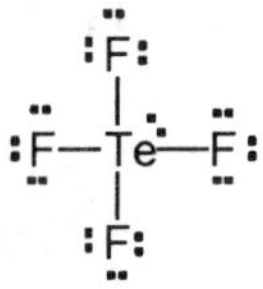 :F—Te—F: / :F:	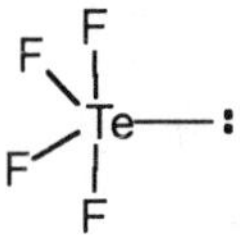 (structure)	tbp ss

9.11

     :F—N—F: / :F: :F—B—F: / :F: :F—Cl—F: / :F:

	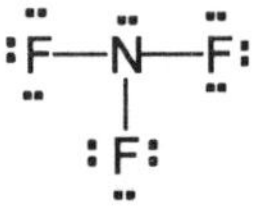	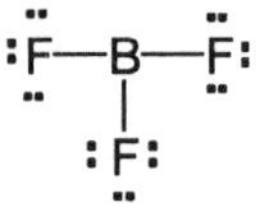	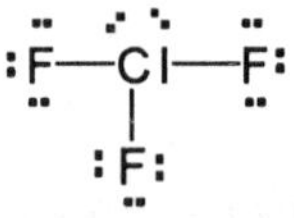
e⁻ pair geometry	td	tr	tbp
molecular geometry	tp	tr	T

The molecular geometries or shapes differ because there are a different number of nonbonding electron pairs around the central atom in each case.

9.12

     :F: / :F—Si—F: / :F: :F: / :F—S—F: / :F: :F: / :F—Xe—F: / :F:

	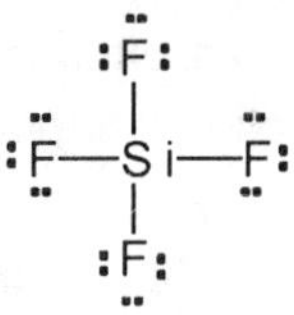	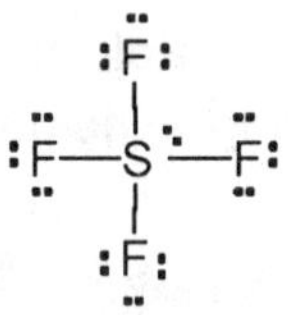	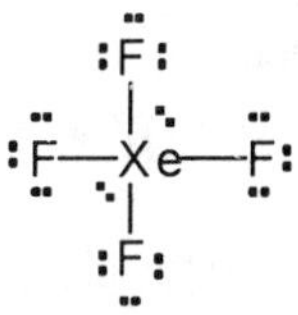
e⁻ pair geometry	td	tbp	octahedral (o)
molecular geometry	td	seesaw (ss)	square planar (s)

Although there are four bonding pairs of electrons in each molecule, the number of nonbonding electrons is different in each case and the molecular shape is influenced by the total number of valence electron pairs.

19.13

$$\left[H-\overset{..}{\underset{..}{N}}-H \right]^{-} \qquad H-\overset{..}{\underset{\underset{H}{|}}{N}}-H \qquad \left[\overset{\overset{H}{|}}{\underset{\underset{H}{|}}{H-N-H}} \right]^{+}$$

Each molecule has 4 pairs of electrons around the N atom, but the number of nonbonded pairs decreases from 2 to 0 going from NH_2^- to NH_4^+. Since lone pairs occupy more space than bonded pairs, the bond angles expand as the number of lone pairs decreases.

9.14

The three nonbonded electron pairs on each F atom have been omitted for clarity.

The F (axial) - A - F (equatorial) angle is largest in PF_5 and smallest in ClF_3. As the number of nonbonded pairs in the equatorial plane increases, they push back the axial A - F bonds, decreasing the F (axial) - A - F (equatorial) bond angles.

9.15   (a)   1 - 109°, 2 - 109°   (b) 3 - 109°, 4 - 109°
       (c)   5 - 180°   (d) 6 - 120°, 7 - 109°, 8 - 109°

9.16   (a)   1 - 109°, 2 - 120°   (b) 3 - 109°, 4 - 120°
       (c)   5 - 109°, 6 - 109°   (d) 7 - 180°, 8 - 109°

Polarity of Molecules

9.17   (a)   A polar molecule has a measurable dipole moment; its centers of positive and negative charge do not coincide. A nonpolar molecule has a zero net dipole moment; its centers of positive and negative charge do coincide.

       (b)   Yes. If X and Y have different electronegativities, they have different attractions for the electrons in the molecule. The electron density around the more electronegative atom will be greater, producing a charge separation or dipole in the molecule.

       (c) $\mu = Qr$. The dipole moment, μ, is the product of the magnitude of the separated charges, Q, and the distance between them, r.

9.18   (a)   A separation of the centers of positive and negative charge is needed to produce a dipole.

(b)   No. Although each X-Y bond is polar, the overall polarity of the molecule depends on the arrangement of bond dipoles, the molecular geometry.

(c)   No. Dipole moment depends on the distance between charge centers (bond distance) **and** the magnitude of the separated charges (electronegativity difference). Two diatomic molecules with the same bond length will probably have different dipole moments because it is unlikely that ΔEN for the two molecules is the same.

9.19   (a)   **Nonpolar**, in a symmetrical tetrahedral structure (Figure 9.1), the bond dipoles cancel.

(b)   **Nonpolar**, the molecule is linear and the bond dipoles cancel.

S=C=S

(c)   **Nonpolar,** in a symmetrical trigonal planar structure (Exercise 9.9 (b)), the bond dipoles cancel.

(d)   **Polar**, in the see-saw molecular geometry (Exercise 9.12), the dipoles don't cancel and there is an unequal charge distribution due to the nonbonded electron pair on S.

(e)   **Polar**, there is an unequal charge distribution due to the nonbonded electron pair on N.

(f)   **Nonpolar**, in a symmetrical trigonal bipyramid, the bond dipoles cancel.

9.20   (a)   **Nonpolar**, symmetrical tetrahedron

(b)   **Polar**, ΔEN > 0

I-F

(c)   **Polar**, bent molecular geometry

(d)   **Polar**, although the bond dipoles are essentially zero, there is an unequal charge distribution due to the non-bonded electron pair on P.

(e)   **Nonpolar**, symmetrical octahedron

(f)   **Polar**, square pyramidal molecular geometry, bond dipoles do not cancel

9.21   In $BeCl_2$ there are just two electron pairs about Be. The structure is linear, so the individual bond dipoles cancel. In SCl_2, there are four electron pairs about S. The two unshared pairs don't cancel with the two S-Cl bonding pairs.

9.22

μ = 1.03 D   μ = 0

BF_3 is a trigonal planar molecule with the central B atom symmetrically surrounded by the three F atoms (Figure 9.11). The individual B-F bond dipoles cancel, and the molecule has a net dipole moment of zero. PF_3 has tetrahedral electron pair geometry with one of the positions in the tetrahedron occupied by a nonbonding electron pair. The individual P-F bond dipoles do not cancel (nothing cancels a nonbonding electron pair except another nonbonding pair), and the molecule is polar.

9.23   Q is the charge at either end of the dipole.

$$Q = \frac{\mu}{r} = \frac{1.82\ D}{0.92\ \text{Å}} \times \frac{1\ \text{Å}}{1 \times 10^{-10}\ m} \times \frac{3.34 \times 10^{-30}\ C \cdot m}{1\ D} \times \frac{1\ e^-}{1.60 \times 10^{-19}\ C} = 0.41\ e^-$$

The calculated charge on H and F is 0.41 e⁻. This can be thought of as the amount of charge "transferred" from H to F.

9.24   Q is the charge at either end of the dipole.

$$Q = \frac{\mu}{r} = \frac{0.82 \, D}{1.41 \, \text{Å}} \times \frac{1 \, \text{Å}}{1 \times 10^{-10} \, m} \times \frac{3.34 \times 10^{-30} \, C \cdot m}{1 \, D} \times \frac{1 \, e^-}{1.60 \times 10^{-19} \, C} = 0.12 \, e^-$$

Comparing this with the result in Exercise 9.23; HF is much more polar than HBr.

9.25

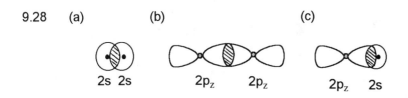

     polar   nonpolar   polar

All three isomers are planar. The molecules on the left and right are polar because the C-Cl bond dipoles do not point in opposite directions. In the middle isomer, the C-Cl bonds and dipoles are pointing in opposite directions (as are the C-H bonds), the molecule is nonpolar and has a measured dipole moment of zero.

9.26   Each C-Cl bond is polar. The question is whether the vector sum of the C-Cl bond dipoles in each molecule will be nonzero. In the *ortho* and *meta* isomers, the C-Cl vectors are at 60° and 120° angles, respectively, and their resultant dipole moments are nonzero. In the *para* isomer, the C-Cl vectors are opposite, at an angle of 180°, with a resultant dipole moment of zero. The *ortho* and *meta* isomers are polar, the *para* isomer is nonpolar.

Orbital Overlap; Hybrid Orbitals

9.27   (a)   "Orbital overlap" occurs when a valence atomic orbital on one atom shares the same region of space with a valence atomic orbital on an adjacent atom.

        (b)   In valence bond theory, overlap of orbitals allows the two electrons in a chemical bond to mutually occupy the space between the bonded nuclei.

        (c)   Valence bond theory is a combination of the atomic orbital concept with the Lewis model of electron pair bonding.

9.28   (a)   (b)   (c)

              2s   2s $2p_z$ $2p_z$ $2p_z$ 2s

9.29   (a) sp^2 -- 120° angles in a plane

        (b) sp^3d -- 90°, 120° and 180° bond angles (trigonal bipyramid)

        (c) sp^3d^2 -- 90° and 180° bond angles (octahedron)

9.30   (a)   sp -- 180°   (b) sp^3 -- 109°   (c) sp^2 -- 120°

        (d) sp^3d^2 -- 90° and 180°   (e) sp^3d -- 90°, 120° and 180°

9.31 (a) B - [He]$2s^2 2p^1$

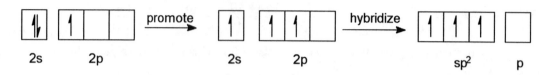

  2s 2p   2s 2p sp^2 p

  (b) The hybrid orbitals are called sp^2. (c)

  (d) A single 2p orbital is unhybridized.
It lies perpendicular to the trigonal
plane of the sp^2 hybrid orbitals.

9.32 (a) P - [Ne]$3s^2 3p^3$

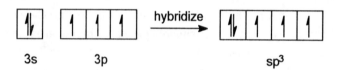

  3s 3p sp^3

  (b) The hybrid orbitals are called sp^3. (c)

  (d) The hybrid orbitals formed in (a) would not be appropriate for PCl_5.
There are 5 equivalent P-Cl bonds in PCl_5, so 5 equivalent hybrid
orbitals are required. A set of 4 sp^3 hybrid orbitals on P could not form
the 5 P-Cl bonds.

9.33 (a) 8 e⁻, 4 e⁻ pairs   (b) 8 e⁻, 4 e⁻ pairs

       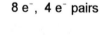

4 e⁻ pairs around B   4 e⁻ pairs around N
tetrahedral e⁻ pair geometry   tetrahedral e⁻ pair geometry
sp^3 hybrid orbitals   geometry sp^3 hybrid orbitals

(c)   22 e⁻, 11 e⁻ pairs   (d)   24 e⁻, 12 e⁻ pairs

:F̈—Ẍe—F̈:   :Ï—B—Ï:
                                 :Ï:

5 e⁻ pairs around Xe   3 e⁻ pairs around B
trigonal bipyramidal e⁻ pair geometry   trigonal planar e⁻ pair geometry
sp^3d^2 hybrid orbitals sp^2 hybrid orbitals

(e)   48 e⁻, 24 e⁻ pairs   Note that the B-I bond is probably too long for π
overlap and formal charge considerations argue
against structures with double bonds. However,
resonance structures that contain a B = I double
bond would use the same hybrid orbitals as the
one given above.

$$\left[\begin{matrix} :F: & :F: \\ :F—P—F: \\ :F: & :F: \end{matrix} \right]^{-}$$

6 e⁻ pairs around P
octahedral e⁻ pair geometry, sp^3d^2

9.34   (a)   8 e⁻, 4 e⁻ pairs   (b)   16 e⁻, 8 e⁻ pairs

$$\left[\begin{matrix} H—\ddot{O}—H \\ | \\ H \end{matrix} \right]^{+}$$ S̈=C=S̈

4 e⁻ pairs around O   2 VSEPR pairs around C
tetrahedral e⁻ pair geometry   linear e⁻ pair geometry
sp^3 hybrid orbitals   sp hybrid orbitals

(c)   32 e⁻, 16 e⁻ pairs   (d)   22 e⁻, 11 e⁻ pairs

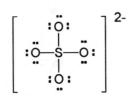

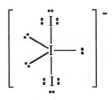

4 e⁻ pairs around S   5 e⁻ pairs around I
tetrahedral e⁻ pair geometry   trigonal bipyramidal e⁻ pair geometry
sp^3 hybrid orbitals sp^3d hybrid orbitals
                                (In a trigonal bipyramid, placing
nonbonding e⁻ pairs in the
equatorial position minimizes
e⁻ pair repulsion.)

(e)   36 e⁻, 18 e⁻ pairs   6 e⁻ pairs around Xe
         (Xe has 8 valence e⁻)   octahedral e⁻ pair geometry

                                             sp^3d^2 hybrid orbitals

                                             (In an octahedron, placing 2 lone pairs
                                             opposite each other minimizes repulsions.)

Multiple Bonds

9.35   (a)

σ

(b)

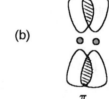

π

(c)   A σ bond is generally stronger than a π bond, because there is more extensive orbital overlap.

9.36   (a)   A double bond is composed of 1 σ and 1 π bond.

       (b)   A triple bond is composed of 1 σ and 2 π bonds.

       (c)   There is free rotation of attached groups around a σ bond, but not around a π bond. Rotation about a π bond would require that the bond be broken; the p orbitals would no longer be in the correct orientation for π overlap. The π overlap that is part of all multiple bonds introduces rigidity into molecules.

9.37

$$\begin{array}{c} H \\ | \\ H-C-H \\ | \\ H \end{array} \qquad \begin{array}{c} O \\ || \\ C \\ H \quad H \end{array}$$

The C atom in CH_4 is sp^3 hybridized; there are no unhybridized p orbitals available for the π overlap required by multiple bonds. In CH_2O, the C atom is sp^2 hybridized, with 1 p atomic orbital available to form the π overlap in the C=O double bond.

9.38 $\begin{array}{c} \ddot{} \quad \ddot{} \\ H-N-N-H \\ | \quad | \\ H \quad H \end{array}$:N≡N:

The N atoms in N_2H_4 are sp^3 hybridized; there are no unhybridized p orbitals available for π bonding. In N_2, the N atoms are sp hybridized, with 2 unhybridized p orbitals on each N atom available to form the 2 π bonds in the N ≡ N triple bond.

9.39 (a) ~109° about the left most C, sp^3; ~120° about the right-hand C, sp^2

(b) The doubly bonded O can be viewed as sp^2, the other as sp^3; the nitrogen is sp^3 with approximately 109° bond angles.

(c) nine σ bonds, one π bond

9.40 (a) 1 - 120°; 2 - 120°; 3 - 109°   (b) 1 - sp^2; 2 - sp^2; 3 - sp^3 (c) 21 σ bonds

9.41 (a) In a localized π bond, the electron density is concentrated strictly between the two atoms forming the bond. In a delocalized π bond, parallel p orbitals on more than two adjacent atoms overlap and the electron density is spread over all the atoms that contribute p orbitals to the network. There are still two regions of overlap, above and below the σ framework of the molecule.

(b) The existence of more than one resonance form is a good indication that a molecule will have delocalized π bonding.

(c)

The existence of more than one resonance form for NO$_2^-$ indicates that the π bond is delocalized. From an orbital perspective, the electron pair geometry around N is trigonal planar, so the hybridization at N is sp^2. This leaves a p orbital on N and one on each O atom perpendicular to the trigonal plane of the molecule, in the correct orientation for delocalized π overlap. Physically, the two N-O bond lengths are equal, indicating that the two N-O bonds are equivalent, rather than one longer single bond and one shorter double bond.

9.42 (a) 24 e$^-$, 12 e$^-$ pairs   (b)

3 VSEPR pairs around S trigonal planar e$^-$ pair geometry, sp^2 hybrid orbitals

(c) The multiple resonance structures indicate delocalized π bonding. All four atoms lie in the trigonal plane of the sp^2 hybrid orbitals. On each atom there is a p atomic orbital perpendicular to this plane in the correct orientation for π overlap. The resulting delocalized π electron cloud is Y-shaped (the shape of the molecule) and has electron density above and below the plane of the molecule.

Molecular Orbitals

9.43 (a) Both atomic and molecular orbitals have a characteristic energy and shape (region where there is a high probability of finding an electron). Each atomic or molecular orbital can hold a maximum of two electrons. Atomic orbitals are localized on single atoms and their energies are the result of interactions between the subatomic particles in a single atom. Molecular orbitals can be delocalized over several or even all the atoms in a molecule and their energies are influenced by interactions between electrons on several atoms.

(b) There is a net stabilization (lowering in energy) that accompanies bond formation because the bonding electrons in H_2 are strongly attracted to both H nuclei.

(c) 2

9.44 (a) According to molecular orbital theory, the electrons in H_2 both reside in the σ bonding molecular orbital.

(b) In the σ anti-bonding molecular orbital, electron density is concentrated away from the nuclei; an electron in this orbital experiences less stabilization by the nucleus than an electron in an isolated atom.

(c) Yes. The Pauli principle, no two electrons can have the same 4 quantum numbers, means that an orbital can hold at most 2 electrons. (Since n, l and m_l are the same for a particular orbital and m_s only has 2 possible values, an orbital can hold at most 2 electrons). This is true for atomic and molecular orbitals.

9.45 (a)

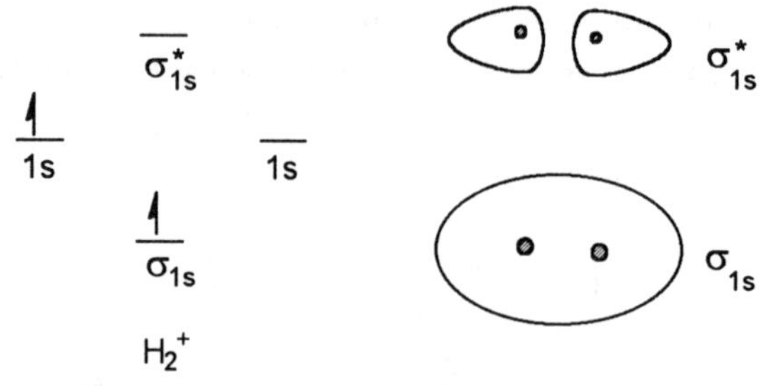

(b) There is 1 electron in H_2^+. The Lewis model of bonding indicates that a bond is a pair of electrons. Since there is only 1 electron in H_2^+, there is no bond by the Lewis definition, so it is not possible to write a Lewis structure.

(c) σ_{1s}^* [empty box] σ_{1s} [box with ↑]

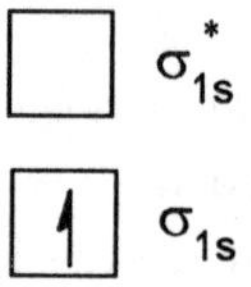

(d) Bond order = 1/2 (1-0) = 1/2

(e)   Yes. The stability of H_2^+ is due to the lower energy state of the σ bonding molecular orbital relative to the energy of a H 1s atomic orbital. If the single electron in H_2^+ is excited to the $\sigma_{1s}{}^*$ orbital, its energy is higher than the energy of a H 1s atomic orbital and H_2^+ will decompose into a hydrogen atom and a hydrogen ion,

$$H_2^+ \xrightarrow{h\nu} H + H^+.$$

9.46   (a)

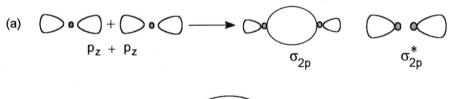

(b)   According to the Duet Rule for H and He, these atoms can accommodate at most 2 electrons in their valence shells. A Lewis structure for H_2^- would violate the duet rule.

(c) $\boxed{\uparrow}$ $\sigma_{1s}{}^*$ (d) Bond order = 1/2 (2-1) = 1/2

                $\boxed{\uparrow\downarrow}$ σ_{1s}

(e)   If 1 electron moves from σ_{1s} to $\sigma_{1s}{}^*$, the bond order becomes -1/2. There is a net increase in energy relative to isolated H atoms, so the molecule will decompose.

$$H_2^- \xrightarrow{h\nu} H + H^-.$$

9.47   In a σ molecular orbital, the electron density is spherically symmetric about the internuclear axis and is concentrated along this axis. In a π molecular orbital, the electron density is concentrated above and below the internuclear axis and zero along it.

(a)

           $p_z + p_z$ σ_{2p} $\sigma_{2p}{}^*$

(b)

           $p_x + p_x$ π_{2p} $\pi_{2p}{}^*$

(c) σ_{2p} is lower in energy than π_{2p} due to greater extent of orbital overlap in the σ molecular orbital. $\sigma_{2p} < \pi_{2p} < \pi^*_{2p} < \sigma_{2p}$

9.48   (a)   zero

(b)   The 2 π_{2p} molecular orbitals are degenerate; they have the same energy, but they have different spatial orientations 90° apart.

(c)   In the bonding molecular orbital the electrons are stabilized by both nuclei. In an antibonding molecular orbital, the electrons are directed away from the nuclei, so π_{2p} is lower in energy than π_{2p}^*.

9.49   (a)   Substances with no unpaired electrons are weakly repelled by a magnetic field. This property is called *diamagnetism*.

(b) H_2 , N_2

(c) O_2^{2-}, Be_2^{2+} (see Figure 9.39)

9.50   (a)   Substances with unpaired electrons are attracted into a magnetic field. This property is called paramagnetism.

(b) B_2 , O_2

(c) N_2^{2-}, Li_2^+ (see Figure 9.39)

9.51   (a)

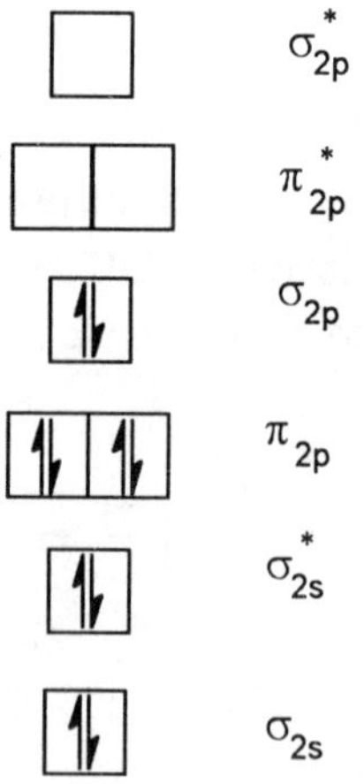

□	σ_{2p}^*
□ □	π_{2p}^*
⇅	σ_{2p}
⇅ ⇅	π_{2p}
⇅	σ_{2s}^*
⇅	σ_{2s}

(a) CO :   B.O. = (8-2) / 2 = 3.0, diamagnetic

(b) NO⁻ :   B.O. = (8-4) / 2 = 2, paramagnetic

(c) CN⁻ :   B.O. = (8-2) / 2 = 3, diamagnetic

(d) OF :   B.O. = (8-5) / 2 = 1.5 paramagnetic

9.52

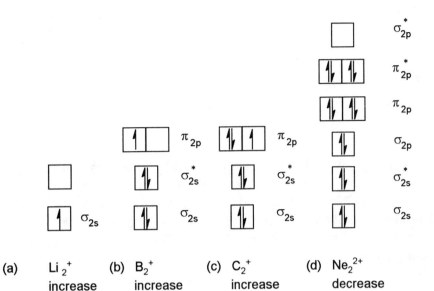

	(a) Li_2^+	(b) B_2^+	(c) C_2^+	(d) Ne_2^{2+}
	increase	increase	increase	decrease

Addition of an electron increases bond order if it occupies a bonding orbital and decreases stability if it occupies an antibonding orbital.

9.53 (a) The separation between related bonding and antibonding orbitals is directly related to the extent of overlap of the atomic orbitals. The Li 2s orbitals are larger than the 1s, the overlap is greater and the energy separation of the resulting σ_{2s} and σ_{2s}^* orbitals is larger than that between σ_{1s} and σ_{1s}^*.

(b) O_2^{2-} has a bond order of 1.0, while O_2^- has a bond order of 1.5. For the same bonded atoms, the greater the bond order the shorter the bond, so O_2^- has the shorter bond.

(c) Interactions between electrons in 2s atomic orbitals on one atom and 2p orbitals on the other has an effect on the energies of the $\sigma_{2s}, \sigma_{2s}^*, \sigma_{2p}$ and σ_{2p}^* molecular orbitals. The energy of the σ_{2s} is lowered and the energy of the σ_{2p} is raised. If the 2s-2p interaction is substantial, as it is in B_2, the energy of the σ_{2p} is actually raised above the energy of the π_{2p}.

9.54 (a) An electron in a σ_{2s} bonding molecular orbital is farther from the bonding nuclei than an electron in a σ_{1s}^* orbital. Thus, the energy of the σ_{2s} bonding orbital is higher.

(b) The extra electron in N_2^- occupies an antibonding orbital, so the bond order is lower in N_2^- than in N_2. This means that N_2^- has a longer, weaker bond than N_2.

(c) In F_2, the interactions between 2s and 2p electrons on the different bonding atoms is not large. The energies of the σ_{2s} and σ_{2p} orbitals are affected, but not enough so that the σ_{2p} is higher in energy than the π_{2p}.

Additional Exercises

9.55   26 e⁻, 13 e⁻ pairs

$$\left[\; :\overset{..}{\underset{..}{O}}-\overset{..}{\underset{..}{Cl}}-\overset{..}{\underset{..}{O}}:\; \right]^{-}$$

The electron pair geometry (four total pairs around Cl) is tetrahedral and molecular shape is trigonal pyramidal. Because there are four valence electron pairs on Cl, sp³ hybrid orbitals are used for bonding. Each Cl-O σ bond is formed by the overlap of an sp³ hybrid orbital on Cl with an atomic or hybrid orbital on O. (Because O is terminal, there is no angular data to indicate the type of orbital used in bonding.) The O-Cl-O bond angles should be approximately 109°.

9.56   (a)   AsF₃; 26 valence e⁻   (b)   OCN⁻; 16 valence e⁻

tetrahedral e⁻ pair geometry
trigonal pyramidal molecular geometry

linear e⁻ pair geometry
linear "molecular" geometry

(c)   H₂CO; 12 valence e⁻   (d)   ICl₂⁻ ; 22 valence e⁻

trigonal planar e⁻ pair geometry
trigonal planar molecular geometry

trigonal bipyramidal e⁻ pair geometry
linear molecular geometry
(In a trigonal bipyramid, nonbonded
pairs lie in the trigonal plane.)

9.57   The σ bond electrons are localized in the region along the internuclear axes. The positions of the atoms and geometry of the molecule are thus closely tied to the locations of these electron pairs. Because the π bond electrons are distributed above and below the plane of the σ bond axis, these electron pairs do not, in effect, influence the geometry of the molecule.

9.58   For any triangle, the law of cosines gives the length of side c as $c^2 = a^2 + b^2 - 2ab \cos\theta$.

Let the edge length of the cube (uy = vy = vz) = X
The length of the face diagonal (uv) is
$(uv)^2 = (uy)^2 + (vy)^2 - 2(uy)(vy) \cos 90$
$(uv)^2 = X^2 + X^2 - 2(X)(X) \cos 90$
$(uv)^2 = 2X^2; \quad uv = \sqrt{2}X$

The length of the body diagonal (uz) is
$(uz)^2 = (vz^2) + (uv)^2 - 2(vz)(uv) \cos 90$
$(uz)^2 = X^2 + (\sqrt{2}X)^2 - 2(X)(\sqrt{2}X) \cos 90$
$(uz)^2 = 3X^2; \quad uz = \sqrt{3}X$

For calculating the characteristic tetrahedral angle, the appropriate triangle has vertices u, v and w. Theta, θ, is the angle formed by sides wu and wv and the hypotenous is side uv.

$wu = wv = uz/2 = \sqrt{3}/2 X; \quad uv = \sqrt{2}X$

$(\sqrt{2}X)^2 = (\sqrt{3}/2X)^2 + (\sqrt{3}/2X)^2 - 2(\sqrt{3}/2X)(\sqrt{3}/2) \cos\theta$

$2X^2 = 3/4 X^2 + 3/4 X^2 - 3/2 X^2 \cos\theta$

$2X^2 = 3/2 X^2 - 3/2 X^2 \cos\theta$

$1/2 X^2 = -3/2 X^2 \cos\theta$

$\cos\theta = -(1/2 X^2) / (3/2 X^2) = -1/3 = -0.3333$

$\theta = 109.47°$

9.59   (a) CO_2, 16 valence e⁻   (b)   NCS⁻, 16 valence e⁻

$$\overset{..}{\underset{..}{O}} = C = \overset{..}{\underset{..}{O}}$$

$2\sigma, \quad 2\pi$

$$\left[\overset{..}{N} = C = \overset{..}{\underset{..}{S}} \right] \quad + \quad \text{two other resonance structures}$$

$2\sigma, \quad 2\pi$ (for any of the resonance structures)

(c) SO_4^{2-}, 32 valence e⁻

(d)   HCO(OH), 18 valence e⁻

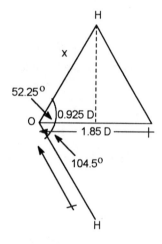

4 σ

(assuming that the best Lewis structure is
the one that does not violate the octet rule)

4 σ, 1 π

9.60   The compound on the right has a dipole moment. In the square planar *trans* structure on the left, all equivalent bond dipoles can be oriented opposite each other, for a net dipole moment of zero.

9.61

(a)   The bond dipoles in H_2O lie along the O-H bonds with the positive end at H and the negative end at O. The dipole moment vector of the H_2O molecule is the resultant (vector sum) of the two bond dipoles. This vector bisects the H-O-H angle and has a magnitude of 1.85 D with the negative end pointing toward O.

(b)   Since the dipole moment vector bisects the H-O-H bond angle, the angle between one H-O bond and the dipole moment vector is 1/2 the H-O-H bond angle, 52.25°. Dropping a perpendicular line from H to the dipole moment vector creates the right triangle pictured. If x = the magnitude of the O-H bond dipole, x cos (52.25) = 0.925 D. **x = 1.51 D.**

(c)   The X-H bond dipoles (Table 9.4) and the electronegativity values of X (Figure 8.7) are

	Electronegativity	Bond dipole
F	4.0	1.82
O	3.5	1.51
Cl	3.0	1.08

Since the electronegativity of O is midway between the values for F and Cl, the O-H bond dipole should be approximately midway between the bond dipoles of HF and HCl. The value of the O-H bond dipole calculated in part (b) is consistent with this prediction.

9.62   The carbons not in the ring all employ sp^3 hybrid orbitals. The ring carbons use an sp^2 hybrid set for the σ bond framework, with a p orbital involved in the delocalized π bonding in the ring. The two oxygen atoms of cumene hydroperoxide are approximately sp^3 hybridized (recall that there are two unshared pairs on each oxygen, not shown.)

9.63

(a)   The molecule is nonplanar. The CH_2 planes at each end are twisted 90° from one another.

(b)   Allene has no dipole moment.

(c)   The bonding in allene would not be described as delocalized. The π electron clouds of the two adjacent C=C are mutually perpendicular. The mechanism for delocalization of π electrons is mutual overlap of parallel p atomic orbitals on adjacent atoms. If adjacent π electron clouds are mutually perpendicular, there is no overlap and no delocalization of π electrons.

9.64 (a)   16 e⁻, 8 e⁻ pairs

$$\left[\ddot{N}=N=\ddot{N} \right]^- \longleftrightarrow \left[:N\equiv N-\ddot{N}: \right]^- \longleftrightarrow \left[:\ddot{N}-N\equiv N: \right]^-$$

(b)   The observed bond length of 1.16 Å is intermediate between the values for N=N, 1.24 Å, and N ≡ N, 1.10 Å. This is consistent with the resonance structures, which indicate contribution from formally double and triple bonds to the true bonding picture in N_3^-.

(c)   In each resonance structure, the central N has 2 VSEPR e⁻ "pairs", so it must be sp hybridized. It is difficult to predict the hybridization of terminal atoms in molecules where there are resonance structures because there are a different number of VSEPR pairs around the terminal atoms in each structure. Since the "true" electronic arrangement is a combination of all resonance structures, we will assume that the terminal N-N bonds have some triple bond character and that the terminal N atoms are sp hybridized. (There is no experimental measure of hybridization at terminal atoms, since there are no bond angles to observe.)

(d)   In each resonance structure, N-N σ bonds are formed by sp hybrids and π bonds are formed by unhybridized p orbitals. Nonbonding e⁻ pairs can reside in sp hybrids or p atomic orbitals.

(e)   Recall that electrons in 2s orbitals are on the average closer to the nucleus than electrons in 2p orbitals. Since sp hybrids have greater s orbital character, it is resonable to expect the radial extension of sp orbitals to be smaller than that of sp^2 or sp^3 orbitals and σ bonds formed by sp orbitals to be slightly shorter than those formed by other hybrid orbitals, assuming the same bonded atoms.

         There are no solitary σ bonds in N_3^-. That is, the two σ bonds in N_3^- are each accompanied by at least one π bond between the bonding pair of atoms. Sigma bonds that are part of a double or triple bond must be shorter so that the p orbitals can overlap enough for the π bond to form. Thus, the observation is not applicable to this molecule. (Comparison of C-H bond lengths in C_2H_2, C_2H_4, C_2H_6 and related molecules would confirm or deny the observation.)

9.65   (a) XeF_6 50 e⁻, 25 e⁻ pairs

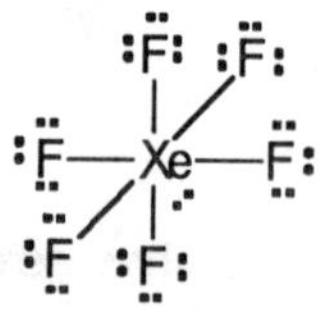

(b)   There are 7 VSEPR e⁻ pairs around Xe, and the maximum number of e⁻ pairs in Table 9.3 is 6.

(c)   Tie 7 balloons together and see what arrangement they adopt (seriously! see Figure 9.3). Alternatively, go to the chemical literature where VSEPR was first proposed and see if there is a preferred orientation for 7 e⁻ pairs.

(d)   Since the hybrid orbitals for 5 VSEPR e⁻ pairs involve one d orbital and for 6 pairs, two d orbitals, a reasonable suggestion would be sp^3d^3.

(e)   One of the 7 VSEPR pairs is a nonbonded pair. The question is whether it occupies an axial or equatorial position. The equatorial plane of a pentagonal bipyramid has F-Xe-F angles of 72°. Placing the nonbonded pair in the equatorial plane would create severe repulsions between it and the adjacent bonded pairs. Thus, the nonbonded pair will reside in the axial position. The molecular structure is a pentagonal pyramid.

9.66   (a)

         To accommodate the π bonding by all 3 O atoms indicated in the resonance structures above, all O atoms are sp^2 hybridized.

(b)   For the first resonance structure, both sigma bonds are formed by overlap of sp^2 hybrid orbitals, the π bond is formed by overlap of atomic p orbitals, one of the nonbonded pairs on the right terminal O atom is in a p atomic orbital, and the remaining 5 nonbonded pairs are in sp^2 hybrid orbirals.

(c)   Only unhybridized p atomic orbitals can be used to form a delocalized π system.

(d)   The unhybridized p orbital on each O atom is used to form the delocalized π system, and in both resonance structures one nonbonded electron pair resides in a p atomic orbital. The delocalized π system then contains 4 electrons, 2 from the π bond and 2 from the nonbonded pair in the p orbital.

9.67   (a)   Each C atom is surrounded by 3 VSEPR electron pairs (2 single and 1 double bonds), so bond angles at each C atom will be approximately 120°.

       Since there is free rotation around the central C-C single bond, other conformations are possible.

(b)   According to Table 8.4, the average C-C length is 1.54 Å, and the average C=C length is 1.34 Å. While the C=C bonds in butadiene appear "normal," the central C-C is significantly shorter than average. Examination of the bonding in butadiene reveals that each C atom is sp^2 hybridized and the π bonds are formed by the remaining unhybridized 2p orbital on each atom. If the central C-C bond is rotated so that all four C atoms are coplanar, the four 2p orbitals are parallel and some delocalization of the π electrons occurs.

9.68   (a) $C_5H_5^-$, 26 valence e^-, 13 e^- pairs

(b)   There are four **other** resonance structures for $C_5H_5^-$, each with the nonbonding electron pair on a different C atom and the double bonds rotated appropriately.

(c)   Each C is surrounded by 3 VSEPR electron pairs, so the hybridization at each C atom is sp^2. Since the ion is planar, the hybridization at the C atom with the nonbonding pair cannot be sp^3. The nonbonding pair must occupy an unhybridized 2p orbital.

(d)   Since the ion is planar, the unhybridized 2p orbitals on the 5 C atoms are parallel and the π bonding is delocalized. The delocalized π cloud adapts the shape of the σ bonding framework and has lobes above and below the plane of the atoms.

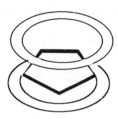

9.69 Ne_2, 16 e⁻ Ne_2^+, 15 e⁻

⇅	σ_{2p}^*	↑	σ_{2p}^*
⇅ ⇅	π_{2p}^*	⇅ ⇅	π_{2p}^*
⇅	σ_{2p}	⇅	σ_{2p}
⇅ ⇅	π_{2p}	⇅ ⇅	π_{2p}
⇅	σ_{2s}^*	⇅	σ_{2s}^*
⇅	σ_{2s}	⇅	σ_{2s}

NO = 1/2 (8 - 8) = 0   BO = 1/2 (8 - 7) = 0.5

Ne_2 has a bond order of zero and is not energetically favored over isolated Ne atoms; it should not exist. Ne_2^+ has a bond order of 0.5 and is slightly lower in energy than isolated Ne atoms, so it will probably exist under special experimental conditions, but be unstable.

9.70   The two possible orbital energy level diagrams are:

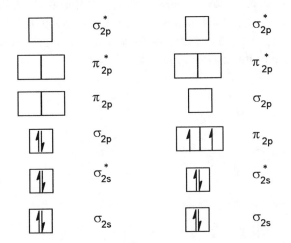

If the σ_{2p} molecular orbital is lower in energy than the π_{2p} orbitals, there are no unpaired electrons, and the molecule is diamagnetic. Switching the order of σ_{2p} and π_{2p} gives 1 unpaired electron in each degenerate π_{2p} orbital and the observed paramagnetism.

9.71   (a)   11 valence electrons, $\sigma_{2s}^{2}\ \sigma_{2s}^{*\,2}\ \pi_{2p}^{4}\ \sigma_{2p}^{2}\ \pi_{2p}^{*\,1}$ (b)   paramagnetic

       (c)   The bond order of NO is $1/2\,(8-3) = 2.5$. The electron that is lost is in an antibonding molecular orbital, so the bond order in NO^{+} is 3.0. The increase in bond order is the driving force for the formation of NO^{+}.

       (d)   To form NO^{-}, an electron is added to an antibonding orbital, and the new bond order is $1/2\,(8-4) = 2$. The order of increasing bond order and bond strength is $NO^{-} < NO < NO^{+}$.

       (e) NO^{+} is isoelectronic with N_{2}, and NO^{-} is isoelectronic with O_{2}.

Integrative Exercises

9.72   (a) PX_{3}, 26 valence e^{-}, 13 e^{-} pairs

$$:\!\overset{\displaystyle ..}{\underset{\displaystyle ..}{X}}\!-\!\overset{\displaystyle ..}{P}\!-\!\overset{\displaystyle ..}{\underset{\displaystyle ..}{X}}\!:$$
$$\underset{\displaystyle ..}{:\!X\!:}$$

                                4 VSEPR e^{-} pairs around P
                                tetrahedral e^{-} pair geometry
                                $109°$ bond angles

       (b)   As electronegativity increases (I < Br < Cl < F), the X-P-X angles decreases.

       (c)   For P-I, ΔEN is $(2.5- 2.1) = 0.4$ and the bond dipole is small. For P-F, ΔEN is $(4.0 - 2.1) = 1.9$ and the bond dipole is large. The greater the ΔEN and bond dipole, the larger the magnitude of negative charge centered on X. The more negative charge centered on X, the greater the repulsion between X and the nonbonding electron pair on P and the smaller the bond angle. (Since electrostatic attractions and repulsions vary as $1/r$, the shorter P-F bond may also contribute to this effect.)

(d) $PBrCl_4$, 40 valence electrons, 20 e⁻ pairs. The molecule will have trigonal bipyramidal electron pair geometry (similar to PCl_5 in Table 9.3.) Based on the argument in part (c), the P-Br bond will have smaller repulsions with P-Cl bonds than P-Cl bonds have with each other. Therefore, the Br will occupy an axial position in the trigonal bipyramid, so that the more unfavorable P-Cl to P-Cl repulsions can be situated in the equatorial plane.

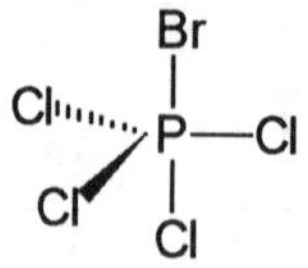

9.73   (a)   Assume 100 g of compound

$$2.1 \text{ g H} \times \frac{1 \text{ mol H}}{1.008 \text{ g H}} = 2.1 \text{ mol H}; \ 2.1 / 2.1 = 1$$

$$29.8 \text{ g N} \times \frac{1 \text{ mol N}}{14.01 \text{ g N}} = 2.13 \text{ mol N}; \ 2.13 / 2.1 \approx 1$$

$$68.1 \text{ g O} \times \frac{1 \text{ mol O}}{16.00 \text{ g O}} = 4.26 \text{ mol O}; \ 4.26 / 2.1 \approx 2$$

The empirical formula is HNO_2; formula weight = 47. Since the approximate molecular weight is 50, the **molecular formula is HNO_2.**

(b)   Assume N is central, since it is unusual for O to be central, and part (d) indicates as much. HNO_2: 18 valence e⁻

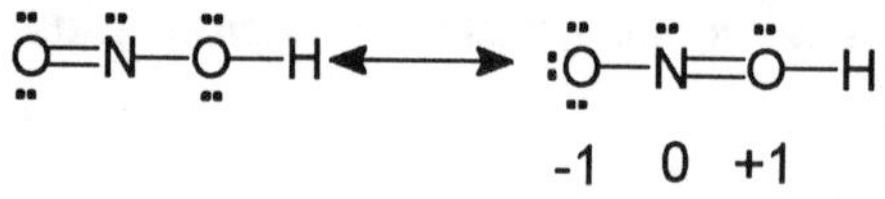

The second resonance form is a minor contributor due to unfavorable formal charges.

(c)   The electron pair geometry around N is trigonal planar; if the resonance structure on the right makes a significant contribution to the molecular structure, all 4 atoms would lie in a plane. If only the left structure contributes, the H could rotate in and out of the molecular plane. The contributions of the two resonance structures could be determined by measuring the O-N-O and N-O-H bond angles.

(d)   3 VSEPR e⁻ pairs around N, sp^2 hybridization

(e)   3 σ, 1 π for both structures (or for H bound to N).

9.74   (a)   3 VSEPR e⁻ pairs around each central C atom, sp^2 hybridization.

(b)   A 180° rotation around the C=C is required to convert the *trans* isomer into the *cis* isomer. A 90° rotation around the C=C double bond eliminates all overlap of the p orbitals that form the π bond and the π bond is broken.

(c)   **average bond enthalpy**

C=C   614 kJ/mol

C-C   348 kJ/mol

The difference in these values, 266 kJ/mol, is the average bond enthalpy of a C-C π bond. This is the amount of energy required to break 1 mol of C-C π bonds. The energy per molecule is

$$266 \text{ kJ/mol} \times \frac{1000 \text{ J}}{1 \text{ kJ}} \times \frac{1 \text{ mol}}{6.022 \times 10^{23} \text{ molecules}} = 4.417 \times 10^{-19}$$

$$= 4.42 \times 10^{-19} \text{ J/molecule}$$

(d) $\lambda = hc/E = \dfrac{6.626 \times 10^{-34} \text{ J} \bullet \text{s} \times 2.998 \times 10^{8} \text{ m/s}}{4.417 \times 10^{-19} \text{ J}} = 4.50 \times 10^{-7} \text{ m} = 450 \text{ nm}$

(e)   Yes, 450 nm light is in the visible portion of the spectrum. A cis-trans isomerization in the retinal portion of the large molecule rhodopsin is the first step in a sequence of molecular transformations in the eye that leads to vision. The sequence of events enables the eye to detect visible photons, in other words, to see.

9.75   (a)   C ≡ C   839 kJ/mol   (1 σ, 2 π)

                C=C   614 kJ/mol   (1 σ, 1 π)

                C-C   348 kJ/mol   (1 σ)

The contribution from 1 π bond would be (614-348) 266 kJ/mol. From a second π bond, (839 - 614), 225 kJ/mol. An average π bond contribution would be (266 + 225)/2 = 246 kJ/mol.

This is $\dfrac{246 \text{ kJ/π bond}}{348 \text{ kJ/σ bond}} \times 100 = 71\%$ of the average enthalpy of a σ bond.

(b)   N ≡ N   941 kJ/mol

            N=N   418 kJ/mol

            N-N   163 kJ/mol

first π = (418 - 163) = 255 kJ/mol

second π = (941 - 418) = 523 kJ/mol

average π bond enthalpy = (255 + 523) / 2 = 389 kJ/mol

This is $\dfrac{389 \text{ kJ/π bond}}{163 \text{ kJ/σ bond}} \times 100 = 240\%$ of the average enthalpy of a σ bond.

N-N σ bonds are weaker than C-C σ bonds, while N-N π bonds are stronger than C-C π bonds. The relative energies of C-C σ and π bonds are similar, while N-N π bonds are much stronger than N-N σ bonds.

(c) N_2H_4, 14 valence e$^-$, 7 e$^-$ pairs N_2H_2, 12 valence e$^-$, 6 e$^-$ pairs

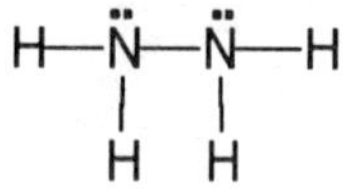

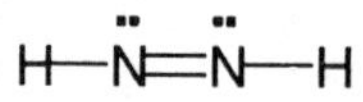

         4 VSEPR e$^-$ pairs around N,   3 VSEPR e$^-$ pairs around N,
         sp^3 hybridization   sp^2 hybridization

         N_2 , 10 valence e$^-$, 5 e$^-$ pairs

            : N≡N :

         2 VSEPR e$^-$ pairs around N,
         sp hybridization

(d)   In the three types of N-N bonds, each N atom has a lone pair of electrons. The lone pair to bond pair repulsions are minimized going from 109° to 120° to 180° bond angles, making the π bonds stronger relative to the σ bond. In the three types of C-C bonds, no lone-pair to bond-pair repulsions exist, and the σ and π bonds have more similar energies.

9.76

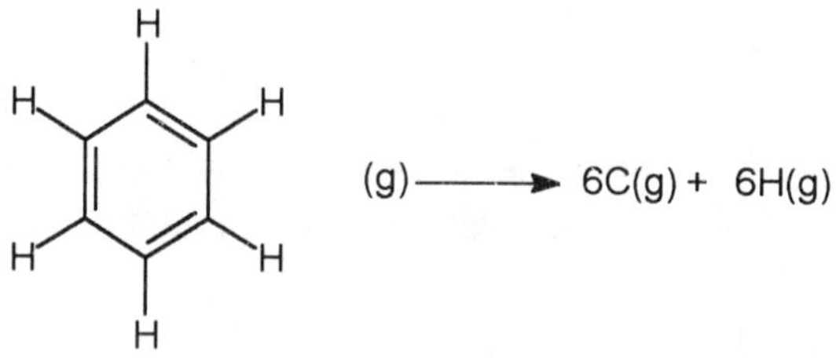

                              (g) $\longrightarrow$ 6C(g) + 6H(g)

ΔH = 6D(C-H) + 3D(C-C) + 3D(C=C) - 0   (The products are isolated atoms;
     = 6(413 kJ) + 3(348 kJ) + 3(614 kJ)   there is no bond making.)
     = 5364 kJ

According to Hess's Law:

$\Delta H°$ = 6$\Delta H_f°$ C(g) + 6$\Delta H_f°$ H(g) - $\Delta H_f°$ C_6H_6(g)
     = 6(718.4 kJ) + 6(217.94 kJ) - (+82.9 kJ)
     = 5535 kJ

The difference in the two results, 171 kJ/mol C_6H_6 is due to the resonance stabilization in benzene. That is, because the π electrons are delocalized, the molecule has a lower overall energy than that predicted for the presence of 3 localized C-C and C=C bonds. Thus, the amount of energy actually required to decompose 1 mole of C_6H_6(g), represented by the Hess's Law calculation, is greater than the sum of the localized bond energies (not taking resonance into account) from the first calculation above.

9.77   (a)   1 eV = 96.485 kJ/mol

H_2: $15.4 \text{ eV} \times \dfrac{96.485 \text{ kJ/mol}}{1 \text{ eV}} = 1486 = 1.49 \times 10^3 \text{ kJ}$

N_2: $15.6 \text{ eV} \times \dfrac{96.485 \text{ kJ/mol}}{1 \text{ eV}} = 1505 = 1.51 \times 10^3 \text{ kJ}$

O_2: $12.1 \text{ eV} \times \dfrac{96.485 \text{ kJ/mol}}{1 \text{ eV}} = 1167 = 1.17 \times 10^3 \text{ kJ}$

F_2: $15.7 \text{ eV} \times \dfrac{96.485 \text{ kJ/mol}}{1 \text{ eV}} = 1515 = 1.52 \times 10^3 \text{ kJ}$

   (b)

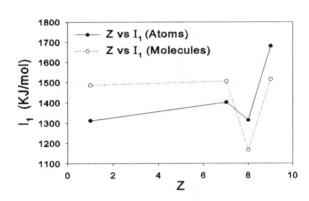

   (c)   In general, I_1 for atoms and molecules increases going across a row of the periodic chart. In both cases, there is a discontinuity at oxygen. The details of the trends are different. The deviation at O is larger for the molecules than the atoms, while the increase at F is much greater for the atoms than the molecules.

   (d)   According to Figures 9.33 and 9.39, H_2 and F_2 are diamagnetic and O_2 is para-magnetic. That is, ionization in H_2, N_2 and F_2 has to overcome spin-pairing energy, while ionization of O_2 removes an already unpaired electron. Thus, the ionization energy of O_2 is much less than I_1 for H_2, N_2 and F_2.

                Despite differences in bond order, bond length and the bonding or antibonding nature of the HOMO in H_2, N_2 and F_2, the ionization energies for these molecules are very similar.

9.78   (a) $3d_{z^2}$

   (b)   Ignoring the donut of the d_{z^2} orbital

(c)   Sc - [Ar] $4s^2 3d^1$ Omitting the core electrons, there are 6 e$^-$ in the energy level diagram.

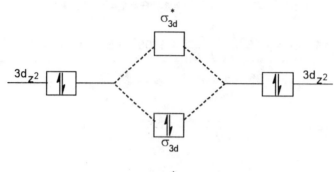

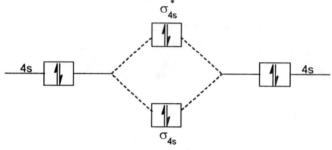

(d)   The bond order in Sc_2 is $1/2 (4 - 2) = 1.0$.

10 Gases

Gas Characteristics; Pressure

10.1   In the gas phase molecules are far apart, while in the liquid they are touching.

    (a)   A gas is much less dense than a liquid because most of the volume of a gas is empty space.

    (b)   A gas is much more compressible because of the distance between molecules.

    (c)   Gaseous molecules are so far apart that there is no barrier to mixing, regardless of the identity of the molecule. All mixtures of gases are homogeneous. Liquid molecules are touching. In order to mix, they must displace one another. Similar molecules displace each other and form homogeneous mixtures. Very dissimilar molecules form heterogeneous mixtures.

10.2   (a)   Because gas molecules are far apart and in constant motion, the gas expands to fill the container. Attractive forces hold liquid molecules together and the volume of the liquid does not change.

    (b) H_2O and CCl_4 molecules are too dissimilar to displace each other and mix in the liquid state. All mixtures of gases are homogeneous. (See Exercise 10.1 (c)).

    (c)   Because gas molecules are far apart, the mass present in 1 mL of a gas is very small. The mass of a gas present in 1 L is on the same order of magnitude as the mass of a liquid present in 1 mL.

10.3   (a) $F = m \times a$. Since both people have the same mass and both experience the acceleration of gravity, the forces they exert on the floor are exactly equal.

    (b) $P = F / A$. The two forces are equal, but the person standing on one foot exerts this force over a smaller area. Thus, the person standing on one foot exerts a greater pressure on the floor.

10.4   The height of the mercury column in a barometer is a measure of atmospheric pressure. The Hg column is steady when the pressure exerted by the atmosphere is equal to the downward pressure due to the mass of the Hg column. Pressure is defined as force **per unit area**. Thus, the larger the diameter of the Hg column, the greater the mass of Hg and the greater the **total force** exerted by the Hg column. However, this force is distributed over a larger cross-sectional area, so the ratio of force/area, the pressure, is the same as that of a smaller diameter column. The same length Hg column is required to balance atmospheric pressure, regardless of the diameter of the column.

Mathematically: $P = \dfrac{F}{A}$; $F = m \times a$; $m = d \times V$ (density)(volume); $V = A \times$ height

Substituting $\quad P = \dfrac{d \times V \times a}{A} = \dfrac{d \times A \times \text{height} \times a}{A}$

$P = d \times \text{height} \times a$

The area (A) terms cancel, and pressure is independent of cross-sectional area of the tube.

10.5   (a) $P_{Hg} = P_{H_2O}$; Using the relationship derived in 10.4: $(d \times h \times a)_{H_2O} = (d \times h \times a)_{Hg}$

Since a, the acceleration due to gravity, is equal in both liquids,

$(d \times h)_{H_2O} = (d \times h)_{Hg}$

$1.00 \text{ g/mL} \times h_{H_2O} = 13.6 \text{ g/mL} \times 760 \text{ mm}$

$h_{H_2O} = \dfrac{13.6 \text{ g/mL} \times 760 \text{ mm}}{1.00 \text{ g/mL}} = 1.034 \times 10^4 = 1.03 \times 10^4 \text{ mm} = 10.3 \text{ m}$

      (b)   Pressure due to H_2O:

$1 \text{ atm} = 1.034 \times 10^4 \text{ mm } H_2O$ (from part (a))

$28 \text{ ft } H_2O \times \dfrac{12 \text{ in}}{1 \text{ ft}} \times \dfrac{2.54 \text{ cm}}{1 \text{ in}} \times \dfrac{10 \text{ mm}}{1 \text{ cm}} \times \dfrac{1 \text{ atm}}{1.034 \times 10^4 \text{ mm}} = 0.826 = 0.83 \text{ atm}$

$P_{total} = P_{atm} + P_{H_2O} = 0.98 \text{ atm} + 0.826 \text{ atm} = 1.81 \text{ atm}$

10.6   Using the relationship derived in Exercise 10.5 for two liquids under the influence of gravity, $(d \times h)_{dbp} = (d \times h)_{Hg}$. At 760 torr, the height of a Hg barometer is 760 mm.

$\dfrac{1.05 \text{ g}}{1 \text{ mL}} \times h_{dbp} = \dfrac{13.6 \text{ g}}{1 \text{ mL}} \times 760 \text{ mm}; \; h_{dbp} = \dfrac{13.6 \text{ g/mL} \times 760 \text{ mm}}{1.05 \text{ g/mL}} = 9.84 \times 10^3 \text{ mm} = 9.84 \text{ m}$

10.7   (a) $0.860 \text{ atm} \times \dfrac{101.325 \text{ kPa}}{1 \text{ atm}} = 87.1 \text{ kPa}$

      (b) $457 \text{ torr} \times \dfrac{1 \text{ atm}}{760 \text{ torr}} = 0.601 \text{ atm}$

      (c) $802 \text{ mm Hg} \times \dfrac{1 \text{ atm}}{760 \text{ mm Hg}} = 1.06 \text{ atm}$

      (d) $0.897 \text{ atm} \times \dfrac{760 \text{ torr}}{1 \text{ atm}} = 682 \text{ torr}$

10.8   (a) $96{,}300 \text{ Pa} \times \dfrac{1 \text{ atm}}{101{,}325 \text{ Pa}} = 0.950 \text{ atm (or 0.095041 atm)}$

                           (96,300 could have 3 or 5 sig figs)

(b) $\quad 1.22 \text{ atm} \times \dfrac{760 \text{ mm Hg}}{1 \text{ atm}} = 927 \text{ mm Hg}$

(c) $\quad 589 \text{ mm Hg} \times \dfrac{1 \text{ torr}}{1 \text{ mm Hg}} = 589 \text{ torr}$

(d) $\quad 8.50 \times 10^{-3} \text{ torr} \times \dfrac{1 \text{ atm}}{760 \text{ torr}} = 1.12 \times 10^{-5} \text{ atm}$

10.9 (a) $\quad 30.45 \text{ in Hg} \times \dfrac{25.4 \text{ mm}}{1 \text{ in}} \times \dfrac{1 \text{ torr}}{1 \text{ mm Hg}} = 773.4 \text{ torr}$

[The result has **4 sig figs** because 25.4 mm/in is considered to be an exact number. (section 1.5)]

(b) The pressure in Chicago is greater than **standard atmospheric pressure**, 760 torr, so it makes sense to classify this weather system as a "high pressure system."

10.10 (a) $\quad \dfrac{1.6 \times 10^5 \text{ Pa}}{1 \text{ Titan atm}} \times \dfrac{1 \text{ Earth atm}}{101{,}325 \text{ Pa}} = \dfrac{1.6 \text{ Earth atm}}{1 \text{ Titan atm}}$

(b) $\quad \dfrac{90 \text{ Earth atm}}{1 \text{ Venus atm}} \times \dfrac{101.3 \text{ kPa}}{1 \text{ Earth atm}} = \dfrac{9.1 \times 10^3 \text{ kPa}}{1 \text{ Venus atm}}$

10.11 $\quad P = \dfrac{m \times a}{A} = \dfrac{135 \text{ lb}}{0.50 \text{ in}^2} \times \dfrac{9.81 \text{ m}}{1 \text{ s}^2} \times \dfrac{0.454 \text{ kg}}{1 \text{ lb}} \times \dfrac{39.4^2 \text{ in}^2}{1 \text{ m}^2} = 1.9 \times 10^3 \text{ kPa}$

10.12 $\quad 1 \text{ Pa} = \dfrac{1 \text{ N}}{\text{m}^2} = \dfrac{1 \text{ kg} \cdot \text{m}}{\text{s}^2} \times \dfrac{1}{\text{m}^2} = \dfrac{1 \text{ kg}}{\text{m} \cdot \text{s}^2}$ Change mass to kg and area to m².

mass: $\dfrac{2.70 \text{ g}}{\text{cm}^3} \times (25.0 \text{ cm})^3 \times \dfrac{1 \text{ kg}}{1000 \text{ g}} = 42.19 = 42.2 \text{ kg}$

area: $(25.0 \text{ cm})^2 \times \dfrac{1 \text{ m}^2}{(100 \text{ cm})^2} = 0.0625 \text{ m}^2$

$P = \dfrac{F}{A} = \dfrac{m \times a}{A} = \dfrac{42.19 \text{ kg}}{0.0625 \text{ m}^2} \times \dfrac{9.81 \text{ m}}{1 \text{ s}^2} = 6.62 \times 10^3 \text{ Pa}$

10.13 (a) It **is not** necessary to know atmospheric pressure to use a closed-end manometer. The gas on one side of the Hg is working against a vacuum on the other side of the Hg. The pressure of the gas is just the difference in the heights of the two arms.

It **is** necessary to know atmospheric pressure to use an open-end manometer. In an open-end manometer the pressure of the sample is balanced against atmospheric pressure. If the Hg level is higher in the open end than in the end exposed to the sample, the pressure of the sample is atmospheric pressure plus the difference in heights of the two sides. If the Hg level is lower in the open end than in the other end, the pressure of the sample is atmospheric pressure minus the difference in heights of the two sides.

(b)   (i)   The Hg level is lower in the open end than the closed end, so the gas pressure is less than atmospheric pressure.

$$P_{gas} = 0.975 \text{ atm} - \left(52 \text{ cm} \times \frac{1 \text{ atm}}{76 \text{ cm}} \right) = 0.29 \text{ atm}$$

        (ii)   The Hg level is higher in the open end, so the gas pressure is greater than atmospheric pressure.

$$P_{gas} = 0.975 \text{ atm} + \left(67 \text{ mm Hg} \times \frac{1 \text{ atm}}{760 \text{ mm Hg}} \right) = 1.063 \text{ atm}$$

        (iii)   This is a closed-end manometer so $P_{gas} = h$.

$$P_{gas} = 10.3 \text{ cm} \times \frac{1 \text{ atm}}{76 \text{ cm}} = 0.136 \text{ atm}$$

10.14   (a)   The gas is exerting 32 mm Hg (torr) more pressure than the atmosphere.

$$P_{gas} = P_{atm} + 32 \text{ torr} = \left(0.967 \text{ atm} \times \frac{760 \text{ torr}}{1 \text{ atm}} \right) + 32 \text{ torr} = 767 \text{ torr}$$

      (b)   The atmosphere is exerting 68 mm Hg (torr) more pressure than the gas.

$$P_{gas} = P_{atm} - 68 \text{ torr} = \left(1.05 \text{ atm} \times \frac{760 \text{ torr}}{1 \text{ atm}} \right) - 68 \text{ torr} = 730 \text{ torr}$$

The Gas Laws; The Ideal-Gas Equation

(In *Solutions to Exercises*, the symbol for molar mass is MM.)

10.15   (a)   If two quantities, X and Y, are **directly proportional**, a change in X causes a proportional change in Y in the **same** direction. Mathematically, if X and Y are directly proportional and C is a constant, Y = CX or Y/X = C.

               If X and Y are **inversely proportional**, a change in X causes a change in Y in the **opposite** direction. Mathematically X • Y = C or Y = C/X.

      (b)   Volume (V) is inversely proportional to pressure (P); volume is directly proportional to absolute temperature (T); volume is directly proportional to quantity of gas (n).

10.16   In the following equations, C indicates a constant, which can have a different value in each relationship.

      (a)   PV = C or $P \propto 1/V$ when n and T are constant

      (b) $V \propto n$ or V = Cn when T and P are constant

      (c) $P \propto T$ or P = CT when n and V are constant

      (d) $PV \propto T$ or PV = CT when n is constant

10.17 $P_1V_1 = P_2V_2$; the proportionality holds true for any pressure or volume units.

$P_1 = 737$ torr, $V_1 = 20.5$ L

(a) $P_2 = 1.80$ atm $\times \dfrac{760 \text{ torr}}{1 \text{ atm}} = 1.368 \times 10^3 = 1.37 \times 10^3$ torr

$V_2 = \dfrac{P_1V_1}{P_2} = \dfrac{737 \text{ torr} \times 20.5 \text{ L}}{1.368 \times 10^3 \text{ torr}} = 11.0$ L

(b) $V_2 = 16.0$ L; $P_2 = \dfrac{P_1V_1}{V_2} = \dfrac{737 \text{ torr} \times 20.5 \text{ L}}{16.0 \text{ L}} = 944$ torr

10.18 $\dfrac{V_1}{T_1} = \dfrac{V_2}{T_2}$; T must be in Kelvins for the relationship to be true.

$V_1 = 8.50$ L, $T_1 = 29.0°C = 302$ K

(a) $T_2 = 125°C = 398$ K; $V_2 = \dfrac{V_1T_2}{T_1} = \dfrac{8.50 \text{ L} \times 398 \text{ K}}{302 \text{ K}} = 11.2$ L

(b) $V_2 = 5.00$ L; $T_2 = \dfrac{T_1V_2}{V_1} = \dfrac{302 \text{ K} \times 5.00 \text{ L}}{8.50 \text{ L}} = 178$ K $= -95°C$

10.19   (a)   Avogadro's hypothesis states that equal volumes of gases at the same temperature and pressure contain equal numbers of molecules. Since molecules react in the ratios of small whole numbers, it follows that the volumes of reacting gases (at the same temperature and pressure) are in the ratios of small whole numbers.

(b)   Since the two gases are at the same temperature and pressure, the ratio of the numbers of atoms is the same as the ratio of volumes. There are 1.5 times as many Xe atoms as Ne atoms.

10.20   According to Avogadro's hypothesis, the mole ratios in the chemical equation will be volume ratios for the gases if they are at the same temperature and pressure.

$$N_2(g) + 3H_2(g) \rightarrow 2NH_3(g)$$

The volumes of H_2 and N_2 are in a stoichiometric $\dfrac{2.1 \text{ L}}{0.70 \text{ L}}$ or $\dfrac{3 \text{ vol } H_2}{1 \text{ vol } N_2}$ ratio, so either can be used to determine the volume of $NH_3(g)$ produced.

0.70 L $N_2 \times \dfrac{2 \text{ mol } NH_3}{1 \text{ mol } N_2} = 1.4$ L $NH_3(g)$ produced.

10.21   (a)   An ideal gas exhibits pressure, volume and temperature relationships which are described by the equation PV = nRT. (An ideal gas obeys the Ideal-Gas Equation.)

(b)   PV = nRT;   P in atmospheres
                               V in liters
                               n in moles
                               T in kelvins

10.22 (a) STP stands for standard temperature, 0°C (or 273 K), and standard pressure, 1 atm.

(b) $V = \dfrac{nRT}{P}$; $V = 1 \text{ mol} \times \dfrac{0.08206 \text{ L} \cdot \text{atm}}{\text{K} \cdot \text{mol}} \times \dfrac{273 \text{ K}}{1 \text{ atm}}$

$V = 22.4$ L for 1 mole of gas at STP

(c) $25°C + 273 = 298$ K

$V = \dfrac{nRT}{P}$; $V = 1 \text{ mol} \times \dfrac{0.08206 \text{ L} \cdot \text{atm}}{\text{K} \cdot \text{mol}} \times \dfrac{298 \text{ K}}{1 \text{ atm}}$

$V \doteq 24.5$ L for 1 mol of gas at 1 atm and 25°C

10.23 (a) $n = 8.25 \times 10^{-2}$ mol, $V = 174$ mL, $T = -15°C = 258$ K

$P = \dfrac{nRT}{V} = 0.0825 \text{ mol} \times \dfrac{0.08206 \text{ L} \cdot \text{atm}}{\text{K} \cdot \text{mol}} \times \dfrac{258 \text{ K}}{0.174 \text{ L}} = 10.0$ atm

(b) $V = 6.38$ L, $T = 35°C = 308$ K, $P = 955 \text{ torr} \times \dfrac{1 \text{ atm}}{760 \text{ torr}} = 1.257 = 1.26$ atm

$n = \dfrac{PV}{RT} = 1.257 \text{ atm} \times \dfrac{\text{K} \cdot \text{mol}}{0.08206 \text{ L} \cdot \text{atm}} \times \dfrac{6.38 \text{ L}}{308 \text{ K}} = 0.317$ mol

(c) $n = 2.95$ mol, $P = 0.76$ atm, $T = 52°C = 325$ K

$V = \dfrac{nRT}{P} = 2.95 \text{ mol} \times \dfrac{0.08206 \text{ L} \cdot \text{atm}}{\text{K} \cdot \text{mol}} \times \dfrac{325 \text{ K}}{0.76 \text{ atm}} = 1.0 \times 10^2$ L

(d) $n = 9.87 \times 10^{-2}$ mol, $V = 164$ mL $= 0.164$ L

$P = 682 \text{ torr} \times \dfrac{1 \text{ atm}}{760 \text{ torr}} = 0.8974 = 0.897$ atm

$T = \dfrac{PV}{nR} = 0.8974 \text{ atm} \times \dfrac{0.164 \text{ L}}{0.0987 \text{ mol}} \times \dfrac{1 \text{ K} \cdot \text{mol}}{0.08206 \text{ L} \cdot \text{atm}} = 18.2$ K

10.24 (a) $n = 1.34$ mol, $V = 3.28$ L, $T = 25°C = 298$ K

$P = \dfrac{nRT}{V} = 1.34 \text{ mol} \times \dfrac{0.08206 \text{ L} \cdot \text{atm}}{\text{K} \cdot \text{atm}} \times \dfrac{298 \text{ K}}{3.28 \text{ L}} = 9.99$ atm

(b) $n = 6.72 \times 10^{-3}$ mol, $T = 145°C = 418$ K

$P = 59.0 \text{ torr} \times \dfrac{1 \text{ atm}}{760 \text{ torr}} = 0.07763 = 0.0776$ atm

$V = \dfrac{nRT}{P} = 6.72 \times 10^{-3} \text{ mol} \times \dfrac{0.08206 \text{ L} \cdot \text{atm}}{\text{K} \cdot \text{mol}} \times \dfrac{418 \text{ K}}{0.07763 \text{ atm}} = 2.97$ L

(c) $V = 2.50$ L, $T = 37°C = 310$ K, $P = 725 \text{ torr} \times \dfrac{1 \text{ atm}}{760 \text{ torr}} = 0.9539 = 0.954$ atm

$n = \dfrac{PV}{RT} = 0.9539 \text{ atm} \times \dfrac{\text{K} \cdot \text{mol}}{0.08206 \text{ L} \cdot \text{atm}} \times \dfrac{2.50 \text{ L}}{310 \text{ K}} = 9.38 \times 10^{-2}$ mol

(d)   n = 0.270 mol, V = 15.0 L, P = 1.88 atm

$$T = \frac{PV}{nR} = \frac{1.88 \text{ atm} \times 15.0 \text{ L}}{0.270 \text{ mol}} \times \frac{K \cdot mol}{0.08206 \text{ L} \cdot atm} = 1.27 \times 10^3 \text{ K}$$

10.25   V = 0.325 L, T = 28°C = 301 K, n = 3.0 g $C_3H_8 \times \dfrac{1 \text{ mol } C_3H_8}{44.1 \text{ g } C_3H_8}$ = 0.0680 = 0.068 mol

$$P = \frac{nRT}{V} = 0.0680 \text{ mol} \times \frac{0.08206 \text{ L} \cdot atm}{K \cdot mol} \times \frac{301 \text{ K}}{0.325 \text{ L}} = 5.2 \text{ atm}$$

10.26   n = g/MM (MM = molar mass)

$$PV = \frac{g \, RT}{MM}; \quad g = \frac{MM \times PV}{RT}; \quad \text{Change m}^3 \text{ to L, then calculate grams (or kg).}$$

$$2.0 \times 10^5 \text{ m}^3 \times \frac{10^3 \text{ dm}^3}{1 \text{ m}^3} \times \frac{1 \text{ L}}{1 \text{ dm}^3} = 2.0 \times 10^8 \text{ L } H_2$$

$$g = \frac{2.02 \text{ g } H_2}{1 \text{ mol } H_2} \times \frac{K \cdot mol}{0.08206 \text{ L} \cdot atm} \times \frac{1 \text{ atm} \times 2.0 \times 10^8 \text{ L}}{300 \text{ K}} = 1.6 \times 10^7 \text{ g} = 1.6 \times 10^4 \text{ kg } H_2$$

10.27   Air is a mixture of N_2 and O_2, but for the purpose of calculating pressure, only the total number of gas molecules is important, not the identity of these molecules.

$$V = 1.05 \text{ L}, \; T = 37°C = 310 \text{ K}, \; P = 740 \text{ mm Hg} \times \frac{1 \text{ atm}}{760 \text{ mm}} = 0.9737 = 0.974 \text{ atm}$$

$$n = \frac{PV}{RT} = 0.9737 \text{ atm} \times \frac{K \cdot mol}{0.08206 \text{ L} \cdot atm} \times \frac{1.05 \text{ L}}{310 \text{ K}} = 0.04019 = 0.0402 \text{ mol of gas}$$

$$0.04019 \text{ mol} \times \frac{6.022 \times 10^{23} \text{ molecules}}{1 \text{ mol}} = 2.42 \times 10^{22} \text{ gas molecules}$$

10.28   Find the volume of the tube in cm³; 1 cm³ = 1 mL; MM = molar mass.

r = d/2 = 2.0 cm/2 = 1.0 cm; h = 4.0 m = 4.0×10^2 cm

$$V = \pi r^2 h = 3.14159 \times (1.0 \text{ cm})^2 \times (4.0 \times 10^2 \text{ cm}) = 1.257 \times 10^3 \text{ cm}^3 = 1.3 \text{ L}$$

$$PV = \frac{g}{MM} RT; \quad g = \frac{MM \times PV}{RT}; \quad P = 1.5 \text{ torr} \times \frac{1 \text{ atm}}{760 \text{ torr}} = 1.97 \times 10^{-3} = 2.0 \times 10^{-3} \text{ atm}$$

$$g = \frac{20.18 \text{ g Ne}}{1 \text{ mol Ne}} \times \frac{K \cdot mol}{0.08206 \text{ L} \cdot atm} \times \frac{1.97 \times 10^{-3} \text{ atm} \times 1.257 \text{ L}}{308 \text{ K}} = 2.0 \times 10^{-3} \text{ g Ne}$$

10.29   (a) $V_2 = \dfrac{P_1 V_1 T_2}{P_2 T_1} = \dfrac{0.880 \text{ atm} \times 0.600 \text{ L} \times 273 \text{ K}}{0.205 \text{ atm} \times 319 \text{ K}} = 2.20 \text{ L}$

(b) $V_2 = \dfrac{0.880 \text{ atm} \times 0.600 \text{ L} \times 273 \text{ K}}{1.00 \text{ atm} \times 319 \text{ K}} = 0.452 \text{ L}$

10.30   (a) $V_2 = \dfrac{P_1 V_1 T_2}{P_2 T_1} = \dfrac{740 \text{ torr} \times 5.62 \text{ L} \times 380 \text{ K}}{680 \text{ torr} \times 306 \text{ K}} = 7.59$ L

      (b) $V_2 = \dfrac{740 \text{ torr} \times 5.62 \text{ L} \times 273 \text{ K}}{760 \text{ torr} \times 306 \text{ K}} = 4.88$ L

      (c) $T_2 = \dfrac{P_2 V_2 T_1}{P_1 V_1} = \dfrac{800 \text{ torr} \times 3.00 \text{ L} \times 306 \text{ K}}{740 \text{ torr} \times 5.62 \text{ L}} = 177$ K

      (d) $P_2 = \dfrac{P_1 V_1 T_2}{V_2 T_1} = \dfrac{740 \text{ torr} \times 5.62 \text{ L} \times 340 \text{ K}}{5.00 \text{L} \times 306 \text{ K}} = 924$ torr

10.31   (a) $g = \dfrac{MM \times PV}{RT}$ (MM = molar mass)

$$\dfrac{32.0 \text{ g O}_2}{1 \text{ mol O}_2} \times \dfrac{K \bullet mol}{0.08206 \text{ L} \bullet atm} \times 18{,}000 \text{ kPa} \times \dfrac{1 \text{ atm}}{101.3 \text{ kPa}} \times \dfrac{42.0 \text{ L}}{296 \text{ K}}$$

$$= 9.83 \times 10^3 \text{ g O}_2$$

      (b) $V_2 = \dfrac{P_1 V_1 T_2}{T_1 P_2} = \dfrac{18{,}000 \text{ kPa} \times 42.0 \text{ L} \times 273 \text{ K}}{296 \text{ K} \times 101.3 \text{ kPa}} = 6.88 \times 10^3$ L

10.32   (a) $g = \dfrac{MM \times PV}{RT}$ (MM = molar mass); $P = 160 \text{ lb/in}^2 \times \dfrac{1 \text{ atm}}{14.7 \text{ lb/in}^2} = 10.884 = 10.9$ atm

$$g = \dfrac{38.0 \text{ g F}_2}{1 \text{ mol F}_2} \times \dfrac{K \bullet mol}{0.08206 \text{ L} \bullet atm} \times \dfrac{10.884 \text{ atm} \times 30.0 \text{ L}}{299 \text{ K}} = 506 \text{ g F}_2$$

      (b) $V_2 = \dfrac{P_1 V_1 T_2}{T_1 P_2} = \dfrac{160 \text{ lb/in}^2 \times 30.0 \text{ L} \times 273 \text{ K}}{299 \text{ K} \times 14.7 \text{ lb/in}^2} = 298$ L

10.33   (a) $5.2 \text{ g} \times 1 \text{ h} \times \dfrac{0.8 \text{ mL O}_2}{1 \text{ g} \bullet hr} = 4.16 = 4 \text{ mL O}_2$ consumed

$$n = \dfrac{PV}{RT} = 1 \text{ atm} \times \dfrac{K \bullet mol}{0.08206 \text{ L} \bullet atm} \times \dfrac{0.00416 \text{ L}}{297 \text{ K}} = 1.71 \times 10^{-4} = 2 \times 10^{-4} \text{ mol O}_2$$

      (b) $1 \text{ qt air} \times \dfrac{0.946 \text{ L}}{1 \text{ qt}} \times 0.21\% \text{ O}_2 \text{ in air} = 0.199 \text{ L O}_2$ available

$$n = 1 \text{ atm} \times \dfrac{K \bullet mol}{0.08206 \text{ L} \bullet atm} \times \dfrac{0.199 \text{ L}}{297 \text{ K}} = 8.16 \times 10^{-3} = 8 \times 10^{-3} \text{ mol O}_2 \text{ available}$$

$$\text{roach uses } \dfrac{1.71 \times 10^{-4} \text{ mol}}{1 \text{ hr}} \times 48 \text{ hr} = 8.19 \times 10^{-3} = 8 \times 10^{-3} \text{ mol O}_2 \text{ consumed}$$

Not only does the roach use 20% of the available O_2, it needs all the O_2 in the jar.

10.34   (a) $P = \dfrac{gRT}{MM \times V}$ (MM = molar mass); mass $= 1800 \times 10^{-9} \text{ g} = 1.8 \times 10^{-6} \text{ g}$;

$$V = 1 \text{ m}^3 = 1 \times 10^3 \text{ L}$$

$$P = \frac{1.8 \times 10^{-6}\,g\,Hg \times 1\,mol\,Hg}{200.6\,g\,Hg} \times \frac{0.08206\,L \cdot atm}{K \cdot mol} \times \frac{283\,K}{1 \times 10^3\,L} = 2.1 \times 10^{-10}\,atm$$

(b) $\quad \dfrac{1.8 \times 10^{-6}\,g\,Hg}{1\,m^3} \times \dfrac{1\,mol\,Hg}{200.6\,g\,Hg} \times \dfrac{6.022 \times 10^{23}\,Hg\,atoms}{1\,mol\,Hg} = 5.4 \times 10^{15}\,Hg\,atoms/m^3$

(c) $\quad 1600\,km^3 \times \dfrac{1000^3\,m^3}{1\,km^3} \times \dfrac{1.8 \times 10^{-6}\,g\,Hg}{1\,m^3} = 2.9 \times 10^6\,g\,Hg/day$

Further Applications of the Ideal-Gas Equation

(In *Solutions to Exercises*, the symbol for molar mass is MM.)

10.35 (a)   Density of a gas $= \dfrac{g}{L}$; $PV = \dfrac{g\,RT}{MM}$; $\dfrac{g}{V} = \dfrac{MM \times P}{RT} = d$

P = 0.96 atm,   T = 35°C = 308 K,   MM of SO_3 = 80.07 g/mol

$$d = \frac{80.07\,g}{mol} \times \frac{K \cdot mol}{0.08206\,L \cdot atm} \times \frac{0.96\,atm}{308\,K} = 3.0\,g/L$$

(b) $MM = \dfrac{gRT}{PV} = \dfrac{4.40\,g}{3.50\,L} \times \dfrac{0.08206\,L \cdot atm}{K \cdot mol} \times \dfrac{314\,K}{560\,torr} \times \dfrac{760\,torr}{1\,atm} = 44.0\,g/mol$

10.36 (a) $d = \dfrac{MM\,P}{RT}$; $P = \dfrac{650\,torr \times 1\,atm}{760\,torr} = 0.8553 = 0.855\,atm$,   T = 100°C = 373 K

MM of SF_4 = 108.1 g/mol

$$d = \frac{108.1\,g}{1\,mol} \times \frac{K \cdot mol}{0.08206\,L \cdot atm} \times \frac{0.8553\,atm}{373\,K} = 3.02\,g/L$$

(b) $MM = \dfrac{dRT}{P} = \dfrac{3.67\,g}{1\,L} \times \dfrac{0.08206\,L \cdot atm}{K \cdot mol} \times \dfrac{288\,K}{825\,torr} \times \dfrac{760\,torr}{1\,atm} = 79.9\,g/mol$

10.37 $MM = \dfrac{gRT}{PV} = \dfrac{1.012\,g}{0.354\,L} \times \dfrac{0.08206\,L \cdot atm}{K \cdot atm} \times \dfrac{372\,K}{742\,torr} \times \dfrac{760\,torr}{1\,atm} = 89.4\,g/mol$

10.38 $MM = \dfrac{gRT}{PV} = \dfrac{0.846\,g}{0.354\,L} \times \dfrac{0.08206\,L \cdot atm}{K \cdot atm} \times \dfrac{373\,K}{752\,torr} \times \dfrac{760\,torr}{1\,atm} = 73.9\,g/mol$

10.39 $mol\,O_2 = \dfrac{PV}{RT} = 3.5 \times 10^{-6}\,torr \times \dfrac{1\,atm}{760\,torr} \times \dfrac{K \cdot mol}{0.08206\,L \cdot atm} \times \dfrac{0.382\,L}{300\,K} = 7.146 \times 10^{-11}$

$$= 7.1 \times 10^{-11}\,mol\,O_2$$

$$7.146 \times 10^{-11}\,mol\,O_2 \times \frac{2\,mol\,Mg}{1\,mol\,O_2} \times \frac{24.3\,g\,Mg}{1\,mol\,Mg} = 3.5 \times 10^{-9}\,g\,Mg$$

10.40 $n_{H_2} = \dfrac{P_{H_2} V}{RT}$; $P = 740 \text{ torr} \times \dfrac{1 \text{ atm}}{760 \text{ torr}} = 0.9737 = 0.974 \text{ atm}$

$n_{H_2} = 0.9737 \text{ atm} \times \dfrac{K \cdot mol}{0.08206 \text{ L} \cdot atm} \times \dfrac{10.0 \text{ L}}{296 \text{ K}} = 0.4009 = 0.401 \text{ mol } H_2$

$0.4009 \text{ mol } H_2 \times \dfrac{1 \text{ mol } CaH_2}{2 \text{ mol } H_2} \times \dfrac{42.10 \text{ g } CaH_2}{1 \text{ mol } CaH_2} = 8.44 \text{ g } CaH_2$

10.41 $kg\ H_2SO_4 \rightarrow g\ H_2SO_4 \rightarrow mol\ H_2SO_4 \rightarrow mol\ NH_3 \rightarrow V\ NH_3$

$150 \text{ kg} \times \dfrac{1000 \text{ g}}{1 \text{ kg}} = 1.50 \times 10^5 \text{ g } H_2SO_4 \times \dfrac{1 \text{ mol}}{98.08 \text{ g}} = 1.529 \times 10^3 = 1.53 \times 10^3 \text{ mol } H_2SO_4$

$1.529 \times 10^3 \text{ mol } H_2SO_4 \times \dfrac{2 \text{ mol } NH_3}{1 \text{ mol } H_2SO_4} = 3.059 \times 10^3 = 3.06 \times 10^3 \text{ mol } NH_3$

$V_{NH_3} = \dfrac{nRT}{P} = 3.059 \times 10^3 \text{ mol} \times \dfrac{0.08206 \text{ L} \cdot atm}{K \cdot mol} \times \dfrac{293 \text{ K}}{25.0 \text{ atm}} = 2.94 \times 10^3 \text{ L } NH_3$

10.42 $g\ glucose \rightarrow mol\ glucose \rightarrow mol\ CO_2 \rightarrow V\ CO_2$

$5.00 \text{ g} \times \dfrac{1 \text{ mol glucose}}{180.1 \text{ g}} \times \dfrac{6 \text{ mol } CO_2}{1 \text{ mol glucose}} = 0.1666 = 0.167 \text{ mol } CO_2$

$V = \dfrac{nRT}{p} = 0.1666 \text{ mol} \times \dfrac{0.08206 \text{ L} \cdot atm}{K \cdot mol} \times \dfrac{310 \text{ K}}{1.00 \text{ atm}} = 4.24 \text{ L } CO_2$

Partial Pressures

10.43   (a)   *Partial pressure* is the pressure exerted by a single component of a gaseous mixture at the same temperature and volume as the mixture.

       (b)   The total pressure of a mixture of gases is equal to the sum of the individual pressures the gases would exert if present in the same container alone.

10.44   No. Only the total pressure of a gaseous mixture can be measured by a manometer, not the partial pressures of the components. The partial pressure of a gaseous component is directly related to the moles of that gas in the mixture. Thus, the moles of a component are measured stoichiometrically and partial pressure is then calculated from moles.

10.45   (a) $P_{He} = \dfrac{nRT}{V} = 0.538 \text{ mol} \times \dfrac{0.08206 \text{ L} \cdot atm}{K \cdot atm} \times \dfrac{298 \text{ K}}{7.00 \text{ L}} = 1.88 \text{ atm}$

$P_{Ne} = \dfrac{nRT}{V} = 0.315 \text{ mol} \times \dfrac{0.08206 \text{ L} \cdot atm}{K \cdot atm} \times \dfrac{298 \text{ K}}{7.00 \text{ L}} = 1.10 \text{ atm}$

$P_{Ar} = \dfrac{nRT}{V} = 0.103 \text{ mol} \times \dfrac{0.08206 \text{ L} \cdot atm}{K \cdot atm} \times \dfrac{298 \text{ K}}{7.00 \text{ L}} = 0.360 \text{ atm}$

      (b) P_t = 1.88 atm + 1.10 atm + 0.360 atm = 3.34 atm

10.46   (a)   4.00 g CH_4 × $\dfrac{1 \text{ mol } CH_4}{16.04 \text{ g } CH_4}$ = 0.2494 = 0.249 mol CH_4

$$P_{CH_4} = \frac{nRT}{V} = 0.2494 \text{ mol} \times \frac{0.08206 \text{ L} \cdot \text{atm}}{K \cdot \text{mol}} \times \frac{273 \text{ K}}{1.50 \text{ L}} = 3.72 \text{ atm}$$

4.00 g C_2H_4 × $\dfrac{1 \text{ mol } C_2H_4}{28.05 \text{ g } C_2H_4}$ = 0.1426 = 0.143 mol C_2H_4

$$P_{C_2H_4} = \frac{nRT}{V} = 0.1426 \text{ mol } C_2H_4 \times \frac{0.08206 \text{ L} \cdot \text{atm}}{K \cdot \text{mol}} \times \frac{273 \text{ K}}{1.50 \text{ L}} = 2.13 \text{ atm}$$

4.00 g C_4H_{10} × $\dfrac{1 \text{ mol } C_4H_{10}}{58.12 \text{ g } C_4H_{10}}$ = 0.06882 = 0.0688 mol C_4H_{10}

$$P_{C_4H_{10}} = \frac{nRT}{V} = 0.06882 \text{ mol} \times \frac{0.08206 \text{ L} \cdot \text{atm}}{K \cdot \text{mol}} \times \frac{273 \text{ K}}{1.50 \text{ L}} = 1.03 \text{ atm}$$

      (b) P_t = 3.72 atm + 2.13 atm + 1.03 atm = 6.88 atm

10.47   The partial pressure of each component is equal to the mole fraction of that gas times the total pressure of the mixture. Find the mole fraction of each component and then its partial pressure.

n_t = 0.55 mol N_2 + 0.20 mol O_2 + 0.10 mol CO_2 = 0.85 mol

χ_{N_2} = $\dfrac{0.55}{0.85}$ = 0.647 = 0.65; P_{N_2} = 0.647 × 1.32 atm = 0.85 atm

χ_{O_2} = $\dfrac{0.20}{0.85}$ = 0.235 = 0.24; P_{O_2} = 0.235 × 1.32 atm = 0.31 atm

χ_{CO_2} = $\dfrac{0.10}{0.85}$ = 0.118 = 0.12; P_{CO_2} = 0.118 × 1.32 atm = 0.16 atm

10.48 n_{N_2} = 3.50 g N_2 × $\dfrac{1 \text{ mol}}{28.02 \text{ g}}$ = 0.125 mol; n_{H_2} = 2.15 g H_2 × $\dfrac{1 \text{ mol}}{2.016 \text{ g}}$ = 1.07 mol

n_{NH_3} = 5.27 g NH_3 × $\dfrac{1 \text{ mol}}{17.03 \text{ g}}$ = 0.309 mol; n_t = 0.125 + 1.07 + 0.309 = 1.50 mol

P_{N_2} = $\dfrac{n_{N_2}}{n_t} \times P_t$ = $\dfrac{0.125}{1.50}$ × 2.50 atm = 0.208 atm

P_{H_2} = $\dfrac{1.07}{1.50}$ × 2.50 atm = 1.78 atm; P_{NH_3} = $\dfrac{0.309}{1.50}$ × 2.50 atm = 0.515 atm

10.49 χ_{O_2} = $\dfrac{P_{O_2}}{P_t}$ = $\dfrac{0.21 \text{ atm}}{8.38 \text{ atm}}$ = 0.025; mole % = 0.025 × 100 = 2.5%

10.50   (a) $n_{O_2} = 5.00 \text{ g } O_2 \times \dfrac{1 \text{ mol}}{32.00 \text{ g}} = 0.156 \text{ mol};\quad n_{N_2} = 7.50 \text{ g } N_2 \times \dfrac{1 \text{ mol}}{28.02 \text{ g}} = 0.268 \text{ mol}$

$n_{H_2} = 1.00 \text{ g } H_2 \times \dfrac{1 \text{ mol}}{2.016 \text{ g}} = 0.496 \text{ mol};\quad n_t = 0.156 + 0.268 + 0.496 = 0.920 \text{ mol}$

$\chi_{O_2} = \dfrac{n_{O_2}}{n_t} = \dfrac{0.156}{0.920} = 0.170;\quad \chi_{N_2} = \dfrac{n_{N_2}}{n_t} = \dfrac{0.268}{0.920} = 0.291$

$\chi_{H_2} = \dfrac{0.496}{0.920} = 0.539$

(b) $P_{O_2} = n \times \dfrac{RT}{V};\quad P_{O_2} = 0.156 \text{ mol} \times \dfrac{0.08206 \text{ L} \cdot \text{atm}}{K \cdot \text{mol}} \times \dfrac{288 \text{ K}}{10.0 \text{ L}} = 0.369 \text{ atm}$

$P_{N_2} = 0.268 \text{ mol} \times \dfrac{0.08206 \text{ L} \cdot \text{atm}}{K \cdot \text{mol}} \times \dfrac{288 \text{ K}}{10.0 \text{ L}} = 0.633 \text{ atm}$

$P_{H_2} = 0.496 \text{ mol} \times \dfrac{0.08206 \text{ L} \cdot \text{atm}}{K \cdot \text{mol}} \times \dfrac{288 \text{ K}}{10.0 \text{ L}} = 1.17 \text{ atm}$

10.51   (a)   The partial pressure of gas A is **not affected** by the addition of gas C. The partial pressure of A depends only on moles of A, volume of container and conditions; none of these factors change when gas C is added.

(b)   The total pressure in the vessel **increases** when gas C is added, because the total number of moles of gas increases.

(c)   The mole fraction of gas B **decreases** when gas C is added. The moles of gas B stay the same, but the total moles increase, so the mole fraction of B (n_B/n_t) decreases.

10.52   (a) $6.00 \text{ g } SO_2 \times \dfrac{1 \text{ mol } SO_2}{64.06 \text{ g } SO_2} = 0.0937 \text{ mol } SO_2;\quad \Delta T = 333 \text{ K} - 291 \text{ K} = 42 \text{ K}$

$\Delta P_{SO_2} = \dfrac{nR\Delta T}{V} = 0.0937 \text{ mol} \times \dfrac{0.08206 \text{ L} \cdot \text{atm}}{K \cdot \text{mol}} \times \dfrac{42 \text{ K}}{5.00 \text{ L}} = +0.065 \text{ atm}$

As the flask is heated, the partial pressure of $SO_2(g)$ increases by 0.065 atm.

(b) $7.50 \text{ g } SO_3 \times \dfrac{1 \text{ mol } SO_3}{80.06 \text{ g } SO_3} = 0.0937 \text{ mol } SO_3;\quad \Delta T = 42 \text{ K}$

$n_t = 0.0937 \text{ mol } SO_2 + 0.0937 \text{ mol } SO_3 = 0.1874 \text{ mol gas}$

$\Delta P_t = \dfrac{n_t R\Delta T}{V} = 0.1814 \text{ mol} \times \dfrac{0.08206 \text{ L} \cdot \text{atm}}{K \cdot \text{mol}} \times \dfrac{42 \text{ K}}{5.00 \text{ L}} = +0.13 \text{ atm}$

The total pressure in the flask increases by 0.13 atm when the flask is heated.

(c)   Neither the moles of each gas nor the total moles of gas are affected by heating, so the mole fraction of $SO_3(g)$ is unchanged. $\chi_{SO_3} = \dfrac{0.0937 \text{ mol}}{0.1874 \text{ mol}} = 0.500$.

10.53 $P_{N_2} = \dfrac{P_1V_1T_2}{V_2T_1} = \dfrac{3.80\ atm \times 1.00\ L \times 293\ K}{10.0\ L \times 299\ K} = 0.372\ atm$

$P_{O_2} = \dfrac{P_1V_1T_2}{V_2T_1} = \dfrac{4.75\ atm \times 5.00\ L \times 293\ K}{10.0\ L \times 299\ K} = 2.33\ atm$

$P_t = 0.372\ atm + 2.33\ atm = 2.70\ atm$

10.54 It is simplest to calculate the partial pressure of each gas as it expands into the total volume, then sum the partial pressures.

$P_2 = P_1V_1 / V_2$

$P_{N_2} = 635\ torr\ (1.0\ L\ /\ 2.5\ L) = 254\ torr$

$P_{Ne} = 212\ torr\ (1.0\ L\ /\ 2.5\ L) = \ \ 85\ torr$

$P_{H_2} = 418\ torr\ (0.5\ L\ /\ 2.5\ L) = \ \ \underline{84\ torr}$

Total pressure $= 423\ torr = 4.2 \times 10^2\ torr$ (two significant figures)

10.55 The gas sample is a mixture of $H_2(g)$ and $H_2O(g)$. Find the partial pressure of $H_2(g)$ and then the moles of $H_2(g)$ and $Zn(s)$.

$P_t = 738\ torr = P_{H_2} + P_{H_2O}$

From Appendix B, the vapor pressure of H_2O at $24°C = 22.38\ torr$

$P_{H_2} = (738\ torr - 22.38\ torr) \times \dfrac{1\ atm}{760\ torr} = 0.9416 = 0.942\ atm$

$n_{H_2} = \dfrac{P_{H_2}V}{RT} = 0.9416\ atm \times \dfrac{K \cdot mol}{0.08206\ L \cdot atm} \times \dfrac{0.159\ L}{297\ K} = 0.006143 = 0.00614\ mol\ H_2$

$0.006143\ mol\ H_2 \times \dfrac{1\ mol\ Zn}{1\ mol\ H_2} \times \dfrac{65.39\ g\ Zn}{1\ mol\ Zn} = 0.402\ g\ Zn$

10.56 $V = \dfrac{nRT}{P}$; find mol O_2 by partial pressures.

P_{H_2O} at $23°C = 21.07\ torr$

$P_{O_2} = (742\ torr - 21.07\ torr) \times \dfrac{1\ atm}{760\ torr} = 0.9486 = 0.949\ atm$

$0.3570\ g\ KClO_3 \times \dfrac{1\ mol\ KClO_3}{122.55\ g} \times \dfrac{3\ mol\ O_2}{2\ mol\ KClO_3} = 0.0043697 = 0.004370\ mol\ O_2$

$V = 0.0043697\ mol \times \dfrac{0.08206\ L \cdot atm}{K \cdot mol} \times \dfrac{296\ K}{0.9486\ atm} = 0.112\ L = 112\ mL$

Kinetic - Molecular Theory; Graham's Law

(In *Solutions to Exercises*, the symbol for molar mass is MM.)

10.57   (a)   They have the same number of molecules (equal volumes of gases at the same temperature and pressure contain equal numbers of molecules).

         (b) N_2 is more dense because it has the larger molar mass. Since the volumes of the samples and the number of molecules are equal, the gas with the larger molar mass will have the greater density.

         (c)   The average kinetic energies are equal (statement 5, section 10.7).

         (d) CH_4 will effuse faster. The lighter the gas molecules, the faster they will effuse (Graham's Law).

10.58   (a) $n \propto P/T$ (V/R is the same for A and B.) Since P is greater and T is smaller for vessel A, it has more molecules.

         (b)   Since vessel A has more molecules and the molar mass of CO_2 is greater than the molar mass of HCl, vessel A contains more mass.

         (c)   Vessel B is at a higher temperature so the average kinetic energy of its molecules is higher.

         (d)   The two factors that affect rms speed are temperature and molar mass. Since the molecules in vessel B have smaller molar mass and higher temperature, they have greater rms speed.

             Mathematically, according to Equation 10.24,

$$\frac{u_A}{u_B} = \sqrt{\frac{T_A/MM_A}{T_B/MM_B}} = \sqrt{\frac{273/44.01}{293/36.46}} = 0.772$$

             The ratio is less than 1, vessel B has the greater rms speed.

10.59   (a) increase in temperature at constant volume, decrease in volume, increase in pressure
          (b) decrease in temperature   (c) increase in volume   (d) increase in temperature

10.60   (a) False. The average kinetic energy per molecule in a collection of gas molecules is the same for all gases at the same temperature.   (b) True.   (c) False. The molecules in a gas sample at a given temperature exhibit a distribution of kinetic energies.   (d) True.

10.61   (a)   In order of increasing speed (and decreasing molar mass):

             $CO_2 \approx N_2O < F_2 < HF < H_2$

(b) $\quad u_{H_2} = \sqrt{\dfrac{3RT}{MM}} = \left(\dfrac{3 \times 8.314 \ kg \cdot m^2/s^2 \cdot K \cdot mol \times 300 \ K}{2.02 \times 10^{-3} \ kg/mol} \right)^{1/2} = 1.92 \times 10^3 \ m/s$

$\quad u_{CO_2} = \left(\dfrac{3 \times 8.314 \ kg \cdot m^2/s^2 \cdot K \cdot mol \times 300 \ K}{44.0 \times 10^{-3} \ kg/mol} \right)^{1/2} = 4.12 \times 10^2 \ m/s$

As expected, the lighter molecule moves at the greater speed.

10.62 (a) The larger the molar mass, the slower the average speed (at constant temperature). In order of increasing speed (and decreasing molar mass):

$SF_6 < HI < Cl_2 < H_2S < CO$

(b) $\quad u = \sqrt{\dfrac{3RT}{MM}} = \left(\dfrac{3 \times 8.314 \ kg \cdot m^2/s^2 \cdot K \cdot mol \times 298 \ K}{28.0 \times 10^{-3} \ kg/mol} \right)^{1/2} = 515 \ m/s$

10.63 The heavier the molecule, the slower the rate of effusion. Thus, the order for increasing rate of effusion is in the order of decreasing mass.

rate $^2H^{37}Cl$ < rate $^1H^{37}Cl$ < rate $^2H^{35}Cl$ < rate $^1H^{35}Cl$

10.64 $\quad \dfrac{rate \ ^{235}U}{rate \ ^{238}U} = \sqrt{\dfrac{238.05}{235.04}} = \sqrt{1.0128} = 1.0064$

There is a slightly greater rate enhancement for $^{235}U(g)$ atoms than $^{235}UF_6(g)$ molecules (1.0043), because ^{235}U is a greater percentage (100%) of the mass of the diffusing particles than in $^{235}UF_6$ molecules. The masses of the isotopes were taken from *The Handbook of Chemistry and Physics*.

10.65 $\quad \dfrac{rate \ (sulfide)}{rate \ (Ar)} = \left[\dfrac{39.9}{MM \ (sulfide)} \right]^{1/2} = 0.28$

MM (sulfide) = $(39.9 / 0.28)^2 = 510 \ g/mol$ (two significant figures)

The empirical formula of arsenic(III) sulfide is As_2S_3 which has a formula mass of 246.1. Twice this is 490 g/mol, close to the value estimated from the effusion experiment. Thus, the formula of the vapor phase molecule is As_4S_6.

10.66 The time required is proportional to the reciprocal of the effusion rate.

$\dfrac{rate \ (X)}{rate \ (O_2)} = \dfrac{28 \ s}{72 \ s} = \left[\dfrac{32 \ g \ O_2}{MM_x} \right]^{1/2}$; $MM_x = 32 \ g \ O_2 \times \left[\dfrac{72}{28} \right]^2 = 210 \ g/mol$ (two sig figs)

Nonideal-Gas Behavior

10.67　(a)　Nonideal gas behavior is observed at very high pressures and/or low temperatures.

(b)　The real volumes of gas molecules and attractive intermolecular forces between molecules cause gases to behave nonideally.

10.68　Ideal gas behavior is most likely to occur at high temperature and low pressure, so the atmosphere on Mercury is more likely to obey the ideal gas law. The higher temperature on Mercury means that the kinetic energies of the molecules will be larger relative to intermolecular attractive forces. Further, the gravitational attractive forces on Mercury are lower because the planet has a much smaller mass. This means that for the same column mass of gas (Figure 10.1), atmospheric pressure on Mercury will be lower.

10.69　The ratio PV/RT is equal to the number of moles of molecules in an ideal-gas sample; this number should be a constant for all pressure, volume and temperature conditions. If the value of this ratio changes with increasing pressure, the gas sample is not behaving ideally (according to the Ideal-Gas Equation).

10.70　(a)　Intermolecular forces cause the molecules to "stick" together and behave as if there are fewer net particles in the sample.

(b)　The real volume of gas molecules causes the amount of free space in the gas sample to be less than the container volume. Using the container volume to calculate PV/RT gives a value larger than that for an ideal gas (which assumes that the total container volume is free space).

(c)　As the temperature of a gas increases, the average kinetic energy of the particles increases. The increased kinetic energy overcomes the attractive forces between molecules and keeps them separate.

10.71　The constants a and b are part of the correction terms in the van der Waals equation. The smaller the values of a and b, the smaller the corrections and the more ideal the gas. Ar (a = 1.34, b = 0.0322) will behave more like an ideal gas than CO_2 (a = 3.59, b = 0.0427) at high pressures.

10.72　The constant a is a measure of the strength of intermolecular attractions among gas molecules; b is a measure of molecular volume. Both increase with increasing molecular mass and structural complexity.

10.73　(a)　$P = 1.00 \text{ mol} \times \dfrac{0.08206 \text{ L} \cdot \text{atm}}{\text{K} \cdot \text{mol}} \times \dfrac{313 \text{ K}}{28.0 \text{ L}} = 0.917 \text{ atm}$

(b)　$P = \dfrac{nRT}{V \cdot nb} - \dfrac{an^2}{V^2} = \dfrac{1.00 \times 0.08206 \times 313}{28.0 - (1.00 \times 0.1383)} - \dfrac{20.4(1.00)^2}{(28.0)^2} = 0.896 \text{ atm}$

10.74 (a)   At STP, the molar volume = 1 mol $\times \dfrac{0.08206 \text{ L} \cdot \text{atm}}{\text{K} \cdot \text{mol}} \times \dfrac{273 \text{ K}}{1 \text{ atm}} = 22.4 \text{ L}$

             Dividing the value for b, 0.0322 L/mol, by 4, we obtain 0.0080 L. Thus, the volume of the Ar atoms is (0.0080/22.4)100 = 0.036% of the total volume.

  (b)   At 100 atm pressure the molar volume is 0.244 L, and the volume of the Ar atoms is 3.6% of the total volume.

Additional Exercises

(In *Solutions to Exercises*, the symbol for molar mass is MM.)

10.75   A mercury barometer with water trapped in its tip would not read the correct pressure. The standard relationship between height of an Hg column and atmospheric pressure (760 mm Hg = 1 atm) assumes that there is a vacuum in the closed end of the barometer and that gravity is the only downward force on the Hg column. Water at the top of the Hg column would establish a vapor pressure which would exert additional downward pressure and partially counterbalance the pressure of the atmosphere. The Hg column would read lower than the actual atmospheric pressure.

10.76 (a) $P = \dfrac{nRT}{V}$; $n = 0.29 \text{ kg O}_2 \times \dfrac{1000 \text{ g}}{1 \text{ kg}} \times \dfrac{1 \text{ mol O}_2}{32.00 \text{ g O}_2} = 9.0625 = 9.1 \text{ mol}$; $V = 2.3 \text{ L}$;

             $T = 273 + 9°C = 282 \text{ K}$

             $P = \dfrac{9.0625 \text{ mol}}{2.3 \text{ L}} \times \dfrac{0.08206 \text{ L} \cdot \text{atm}}{\text{K} \cdot \text{mol}} \times 282 \text{ K} = 91 \text{ atm}$

  (b) $V = \dfrac{nRT}{P}$; $= \dfrac{9.0625 \text{ mol}}{0.95 \text{ atm}} \times \dfrac{0.08206 \text{ L} \cdot \text{atm}}{\text{K} \cdot \text{mol}} \times 299 \text{ K} = 2.3 \times 10^2 \text{ L}$

10.77 $P = \dfrac{nRT}{V}$; $n = 1 \times 10^{-5} \text{ mol}$, $V = 0.200 \text{ L}$, $T = 23°C = 296 \text{ K}$

         $P = 1 \times 10^{-5} \text{ mol} \times \dfrac{0.08206 \text{ L} \cdot \text{atm}}{\text{K} \cdot \text{mol}} \times \dfrac{296 \text{ K}}{0.200 \text{ L}} \times \dfrac{760 \text{ mm Hg}}{1 \text{ atm}} = 0.9 \text{ mm Hg}$

10.78 $\dfrac{P_1}{T_1} = \dfrac{P_2}{T_2}$; $P_1 = 2.2 \text{ atm}$, $T_1 = 297 \text{ K}$, $P_2 = 3.0 \text{ atm}$, $T_2 = ?$

         (T must be in kelvins for the P, T relationship to hold true.)

         $T_2 = \dfrac{P_2 T_1}{P_1} = \dfrac{3.0 \text{ atm} \times 297 \text{ K}}{2.2 \text{ atm}} = 405 \text{ K or } 132 °C$

10.79 (a) $n = \dfrac{PV}{RT} = 3.00 \text{ atm} \times \dfrac{K \cdot mol}{0.08206 \text{ L} \cdot atm} \times \dfrac{1.00 \text{ L}}{300 \text{ K}} = 0.122 \text{ mol } C_3H_8(g)$

(b) $\dfrac{0.590 \text{ g } C_3H_8(l)}{1 \text{ mL}} \times 1.00 \times 10^3 \text{ mL} \times \dfrac{1 \text{ mol } C_3H_8}{44.094 \text{ g}} = 13.4 \text{ mol } C_3H_8 \text{ (l)}$

(c) A 1.00 L container holds many more moles (molecules) of $C_3H_8(l)$ because in the liquid phase the molecules are touching. In the gas phase, the molecules are far apart (statement 2, section 10.7) and many fewer molecules will fit in the 1.00 L container.

10.80 Calculate mol O_3/L, then molecules/L.

$PV = nRT; \quad \dfrac{n}{V} = \dfrac{P}{RT} = \dfrac{3.0 \times 10^{-3} \text{ atm}}{250 \text{ K}} \times \dfrac{K \cdot mol}{0.08206 \text{ L} \cdot atm} = 1.462 \times 10^{-4}$

$= 1.46 \times 10^{-4} \text{ mol } O_3/L$

$\dfrac{1.462 \times 10^{-4} \text{ mol } O_3}{L} \times \dfrac{6.022 \times 10^{23} \text{ molecules}}{mol \; O_3} = 8.8 \times 10^{19} \; O_3 \text{ molecules/L}$

10.81 Volume of laboratory $= 110 \text{ m}^2 \times 2.7 \text{ m} \times \dfrac{1000 \text{ L}}{1 \text{ m}^3} = 2.97 \times 10^5 = 3.0 \times 10^5 \text{ L}$

Calculate the **total** moles of gas in the laboratory at the conditions given.

$n_t = \dfrac{PV}{RT} = 1.00 \text{ atm} \times \dfrac{K \cdot mol}{0.08206 \text{ L} \cdot atm} \times \dfrac{2.97 \times 10^5 \text{ L}}{297 \text{ K}} = 1.22 \times 10^4 = 1.2 \times 10^4 \text{ mol gas}$

An $Ni(CO)_4$ concentration of 1 part in 10^9 means 1 mol $Ni(CO)_4$ in 1×10^9 total moles of gas.

$\dfrac{x \text{ mol } Ni(CO)_4}{1.22 \times 10^4 \text{ mol gas}} = \dfrac{1}{10^9} = 1.22 \times 10^{-5} \text{ mol } Ni(CO)_4$

$1.22 \times 10^{-5} \text{ mol } Ni(CO)_4 \times \dfrac{170.74 \text{ g } Ni(CO)_4}{1 \text{ mol } Ni(CO)_4} = 2.1 \times 10^{-3} \text{ g } Ni(CO)_4$

10.82 (a) $n = \dfrac{PV}{RT} = 0.980 \text{ atm} \times \dfrac{K \cdot mol}{0.08206 \text{ L} \cdot atm} \times \dfrac{0.600 \text{ L}}{353 \text{ K}} = 0.2030 = 0.203 \text{ mol air}$

$\text{mol } O_2 = 0.203 \text{ mol air} \times \dfrac{0.2095 \text{ mol } O_2}{1 \text{ mol air}} = 0.04253 = 0.0425 \text{ mol } O_2$

(b) $C_8H_{18}(l) + 25/2 \, O_2(g) \rightarrow 8CO_2(g) + 9H_2O(g)$

$0.04253 \text{ mol } O_2 \times \dfrac{1 \text{ mol } C_8H_{18}}{12.5 \text{ mol } O_2} \times \dfrac{114.2 \text{ g } C_8H_{18}}{1 \text{ mol } C_8H_{18}} = 0.389 \text{ g } C_8H_{18}$

10.83   (a) $5.00 \text{ g HCl} \times \dfrac{1 \text{ mol HCl}}{36.46 \text{ g HCl}} = 0.1371 = 0.137 \text{ mol HCl}$

$5.00 \text{ g NH}_3 \times \dfrac{1 \text{ mol NH}_3}{17.03 \text{ g NH}_3} = 0.2936 = 0.294 \text{ mol}$

The gases react in a 1:1 mole ratio, HCl is the limiting reactant and is completely consumed. $(0.2936 \text{ mol} - 0.1371 \text{ mol}) = 0.1565 = 0.157 \text{ mol NH}_3$ remain in the system.

$NH_3(g)$ is the only gas remaining after reaction. $V_t = 4.00 \text{ L}$

(b) $P = \dfrac{nRT}{V} = 0.1565 \text{ mol} \times \dfrac{0.08206 \text{ L} \bullet \text{atm}}{\text{K} \bullet \text{mol}} \times \dfrac{298 \text{ K}}{4.00 \text{ L}} = 0.957 \text{ atm}$

10.84 $MM_{avg} = \dfrac{dRT}{P} = \dfrac{1.104 \text{ g}}{1 \text{ L}} \times \dfrac{0.08206 \text{ L} \bullet \text{atm}}{\text{K} \bullet \text{mol}} \times \dfrac{300 \text{ K}}{435 \text{ mm Hg}} \times \dfrac{760 \text{ mm Hg}}{1 \text{ atm}}$

$= 47.48 = 47.5 \text{ g/mol}$

χ = mole fraction O_2; $1 - \chi$ = mole fraction Kr

$47.48 \text{ g} = \chi(32.00) + (1-\chi)(83.80)$

$36.3 = 51.8 \chi; \chi = 0.701; 70.1\% \text{ O}_2$

10.85   Calculate the number of moles of Ar in the vessel:

$n = (339.712 - 337.428)/39.948 = 0.05717 \text{ mol}$

The total number of moles of the mixed gas is the same (Avogadro's Law). Thus, the average atomic weight is $(339.218 - 337.428)/0.05717 = 31.31$. Let the mole fraction of Ne be χ. Then,

$\chi (20.183) + (1 - \chi) (39.948) = 31.31; 8.638 = 19.765 \chi; \chi = 0.437$

Neon is thus 43.7 mole percent of the mixture.

10.86   The balloon will expand; H_2 (MM = 2 g/mol) will effuse in through the walls of the balloon faster than He (MM = 4 g/mol) will effuse out, because the gas with the smaller molar mass effuses more rapidly.

10.87   (a)   The quantity d/P = MM/RT should be a constant at all pressures for an ideal gas. It is not, however, because of nonideal behavior. If we graph d/P vs P, the ratio should approach ideal behavior at low P. At P = 0, d/P = 2.2525. Using this value in the formula MM = d/P × RT, MM = 50.46 g/mol.

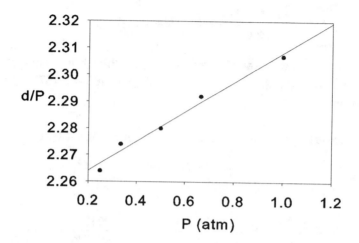

(b) The ratio d/P varies with pressure because of the finite volumes of gas molecules and attractive intermolecular forces.

10.88 In the expanded container volume, there are fewer collisions with the walls and with other gas particles. Attractive and repulsive forces which might change the speed of the particles become less important, and the particles maintain essentially the same average speed, kinetic energy and temperature after the expansion.

10.89

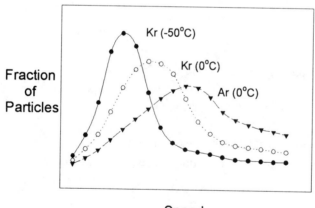

10.90 (a) $80.00 \text{ kg N}_2(g) \times \dfrac{1000 \text{ g}}{1 \text{ kg}} \times \dfrac{1 \text{ mol N}_2}{28.02 \text{ g N}} = 2855 \text{ mol N}_2$

$P = \dfrac{nRT}{V} = 2855 \text{ mol} \times \dfrac{0.08206 \text{ L} \cdot \text{atm}}{\text{K} \cdot \text{mol}} \times \dfrac{573 \text{ K}}{1000.0 \text{ L}} = 134.2 \text{ atm}$

(b)   According to Equation 10.26,

$$P = \frac{nRT}{V-nb} - \frac{n^2a}{V^2}$$

$$P = \frac{(2855 \text{ mol}) (0.08206 \text{ L} \cdot \text{atm/K} \cdot \text{mol}) (573 \text{ K})}{1000.0 \text{ L} - (2855 \text{ mol}) (0.0391 \text{ L/mol})} - \frac{(2855 \text{ mol})^2 (1.39 \text{ L}^2 \cdot \text{atm/mol}^2)}{(1000.0 \text{ L})^2}$$

$$P = \frac{134{,}243 \text{ L} \cdot \text{atm}}{1000.0 \text{ L} - 111.6 \text{ L}} - 11.3 \text{ atm} = 151.1 \text{ atm} - 11.3 \text{ atm} = 139.8 \text{ atm}$$

(c)   The pressure corrected for the real volume of the N_2 molecules is 151.1 atm, 16.9 atm higher than the ideal pressure of 134.2 atm. The 11.3 atm correction for intermolecular forces reduces the calculated pressure somewhat, but the "real" pressure is still higher than the ideal pressure. The correction for the real volume of molecules dominates. Even though the value of b is small, the number of moles of N_2 is large enough so that the molecular volume correction is larger than the attractive forces correction.

Integrative Exercises

(In *Solutions to Exercises*, the symbol for molar mass is MM.)

10.91 $MM = \dfrac{gRT}{VP} = \dfrac{1.05 \text{ g}}{0.500 \text{ L}} \times \dfrac{0.08206 \text{ L} \cdot \text{atm}}{\text{K} \cdot \text{mol}} \times \dfrac{298 \text{ K}}{750 \text{ torr}} \times \dfrac{760 \text{ torr}}{1 \text{ atm}} = 52.0 \text{ g/mol}$

$0.462 \text{ g C} \times \dfrac{1 \text{ mol C}}{12.01 \text{ g C}} = 0.0385 \text{ mol C}$

$0.538 \text{ g N} \times \dfrac{1 \text{ mol}}{14.01 \text{ g N}} = 0.0384 \text{ mol N}$

The mole ratio is 1C:1N and the empirical formula is CN; the formula weight of CN = 12 + 14 = 26 g. Since molar mass of 52.0 g is twice the empirical formula weight, the molecular formula is C_2N_2.

10.92 $MM = \dfrac{gRT}{VP} = \dfrac{1.56 \text{ g}}{1.00 \text{ L}} \times \dfrac{0.08206 \text{ L} \cdot \text{atm}}{\text{K} \cdot \text{mol}} \times \dfrac{323 \text{ K}}{0.984 \text{ atm}} = 42.0 \text{ g/mol}$

Assume 100 g cyclopropane

$100 \text{ g} \times 0.857\% \text{ C} = 85.7 \text{ g C} \times \dfrac{1 \text{ mol C}}{12.01 \text{ g}} = \dfrac{7.136 \text{ mol C}}{7.136} = 1 \text{ mol C}$

$100 \text{ g} \times 0.143\% \text{ H} = 14.3 \text{ g H} \times \dfrac{1 \text{ mol H}}{1.008 \text{ g}} = \dfrac{14.19 \text{ mol H}}{7.136} = 2 \text{ mol H}$

The empirical formula of cyclopropane is CH_2 and the empirical formula weight is 12 + 2 = 14 g. The ratio of molar mass to empirical formula weight, 42.0 g / 14 g, is 3; therefore, there are three empirical formula units in one cyclopropane molecule. The molecular formula is $3 \times (CH_2) = C_3H_6$.

10.93   Strategy: Calculate ΔH_{rxn}° using Hess's Law and data from Appendix C. Using the Ideal-Gas Equation, calculate moles CO_2 produced. Use stoichiometry to calculate the overall enthalpy change, ΔH.

$$\Delta H_{rxn}^{\circ} = \Delta H_f^{\circ} Na^+(aq) + \Delta H_f^{\circ} CO_2(g) + \Delta H_f^{\circ} H_2O(l) - \Delta H_f^{\circ} NaHCO_3(s) - \Delta H_f^{\circ} H^+(aq)$$

$$= (-240.1 \text{ kJ} - 393.5 \text{ kJ} - 285.83 \text{ kJ}) - (-947.7 \text{ kJ} + 0)$$

$$= 28.27 = 28.3 \text{ kJ}$$

$$n = \frac{PV}{RT} = 715 \text{ torr} \times \frac{1 \text{ atm}}{760 \text{ torr}} \times \frac{K \cdot mol}{0.08206 \text{ L} \cdot atm} \times \frac{10.0 \text{ L}}{292 \text{ K}} = 0.3926 = 0.393 \text{ mol } CO_2$$

$$\Delta H = \frac{28.27 \text{ kJ}}{1 \text{ mol } CO_2} \times 0.3926 \text{ mol } CO_2 = 11.1 \text{ kJ}$$

10.94   The number of moles of CO_2 is equal to the total number of moles of carbon atoms in the CH_4/C_2H_2 mixture. Let x = moles CH_4, y = moles C_2H_4. Then, total moles gas in the mixture = x + y. Total moles CO_2 = x + 2y.

$$\frac{x+y}{x+2y} = \frac{70.5 \text{ mm Hg}}{96.4 \text{ mm Hg}}; \frac{x}{y} = 1.72; \ x = 1.72 \ y$$

Let x + y = 1; y = 0.37 = the fraction that is acetylene, C_2H_2.

Notice that the only property of gases needed here is that equal numbers of moles of gas exert equal pressures under the same conditions of volume and temperature.

10.95   After reaction, the flask contains $IF_5(g)$ and whichever reactant is in excess. Determine the limiting reactant, which regulates the moles of IF_5 produced and moles of excess reactant.

$$I_2(s) + 5F_2(g) \rightarrow 2 \ IF_5(g)$$

$$10.0 \text{ g } I_2 \times \frac{1 \text{ mol } I_2}{253.8 \text{ g } I_2} \times \frac{5 \text{ mol } F_2}{1 \text{ mol } I_2} = 0.1970 = 0.197 \text{ mol } F_2$$

$$10.0 \text{ g } F_2 \times \frac{1 \text{ mol } F_2}{38.00 \text{ g } F_2} = 0.2632 = 0.263 \text{ mol } F_2 \text{ available}$$

I_2 is the limiting reactant; F_2 is in excess.

0.263 mol F_2 available - 0.197 mol F_2 reacted = 0.066 mol F_2 remain.

$$10.0 \text{ g } I_2 \times \frac{1 \text{ mol } I_2}{253.8 \text{ g } I_2} \times \frac{2 \text{ mol } IF_5}{1 \text{ mol } I_2} = 0.0788 \text{ mol } IF_5 \text{ produced}$$

(a) $$P_{IF_5} = \frac{nRT}{V} = 0.0788 \text{ mol} \times \frac{0.08206 \text{ L} \cdot atm}{K \cdot mol} \times \frac{398 \text{ K}}{5.00 \text{ L}} = 0.515 \text{ atm}$$

(b) $$\chi_{IF_5} = \frac{\text{mol } IF_5}{\text{mol } IF_5 + \text{mol } F_2} = \frac{0.0788}{0.0788 + 0.066} = 0.544$$

10.96   (a) $MgCO_3(s) + 2HCl(aq) \rightarrow MgCl_2(aq) + H_2O(l) + CO_2(g)$

               $CaCO_3(s) + 2HCl(aq) \rightarrow CaCl_2(aq) + H_2O(l) + CO_2(g)$

     (b) $n = \dfrac{PV}{RT} = 735 \text{ torr} \times \dfrac{1 \text{ atm}}{760 \text{ torr}} \times \dfrac{K \bullet mol}{0.08206 \text{ L} \bullet \text{atm}} \times \dfrac{2.26 \text{ L}}{296 \text{ K}}$

                                      $= 0.08998 = 0.0900 \text{ mol } CO_2$

     (c) $x = g \text{ } MgCO_3$, $y = g \text{ } CaCO_3$, $x + y = 8.05$ g

             $mol \text{ } MgCO_3 + mol \text{ } CaCO_3 = mol \text{ } CO_2$ total

             $\dfrac{x}{84.32} + \dfrac{y}{100.09} = 0.08998$; $y = 8.05 - x$

             $\dfrac{x}{84.32} + \dfrac{8.05 - x}{100.09} = 0.08998$

             $100.09x - 84.32x + 84.32(8.05) = 0.08998 \,(84.32)(100.09)$

             $15.77x + 678.776 = 759.417$; $x = 5.11$ g $MgCO_3$

             mass % $MgCO_3 = \dfrac{5.11 \text{ g } MgCO_3}{8.05 \text{ g sample}} \times 100 = 63.5\%$

             [By strict sig fig rules, the answer has 2 sig figs: $15.77x + 679$ (3 digits from 8.05) = 759.4; $759.4 - 679 = 80$ (no decimal places, 2 sig figs) leads to 5.1 g $MgCO_3$.]

11 Intermolecular Forces, Liquids and Solids

Kinetic-Molecular Theory

11.1   (a)   solid < liquid < gas   (b)   gas < liquid < solid

11.2   Liquids are fluid and take the shape of their container. Solids are rigid, they have their own volume and do not flow. Solids are usually denser than liquids, indicating less empty space and a higher degree of order. Water is a notable exception.

11.3   (a)   liquid   (b)   gas

11.4   Density is the ratio of the mass of a substance to the volume it occupies. For the same substance in different states, mass will be the same. In the liquid and solid states, the particles are touching and there is very little empty space, so the volumes occupied by a unit mass are very similar and the densities are similar. In the gas phase, the molecules are far apart, so a unit mass occupies a much greater volume than the liquid or solid, and the density of the gas phase is much less.

11.5   As the temperature of a substance is increased, the average kinetic energy of the particles increases. In a collection of particles (molecules), the state is determined by the strength of interparticle forces relative to the average kinetic energy of the particles. As the average kinetic energy increases, more particles are able to overcome intermolecular attractive forces and move to a less ordered state, from solid to liquid to gas.

11.6   At constant temperature, the average kinetic energy of a collection of particles is constant. Compression brings particles closer together and increases the number of particle-particle collisions. With more collisions, the likelihood of intermolecular attractions causing the particles to coalesce (liquefy) is greater.

Intermolecular Forces

11.7   (a)   London dispersion forces   (b) dipole-dipole forces
        (c)   dipole-dipole or in certain cases hydrogen bonding

11.8   (a)   Hydrogen bonds   (b) ionic bonds   (c) London dispersion forces
        (d)   dipole-dipole forces

11.9   CO is a polar covalent molecule (ΔEN = 1.0) and N_2 is nonpolar. Dipole-dipole forces between CO molecules are stronger than London dispersion forces between N_2 molecules. A higher temperature and greater average kinetic energy is required to overcome the dipole-dipole forces in CO and separate (vaporize) the molecules.

11.10 (a)   HCl has stronger dipole-dipole forces because it is a more polar molecule. (Strictly speaking, it does not have hydrogen bonding.)

(b)   HI has stronger London dispersion forces because it is a larger molecule with a more polarizable electron cloud.

11.11 (a)   *Polarizability* is the ease with which the charge distribution (electron cloud) in a molecule can be distorted to produce a transient dipole.

(b)   Te is most polarizable because its valence electrons are farthest from the nucleus and least tightly held.

(c)   Polarizability increases as molecular size (and thus molecular weight) increases. In order of increasing polarizability: $CH_4 < SiH_4 < SiCl_4 < GeCl_4 < GeBr_4$

11.12 (a)   A more polarizable molecule can develop a larger transient dipole, increasing the strength of electrostatic attractions among polarized molecules.

(b)   Helium is a collection of individual isolated atoms with no tendency to form chemical bonds. In order to liquefy He(g), there must be attractive forces between He atoms strong enough to overcome their kinetic energies. Since He(g) can be liquefied, some kind of attractive interaction between particles with normally uniform charge distributions (dispersion forces) must exist.

(c)   The strengths of dispersion forces increase with increasing molecular size, because the relatively diffuse electron clouds are less tightly held by distant nuclei.

11.13 (a) C_6H_{14} - dispersion; C_8H_{18} - dispersion. C_8H_{18} has the higher boiling point due to greater molar mass and similar strength of forces.

(b) C_3H_8 - dispersion; CH_3OCH_3 - dipole-dipole and dispersion. CH_3OCH_3 has the higher boiling point due to stronger intermolecular forces and similar molar mass.

(c) CH_3OH - hydrogen bonding, dipole-dipole and dispersion; CH_3SH - dipole-dipole and dispersion. CH_3OH has the higher boiling point due to the influence of hydrogen bonding (Figure 11.7).

(d) NH_2NH_2 - hydrogen bonding, dipole-dipole and dispersion; CH_3CH_3 - dispersion. NH_2NH_2 has the higher boiling point due to much stronger intermolecular forces.

11.14 (a)   HF has the higher boiling point because hydrogen bonding is stronger than dipole-dipole forces.

(b) $CHBr_3$ has the higher boiling point because it has the higher molar mass which leads to greater polarizability and stronger dispersion forces.

(c)   ICl has the higher boiling point because it is a polar molecule. For molecules with similar structures and molar masses, dipole-dipole forces are stronger than dispersion forces.

11.15   Both hydrocarbons experience dispersion forces. Rodlike butane molecules can contact each other over the length of the molecule, while spherical 2-methylpropane molecules can only touch tangentially. The larger contact surface of butane produces greater polarizability and a higher boiling point.

11.16   Both molecules experience hydrogen bonding through their -OH groups and dispersion forces between their hydrocarbon portions. The position of the -OH group in isopropyl alcohol shields it somewhat from approach by other molecules and slightly decreases the extent of hydrogen bonding. Also, isopropyl alcohol is less rodlike (it has a shorter chain) than n-propyl alcohol, so dispersion forces are weaker. Since hydrogen bonding and dispersion forces are weaker in isopropyl alcohol, it has the lower boiling point.

11.17   Surface tension (Section 11.3), high boiling point (relative to H_2S, H_2Se, H_2Te, Figure 11.7), high heat capacity per gram, high enthalpy of vaporization; the solid is less dense than the liquid; it is a liquid at room temperature despite its low molar mass.

11.18   (a)   Water expands when it freezes to maximize the number of hydrogen bonding interactions in the structure. In the solid state, each H atom is involved in one hydrogen bond and each O atom participates in two hydrogen bonds. The H_2O molecules must be far enough apart to allow for the steric requirements of these four interactions.

(b)   Since ice, $H_2O(s)$, floats on water, it insulates the $H_2O(l)$ from extreme cold, allowing aquatic life to exist. If $H_2O(s)$ sank to the bottom of natural waters, temperatures of the $H_2O(l)$ (unfrozen water) would be too low to support living organisms.

Viscosity and Surface Tension

11.19   Viscosities and surface tensions of liquids both increase as intermolecular forces become stronger.

11.20   As temperature increases, the average kinetic energy of the molecules increases and intermolecular attractions are more easily overcome. Surface tensions and viscosities decrease.

11.21   (a)   Ethanol molecules experience hydrogen bonding, a stronger intermolecular force than the weak dipole-dipole forces between ether molecules. The stronger forces between ethanol molecules make the liquid more resistant to flow and thus more viscous.

(b)   The shape of a meniscus depends on the strength of the cohesive forces within a liquid relative to the adhesive forces between the walls of the capillary and the liquid. Because polar water molecules have strong cohesive intermolecular interactions relative to the weak adhesive forces between the polar water molecules and nonpolar polyethylene, the meniscus is concave-downward.

11.22   (a) $CHBr_3$ has a higher molar mass, is more polarizable and has stronger dispersion forces, so the surface tension is greater (see Exercise 11.14(b)).

(b)   As temperature increases, the viscosity of the oil decreases because the average kinetic energies of the molecules increase (Exercise 11.20).

(c)   Adhesive forces between polar water and nonpolar car wax are weak, so the large surface tension of water draws the liquid into the shape with the smallest surface area, a sphere.

Changes of State

11.23   Endothermic: melting (s → l), vaporization (l → g), sublimation (s → g)
Exothermic: condensation (g → l), freezing (l → s), deposition (g → s)

11.24   (a) Condensation   (b) sublimation   (c) vaporization (evaporation)   (d) freezing

11.25   (a)   Ice, $H_2O(s)$, sublimes to water vapor, $H_2O(g)$.

(b)   The heat energy required to increase the kinetic energy of molecules enough to melt the solid does not produce a large separation of molecules. The specific order is disrupted, but the molecules remain close together. On the other hand, when a liquid is vaporized, the intermolecular forces which maintain close molecular contacts must be overcome. Because molecules are being separated, the energy requirement is higher than for melting.

11.26   (a)   Liquid ethyl chloride at room temperature is far above its boiling point. When the liquid boils, heat is required to overcome the intermolecular forces between molecules. This heat is extracted from the surface that contacts the liquid.

(b)   At 279°C, CS_2 is above its critical temperature. The molecules have too much kinetic energy to be liquefied, regardless of pressure.

11.27   Evaporation of 10 g of water requires:

$$10.0 \text{ g } H_2O \times \frac{2.4 \text{ kJ}}{1 \text{ g } H_2O} = 24.0 \text{ kJ or } 2.4 \times 10^4 \text{ J}$$

Cooling a certain amount of water by 13°C:

$$2.40 \times 10^4 \text{ J} \times \frac{1 \text{ g} \cdot \text{K}}{4.18 \text{ J}} \times \frac{1}{13°C} = 442 = 4.4 \times 10^2 \text{ g } H_2O$$

11.28 Energy released when 100 g of H_2O is cooled from 18°C to 0°C:

$$\frac{4.184\ J}{g \bullet K} \times 100\ g\ H_2O \times 18°C = 7.53 \times 10^3\ J = 7.5\ kJ$$

Energy released when 100 g of H_2O is frozen (there is no change in temperature during a change of state):

$$\frac{334\ J}{g} \times 100\ g\ H_2O = 3.34 \times 10^4\ J = 33.4\ kJ$$

Total energy released = 7.53 kJ + 33.4 kJ = 40.93 = 40.9 kJ

Mass of freon which will absorb 40.9 kJ when vaporized:

$$40.93\ kJ \times \frac{1 \times 10^3\ J}{1\ kJ} \times \frac{1\ g\ CCl_2F_2}{289\ J} = 142\ g\ CCl_2F_2$$

11.29 Consider the process in steps, using the appropriate thermochemical constant.

Heat the liquid from -50°C to 23.8°C (223 K to 296.8 K), using the specific heat of the liquid.

$$10.0\ g\ CCl_3F \times \frac{0.87\ J}{g \bullet K} \times 73.8\ K \times \frac{1\ kJ}{1000\ J} = 0.642 = 0.64\ kJ$$

Boil the liquid at 23.8°C (296.8 K), using the enthalpy of vaporization.

$$10.0\ g\ CCl_3F \times \frac{1\ mol\ CCl_3F}{137.4\ g\ CCl_3F} \times \frac{24.75\ kJ}{mol} = 1.801 = 1.80\ kJ$$

Heat the gas from 23.8°C to 50°C (296.8 K to 323 K), using the specific heat of the gas.

$$10.0\ g\ CCl_3F \times \frac{0.59\ J}{g \bullet K} \times 26.2\ K \times \frac{1\ kJ}{1000\ J} = 0.1546 = 0.15\ kJ$$

The total energy required is 0.642 kJ + 1.801 kJ + 0.1546 kJ = 2.60 kJ.

11.30 Heat the solid from -150°C to -114°C (123 K to 149 K), using the specific heat of the solid.

$$50.0\ g\ C_2H_5OH \times \frac{0.97\ J}{g \bullet K} \times 36\ K \times \frac{1\ kJ}{1000\ J} = 1.746 = 1.7\ kJ$$

At -114°C (159 K), melt the solid, using its enthalpy of fusion.

$$50.0\ g\ C_2H_5OH \times \frac{1\ mol\ C_2H_5OH}{46.07\ g\ C_2H_5OH} \times \frac{5.02\ kJ}{1\ mol} = 5.448 = 5.45\ kJ$$

Heat the liquid from -114°C to 78°C (159 K to 351 K), using the specific heat of the liquid.

$$50\ g\ C_2H_5OH \times \frac{2.3\ J}{g \bullet K} \times 192\ K \times \frac{1\ kJ}{1000\ J} = 22.1 = 22\ kJ$$

At 78°C (351 K), vaporize the liquid, using its enthalpy of vaporization.

$$50.0 \text{ g C}_2\text{H}_5\text{OH} \times \frac{1 \text{ mol C}_2\text{H}_5\text{OH}}{46.07 \text{ g C}_2\text{H}_5\text{OH}} \times \frac{38.56 \text{ kJ}}{1 \text{ mol}} = 41.849 = 41.85 \text{ kJ}$$

The total energy required is 1.746 kJ + 5.448 kJ + 22.1 kJ + 41.849 kJ = 71.143 = 71 kJ. (The result has zero decimal places, from 22 kJ required to heat the liquid.)

11.31   (a)   The critical temperature is the highest temperature at which a gas can be liquefied, regardless of pressure.

     (b)   As the force of attraction between molecules increases, the critical temperature of the compound increases.

     (c)   The critical pressure is the pressure required to cause liquefaction at the critical temperature.

11.32   (a)   Those gases whose critical temperatures are greater than 298 K can be liquefied at room temperature. The gases in Table 11.4 that meet this criterion are: NH_3, CO_2, H_2O.

     (b)   The temperature of $N_2(l)$ is 77 K. All of the gases in Table 11.4 have critical temperatures higher than 77 K, so all of them can be liquefied at this temperature, given sufficient pressure.

Vapor Pressure and Boiling Point

11.33 The boiling point is the temperature at which the vapor pressure of a liquid equals the external pressure acting on the surface of the liquid. If the external pressure is changed, then the temperature required to produce that vapor pressure will also change.

Melting of a solid occurs when the vapor pressures of the solid and liquid phases are equal. Both of these vapor pressures will vary to a degree with the external pressure acting on the solid or liquid, but the variation is not great, and both solid and liquid vapor pressures tend to change in the same way with applied pressure.

11.34   (a)   No effect.

     (b)   Vapor pressure increases with increasing temperature because average kinetic energies of molecules increase.

     (c)   Vapor pressure would decrease with increasing intermolecular attractive forces because fewer molecules would have sufficient kinetic energy to overcome the attractive forces and escape to the vapor phase.

     (d)   No effect.

(e)   No effect. The pressure of the air above the liquid determines how quickly liquid-vapor equilibrium is reached but not the equilibrium vapor pressure.

11.35 $CBr_4 < CHBr_3 < CH_2Br_2 < CH_2Cl_2 < CH_3Cl < CH_4$

The weaker the intermolecular forces, the higher the vapor pressure, the more volatile the compound. The order of increasing volatility is the order of decreasing strength of intermolecular forces. By analogy to the boiling points of HCl and HBr (Section 11.2), the trend will be dominated by dispersion forces, even though four of the molecules ($CHBr_3$, CH_2Br_2, CH_2Cl_2 and CH_3Cl) are polar. Thus, the order of increasing volatility is the order of decreasing molar mass and decreasing strength of dispersion forces.

11.36   (a)   The less volatile compound, $AsCl_3$, has the stronger intermolecular forces. In the more volatile one, PCl_3, more molecules have sufficient kinetic energy to overcome intermolecular attractive forces and escape to the gas phase. Since both compounds have the same distribution of kinetic energies at 25°C, the attractive forces must be stronger and harder to overcome in $AsCl_3$. This makes sense, because $AsCl_3$ is heavier and has stronger dispersion forces than PCl_3. $AsCl_3$ is also slightly more polar, but dispersion forces account for most of the volatility difference between these two compounds.

(b) PCl_3; the more volatile compound has the higher vapor pressure.

(c) $AsCl_3$; the less volatile compound requires a greater increase in temperature (kinetic energy) to promote enough molecules to the gas phase so that vapor pressure is equal to atmospheric pressure.

11.37   The water in the two pans is at the same temperature, the boiling point of water at the atmospheric pressure of the room. During a phase change, the temperature of a system is constant. All energy gained from the surroundings is used to accomplish the transition, in this case to vaporize the liquid water. The pan of water that is boiling vigorously is gaining more energy and the liquid is being vaporized more quickly than in the other pan, but the temperature of the phase change is the same.

11.38   (a)   On a humid day, there are more gaseous water molecules in the air and more are recaptured by the surface of the liquid, making evaporation slower.

(b)   At high altitude, atmospheric pressure is lower and water boils at a lower temperature. The eggs must be cooked longer at the lower temperature.

11.39   The boiling point is the temperature at which the vapor pressure of a liquid equals atmospheric pressure.

(a)   The boiling point of ethanol at 300 torr is ~56°C, or, at 56°C, the vapor pressure of ethanol is 300 torr.

(b)   At a pressure of 15 torr, water would boil at ~17.5°C, or, the vapor pressure of water at 17.5°C is 15 torr.

11.40   (a)   A pressure of 1.3 atm corresponds to $1.3 \text{ atm} \times \dfrac{760 \text{ torr}}{1 \text{ atm}} = 988$ torr. According to Appendix B, the vapor pressure of water reaches 988 torr somewhere between

               106-108°C. By linear interpolation, the boiling point should be near

$$106°C + \left[\frac{(988 - 938)\text{torr}}{(1004 - 938)\text{torr}} \times 2°C \right] = 107.5°.$$

      (b)   The vapor pressure of diethyl ether at 20°C is approximately 450 torr. Thus, at 20°C diethyl ether would boil at an external pressure of 450 torr.

11.41   From Appendix B, the temperature at which the vapor pressure of water is 350 torr is approximately 80°C.

11.42   The boiling point of a liquid is the temperature at which its vapor pressure equals atmospheric pressure. According to Appendix B, the vapor pressure of water is 680 torr at approximately 97°C.

Phase Diagrams

11.43   The liquid/gas line of a phase diagram ends at the critical point, the temperature and pressure beyond which the gas and liquid phases are indistinguishable. At temperatures higher than the critical temperature, a gas cannot be liquefied, regardless of pressure.

11.44   The triple point on a phase diagram represents the temperature at which the gas, liquid and solid phases are in equilibrium.

11.45   (a)   The water vapor would condense to form a solid at a pressure of around 4 torr. At higher pressure, perhaps 5 atm or so, the solid would melt to form liquid water. This occurs because the melting point of ice, which is 0°C at 1 atm, decreases with increasing pressure.

      (b)   In thinking about this exercise, keep in mind that **total** pressure is being maintained constant at 0.3 atm. That pressure is made up of water vapor pressure and some other pressure, which could come from an inert gas. The water at -1.0°C and 0.30 atm is in the solid form. Upon heating, it melts to form liquid water slightly above 0°C. The liquid converts to a vapor when the temperature reaches the point at which the vapor pressure of water reaches 0.3 atm (228 torr). From Appendix B we see that this occurs just below 70°C.

11.46   (a)   Solid CO_2 sublimes to form $CO_2(g)$ at a temperature of about -60°C.

      (b)   Solid CO_2 melts to form $CO_2(l)$ at a temperature of about -50°C. The CO_2 (l) boils when the temperature reaches approximately -40°C.

11.47   (a)

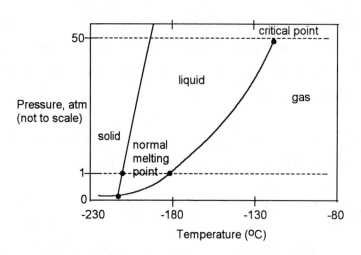

(b) $O_2(s)$ is denser than $O_2(l)$ because the solid-liquid line on the phase diagram is normal. That is, as pressure increases, the melting temperature increases. [Note that the solid-liquid line for O_2 is nearly vertical, indicating a small difference in the densities of $O_2(s)$ and $O_2(l)$].

(c) $O_2(s)$ will melt when heated at a pressure of 1 atm, since this is a much greater pressure than the pressure at the triple point.

11.48   (a)

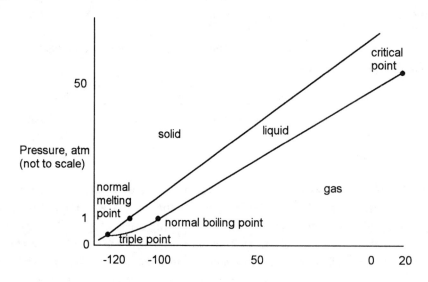

(b)   Xe(s) will not float in Xe(l). The solid-liquid line on the phase diagram is normal, the melting point of Xe(s) increases with increasing pressure, which means that Xe(s) is denser than Xe(l).

(c)   Cooling Xe(g) at 100 torr will cause deposition of the solid. A pressure of 100 torr is below the pressure of the triple point, so the gas will change directly to the solid upon cooling.

Structures of Solids

11.49 In a crystalline solid, the component particles (ions or molecules) are arranged in an ordered repeating pattern. In an amorphous solid, there is no orderly structure.

11.50 In amorphous silica (SiO_2) the regular structure of quartz is disrupted; the loose, disordered structure, Figure 11.28(b), has many vacant "pockets" throughout. There are fewer SiO_2 groups per volume in the amorphous solid; the packing is less efficient and less dense.

11.51 The unit cell is the building block of the crystal lattice. When repeated in three dimensions, it produces the crystalline solid. It is a parallelepiped with characteristic distances and angles. Unit cells can be primitive (lattice points only at the corners of the parallelepiped) or centered (lattice points at the corners and at the middle of faces or the middle of the parallelepiped).

11.52 (a)   8 corners × 1/8 sphere/corner = 1 sphere

     (b)   8 corners × 1/8 sphere/corner + 1 center × 1 sphere/center = 2 spheres

     (c)   8 corners × 1/8 sphere/corner + 6 faces × ½ sphere/face = 4 spheres

11.53 (a)   4 [See Exercise 11.52(c).]

     (b)   Each sphere is in contact with 12 nearest neighbors; its coordination number is thus 12.

     (c)   The length of the face diagonal of a face-centered cubic unit cell is four times the radius of the metal and $\sqrt{2}$ times the unit cell dimension (usually designated a for cubic cells).

$$4 \times 1.24 \text{ Å} = \sqrt{2} \times a$$

$$a = \frac{4 \times 1.24 \text{Å}}{\sqrt{2}} = 3.507 = 3.51 \times 10^{-8} \text{ cm}$$

     (d)   The density of the metal is the mass of the unit cell contents divided by the volume of the unit cell.

$$\text{density} = \frac{4 \text{ Ni atoms}}{(3.507 \times 10^{-8} \text{ cm})^3} \times \frac{58.69 \text{ g Ni}}{6.022 \times 10^{23} \text{ Ni atoms}} = 9.04 \text{ g/cm}^3$$

11.54 (a)   The face diagonal length is just $\sqrt{2}$ times the unit cell length. Because the atoms are in contact, this diagonal has a length equal to four atomic radii.

$$r = (\sqrt{2} \times 3.61 \text{ Å})/4 = 1.276 = 1.28 \text{ Å}$$

     (b)   The density of the metal is the mass of the unit cell contents divided by the volume of the unit cell. There are four Cu atoms in a face-centered cubic unit cell.

$$\text{density} = \frac{4 \text{ Cu atoms}}{(3.61 \text{ Å})^3} \times \frac{63.546 \text{ g Cu}}{6.022 \times 10^{23} \text{ Cu atoms}} \times \frac{(1 \times 10^8 \text{ Å})^3}{1 \text{ cm}^3} = 8.97 \text{ g/cm}^3$$

11.55 (a) Each sphere is in contact with 12 nearest neighbors; its coordination number is thus 12.

(b) Each sphere has a coordination number of six.

(c) Each sphere has a coordination number of eight.

11.56 (a) Na^+, 6 (b) Zn^{2+}, 4 (c) Ca^{2+}, 8

11.57 In the face-centered cubic structure, there are four NiO units in the unit cell. Density is the mass of the unit cell contents divided by the unit cell volume (a^3).

$$\text{density} = \frac{4 \text{ NiO units}}{(4.18 \text{ Å})^3} \times \frac{74.7 \text{ g NiO}}{6.022 \times 10^{23} \text{ NiO units}} \times \left(\frac{1 \text{ Å}}{1 \times 10^{-8} \text{ cm}}\right)^3 = \frac{6.79 \text{ g}}{cm^3}$$

11.58 There are four PbSe units in the unit cell. The unit cell edge is designated a.

$$8.27 \text{ g/cm}^3 = \frac{4 \text{ PbSe units}}{a^3} \times \frac{286.2 \text{ g}}{6.022 \times 10^{23} \text{ PbSe units}} \times \left(\frac{1 \text{ Å}}{10^{-8} \text{ cm}}\right)^3$$

$a^3 = 229.87 \text{ Å}^3$, $a = 6.13$ Å

11.59 The volume of the unit cell is $(2.86 \times 10^{-8} \text{ cm})^3$. The mass of the unit cell is:

$$\frac{7.92 \text{ g}}{cm^3} \times \frac{(2.86 \times 10^{-8})^3 \text{ cm}^3}{\text{unit cell}} = 1.853 \times 10^{-22} \text{ g/unit cell}$$

There are two atoms of the element present in the body-centered cubic unit cell. Thus the atomic weight is:

$$\frac{1.853 \times 10^{-22} \text{ g}}{\text{unit cell}} \times \frac{1 \text{ unit cell}}{2 \text{ atoms}} \times \frac{6.022 \times 10^{23} \text{ atoms}}{1 \text{ mol}} = 55.8 \text{ g/mol}$$

11.60 Avogadro's number is the number of KCl formula units in 74.55 g of KCl.

$$74.55 \text{ g KCl} \times \frac{1 \text{ cm}^3}{1.984 \text{ g}} \times \frac{(1 \times 10^{10} \text{ pm})^3}{1 \text{ cm}^3} \times \frac{4 \text{ KCl units}}{628^3 \text{ pm}^3} = 6.07 \times 10^{23} \text{ KCl formula units}$$

Bonding in Solids

11.61 (a) Hydrogen bonding, dipole-dipole forces, London dispersion forces

(b) covalent chemical bonds (mainly)

(c) ionic bonds (mainly)

(d) metallic bonds

11.62 (a) Molecular (b) molecular (c) metallic (d) ionic

(e) network covalent (f) molecular

11.63 Carbon (graphite and diamond) and SiO_2 (quartz)

11.64   In molecular solids, relatively weak intermolecular forces (hydrogen bonding, dipole-dipole, dispersion) bind the molecules in the lattice, so relatively little energy is required to disrupt these forces. In network-covalent solids, covalent bonds join atoms into an extended network. Melting or deforming a network-covalent solid means breaking these covalent bonds, which requires a large amount of energy.

11.65   (a)   KBr - strong ionic versus weak dispersion forces

        (b) SiO_2 - covalent bonds establish the network structure of the lattice versus weak dispersion forces holding CO_2 molecules together

        (c)   Se - network covalent lattice versus weak dispersion forces

        (d) MgF_2 - due to higher charge on Mg^{2+} than Na^+

11.66   (a) C_6Cl_6 - both are influenced by dispersion forces, C_6Cl_6 has the higher molar mass.

        (b)   HF - it has hydrogen bonding and HCL does not

        (c) SiO_2 - both have regular lattices, but the covalent bonds in the SiO_2 lattice require more energy to break than the ionic bonds in KO_2

        (d)   Xe - greater atomic weight, stronger dispersion forces

11.67   According to Table 11.6, the solid could be either ionic with low water solubility or network covalent. Due to the extremely high sublimation temperature, it is probably network covalent.

11.68   Because of its relatively high melting point and properties as a conducting solution, the solid must be ionic.

11.69   (a)   According to Figure 11.44(b), there are 4 Ag I units in the "zinc blende" unit cell [4 complete Ag^+ spheres, 6(½) + 8(1/8) I^- sphere's.]

$$5.69 \frac{g}{cm^3} = \frac{4\,AgI\,units}{a^3} \times \frac{234.8\,g}{6.022 \times 10^{23}\,AgI\,units} \times \left(\frac{1\,Å}{1 \times 10^{-8}\,cm}\right)^3$$

$$a^3 = 274.10\,Å^3,\ a = 6.50\,Å$$

        (b)   In an orthonormal coordinate system, the distance between two points (x_1, y_1, z_1) and (x_2, y_2, z_2) is $\sqrt{(x_1 - x_2)^2 + (y_1 - y_2)^2 + (z_1 - z_2)^2}$.

On Figure 11.44(b), select a right-handed coordinate system and select a bonded pair of ions. One possibility is shown below.

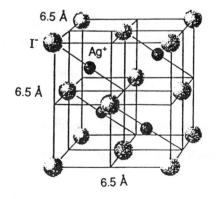
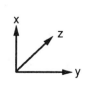

In a cubic unit cell, the lengths of all three cell edges are the same, in this case 6.50 Å.

For Ag^+: $x = 0.75(6.50 \text{ Å})$, $y = 0.25(6.50 \text{ Å})$, $z = 0.25(6.50 \text{ Å})$

$\qquad x = 4.875$, $y = 1.625$, $z = 1.625$

I^-: $x = 6.50$, $y = 0$, $z = 0$

The Ag-I distance is then $\sqrt{(4.875 - 6.50)^2 + (1.625 - 0)^2 + (1.625 - 0)^2}$.

Ag-I = $\sqrt{(1.625)^2 + (1.625)^2 + (1.625)^2}$ = 2.81 Å

11.70 (a) The U atoms in UO_2 are represented by the smaller spheres in Figure 11.44(c). The chemical formula requires twice as many O^{2-} ions as U^{4+} ions. There are eight complete large spheres and four total ($8 \times 1/8 + 6 \times \frac{1}{2}$) small spheres, so the small ones must represent U^{4+}. (It is probably true that O^{2-} has a physically larger radius than U^{4+}, but the elements' large separation on the periodic chart makes the relative radii difficult to estimate from trends.)

(b) According to Figure 11.44(c), there are four UO_2 units in the "fluorite" unit cell.

$$\frac{4 \text{ } UO_2 \text{ units}}{(5.468 \text{ Å})^3} \times \frac{270.03 \text{ g}}{6.022 \times 10^{23} \text{ } UO_2 \text{ units}} \times \left(\frac{1 \text{ Å}}{1 \times 10^{-8} \text{ cm}} \right)^3 = 10.97 \text{ g/cm}^3$$

Additional Exercises

11.71 (a) An *instantaneous dipole* is the temporary, small charge separation that occurs when negatively charged electron clouds bump into each other. It is sometimes called an induced dipole.

(b) *Polarizability* is the tendency of a normally symmetrical electron cloud to adopt an instantaneous dipole upon contact with another molecule or outside force.

(c) *Intermolecular attractive forces* are electrostatic attractions between neutral molecules owing to either permanent or induced dipole moments. Portions of molecules with partial charges of opposite sign are attracted to each, resulting in an overall lowering in energy of the system.

11.72 (a) *Volatility* is the tendency to evaporate. Molecules with low boiling points and high room temperature vapor pressures are volatile.

(b) *Capillary action* is the tendency of a liquid to move up a thin tube due to adhesive forces between the liquid and the tube and cohesive forces between liquid molecules.

(c) The *meniscus* is the curved upper surface of a liquid in a tube.

(d) *Sublimation* is the transformation of a solid directly to the gas phase, without melting.

11.73 (a)   Dipole-dipole attractions (polar covalent molecules): SO_2, IF, HBr

        (b)   Hydrogen bonding (O-H, N-H or F-H bonds): CH_3NH_2, HCOOH

11.74 (a)   The *cis* isomer has stronger dipole-dipole forces; the *trans* isomer is nonpolar. The higher boiling point of the *cis* isomer supports this conclusion.

        (b)   While boiling points are primarily a measure of strength of intermolecular forces, melting points are influenced by crystal packing efficiency as well as intermolecular forces. Since the nonpolar *trans* isomer with weaker intermolecular forces has the higher melting point, it must pack more efficiently.

11.75 (a)   In dibromethane, CH_2Br_2, the dispersion force contribution will be larger than for CH_2Cl_2, because bromine is more polarizable than the lighter element chlorine. At the same time, the dipole-dipole contribution for CH_2Cl_2 is greater than for CH_2Br_2 because CH_2Cl_2 has a larger dipole moment.

        (b)   Just the opposite comparisons apply to CH_2F_2, which is less polarizable and has a higher dipole moment than CH_2Cl_2.

11.76   When a halogen atom (Cl or Br) is substituted for H in benzene, the molecule becomes polar. These molecules experience dispersion forces similar to those in benzene plus dipole-dipole forces, so they have higher boiling points than benzene. C_6H_5Br has a higher molar mass and is more polarizable than C_6H_5Cl so it has the higher boiling point. C_6H_5OH experiences hydrogen bonding, the strongest force between neutral molecules, so it has the highest boiling point.

11.77   The more carbon atoms in the hydrocarbon, the longer the chain, the more polarizable the electron cloud, the higher the boiling point. A plot of the number of carbon atoms versus boiling point is shown below. For 8 C atoms, the boiling point is approximately 130°C.

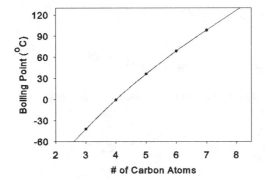

11.78   (a) Viscosity   (b) boiling point   (c) surface tension   (d) enthalpy of vaporization, ΔH_{vap}

11.79   Propylamine experiences hydrogen bonding interactions while trimethylamine, with no N-H bonds, does not. Also, the rod-like shape of propylamine (see Exercise 11.15) leads to stronger dispersion forces than in trimethylamine. The stronger intermolecular forces in propylamine lead to the lower vapor pressure.

11.80   The two O-H groups in ethylene glycol are involved in many hydrogen bonding interactions, leading to its high boiling point and viscosity, relative to pentane, which experiences only dispersion forces.

11.81   (a)   If the Clausius-Clapeyron equation is obeyed, a graph of ln VP vs 1/T(K) should be linear. Here are the data in a form for graphing.

T(K)	1/T	VP(torr)	ln(VP)
280.0	3.571×10^{-3}	32.42	3.479
300.0	3.333×10^{-3}	92.47	4.527
320.0	3.125×10^{-3}	225.1	5.417
330.0	3.030×10^{-3}	334.4	5.812
340.0	2.941×10^{-3}	482.9	6.180

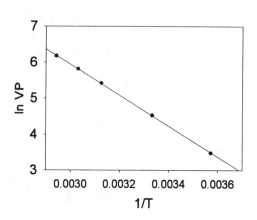

According to the graph, the Clausius-Clapeyron equation is obeyed, to a first approximation.

$$\Delta H_{vap} = -\text{slope} \times R; \quad \text{slope} = \frac{3.479 - 6.180}{(3.571 - 2.941) \times 10^{-3}} = -\frac{2.701}{0.630 \times 10^{-3}} = -4.29 \times 10^{3}$$

$$\Delta H_{vap} = -(-4.29 \times 10^{3}) \times 8.314 \text{ J/K} \cdot \text{mol} = 35.7 \text{ kJ/mol}$$

        (b)   The normal boiling point is the temperature at which the vapor pressure of the liquid equals atmospheric pressure, 760 torr. From the graph,

            ln 760 = 6.63, 1/T for this VP = 2.828×10^{-3}; T = 353.6 K

11.82   (a)   The Clausius-Clapeyron equation is $\ln P = \dfrac{-\Delta H_{vap}}{RT} + C.$

            For two vapor pressures, P_1 and P_2, measured at corresponding temperatures T_1 and T_2, the relationship is

$$\ln P_1 - \ln P_2 = \left(\frac{-\Delta H_{vap}}{RT_1} + C \right) - \left(\frac{-\Delta H_{vap}}{RT_2} + C \right)$$

$$\ln P_1 - \ln P_2 = \frac{-\Delta H_{vap}}{R} \left(\frac{1}{T_1} - \frac{1}{T_2} \right) + C - C; \quad \ln \frac{P_1}{P_2} = \frac{-\Delta H_{vap}}{R} \left(\frac{1}{T_1} - \frac{1}{T_2} \right)$$

(b) $P_1 = 10.00$ torr, $T_1 = 716$ K; $P_2 = 400.0$ torr, $T_2 = 981$ K

$$\ln\frac{10.00}{400.0} = \frac{-\Delta H_{vap}}{8.314\ \text{J/K}\cdot\text{mol}}\left(\frac{1}{716} - \frac{1}{981}\right)$$

$-3.6889\ (8.314\ \text{J/K}\cdot\text{mol}) = -\Delta H_{vap}(3.773 \times 10^{-4}/\text{K})$

$\Delta H_{vap} = 8.129 \times 10^4 = 8.13 \times 10^4$ J/mol $= 81.3$ kJ/mol

(c)   The normal boiling point of a liquid is the temperature at which the vapor pressure of the liquid is 760 torr.

$P_1 = 400.0$ torr, $T_1 = 981$ K; $P_2 = 760$ torr, $T_2 = $ b.p. of potassium

$$\ln\left(\frac{400.0}{760.0}\right) = \frac{-8.129 \times 10^4\ \text{J/mol}}{8.314\ \text{J/K}\cdot\text{mol}}\left(\frac{1}{981\ \text{K}} \times \frac{1}{T_2}\right)$$

$$\frac{-0.64185}{-9.7775 \times 10^3} = 1.0194 \times 10^{-3} - \frac{1}{T_2};\ \frac{1}{T_2} = 1.0194 \times 10^{-3} - 6.565 \times 10^{-5}$$

$\frac{1}{T_2} = 9.5375 \times 10^{-4};\ T_2 = 1048$ K $(775°C)$

(d) $P_1 = $ VP of K(l) at $100°C$, $T_1 = 373$ K; $P_2 = 10.00$ torr, $T_2 = 716$ K

$$\ln\frac{P_1}{10.00\ \text{torr}} = \frac{-8.129 \times 10^4\ \text{J/mol}}{8.314\ \text{J/K}\cdot\text{mol}}\left(\frac{1}{373} - \frac{1}{716}\right)$$

$$\ln\frac{P_1}{10.00\ \text{torr}} = \frac{-8.129 \times 10^4\ \text{J/mol}}{8.314\ \text{J/K}\cdot\text{mol}} \times 1.284 \times 10^{-3} = -12.5543$$

$$\frac{P_1}{10.00\ \text{torr}} = e^{-12.5543} = 3.530 \times 10^{-6};\ P_1 = 3.53 \times 10^{-5}\ \text{torr}$$

11.83   Physical data for the two compounds from the *Handbook of Chemistry and Physics*:

	MM	dipole moment	boiling point
CH_2Cl_2	85 g/mol	1.60 D	40.0°C
CH_3I	142 g/mol	1.62 D	42.4°C

(a)   The two substances have very similar molecular structures; each is an unsymmetrical tetrahedron with a single central carbon atom and no hydrogen bonding. Since the structures are very similar, the magnitudes of the dipole-dipole forces should be similar. This is verified by their very similar dipole moments. The heavier compound, CH_3I, will have slightly stronger London dispersion forces. Since the nature and magnitude of the intermolecular forces in the two compounds is nearly the same, it is very difficult to predict which will be more volatile (or which will have the higher boiling point as in part (b)).

(b)   Given the structural similarities discussed in part (a), one would expect the boiling points to be very similar, and they are. Based on its larger molar mass (and dipole-dipole forces being essentially equal) one might predict that CH_3I would have a slightly higher boiling point; this is verified by the known boiling points.

(c)   According to Equation 11.1, $\ln P = \dfrac{-\Delta H_{vap}}{RT} + C$

A plot of ln P vs. 1/T for each compound is linear. Since the order of volatility changes with temperature for the two compounds, the two lines must cross at some temperature; the slopes of the two lines, ΔH_{vap} for the two compounds, and the y-intercepts, C, must be different.

(d)   **CH_2Cl_2**

lnVP	T(K)	1/T
2.303	229.9	4.351×10^{-3}
3.689	250.9	3.986×10^{-3}
4.605	266.9	3.747×10^{-3}
5.991	297.3	3.364×10^{-3}

CH_3I

lnVP	T(K)	1/T
2.303	227.4	4.398×10^{-3}
3.689	249.0	4.016×10^{-3}
4.605	266.2	3.757×10^{-3}
5.991	298.5	3.350×10^{-3}

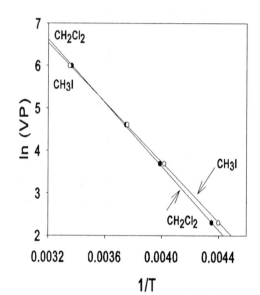

For CH_2Cl_2, $-\Delta H_{vap}/R$ = slope =

$$\frac{(5.991 - 2.303)}{(3.364 \times 10^{-3} - 4.350 \times 10^{-3})} = \frac{-3.688}{0.987 \times 10^{-3}} = -3.74 \times 10^3 = -\Delta H_{vap}/R$$

$\Delta H_{vap} = 8.314 (3.74 \times 10^3) = 3.107 \times 10^4$ J/mol = 31.1 kJ/mol

For CH_3I, $-\Delta H_{vap}/R$ = slope =

$$\frac{(5.991 - 2.303)}{(3.350 \times 10^{-3} - 4.398 \times 10^{-3})} = \frac{-3.688}{1.048 \times 10^{-3}} = -3.519 \times 10^3 = -\Delta H_{vap}/R$$

$\Delta H_{vap} = 8.314 (3.519 \times 10^3) = 2.926 \times 10^4$ J/mol = 29.3 kJ/mol

11.84   The length of the body diagonal in a body-centered cubic cell is 4r, where r is the radius of a Cr atom. To deduce the value of r, look at the triangle formed by the cube side (length = a), a face diagonal (length $= \sqrt{2}\ a$) and the body diagonal (length = 4r).

The angle θ is arctan $1/\sqrt{2} = 35.264°$.

Sin $\theta = a/4r$; r = 2.884 Å/4 sin 35.264° = 1.249 Å.

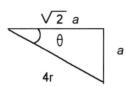

11.85   The most effective diffraction of light by a grating occurs   when the wavelength of light and the separation of the slits in the grating are similar. When X-rays are diffracted by a crystal, layers of atoms serve as the "slits." The most effective diffraction occurs when the distances between layers of atoms are similar to the wavelength of the X-rays. Typical interlayer distances in crystals range from 2 Å to 20 Å. Visible light, 400-700 nm or 4,000 to 7,000 Å, is too long to be diffracted effectively by crystals. Molybdenum x-rays of 0.71 Å are on the same order of magnitude as interlayer distances in crystals and are diffracted.

11.86   (a)   Both diamond (d = 3.5 g/cm^3) and graphite (d = 2.3 g/cm^3) are network covalent solids with efficient packing arrangements in the solid state; there is relatively little empty space in their respective crystal lattices. Diamond, with bonded C-C distances of 1.54 Å in all directions, is more dense than graphite, with shorter C-C distances within carbon sheets but longer 3.41 Å separations between sheets (Figure 11.42). Buckminsterfullerene has much more empty space, both inside each C$_{60}$ "ball" and between balls, than either diamond or graphite, so its density will be considerably less than 2.3 g/cm^3.

(b)   In a face-centered-cubic unit cell, there are 4 complete C$_{60}$ units.

$$\frac{4\,C_{60}\ units}{(14.2\ \text{Å})^3} \times \frac{729.66\ g}{6.022 \times 10^{23}\ C_{60}\ units} \times \left(\frac{1\ \text{Å}}{1 \times 10^{-8}\ cm}\right)^3 = 1.67\ g/cm^3$$

(1.67 g/cm^3 is the smallest density of the three allotropes, diamond, graphite and buckminsterfullerene.)

Integrative Exercises

11.87   (a)   The greater dipole moment of HCl (1.08 D vs 0.82 D for HBr) indicates that it will experience greater dipole-dipole forces.

(b)   The longer bond length in HBr (1.41 Å vs 1.27 Å for HCl) indicates a more diffuse and therefore more polarizable electron cloud in HBr. It experiences stronger London dispersion forces.

(c)   For molecules with similar structures, the compound with the higher boiling point experiences stronger intermolecular forces. Since HBr has the higher boiling point, weaker dipole-dipole forces but stronger dispersion forces, dispersion forces must determine the boiling point.

(d)   HF experiences hydrogen bonding, a much stronger force than the dipole-dipole and dispersion forces operating in the other hydrogen halides, which results in the high boiling point.

HI, with the longest bond length and greatest polarizability in the series, has the strongest dispersion forces and highest boiling point of the compounds that do not experience hydrogen bonding.

11.88 $n = \dfrac{PV}{RT} = 735 \text{ torr} \times \dfrac{1 \text{ atm}}{760 \text{ torr}} \times \dfrac{K \cdot mol}{0.08206 \text{ L} \cdot atm} \times \dfrac{3.00 \text{ L}}{290 \text{ K}} = 0.1219 = 0.122 \text{ mol } C_4H_{10}$

$0.1219 \text{ mol } C_4H_{10} \times \dfrac{21.3 \text{ kJ}}{1 \text{ mol } C_4H_{10}} = 2.60 \text{ kJ}$

11.89   Strategy:

i)   Using thermochemical data from Appendix B, calculate the energy (enthalpy) required to melt and heat the H_2O.

ii)   Using Hess's Law, calculate the enthalpy of combustion, ΔH_{comb}, for C_3H_8.

iii)   Solve the stoichiometry problem.

i)   Heat $H_2O(s)$ from -10.0°C to 0.0°C; $1000 \text{ g } H_2O \times \dfrac{2.092 \text{ J}}{g \cdot °C} \times 10.0 °C = 20.92 = 20.9 \text{ kJ}$

Melt $H_2O(s)$; $1000 \text{ g } H_2O \times \dfrac{6.008 \text{ kJ}}{mol \, H_2O} \times \dfrac{1 \text{ mol } H_2O}{18.02 \text{ g } H_2O} = 333.41 = 333.4 \text{ kJ}$

Heat $H_2O(l)$ from 0.0°C to 75.0°C; $1000 \text{ g } H_2O \times \dfrac{4.184 \text{ J}}{g \cdot °C} \times 75.0 °C = 313.8 = 314 \text{ kJ}$

Total energy = 20.92 kJ + 333.4 kJ + 313.8 kJ = 668.1 = 668 kJ

(The result has zero decimal places because 314 kJ has zero decimal places.)

ii) $C_3H_8(g) + 5O_2(g) \rightarrow 3CO_2(g) + 4H_2O(l)$

Assume that one product is $H_2O(l)$, since this leads to a more negative ΔH_{comb} and fewer grams of $C_3H_8(g)$ required.

$\Delta H_{comb} = 3\Delta H^°_f \, CO_2(g) + 4\Delta H^°_f \, H_2O(l) - \Delta H^°_f \, C_3H_8(g) - \Delta H^°_f \, O_2(g)$

$= 3(-393.5 \text{ kJ}) + 4(-285.83 \text{ kJ}) - (-103.85 \text{ kJ}) - 0 = -2219.97 = -2220 \text{ kJ}$

iii) $668.1 \text{ kJ required} \times \dfrac{1 \text{ mol } C_3H_8}{2219.97 \text{ kJ}} \times \dfrac{44.094 \text{ g } C_3H_8}{1 \text{ mol } C_3H_8} = 13.3 \text{ g } C_3H_8$

(668 kJ required has 3 sig figs and so does the result)

11.90 $P = \dfrac{nRT}{V} = \dfrac{g\,RT}{MM \bullet V}$; T = 273 + 26°C = 299 K; V = 5.00 L

g C_6H_6(g) = 7.2146 − 5.1493 = 2.0653 g C_6H_6(g)

P (vapor) = $\dfrac{2.0653\ g}{78.11\ g/mol} \times \dfrac{299\ K}{5.00\ L} \times \dfrac{0.08206\ L \bullet atm}{K \bullet mol} \times \dfrac{760\ torr}{1\ atm}$ = 98.6 torr

11.91 $PV = \dfrac{g}{MM} \times RT$; $g = \dfrac{PV \bullet MM}{RT}$; T = 313 K; MM = 18.02 g/mol

P (the vapor pressure of H_2O at 40°C) = 55.3 torr $\times \dfrac{1\ atm}{760\ torr}$ = 0.07276 = 0.0728 atm

V = 4.0 m × 4.0 m × 3.0 m $\times \dfrac{10^3\ dm^3}{1\ m^3}$ = 4.8 × 10⁴ L

$g = \dfrac{0.07276\ atm \times 4.8 \times 10^4\ L \times 18.02\ g}{313\ K \times mol} \times \dfrac{K \bullet mol}{0.08206\ L \bullet atm}$ = 2.450 × 10³ g = 2.5 kg H_2O

11.92   (a)   There are 90 bonds in the C_{60} molecule. [Hint: Use a soccer ball to count.]

(b)   If 60 C atoms each share four pairs of electrons with atoms other than C, a total of 240 bonds are required. If C is bound only to C, as is the case in C_{60}, two C atoms participate in each single bond, so 120 single bonds are required to satisfy each C atom. Since there are only 90 bonds in C_{60}, some of these must be double bonds. In a double bond, two C atoms each share two pairs of electrons; each double bond takes care of four of the required bonds. Thus,

> 60 single bonds × 2 C atoms
> + 30 double bonds × 2 C atoms × 2 pairs of electrons
> 240 total "bonds"

Of the 90 bonds in C_{60}, 30 are formally double bonds.

12 Modern Materials

Liquid Crystals

12.1   Both an ordinary liquid and a nematic liquid crystal phase are fluids; they are converted directly to the solid phase upon cooling. The nematic phase is cloudy and more viscous than an ordinary liquid. Upon heating, the nematic phase is converted to an ordinary liquid.

12.2   In an ordinary liquid, molecules are oriented randomly and their relative orientations are continuously changing. In liquid crystals, the molecules are aligned in at least one dimension. The relative orientations in the other two dimensions may change, but alignment in the oriented direction is maintained.

12.3   Reinitzer observed that cholesteryl benzoate has a phase that exhibits properties intermediate between those of the solid and liquid phases. This "liquid-crystalline" phase, formed by melting at 145°C, is opaque, changes color from red to blue as the temperature is increased, and becomes clear at 179°C.

12.4   In the solid state, there is three-dimensional order; the relative orientation of the molecules is fixed and repeating in all three dimensions. Essentially no translational or rotational motion is allowed. When a substance changes to the nematic liquid crystalline base, the molecules remain aligned in one dimension (the long dimension of the molecule). Translational motion is allowed, but rotational motion is restricted. Transformation to the isotropic liquid phase destroys the one-dimensional order. Free translational and rotational motion result in random molecular orientations that change continuously.

12.5   Because order is maintained in at least one dimension, the molecules in a liquid crystalline phase are not totally free to change orientation. This makes the liquid crystalline phase more resistant to flow, more viscous, than the isotropic liquid.

12.6   The presence of polar groups or nonbonded pairs of electrons leads to relatively strong dipole-dipole interactions between molecules. These are a significant part of the orienting forces necessary for liquid crystal formation.

12.7   In the nematic phase, the long axes of the molecules are aligned. Translational motion is allowed, but rotational motion is restricted. In the smectic phase, both the long axes and the ends of the molecules are aligned; the molecules are organized into sheets. Both translational and rotational motion are restricted.

12.8 All of the molecules in Figure 12.5 are long, rod-like and somewhat rigid. The rigid rods can have flexible tails, as in cholesteric phases (Figure 12.7). Functional groups that contribute to the rigidity of the rod and increase the strength of the dispersion forces among molecules are benzene rings, C=N and N=N bonds.

12.9 nematic - one dimension; smectic - at least two dimensions.

12.10 The three major classes of liquid-crystalline materials are nematic, smectic and chlolesteric. Nematic and smetic phases are formed by rod-like molecules aligned with their long axes parallel. Smectic phases have some additional ordering, which often involves alignment of the ends of the molecular rods to form sheets. In cholesteric phases, the molecules are stacked in layers. Within a layer, the molecules are aligned, but they are usually twisted with respect to molecules in adjacent layers (Figure 12.6).

Polymers

12.11 *n*-decane does not have a sufficiently high chain length or molecular mass to be considered a polymer.

12.12 Monomers are small molecules with low molecular mass that are joined together to form polymers. They are the repeating units of a polymer. Three monomers mentioned in this chapter are

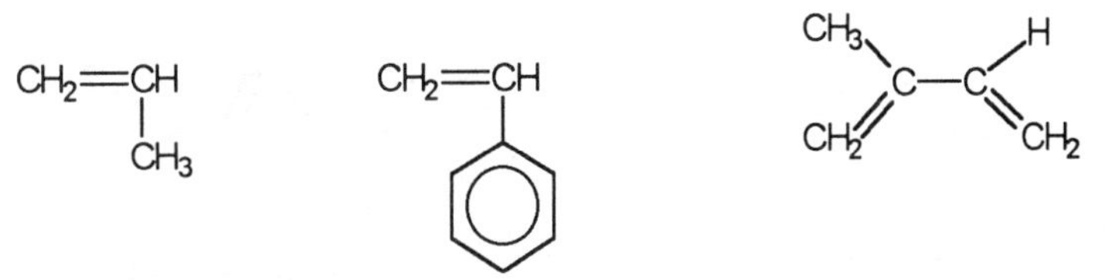

propylene
(propene)

styrene
(phenyl ethene)

isoprene
(2-methyl-1,3-butadiene)

12.13

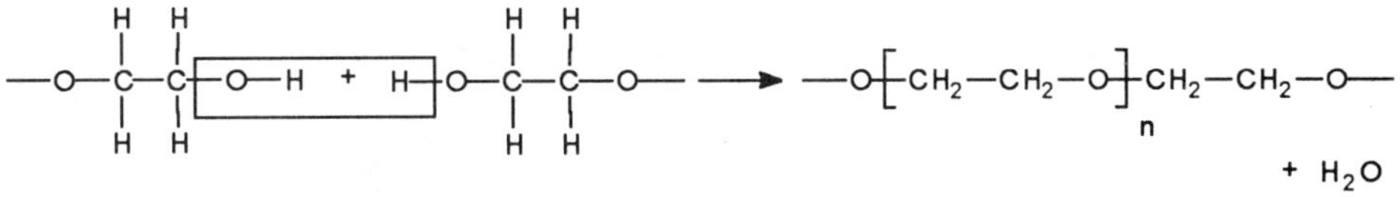

12.14 Condensation reaction to form an ester:

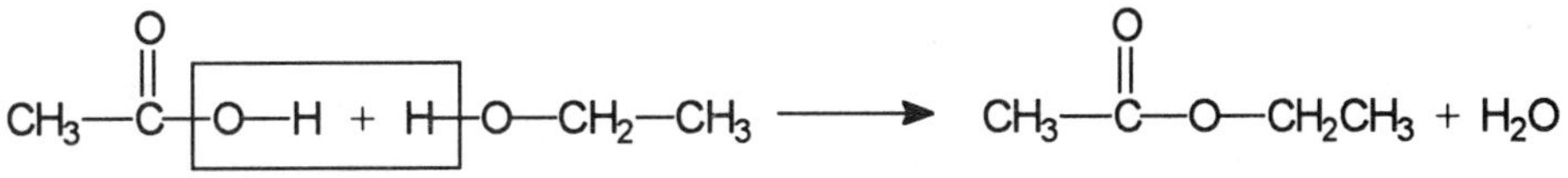

acetic acid   ethanol   ethyl acetate

If a dicarboxylic acid (two -COOH groups, usually at opposite ends of the molecule) and a dialcohol (two -OH groups, usually at opposite ends of the molecule) are combined, there is the potential for propagation of the polymer chain at both ends of both monomers. Polyethylene terephthalate (Table 12.1) is an example of a polyester formed from the monomers ethylene glycol and terephthalic acid.

12.15   (a)

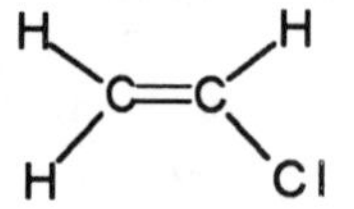

vinyl chloride (chloroethylene or chloroethene)

(b)

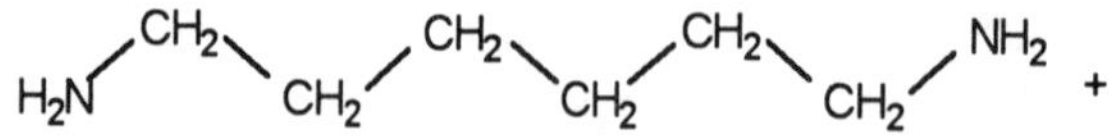

hexanediamine

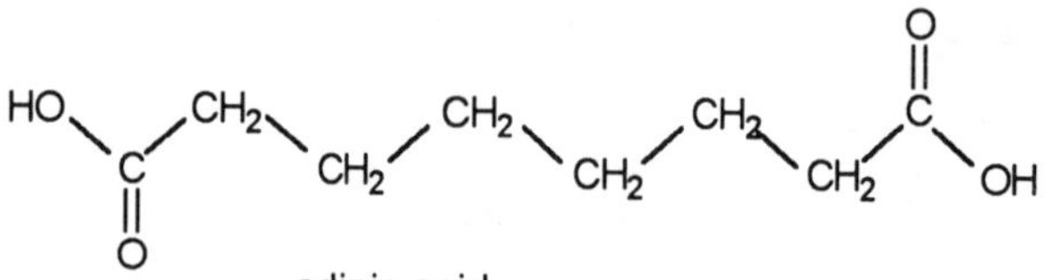

adipic acid

(Formulas given in Equation 12.3.)

(c)

ethylene glycol   terephthalic acid

12.16   (a)   By analogy to polyisoprene, Table 12.1,

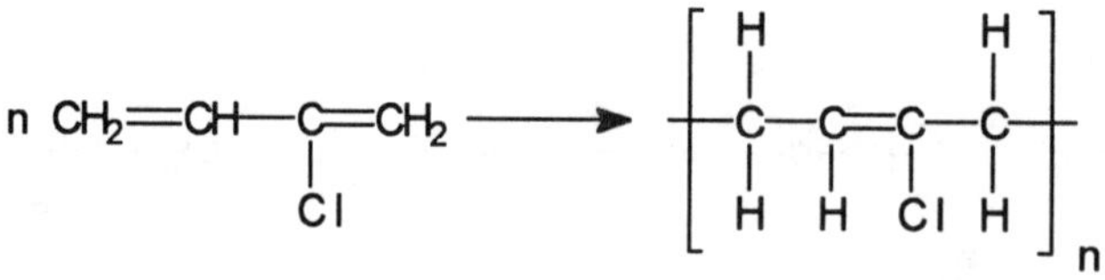

(b)

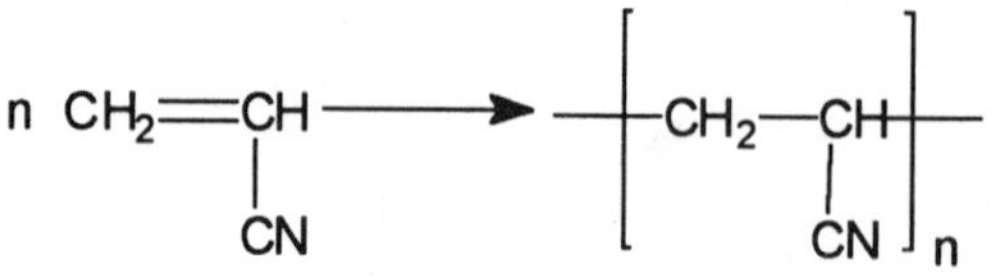

12.17

12.18   When nylon polymers are made, H_2O is produced as the C-N bonds are formed. Reversing this process (adding H_2O across the C-N bond), we see that the monomers used to produce Nomex are:

                                   and

12.19   High density polyethylene (HDPE) has a higher molecular mass, melting point, density and mechanical strength than low density polyethylene (LDPE). At the molecular level, the longer, unbranched chains of HDPE fit closer together and have more crystalline (ordered, aligned) regions than the shorter, branched chains of LDPE. Closer packing leads to higher density. The stronger dispersion forces that result from tighter packing and more alignment lead to higher melting point and greater mechanical strength.

12.20   Most of a polymer backbone is composed of σ bonds. The geometry around individual atoms is tetrahedral with bond angles of $109°$, so the polymer is not flat, and there is relatively free rotation around the σ bonds. The flexibility of the molecular chains causes flexibility of the bulk material. Flexibility is enhanced by molecular features that inhibit order, such as branching, and diminished by features that encourager order, such as crosslinking or delocalized π electron density.

         Crosslinking is the formation of chemical bonds between polymer chains. It reduces flexibility of the molecular chains and increases the hardness of the material. Crosslinked polymers are less chemically reactive because of the links.

12.21   The function of the material (polymer) determines whether high molecular mass and high degree of crystallinity are desirable properties. If the material will be formed into containers or pipes, rigidity and structural strength are required. If the polymer will be used as a flexible wrapping or as a garment material, rigidity is an undesirable property.

12.22   (a)   An *elastomer* is a polymer material that recovers its shape when released from a distorting force. A typical elastomeric polymer can be stretched to at least twice its original length and return to its original dimensions upon release.

(b)   A *thermoplastic* material can be shaped and reshaped by application of heat and/or pressure.

(c)   A *thermosetting plastic* can be shaped once, through chemical reaction in the shape-forming process, but cannot easily be reshaped, due to the presence of chemical bonds that crosslink the polymer chains.

(d)   A *plasticizer* is a substance of relatively low molar mass added to a polymer material to soften it.

Ceramics

12.23   Structurally, polymers are formed from organic monomers held together by covalent bonds, whereas ceramics are formed from inorganic materials linked by ionic or highly polar covalent bonds. Ceramics are often stabilized by a three-dimensional bonding network, whereas in polymers, covalent bonds link atoms into a long, chain-like molecule with only weak interactions between molecules. Polymers are nearly always amorphous, and though ceramics can be amorphous, they are often crystalline.

In terms of physical properties, ceramics are generally much harder, more heat-stable and higher melting than polymers. (These properties are true in general for network solids relative to molecular solids, as described in Table 11.6.)

12.24   The rigid, three-dimensional network structure of ceramics explains these physical properties. They are high-melting, because melting requires that ionic or covalent bonds be broken. They are hard and brittle because of the rigidity of the network. There is no mechanism for deformation without destruction of the network structure.

12.25   Since Zr and Ti are in the same family, assume that the stoichiometry of the compounds in a sol-gel process will be the same for the two metals.

   i.   Alkoxide formation - oxidation-reduction reaction
          $Zr(s) + 4CH_3CH_2OH(l) \rightarrow Zr(OCH_2CH_3)_4(s) + 2H_2(g)$
                               alkoxide

   ii.   Sol formation - metathesis reaction
          $Zr(OCH_2CH_3)_4(soln) + 4H_2O(l) \rightarrow Zr(OH)_4(s) + 4CH_3CH_2OH(l)$
                                              "precipitate"   nonelectrolyte
                                                  sol

   $Zr(OCH_2CH_3)_4(s)$ is dissolved in an alcohol solvent and then reacted with water. In general, reaction with water is called *hydrolysis*. The alkoxide anions ($CH_3CH_2O^-$) combine with H^+ from H_2O to form the nonelectrolyte $CH_3CH_2OH(l)$, and Zr^{2+} cations combine with OH^- to form the $Zr(OH)_4$ solid. The product $Zr(OH)_4(s)$ is not a traditional coagulated precipitate, but finely divided evenly dispersed particles called a sol.

iii.　　Gel formation - condensation reaction

$(OH)_3Zr-O-H(s) + H-O-Zr(OH)_3(s) \rightarrow (HO)_3Zr-O-Zr(OH)_3(s) + H_2O(l)$

　　　　　　　　　　　　　　　　　gel

Adjusting the acidity of the $Zr(OH)_4$ sol initiates condensation, the splitting-out of $H_2O(l)$ and formation of a zirconium-oxide network solid. The solid remains suspended in the solvent mixture and is called a gel.

iv.　　Processing - physical changes

The gel is heated to drive off solvent and the resulting solid consists of dry, uniform and finely divided ZrO_2 particles.

12.26　Very small, uniformly sized and shaped particles are required for the production of a strong ceramic object by sintering. During sintering, the small ceramic particles are heated to a high temperature below the melting point of the solid. This high temperature initiates condensation reactions between atoms at the surfaces of the spheres; the spheres are then connected by chemical bonds between atoms in different spheres. The more uniform the particle size and the greater the total surface area of the solid, the more chemical bonds are formed, and the stronger the ceramic object.

12.27　Concrete is a typically brittle ceramic and susceptible to catastrophic fracture. Steel reinforcing rods are added to resist stress applied along the long direction of the rod. By analogy, the shape of the reinforcing material in the ceramic composite should be rod-like, with a length much greater than its diameter. This is the optimal shape because rods have great strength when the load or stress is applied parallel to the long direction of the rod. Rods can be oriented in many directions, so that the material (concrete or ceramic composite) is strengthened in all directions.

12.28　Ceramic materials typically have rigid three-dimensional lattices; movements of atoms with respect to one another require the breaking of chemical bonds. For this reason, most ceramics are not very flexible, especially compared to an organic polymer solid. As bonds in a ceramic are broken under great stress, the break tends to propagate itself by applying stress to adjacent atoms.

12.29　By analogy to the ZnS structure, the C atoms form a face-centered cubic array with Si atoms occupying **alternate** tetrahedral holes in the lattice. This means that the coordination numbers of both Si and C are 4; each Si is bound to 4 C atoms in a tetrahedral arrangement, and each C is bound to 4 Si atoms in a tetrahedral arrangement, producing an extended three-dimensional network. ZnS, an ionic solid, sublimes at 1185° and 1 atm pressure and melts at 1850° and 150 atm pressure. The considerably higher melting point of SiC, 2800° at 1 atm, indicates that SiC is probably not a purely ionic solid and that the Si-C bonding network has significant covalent character. This is reasonable, since the electronegativities of Si and C are similar (Figure 8.7). SiC is high-melting because a great deal of chemical energy is stored in the covalent Si-C bonds, and it is hard because the three-dimensional lattice resists any change that would weaken the Si-C bonding network.

12.30   The four ceramics in Table 12.4 are Al_2O_3, SiC, ZrO_2 and BeO. All are network solids with some ionic and some covalent bonding character. They have a higher modulus of elasticity, which is to say they are more rigid than the metals listed. The high melting points, hardness and rigidity of the these four solids are due to the rigid, highly ordered, three-dimensional network structure of the solids. These network structures resist deformation (hardness) and require a great deal of energy (high melting point) to break the chemical bonds that form the network.

12.31   A superconducting material offers no resistance to the flow of electrical current; it is the frictionless flow of electrons. Superconductive materials could transmit electricity with no heat loss and therefore much greater efficiency than current carriers. Because of the Meisner effect, they are also potential materials for magnetically levitated trains.

12.32   Because they are brittle ceramics, it is difficult to mold superconductors into useful shapes such as wires and these wires would be fragile at best. For presently known superconductors, the amount of current per cross-sectional area that can be carried by these wires is limited. The low temperatures required for superconductivity renders today's superconducting ceramics impractical for widespread use.

Thin Films

12.33   In general, a useful thin film should:

(a)   be chemically stable in its working environment
(b)   adhere to its substrate
(c)   have a uniform thickness
(d)   have an easily controllable composition
(e)   be nearly free of imperfections

12.34   Adhesion is due to attractive intermolecular forces. These include ion-dipole, dipole-dipole, dispersion and hydrogen bonding. Adhesive interactions will be strongest between substances with similar intermolecular forces, so the bonding characteristics (that determine intermolecular forces) of the thin film material should be matched to those of the substrate.

12.35   There are three major methods of producing thin films.

(i)   In *vacuum deposition*, a substance is vaporized or evaporated by heating under vacuum and then deposited on the desired substrate. No chemical change occurs.

(ii)   In *sputtering*, ions accelerated to high energies by applying a high voltage are allowed to strike the target material, knocking atoms from its surface. These target material atoms are further accelerated toward the substrate, forming a thin film. No net chemical change occurs, because the material in the film is the target material.

(iii) In *chemical vapor deposition*, two gas phase substances react at the substrate surface to form a stable product which is deposited as a thin film. This involves a net chemical change.

12.36   The coating in Figure 12.27 is a metallic film that reflects most of the incident sunlight. The exclusion of sunlight from the interior of the building reduces glare and cooling load. The opacity of the film provides privacy.

Additional Exercises

12.37   A dipole moment (permanent, partial charge separation) roughly parallel to the long dimension of the molecule would cause the molecules to reorient when an electric field is applied perpendicular to the usual direction of molecular orientation.

12.38

   Teflon is formed by addition polymerization.

12.39   (a) Polymer   (b) ceramic   (c) ceramic   (d) polymer   (e) liquid crystal (an organic molecule with a characteristic long axis and the kinds of functional groups often found in compounds with liquid-crystalline phases (Figure 12.5); not enough repeating units to be a polymer)

12.40   At the temperature where a substance changes from the solid to the liquid-crystalline phase, kinetic energy sufficient to overcome most of the long range order in the solid has been supplied. A few van der Waals forces have sufficient attractive energy to impose the one-dimensional order characteristic of the liquid-crystalline state. Very little additional kinetic energy (and thus a relatively small increase in temperature) is required to overcome these aligning forces and produce an isotropic liquid.

12.41   In a liquid crystal display (Figure 12.9), the molecules must be free to rotate by 90°. The long directions of molecules remain aligned but any attractive forces between the ends of molecules are disrupted. At low Antarctic temperatures, the liquid crystalline phase is closer to its freezing point. The molecules have less kinetic energy due to temperature and the applied voltage may not be sufficient to overcome orienting forces among the ends of molecules. If some or all of the molecules do not rotate when the voltage is applied, the display will not function properly.

12.42   This phenomenon is similar to supercooling, Section 11.4. When the isotropic liquid is cooled below the liquid crystal-liquid transition temperature, the kinetic energy of the molecules has been decreased enough so that formation of the liquid crystalline phase is energetically favorable. However, the molecules may not be correctly organized so that long range ordering can take place.

12.43   Ceramics are usually three-dimensional network solids, whereas plastics most often consist of large, chain-like molecules (the chain may be branched) held loosely together by relatively weak van der Waals forces. Ceramics are rigid precisely because of the many strong bonding interactions intrinsic to the network. Once a crack forms, atoms near the defect are subject to great stress, and the crack is propagated. They are stable to high temperatures because tremendous kinetic energy (temperature) is required for an atom to break free from the bonding network. On the other hand, plastics are flexible because the molecules themselves are flexible (free rotation around the sigma bonds in the polymer chain), and it is easy for the molecules to move relative to one another (weak intermolecular forces). (However, recall that rigidity of the plastic increases as crosslinking of the polymer chain increases. The melamine-formaldehyde polymer in Figure 12.20 is a very rigid, brittle polymer.) Plastics are not thermally stable because their largely organic molecules are subject to oxidation and/or bond breaking at high temperatures.

12.44   Both metals and ceramics have highly ordered three-dimensional structures or lattices, but the bonding characteristics of these lattices are different. Many metals adopt close-packed structures where each metal has 12 nearest neighbors (Sections 11.8 and 23.5). Since each metal atom has too few electrons to form a covalent (electron-pair) bond with each of its nearest neighbors, the bonding electrons are delocalized and free to move throughout the lattice. This provides a mechanism for stress relief without catastrophic fracture. Metal atoms or planes of metal atoms can slip relative to each other without significant damage to the metallic bonding network.

In the three dimensional lattices of ceramics, bonding electrons are localized. In ionic solids such as Al_2O_3 they are localized on the ions; in network covalent solids such as SiC (Exercise 12.25), the electrons are localized in the covalent bonds between the atoms. In either case, atoms or ions cannot move without breaking chemical bonds. Localized (ionic or covalent) bonding in ceramics renders them harder than metals, which have delocalized bonding throughout the crystal lattice.

12.45   HDPE (Table 12.3, 0.96 g/cm^3, mp = $133\,°$C) is less dense, much lower melting, softer and more elastic than silica glass (2.66 g/cm^3, mp = $2661\,°$C). HDPE consists of long, flexible, unbranched chains of $-CH_2-$ groups with an average molecular mass of 1×10^6 amu. Interactions among chains are weak dispersion forces. The solid state structure resembles a bowl of tangled strands of spaghetti with no long range order. Silica (SiO_2) glass (Figure 11.28) is a network of SiO_2 units held together by strong polar covalent Si-O bonds. The SiO_2 units are bound into rings of varying sizes that result in a rigid three-dimensional network with no long range order or crystallinity.

The higher density of SiO_2 is due to the strict steric requirements of the Si-O covalent bonds. In HDPE no strong forces hold polymer chains close together. Twists and bends in the flexible chains often prevent close contact and result in a low density for the material. SiO_2 is higher melting because covalent bonds must be broken before the SiO_2 units can move relative to each other. Little thermal energy is required to overcome the dispersion forces among polymer chains and thus melt HDPE. Silica glass is harder and more rigid because of the rigid covalent network of SiO_2 groups. There is no mechanism for the SiO_2 units to move relative to each other without breaking covalent bonds, which results in catastrophic fracture. In HDPE, the polymer chains can move relative to each other and there is flexibility within the chains, rendering the material soft and flexible.

Chemically, HDPE is combustible to $CO_2(g)$ + $H_2O(g)$, because C atoms in the polymer are in a low oxidation state. SiO_2 glass is unreactive because Si atoms in the ceramic network are already in their maximum oxidation state.

12.46   (a)

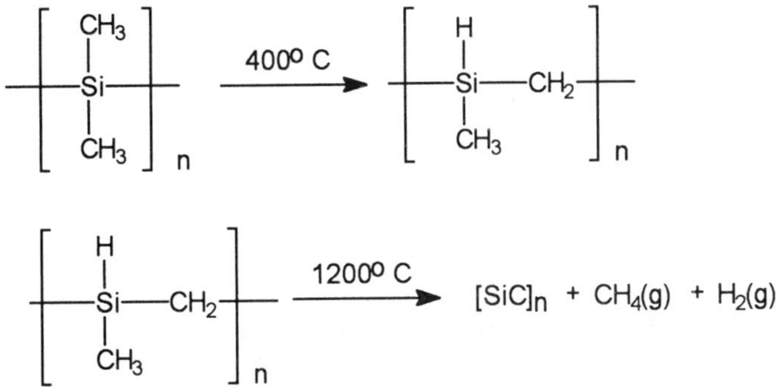

        (b) $2NbBr_5(g) + 5H_2(g) \rightarrow 2Nb(s) + 10HBr(g)$

        (c) $SiCl_4(l) + C_2H_5OH(l) \rightarrow Si(OC_2H_5)_4(s) + 4HCl(g)$

        (d)

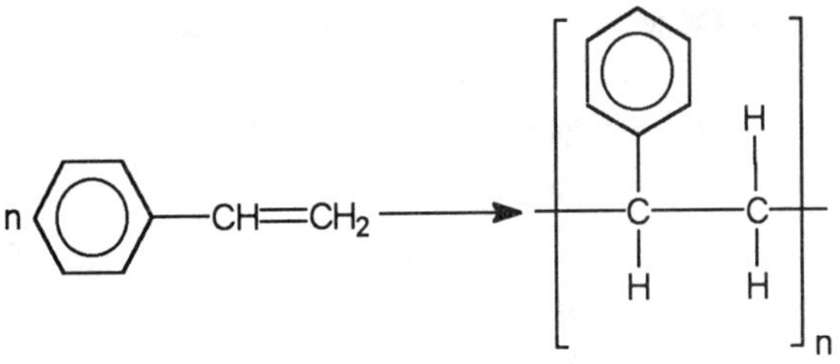

12.47   Both Kevlar and Nylon 6,6 contain the

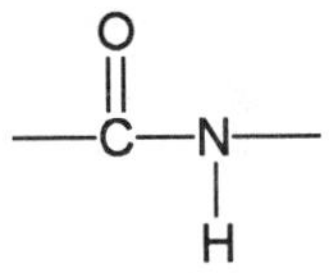

functional group. According to Section 25.3, this is called an **amide** group. Both are formed by the condensation reaction of a diacid and a diamine. Both have strong -N-H----O hydrogen bonds between polymer chains. They differ in the "spacer" between amide groups. In Kevlar, the spacer is a rigid, planar benzene ring. In Nylon 6,6, the spacer is a flexible $-(CH_2)_6-$ chain. The rigid, planar benzene groups in the Kevlar backbone, coupled with strong interchain hydrogen bonding causes Kevlar to form ordered, planar sheets. Because the sheets are planar and there are relatively strong dispersion forces between benzene rings, the arrangement of sheets is also regular. The flexibility in the Nylon 6,6 backbone inhibits sheet formation and produces the more flexible material. Both can be drawn into fibers, but nylon fibers are used for textiles and fishing line, while Kevlar fibers are used to make bullet-proof vests and cables stronger than steel.

233

12.48

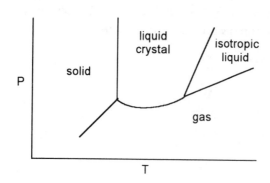

12.49 $TiCl_4(g) + 2SiH_4(g) \rightarrow TiSi_2(s) + 4HCl(g) + 2H_2(g)$

As a ceramic, $TiSi_2$ will have a three dimensional network structure similar to that of Si. (Si(s) has a diamond-like network covalent structure, Figure 11.42). At the surface of the thin film there will be Ti atoms and Si atoms with incomplete valences that can and will chemically bond with Si atoms on the surface of the substrate. This kind of bonding would not be possible with a Cu thin film. Strong adherence to the surface is an essential component of thin film performance.

12.50 The formula of the compound deposited as a thin film is indicated by boldface type.

(a) $SiH_4(g) + 2H_2(g) + 2CO_2(g) \rightarrow \mathbf{SiO_2(s)} + 2H_2(g) + 2CO(g)$

(b) $TiCl_4(g) + 2H_2O(g) \rightarrow \mathbf{TiO_2(s)} + 4HCl(g)$

(c) $GeCl_4(g) \rightarrow \mathbf{Ge(s)} + 2Cl_2(g)$ The H_2 carrier gas dilutes the $GeCl_4(g)$ so the reaction occurs more evenly and at a controlled rate; it does not participate in the reaction.

12.51 If the expected oxidation states on Y and Ba are +3 and +2, respectively, the **average** oxidation state of Cu is +2 1/3. That is, two Cu ions are in the +2 state and one is in the +3 state. Y^{3+} and Ba^{2+} have the stable electron configurations of their nearest noble gases, while Cu has an incomplete d orbital set. Cu(II) is d^9, Cu(III) is d^8 and both have unpaired electrons. Although the mechanism by which the copper 3d electrons interact through the bridging oxygen atoms to form a superconducting state is still not clear, it is evident that the electronic structure of the copper ions is essential to the observed superconductivity.

12.52 These compounds cannot be vaporized without destroying their chemical identities. Anions with names ending in "ite" or "ate" are oxyanions with nonmetallic elements as central atoms. Under the conditions of vacuum deposition, high temperatures and low pressures, these anions tend to chemically decompose to form gaseous nonmetal oxides. For example:

$$MnSO_4(s) \rightarrow MnO(s) + SO_3(g)$$

12.53   Application of an electric field in the direction perpendicular to the planes of the plates would cause the molecules to orient along the direction of the applied field, vertical in the diagram below.

Because the molecules have thermal energy which causes them to move randomly, alignment parallel to the applied field is not complete, but some net reorientation does take place. This happens because the molecules in a nematic substance generally have a dipole moment (partial charge separation) oriented roughly parallel to the long direction of the molecule. Also, the molecules are more polarizable in their long direction, so London dispersion forces between molecules are stronger when the molecules are aligned in the direction of the applied field.

Integrative Exercises

12.54   (a)

HDPE

$$\Delta H = D(C=C) - 2D(C-C) = 614 - 2(348) = 82 \text{ kJ/mol } C_2H_4$$

(b) $(n + 1)$ HOOC—$(CH_2)_6$—COOH + $(n + 1)$ H_2N—$(CH_2)_6$—NH_2 ⟶

Nylon 6,6

$$\Delta H = 2D(C-O) + 2D(N-H) - 2D(C-N) - 2D(H-O)$$

$$\Delta H = 2(358) + 2(391) - 2(293) - 2(463) = -14 \text{ kJ/mol}$$

(This is -14 kJ/mol of either reactant.)

(c)

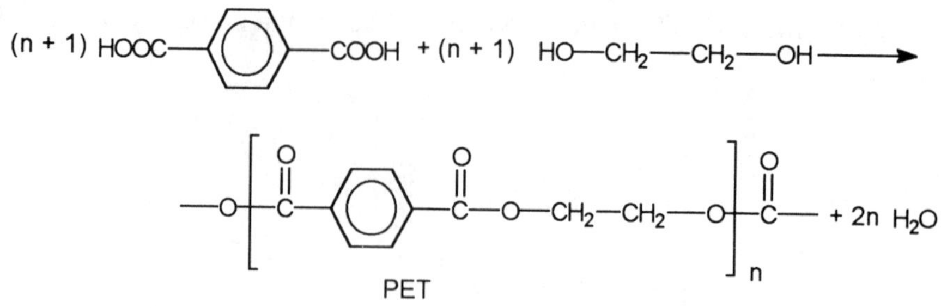

$$\Delta H = 2D(\text{C-O}) + 2D(\text{O-H}) - 2D(\text{C-O}) - 2D(\text{O-H}) = 0 \text{ kJ}$$

12.55 (a) sp^3 hybrid orbitals at C, 109° bond angles around C

(b)

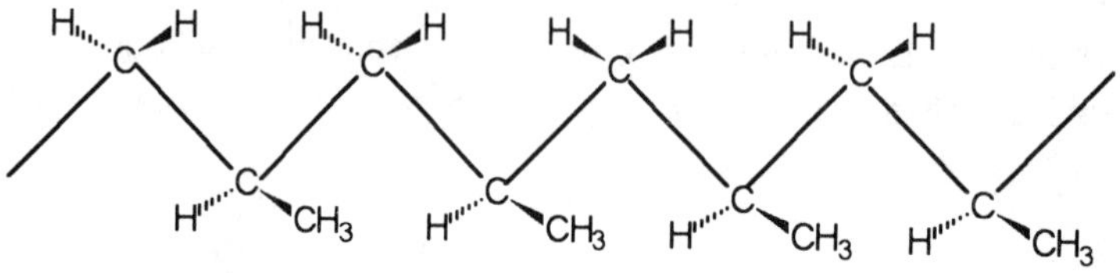

isotactic

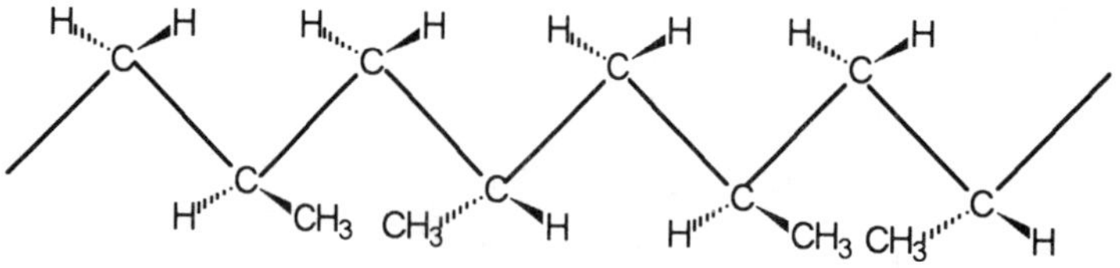

syndiotactic

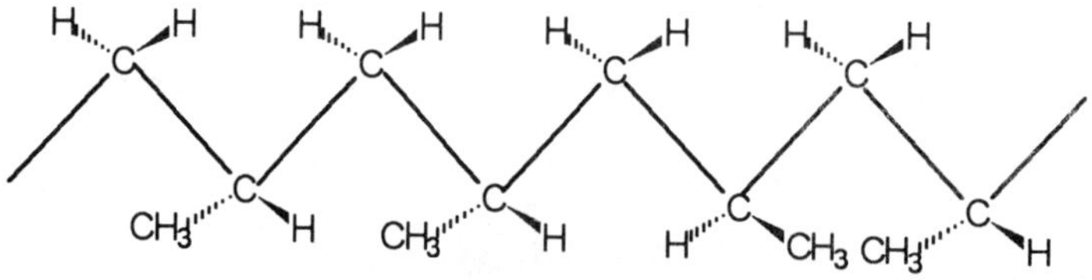

atactic

Isotactic polypropylene has the highest degree of crystallinity and highest melting point. The regular shape of the polymer backbone allows for close, orderly (almost zipper like) contact between chains. This maximizes dispersion forces between chains and produces higher order (crystallinity) and melting point. Atacticpolypropylene has the least order and the lowest melting point.

(c)   Cotton, with $-\overset{\displaystyle |}{\underset{\displaystyle OH}{C}}-$ groups and polyester, with $-\overset{\displaystyle O}{\overset{\displaystyle \|}{C}}-O-C-$

groups, both participate in hydrogen bonding interactions with H_2O molecules. These are strong intermolecular forces that hold the "moisture" at the surface of the fabric next to the skin. Polypropylene has no strong interactions with water and capillary action "wicks" the moisture away from the skin.

12.56   (a)   The average molar mass (MM_{avg}) of the gas is the average of the individual molar masses weighted for their relative abundances.

$MM_{avg} = (40/41)(16.04) + (1/41)(2.016) = 15.70$ g $= 16$ g

gas density (g/L) $= MM_{avg}$ P/RT; $P = 90$ torr $\times \dfrac{1 \text{ atm}}{760 \text{ torr}} = 0.1184$ atm $= 0.12$ atm

$\rho = 15.70$ g/mol $\times 0.1184$ atm $\times \dfrac{K \cdot mol}{0.08206 \, L \cdot atm} \times \dfrac{1}{1123 \, K} = 0.020$ g/L

(b)   The density of diamond, C(s,d), is 3.51 g/mL. Calculate the volume of the diamond film, and then its mass.

$V = 1$ cm$^2 \times 2.5$ μm $\times \dfrac{1 \times 10^{-6} \text{ m}}{1 \, \mu m} \times \dfrac{100 \text{ cm}}{m} \times \dfrac{1 \text{ mL}}{1 \text{ cm}^3} = 2.5 \times 10^{-4}$ mL

$V = 3 \times 10^{-4}$ mL (to 1 sig fig)

$\dfrac{3.51 \text{ g C (s,d)}}{1 \text{ mL}} \times 2.5 \times 10^{-4}$ mL $= 8.775 \times 10^{-4}$ g $= 9 \times 10^{-4}$ g C (s, d)

Calculate how many grams of gas are needed to deposit 9×10^{-4} g of film, and the volume of this amount of gas.

8.775×10^{-4} g C (s, d) $\times \dfrac{15.70 \text{ g gas}}{12.01 \text{ g C (s,d)}} = 1.147 \times 10^{-3}$ g gas contains the amount of C(s,d) in the film.

But the process is only 0.5% efficient.

$\dfrac{1.147 \times 10^{-3} \text{ g gas needed}}{\text{total of gas}} \times 100 = 0.5$; total g gas $= 0.2294$ g $= 0.23$ g gas

$V = \dfrac{g \, R \, T}{MM_{avg} \, P} = \dfrac{0.2294 \text{ g}}{15.70 \text{ g/mol}} \times \dfrac{0.08206 \, L \cdot atm}{K \cdot mol} \times \dfrac{1123 \, K}{0.1184 \text{ atm}} = 11.37$ L $= 11$ L

12.57 (a)   The data (14.99%) has 4 sig figs, so use molar masses to 5 sig figs.

$$\text{mass \% O} = 14.99 = \frac{(8+x)\,15.999}{746.04 + (8+x)\,15.999} \times 100$$

rounded (to show sig figs)

(8+x)15.999
   = 0.1499 [746.04 + (8+x) 15.999]
127.99 + 15.999x
   = 0.1499(874.04 + 15.999x)
15.999x - 2.398x
   = 131.0 - 127.99
13.601x = 3.0; x = 0.22

unrounded

(8+x)15.999
   = 0.1499 [746.04 + (8+x) 15.999]
127.992 + 15.999x
   = 0.1499 (874.036 + 15.999x
15.999x - 2.3983x
   = 131.018 - 127.992
13.6007x = 3.026; x = 0.2225

(b)   **Hg** and **Cu** both have more than one stable oxidation state. If different Cu ions (or Hg ions) in the solid lattice have different charges, then the average charge is a noninteger value. Ca and Ba are stable only in the +2 oxidation state; they are unlikely to have noninteger average charge.

(c) Ba^{2+} is largest; Cu^{2+} is smallest. For ions with the same charge, size decreases going up or across the periodic table. In the +2 state, Hg is smaller than Ba. If Hg has an average charge greater than 2+, it will be smaller yet. The same argument is true for Cu and Ca.

12.58 (a)

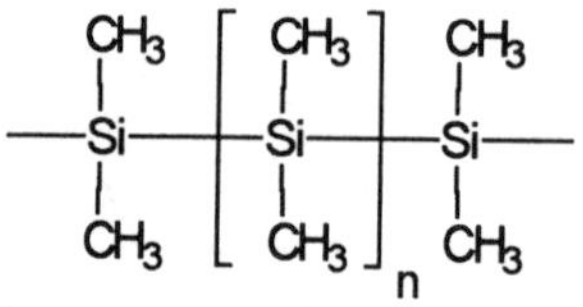

Si-Si   226 kJ ◄— lowest
Si-C   301 kJ
C-H   413 kJ

(b)   Si-Si bonds are weakest, so they are most likely to break upon heating.

(c)   At this point, a C-H bond must be broken, because the product polymer has $-CH_2-$ in the backbone, while only terminal $-CH_3$ groups are present in the reactant polymer.

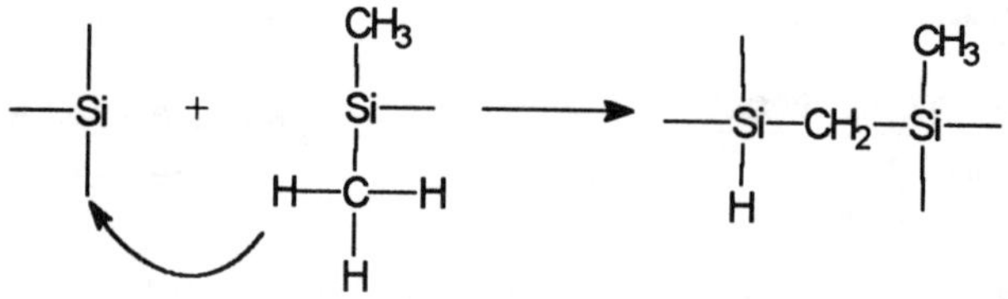

ΔH = D(C-H) - D(Si-C) - D(Si-H) = 413 - 301 - 323 = -211 kJ/mol

12.59 (a)

There are several other resonance structures involving alternate placement of the double bonds in the benzene rings.

(b) Both N atoms are surrounded by 3 VSEPR "pairs" of electrons, so the hybridization at both atoms is sp^2. The bond angles around the N attached to O will be approximately $120°$.

(c) When the -OCH$_3$ group is replaced by the -CH$_2$CH$_2$CH$_2$CH$_3$ group, a small rather compact group with some polarity is replaced by a larger, more flexible, nonpolar group. Thus, the molecules don't line up as well in the solid, and the melting point and liquid crystal temperature range are lower.

(d) The density decreases going from solid to nematic liquid crystal to isotropic liquid. In the nematic liquid crystal, most of the long range order of the solid state is lost. The molecules are moving relative to one another with their long axes more or less aligned. The result is more empty space and a lower density than the solid. There is a further small decrease in density when the last degree of order is lost and substance becomes an isotropic liquid.

13 Properties of Solutions

The Solution Process

13.1   If the enthalpy released due to solute-solvent attractive forces (ΔH_3) is at least as large as the enthalpy required to separate the solute particles (ΔH_1), the overall enthalpy of solution (ΔH_{soln}) will be either slightly endothermic (owing to $+\Delta H_2$) or exothermic. Even if ΔH_{soln} is slightly endothermic, the increase in disorder due to mixing will cause a significant amount of solute to dissolve. If the magnitude of ΔH_3 is small relative to the magnitude of ΔH_1, ΔH_{soln} will be large and endothermic (energetically unfavorable) and not much solute will dissolve.

13.2   From weakest to strongest solvent-solute interactions:
(b), dispersion forces < (c), hydrogen bonding < (a), ion-dipole

13.3   (a)   dispersion   (b) ion-dipole   (c) hydrogen bonding   (d) dipole-dipole

13.4   (a)   ion-dipole   (HCl is a strong electrolyte that dissociates into H^+ and Cl^- in aqueous solution)

(b)   dispersion   (c) ion-dipole   (d) ion-ion

13.5   Water, H_2O, is a polar solvent that forms hydrogen bonds with other H_2O molecules. The more soluble solute in each case will have intermolecular interactions that are most similar to the hydrogen bonding in H_2O.

(a) CH_3CH_2OH is more soluble because it has a shorter nonpolar hydrocarbon chain. The longer nonpolar hydrocarbon chain of $CH_3CH_2CH_2CH_2OH$ is capable only of dispersion forces and interferes with hydrogen bonding interactions between the -OH group and water.

(b)   Ionic $CaCl_2$ is more soluble because ion-dipole solute-solvent interactions are more similar to ionic solute-solute and hydrogen bonding solvent-solvent interactions than the weak dispersion forces between CCl_4 and H_2O.

(c) C_6H_5OH is more soluble because it is capable of hydrogen bonding. Nonpolar C_6H_6 is capable only of dispersion force interactions and does not have strong intermolecular interactions with polar (hydrogen bonding) H_2O.

13.6   Hexane is a nonpolar hydrocarbon that experiences dispersion forces with other C_6H_{14} molecules. Solutes that primarily experience dispersion forces will be more soluble in hexane.

(a)   Cyclohexane, C_6H_{12}, is also a nonpolar hydrocarbon and will be more soluble in hexane. Glucose experiences hydrogen bonding with itself; these solute-solute interactions are less likely to be overcome by weak solute-solvent interactions.

(b)   Propionic acid, which experiences hydrogen bonding, will be more soluble than sodium propionate, which experiences ion-ion forces. The hydrogen bonding is weaker (relatively speaking) and more likely to be overcome by dispersion forces with hexane.

(c)   Ethyl chloride, which has a -CH_2CH_3 group capable of dispersion forces, will be more soluble. HCl experiences dipole-dipole forces less likely to be disrupted by dispersion forces with hexane.

13.7   The overall energy change associated with dissolution depends on the relative magnitudes of the solute-solute, solvent-solvent, and solute-solvent interactions. If disrupting the "old" solute-solute and solvent-solvent interactions requires more energy than is released by the "new" solute-solvent interactions, the process is endothermic. When $NH_4NO_3(s)$ dissolves in water, disrupting the ion-ion interactions among NH_4^+ and NO_3^- ions of the solute (ΔH_1) and the hydrogen bonding among H_2O molecules of the solvent (ΔH_2) requires more energy than is released by subsequent ion-dipole and hydrogen bonding interactions among NH_4^+ and H_2O and NO_3^- and H_2O (ΔH_3).

13.8   Separation of solvent molecules, ΔH_2, will be smallest in this case, because hydrogen bonding is the weakest of the intermolecular forces involved. ΔH_1 involves breaking ionic bonds, and ΔH_3 involves formation of ion-dipole interactions, both stronger forces than hydrogen bonding.

13.9   (a) ΔH_{soln} is determined by the relative magnitudes of the "old" solute-solute (ΔH_1) and solvent-solvent (ΔH_2) interactions and the new solute-solvent interactions (ΔH_3); $\Delta H_{soln} = \Delta H_1 + \Delta H_2 + \Delta H_3$. Since the solute and solvent in this case experience very similar London dispersion forces, the energy required to separate them individually and the energy released when they are mixed are approximately equal. $\Delta H_1 + \Delta H_2 \approx -\Delta H_3$. Thus, ΔH_{soln} is nearly zero.

(b)   Mixing hexane and heptane produces a homogeneous solution from two pure substances, and the randomness of the system increases. Since no strong intermolecular forces prevent the molecules from mixing, they do so spontaneously due to the increase in disorder.

13.10   KBr is quite soluble in water because of the sizeable increase in disorder of the system (ordered KBr lattice → freely moving hydrated ions) associated with the dissolving process. An increase in disorder or randomness in a process tends to make that process spontaneous.

Concentrations of Solutions

13.11   (a)   mass % = $\dfrac{\text{mass solute}}{\text{total mass solution}} \times 100 = \dfrac{16.5\,\text{g CaCl}_2}{16.5\,\text{g CaCl}_2 + 456\,\text{g H}_2\text{O}} \times 100 = 3.49\%$

(b) $ppm = \dfrac{\text{mass solute}}{\text{total mass solution}} \times 10^6; \dfrac{83.5\,g\,Ag}{1\,\text{ton ore}} \times \dfrac{1\,\text{ton}}{2000\,lb} \times \dfrac{1\,lb}{453.6\,g} \times 10^6$

$= 92.0\,ppm$

13.12   (a) $\text{mass \%} = \dfrac{\text{mass solute}}{\text{total mass solution}} \times 100$

mass solute $= 0.065\,mol\,I_2 \times \dfrac{253.8\,g\,I_2}{1\,mol\,I_2} = 16.497 = 16\,g\,I_2$

mass % $I_2 = \dfrac{16.497\,g\,I_2}{16.497\,g\,I_2 + 120.0\,g\,CCl_4} \times 100 = 12\%\,I_2$

(b) $\dfrac{0.412\,g\,Ca^{2+}}{1\,kg\,H_2O} \times \dfrac{1\,mol\,Ca^{2+}}{40.08\,g\,Ca^{2+}} = 0.0103\,mol\,Ca^{2+}/kg\,H_2O\ (0.0103\,m\ Ca^{2+})$

$ppm = \dfrac{\text{mass solute}}{\text{total mass solution}} \times 10^6 = \dfrac{0.412\,g\,Ca^{2+}}{1 \times 10^3\,g\,H_2O} \times 10^6 = 412\,ppm\,Ca^{2+}$

13.13   (a) $8.5\,g\,CH_3OH \times \dfrac{1\,mol\,CH_3OH}{32.04\,g\,CH_3OH} = 0.265 = 0.27\,mol\,CH_3OH$

$224\,g\,H_2O \times \dfrac{1\,mol\,H_2O}{18.02\,g\,H_2O} = 12.43 = 12.4\ mol\,H_2O$

$\chi_{CH_3OH} = \dfrac{0.265}{0.265 + 12.43} = 0.021$

(b) $\dfrac{65.2\,g\,CH_3OH}{32.04\,g/mol} = 2.035 = 2.04\,mol\,CH_3OH;\quad \dfrac{144\,g\,CCl_4}{153.8\,g/mol} = 0.9363 = 0.936\,mol\,CCl_4$

$\chi_{CH_3OH} = \dfrac{2.035}{2.035 + 0.9363} = 0.685$

13.14   (a) $\dfrac{4.5\,g\,C_6H_5OH}{94.11\,g/mol} = 0.0478 = 0.048\,mol\,C_6H_5OH$

$\dfrac{855\,g\,H_2O}{18.02\,g/mol} = 47.447 = 47.4\,mol\,H_2O \quad \chi_{C_6H_5OH} = \dfrac{0.0478}{0.0478 + 47.447} = 1.0 \times 10^{-3}$

(b) $\dfrac{44.0\,g\,C_6H_5OH}{94.11\,g/mol} = 0.4675 = 0.468\,mol\,C_6H_5OH$

$\dfrac{550\,g\,CH_3CH_2OH}{46.07\,g/mol} = 11.94 = 11.9\,mol\,CH_3CH_2OH$

$\chi_{C_6H_5OH} = \dfrac{0.4675}{0.4675 + 11.94} = 0.0377$

13.15   (a) $M = \dfrac{\text{mol solute}}{\text{L soln}}; \dfrac{10.5\,g\,NaCl}{0.350\,L\,soln} \times \dfrac{1\,mol\,NaCl}{58.44\,NaCl} = 0.513\,M$

(b) $\dfrac{40.7 \text{ g LiClO}_4 \cdot 3H_2O}{0.125 \text{ L soln}} \times \dfrac{1 \text{ mol LiClO}_4 \cdot 3H_2O}{160.4 \text{ g LiClO}_4 \cdot 3H_2O} = 2.03 \ M$

(c) $M_c \times L_c = M_d \times L_d$; 1.50 M HNO$_3$ × 0.0400 L = ?M HNO$_3$ × 0.500 L

500 mL of 0.120 M HNO$_3$

13.16 (a) $M = \dfrac{\text{mol solute}}{\text{L soln}}$; $\dfrac{25.0 \text{ g MgBr}_2}{0.355 \text{ L soln}} \times \dfrac{1 \text{ mol MgBr}_2}{184.1 \text{ g MgBr}_2} = 0.3825 = 0.383 \ M$

(b) $\dfrac{5.66 \text{ g Mn(NO}_3)_2 \cdot 2H_2O}{0.185 \text{ L soln}} \times \dfrac{1 \text{ mol Mn(NO}_3)_2 \cdot 2H_2O}{215.0 \text{ g Mn(NO}_3)_2 \cdot 2H_2O} = 0.142 \ M$

(c) $M_c \times L_c = M_d \times L_d$; 3.00 M H$_2SO_4$ × 0.085 L = ?M H$_2$SO$_4$ × 0.500 L

500 mL of 0.510 M H$_2$SO$_4$

13.17 (a) $m = \dfrac{\text{mol solute}}{\text{kg solvent}}$; $\dfrac{13.0 \text{ g C}_6H_6}{17.0 \text{ g CCl}_4} \times \dfrac{1 \text{ mol C}_6H_6}{78.11 \text{ g C}_6H_6} \times \dfrac{1000 \text{ g CCl}_4}{1 \text{ kg CCl}_4} = 9.79 \ m \text{ C}_6H_6$

(b) The density of H$_2$O = 0.997 g/mL = 997 kg/L.

$\dfrac{5.85 \text{ mol NaCl}}{0.250 \text{ L H}_2O} \times \dfrac{1 \text{ mol NaCl}}{58.44 \text{ g NaCl}} \times \dfrac{1 \text{ L H}_2O}{0.997 \text{ kg H}_2O} = 0.402 \ M \text{ NaCl}$

13.18 (a) $m = \dfrac{\text{mol solute}}{\text{kg solvent}}$; mol S$_8$ = m × kg C$_{10}$H$_8$ = 0.16 m × 0.1000 kg C$_{10}$H$_8$ = 0.016 mol

0.016 mol S$_8$ × $\dfrac{256.5 \text{ g S}_8}{1 \text{ mol S}_8}$ = 4.104 = 4.1 g S$_8$

(b) 16.0 mol H$_2$O × $\dfrac{18.02 \text{ g H}_2O}{1 \text{ mol H}_2O}$ = 288.3 g H$_2$O = 0.288 kg H$_2$O

$m = \dfrac{1.80 \text{ mol KCl}}{0.2883 \text{ kg H}_2O} = 6.2435 = 6.24 \ m$

13.19 Given: 100.0 mL of CH$_3$CN(l), 0.786 g/mL; 20.0 mL CH$_3$OH, 0.791 g/mL

(a) mol CH$_3$CN = $\dfrac{0.786 \text{ g}}{1 \text{ mL}}$ × 100.0 mL × $\dfrac{1 \text{ mol CH}_3CN}{41.05 \text{ g CH}_3CN}$ = 1.9147 = 1.91 mol

mol CH$_3$OH = $\dfrac{0.791 \text{ g}}{1 \text{ mL}}$ × 20.0 mL × $\dfrac{1 \text{ mol CH}_3OH}{32.04 \text{ g CH}_3OH}$ = 0.4938 = 0.494 mol

$\chi_{\text{CH}_3\text{OH}} = \dfrac{0.4938 \text{ mol CH}_3OH}{1.9147 \text{ mol CH}_3CN + 0.4938 \text{ mol CH}_3OH} = 0.205$

(b)   Assuming CH_3OH is the solute and CH_3CN is the solvent,

$$100.0 \text{ mL } CH_3CN \times \frac{0.786 \text{ g}}{1 \text{ mL}} \times \frac{1 \text{ kg}}{1000 \text{ g}} = 0.0786 \text{ kg } CH_3CN$$

$$m_{CH_3OH} = \frac{0.4938 \text{ mol } CH_3OH}{0.0786 \text{ kg } CH_3CN} = 6.28 \ m \ CH_3OH$$

(c)   The total volume of the solution is 120.0 mL, assuming volumes are additive.

$$M = \frac{0.4938 \text{ mol } CH_3OH}{0.1200 \text{ L solution}} = 4.12 \ M \ CH_3OH$$

13.20   Given: 15.0 g C_4H_4S, 1.065 g/mL; 250.0 mL C_7H_8, 0.897 g/mL

(a) $$\text{mol } C_4H_4S = 15.0 \text{ g } C_4H_4S \times \frac{1 \text{ mol } C_4H_4S}{84.15 \text{ g } C_4H_4S} = 0.1783 = 0.178 \text{ mol } C_4H_4S$$

$$\text{mol } C_7H_8 = \frac{0.867 \text{ g}}{1 \text{ mL}} \times 250.0 \text{ mL} \times \frac{1 \text{ mol } C_7H_8}{92.14 \text{ g } C_7H_8} = 2.352 = 2.35 \text{ mol}$$

$$\chi_{C_4H_4S} = \frac{0.1783 \text{ mol } C_4H_4S}{0.1783 \text{ mol } C_4H_4S + 2.352 \text{ mol } C_7H_8} = 0.07047 = 0.0705$$

(b) $$m_{C_4H_4S} = \frac{\text{mol } C_4H_4S}{\text{kg } C_7H_8}; \ 250.0 \text{ mL} \times \frac{0.867 \text{ g}}{1 \text{ mL}} \times \frac{1 \text{ kg}}{1000 \text{ g}} = 0.2168 = 0.217 \text{ kg } C_7H_8$$

$$m_{C_4H_4S} = \frac{0.1783 \text{ mol } C_4H_4S}{0.2168 \text{ kg } C_7H_8} = 0.822 \ m \ C_4H_4S$$

(c) $$15.0 \text{ g } C_4H_4S \times \frac{1 \text{ mL}}{1.065 \text{ g}} = 14.08 = 14.1 \text{ mL } C_4H_4S;$$

$$V_{soln} = 14.1 \text{ mL } C_4H_4S + 250.0 \text{ mL } C_7H_8 = 264.1 \text{ mL}$$

$$M_{C_4H_4S} = \frac{0.1783 \text{ mol } C_4H_4S}{0.2641 \text{ L soln}} = 0.675 \ M \ C_4H_4S$$

13.21   (a) $$\text{mass \%} = \frac{\text{mass } C_6H_8O_6}{\text{total mass solution}} \times 100;$$

$$\frac{80.5 \text{ g } C_6H_8O_6}{80.5 \text{ g } C_6H_8O_6 + 210 \text{ g } H_2O} \times 100 = 27.71 = 27.7\%$$

(b) $$\text{mol } C_6H_8O_6 = \frac{80.5 \text{ g } C_6H_8O_6}{176.1 \text{ g/mol}} = 0.4571 = 0.457 \text{ mol } C_6H_8O_6$$

$$\text{mol } H_2O = \frac{210 \text{ g } H_2O}{18.02 \text{ g/mol}} = 11.654 = 11.7 \text{ mol } H_2O$$

$$\chi_{C_6H_8O_6} = \frac{0.4571 \text{ mol } C_6H_8O_6}{0.4571 \text{ mol } C_6H_8O_6 + 11.654 \text{ mol } H_2O} = 0.0377$$

(c) $m = \dfrac{0.4571 \text{ mol } C_6H_8O_6}{0.210 \text{ kg } H_2O} = 2.18\ m\ C_6H_8O_6$

(d) $M = \dfrac{\text{mol } C_6H_8O_6}{\text{L solution}}$; $290.5 \text{ g soln} \times \dfrac{1 \text{ mL}}{1.22 \text{ g}} \times \dfrac{1 \text{ L}}{1000 \text{ mL}} = 0.2381 = 0.238 \text{ L}$

$M = \dfrac{0.4571 \text{ mol } C_6H_8O_6}{0.2381 \text{ L soln}} = 1.92\ M$

13.22   (a) $\dfrac{571.6 \text{ g } H_2SO_4}{1 \text{ L soln}} \times \dfrac{1 \text{ L soln}}{1329 \text{ g soln}} = 0.430098 \text{ g } H_2SO_4/\text{g soln}$

mass percent is thus $0.4301 \times 100 = 43.01\%$

(b)   In a liter of solution there are $1329 - 571.6 = 757.4 = 757 \text{ g } H_2O$.

$\dfrac{571.6 \text{ g } H_2SO_4}{98.09 \text{ g/mol}} = 5.827 \text{ mol } H_2SO_4$; $\dfrac{757.4 \text{ g } H_2O}{18.02 \text{ g/mol}} = 42.03 = 42.0 \text{ mol } H_2O$

$\chi_{H_2SO_4} = \dfrac{5.827}{42.03 + 5.827} = 0.122$

(The result has 3 sig figs because 42.0 mol H_2O limits the denominator to 3 sig figs.)

(c) $\text{molality} = \dfrac{5.827 \text{ mol } H_2SO_4}{0.7574 \text{ kg } H_2O} = 7.693 = 7.69\ m$

(d) $\text{molarity} = \dfrac{5.827 \text{ mol } H_2SO_4}{1 \text{ L soln}} = 5.827\ M$

13.23   Assume a solution volume of 1.00 L. Calculate the mass of 1.00 L of solution and the mass of HNO_3 in 1.00 L of solution.

$1.00 \text{ L} \times \dfrac{1000 \text{ mL}}{1 \text{ L}} \times \dfrac{1.42 \text{ g soln}}{\text{mL soln}} = 1.42 \times 10^3 \text{ g soln}$

$16\ M = \dfrac{16 \text{ mol } HNO_3}{1 \text{ L soln}} \times \dfrac{63.02 \text{ g } HNO_3}{1 \text{ mol } HNO_3} = 1008 = 1.0 \times 10^3 \text{ g } HNO_3$

$\text{mass \%} = \dfrac{1008 \text{ g } HNO_3}{1.42 \times 10^3 \text{ g soln}} \times 100 = 71\%$

13.24   Assume 1.00 L of solution. Calculate mol NH_3 in 1.00 L.

$1.00 \text{ L soln} \times \dfrac{1000 \text{ mL}}{1 \text{ L}} \times \dfrac{0.90 \text{ g soln}}{1 \text{ mL soln}} = 9.0 \times 10^2 \text{ g soln/L}$

$\dfrac{900 \text{ g soln}}{1.00 \text{ L soln}} \times \dfrac{28 \text{ g } NH_3}{100 \text{ g soln}} \times \dfrac{1 \text{ mol } NH_3}{17.03 \text{ g } NH_3} = 14.80 = 15 \text{ mol } NH_3/\text{L soln} = 15\ M$

13.25 (a) $\dfrac{0.240\ \text{mol MgBr}_2}{1\ \text{L soln}} \times 0.400\ \text{L} = 9.60 \times 10^{-2}\ \text{mol MgBr}_2$

(b) $\dfrac{0.460\ \text{mol glucose}}{1\ \text{L soln}} \times 80.0\ \mu\text{L} \times \dfrac{1 \times 10^{-6}\ \text{L}}{1\ \mu\text{L}} = 3.68 \times 10^{-5}\ \text{mol glucose}$

(c) $\dfrac{0.040\ \text{mol Na}_2\text{CrO}_4}{1\ \text{L soln}} \times 3.00\ \text{L} = 0.12\ \text{mol Na}_2\text{CrO}_4$

13.26 (a) $75.0\ \text{g soln} \times \dfrac{2.50\ \text{g C}_{12}\text{H}_{22}\text{O}_{11}}{100\ \text{g soln}} \times \dfrac{1\ \text{mol C}_{12}\text{H}_{22}\text{O}_{11}}{342.3\ \text{g C}_{12}\text{H}_{22}\text{O}_{11}} = 5.48 \times 10^{-3}\ \text{mol C}_{12}\text{H}_{22}\text{O}_{11}$

(b) $300\ \text{g soln} \times \dfrac{0.460\ \text{g NaCl}}{100\ \text{g soln}} \times \dfrac{1\ \text{mol NaCl}}{58.44\ \text{g NaCl}} = 2.36 \times 10^{-2}\ \text{mol NaCl}$

(c) $\dfrac{2.55\ \text{mol HNO}_3}{1\ \text{L soln}} \times 1.20\ \text{L} = 3.06\ \text{mol HNO}_3$

13.27 (a) $\dfrac{1.50 \times 10^{-2}\ \text{mol KBr}}{1\ \text{L soln}} \times 0.85\ \text{L} \times \dfrac{119.0\ \text{g KBr}}{1\ \text{mol KBr}} = 1.5\ \text{g KBr}$

Weigh out 1.5 g KBr, dissolve in water, dilute with stirring to 0.85 L (850 mL).

(b) Determine the mass % of KBr:

$\dfrac{0.180\ \text{mol KBr}}{1000\ \text{g H}_2\text{O}} \times \dfrac{119.0\ \text{g KBr}}{1\ \text{mol KBr}} = 21.42 = 21.4\ \text{g KBr/kg H}_2\text{O}$

Thus, mass fraction $= \dfrac{21.42\ \text{g KBr}}{1000 + 21.42} = 0.02097 = 0.0210$

In 165 g of the 0.180 *m* solution, there are

$(165\ \text{g soln}) \times \dfrac{0.02097\ \text{g KBr}}{1\ \text{g soln}} = 3.460 = 3.46\ \text{g KBr}$

Weigh out 3.46 g KBr, dissolve it in 165 - 3.46 = 162 g H_2O to make exactly 165 g of 0.180 *m* solution.

(c) Calculate the total mass of 1.85 L of solution, and from the mass % of KBr, the mass of KBr required.

$1.85\ \text{L soln} \times \dfrac{1000\ \text{mL}}{1\ \text{L}} \times \dfrac{1.10\ \text{g soln}}{1\ \text{mL}} = 2035 = 2.04 \times 10^{3}\ \text{g soln}$

0.120 (2035 g soln) = 244.2 = 244 g KBr

Dissolve 224 g KBr in water, dilute with stirring to 1.85 L.

(d)   Calculate moles KBr needed to precipitate 16.0 g AgBr.

$$16.0 \text{ g AgBr} \times \frac{1 \text{ mol AgBr}}{187.8 \text{ g AgBr}} \times \frac{1 \text{ mol KBr}}{1 \text{ mol AgBr}} = 0.08520 = 0.0852 \text{ mol KBr}$$

$$0.0852 \text{ mol KBr} \times \frac{1 \text{ L soln}}{0.150 \text{ mol KBr}} = 0.568 \text{ L soln}$$

Weigh out 0.0852 mol KBr (10.1 g KBr), dissolve it in a small amount of water and dilute to 0.568 L.

13.28   (a) $$\frac{0.110 \text{ mol Na}_2\text{CO}_3}{1 \text{ L soln}} \times 1.60 \text{ L} \times \frac{106.0 \text{ g Na}_2\text{CO}_3}{1 \text{ mol Na}_2\text{CO}_3} = 18.66 = 18.7 \text{ g Na}_2\text{CO}_3$$

Weigh 18.7 g Na_2CO_3, dissolve in a small amount of water, continue adding water with thorough mixing up to a total solution volume of 1.60 L.

(b)   Determine the mass % of $(NH_4)_2SO_4$ in the solution:

$$\frac{0.65 \text{ mol (NH}_4)_2\text{SO}_4}{1000 \text{ g H}_2\text{O}} \times \frac{132 \text{ g (NH}_4)_2\text{SO}_4}{1 \text{ mol (NH}_4)_2\text{SO}_4} = \frac{85.8 \text{ g (NH}_4)_2\text{SO}_4}{1000 \text{ g H}_2\text{O}}$$

$$\text{mass \%} = \frac{85.8 \text{ g (NH}_4)_2\text{SO}_4}{1000 \text{ g H}_2\text{O} + 85.8 \text{ g (NH}_4)_2\text{SO}_4} \times 100 = 7.90 = 7.9\%$$

In 120 g of solution, there are 0.079(120) = 9.48 = 9.5 g $(NH_4)_2SO_4$.

Weigh out 9.5 g $(NH_4)_2SO_4$ and dissolve it in 120 - 9.5 = 110.5 g H_2O to make exactly 150 g of solution.

(110.5 g H_2O/0.997 g H_2O/mL @ 25° = 110.8 mL H_2O)

(c) $$1.20 \text{ L} \times \frac{1000 \text{ mL}}{1 \text{ L}} \times \frac{1.20 \text{ g}}{1 \text{ mL}} = 1440 \text{ g solution}; \ 0.200(1440 \text{ g soln}) = 288 \text{ g Pb(NO}_3)_2$$

Weigh 288 g $Pb(NO_3)$ and add (1440 - 288) = 1152 g H_2O to make exactly 1400 g or 1.20 L of solution.

(1152 g H_2O/0.997 g/mL @ 25°C = 1155 mL H_2O)

(d)   Calculate the mol HCl necessary to neutralize 6.6 g $Ba(OH)_2$.

$$Ba(OH)_2(s) + 2HCl(aq) \rightarrow BaCl_2(aq) + 2H_2O(l)$$

$$6.6 \text{ g Ba(OH)}_2 \times \frac{1 \text{ mol Ba(OH)}_2}{171 \text{ g Ba(OH)}_2} \times \frac{2 \text{ mol HCl}}{1 \text{ mol Ba(OH)}_2} = 0.0772 = 0.077 \text{ mol HCl}$$

$$M = \frac{\text{mol}}{\text{L}}; \ L = \frac{\text{mol}}{M} = \frac{0.0772 \text{ mol HCl}}{0.50 \ M \text{ HCl}} = 0.1544 = 0.15 \text{ L} = 150 \text{ mL}$$

150 mL of 0.50 *M* HCl are needed.

$M_c \times L_c = M_d \times L_d$; 6.0 *M* × L_c = 0.50 *M* × 0.1544 L; L_c = 0.01290 L = 13 mL

Using a pipette, measure exactly 13 mL of 6.0 *M* HCl and dilute with water to a total volume of 150 mL.

Saturated Solutions: Factors Affecting Solubility

13.29   (a)   Supersaturated

       (b)   Add a seed crystal. Supersaturated solutions exist because not enough solute molecules are properly aligned for crystallization to occur. A seed crystal provides a nucleus of already aligned molecules, so that ordering of the dissolved particles is more facile.

13.30   (a) $\dfrac{1.22 \text{ mol MnSO}_4\cdot\text{H}_2\text{O}}{1 \text{ L soln}} \times \dfrac{169.0 \text{ g MnSO}_4\cdot\text{H}_2\text{O}}{1 \text{ mol}} \times 0.100 \text{ L}$

                      = 20.6 g $MnSO_4\cdot H_2O$/100 mL

            The 1.22 *M* solution is unsaturated.

       (b)   Add a known mass, say 5.0 g, of $MnSO_4\cdot H_2O$ to the unknown solution. If the solid dissolves, the solution is unsaturated. If there is undissolved $MnSO_4\cdot H_2O$, filter the solution and weigh the solid. If there is less than 5.0 g of solid, some of the added $MnSO_4\cdot H_2O$ dissolved and the unknown solution is unsaturated. If there is exactly 5.0 g, no additional solid dissolved and the unknown is saturated. If there is more than 5.0 g, excess solute has precipitated and the solution is supersaturated.

13.31   (a)   unsaturated   (b) saturated   (c) saturated   (d) unsaturated

13.32   (a)   at 30°C, $\dfrac{10 \text{ g KClO}_3}{100 \text{ g H}_2\text{O}} \times 250 \text{ g H}_2\text{O} = 25 \text{ g KClO}_3$

       (b) $\dfrac{66 \text{ g Pb(NO}_3)_2}{100 \text{ g H}_2\text{O}} \times 250 \text{ g H}_2\text{O} = 165 = 1.7 \times 10^2 \text{ g Pb(NO}_3)_2$

       (c) $\dfrac{3 \text{ g Ce}_2(\text{SO}_4)_3}{100 \text{ g H}_2\text{O}} \times 250 \text{ g H}_2\text{O} = 7.5 = 8 \text{ g Ce}_2(\text{SO}_4)_3$

13.33   In order for the dissolving process to be energetically favorable, the energy required to disrupt solute-solute and solvent-solvent intermolecular interactions ($\Delta H_1 + \Delta H_2$) must be similar to the energy released when solute-solvent interactions form (ΔH_3). This can only occur if the same types of interactions are being disrupted and formed. That is, if the types of intermolecular interactions in the pure solute and solvent are similar.

13.34 C_{He} = 3.7 ×10^{-4} M/atm × 2.5 atm = 9.2 × 10^{-4} M

        C_{N_2} = 6.0 × 10^{-4} M/atm × 2.5 atm = 1.5 × 10^{-3} M

13.35 $\dfrac{1.38 \times 10^{-3} M}{1 \text{ atm O}_2 \text{ pressure}}$ × 0.21 atm O_2 pressure = 2.9 × 10^{-4} M

13.36 $\dfrac{1.23 \times 10^{-3} M}{1 \text{ atm CH}_4 \text{ pressure}}$ × 2.1 × 10^{-6} atm = 2.6 × 10^{-6} M

Colligative Properties

13.37   Freezing point depression, boiling point elevation, osmotic pressure, vapor pressure lowering

13.38   Any concentration unit that expresses mole fraction, for example mole fraction or molality, is appropriate.

13.39   For a solution with a nonvolatile solute, the vapor pressure above the solution is equal to the mole fraction of the *solvent* times the vapor pressure of the pure solvent. Nonvolatile solute particles at the surface of a solution physically limit the number of chances for a solvent molecule to escape to the vapor phase. This causes the vapor pressure above the solution to be less than the vapor pressure of the pure solvent.

13.40   Ideal solutions obey Raoult's law. Calculate the vapor pressure predicted by Raoult's law and compare it to the experimental vapor pressure. Assume ethylene glycol (eg) is the solute.

        χ_{H_2O} = χ_{eg} = 0.500; $P_A = \chi_A P_A^{\circ}$ = 0.500(149) mm Hg = 74.5 mm Hg

        The vapor pressure of the solution (P_A), 67 mm Hg, is less than the value predicted by Raoult's law for an ideal solution. The solution is not ideal.

13.41   For this problem, it will be convenient to express Raoult's law in terms of the lowering of the vapor pressure of the solvent, ΔP_A.

        $\Delta P_A = P_A^{\circ} - \chi_A P_A^{\circ} = P_A^{\circ}(1 - \chi_A)$

        $1 - \chi_A = \chi_B$, the mole fraction of the *solute* particles

        $\Delta P_A^{\circ} = \chi_B P_A^{\circ}$; the vapor pressure of the solvent (A) is lowered according to the mole fraction of solute (B) particles present.

      (a)   Calculate χ_B by vapor pressure lowering; $\chi_B = \Delta P_A / P_A^{\circ}$. Given moles solvent, calculate moles solute from the definition of mole fraction.

             $\chi_{C_2H_6O_2} = \dfrac{13.2 \text{ torr}}{100 \text{ torr}}$ = 0.132

             $\dfrac{1.00 \times 10^3 \text{ g C}_2\text{H}_5\text{OH}}{46.07 \text{ g/mol}}$ = 21.71 = 21.7 mol C_2H_5OH; let y = mol $C_2H_6O_2$

$$\chi_{C_2H_6O_2} = \frac{y\ mol\ C_2H_6O_2}{y\ mol\ C_2H_6O_2 + 21.71\ mol\ C_2H_5OH} = 0.132 = \frac{y}{y + 21.71}$$

0.132 y + 2.866 = y; 0.868 y = 2.866; y = 3.302 = 3.30 mol $C_2H_6O_2$

3.302 mol $C_2H_6O_2 \times \dfrac{62.07\ g}{1\ mol} = 205\ g\ C_2H_6O_2$

(b) $\Delta P_A = \chi_B P_A°$; vapor pressure of the solvent is lowered according to the mole fraction of solute particles present.

P_{H_2O} at 40°C = 55.3 torr; $\dfrac{500\ g\ H_2O}{18.02\ g/mol} = 27.747 = 27.7\ mol\ H_2O$

$\chi_{ions} = \dfrac{4.60\ torr}{55.3\ torr} = \dfrac{y\ mol\ ions}{y\ mol\ ions + 27.747\ mol\ H_2O} = 0.08318 = 0.0832$

$0.08318 = \dfrac{y}{y + 27.747}$; 0.08318 y + 2.308 = y; 0.9168 y = 2.308,

y = 2.517 = 2.52 mol of ions

This result has 3 sig figs because (27.7 × 0.0832 = 2.31) has 3 sig figs.

2.517 mol ions $\times \dfrac{1\ mol\ KBr}{2\ mol\ ions} \times \dfrac{119.0\ g\ KBr}{mol\ KBr} = 149.8 = 150\ g\ KBr$

13.42 H_2O vapor pressure will be determined by the mole fraction of H_2O in the solution. The vapor pressure of pure H_2O at 338 K (65°C) = 187.5 torr; at 343 K (70°C) = 233.7 torr.

(a) $\dfrac{16.2\ g\ C_{12}H_{22}O_{11}}{342.3\ g/mol} = 0.04733 = 0.0473\ mol$; $\dfrac{105.7\ g\ H_2O}{18.02\ g/mol} = 5.8657 = 5.866\ mol$

$P_{H_2O} = \chi_{H_2O} P_{H_2O}° = \dfrac{5.8657\ mol\ H_2O}{5.8657 + 0.04733} \times 187.5\ torr = 186.0\ torr$

(b) $\dfrac{5.00\ g\ Na_2SO_4}{142.1\ g\ Na_2SO_4/mol} \times \dfrac{3\ mol\ ions}{1\ mol\ Na_2SO_4} = 0.1056 = 0.106\ mol\ ions$;

$\dfrac{92.0\ g\ H_2O}{18.02\ g/mol} = 5.105 = 5.11\ mol\ H_2O$

$P_{H_2O} = \dfrac{5.105\ mol\ H_2O}{5.105 + 0.1056} \times 187.5\ torr = 183.7 = 184\ torr$

(c) $\dfrac{32.5\ g\ C_3H_8O_3}{92.10\ g/mol} = 0.3529 = 0.353\ mol$; $\dfrac{140\ g\ H_2O}{18.02\ g/mol} = 7.769 = 7.77\ mol$

$P_{H_2O} = \dfrac{7.769\ mol\ H_2O}{7.769 + 0.3529} \times 233.7\ torr = 223.5 = 224\ torr$

13.43 At 63.5°C, $P°_{H_2O} = 175$ torr, $P°_{Eth} = 400$ torr.

Let G = the mass of H_2O and/or C_2H_5OH.

(a) $$\chi_{Eth} = \frac{\dfrac{G}{46.07 \text{ g } C_2H_5OH}}{\dfrac{G}{46.07 \text{ g } C_2H_5OH} + \dfrac{G}{18.02 \text{ g } H_2O}}$$

Multiplying top and bottom of the right side of the equation by 1/G gives:

$$\chi_{Eth} = \frac{1/46.07}{1/46.07 + 1/18.02} = \frac{0.02171}{0.02171 + 0.05549} = 0.2812$$

(b) $P_T = P_{Eth} + P_{H_2O}$; $P_{Eth} = \chi_{Eth} P°_{Eth}$; $P_{H_2O} = \chi_{H_2O} P°_{H_2O}$

$\chi_{Eth} = 0.2812$, $P_{Eth} = 0.2812 (400 \text{ torr}) = 112.48 = 112$ torr

$\chi_{H_2O} = 1 - 0.2812 = 0.7188$; $P_{H_2O} = 0.7188(175 \text{ torr}) = 125.8 = 126$ torr

$P_T = 112.5$ torr + 125.8 torr = 238.3 = 238 torr

(c) χ_{Eth} in vapor $= \dfrac{P_{Eth}}{P_{total}} = \dfrac{112.5 \text{ torr}}{238.3 \text{ torr}} = 0.4721 = 0.472$

13.44 (a) Since C_6H_6 and C_7H_8 form an ideal solution, we can use Raoult's Law. Since both components are volatile, both contribute to the total vapor pressure of 40 torr.

$P_T = P_{C_6H_6} + P_{C_7H_8}$; $P_{C_6H_6} = \chi_{C_6H_6} P°_{C_6H_6}$; $P_{C_7H_8} = \chi_{C_7H_8} P°_{C_7H_8}$

$\chi_{C_7H_8} = 1 - \chi_{C_6H_6}$; $P_T = \chi_{C_6H_6} P°_{C_6H_6} + (1 - \chi_{C_6H_6}) P°_{C_7H_8}$

$40 \text{ torr} = \chi_{C_6H_6}(75 \text{ torr}) + (1 - \chi_{C_6H_6})22 \text{ torr}$

$18 \text{ torr} = 53 \text{ torr } (\chi_{C_6H_6})$; $\chi_{C_6H_6} = \dfrac{18 \text{ torr}}{53 \text{ torr}} = 0.34$; $\chi_{C_7H_8} = 0.66$

(b) $P_{C_6H_6} = 0.34(75 \text{ torr}) = 25.5$ torr; $P_{C_7H_8} = 0.66(22 \text{ torr}) = 14.5$ torr

In the vapor, $\chi_{C_6H_6} = \dfrac{P_{C_6H_6}}{P_T} = \dfrac{25.5 \text{ torr}}{40 \text{ torr}} = 0.64$; $\chi_{C_7H_8} = 0.36$

13.45 $\Delta T = K (m)$; first, calculate the **molality** of each solution

(a) 0.17 m (b) $16.8 \text{ mol CHCl}_3 \times \dfrac{119.4 \text{ g CHCl}_3}{\text{mol CHCl}_3} = 2.006 = 2.01$ kg;

$\dfrac{1.92 \text{ mol } C_{10}H_8}{2.006 \text{ kg CHCl}_3} = 0.9571 = 0.957 \ m$

(c) $5.44 \text{ g KBr} \times \dfrac{1 \text{ mol KBr}}{119.0 \text{ g KBr}} \times \dfrac{2 \text{ mol particles}}{1 \text{ mol KBr}} = 0.09143 = 0.0914 \text{ mol particles}$

$6.35 \text{ g } C_6H_{12}O_6 \times \dfrac{1 \text{ mol } C_6H_{12}O_6}{180.2 \text{ g } C_6H_{12}O_6} = 0.03524 = 0.0352 \text{ mol particles}$

$m = \dfrac{(0.09143 + 0.03524) \text{ mol particles}}{0.200 \text{ kg } H_2O} = 0.63335 = 0.633 \, m$

Then, f.p. = T_f - $K_f(m)$; b.p. = T_b + $K_b(m)$; T in °C

	m	T_f	$-K_f(m)$		f.p.	T_b	$+K_b(m)$		b.p.
(a)	0.17	-114.6	-1.99(0.17)	= -0.34	-114.9	78.4	1.22(0.17)	= 0.21	78.6
(b)	0.957	-63.5	-4.68(0.957)	= -4.48	-68.0	61.2	3.63(0.957)	= 3.47	64.7
(c)	0.633	0.0	-1.86(0.633)	= -1.20	-1.2	100.0	0.52(0.633)	= 0.33	100.3

13.46 $\Delta T = K(m)$; first calculate the **molality** of the solute particles.

(a) 0.60 m (b) $\dfrac{16.0 \text{ g } C_{10}H_{22}}{0.420 \text{ kg } CHCl_3} \times \dfrac{1 \text{ mol } C_{10}H_{22}}{142.3 \text{ g } C_{10}H_{22}} = 0.2677 = 0.268 \, m$

(c) $m = \dfrac{0.40 \text{ mol eg} + 2(0.12) \text{ mol KBr}}{0.140 \text{ kg } H_2O} = \dfrac{0.64 \text{ mol particles}}{0.140 \text{ kg } H_2O} = 4.57 = 4.6 \, m$

Then, f.p. = T_f - $K_f(m)$; b.p. = T_b + $K_b(m)$; T in °C

	m	T_f	$-K_f(m)$		f.p.	T_b	$+K_b(m)$		b.p.
(a)	0.60	-114.6	-1.99(0.60)	= -1.2	-115.8	78.4	1.22(0.60)	= 0.73	79.1
(b)	0.268	-63.5	-4.68(0.268)	= -1.25	-64.8	61.2	3.63(0.268)	= 0.973	62.2
(c)	4.6	0.0	-1.86(4.6)	= -8.6	-8.6	100.0	0.52(4.6)	= 2.4	102.4

13.47 0.030 m phenol < 0.040 m glycerin = 0.020 m KBr. Phenol is very slightly ionized in water, but not enough to match the number of particles in a 0.040 m glycerin solution. The KBr solution is 0.040 m in particles, so it has the same boiling point as 0.040 glycerin, which is a nonelectrolyte.

13.48 For dilute aqueous solutions such as these, M is essentially equal to m. For the purposes of comparison, assume we can use M.

The more solute particles, the lower the freezing point. Since LiBr and $Zn(NO_3)_2$ are electrolytes, the particle concentrations in these solutions are 0.150 M and 0.090 M, respectively (although ion-ion attractive forces may decrease the real concentrations somewhat). Thus, the order of decreasing freezing points is:
(a) 0.075 M glucose > (c) 0.030 M $Zn(NO_3)_2$ > (b) 0.075 M LiBr.

13.49 $\pi = MRT$; $T = 20°C + 273 = 293\ K$

$$M = \frac{\text{mol urea}}{\text{L soln}} = \frac{2.02\ \text{g }(NH_2)_2CO}{0.145\ \text{L soln}} \times \frac{1\ \text{mol urea}}{60.06\ \text{g }(NH_2)_2CO} = 0.2320 = 0.232\ M$$

$$\pi = \frac{0.232\ \text{mol}}{L} \times \frac{0.08206\ \text{L}\cdot\text{atm}}{K\cdot\text{mol}} \times 293\ K = 5.557 = 5.58\ \text{atm}$$

13.50 $\pi = MRT$; $T = 20°C + 273 = 293\ K$

$$M\ (\text{of ions}) = \frac{\text{mol NaCl} \times 2}{\text{L soln}} = \frac{3.4\ \text{g NaCl}}{1\ \text{L soln}} \times \frac{1\ \text{mol NaCl}}{58.4\ \text{g NaCl}} \times \frac{2\ \text{mol ions}}{1\ \text{mol NaCl}} = 0.116 = 0.12\ M$$

$$\pi = \frac{0.116\ \text{mol}}{L} \times \frac{0.0821\ \text{L}\cdot\text{atm}}{K\cdot\text{mol}} \times 293\ K = 2.8\ \text{atm}$$

13.51 $\Delta T_b = K_b m$; $m = \dfrac{\Delta T_b}{K_b} = \dfrac{+0.49}{5.02} = 0.0976 = 0.098\ m$ adrenaline

$$m = \frac{\text{mol adrenaline}}{\text{kg CCl}_4} = \frac{\text{g adrenaline}}{\text{MM adrenaline} \times \text{kg CCl}_4}$$

$$\text{MM adrenaline} = \frac{\text{g adrenaline}}{m \times \text{kg CCl}_4} = \frac{0.64\ \text{g adrenaline}}{0.0976\ m \times 0.0360\ \text{kg CCl}_4} = 1.8 \times 10^2\ \text{g/mol adrenaline}$$

13.52 $\Delta T_f = 5.5 - 4.1 = 1.4$; $m = \dfrac{\Delta T_f}{K_f} = \dfrac{1.4}{5.12} = 0.273 = 0.27\ m$

$$\text{MM lauryl alcohol} = \frac{\text{g lauryl alcohol}}{m \times \text{kg C}_6H_6} = \frac{5.00\ \text{g lauryl alcohol}}{0.273 \times 0.100\ \text{kg C}_6H_6}$$

$$= 1.8 \times 10^2\ \text{g/mol lauryl alcohol}$$

13.53 $\pi = MRT$; $M = \dfrac{\pi}{RT}$; $T = 25°C + 273 = 298\ K$

$$M = 0.953\ \text{torr} \times \frac{1\ \text{atm}}{760\ \text{torr}} \times \frac{K\cdot\text{mol}}{0.08206\ \text{L}\cdot\text{atm}} \times \frac{1}{298\ K} = 5.128 \times 10^{-5} = 5.13\ M$$

$$\text{mol} = M \times L = 5.128 \times 10^{-5} \times 0.210\ L = 1.077 \times 10^{-5} = 1.08 \times 10^{-5}\ \text{mol lysozyme}$$

$$\text{MM} = \frac{g}{\text{mol}} = \frac{0.150\ g}{1.077 \times 10^{-5}\ \text{mol}} = 1.39 \times 10^4\ \text{g/mol lysozyme}$$

13.54 $M = \pi/RT = \dfrac{2.12\ \text{atm}}{298\ K} \times \dfrac{\text{mol}\cdot K}{0.08206\ \text{L}\cdot\text{atm}} = 0.08669 = 0.0867\ M$

$$\text{MM} = \frac{g}{\text{mol}} = \frac{g}{M \times L} = \frac{22.3\ g}{0.08669\ M \times 0.250\ L} = 1.03 \times 10^3\ \text{g/mol}$$

13.55   Attractive forces between ions in electrolyte solutions lead to ion-pairing, which reduces the effective number of particles in solution. Fewer effective particles in solution leads to smaller ΔT_f and ΔT_b values, higher freezing points and lower boiling points than predicted for ideal solutions.

13.56   If these were ideal solutions, they would have equal ion concentrations and equal ΔT_f values. Data in Table 13.6 indicates that the van't Hoff factors (i) for both salts are less than the ideal values. For 0.030 *m* NaCl, *i* is between 1.87 and 1.94, about 1.92. For 0.020 *m* K_2SO_4, *i* is between 2.30 and 2.70, about 2.66. From Equation 13.14,

ΔT_f (measured) = *i* × ΔT_f (calculated for nonelectrolyte)

NaCl: ΔT_f (measured) = 1.92 × 0.030 *m* × 1.86 °C/*m* = 0.11 °C

K_2SO_4: ΔT_f (measured) = 2.66 × 0.020 *m* × 1.86 °C/*m* = 0.099 °C

0.030 *m* NaCl would have the larger ΔT_f.

The deviations from ideal behavior are due to ion-pairing in the two electrolyte solutions. K_2SO_4 has more extensive ion pairing and a larger deviation from ideality because of the higher charge on SO_4^{2-} relative to Cl^-.

Colloids

13.57   The outline of a light beam passing through a colloid is visible, whereas light passing through a true solution is invisible unless collected on a screen. This is the Tyndall effect. To determine whether Faraday's (or anyone's) apparently homogeneous dispersion is a true solution or a colloid, shine a beam of light on it and see if the light is scattered.

13.58   Suspensions are classified as solutions or colloids according to the size of the dispersed particles. Solute particles have diameters less than 10 Å. Clearly a protein with a molecular mass of 30,000 amu will be longer than 10 Å. The aqueous suspensions are colloids because of the size of protein molecules.

13.59   (a)   hydrophobic   (b) hydrophilic   (c) hydrophobic

13.60   (a)   When the colloid particle mass becomes large enough so that gravitational and interparticle attractive forces are greater than the kinetic energies of the particles, settling and aggregation can occur.

        (b)   Hydrophobic colloids do not attract a sheath of water molecules around them and thus tend to aggregate from aqueous solution. They can be stabilized as colloids by adsorbing charges on their surfaces. The charged particles interact with solvent water, stabilizing the colloid.

        (c)   Charges on colloid particles can stabilize them against aggregation. Particles carrying like charges repel one another and are thus prevented from aggregating and settling out.

13.61 Colloid particles are stabilized by attractive intermolecular forces with the dispersing medium (solvent) and do not coalesce because of electrostatic repulsions between groups at the surface of the dispersed particles. Colloids can be coagulated by heating (more collisions, greater chance that particles will coalesce); hydrophilic colloids can be coagulated by adding electrolytes, which neutralize surface charges allowing the colloid particles to collide more freely.

13.62 (a) The nonpolar hydrophobic tails of soap particles (the hydrocarbon chain of stearate ions) establish attractive intermolecular dispersion forces with the nonpolar oil molecules, while the charged hydrophilic head of the soap particles interacts with H_2O to keep the oil molecules suspended. (This is the mechanism by which laundry detergents remove greasy dirt from clothes.)

     (b) The electrolytes from the acid neutralize the surface charges of the suspended particles in milk, causing the colloid to coagulate.

Additional Exercises

13.63 Assume 1.00 L of solution. 1000 mL soln × 0.945 g/mL = 945 g solution

$$\frac{945 \text{ g soln}}{1.0 \text{ L soln}} \times \frac{32.0 \text{ g } C_3H_7OH}{100 \text{ g soln}} \times \frac{1 \text{ mol } C_3H_7OH}{60.09 \text{ g } C_3H_7OH} = 5.032 = 5.03 \text{ } M$$

$$945 \text{ g soln} \times \frac{32.0 \text{ g } C_3H_7OH}{100 \text{ g soln}} = 302.4 = 302 \text{ g } C_3H_7OH; \text{ } 945 \text{ g} - 302.4 \text{ g} = 642.6 = 643 \text{ g } H_2O$$

$$\frac{302.4 \text{ g } C_3H_7OH}{0.6426 \text{ kg } H_2O} \times \frac{1 \text{ mol } C_3H_7OH}{60.09 \text{ g } C_3H_7OH} = 7.831 = 7.83 \text{ } m$$

13.64 (a)

$$\frac{1.80 \text{ mol } CH_3CN}{1 \text{ L soln}} \times \frac{86.85 \text{ g LiBr}}{1 \text{ mol LiBr}} = 156.3 = 156 \text{ g LiBr}$$

1 L soln = 826 g soln; g CH_3CN = 826 - 156.3 = 669.7 = 670 g CH_3CN

$$m \text{ LiBr} = \frac{1.80 \text{ mol LiBr}}{0.6697 \text{ kg } CH_3CN} = 2.69 \text{ } m$$

     (b)

$$\frac{669.7 \text{ g } CH_3CN}{41.05 \text{ g/mol}} = 16.31 = 16.3 \text{ mol } CH_3CN; \text{ } \chi_{LiBr} = \frac{1.80}{1.80 + 16.31} = 0.0994$$

     (c)

$$\text{mass \%} = \frac{669.7 \text{ g } CH_3CN}{826 \text{ g soln}} \times 100 = 81.1\% \text{ } CH_3CN$$

13.65 $8.00 \text{ L soln} \times \dfrac{1.208 \text{ g}}{1 \text{ cm}^3} \times \dfrac{1000 \text{ cm}^3}{1 \text{ L}} \times \dfrac{0.180 \text{ g } CuSO_4}{1 \text{ g soln}} = 1.7395 \times 10^3 = 1.74 \times 10^3 \text{ g } CuSO_4$

$$1.74 \times 10^3 \text{ g } CuSO_4 \times \frac{249.6 \text{ g } CuSO_4 \cdot 5H_2O}{159.6 \text{ g } CuSO_4} = 2.721 \times 10^3 = 2.72 \times 10^3 \text{ g } CuSO_4 \cdot 5H_2O$$

$$\frac{2.721 \times 10^3 \text{ g } CuSO_4 \cdot 5H_2O}{8.00 \text{ L}} \times \frac{1 \text{ mol } CuSO_4 \cdot 5H_2O}{249.6 \text{ g } CuSO_4 \cdot 5H_2O} = 1.36 \text{ } M$$

13.66 $\dfrac{2.16 \text{ g } C_7H_6O_2}{122.1 \text{ g/mol}} = 0.01769 = 0.0177 \text{ mol } C_7H_6O_2$

180 mL $CCl_4 \times \dfrac{1.59 \text{ g}}{1 \text{ mL}} = 286.2 = 286 \text{ g } CCl_4$; $\dfrac{286.2 \text{ g } CCl_4}{153.8 \text{ g/mol}} = 1.861 = 1.86 \text{ mol } CCl_4$

180 mL $C_2H_5OH \times \dfrac{0.782 \text{ g}}{1 \text{ mL}} = 140.8 = 141 \text{ g } C_2H_5OH$;

$\dfrac{140.8 \text{ g } C_2H_5OH}{46.06 \text{ g/mol}} = 3.057 = 3.06 \text{ mol } C_2H_5OH$

CCl_4   **C_2H_5OH**

(a) $\chi = \dfrac{0.01769}{0.01769 + 1.861} = 9.42 \times 10^{-3}$ $\chi = \dfrac{0.01769}{0.01769 + 3.057} = 5.75 \times 10^{-3}$

$m = \dfrac{0.01769 \text{ mol } C_7H_6O_2}{0.02862 \text{ kg } CCl_4} = 0.0618 \, m$ $m = \dfrac{0.01769 \text{ mol } C_7H_6O_2}{0.1408 \text{ kg } C_2H_5OH} = 0.126 \, m$

(b) CCl_4: 2.16 g $C_7H_6O_2$ + 286.2 g CCl_4 = 288.36 = 288 g soln

288.36 g soln $\times \dfrac{1 \text{ mL}}{1.59 \text{ g}} = 181.4 = 181 \text{ mL soln}$; $M = \dfrac{0.01769 \text{ mol } C_7H_6O_2}{0.1814 \text{ L soln}} = 0.0975 \, M$

C_2H_5OH: 2.16 g $C_7H_6O_2$ + 140.8 g C_2H_5OH = 142.96 = 143 g soln

142.96 g soln $\times \dfrac{1 \text{ mL}}{0.782 \text{ g}} = 182.8 = 183 \text{ mL soln}$; $M = \dfrac{0.01769 \text{ mol } C_7H_6O_2}{0.1828 \text{ L soln}} = 0.0968 \, M$

(c)   The molarities vary slightly because of the assumption that density of the solution equals density of the solvent; nonetheless, the molarities are close. The molalities differ by a factor of about two, because the densities of the two solvents are not the same. When equal volumes of the solvents are used, the masses of solvent are very different and the molalities of the solutions are different.

13.67   (a)   Assume 1000 g of solution. 1000 g soln $\times \dfrac{1 \text{ mL}}{1.22 \text{ g}} = 819.7 = 820 \text{ mL}$

1000 g soln $\times \dfrac{40.0 \text{ g KSCN}}{100 \text{ g soln}} = 400 \text{ g KSCN}$; 1000 g soln - 400 g KSCN = 600 g H_2O

$\dfrac{400 \text{ g KSCN}}{97.19 \text{ g/mol}} = 4.116 = 4.12 \text{ mol KSCN}$; $\dfrac{600 \text{ g } H_2O}{18.02 \text{ g/mol}} = 33.30 = 33.3 \text{ mol } H_2O$

$\chi_{KSCN} = \dfrac{4.116}{4.116 + 33.30} = 0.110$; $m = \dfrac{4.116 \text{ mol KSCN}}{0.600 \text{ kg } H_2O} = 6.86 \, m$

$M = \dfrac{4.116 \text{ mol KSCN}}{0.8197 \text{ L}} = 5.02 \, M$

(b)   If there are 4.12 mol of KSCN, there are 8.24 moles of ions. There are then 33.3 mol H_2O/8.24 mol ions $\approx$ 4 mol H_2O for each mol of ions, or 4 water molecules for each ion. This is too few water molecules to completely hydrate the anions and cations in the solution.

For a solution that is this concentrated, one would expect significant ion-pairing, because the ions are not completely surrounded and separated by H_2O molecules. Because of ion-pairing, the effective number of particles will be less than that indicated by m and M, so the observed colligative properties will be significantly different from those predicted by formulas for ideal solutions. The observed freezing point will be higher, the boiling point lower and the osmotic pressure lower than predicted.

13.68 $P_{Rn} = \chi_{Rn}P_{total}$; $P_{Rn} = 3.5 \times 10^{-6}$ (36 atm) $= 1.26 \times 10^{-4} = 1.3 \times 10^{-4}$ atm

$C_{Rn} = k\, P_{Rn}$; $C_{Rn} = \dfrac{7.27 \times 10^{-3}\,M}{1\text{ atm}} \times 1.26 \times 10^{-4}$ atm $= 9.2 \times 10^{-7}\,M$

13.69   (a)   0.100 m K_2SO_4 is 0.300 m in particles. H_2O is the solvent.

               $\Delta T_f = K_f\, m = -1.86(0.300) = -0.558$; $T_f = 0.000 - 0.558 = -0.558°C = -0.6°C$

      (b) ΔT_f (nonelectrolyte) $= -1.86(0.100) = -0.186$; $T_f = 0.000 - 0.186 = -0.186°C = -0.2°C$

               T_f (measured) $= i \times T_f$ (nonelectrolyte)

               From Table 13.6, i for 0.100 m $K_2SO_4 = 2.32$

               T_f (measured) $= 2.32(-0.186°C) = -0.432°C$

13.70   The compound with the larger i value is the stronger electrolyte.

       $i = \dfrac{\Delta T_f\text{(measured)}}{\Delta T_f\text{(calculated)}}$ The idealized value is 3 for both salts.

       $Hg(NO_3)_2$; $m = \dfrac{10.0\text{ g }Hg(NO_3)_2}{1.00\text{ kg }H_2O} \times \dfrac{1\text{ mol }Hg(NO_3)_2}{324.6\text{ g }Hg(NO_3)_2} = 0.0308\,m$

            ΔT_f (nonelectrolyte) $= -1.86(0.0308) = -0.0573°C$

            $i = \dfrac{-0.162°C}{-0.0573°C} = 2.83$

       $HgCl_2$; $m = \dfrac{10.0\text{ g }HgCL_2}{1.00\text{ kg }H_2O} \times \dfrac{1\text{ mol }HgCl_2}{271.5\text{ g }HgCl_2} = 0.0368\,m$

            ΔT_f (nonelectrolyte) $= -1.86(0.0368) = -0.0685°C$

            $i = \dfrac{-0.0685}{-0.0685} = 1.00$

With an i value of 2.83, $Hg(NO_3)_2$ is almost completely dissociated into ions; with an i value of 1.00, the $HgCl_2$ behaves essentially like a nonelectrolyte. Clearly, $Hg(NO_3)_2$ is the stronger electrolyte.

13.71 (a) $m = \dfrac{\text{mol Na(s)}}{\text{kg Hg(l)}}$; $1.0 \text{ cm}^3 \text{ Na(s)} \times \dfrac{0.97 \text{ g}}{1 \text{ cm}^3} \times \dfrac{1 \text{ mol}}{23.0 \text{ g Na}} = 0.0422 = 0.042 \text{ mol Na}$

$20.0 \text{ cm}^3 \text{ Hg(l)} \times \dfrac{13.6 \text{ g}}{1 \text{ cm}^3} \times \dfrac{1 \text{ kg}}{1000 \text{ g}} = 0.272 \text{ kg Hg(l)}$;

$m = \dfrac{0.0422 \text{ mol Na}}{0.272 \text{ Kg Hg(l)}} = 0.155 = 0.16 \; m$

(b) $M = \dfrac{\text{mol Na(s)}}{\text{L soln}} = \dfrac{0.0422 \text{ mol Na}}{0.021 \text{ L soln}} = 2.01 = 2.0 \; M$

(c)   Clearly, molality and molarity are not the same for this amalgam. Only in the instance that one kg solvent and the mass of one liter solution are nearly equal do the two concentration units have similar values. In this example, one kg Hg has a volume much less than one liter.

13.72 In this equilibrium system, molecules move from the surface of the solid into solution, while molecules in solution are deposited on the surface of the solid. As molecules leave the surface of the small particles of powder, the reverse process preferentially deposits other molecules on the surface of a single crystal. Eventually, all molecules that were present in the 50 g of powder are deposited on the surface of a 50 g crystal; this can only happen if the dissolution and deposition processes are ongoing.

13.73 Assume the radiator solution is prepared by mixing 1.00 L of each of the liquids. To calculate the freezing and boiling points, we need the molality of this solution. Assume H_2O is the solvent.

$m = \dfrac{\text{mol } C_2H_6O_2}{\text{kg } H_2O}$; $\text{kg } H_2O = 1.00 \text{ L} \times \dfrac{1000 \text{ mL}}{1 \text{ L}} \times \dfrac{1.00 \text{ g}}{\text{mL}} = 1.00 \text{ kg}$

$\text{mol } C_2H_6O_2 = 1.00 \text{ L} \times \dfrac{1000 \text{ mL}}{1 \text{ L}} \times \dfrac{1.12 \text{ g}}{1 \text{ mL}} \times \dfrac{1 \text{ mol } C_2H_6O_2}{62.07 \text{ g } C_2H_6O_2} = 18.04 = 18.0 \text{ mol } C_2H_6O_2$

$m = \dfrac{18.04 \text{ mol } C_2H_6O_2}{1.00 \text{ kg } H_2O} = 18.04 \; m$

$\Delta T_f = K_f m = -1.86(18.04) = -33.6°C$; f.p. $= 0.0 - 33.6 = -33.6°C$

$\Delta T_b = K_b m = 0.52(18.04) = 9.4°C$; b.p. $= 100.0 + 9.4 = +109.4°C$

13.74 (a) $K_b = \dfrac{\Delta T_b}{m}$; $\Delta T_b = 47.46°C - 46.30°C = 1.16°C$

$m = \dfrac{\text{mol solute}}{\text{kg } CS_2} = \dfrac{0.250 \text{ mol}}{400.0 \text{ mL } CS_2} \times \dfrac{1 \text{ mL } CS_2}{1.261 \text{ g } CS_2} \times \dfrac{1000 \text{ g}}{1 \text{ kg}} = 0.4956 = 0.496 \; m$

$K_b = \dfrac{1.16°C}{0.4956 \; m} = 2.34°C/m$

(b) $m = \dfrac{\Delta T_b}{K_b} = \dfrac{(47.08 - 46.30)°C}{2.34°C/m} = 0.333 = 0.33\ m$

$m = \dfrac{\text{mol unknown}}{\text{kg } CS_2}$; $m \times \text{kg } CS_2 = \dfrac{\text{g unknown}}{\text{MM unknown}}$; $MM = \dfrac{\text{g unknown}}{m \times \text{kg } CS_2}$

$50.0\ \text{mL } CS_2 \times \dfrac{1.261\ \text{g } CS_2}{1\ \text{mL}} \times \dfrac{1\ \text{kg}}{1000\ \text{g}} = 0.06305 = 0.0631\ \text{kg } CS_2$

$MM = \dfrac{5.39\ \text{g unknown}}{0.333\ m \times 0.06305\ \text{kg } CS_2} = 257 = 2.6 \times 10^2\ \text{g/mol}$

13.75   Mole fraction ethylalcohol, $\chi_{C_2H_5OH} = \dfrac{P_{C_2H_5OH}}{P°_{C_2H_5OH}} = \dfrac{8\ \text{torr}}{100\ \text{torr}} = 0.08$

$\dfrac{620 \times 10^3\ \text{g } C_{24}H_{50}}{338.6\ \text{g/mol}} = 1.83 \times 10^3\ \text{mol } C_{24}H_{50}$;   let y = mol C_2H_5OH

$\chi_{C_2H_5OH} = 0.08 = \dfrac{y}{y + 1.83 \times 10^3}$; $0.92\,y = 146.4$; $y = 1.6 \times 10^2\ \text{mol } C_2H_5OH$

(Strictly speaking, y should have 1 sig fig because 0.08 has 1 sig fig, but this severely limits the calculation.)

$1.6 \times 10^2\ \text{mol } C_2H_5OH \times \dfrac{46\ \text{g } C_2H_5OH}{1\ \text{mol}} = 7.4 \times 10^3\ \text{g or } 7.4\ \text{kg } C_2H_5OH$

13.76   The solvent vapor pressure over each solution is determined by the total particle concentrations present in the solutions. When the particle concentrations are equal, the vapor pressures will be equal and equilibrium established. The particle concentration of the nonelectrolyte is just 0.065 M; the ion concentration of the NaCl is 2 × 0.035 M = 0.070 M. Solvent will diffuse from the less concentrated nonelectrolyte solution. Let x = volume of solvent transferred.

$\dfrac{0.065\ M \times 20.0\ \text{mL}}{(20.0 - x)\ \text{mL}} = \dfrac{0.070\ M \times 20.0\ \text{mL}}{(20.0 + x)\ \text{mL}}$; $1.3(20.0 + x) = 1.4(20.0 - x)$

$2.7x = 2$, $x = 0.74 = 0.7$ mL transferred

The nonelectrolyte beaker contains 20.0 - 0.74 = 19.3 mL solution; the NaCl beaker contains 20.0 + 0.74 = 20.7 mL solution.

13.77 $M = \dfrac{\pi}{RT} = \dfrac{6.41\ \text{atm}}{298\ \text{K}} \times \dfrac{\text{K} \cdot \text{mol}}{0.08206\ \text{L} \cdot \text{atm}} = 0.2621 = 0.262\ M$

There are 0.262 mol of **particles** per liter of solution and 0.131 mol particles in 500 mL. Each mole of sucrose provides 1 mol of particles and each mole of NaCl provides 2.

Let x = g $C_{12}H_{22}O_{11}$, 15.0 - x = g NaCl

$$\frac{x \text{g } C_{12}H_{22}O_{11}}{342.3 \text{ g } C_{12}H_{22}O_{11}} + \frac{2(15.0 - x)}{58.44 \text{ g NaCl}} = 0.131$$

$58.44x + 2(342.3)(15.0 - x) = 0.131(58.44)(342.3)$

$58.44x - 684.6x + 10,269 = 2,621$

$-626.2x = -7648$; $x = 12.2$ g $C_{12}H_{22}O_{11}$

$$\frac{12.2 \text{ g } C_{12}H_{22}O_{11}}{15.0 \text{ g mixture}} \times 100 = 81.3\% \text{ sucrose, } 18.7\% \text{ NaCl}$$

13.78 $M = \dfrac{\pi}{RT} = \dfrac{57.1 \text{ torr}}{298 \text{ K}} \times \dfrac{1 \text{ atm}}{760 \text{ torr}} \times \dfrac{K \bullet mol}{0.08206 \text{ L} \bullet atm} = 3.072 \times 10^{-3} = 3.07 \times 10^{-3} \ M$

$\dfrac{0.036 \text{ g solute}}{100 \text{ g } H_2O} \times \dfrac{1000 \text{ g } H_2O}{1 \text{ kg } H_2O} = 0.36$ g solute/kg H_2O

Assuming molarity and molality are the same in this dilute solution, we can then say 0.36 g solute = 3.072×10^{-3} mol; MM = 117 g/mol. Because the salt is completely ionized, the formula weight of the lithium salt is **twice** this calculated value, or **234 g/mol**. The organic portion, $C_nH_{2n-1}O_2^-$, has a formula weight of 234-7 = 227 g. Subtracting 32 for the oxygens, and adding 1 to make the formula C_nH_{2n}, we have C_nH_{2n}, MM = 196 g/mol. Since each CH_2 unit has a mass of 14, n = 196/14 = 14. Thus the formula for our salt is $LiC_{14}H_{27}O_2$.

Integrative Exercises

13.79   Liquids and solids are condensed phases; molecules in these two phases are touching and there is little empty space in the pure substances. Changes in pressure have little effect on the number or nature of solute-solvent interactions. In the gas phase, molecules are far apart and constantly moving. An increase in gas pressure above a solution means an increase in the number of gas molecules per unit volume. Either way, the number of collisions of the gas with the surface of the solution increases, which increases the number of gas molecules that dissolve. An increase in pressure of the gas above the solution increases the solubility of the gas.

13.80   (a) $\Delta T_f = K_f m = K_f \times \dfrac{mol\ C_7H_6O_2}{kg\ C_6H_6} = K_f \times \dfrac{g\ C_7H_6O_2}{kg\ C_6H_6 \times MM\ C_7H_6O_2}$

MM = $\dfrac{K_f \times g\ C_7H_6O_2}{\Delta T_f \times kg\ C_6H_6} = \dfrac{5.12 \times 0.55}{0.360 \times 0.032} = 244.4 = 2.4 \times 10^2$ g/mol

(b)   The formula weight of $C_7H_6O_2$ is 122 g/mol. The experimental molar mass is twice this value, indicating that benzoic acid is associated into dimers in benzene solution. This is reasonable, since the carboxyl group is capable of strong hydrogen bonding with itself. Many carboxylic acids exist as dimers in solution. The structure of benzoic acid dimer in benzene solution is:

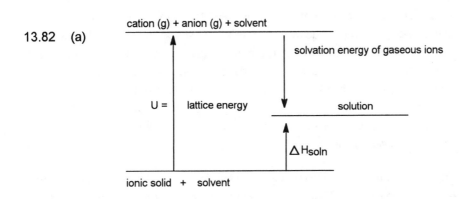

13.81   For ionic solids, the exothermic part of the solution process is step (3), surrounding the separated ions by solvent molecules. The released energy comes from the attractive interaction of the solvent with the separated ions. In the hydrates, one water molecule is already associated with the cation, reducing the total energy released during solvation.

13.82   **(a)**

cation (g) + anion (g) + solvent

solvation energy of gaseous ions

U =   lattice energy   solution

ΔH_{soln}

ionic solid + solvent

(b)   If the lattice energy (U) of the ionic solid (ion-ion forces) is too large relative to the solvation energy of the gaseous ions (ion-dipole forces), ΔH_{soln} will be too large and positive (endothermic) for solution to occur. This is the case for solutes like NaBr.

Lattice energy is inversely related to the distance between ions, so salts with large cations like $(CH_3)_4N^+$ have smaller lattice energies than salts with simple cations like Na^+. The smaller lattice energy of $(CH_4)_3NBr$ causes it to be more soluble in nonaqueous polar solvents. Also, the $-CH_3$ groups in the large cation are capable of dispersion interactions with the $-CH_3$ (or other nonpolar groups) of the solvent molecules. This produces a more negative solvation energy for the salts with large cations.

Overall, for salts with larger cations, U is smaller (less positive), the solvation energy of the gaseous ions is more negative, and ΔH_{soln} is less endothermic. These salts are more soluble in polar nonaqueous solvents.

13.83   **(a)**   The outer periphery of the BHT molecule is mostly hydrocarbon-like groups, such as $-CH_3$. The one $-OH$ group is rather buried inside, and probably does little to enhance solubility in water. Thus, BHT accumulates in the essentially nonpolar fats, and has a fairly long residence time in the body.

(b)   Put a known amount of BHT, water and oil in the test tube and shake. The BHT will dissolve in one or both of the liquids. The nonpolar vegetable oil is immiscible with and less dense than polar H_2O. The oil will collect on top. Pour off the oil layer and measure the concentration of BHT. Measure the concentration of BHT in the H_2O remaining in the tube.

13.84 $\chi_{CHCl_3} = \chi_{C_3H_6O} = 0.500$

(a)   For an ideal solution, Raoult's Law is obeyed.

$P_T = P_{CHCl_3} + P_{C_3H_6O}; \ P_{CHCl_3} = 0.5(300 \text{ torr}) = 150 \text{ torr}$

$P_{C_3H_6O} = 0.5(360 \text{ torr}) = 180 \text{ torr}; \ P_T = 150 \text{ torr} + 180 \text{ torr} = 330 \text{ torr}$

(b)   The real solution has a lower vapor pressure, 250 torr, than an ideal solution of the same composition, 330 torr. Thus, fewer molecules escape to the vapor phase from the liquid. This means that fewer molecules have sufficient kinetic energy to overcome intermolecular attractions. Clearly, even weak hydrogen bonds such as this one are stronger attractive forces than dipole-dipole or dispersion forces. These hydrogen bonds prevent molecules from escaping to the vapor phase and result in a lower than ideal vapor pressure for the solution. There is essentially no hydrogen bonding in the individual liquids.

(c)   According to Coulomb's law, electrostatic attractive forces lead to an overall lowering of the energy of the system. Thus, when the two liquids mix and hydrogen bonds are formed, the energy of the system is decreased and $\Delta H_{soln} < 0$; the solution process is exothermic.

13.85   Since these are very dilute solutions, assume that the density of the solution $\approx$ the density of $H_2O \approx 1.0 \text{ g/mL}$ at $25°C$. Then, 100 g solution = 100 g H_2O = 0.100 kg H_2O.

(a) CF_4: $\dfrac{0.0015 \text{ g } CF_4}{0.100 \text{ kg } H_2O} \times \dfrac{1 \text{ mol } CF_4}{88.00 \text{ g } CF_4} = 1.7 \times 10^{-4} \ m$

$CCIF_3$: $\dfrac{0.009 \text{ g } CCIF_3}{0.100 \text{ kg } H_2O} \times \dfrac{1 \text{ mol } CCIF_3}{104.46 \text{ g } CCIF_3} = 8.6 \times 10^{-4} \ m = 9 \times 10^{-4} \ m$

CCl_2F_2: $\dfrac{0.028 \text{ g } CCl_2F_2}{0.100 \text{ kg } H_2O} \times \dfrac{1 \text{ mol } CCl_2F_2}{120.9 \text{ g } CCl_2F_2} = 2.3 \times 10^{-2} \ m$

$CHCIF_2$: $\dfrac{0.30 \text{ g } CHCIF_2}{0.100 \text{ kg } H_2O} \times \dfrac{1 \text{ mol } CHCIF_2}{86.47 \text{ g } CHCIF_2} = 3.5 \times 10^{-2} \ m$

(b) $m = \dfrac{\text{mol solute}}{\text{kg solvent}}$; $M = \dfrac{\text{mol solute}}{\text{L solution}}$

Molality and molarity are numerically similar when kilograms solvent and liters solution are nearly equal. This is true when solutions are dilute, so that the density of the solution is essentially the density of the solvent, and when the density of the solvent is nearly 1 g/mL. That is, for dilute aqueous solutions such as the ones in this problem, $M \approx m$.

(c)   Water is a polar solvent; the solubility of solutes increases as their polarity increases. All the fluorocarbons listed have tetrahedral molecular structures. CF_4, a symmetrical tetrahedron, is nonpolar and has the lowest solubility. As more different atoms are bound to the central carbon, the electron density distribution in the molecule becomes less symmetrical and the molecular polarity increases. The most polar fluorocarbon, $CHClF_2$, has the greatest solubility in H_2O. It may experience weak hydrogen bonding with water.

(d) $C_g = k\,P_g$. Assume $M = m$ for $CHClF_2$. $P_g = 1$ atm

$k = \dfrac{C_g}{P_g} = \dfrac{M}{P_g}$; $k = \dfrac{3.5 \times 10^{-2}\,M}{1.0\,\text{atm}} = 3.5 \times 10^{-2}$ mol/L • atm

This value is greater than the Henry's law constant for $N_2(g)$, because $N_2(g)$ is nonpolar and of lower molecular mass than $CHClF_2$. In fact, the Henry's law constant for nonpolar CF_4, 1.7×10^{-4} mol/L • atm is similar to the value for N_2, 6.8×10^{-4} mol L • atm.

13.86 (a) $\dfrac{1.3 \times 10^{-3}\,\text{mol }CH_4}{\text{L soln}} \times 4.0\,\text{L} = 5.2 \times 10^{-3}\,\text{mol }CH_4$

$V = \dfrac{nRT}{P} = \dfrac{5.2 \times 10^{-3}\,\text{mol} \times 273\,\text{K}}{1.0\,\text{atm}} \times \dfrac{0.08206\,\text{L•atm}}{\text{K•mol}} = 0.12\,\text{L}$

(b)   All three hydrocarbons are nonpolar; they have zero net dipole movement. In CH_4 and C_2H_6, the C atoms are tetrahedral and all bonds are σ bonds. C_2H_6 has a higher molar mass than CH_4, which leads to stronger dispersion forces and greater water solubility. In C_2H_4, the C atoms are trigonal planar and the π electron cloud is symmetric above and below the plane that contains all the atoms. The π cloud in C_2H_4 is an area of concentrated electron density that experiences attractive forces with the positive ends of H_2O molecules. These forces increase the solubility of C_2H_4 relative to the other hydrocarbons.

(c)   The molecules have similar molar masses. NO is most soluble because it is polar. The triple bond in N_2 is shorter than the double bond in O_2. It is more difficult for H_2O molecules to surround the smaller N_2 molecules, so they are less soluble than O_2 molecules.

(d) H_2S and SO_2 are polar molecules capable of hydrogen bonding with water. Hydrogen bonding is the strongest force between neutral molecules and causes the much greater solubility. H_2S is weakly acidic in water. SO_2 reacts with water to form H_2SO_3, a weak acid. The large solubility of SO_2 is a sure sign that a chemical process has occurred.

(e) N_2 and C_2H_4. The dispersion forces due to the π cloud in C_2H_4 cause it to be more soluble in H_2O.

NO (31) and C_2H_6 (30). The structures of these two molecules are very different, yet they have similar solubilities. NO is slightly polar, but too small to be easily hydrated. The larger C_2H_6 is nonpolar, but more polarizable (stronger dispersion forces).

NO (31) and O_2 (32). The slightly polar NO is more soluble than the slightly larger (longer O=O bond than $N \equiv O$ bond) but nonpolar O_2.

13.87   The stronger the intermolecular forces, the higher the heat (enthalpy) of vaporization.

(a)   None of the substances are capable of hydrogen bonding in the pure liquid, and they have similar molar masses. All intermolecular forces are van der Waals forces, dipole-dipole and dispersion forces. In decreasing order of strength of forces: acetonitrile > acetone > acetaldehyde > ethylene oxide > cyclopropane > propane

The first four compounds have dipole-dipole and dispersion forces, the last two only dispersion forces.

(b)   The order of solubility in hexane should be the reverse of the order above. The least polar substance, propane, will be most soluble in hexane. Ethanol, CH_3CH_2OH, is capable of hydrogen bonding with the four polar compounds. Thus, acetonitrile, acetaldehyde, acetone and ethylene oxide should be more soluble than propane and cyclopropane, but without further information we cannot distinguish among the polar molecules.

14 Chemical Kinetics

Reaction Rates

14.1 (a) *Reaction rate* is the change in the amount of products or reactants in a given amount of time; it is the speed of a chemical reaction.

(b) Rates depend on concentration of reactants, surface area of reactants, temperature and presence of catalyst.

(c) The stoichiometry of the reaction (mole ratios of reactants and products) must be known to relate rate of disappearance of products to rate of appearance of reactants.

14.2 (a) M/s

(b) The hotter the oven, the faster the cake bakes. Milk sours faster in hot weather than cool weather.

(c) The *average rate* is the rate over a period of time, while the *instantaneous rate* is the rate at a particular time.

14.3

Time(s)	Mol A	(a) Mol B	Δ Mol A	(b) Rate (Δ mol A/s)
0	0.100	0		
30	0.0595	0.0405	-0.0405	13.5×10^{-4}
60	0.0354	0.0646	-0.0241	8.03×10^{-4}
90	0.0210	0.0790	-0.0144	4.80×10^{-4}
120	0.0125	0.0875	-0.0085	2.8×10^{-4}

(c) The volume of the container must be known to report the rate in units of concentration (mol/L) per time.

14.4

Time(min)	Mol A	(a) Mol B	Δ Mol A	(b) Rate (Δ mol A/s)
0	0.065	0		
10	0.051	0.028	-0.014	2.3×10^{-5}
20	0.042	0.046	-0.011	1.8×10^{-5}
30	0.036	0.058	-0.006	1.0×10^{-5}
40	0.031	0.068	-0.005	0.8×10^{-5}

(c) $$\frac{\Delta M_B}{\Delta t} = \frac{(0.058 - 0.028) \text{ mol}/0.100 \text{ L}}{(30 - 10) \text{ min}} \times \frac{1 \text{ min}}{60 \text{ s}} = 2.5 \times 10^{-4} \ M/s$$

14.5

Time (sec)	Time Interval (sec)	Concentration (M)	ΔM	Rate (M/s)
0		0.0165		
2,000	2,000	0.0110	-0.0055	28×10^{-7}
5,000	3,000	0.00591	-0.0051	17×10^{-7}
8,000	3,000	0.00314	-0.00277	9.23×10^{-7}
12,000	4,000	0.00137	-0.00177	4.43×10^{-7}
15,000	3,000	0.00074	-0.00063	2.1×10^{-7}

14.6

Time (min)	Time Interval (min)	Concentration (M)	ΔM	Rate (M/s)
0		1.85		
79	79	1.67	-0.18	3.8×10^{-5}
158	79	1.52	-0.15	3.2×10^{-5}
316	158	1.30	-0.22	2.3×10^{-5}
632	316	1.00	-0.30	1.6×10^{-5}

14.7   From the slopes of the lines in the figure at right, the rates are 1.5×10^{-6} M/s at 4000 s, 4.3×10^{-7} M/s at 10,000 s.

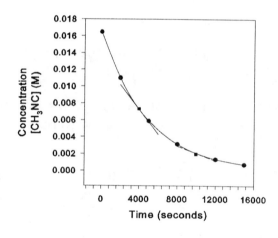

14.8   From the slopes of the lines in the figure at right, the rates are: at t = 100 min, 1.9×10^{-3} M/min or 3.2×10^{-3} M/s; at t = 500 min, 9.4×10^{-4} M/min or 1.6×10^{-3} M/s.

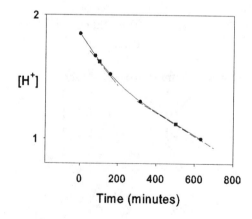

14.9   (a) $-\Delta[H_2O_2]/\Delta t = \Delta[H_2]/\Delta t = \Delta[O_2]/\Delta t$

       (b) $-\Delta[N_2O]/2\Delta t = \Delta[N_2]/2\Delta t = \Delta[O_2]/\Delta t$

              $-\Delta[N_2O]/\Delta t = \Delta[N_2]/\Delta t = 2\Delta[O_2]/\Delta t$

       (c) $-\Delta[N_2]/\Delta t = \Delta[NH_3]/2\Delta t$

              $-\Delta[H_2]/3\Delta t = \Delta[NH_3]/2\Delta t; \ -\Delta[H_2]/\Delta t = 3\Delta[NH_3]/2\Delta t$

14.10   (a) $-\Delta[NO]/\Delta t = \Delta[NOBr]/\Delta t; \ -\Delta[Br_2]/\Delta t = \Delta[NOBr]/2\Delta t$

       (b) $-\Delta[CH_4]/\Delta t = \Delta[CO_2]/\Delta t = \Delta[H_2]/4\Delta t$

              $-\Delta[H_2O]/2\Delta t = \Delta[CO_2]/\Delta t = \Delta[H_2]/4\Delta t; \ -\Delta[H_2O]/\Delta t = 2\Delta[CO_2]/\Delta t = \Delta[H_2]/2\Delta t$

       (c) $-\Delta[C_2H_2]/3\Delta t = \Delta[C_6H_6]/\Delta t; \ -\Delta[C_2H_2]/\Delta t = 3\Delta[C_6H_6]/\Delta t$

14.11   (a) $\dfrac{\Delta[H_2O]}{2\Delta t} = \dfrac{-\Delta[H_2]}{2\Delta t} = \dfrac{-\Delta[O_2]}{\Delta t}$

              H_2 is burning, $\dfrac{-\Delta[H_2]}{\Delta t} = 4.6$ mol/s

              O_2 is consumed, $\dfrac{-\Delta[O_2]}{\Delta t} = \dfrac{-\Delta[H_2]}{2\Delta t} = \dfrac{4.6 \text{ mol/s}}{2} = 2.3$ mol/s

              H_2O is produced, $\dfrac{+\Delta[H_2O]}{\Delta t} = \dfrac{-\Delta[H_2]}{\Delta t} = 4.6$ mol/s

       (b)   The change in total pressure is the sum of the changes of each partial pressure. NO and Cl_2 are disappearing and NOCl is appearing.

              $-\Delta P_{NO}/\Delta t = -30$ torr/min

              $-\Delta P_{Cl_2}/\Delta t = \Delta P_{NO}/2\Delta t = -15$ torr/min

              $+\Delta P_{NOCl}/\Delta t = -\Delta P_{NO}/\Delta t = +30$ torr/min

              $\Delta P_T/\Delta t = -30$ torr/min $- 15$ torr/min $+ 30$ torr/min $= -15$ torr/min

14.12   (a) $\dfrac{-\Delta[CH_4]}{\Delta t} = \dfrac{+\Delta[CO_2]}{\Delta t} = \dfrac{\Delta[H_2O]}{2\Delta t}$

              If CH_4 is burning, $-\Delta[CH_4]/\Delta t$, at 0.40 M/s, CO_2 is being produced, $+\Delta[CO_2]/\Delta t$, at 0.40 M/s and H_2O is being produced, $+\Delta[H_2O]/\Delta t$, at 0.80 M/s.

       (b) $\dfrac{+\Delta[NH_3]}{\Delta t} = \dfrac{-2\Delta[N_2]}{\Delta t} = \dfrac{-2\Delta[H_2]}{3\Delta t}; \ \dfrac{+\Delta[NH_3]}{\Delta t} = 100$ torr/hr

              $\dfrac{-\Delta[N_2]}{\Delta t} = 50$ torr/hr; $\ \dfrac{-\Delta[H_2]}{\Delta t} = 150$ torr/hr

              $\Delta P_{total} = (+100$ torr $- 50$ torr $- 150$ torr$) = -100$ torr/hr

Rate Laws

14.13 (a) A *rate law* is an algebraic expression that shows how the rate of a reaction varies with concentrations of reactants. The *rate constant*, k, is a number, the proportionality constant in the rate law; it depends on temperature and the system.

(b) The *reaction order* for a specific reactant is the exponent of that reactant in the rate law. It indicates how a change in that reactant affects the reaction rate. The *overall reaction order* is the sum of the exponents in the rate law.

(c) Units of k = $\dfrac{\text{units of rate}}{(\text{units of concentration})^2} = \dfrac{M/s}{M^2} = M^{-1}s^{-1}$

14.14 (a) If [A] changes, the rate will change, but the rate constant, k, will remain the same. The rate law is a general algebraic expression for the rate of a reaction (dependent variable) at any reactant concentration (independent variables); there are an infinite number of combinations of [A] and rate. The rate constant, k, is the proportionality constant that does not change (unless temperature changes).

(b) The reaction is first order in A, second order in B, and third order overall.

(c) Units of k = $\dfrac{M/s}{M^3} = M^{-2}s^{-1}$

14.15 (a) rate = $k[N_2O_5] = 6.08 \times 10^{-4}\,s^{-1}\,[N_2O_5]$

(b) rate = $6.08 \times 10^{-4}\,s^{-1}\,(0.100\,M) = 6.08 \times 10^{-5}\,M/s$

(c) rate = $6.08 \times 10^{-4}\,s^{-1}\,(0.200\,M) = 12.16 \times 10^{-5} = 1.22 \times 10^{-4}\,M/s$

When the concentration of N_2O_5 doubles, the rate of the reaction doubles.

14.16 (a) rate = $k[H_2][NO]^2$

(b) rate = $(6.0 \times 10^4\,M^{-2}s^{-1})(0.050\,M)^2(0.010\,M) = 1.5\,M/s$

(c) rate = $(6.0 \times 10^4\,M^{-2}s^{-1})(0.10\,M)^2(0.010\,M) = 6.0\,M/s$

(Note that doubling [NO] causes a quadrupling in rate.)

14.17 (a,b) rate = $k[CH_3Br][OH^-]$; k = $\dfrac{\text{rate}}{[CH_3Br][OH^-]}$

at 298 K, k = $\dfrac{0.28\,M/s}{(0.010\,M)(0.10\,M)} = 2.8 \times 10^2\,M^{-1}s^{-1}$

(c) Since the rate law is first order in [OH⁻], if [OH⁻] is tripled, the rate triples.

14.18 (a,b) rate = $k[NO]^2[O_2]$; k = rate/$[NO]^2[O_2]$

k = $\dfrac{7.2 \times 10^{-5}\,M/s}{(0.030\,M)^2(0.040\,M)} = 2.0\,M^{-2}s^{-1}$

(c)   Since the reaction is second order in NO, if the [NO] is decreased by a factor of 2, the rate would decrease by a factor of 2^2, or 4.

14.19 (a)   rate = $k[A][C]^2$

       (b)   new rate = $(1/2)(1/2)^2 = 1/8$ of old rate

14.20 (a)   rate = $k[A][B]$ (b) rate = $k[Y]^2$

14.21 (a)   Doubling [NO] while holding $[O_2]$ constant increases the rate by a factor of 4 (experiments 1 and 3). Reducing $[O_2]$ by a factor of 2 while holding [NO] constant reduces the rate by a factor of 2 (experiments 2 and 3). The rate is second order in [NO] and first order in $[O_2]$. rate = $k[NO]^2[O_2]$

     (b,c)   From experiment 1: $k = \dfrac{1.41 \times 10^{-2}\ M/s}{(0.0126\ M)^2(0.0125\ M)} = 7.11 \times 10^3\ M^{-2}s^{-1}$

           (Any of the three sets of initial concentrations and rates could be used to calculate the rate constant k.)

14.22 (a)   Doubling $[NH_3]$ while holding $[BF_3]$ constant doubles the rate (experiments 1 and 2). Doubling $[BF_3]$ while holding $[NH_3]$ constant doubles the rate (experiments 4 and 5). Thus, the reaction is first order in both BF_3 and NH_3; rate = $k[BF_3][NH_3]$.

       (b)   The reaction is second order overall.

       (c)   At t = 0, $k = \dfrac{0.2130\ M/s}{(0.250\ M)(0.250\ M)} = 3.41\ M^{-1}s^{-1}$

           (Any of the five sets of initial concentrations and rates could be used to calculate the rate constant k.)

14.23 (a)   Increasing [NO] by a factor of 2.5 while holding $[Br_2]$ constant (experiments 1 and 2) increases the rate by a factor 6.25 or $(2.5)^2$. Increasing $[Br_2]$ by a factor of 2.5 while holding [NO] constant increases the rate by a factor of 2.5. The rate law for the appearance of NOBr is: rate = $\Delta[NOBr]/\Delta t = k[NO]^2[Br_2]$.

       (b)   From experiment 1: $k = \dfrac{24\ M/s}{(0.10\ M)^2(0.20\ M)} = 1.2 \times 10^4\ M^{-2}s^{-1}$

       (c) $\Delta[NOBr]/2\Delta t = -\Delta[Br_2]/\Delta t$; the rate of disappearance of Br_2 is half the rate of appearance of NOBr.

       (d) $\dfrac{-\Delta[Br_2]}{\Delta t} = \dfrac{k[NO]^2[Br_2]}{2} = \dfrac{1.2 \times 10^4}{2\ M^2 s} \times (0.075\ M)^2 \times (0.185\ M) = 6.2\ M/s$

14.24 **(a)** Changing $[S_2O_8^{2-}]$ from 0.023 M to 0.054 M, a factor of 2.35, increases the rate by this same factor (experiments 1 and 2). Changing $[I^-]$ from 0.048 M to 0.019 M, a factor of 2.53, decreases rate by the same factor (experiments 2 and 3). The rate law is: rate = $k[S_2O_8^{2-}][I^-]$.

(b) From experiment 1: $k = \dfrac{6.8 \times 10^{-6}\ M/s}{(0.023\ M)(0.048\ M)} = 6.16 \times 10^{-3}\ M^{-1}s^{-1}$

(c) $-\Delta[S_2O_8^{2-}]/\Delta t = -\Delta[I^-]/3\Delta t$; the rate of disappearance of $S_2O_8^{2-}$ is one-third the rate of disappearance of I^-.

(d) $\dfrac{-\Delta[I^-]}{\Delta t} = \dfrac{-3\Delta[S_2O_8^{2-}]}{\Delta t} = 3(6.16 \times 10^{-3}\ M^{-1}s^{-1})(0.075\ M)(0.060\ M) = 8.3 \times 10^{-5}\ M/s$

Change of Concentration with Time

14.25 **(a)** A *first-order reaction* depends on the concentration, raised to the first power, of only one reactant; rate = $k[A]^1$.

(b) A graph of ln[A] vs time yields a straight line for a first-order reaction.

(c) The half-life of a first-order reaction **does not** depend on initial concentration; it is determined by the value of the rate constant, k.

14.26 **(a)** A *second order reaction* depends on the concentration of a single reactant raised to the second power, or the concentrations of two reactants raised to the first power; rate = $k[A]^2$ or $k[A][B]$.

(b) A graph of 1/[A] vs time yields a straight line for a second-order reaction.

(c) The half-life of a second-order reaction **does** depend on initial reactant concentration; $t_{1/2} = 1/[A]_o$. When solving the general rate equation for $[A]_t = 1/2\ [A]_o$, the $[A]_o$ terms do not drop out. The greater $[A]_o$, the shorter the half-life.

14.27 For a first order reaction, $t_{1/2} = 0.693/k$, $k = 5.1 \times 10^{-4}s^{-1}$

$t_{1/2} = \dfrac{0.693}{5.1 \times 10^{-4}\ s^{-1}} = 1.4 \times 10^3\ s$ or 23 min

14.28 $t_{1/2} = 0.91$ s; $t_{1/2} = 0.693/k$, $k = 0.693/t_{1/2}$

$k = \dfrac{0.693}{0.91\ s} = 0.76\ s^{-1}$

14.29 **(a)** Rearranging Equation [14.13] for a first order reaction:

$\ln[A]_t = -kt + \ln[A]_0$

1.5 min = 90 s; $[N_2O_5]_0 = (0.300\ mol/0.500\ L) = 0.600\ M$

$\ln[N_2O_5]_{90} = -(6.82 \times 10^{-3}s^{-1})(90\ s) + \ln(0.600)$

$\ln[N_2O_5]_{90} = -0.6138 + (-0.5108) = -1.1246 = -1.125$

$[N_2O_5]_{90} = 0.3248 = 0.325\ M;\ \text{mol}\ N_2O_5 = 0.3248\ M \times 0.500\ L = 0.162\ \text{mol}$

(b) $[N_2O_5]_t = 0.030\ \text{mol}/0.500\ L = 0.060\ M;\ [N_2O_5]_0 = 0.600\ M$

$\ln(0.060) = -(6.82 \times 10^{-3}s^{-1})\ (t) + \ln(0.600)$

$t = \dfrac{-[\ln(0.060) - \ln(0.600)]}{(6.82 \times 10^{-3}s^{-1})} = 337.6 = 338\ s \times \dfrac{1\ \text{min}}{60\ s} = 5.63\ \text{min}$

(c) $t_{1/2} = 0.693/k = 0.693/6.82 \times 10^{-3}s^{-1} = 101.6 = 102\ s\ \text{or}\ 1.69\ \text{min}$

14.30 (a)   For a first order reaction:

$\ln[A]_t = -kt + \ln[A]_0;\ k = 1.65\ yr^{-1},\ t = 3\ \text{months} = 0.250\ yr$

$\ln[A]_{0.25} = -1.65\ yr^{-1} \times 0.250\ yr + \ln(6.0 \times 10^{-3})$

$\ln[A]_{0.25} = -5.529 = -5.53;\ [A]_{0.25} = 3.97 \times 10^{-3}\ M$

$\text{At} = 1.0\ yr,\ \ln[A]_{1.0} = (-1.65\ yr^{-1} \times 1.0\ yr) + \ln(6.0 \times 10^{-3})$

$\ln[A]_{1.0} = -6.766 = -6.77;\ [A]_{1.0} = 1.15 \times 10^{-3}\ M$

(b) $\ln[A]_t - \ln[A]_0 = -kt;\ t = (\ln[A]_0 - \ln[A]_t)/k$

$t = \dfrac{[\ln(6.0 \times 10^{-3}) - \ln(1.0 \times 10^{-3})]}{1.65\ yr^{-1}} = \dfrac{1.792}{1.65\ yr^{-1}} = 1.09 = 1.1\ yr$

(c) $t_{1/2} = 0.693/k = 0.693/1.65\ yr^{-1} = 0.420\ yr = 5.04\ \text{mo} = 153\ \text{days}$

14.31

t(s)	$P_{SO_2Cl_2}$	$\ln P_{SO_2Cl_2}$
0	1.000	0
2500	0.947	-0.0545
5000	0.895	-0.111
7500	0.848	-0.165
10000	0.803	-0.219

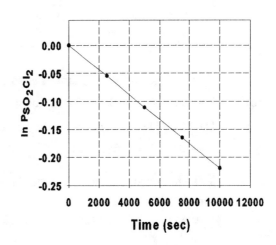

Graph $\ln P_{SO_2Cl_2}$ vs. time. (Pressure is a satisfactory concentration unit for a gas, since the concentration in moles/liter is proportional to P.) The graph is linear with slope $-2.19 \times 10^{-5}\ s^{-1}$ as shown on the figure. The rate constant k = -slope = $2.19 \times 10^{-5}\ s^{-1}$.

14.32

t(s)	P_{CH_3NC}	$\ln P_{CH_3NC}$
0	502	6.219
2000	335	5.814
5000	180	5.193
8000	95.5	4.559
12000	41.7	3.731
15000	22.4	3.109

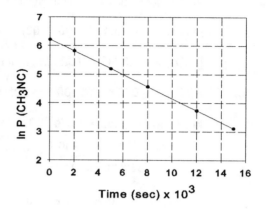

A graph of ln P vs t is linear with a slope of $-2.08 \times 10^{-4}\,s^{-1}$. The rate constant k = -slope $= 2.08 \times 10^{-4}\,s^{-1}$. Half-life $= t_{1/2} = 0.693/k = 3.33 \times 10^{3}$ s.

14.33   (a)   Make both first- and second-order plots to see which is linear. Moles is a satisfactory concentration unit, since volume is constant.

time(s)	mol A	ln (mol A)	1/mol A
0	0.1000	-2.303	10.00
30	0.0595	-2.822	16.8
60	0.0354	-3.341	28.2
90	0.0210	-3.863	47.6
120	0.0125	-4.382	80.0

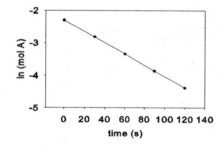 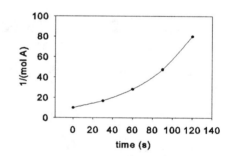

The plot of ln (mol A) vs time is linear, so the reaction is first-order in A.

  (b)   k = -slope = -[-4.382 - (-2.822)]/90 s = 0.01733 = 0.017 s^{-1}

(The best fit to this line yields the same value for the slope.

  (c) $t_{1/2} = 0.693/k = 0.693/0.0173\,s^{-1} = 40$ s

14.34   (a)   Make both first- and second-order plots to see which is linear.

time(min)	mol A	[A] (*M*)	ln[A]	1/mol A
0	0.065	0.65	-0.43	1.5
10	0.051	0.51	-0.67	2.0
20	0.042	0.42	-0.87	2.4
30	0.036	0.36	-1.02	2.8
40	0.031	0.31	-1.17	3.2

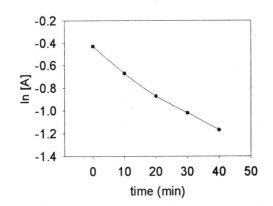 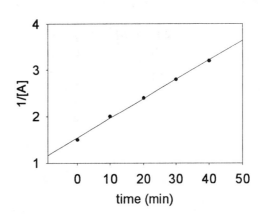

The plot of 1/[A] vs time is linear, so the reaction is second-order in [A].

(b) k = slope = (3.2 - 2.0) M^{-1} / 30 min = 0.040 M^{-1} min^{-1}

(The best fit to the line yields slope =0.042 $M^{-1}min^{-1}$.)

(c) $t_{1/2}$ = 1/k[A]$_o$ = 1/(0.040 M^{-1} min^{-1})(0.65 M) = 38.46 = 38 min

(Using the "best-fit" slope, $t_{1/2}$ = 37 min.)

14.35 (a)   Make both first and second order plots to see which is linear.

time(s)	[NO$_2$](M)	ln[NO$_2$]	1/[NO$_2$]
0.0	0.100	-2.303	10.0
5.0	0.017	-4.08	59
10.0	0.0090	-4.71	110
15.0	0.0062	-5.08	160
20.0	0.0047	-5.36	210

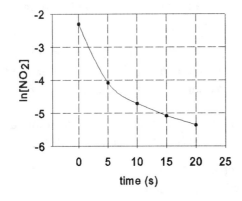 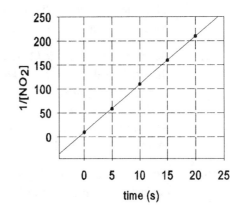

The plot of 1/[NO$_2$] vs time is linear, so the reaction is second order in NO$_2$.

(b)   The slope of the line is (210 - 59) M^{-1} / 15.0 s = 10.07 = 10 $M^{-1}s^{-1}$ = k.
       (The slope of the best-fit line is 10.02 = 10 $M^{-1}s^{-1}$.)

14.36 (a) Make both first- and second-order plots to see which is linear.

time(min)	$[C_{12}H_{22}O_{11}]$ (M)	$\ln[C_{12}H_{22}O_{11}]$	$1/[C_{12}H_{22}O_{11}]$
0	0.316	-1.152	3.16
39	0.274	-1.295	3.65
80	0.238	-1.435	4.20
140	0.190	-1.661	5.26
210	0.146	-1.924	6.85

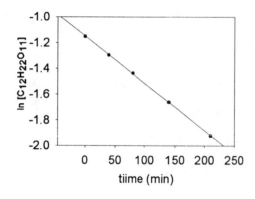 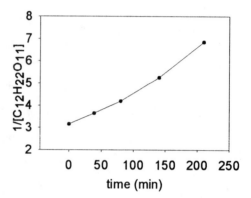

The plot of $\ln [C_{12}H_{22}O_{11}]$ in linear, so the reaction is first order in $C_{12}H_{22}O_{11}$.

(b) $k = -$slope $= - (-1.924 - (-1.295)) / 171$ min $= 3.68 \times 10^{-3}$ min^{-1}

(The slope of the best-fit line is -3.67×10^{-3} min^{-1}.)

Temperature and Rate

14.37 (a) The central idea of the *collision model* is that molecules must collide to react.

(b) The energy of the collision and the orientation of the molecules when they collide determine whether a reaction will occur.

(c) According to the Kinetic Molecular Theory (Chapter 10), the higher the temperature, the greater the speed and kinetic energy of the molecules. Therefore, at a higher temperature, there are more total collisions and each collision is more energetic.

14.38 (a) *Activation energy* is the minimum amount of energy required to initiate a chemical reaction. It is provided by the kinetic energies of the reacting particles.

(b) No. In addition to having sufficient kinetic energy, the particles must collide in the correct **orientation** to form the activated complex.

(c) The collision model is based on the kinetic theory of gases. The kinetic theory states that the higher the temperature, the greater the average kinetic energy of a collection of particles. The collision theory uses this idea to explain the effect of temperature on reaction rate. The higher the temperature, the greater the average kinetic energy of the particles, the more total collisions and the more collisions that have sufficient energy to react.

14.39 (a)

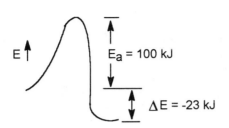

                                                (b)   123 kJ/mol

14.40 (a)

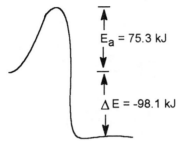

                                                (b)   173.4 kJ/mol

14.41 Assuming all collision factors (A) to be the same, reaction rate depends only on E_a; it is independent of ΔE. Based on the magnitude of E_a, reaction (c) is fastest and reaction (b) is slowest.

14.42 E_a for the reverse reactions is

    (a)   30 - 10 = 20 kJ   (b) 45 - (-10) = 55 kJ   (c) 20 - (-20) = 40 kJ
Based on the magnitude of E_a, the reverse of reaction (a) is fastest and the reverse of reaction (b) is slowest.

14.43 No. The value of A, which is related to frequency and effectiveness of collisions, can be different for each reaction and k is proportional to A.

14.44 From Equation [14.19], reactions with different variations of k with respect to temperature have different activation energies, E_a. The fact that k for the two reactions is the same at a certain temperature is accidental. The reaction with the higher rate at 35° has the larger activation energy, because it was able to use the increase in energy more effectively.

14.45 $T_1 = 20°C + 273 = 293$ K; $T_2 = 60°C + 273 = 333$ K; $k_1 = 1.75 \times 10^{-1} s^{-1}$

    (a) $\ln\left(\dfrac{k_1}{k_2}\right) = \dfrac{E_a}{R}\left(\dfrac{1}{333} - \dfrac{1}{293}\right) = \dfrac{55.5 \times 10^3 \text{ J/mol}}{8.314 \text{ J/mol}}(-4.100 \times 10^{-4})$

          $\ln(k_1/k_2) = -2.7367 = -2.74$; $k_1/k_2 = 0.0648 = 0.065$; $k_2 = \dfrac{0.175 \text{ s}^{-1}}{0.0648} = 2.7 \text{ s}^{-1}$

(b) $\ln\left(\dfrac{k_1}{k_2}\right) = \dfrac{121 \times 10^3 \text{ J/mol}}{8.314 \text{ J/mol}}\left(\dfrac{1}{333} - \dfrac{1}{293}\right) = -5.9666 = -5.97$

$k_1/k_2 = 2.563 \times 10^{-3} = 2.6 \times 10^{-3}$; $k_2 = \dfrac{0.175 \text{ s}^{-1}}{2.563 \times 10^{-3}} = 68 \text{ s}^{-1}$

14.46   According to Equation 14.20, $\ln\left(\dfrac{k_1}{k_2}\right) = \dfrac{E_a}{R}\left[\dfrac{1}{T_2} - \dfrac{1}{T_1}\right]$

(a) $T_2 = 700°C = 973 \text{ K}$, $T_1 = 600°C = 873 \text{ K}$

$\ln\left(\dfrac{k_{873}}{k_{973}}\right) = \dfrac{182 \times 10^3 \text{ J/mol}}{8.314 \text{ J/mol}} \times \left[\dfrac{1}{973} - \dfrac{1}{873}\right] = -2.5771$; $\dfrac{k_{873}}{k_{973}} = 0.07599$

$k_{873} = 0.07599 (1.57 \times 10^{-5} M^{-1} s^{-1}) = 1.19 \times 10^{-6} M^{-1} s^{-1}$

(b) $T_2 = 700°C = 973 \text{ K}$, $T_1 = 800°C = 1073 \text{ K}$

$\ln\left(\dfrac{k_{1073}}{k_{973}}\right) = \dfrac{182 \times 10^3 \text{ J/mol}}{8.314 \text{ J/mol}} \times \left[\dfrac{1}{973} - \dfrac{1}{1073}\right] = 2.0968$; $\dfrac{k_{1073}}{k_{973}} = 8.1398$

$k_{1073} = 8.1398 (1.57 \times 10^{-5} M^{-1} s^{-1}) = 1.28 \times 10^{-4} M^{-1} s^{-1}$

14.47

k	ln k	T(K)	1/T($\times 10^3$)
0.0521	-2.995	288	3.47
0.101	-2.293	298	3.36
0.184	-1.693	308	3.25
0.332	-1.103	318	3.14

The slope, -5.7×10^3, equals $-E_a/R$. Thus,
$E_a = 5.7 \times 10^3 \times 8.314 \text{ J/mol} = 47 \text{ kJ/mol}$.

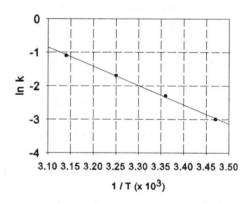

14.48

k	ln k	T(K)	1/T($\times 10^3$)
0.028	-3.58	600	1.67
0.22	-1.51	650	1.54
1.3	0.26	700	1.43
6.0	1.79	750	1.33
23	3.14	800	1.25

Using the relationship $\ln k = \ln A - E_a/RT$,
the slope, $-15.94 \times 10^3 = -16 \times 10^3$, is $-E_a/R$.
$E_a = 15.94 \times 10^3 \times 8.314 \text{ J/mol} =$
$1.3 \times 10^2 \text{ kJ/mol}$. To calculate A, we will
use the rate data at 700 K. From the equation
given above, $0.262 = \ln A - 15.94 \times 10^3/700$;
$\ln A = 0.262 + 22.771$. $A = 1.0 \times 10^{10}$.

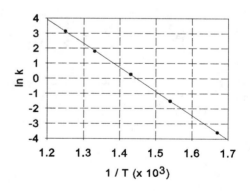

14.49 $T_1 = 50°C + 273 = 323\ K;\ T_2 = 0°C + 273 = 273\ K$

$$\ln\left(\frac{k_1}{k_2}\right) = \frac{E_a}{R}\left[\frac{1}{T_2} - \frac{1}{T_1}\right] = \frac{76.7\ kJ/mol}{8.314\ J/mol} \times \frac{1000\ J}{1\ kJ}\left[\frac{1}{273} - \frac{1}{323}\right]$$

$\ln(k_1/k_2) = 9.225 \times 10^3\ (5.670 \times 10^{-4}) = 5.231 = 5.23;\ k_1/k_2 = 187 = 1.9 \times 10^2$

The reaction will occur 190 times faster at 50°C, assuming equal initial concentrations.

14.50 $T_1 = 20°C = 293\ K;\ T_2 = 40°C = 313\ K;\ k_2 = 4k_1,\ k_1/k_2 = 1/4.$

$$\ln\left(\frac{k_1}{k_2}\right) = \frac{E_a}{R}\left[\frac{1}{T_2} - \frac{1}{T_1}\right];\ \ln(0.25) = \frac{E_a}{8.314\ J/mol}\left[\frac{1}{313} - \frac{1}{293}\right]$$

$$E_a = \frac{\ln(0.25)\,(8.314\ J/mol)}{(-2.181 \times 10^{-4})} = \frac{+5.29 \times 10^4\ J}{mol} = 53\ kJ/mol$$

Reaction Mechanisms

14.51   (a)   An *elementry step* is a process that occurs in a single event; the order is given by the coefficients in the balanced equation for the step.

         (b)   A *unimolecular* elementary step involves only one reactant molecule; the activated complex is derived from a single molecule. A *bimolecular* elementary step involves two reactant molecules in the activated complex and the overall process.

         (c)   A *reaction mechanism* is a series of elementary steps that describe how an overall reaction occurs and explain the experimentally determined rate law.

14.52   (a)   The *molecularity* of a process indicates the number of molecules that participate as reactants in the process. A unimolecular process has one reactant molecule, a bimolecular process has two reactant molecules and a termolecular process has three reactant molecules.

         (b)   Termolecular processes are rare because it is highly unlikely that three molecules will simultaneously collide with the correct energy and orientation to form an activated complex.

         (c)   An *intermediate* is a substance that is produced and then consumed during a chemical reaction. It does not appear in the balanced equation for the overall reaction.

14.53   (a)   bimolecular, rate = $k[N_2O][Cl]$
         (b)   unimolecular, rate = $k[Cl_2]$
         (c)   bimolecular, rate = $k[NO][Cl_2]$

14.54   (a)   bimolecular, rate = $k[NO][O_3]$
         (b)   bimolecular, rate = $k[CO][Cl_2]$
         (c)   unimolecular, rate = $k[O_3]$

14.55 (a)

$$NO(g) + NO(g) \rightarrow N_2O_2(g)$$
$$N_2O_2(g) + H_2(g) \rightarrow N_2O(g) + H_2O(g)$$

$$2NO(g) + N_2O_2(g) + H_2(g) \rightarrow N_2O_2(g) + N_2O(g) + H_2O(g)$$
$$2NO(g) + H_2(g) \rightarrow N_2O(g) + H_2O(g)$$

(b) First step: $-\Delta[NO]/\Delta t = k[NO][NO] = k[NO]^2$
Second step: $-\Delta[H_2]/\Delta t = k[H_2][N_2O_2]$

(c) N_2O_2 is the intermediate; it is produced in the first step and consumed in the second.

(d) Since $[H_2]$ appears in the rate law, the second step must be slow relative to the first.

14.56 (a)   i. $HBr + O_2 \rightarrow HOOBr$
ii. $HOOBr + HBr \rightarrow 2HOBr$
iii. $2HOBr + 2HBr \rightarrow 2H_2O + 2Br_2$

$$4HBr + O_2 \rightarrow 2H_2O + 2Br_2$$

(b)   i. $\dfrac{-\Delta[HBr]}{\Delta t} = k[HBr][O_2]$ ii. $-\Delta[HOOBr] = k[HOOBr][HBr]$

iii. $\dfrac{-\Delta[HOBr]}{\Delta t} = k[HOBr][HBr]$

(c) HOOBr and HOBr are both intermediates; HOOBr is produced in i and consumed in ii and HOBr is produced in ii and consumed in iii.

(d) The first order dependence of the reaction on HBr means that the second and third steps cannot be slow or rate-determining. It must be that the first step is rate-determining; thus neither HOOBr or HOBr accumulates enough to be determined.

14.57 (a)   rate = $k[NO][Cl_2]$

(b) Since the observed rate law is second-order in [NO], the second step must be slow relative to the first step; the second step is rate determining.

14.58 Both (b) and (d) are consistent with the observed rate law. For a multistep reaction, the rate law is determined by the slow step and any preceeding steps. In both (b) and (d), the first step is the slow step and the rate law is $k[H_2][ICl]$. The subsequent fast steps do not influence the rate law. Mechanisms (a) and (c) would lead to a rate law which is second order in ICl.

Catalysis

14.59 (a) A catalyst increases the rate of reaction by decreasing the activation energy, E_a, or increasing the frequency factor A. Lowering the activation energy is more common and more dramatic.

(b) A homogeneous catalyst is in the same phase as the reactants; a hetereogeneous catalyst is in a different phase and is usually a solid.

14.60   (a)   No, the presence of a catalyst does not affect ΔH. ΔH is a state function; the difference in enthalpy between reactants and products is independent of reaction path. A catalyst affects the path of the reaction (the reaction mechanism) and thus the rate, but not the overall difference in enthalpy between reactants and products.

        (b)   The activity of a heterogeneous catalyst depends on the total surface area and number of active sites per unit amount of catalyst, which are influenced by method of preparation and prior treatment. The total surface area depends on particle size and processing techniques and the number of active sites depends on the presence or absence of impurities and possible exposure to contaminants.

14.61   (a) $2[NO_2(g) + SO_2(g) \rightarrow NO(g) + SO_3(g)]$
              $\underline{2NO(g) + O_2(g) \rightarrow 2NO_2(g)}$

              $2SO_2(g) + O_2(g) \rightarrow 2SO_3(g)$

        (b) $NO_2(g)$ is a catalyst because it is consumed and then reproduced in the reaction sequence. ($NO(g)$ is an intermediate because it is produced and then consumed.)

        (c)   Since NO_2 is in the same state as the other reactants, this is homogeneous catalysis.

14.62   (a) $2[NO(g) + N_2O(g) \rightarrow N_2(g) + NO_2(g)]$
              $\underline{2NO_2(g) \rightarrow 2NO(g) + O_2(g)}$

              $2N_2O(g) \rightarrow 2N_2(g) + O_2(g)$

        (b)   An intermediate is produced and then consumed during the course of the reaction. A catalyst is consumed and then reproduced. In other words, the catalyst is present when the reaction sequence begins and after the last step is completed. In this reaction, NO is the catalyst and NO_2 in an intermediate.

        (c)   Since NO is in the same state as the reactant, N_2O, the catalysis in this reaction is homogeneous.

        (d)   rate = $k[N_2O][NO]$ (Yes, even though NO is a catalyst, it appears in the rate law because it participates in the rate determining step.)

14.63   Use of chemically stable supports such as alumina and silica makes it possible to obtain very large surface areas per unit mass of the precious metal catalyst. This is so because the metal can be deposited in a very thin, even monomolecular, layer on the surface of the support.

14.64   Catalytic converters are heterogeneous catalysts that adsorb gaseous CO and hydrocarbons and speed up their oxidation to $CO_2(g)$ and $H_2O(g)$. They also adsorb nitrogen oxides, $(NO)_x$, and speed up their reduction to $N_2(g)$ and $O_2(g)$. Unfortunately, lead-containing compounds in leaded gasoline are also *adsorbed* onto the *active sites* of the catalysts, blocking uptake of CO, C_xH_y and $(NO)_x$ *substrates* and *poisoning* the catalysts. In order for catalytic converters to work efficiently, leaded gasoline cannot be burned in the engine.

14.65 As illustrated in Figure 14.23, the two C-H bonds that exist on each carbon of the ethylene molecule before adsorption are retained in the process in which a D atom is added to each C (assuming we use D_2 rather than H_2). To put two deuteriums on a single carbon, it is necessary that one of the already existing C-H bonds in ethylene be broken while the molecule is adsorbed, so the H atom moves off as an adsorbed atom, and is replaced by a D. This requires a larger activation energy than simply adsorbing C_2H_4 and adding one D atom to each carbon.

14.66 Replacement of hydrogen by -CH_3 or other organic groups would reduce the tendency to undergo hydrogenation. It would be difficult for the C=C bond to get close enough to the metal so that the π electrons of the alkene could interact with the metal orbitals to form a stable binding arrangement. With a -CH_3 or other bulky group blocking the way, it would also be difficult for an adsorbed H to find its way into the vicinity of the C=C.

14.67 (a) In the laboratory, if a reaction is too slow, we usually heat it to speed it up.

(b) Living organisms operate efficiently in a very narrow temperature range; heating to increase reaction rate is not an option. Therefore, the role of enzymes as homogeneous catalysts that speed up desirable reactions without heating and undesirable side-effects is crucial for biological systems.

(c) *catalase*: $2H_2O_2 \rightarrow 2H_2O + O_2$; *nitrogenase*: $N_2 \rightarrow 2NH_3$ (nitrogen fixation)

14.68 (a) An enzyme is a very large biological molecule that acts as a catalyst for a specific reaction or reactions. The active site is the specific location on the large enzyme molecule where the reacting substance or substrate is bound. Binding to the active site can activate the substrate by distorting its molecular structure or electron density distribution, and enables the substrate to react very quickly.

(b) An enzyme catalyzes a reaction by binding the substrate onto its active site and somehow activating the substrate. The enzyme is specific because the electronic and geometric characteristics of the active site (the lock) can only accommodate a certain substrate (the key). If the substrate doesn't fit the active site, the reaction is not catalyzed.

(c) An enzyme inhibitor prevents the substrate from binding to the active site, by either occupying the active site itself or attaching to some other area of the protein so that the shape of the active site is changed significantly, and the substrate no longer fits the "lock."

14.69 Let k = the rate constant for the uncatalyzed reaction,
k_c = the rate constant for the catalyzed reaction

According to Equation 14.19, $\ln k = -E_a/RT + \ln A$

Subtracting ln k from ln k_c,

$$\ln k_c - \ln k = -\left[\frac{55 \text{ kJ/mol}}{RT} + \ln A\right] - \left[-\frac{85 \text{ kJ/mol}}{RT} + \ln A\right]$$

(a)   RT = 8.314 J/K•mol × 298 k × 1 kJ/1000 J = 2.478 kJ/mol; ln A is the same for both reactions.

$$\ln (k_c /k) = \frac{85 \text{ kJ/mol} - 55 \text{kJ/mol}}{2.478 \text{ kJ/mol}}; \quad k_c /k = 1.8 \times 10^5$$

The catalyzed reaction is approximately 180,000 times faster at 25°C.

(b)   RT = 8.314 J/K•mol × 398 k × 1 kJ/1000 J = 3.309 kJ/mol

$$\ln (k_c /k) = \frac{30 \text{ kJ/mol}}{3.309 \text{ kJ/mol}}; \quad k_c /k = 8.7 \times 10^3$$

The catalyzed reaction is 8700 times faster at 125°C.

14.70   Let k and E_a equal the rate constant and activation energy for the uncatalyzed reaction. Let k_c and E_{ac} equal the rate constant and activation energy of the catalyzed reaction. A is the same for the uncatalyzed and catalyzed reactions. $k_c / k = 500$, T = 37°C = 310 K.

According to Equation 14.19, ln k = E_a/RT + ln A. Subtracting ln k from ln k_c

$$\ln k_c - \ln k = \left[\frac{-E_{ac} + A}{RT}\right] - \left[\frac{-E_a + A}{RT}\right]$$

$$\ln (k_c/k) = \frac{E_a - E_{ac}}{RT}; \quad E_a - E_{ac} = RT \ln (k_c / k)$$

$$E_a - E_{ac} = \frac{8.314 \text{ J}}{K•mol} \times 310 \text{ K} \times \ln (500) = 1.602 \times 10^4 \text{ J} = 16.0 \text{ kJ}$$

The enzyme must lower the activation energy by 16.0 kJ in order to achieve a 500-fold increase in reaction rate.

Additional Exercises

14.71   rate = $\dfrac{-\Delta[H_2S]}{\Delta t} = \dfrac{\Delta[Cl^-]}{2\Delta t} = k[H_2S][Cl_2]$

$$\frac{-\Delta[H_2S]}{\Delta t} = (3.5 \times 10^{-2} \ M^{-1}s^{-1})(1.6 \times 10^{-4} \ M)(0.070 \ M) = 3.92 \times 10^{-7} = 3.9 \times 10^{-7} \ M/s$$

$$\frac{\Delta[Cl^-]}{\Delta t} = \frac{2\Delta[H_2S]}{\Delta t} = 2(3.92 \times 10^{-7} \ M/s) = 7.8 \times 10^{-7} \ M/s$$

14.72   (a)   Note that the rate increases by a factor of four when $[C_2O_4^{2-}]$ doubles (compare experiment 3 with 4, or 2 with 1). The rate doubles when $[HgCl_2]$ doubles (compare experiment 1 with 4, or 2 with 3). The rate law is apparently: rate = $k[HgCl_2][C_2O_4^{2-}]^2$

(b) $k = \dfrac{\text{rate}}{[HgCl_2][C_2O_4^{2-}]^2}$ Using the data for Experiment 3,

        $k = \dfrac{(3.5 \times 10^{-5}\ M/s)}{[0.052\ M][0.30\ M]^2} = 7.5 \times 10^{-3}\ M^{-2}s^{-1}$

(c)   rate $= (7.5 \times 10^{-3}\ M^{-2}s^{-1})(0.080\ M)(0.10\ M)^2 = 6.0 \times 10^{-6}\ M/s$

14.73   (a) $k = (8.56 \times 10^{-5}\ M/s)/(0.200\ M) = 4.28 \times 10^{-4}\ s^{-1}$

      (b)   ln [urea] $= -(4.28 \times 10^{-4}s^{-1} \times 5.00 \times 10^3\ s) + \ln(0.500)$

               ln [urea] $= -2.14 - 0.693 = -2.833 = -2.83$;   [urea] $= 0.0588 = 0.059\ M$

      (c) $t_{1/2} = 0.693/k = 0.693/4.28 \times 10^{-4}\ s^{-1} = 1.62 \times 10^3\ s$

14.74   (a)   Because a plot of ln[SO_2Cl_2] vs time is linear, the reaction is first order. The rate law is rate $= k[SO_2Cl_2]$. For a first order reaction, Equation [14.13] is appropriate.

             $\ln([A]_t/[A]_o) = -kt;\ k = -\ln[[A]_t/[A]_o]/t$

             $k = \dfrac{-\ln(0.280/0.400)}{240\ s} = \dfrac{-(-0.3567)}{240\ s} = 1.486 \times 10^{-3} = 1.49 \times 10^{-3}\ s^{-1}$

      (b)   For a first order reaction, $t_{1/2} = 0.693/k$ (Equation [14.15]).

             $t_{1/2} = 0.693 / 1.486 \times 10^{-3}s^{-1} = 466\ s = 7.77\ min$

14.75

Time (s)	[C_5H_6] (M)	ln[C_5H_6]	1/ [C_5H_6]
0	0.0400	-3.219	25.0
50	0.0300	-3.507	33.3
100	0.0240	-3.729	41.7
150	0.0200	-3.912	50.0
200	0.0174	-4.051	57.5

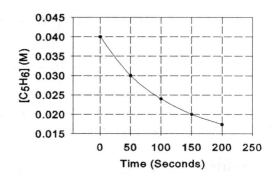

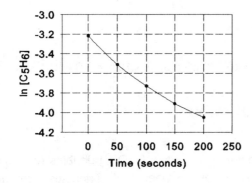

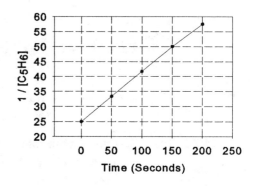

The plot of $1/[C_5H_6]$ vs time is linear and the reaction is second order.

The slope of this line is k. k = slope = $(50.0 - 25.0)$ M^{-1}/ $(150-0)$s = 0.167 $M^{-1}s^{-1}$

(The best-fit slope and k value is 0.163 $M^{-1}s^{-1}$.)

14.76

ln k	1/T
-11.51	3.33×10^{-3}
- 9.90	3.13×10^{-3}
-8.52	2.94×10^{-3}
-7.60	2.82×10^{-3}

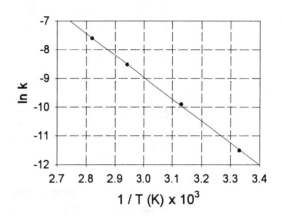

The activation energy E_a, equals $(-slope) \times (8.314$ J/mol$)$. Thus,
$E_a = 7.8 \times 10^3 (8.314)$
$= 6.5 \times 10^4$ J/mol = 65 kJ/mol.
(The best-fit slope is $-7.63 \times 10^3 =$
-7.6×10^3 and the value of E_a is
63 kJ/mol.)

14.77   (a) $T_1 = 37°C = 310$ K; $T_2 = 50°C = 323$ K; $t_{1/2}(50) = 2$ min

$t_{1/2} = 0.693/k$; $k_2 = 0.693 / t_{1/2} = 0.693 / 2$ min $= 0.347 = 0.3$ min^{-1}

$\ln (k_1 / k_2) = E_a /R \left(\dfrac{1}{T_2} - \dfrac{1}{T_1} \right)$; $\ln (k_1 / k_2) = \dfrac{400 \times 10^3 \text{ J/mol}}{8.314 \text{ J/mol}} \left(\dfrac{1}{323} - \dfrac{1}{310} \right)$

$\ln (k_1 / k_2) = -6.2464$; $k_1 / 0.347$ min$^{-1} = 1.937 \times 10^{-3}$;

$k_1 = 6.72 \times 10^{-4} = 7 \times 10^{-4}$ min^{-1}

$t_{1/2} (37) = 0.693 / 6.72 \times 10^{-4}$ min$^{-1} = 1 \times 10^3$ min (~ 17 hr)

(b)   A 13° increase in temperature causes the half-life of coiled DNA to decrease by a factor of 500. Clearly temperature increases (fevers) that are sustained over a period of time can have negative effects on human DNA. To prevent these effects, the human body needs (and has developed) a robust temperature regulation system.

14.78   (a)   rate = $k[H_2O_2][I^-]$

(b) $2H_2O_2(aq) \rightarrow 2H_2O(l) + O_2(g)$

(c) $IO^-(aq)$ is the intermediate.

14.79   (a)
$$Cl_2(g) \rightleftharpoons 2Cl(g)$$
$$Cl(g) + CHCl_3(g) \rightarrow HCl(g) + CCl_3(g)$$
$$Cl(g) + CCl_3(g) \rightarrow CCl_4(g)$$

$$Cl_2(g) + 2Cl(g) + CHCl_3(g) + CCl_3(g) \rightarrow 2Cl(g) + HCl(g) + CCl_3(g) + CCl_4(g)$$
$$Cl_2(g) + CHCl_3(g) \rightarrow HCl(g) + CCl_4(g)$$

(b) $Cl(g)$, $CCl_3(g)$

(c)   Step 1 - unimolecular, Step 2 - bimolecular, Step 3 - bimolecular

(d)   Step 2, the slow step, is rate determining.

(e)   If Step 2 is rate determining, rate = $k_2[CHCl_3][Cl]$. Cl is an intermediate formed in Step 1, an equilibrium. By definition, the rates of the forward and reverse processes are equal; $k_1 [Cl_2] = k_{-1} [Cl]^2$. Solving for [Cl] in terms of $[Cl_2]$,

$$[Cl]^2 = \frac{k_1}{k_{-1}} [Cl_2]; \; [Cl] = \left(\frac{k_1}{k_{-1}} [Cl_2] \right)^{1/2}$$

Substituting into the overall rate law

$$\text{rate} = k_2 \left(\frac{k_1}{k_{-1}} \right)^{1/2} [CHCl_3][Cl_2]^{1/2} = k[CHCl_3][Cl_2]^{1/2} \quad \text{(The overall order is 3/2.)}$$

14.80   Ice crystals in the upper atmosphere act as **heterogeneous** catalysts for the destruction of ozone. The system is similar to the one pictured in Figure 14.23, where ice crystals act as the solid support and $O_3(g)$ and $Cl(g)$ are adsorbed onto the surface.

14.81   (a) $(CH_3)_3AuPH_3 \rightarrow C_2H_6 + (CH_3)AuPH_3$

(b) $(CH_3)_3Au$, $(CH_3)Au$ and PH_3 are intermediates.

(c)   Step 1 is unimolecular, Step 2 is unimolecular, Step 3 is bimolecular.

(d)   Step 2, the slow step, is rate determining.

(e)   If Step 2 is rate determining, rate = $k_2[(CH_3)_3Au]$.

$(CH_3)_3Au$ is an intermediate formed in Step 1, an equilibrium. By definition, the rates of the forward and reverse processes in Step 1 are equal:

$k_1[(CH_3)_3 AuPH_3] = k_{-1} [(CH_3)_3Au][PH_3]$; solving for $[(CH_3)_3Au]$,

$$[(CH_3)_3 Au] = \frac{k_1[(CH_3)_3 AuPH_3]}{k_{-1} [PH_3]}$$

Substituting into the rate law

$$\text{rate} = \left(\frac{k_2 k_1}{k_{-1}} \right) \frac{[(CH_3)_3 AuPH_3]}{[PH_3]} = \frac{k[(CH_3)_3 AuPH_3]}{[PH_3]}$$

(f)   The rate is inversely proportional to $[PH_3]$, so adding PH_3 to the $(CH_3)_3 AuPH_3$ solution would decrease the rate of the reaction.

14.82   Catalysts generally <u>do</u> take part in the reaction, but there is no net consumption of the catalyst. A good example is the role of Br_2 and Br^- in catalyzing the decomposition of H_2O_2, Equations [14.29] and [14.30].

14.83   Enzyme: carbonic anhydrase; substrate: carbonic acid (H_2CO_3); turnover number: 1×10^7 molecules/s.

14.84   The fact that the rate doubles with a doubling of the concentration of sugar tells us that the fraction of enzyme tied up in the form of an enzyme-substrate complex is small. A doubling of the substrate concentration leads to a doubling of the concentration of enzyme-substrate complex, because most of the enzyme molecules are available to bind substrates. The behavior of inositol suggests that it acts as a competitor with sucrose for binding at the active sites of the enzyme system. Such a competition results in a lower effective concentration of active sites for binding of sucrose, and thus results in a lower reaction rate.

Integrative Exercises

14.85   (a)   rate = $k[H_2O_2][Br^-][H^+]$

       (b)   The units of k are $M^{-2}s^{-1}$.

       (c)   Calculate the ratio of $[H^+]$ after the addition of acid to $[H^+]$ of the neutral solution. Since the reaction is first order in $[H^+]$, the rate of the reaction will increase by this ratio.

               $[H^+]$ of the neutral solution is $1.0 \times 10^{-7}\ M$.

$$[H^+]_f = \frac{6.0\ M \times 0.0015\ L}{0.2015\ L} = 4.47 \times 10^{-2} = 4.5 \times 10^{-2}\ M$$

               The ratio is $4.47 \times 10^{-7}\ M / 1.0 \times 10^{-7}\ M = 4.5 \times 10^5$. The rate will increase by a factor of 4.5×10^5.

14.86   (a)   Use an apparatus such as the one pictured in Figure 10.3 (c) (an open-end manometer), a clock, a rule and a constant temperature bath. Since P = (n/V)RT, $\Delta P/\Delta t$ at constant temperature is an acceptable measure of reaction rate.

               Load the flask with HCl(aq) and read the height of the Hg in both arms of the manometer. Quickly add Zn(s) to the flask and record time = 0 when the Zn(s) contacts the acid. Record the height of the Hg in one arm of the manometer at convenient time intervals such as 5 sec. (The decrease in the short arm will be the same as the increase in the tall arm). Calculate the pressure of $H_2(g)$ at each time.

       (b)   Keep the amount of Zn(s) constant and vary the concentration of HCl(aq) to determine the reaction order for H^+ and Cl^-. Keep the concentration of HCl(aq) constant and vary the amount of Zn(s) to determine the order for Zn(s). Combine this information to write the rate law.

       (c) $-\Delta[H^+]/2\Delta t = \Delta[H_2]/\Delta t$; $-\Delta[H^+]/\Delta t = 2\Delta[H_2]/\Delta t$

               $[H_2]$ = mol H_2/L H_2 = n/V; $[H_2]$ = P (in atm)/RT

               Then, the rate of disappearance of H^+ is twice the rate of appearance of $H_2(g)$.

  (d)   By changing the temperature of the constant temperature bath, measure the rate data at several (at least three) temperatures and calculate the rate constant k at these temperatures. Plot ln k vs 1/T. The slope of the line is $-E_a/R$ and E_a = -slope (R).

  (e)   Measure rate data at constant temperature, HCl concentration and mass of Zn(s), varying only the form of the Zn(s). Compare the rate of reaction for metal strips and granules.

14.87   Enzymes and proteins are biopolymers, with much of the same structural flexibility as synthetic polymers (Chapter 12). The three dimensional shape of the protein is determined by many relatively weak intermolecular interactions and is sensitive to changes in local environment. Changes in temperature change the kinetic energy of the various groups on the enzyme and their tendency to form intermolecular associations or break free from them. Thus, changing the temperature changes the overall shape of the protein and specifically the shape of the active site. At body temperature, the competition between kinetic energy driving groups apart and intermolecular attraction pulling them together forms an active site that is optimum for a specific substrate. At other temperatures, a different structural equilibrium is reached, the shape of the active site is slightly different and the enzyme is less active.

14.88 (a)   If the reaction proceeds in a single elementary step, the coefficients in the balanced equation are the reaction orders for the respective reactants.

            rate $= k[Ce^{4+}]^2 [Tl^+]$

  (b)   If the uncatalyzed reaction occurs in a single step, it is *termolecular*. The activated complex requires collision of three particles with the correct energy and orientation for reaction. The probability of an effective three-particle collision is low and the rate is slow.

  (c)   The first step is rate-determining.

  (d)   The ability of Mn to adopt every oxidation state from +2 to +7 makes it especially suitable to catalyze this (and many other) reactions.

14.89 (a)   The two figures both have the same basic shape and represent the energies of the reactants, intermediates and products of a chemical reaction.

  (b)   Activation energies are lower than the sum of bond dissociation enthalpies for net bonds broken, because the transition state often involves partially broken old bonds as well as partially formed new ones.

  (c) E_a (max) = D(C=O) + D(C-C) + D(C-H)
                  = 799 + 348 +413 =1560 kJ/mol CH_3CHO

(d)   Assuming that the stretching of the C-C bond and the partial formation of a new C-H bond roughly cancel, the net change for the production of this transition state is

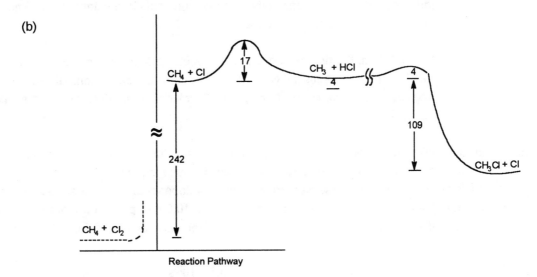

$E_a = D(C=O) - D(C \equiv O) + D(C-H) = 799 - 1072 + 413 = 140$ kJ/mol.

14.90   (a) $D(Cl-Cl) = 242$ kJ/mol Cl_2

$$\frac{242 \text{ kJ}}{\text{mol } Cl_2} \times \frac{1000 \text{ J}}{\text{kJ}} \times \frac{1 \text{ mol}}{6.022 \times 10^{23} \text{ molecules}} = 4.019 \times 10^{-19} = 4.02 \times 10^{-19} \text{ J}$$

$$\lambda = hc/E = \frac{6.626 \times 10^{-34} \text{ J} \cdot \text{s} \times 2.998 \times 10^8 \text{ m/s}}{4.019 \times 10^{-19} \text{ J}} = 4.94 \times 10^{-7} \text{ m}$$

This wavelength, 494 nm, is in the visible portion of the spectrum.

(b)

(c)   Since $D(Cl-Cl)$ is 242 kJ/mol, $CH_4(g) + Cl_2(g)$ should be about 242 kJ below the starting point on the diagram. For the reaction $CH_4(g) + Cl_2(g) \rightarrow CH_3(g) + HCl(g) + Cl(g)$, E_a is $242 + 17 = 259$ kJ. (From bond dissociation enthalpies, ΔH for the overall reaction $CH_4(g) + Cl_2(g) \rightarrow CH_3Cl(g) + Cl(g)$ is -104 kJ, so the graph above is simply a sketch of the relative energies of some of the steps in the process.)

(d) CH_3, 7 valence e^-, odd electron species   H—C̈—H
                                                               |
                                                               H

(e)   This sequence is called a chain reaction because Cl• radicals are regenerated in Reaction 4, perpetuating the reaction. Absence of Cl• terminates the reaction, so Cl•+Cl• → Cl_2 is a termination step.

15 Chemical Equilibrium

The Concept of Equilibrium; Equilibrium Expressions

15.1 (a) At equilibrium the forward and reverse reactions proceed at equal rates. Reactants are continually transformed into products, but products are also transformed into reactants at the same rate, so the net concentrations of reactants and products are constant at equilibrium.

(b) At equilibrium, it is the forward and reverse rates, not the rate constants, that are equal. It is true that the ratio of the two rate constants is constant at equilibrium, but this ratio is not constrained to have a value of 1.

(c) At equilibrium, the net concentrations of reactants and products are *constant* (see part (a)), but not necessarily equal. The relative concentrations of reactants and products at equilibrium are determined by their initial concentrations and the value of the equilibrium constant.

15.2 In this equilibrium system, ions move from the surface of the solid into solution, while ions in solution are deposited on the surface of the solid. As ions leave the surface of the small particles of powder, the reverse process preferentially deposits ions on the surface of a single crystal. Eventually, all ions that were present in the 10.0 g of powder are deposited on the surface of a 10.0 g crystal; this can only happen if the dissolution and deposition processes are ongoing.

15.3 (a) $K = \dfrac{k_f}{k_r}$, Equation [15.3]; $K = \dfrac{9.6 \times 10^2 \, s^{-1}}{3.8 \times 10^4 \, s^{-1}} = 2.5 \times 10^{-2}$

(b) $rate_f = rate_r$; $k_f [A] = k_r [B]$

Since $k_r > k_f$, in order for the two rates to be equal, [A] must be greater than [B].

15.4 $rate_f = k_f [A][B]$; $rate_r = k_r [C][D]$

At equilibrium, $rate_f = rate_r$: $k_f [A][B] = k_r [C][D]$; $\dfrac{k_f}{k_r} = \dfrac{[C][D]}{[A][B]} = K$

15.5 (a) The *law of mass action* expresses the relationship between the concentrations of reactants and products at equilibrium for any reaction. The law of mass action is a generic equilibrium expression.

$$K_c = \dfrac{[H_2O_2]}{[H_2][O_2]}$$

288

(b)   The *equilibrium expression* is an algebraic equation where the variables are the equilibrium concentrations of the reactants and products for a specific chemical reaction. The *equilibrium constant* is a number; it is the ratio calculated from the equilibrium expression for a particular chemical reaction. For any reaction, there are an infinite number of sets of equilibrium concentrations, depending on initial concentrations, but there is only one equilibrium constant.

(c)   Introduce a known quantity of $H_2O_2(g)$ into a vessel of known volume at constant (known) temperature. After equilibrium has been established, measure the total pressure in the flask. Using an equilibrium table, such as the one in Sample Exercise 15.7, calculate equilibrium pressures and concentrations of $H_2(g)$, $O_2(g)$ and $H_2O_2(g)$ and calculate K_c.

15.6   (a)   Yes. The algebraic form of the law of mass action depends only on the coefficients of a chemical equation, not on the reaction mechanism.

(b) $N_2(g) + 3H_2(g) \rightleftharpoons 2NH_3(g)$. The *Haber process* is the primary industrial method of nitrogen fixation, that is, of converting $N_2(g)$ into usable forms. The major use of $NH_3(g)$ from the Haber process is for fertilizer.

(c) $K_c = \dfrac{[NH_3]^2}{[N_2][H_2]^3}$; $K_p = \dfrac{P_{NH_3}^2}{P_{N_2} \times P_{H_2}^3}$

15.7   (a) $K_c = \dfrac{[NO]^2}{[N_2][O_2]}$ (b) $K_c = \dfrac{[N_2][H_2]^2}{[N_2H_4]}$ (c) $K_c = \dfrac{[C_2H_6]^2[O_2]}{[C_2H_4]^2[H_2O]^2}$

(d) $K_c = \dfrac{[H_2O]}{[H_2]}$ (e) $K_c = \dfrac{1}{[Cl_2]^2}$

homogeneous: (a), (b), (c); heterogeneous: (d), (e)

15.8   (a) $K_c = \dfrac{[N_2O][NO_2]}{[NO]^3}$; $K_p = \dfrac{P_{N_2O} \times P_{NO_2}}{P_{NO}^3}$ (b) $K_c = \dfrac{[CS_2][H_2]^4}{[CH_4][H_2S]^2}$; $K_p = \dfrac{P_{CS_2} \times P_{H_2}^4}{P_{CH_4} \times P_{H_2S}^2}$

(c) $K_c = \dfrac{[CO]^4}{[Ni(CO)_4]}$; $K_p = \dfrac{P_{CO}^4}{P_{Ni(CO)_4}}$ (d) $K_c = \dfrac{[H_2O]^3}{[H_2]^3}$; $K_p = \dfrac{P_{H_2O}^3}{P_{H_2}^3}$

(e) $K_c = \dfrac{[NO_2]^4[O_2]}{[N_2O_5]^2}$; $K_p = \dfrac{P_{NO_2}^4 \times P_{O_2}}{P_{N_2O_5}^2}$

Homogeneous: (a), (b), (e); heterogeneous: (c), (d)

15.9   (a) $K_c' = (K_c)^{-1} = 1/2.4 \times 10^3 = 4.2 \times 10^{-4}$

(b)   Since $K_c > 1$ when N_2 and O_2 are products and $K_c' < 1$ when N_2 and O_2 are reactants, the equilibrium favors N_2 and O_2 at this temperature.

15.10 (a) $2SO_2(g) + O_2(g) \rightleftharpoons 2SO_3(g)$ is the reverse of the reaction given.

$$K_c' = (K_c)^{-1} = 1/2.4 \times 10^{-3} = 4.2 \times 10^2$$

(b) Since $K_c < 1$ (when SO_3 is the reactant) and $K_c' > 1$ (when SO_3 is the product), the equilibrium favors SO_3 at this temperature.

15.11 (a) $$K_c = \frac{[Na_2O][SO_2]}{[Na_2SO_3]}$$

(b) The molar concentration, the ratio of moles of a substance to volume occupied by the substance, is a constant for pure solids and liquids.

(c) constant 1 = $[Na_2O]$; constant 2 = $[Na_2SO_3]$

$$K_c = \frac{\text{constant 1 } [SO_2]}{\text{constant 2}}; \quad K_c' = K_c \frac{\text{constant 2}}{\text{constant 1}} = [SO_2]$$

15.12 (a) $$K_c = \frac{[Hg]^4[O_2]}{[Hg_2O]^2}$$

(b) The molar concentration, the ratio of moles of a substance to volume occupied by the substance, is a constant for pure solids and liquids.

(c) constant 1 = $[Hg]^4$; constant 2 = $[Hg_2O]^2$

$$K_c = \frac{\text{constant 1 } [O_2]}{\text{constant 2}}; \quad K_c' = K_c \frac{\text{constant 2}}{\text{constant 1}} = [O_2]$$

Calculating Equilibrium Constants

15.13 $$K_c = \frac{[H_2][I_2]}{[HI]^2} = \frac{(4.79 \times 10^{-4})(4.79 \times 10^{-4})}{(3.53 \times 10^{-3})^2} = 1.84 \times 10^{-2}$$

15.14 $$K_c = \frac{[H_2][CO]}{[H_2O]} = \frac{(4.0 \times 10^{-2})(4.0 \times 10^{-2})}{1.0 \times 10^{-2}} = 0.16$$

15.15 $2NO(g) + Cl_2(g) \rightleftharpoons 2NOCl$

$$K_p = \frac{P_{NOCl}^2}{P_{NO}^2 \times P_{Cl_2}} = 52.0; \quad P_{NOCl}^2 = 52.0 \left(P_{NO}^2 \times P_{Cl_2} \right)$$

$$P_{NOCl} = \left[52.0 \left(P_{NO}^2 \times P_{Cl_2} \right) \right]^{1/2} = [52.0 ((0.095)^2 \times 0.171)]^{1/2}$$

$$P_{NOCl} = 0.28 \text{ atm}$$

15.16 (a) $K_p = \dfrac{P_{PCl_5}}{P_{PCl_3} \times P_{Cl_2}} = \dfrac{1.30 \text{ atm}}{0.124 \text{ atm} \times 0.157 \text{ atm}} = 66.8$

       (b)   Since $K_p > 1$, products (the numerator of the K_p expression) are favored over reactants (the demonimator of the K_p expression).

15.17 (a)   Since the reaction is carried out in a 1.00 L vessel, the moles of each component are equal to the molarity.

	2NO	+	2H$_2$	$\rightleftharpoons$	N$_2$	+	2H$_2$O
initial	0.100 *M*		0.050 *M*		0 *M*		0.100 *M*
change	-0.038 *M*		-0.038 *M*		+0.019 *M*		+0.038 *M*
equil.	0.062 *M*		0.012 *M*		0.019 *M*		0.138 *M*

First calculate the change in [NO], 0.100 - 0.062 = 0.038 *M*. From the stoichiometry of the reaction, calculate the change in the other concentrations. Finally, calculate the equilibrium concentrations.

       (b) $K_c = \dfrac{[N_2][H_2O]^2}{[NO]^2[H_2]^2} = \dfrac{(0.019)(0.138)^2}{(0.062)^2(0.012)^2} = 6.5 \times 10^2$

15.18 (a)   Calculate the initial concentrations of H$_2$(g) and Br$_2$(g) and the equilibrium concentration of H$_2$.

$\dfrac{1.374 \text{ H}_2}{2.00 \text{ L}} \times \dfrac{1 \text{ mol H}_2}{2.016 \text{ g H}_2} = 0.341 \text{ } M \text{ H}_2; \quad \dfrac{70.31 \text{ g Br}_2}{2.00 \text{ L}} \times \dfrac{1 \text{ mol Br}_2}{159.8 \text{ g Br}_2} = 0.220 \text{ } M \text{ Br}_2$

$\dfrac{0.566 \text{ g H}_2}{2.00 \text{ L}} \times \dfrac{1 \text{ mol H}_2}{2.016 \text{ g H}_2} = 0.140 \text{ } M \text{ H}_2$

	H$_2$(g)	+	Br$_2$(g)	$\rightleftharpoons$	2HBr(g)
initial	0.341 *M*		0.220 *M*		0
change	-0.201 *M*		-0.201 *M*		+2(0.201) *M*
equil.	0.140 *M*		0.019 *M*		0.402 *M*

The change in [H$_2$] is (0.341 *M* - 0.140 *M* = 0.201 *M*). The changes in [Br$_2$] and [HBr] are set by stoichiometry, resulting in the equilibrium concentrations shown in the table.

       (b) $K_c = \dfrac{[HBr]^2}{[H_2][Br_2]} = \dfrac{(0.402)^2}{(0.140)(0.019)} = 61$

Without intermediate rounding, equilibrium concentrations are [H$_2$] = 0.1404, [Br$_2$] = 0.0196, [HBr] = 0.4008 and K_c = 58. Note that this is a difference of less than 5 in the last significant figure of K_c.

15.19 (a) $K_c' = 1/K_c = 1/20.4 = 0.0490$

(b) $K_c' = K_c^2 = (20.4)^2 = 416.16 = 416$

(c) $K_p = K_c(RT)^{\Delta n}$; $\Delta n = -1$; $T = 700 + 273 = 973$

$K_p = 416.16/(0.08206)(973) = 5.21$

15.20 (a) $K_p' = 1/K_p = 1/0.0752 = 13.3$

(b) $K_p' = (K_p)^{1/2} = (0.0752)^{1/2} = 0.274$

(c) $K_c = K_p/RT^{\Delta n}$; $\Delta n = +1$; $T = 480 + 273 = 753K$
$K_c = 0.0752/(0.08206)(753) = 1.22 \times 10^{-3}$

15.21 (a) Since the reaction is carried out in a 1.00 L vessel, the moles of each component are equal to the molarity.

	CO_2	+	H_2	⇌	CO	+	H_2O
initial	0.1000 *M*		0.0500 *M*		0 *M*		0.1000 *M*
change	-0.0046 *M*		-0.0046 *M*		+0.0046 *M*		+0.0046 *M*
equil	0.0954 *M*		0.0454 *M*		0.0046 *M*		0.1046 *M*

First calculate the change in $[CO_2]$, $0.1000 - 0.0954 = 0.0046$ *M*. From the stoichiometry of the reaction, calculate the change in the other concentrations. Finally, calculate the equilibrium concentrations.

(b) $K_c = \dfrac{[CO][H_2O]}{[CO_2][H_2]} = \dfrac{(0.0046)(0.1046)}{(0.0954)(0.0454)} = 0.1111 = 0.11$

(c) No. In order to calculate K_p from K_c, the temperature of the reaction must be known.

15.22 (a) $N_2O_4(g) \rightleftharpoons 2NO_2(g)$

initial	1.500 atm	1.000 atm
change	+0.244 atm	-0.488 atm
equil	1.744 atm	0.512 atm

The change in P_{NO_2} is $(1.000 - 0.512) = -0.488$ atm, so the change in $P_{N_2O_4}$ is $+(0.488/2) = +0.244$ atm.

(b) $K_p = \dfrac{P_{NO_2}^2}{P_{N_2O_4}} = \dfrac{(0.512)^2}{(1.744)} = 0.1503 = 0.150$

(c) Yes, because temperature is specified. $K_p = K_c/(RT)^{\Delta n}$; $\Delta n = +1$; $T = 298$ K
$K_c = 0.1503/(0.08206)(298) = 6.15 \times 10^{-3}$

Applications of Equilibrium Constants

15.23 **(a)** A *reaction quotient* is the result of the law of mass action for a general set of concentrations, whereas the equilibrium constant requires equilibrium concentrations.

(b) In the direction of more products, to the right.

(c) If Q = K, the system is at equilibrium; the concentrations used to calculate Q must be equilibrium concentrations.

15.24 **(a)** If the value of Q equals the value of K, the system is at equilibrium.

(b) In the direction of less products (more reactants), to the left.

(c) Q = 0 if the concentration of any product is zero.

15.25 $K_c = \dfrac{[CO][Cl_2]}{[COCl_2]} = 2.19 \times 10^{-10}$ at 100°C

(a) $Q = \dfrac{(3.31 \times 10^{-6})(3.31 \times 10^{-6})}{(5.00 \times 10^{-2})} = 2.19 \times 10^{-10}$; $Q = K_c$

The mixture is at equilibrium.

(b) $Q = \dfrac{(1.11 \times 10^{-5})(3.25 \times 10^{-6})}{(3.50 \times 10^{-3})} = 1.03 \times 10^{-8}$; $Q > K_c$

The reaction will proceed to the left to attain equilibrium.

(c) $Q = \dfrac{(1.56 \times 10^{-6})(1.56 \times 10^{-6})}{(1.45)} = 1.68 \times 10^{-12}$, $Q < K_c$

The reaction will proceed to the right to attain equilibrium.

15.26 Calculate the reaction quotient in each case, compare with $K_p = \dfrac{P^2_{NH_3}}{P_{N_2} \times P^3_{H_2}} = 4.51 \times 10^{-5}$

(a) $Q = \dfrac{(105)^2}{(35)(495)^3} = 2.6 \times 10^{-6}$

Since $Q < K_p$, reaction will shift to the right to attain equilibrium.

(b) $Q = \dfrac{(35)^2}{(0)(595)^3} = \infty$

Since $Q > K_p$, reaction must shift to the left to attain equilibrium. There must be **some** N_2 present to attain equilibrium. In this example, the only source of N_2 is the decomposition of NH_3.

(c) $Q = \dfrac{(26)^2}{(42)^3(202)} = 4.52 \times 10^{-5}$; $Q = K_p$ Reaction is at equilibrium.

(d) $Q = \dfrac{(105)^2}{(5.0)(55)^3} = 1.3 \times 10^{-2}$; $Q > K_p$

Reaction will proceed to the left to attain equilibrium.

15.27 $K_c = \dfrac{[SO_2][Cl_2]}{[SO_2Cl_2]}$; $[Cl_2] = \dfrac{K_c[SO_2Cl_2]}{[SO_2]} = \dfrac{(0.078)(0.136)}{(0.072)} = 0.15\ M$

15.28 $K_p = \dfrac{P_{SO_3}^2}{P_{SO_2}^2 \times P_{O_2}}$; $P_{SO_3} = \left(K_p \times P_{SO_2}^2 \times P_{O_2}\right)^{1/2} = [0.345(0.215)^2(0.679)]^{1/2} = 0.104\ \text{atm}$

15.29 $K_c = 1.04 \times 10^{-3} = \dfrac{[Br]^2}{[Br_2]}$; $[Br_2] = \dfrac{0.245\ \text{g Br}_2}{159.8\ \text{g Br}_2/\text{mol} \times 0.200\ \text{L}} = 7.666 \times 10^{-3}$

$$= 7.67 \times 10^{-3}\ M$$

$[Br] = (1.04 \times 10^{-3}\ [Br_2])^{1/2} = (1.04 \times 10^{-3}\ (7.666 \times 10^{-3}))^{1/2} = 2.824 \times 10^{-3} = 2.82 \times 10^{-3}\ M$

$\text{g Br} = \dfrac{2.824 \times 10^{-3}\ \text{mol Br}}{1\ \text{L}} \times \dfrac{79.90\ \text{g Br}}{1\ \text{mol Br}} \times 0.200\ \text{L} = 0.04513 = 0.0451\ \text{g Br}$

$[Br_2] = 7.67 \times 10^{-3}\ M$; $[Br] = 2.82 \times 10^{-3}\ M$;   0.0451 g Br

15.30 $PV = nRT$; $P = gRT/(MM \times V)$

$P_{H_2} = \dfrac{0.056\ \text{g H}_2}{2.016\ \text{g/mol}} \times \dfrac{0.08206\ \text{L}\cdot\text{atm}}{\text{K}\cdot\text{mol}} \times \dfrac{700\ \text{K}}{2.000\ \text{L}} = 0.7978 = 0.80\ \text{atm}$

$P_{I_2} = \dfrac{4.36\ \text{g I}_2}{253.8\ \text{g/mol}} \times \dfrac{0.08206\ \text{L}\cdot\text{atm}}{\text{K}\cdot\text{mol}} \times \dfrac{700\ \text{K}}{2.000\ \text{L}} = 0.4934 = 0.494\ \text{atm}$

$K_p = 55.3 = \dfrac{P_{HI}^2}{P_{H_2} \times P_{I_2}}$; $P_{HI} = [55.3\,(P_{H_2})\,(P_{I_2})]^{1/2} = [55.3(0.7978)(0.4934)]^{1/2} = 4.666$

$$= 4.7\ \text{atm}$$

$g_{HI} = \dfrac{MM_{HI}\,P_{HI}\,V}{RT} = \dfrac{128.0\ \text{g HI}}{\text{mol HI}} \times \dfrac{\text{K}\cdot\text{mol}}{0.08206\ \text{L}\cdot\text{atm}} \times \dfrac{4.666\ \text{atm} \times 2.000\ \text{L}}{700\ \text{K}} = 20.79 = 21\ \text{g HI}$

15.31

	2NO(g)	$\rightleftharpoons$	N_2(g)	+	O_2(g)	$K_c = \dfrac{[N_2][O_2]}{[NO]^2} = 2.4 \times 10^3$
initial	0.500 *M*		0		0	
change	-2x		+x		+x	
equil.	0.500-2x		+x		+x	

$2.4 \times 10^3 = \dfrac{x^2}{(0.500-2x)^2}$; $(2.4 \times 10^3)^{1/2} = \dfrac{x}{0.500 - 2x}$

$x = (2.4 \times 10^3)^{1/2}\,(0.500 - 2x)$; $97.98x + x = 24.495$

$98.98x = 24.495$, $x = 0.2474 = 0.25$

$[N_2] = [O_2] = 0.25\ M$; $[NO] = 0.500 - 2(0.2474) = 0.0051 = 0.005\ M$

15.32 $Br_2(g)$ + $Cl_2(g)$ $\rightleftharpoons$ $2BrCl(g)$ $K_c = \dfrac{[BrCl]^2}{[Br_2][Cl_2]} = 7.0$

initial	0.50 *M*	0.50 *M*	0
change	-x	-x	+2x
equil.	(0.50-x) *M*	(0.50-x) *M*	+2x

$7.0 = \dfrac{(2x)^2}{(0.50-x)^2}$ (Assuming x is small leads to x = 0.66 *M*, [BrCl] = 1.3 *M*. Clearly x is not small compared to 0.50 *M*.)

$(7.0)^{1/2} = \dfrac{2x}{0.50-x}$,   2.646(0.50-x) = 2x, 1.323 = 4.646x, x = 0.2848 = 0.28 *M*

[BrCl] = 2x = 0.5696 = 0.57 *M*

15.33 $K_p = \dfrac{P_{NO}^2 P_{Br_2}}{P_{NOBr}^2}$

When $P_{NOBr} = P_{NO}$, these terms cancel and $P_{Br_2} = K_p = 0.416$ atm. This is true for all cases where $P_{NOBr} = P_{NO}$.

15.34 $K_c = [H_2S][NH_3] = 1.2 \times 10^{-4}$. The concentrations of H_2S and NH_3 will be the same; call this quantity y. then, $y^2 = 1.2 \times 10^{-4}$, $y = 1.1 \times 10^{-2}$ *M*.

15.35   **(a)**   Starting with only $PH_3BCl_3(s)$, the equation requires that the equilibrium concentrations of $PH_3(g)$ and $BCl_3(g)$ are equal.

         $K_c = [PH_3][BCl_3]$; $1.87 \times 10^{-3} = x^2$; x = 0.04324 = 0.0432 *M* PH_3 and BCl_3

  (b)   Since the mole ratios are 1:1:1, mol $PH_3BCl_3(s)$ required = mol PH_3 or BCl_3 produced.

         $\dfrac{0.04324 \text{ mol } PH_3}{1 \text{ L}} \times 0.500 \text{ L} \times \dfrac{151.2 \text{ g } PH_3BCl_3}{1 \text{ mol } PH_3BCl_3} = 3.27$ g PH_3BCl_3

         In fact, some $BH_3BCl_3(s)$ must remain for the system to be in equilibrium, so a bit more than 3.27 g PH_3BCl_3 is needed.

15.36   **(a)**   At equilibrium, $P_{NH_3} = P_{H_2S} = x$. $K_p = 0.070 = P_{NH_3} \times P_{H_2S} = x^2$; x = 0.2646 = 0.26 atm

  (b)   Calculate moles NH_3 or H_2S that would produce 0.26 atm pressure.

         $n = \dfrac{PV}{RT} = 0.2646 \text{ atm} \times \dfrac{15.0 \text{ L}}{295 \text{ K}} \times \dfrac{K \bullet mol}{0.08206 \text{ L} \bullet atm} = 0.1640 = 0.16$ mol NH_3

         Since some $NH_4SH(s)$ must remain for the system to be at equilibrium, a bit more than 0.1640 mol $NH_4SH(s)$ is required.

         $0.1640 \text{ mol } NH_4SH(s) \times \dfrac{51.12 \text{ g } NH_4SH}{1 \text{ mol } NH_4SH} = 8.4$ g NH_4SH

15.37 $K_c = \dfrac{[PCl_3][Cl_2]}{[PCl_5]}$; $[PCl_5]$ initial $= \dfrac{0.100 \text{ mol}}{5.00 \text{ L}} = 0.0200\ M$

$$PCl_2 \rightleftharpoons PCl_3 + Cl_2$$

	PCl_2	PCl_3	Cl_2
initial	0.0200 M	0	0
change	-x	+x	+x
equil.	0.0200-x	x	x

$K_c = 1.80 = \dfrac{x^2}{0.0200 - x}$ Assume x is small compared to 0.0200.

$1.80 \approx \dfrac{x^2}{0.0200}$; $x^2 = 0.0360$; $x = 0.190\ M$

Clearly, our assumption is not true. Solve the quadratic formula to obtain the value of x.

$x^2 = 1.80(0.0200-x)$; $x^2 + 1.80x - 0.0360 = 0$; for equations in the form $ax^2 + bx + c = 0$,

$$x = \frac{-b + \sqrt{b^2 - 4ac}}{2} = \frac{-1.80 + \sqrt{(1.80)^2 + 4(0.0360)}}{2}$$

$$x = \frac{-1.80 + \sqrt{3.384}}{2} = \frac{0.03957}{2} = 0.0198 = 0.02\ M$$

$[PCl_3] = [Cl_2] = 0.0198\ M$; $[PCl_5] = 0.0200\ M - 0.0198\ M = 2 \times 10^{-4}\ M$

(For this initial condition, the reaction essentially goes to completion.)

15.38 $K_c = 280 = \dfrac{[IBr]^2}{[I_2][Br_2]}$; $[Br]$ initial $= \dfrac{0.500 \text{ mol}}{1.000 \text{ L}} = 0.500\ M$

$$I_2 + Br_2 \rightleftharpoons 2\ IBr$$

	I_2	Br_2	$2\ IBr$
initial	0 M	0 M	0.500 M
change	+x	+x	-2x
equil.	x	x	0.500-2x

Since no I_2 or Br_2 were present initially, the amounts present at equilibrium are produced by the reverse reaction and stoichiometrically equal. Let these amounts equal x. The amount of HBr that reacts is then 2x. Substitute the equilibrium concentrations (in terms of x) into the equilibrium expression and solve for x.

$K_c = 280 = \dfrac{(0.500 - 2x)^2}{x^2}$; taking the square root of both sides

$16.733 = \dfrac{0.500 - 2x}{x}$; $16.733x + 2x = 0.500$; $18.733x = 0.500$

$x = 0.0267\ M$; $[I_2] = 0.0267\ M$, $[Br_2] = 0.0267\ M$

$[IBr] = 0.500 - 2x = 0.500 - 0.0534 = 0.447\ M$

15 Chemical Equilibrium

LeChatelier's Principle

15.39 (a) Shift equilibrium to the right; more $SO_3(g)$ is formed, the amount of $SO_2(g)$ decreases.

(b) Heating an exothermic reaction decreases the value of K. More SO_2 and O_2 will form, the amount of SO_3 will decrease.

(c) Since, $\Delta n = -1$, a change in volume will affect the equilibrium position and favor the side with more moles of gas. The amounts of SO_2 and O_2 increase and the amount of SO_3 decreases.

(d) No effect. Speeds up the forward and reverse reactions equally.

(e) No effect. Does not appear in the equilibrium expression.

(f) Shift equilibrium to the right; amounts of SO_2 and O_2 decrease.

15.40 (a) Increase   **(b)** increase   **(c)** decrease   **(d)** no effect   **(e)** no effect   **(f)** no effect

15.41 (a) No effect   **(b)** no effect   **(c)** increase equilibrium constant   **(d)** no effect

15.42 (a) The reaction must be endothermic ($+\Delta H$) if heating increases the fraction of products.

(b) There must be more moles of gas in the products if increasing the volume of the vessel increases the fraction of products.

15.43 (a) $\Delta H° = \Delta H_f° \, NO_2(g) + \Delta H_f° \, N_2O(g) - 3\Delta H_f° \, NO(g)$

$\Delta H° = 33.84 \text{ kJ} + 81.6 \text{ kJ} - 3(90.37 \text{ kJ}) = -155.7 \text{ kJ}$

(b) The reaction is exothermic ($-\Delta H°$), so the equilibrium constant will decrease with increasing temperature.

(c) Δn does not equal zero, so a change in volume at constant temperature will affect the fraction of products in the equilibrium mixture. An increase in container volume would favor reactants, while a decrease in volume would favor products.

15.44 (a) $\Delta H° = \Delta H_f° \, CH_3OH(g) - \Delta H_f° \, CO(g) - 2\Delta H_f° \, H_2(g)$

$= -201.2 \text{ kJ} - (-110.5 \text{ kJ}) - 0 \text{ kJ}$

$= -90.7 \text{ kJ}$

(b) The reaction is exothermic; an increase in temperature would decrease the value of K_c and decrease the yield. A *low temperature* is needed to maximize yield.

(c) Assuming equal pressures of CO and H_2, increasing total pressure would increase the concentration of each gas, shifting the equilibrium toward products. The extent of conversion to CH_3OH increases as the total pressure increases.

Additional Exercises

15.45   **(a)**   Since both the forward and reverse processes are elementary steps, we can write the rate laws directly from the chemical equation.

$$rate_f = k_f\,[CO][Cl_2] = rate_r = k_r\,[COCl][Cl]$$

$$\frac{k_f}{k_r} = \frac{[COCl][Cl]}{[CO][Cl_2]} = K$$

$$K = \frac{k_f}{k_r} = \frac{1.4 \times 10^{-28}\ M^{-1}\,s^{-1}}{9.3 \times 10^{10}\ M^{-1}\,s^{-1}} = 1.5 \times 10^{-39}$$

   (b)   Since the K is quite small, reactants are much more plentiful than products at equilibrium.

15.46 $[CO] = \dfrac{8.62\ g\ CO}{5.00\ L} \times \dfrac{1\ mol\ CO}{28.01\ g\ CO} = 0.06155 = 0.0616\ M$

$[H_2] = \dfrac{2.60\ g\ H_2}{5.00\ L} \times \dfrac{1\ mol\ H_2}{2.016\ g\ H_2} = 0.2579 = 0.258\ M$

$[CH_4] = \dfrac{43.0\ g\ CH_4}{5.00\ L} \times \dfrac{1\ mol\ CH_4}{16.04\ g\ CH_4} = 0.5362 = 0.536\ M$

$[H_2O] = \dfrac{48.4\ g\ H_2O}{5.00\ L} \times \dfrac{1\ mol\ H_2O}{18.02\ g\ H_2O} = 0.5372 = 0.537\ M$

$K_c = \dfrac{[CO][H_2]^3}{[CH_4][H_2O]} = \dfrac{(0.06155)(0.2579)^3}{(0.5362)(0.5372)} = 3.67 \times 10^{-3}$

15.47   **(a)** $H_2(g) + S(s) \rightleftharpoons H_2S(g)$

$$K_c = \frac{[H_2S]}{[H_2]}$$

   (b)   Calculate the molarity of H_2S and H_2. $M = mol/L$

$$[H_2S] = 0.46\ g\ H_2S \times \frac{1\ mol\ H_2S}{34.1\ g\ H_2S} \times \frac{1}{1.00\ L} = 0.0135\ M$$

$$[H_2] = 0.40\ g\ H_2 \times \frac{1\ mol\ H_2}{2.02\ g\ H_2} \times \frac{1}{1.00\ L} = 0.198\ M$$

$$K_c = \frac{(0.0135)}{(0.198)} = 0.068$$

   (c)   Since S is a pure solid, its concentration doesn't change during the reaction, and [S] does not appear in the equilibrium expression.

15.48 First, calculate the number of moles of each component present.

$$\frac{3.22 \text{ g NOBr}}{109.9 \text{ g/mol}} = 0.02930 = 0.0293 \text{ mol NOBr}; \quad \frac{3.08 \text{ g NO}}{30.01 \text{ g/mol}} = 0.1026 = 0.103 \text{ mol NO}$$

$$\frac{4.19 \text{ g Br}_2}{159.8 \text{ g/mol}} = 0.02622 = 0.0262 \text{ mol Br}_2$$

(a) In calculating K_c, divide each number of moles by 5.00 L to convert to moles/L, then insert into the expression for K_c to obtain:

$$K_c = \frac{[\text{Br}_2][\text{NO}]^2}{[\text{NOBr}]^2} = \frac{(5.244 \times 10^{-3})(2.053 \times 10^{-2})^2}{(5.860 \times 10^{-3})^2} = 6.436 \times 10^{-2} = 6.44 \times 10^{-2}$$

(b) $K_p = K_c(RT)^{\Delta n} = (6.436 \times 10^{-2})(0.08206 \times 373) = 1.97$

(c) The total moles of gas present is $0.02930 + 0.1026 + 0.02622 = 0.1581 = 0.158$

$$P = (0.1581 \text{ mol}) \times \frac{0.08206 \text{ L} \cdot \text{atm}}{1 \text{ mol} \cdot \text{K}} \times \frac{373 \text{ K}}{5.00 \text{ L}} = 0.968 \text{ atm}$$

15.49 (a)

	A(g)	$\rightleftharpoons$	2B(g)
initial	0.75 atm		0
change	-0.25 atm		+0.50 atm
equil.	0.50 atm		0.50 atm

$$P_T = P_A + P_B = 0.50 \text{ atm} + 0.50 \text{ atm} = 1.00 \text{ atm}$$

(b) $$K_p = \frac{(P_B)^2}{P_A} = \frac{(0.50)^2}{0.50} = 0.50$$

(c) $$K_c = \frac{K_p}{(RT)^{\Delta n}}; \quad \Delta n = +1, \quad T = 0°C + 273 = 273 \text{ K}$$

$$K_c = \frac{0.50}{(0.08206 \times 273)^{+1}} = 0.022$$

15.50 (a) $$K_p = \frac{P^2_{\text{NH}_3}}{P_{\text{N}_2} \times P^3_{\text{H}_2}} = 4.34 \times 10^{-3}$$

$$P_{\text{NH}_3} = \frac{gRT}{MM \times V} = \frac{0.753 \text{ g}}{17.03 \text{ g/mol}} \times \frac{0.08206 \text{ L} \cdot \text{atm}}{\text{K} \cdot \text{mol}} \times \frac{573 \text{ K}}{1.000 \text{ L}} = 2.079 = 2.08 \text{ atm}$$

	N$_2$(g)	+	3H$_2$(g)	$\rightleftharpoons$	2NH$_3$(g)
initial	0 atm		0 atm		?
change	x		3x		-2x
equil.	x atm		3x atm		2.079 atm

(Remember, only the change line reflects the stoichiometry of the reaction.)

$$K_p = \frac{(2.079)^2}{(x)(3x)^3} = 4.34 \times 10^{-3}; \quad 27x^4 = \frac{(2.079)^2}{4.34 \times 10^{-3}}; \quad x^4 = 36.885$$

$$x = 2.464 = 2.46 \text{ atm} = P_{N_2}; \quad P_{H_2} = 3x = 7.393 = 7.39 \text{ atm}$$

$$g_{N_2} = \frac{MM \times PV}{RT} = \frac{28.02 \text{ g N}_2}{\text{mol N}_2} \times \frac{K \cdot mol}{0.08206 \text{ L} \cdot atm} \times \frac{2.464 \text{ atm} \times 1.000 \text{ L}}{573 \text{ K}} = 1.47 \text{ g N}_2$$

$$g_{H_2} = \frac{2.016 \text{ g H}_2}{\text{mol H}_2} \times \frac{K \cdot mol}{0.08206 \text{ L} \cdot atm} \times \frac{7.393 \text{ atm} \times 1.00 \text{ L}}{573 \text{ K}} = 0.317 \text{ g N}_2$$

(b)   The initial $P_{NH_3} = 2.079 \text{ atm} + 2(2.464 \text{ atm}) = 7.007 = 7.01 \text{ atm}$

$$g_{NH_3} = \frac{MM \times PV}{RT} = \frac{17.03 \text{ g NH}_3}{\text{mol NH}_3} \times \frac{K \cdot mol}{0.08206 \text{ L} \cdot atm} \times \frac{7.007 \text{ atm} \times 1.000 \text{ L}}{573 \text{ K}}$$
$$= 2.54 \text{ g NH}_3$$

(c) $P_t = P_{N_2} + P_{H_2} + P_{NH_3} = 2.464 \text{ atm} + 7.393 \text{ atm} + 2.079 \text{ atm} = 11.94 \text{ atm}$

15.51 $K_c = \dfrac{[I_2][Br_2]}{[IBr]^2}$;   initial $[Br] = \dfrac{0.040 \text{ mol}}{1.00 \text{ L}} = 0.040 \text{ } M$

	2IBr	⇌	I_2	+	Br_2
initial	0.040 *M*		0		0
change	-2x		x		x
equil.	0.040-2x		x		x

$$K_c = 8.5 \times 10^{-3} = \frac{x^2}{(0.040-2x)^2}; \quad \text{Taking the square root of both sides}$$

$$\frac{x}{0.040-2x} = \sqrt{8.5 \times 10^{-3}} = 0.0922; \quad x = 0.0922(0.040-2x)$$

$x + 0.1844x = 0.003688; \quad 1.1844x = 0.003688, \quad x = 0.003114 = 0.0031$

$[IBr] = 0.040 - 2(0.0031) = 0.034 \text{ } M$

15.52   (a) $K_p = K_c(RT)^{\Delta n}; \quad K_c = \dfrac{K_p}{(RT)^{\Delta n}} = \dfrac{0.052}{(0.08206 \times 333)^2} = 6.964 \times 10^{-5} = 7.0 \times 10^{-5}$

(b) $[BCl_3] = 0.0216 \text{ mol}/0.500 \text{ L} = 0.0432 \text{ } M$

        PH_3BCl_3 is a solid and its concentration is taken as a constant, C.

	PH_3BCl_3	⇌	PH_3	+	BCl_3
initial	C		0 *M*		0.0432 *M*
change			+x *M*		+x *M*
equil.	C		x *M*		0.0432+x *M*

$$K_c = [PH_3][BCl_3]; \quad 7.0 \times 10^{-5} = x(0.0432 + x)$$

$$x^2 + 0.0432x - 7.0 \times 10^{-5} = 0$$

$$x = \frac{-0.0432 \pm [(0.0432)^2 - 4(-7.0 \times 10^{-5})]^{1/2}}{2} = 0.00156 = 1.6 \times 10^{-3} \, M = [PH_3]$$

Check: $(1.56 \times 10^{-3} + 0.0432)(1.56 \times 10^{-3}) = 7.0 \times 10^{-5}$; the solution is correct to two significant figures.

15.53 $K_p = P_{NH_3} \times P_{H_2S}; \quad P_t = 0.614$ atm

If the equilibrium amounts of NH_3 and H_2S are due solely to the decomposition of $NH_4HS(s)$, the equilibrium pressures of the two gases are equal, and each is 1/2 of the total pressure.

$$P_{NH_3} = P_{H_2S} = 0.614 \text{ atm}/2 = 0.307 \text{ atm}$$

$$K_p = (0.307)^2 = 0.0943$$

15.54   First find the initial moles of SO_3 and then the total moles of gas at equilibrium.

$$\frac{0.831 \text{ g } SO_3}{80.07 \text{ g/mol}} = 0.01038 = 0.0104 \text{ mol } SO_3; \quad n_t = \frac{1.30 \text{ atm} \times 1.00 \text{ L}}{1100 \text{ K}} \times \frac{K \cdot mol}{0.08206 \text{ L} \cdot atm}$$

$$= 0.01440 = 0.0144 \text{ mol}$$

	$2SO_3$	$\rightleftharpoons$	$2SO_2$	$+$	O_2
initial	0.01038		0		0
change	-2x		+2x		+x
equil.	0.01038-2x		2x		x
[equil.]	0.00234 *M*		0.00804 *M*		0.00402 *M*

$n_t = 0.0144 = 0.01038 - 2x + 2x + x; \quad x = 0.00402 = 0.0040 \text{ mol } O_2$

Since the volume is 1 L, the equilibrium molar concentrations are equal to the moles of each component.

$$K_c = \frac{[SO_2]^2[O_2]}{[SO_3]^2} = \frac{(0.00804)^2(0.00402)}{(0.00234)^2} = 0.04746 = 0.047$$

$$K_p = K_c(RT)^{\Delta n} = 4.746 \times 10^{-2}(0.08206 \times 1100)^1 = 4.3$$

15.55   In general, the reaction quotient is of the form $Q = \dfrac{[NOCl]^2}{[NO]^2[Cl_2]}$.

(a) $Q = \dfrac{(0.11)^2}{(0.15)^2(0.31)} = 1.7$

$Q > K_p$. Therefore, the reaction will shift toward reactants, to the left, in moving toward equilibrium.

(b) $Q = \dfrac{(0.050)^2}{(0.12)^2 (0.10)} = 1.7$

$Q > K_p$. Therefore, the reaction will shift toward reactants, to the left, in moving toward equilibrium.

(c) $Q = \dfrac{(5.10 \times 10^{-3})^2}{(0.15)^2 (0.20)} = 5.8 \times 10^{-3}$

$Q < K_p$. Therefore, the reaction mixture will shift in the direction of more product, to the right, in moving toward equilibrium.

15.56 $K_c = [CO_2] = 0.0108$

(a) $[CO_2] = 15.0 \text{ g } CO_2 \times \dfrac{1 \text{ mol } CO_2}{44.01 \text{ g } CO_2} = \dfrac{0.3408 \text{ mol}}{10.0 \text{ L}} = 0.0341 \; M$

$Q = 0.0341 > K_c$. The reaction proceeds to the left to achieve equilibrium and the amount of $CaCO_3(s)$ increases.

(b) $[CO_2] = 4.75 \text{ g } CO_2 \times \dfrac{1 \text{ mol } CO_2}{44.01 \text{ g } CO_2} \times \dfrac{1}{10.0 \text{ L}} = 0.0108 \; M$

$Q = 0.0108 = K_c$. The mixture is at equilibrium and the amount of $CaCO_3(s)$ remains constant.

(c) $[CO_2] = 2.50 \text{ g } CO_2 \times \dfrac{1 \text{ mol } CO_2}{44.01 \text{ g } CO_2} \times \dfrac{1}{10.0 \text{ L}} = 0.00568 \; M$

$Q = 0.00568 < K_c$. The reaction proceeds to the right to achieve equilibrium and the amount of $CaCO_3(s)$ decreases.

15.57 $K_c = K_p = \dfrac{P_{CO_2}}{P_{CO}} = 600$

If P_{CO} is 150 torr, P_{CO_2} can never exceed 760 − 150 = 610 torr. Then Q = 610/150 = 4.1. Since this is far less than K, the reaction will shift in the direction of more product. Reduction will therefore occur.

15.58   (a) $K_c = \dfrac{[Ni(CO)_4]}{[CO]^4}$

(b)   Increasing the temperature to 200°C favors the reverse process (decomposition of $Ni(CO)_4(g)$) and thus the value of K is smaller at the higher temperature. This is the behavior expected from an <u>exothermic</u> reaction (heat is a product).

(c)   At the temperature of the exhaust pipe, the $Ni(CO)_4$ product is a gas and is carried into the atmosphere with other exhaust gases. Thus, equilibrium is never established (we do not have a closed system) and the reaction proceeds to the right as $Ni(CO)_4$ product is removed.

15.59 (a) $K_c = \dfrac{K_p}{(RT)^{\Delta n}}$; $\Delta n = +1$, $T = 700$ K

$K_c = \dfrac{0.76}{(0.08206)(700)^{+1}} = 0.01323 = 0.013$

(b)
$$CCl_4(g) \; \rightleftharpoons \; C(s) \; + \; 2Cl_2(g)$$

initial	2.00 atm	0 atm
change	-x atm	+2x atm
equil.	(2.00-x) atm	2x atm

$K_p = 0.76 = \dfrac{P_{Cl_2}^2}{P_{CCl_4}} = \dfrac{(2x)^2}{(2.00-x)}$

$1.52 - 0.76x = 4x^2; \quad 4x^2 + 0.76x - 1.52 = 0$

Using the quadratic formula, $a = 4$, $b = 0.76$, $c = -1.52$

$x = \dfrac{-0.76 \pm \sqrt{(0.76)^2 - 4(4)(-1.52)}}{2(4)} = \dfrac{-0.76 + 4.99}{8} = 0.5287 = 0.53$ atm

Fraction CCl_4 reacted $= \dfrac{x \, atm}{2.00 \, atm} = \dfrac{0.53}{2.00} = 0.264 = 26\%$

(c) $P_{Cl_2} = 2x = 2(0.5287) = 1.06$ atm

$P_{CCl_4} = 2.00 - x = 2.00 - 0.5287 = 1.47$ atm

15.60 (a) $Q = \dfrac{P_{PCl_5}}{P_{PCl_3} \times P_{Cl_2}} = \dfrac{(0.10)}{(0.30)(0.60)} = 0.56$

0.56 (Q) > 0.0870 (K), the reaction proceeds to the left.

(b)
$$PCl_3(g) \; + \; Cl_2(g) \; \rightleftharpoons \; PCl_5(g)$$

initial	0.30 atm	0.60 atm	0.10 atm
change	+x atm	+x atm	-x atm
equil.	(0.30 + x) atm	(0.60 + x) atm	(0.10 - x) atm

(Since the reaction proceeds to the left, P_{PCl_5} must decrease and P_{PCl_3} and P_{Cl_2} must increase.)

$K_p = 0.0870 = \dfrac{(0.10 - x)}{(0.30 + x)(0.60 + x)}$; $0.0870 = \dfrac{(0.10 - x)}{(0.18 + 0.90\,x + x^2)}$

$0.0870\,x^2 + 0.0783\,x + 0.0157 = 0.10 - x; \quad 0.0870\,x^2 + 1.0783\,x - 0.0843 = 0$

$x = \dfrac{-1.0783 \pm \sqrt{(1.0783)^2 - 4(0.0870)(-0.0843)}}{2(0.0870)} = \dfrac{-1.0783 + 1.0918}{0.174} = 0.078$

$P_{PCl_3} = (0.30 + 0.078)$ atm $= 0.378$ $P_{Cl_2} = (0.60 + 0.078)$ atm $= 0.678$ atm

$P_{PCl_5} = (0.10 - 0.078)$ atm $= 0.022$ atm

To 2 decimal places, the pressures are 0.38, 0.68 and 0.02 atm, respectively. When substituting into the K_p expression, pressures to 3 decimal places yield a result much closer to 0.0870.

(c)   Increasing the volume of the container favors the process where more moles of gas are produced, so the reverse reaction is favored and the equilibrium shifts to the left; the mole fraction of P_{PCl_5} decreases.

(d)   For an exothermic reaction, increasing the temperature decreases the value of K; more reactants and fewer products are present at equilibrium and the partial pressure of P_{PCl_5} decreases.

15.61   First calculate K_c for the equilibrium

$$H_2 + I_2 \rightleftharpoons 2HI$$

$$K_c = \frac{[HI]^2}{[H_2][I_2]} = \frac{(0.155)^2}{(2.24 \times 10^{-2})(2.24 \times 10^{-2})} = 47.88 = 47.9$$

The added HI represents a concentration of $\dfrac{0.100 \text{ mol}}{5.00 \text{ L}} = 0.0200$ M.

	H_2	+	I_2	$\rightleftharpoons$	2HI
initial	2.24×10^{-2} M		2.24×10^{-2} M		$0.155 + 0.0200$ M
change	$+ x$ M		$+ x$ M		$-2x$ M
equil.	2.24×10^{-2} M $+ x$ M		2.24×10^{-2} M $+ x$ M		$0.175 - 2x$ M

$\dfrac{(0.175 - 2x)^2}{(2.24 \times 10^{-2} + x)^2} = 47.88$. Take the square root of both sides:

$\dfrac{0.175 - 2x}{2.24 \times 10^{-2} + x} = (47.88)^{1/2} = 6.920 = 6.92$

$0.175 - 2x = 0.155 + 6.92x$; $x = 2.242 \times 10^{-3} = 2.24 \times 10^{-3}$

$[I_2] = [H_2] = 2.24 \times 10^{-2} + 2.24 \times 10^{-3} = 2.464 \times 10^{-2} = 2.46 \times 10^{-2}$ M

$[HI] = 0.175 - 2x = 0.1705 = 0.171$ M

15.62 (a)   Since the volume of the vessel = 1.00 L, mol = M. The reaction will proceed to the left to establish equilibrium.

$$A(g) \; + \; 2B(g) \; \rightleftharpoons \; 2C(g)$$

	A(g)	2B(g)	2C(g)
initial	0 M	0 M	1.00 M
change	+x M	+2x M	-2x M
equil.	x M	2x M	(1.00 - 2x) M

At equilibrium, [C] = (1.00 - 2x) M, [B] = 2x M.

(b)   x must be less than 0.50 M (so that [C], 1.00 -2x, is not less than zero).

(c) $K_c = \dfrac{[C]^2}{[A][B]^2}$; $\dfrac{(1.00 - 2x)^2}{(x)(2x)^2} = 0.25$

$1.00 - 4x + 4x^2 = 0.25(4x)^3$; $x^3 - 4x^2 + 4x - 1 = 0$

(d)

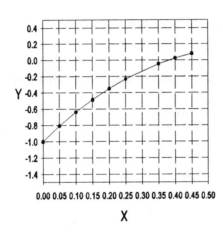

x	y
0.0	-1.000
0.05	-0.810
0.10	-0.639
0.15	-0.487
0.20	-0.352
0.25	-0.234
0.35	-0.047
0.40	+0.024
0.45	+0.081
~0.383	0.00

(e)   From the plot, x ≈ 0.383 M

[A] = x = 0.383 M; [B] = 2x = 0.766 M

[C] = 1.00 - 2x = 0.234 M

Using the K_c expression as a check:

$K_c = 0.25$; $\dfrac{(0.234)^2}{(0.383)(0.766)^2} = 0.24$;   the estimated values are reasonable.

15.63 $K_p = \dfrac{P_{O_2} \times P_{CO}^2}{P_{CO_2}^2}$; $P_{O_2} = (0.03)(1 \text{ atm}) = 0.03$ atm

$P_{CO} = (0.002)(1 \text{ atm}) = 0.002$ atm; $P_{CO_2} = (0.12)(1 \text{ atm}) = 0.12$ atm

$Q = \dfrac{(0.03)(0.002)^2}{(0.12)^2} = 8.3 \times 10^{-6} = 8 \times 10^{-6}$

Since $Q > K_p$, the system will shift to the left to attain equilibrium. Thus a catalyst that promoted the attainment of equilibrium would result in a lower CO content in the exhaust.

15.64   The patent claim is false. A catalyst does not alter the position of equilibrium in a system, only the rate of approach to the equilibrium condition.

Integrative Exercises

15.65   (a)   (i) $K_c = [Na^+]/[Ag^+]$ (ii) $K_c = [Hg^{2+}]^3 / [Al^{3+}]^2$

(iii) $K_c = [Zn^{2+}][H_2] / [H^+]^2$

(b)   According to Table 4.5, the activity series of the metals, a metal can be oxidized by any metal cation below it on the table.

(i) Ag^+ is far below Na, so the reaction will prceed to the right and K_c will be large.

(ii) Al^{3+} is above Hg, so the reaction will not proceed to the right and K_c will be small.

(iii) H^+ is below Zn, so the reaction will proceed to the right and K_c will be large.

(c) $K_c < 1$ for this reaction, so Fe^{2+} (and thus Fe) is above Cd on the table. In other words, Cd is below Fe. The value of K_c, 0.06, is small but not extremely small, so Cd will be only a few rows below Fe.

15.66   (a) $AgCl(s) \rightarrow Ag^+(aq) + Cl^-(aq)$

(b) $K_c = [Ag^+][Cl^-]$

(c)   Using thermodynamic data from Appendix C, calculate ΔH for the reaction in part (a).

$\Delta H^\circ = \Delta H^\circ_f \, Ag^+(aq) + \Delta H^\circ_f \, Cl^-(aq) - \Delta H^\circ_f \, AgCl(s)$
$\Delta H^\circ = +105.90 \, kJ - 167.2 \, kJ - (-127.0 \, kJ) = +65.7 \, kJ$

The reaction is endothermic (heat is a reactant), so the solubility of AgCl(s) in $H_2O(l)$ will increase with increasing temperature.

15.67   (a)   At equilibrium, the forward and reverse reactions occur at <u>equal</u> rates.

(b)   One expects the reactants to be favored at equilibrium since they are lower in energy.

(c)   A catalyst lowers the activation energy for both the forward and reverse reactions; the "hill" would be lower.

(d)   Since the activation energy is lowered for both processes, the new rates would be equal and the ratio of the rate constants, k_f /k_r , would remain unchanged.

(e)   Since the reaction is endothermic (the energy of the reactants is lower than that of the products, ΔE is positive), the value of K should increase with increasing temperature.

15.68   Consider the energy profile for an exothermic reaction.

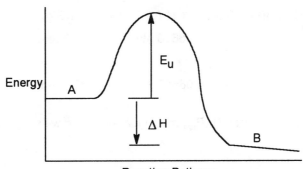

The activation energy in the forward direction, E_{af}, equals E_u, and the activation energy in the reverse reaction, E_{ar}, equals $E_u - \Delta H$. (The same is true for an endothermic reaction because the sign of ΔH is the positive and $E_{ar} < E_{af}$). For the reaction in question,

$$K = \frac{k_f}{k_r} = \frac{A_f e^{-E_{af}/RT}}{A_r e^{-E_{ar}/RT}}$$

Since the ln form of the Ahrrenius equation is easier to manipulate, we will consider ln K.

$$\ln K = \ln \left(\frac{k_f}{k_r} \right) = \ln k_f - \ln k_r = \frac{-E_{af}}{RT} + \ln A - \left[\frac{-E_{ar}}{RT} + \ln A_r \right]$$

Substituting E_u for E_{af} and $(E_u - \Delta H)$ for E_{ar}

$$\ln K = \frac{-E_u}{RT} + \ln A_f - \left[\frac{-(E_u - \Delta H)}{RT} + \ln Ar \right]; \quad \ln K = \frac{-E_u + (E_u - \Delta H)}{RT} + \ln A_f - \ln A_r$$

$$\ln K = \frac{-\Delta H}{RT} + \ln \frac{A_f}{A_r}$$

For the catalyzed reaction, $E_c < E_u$ and $E_{af} = E_c$, $E_{ar} = E_c - \Delta H$. The catalyst does not change the value of ΔH.

$$\ln K_c = \frac{-E_c + (E_c - \Delta H)}{RT} + \ln A_f - \ln A_r$$

$$\ln K_c = \frac{-\Delta H}{RT} + \frac{\ln A_f}{A_r}$$

Thus, assuming A_f and A_f are not changed by the catalyst, ln K = ln K_c and K = K_c.

15.69   (a) $P = \dfrac{g\,RT}{MM \times V} = \dfrac{0.300\ g\ NH_3}{17.03\ g/mol\ NH_3} \times \dfrac{298\ K}{5.00\ L} \times \dfrac{0.08206\ L \cdot atm}{mol \cdot K} = 0.08616$

                                                                                                    $= 0.0862\ atm$

(b) $K_p = P_{NH_3} \times P_{H_2S}$. Before solid is added, $Q = P_{NH_3} \times P_{H_2S} = 0.0862$ atm $\times\ 0\ = 0$.

Q < K and the reaction will proceed to the right. However, no $NH_4SH(s)$ is present to produce $H_2S(g)$, so the reaction cannot proceed.

(c)

	$NH_4SH(s)$	$\rightleftharpoons$	$NH_3(g)$	+	$H_2S(g)$
initial			0.08616 atm		0 atm
change			+x atm		+x atm
equil.			0.08616 + x atm		x atm

Since Q < K initially (part (a)), P_{NH_3} must increase along with P_{H_2S} until equilibrium is established.

$K_p = P_{NH_3} \times P_{H_2S}$; $0.120 = (0.08616 + x)(x)$; $0 = x^2 + 0.08616\,x - 0.120$

Solve for x using the quadratic formula.

$$x = \frac{-0.08616 \pm \sqrt{(0.08616)^2 - 4(1)(-0.120)}}{2}\ ;\quad x = 0.3060 = 0.306\ \text{atm}$$

$P_{H_2S} = 0.306$ atm; $P_{NH_3} = (0.08616 + 0.3060)$ atm $= 0.3922 = 0.392$ atm

(d) $\chi_{NH_3} = \dfrac{P_{NH_3}}{P_T} = \dfrac{0.3922\ \text{atm}}{(0.3060 + 0.3922)\ \text{atm}} = 0.562$

(e)   The minimum amount of $NH_4SH(s)$ required is slightly greater than the number of moles H_2S present at equilibrium. We can calculate the mol H_2S present at equilibrium using the ideal-gas equation.

$$nH_2S = \frac{P_{H_2S}V}{RT} = 0.306\ \text{atm} \times \frac{K\cdot mol}{0.08206\ L\cdot atm} \times \frac{5.00\ L}{298\ K} = 0.06257$$

$$= 0.0626\ \text{mol}\ H_2S$$

$$0.06257\ \text{mol}\ H_2S \times \frac{1\ mol\ NH_4SH}{1\ mol\ H_2S} \times \frac{51.12\ g\ NH_4SH}{1\ mol\ NH_4SH} = 3.20\ g\ NH_4SH$$

The minimum amount is slightly greater than 3.20 g NH_4SH.

15.70 $K_c = \dfrac{[CO]^2}{[CO_2]}$

To calculate K_c, find concentrations in units of moles/L. From the ideal-gas equation, $n/V = P/RT$. Since the total pressure is 1 atm in all cases, n/V for CO_2 and CO at each temperature can be calculated. For example, at 850° (1123 K):

$$[CO_2] = \frac{0.0623\ \text{atm}}{\left[\dfrac{0.08206\ L\cdot atm}{mol\cdot K}\right](1123\ K)} = 6.76 \times 10^{-4}\ M$$

$$[CO] = \frac{0.9377 \text{ atm}}{\left[\dfrac{0.08206 \text{ L} \cdot \text{atm}}{\text{mol} \cdot \text{K}}\right](1123 \text{ K})} = 1.018 \times 10^{-2} \text{ M}$$

Temp (K)	$[CO_2]$	$[CO]$	K_c
1123	6.76×10^{-4}	1.018×10^{-2}	0.153
1223	1.32×10^{-4}	9.833×10^{-3}	0.732
1323	3.4×10^{-5}	9.177×10^{-3}	2.5
1473	5×10^{-6}	8.268×10^{-3}	14 (1×10^1 to 1 sig fig)

Because K_c grows larger with increasing temperature, the reaction must be endothermic in the forward direction.

15.71   (a) $H_2O(l) \rightleftharpoons H_2O(g)$; $K_p = P_{H_2O}$

      (b)   At 30°C, the vapor pressure of $H_2O(l)$ is 31.82 torr.

             $K_p = P_{H_2O} = 31.82$ torr $= 0.041868 = 0.04187$ atm

      (c) $K_c = K_p/(RT)^{\Delta n}$; $\Delta n = +1$; $K_c = 0.041868/(0.08206)(303) = 1.684 \times 10^{-3}$

      (d)   From part (b), the value of K_p is the vapor pressure of the liquid at that temperature. By definition, vapor pressure = atmospheric pressure = 1 atm at the normal boiling point. $K_p = 1$ atm

15.72   (a)

           C-C B.O. = 1   C=C B.O. = 2

      (b) $\Delta H = D$(bond breaking) $- D$(bond making)

           E1: $\Delta H = D(C=C) + D(Cl-Cl) - D(C-C) - 2D (C-Cl)$
                    $\Delta H = 614 + 242 - 348 - 2(328) = -148$ kJ

           E2: $\Delta H = D(C-C) + D(C-H) + D(C-Cl) - D(C=C) - D(H-Cl)$
                   $= 348 + 413 + 328 - 614 - 431 = +44$ kJ

      (c)   E1 is exothermic with $\Delta n = -1$. The yield of $C_2H_4Cl_2(g)$ would decrease with increasing temperature and with increasing container volume.

      (d)   E2 is endothermic with $\Delta n = 1$. The yield of C_2H_3Cl would increase with increasing temperature and with increasing container volume.

      (e)   The boiling points of the reactants and products are C_2H_4, -103.7°C; Cl_2, -34.6°C, $C_2H_4Cl_2$, +83.5°C; C_2H_3Cl, -13.4°C; HCl, -84.9°C.

Because the products of E1 and E2 are optimized by different conditions, carry out the two equilibria in separate reactors. Since E1 is exothermic and $\Delta n = -1$, the reactor on the left should be as small and cold as possible to maximize yield of $C_2H_4Cl_2$. At temperatures below $83.5°C$, $C_2H_4Cl_2$ will condense to the liquid and it can be easily transferred to the second (right) reactor.

Since E_2 is endothermic, the reactor on the right should be as large and hot as possible to optimize production of C_2H_3Cl. The outlet stream will be a mixture of $C_2H_3Cl(g)$, $C_2H_4Cl_2(g)$ and $HCl(g)$. This mixture could be run through a heat exchanger to condense and subsequently recycle $C_2H_4Cl_2(l)$. Since HCl has a lower boiling point than C_2H_3Cl, it cannot be removed by condensation. The $C_2H_3Cl(g)$ / $HCl(g)$ mixture could be bubbled through a basic aqueous solution such as $NaHCO_3(aq)$ or $NaOH(aq)$ to remove $HCl(g)$, leaving pure $C_2H_3Cl(g)$.

Other details of reactor design such as the use of catalysts to speed up these reactions, the exact costs and benefits of heat exchange, recycling unreacted components and separation and recovery of products are issues best resolved by chemical engineers.

16 Acid-Base Equilibria

Acid-Base Equilibria

16.1   Solutions of HCl and H_2SO_4 taste sour, turn litmus paper red (are acidic), neutralize solutions of bases, react with active metals to form $H_2(g)$ and conduct electricity. The two solutions have these properties in common because both solutes are strong acids. That is, they both dissociate completely in H_2O to form $H^+(aq)$ and an anion. (The first dissociation step for H_2SO_4 is complete, but HSO_4^- is not completely dissociated.) The presence of ions enables the solutions to conduct electricity; the presence of $H^+(aq)$ in excess of 1×10^{-7} M accounts for all the other properties listed.

16.2   When NaOH dissolves in water, it completely dissociates to form $Na^+(aq)$ and $OH^-(aq)$. CaO is the oxide of a metal; it dissolves in water according to the following process: $CaO(s) + H_2O(l) \rightarrow Ca^{2+}(aq) + 2OH^-(aq)$. Thus, the properties of both solutions are dominated by the presence of $OH^-(aq)$. Both solutions taste bitter, turn litmus paper blue (are basic), neutralize solutions of acids and conduct electricity.

16.3   (a)   *Autoionization* is the ionization of a neutral molecule (in the absence of any other reactant) into an anion and a cation. The equilibrium expression for the autoionization of water is $H_2O(l) \rightleftharpoons H^+(aq) + OH^-(aq)$.

      (b)   Pure water is a poor conductor of electricity because it contains very few ions. Ions, mobile charged particles, are required for the conduction of electricity in liquids.

      (c)   If a solution is *acidic*, it contains more H^+ than OH^- ($[H^+] > [OH^-]$).

16.4   (a) $H_2O(l) \rightleftharpoons H^+(aq) + OH^-(aq)$

      (b) $K_w = [H^+][OH^-]$. The $[H_2O(l)]$ is omitted because water is a pure liquid. The molarity (mol/L) of pure solids or liquids does not change as equilibrium is established, so it is usually omitted from equilibrium expressions.

      (c)   If a solution is *basic*, it contains more OH^- than H^+ ($[OH^-] > [H^+]$).

16.5   In pure water at 25°C, $[H^+] = [OH^-] = 1 \times 10^{-7}$ M. If $[H^+] > 1 \times 10^{-7}$ M, the solution is acidic; if $[H^+] < 1 \times 10^{-7}$ M, the solution is basic.

      (a) $[H^+] = \dfrac{K_w}{[OH^-]} = \dfrac{1.0 \times 10^{-14}}{7 \times 10^{-4}\,M} = 1.43 \times 10^{-11} = 1 \times 10^{-11}$ $M < 1 \times 10^{-7}$ M; basic

(b)　　$[H^+] = \dfrac{K_w}{[OH^-]} = \dfrac{1.0 \times 10^{-14}}{8.2 \times 10^{-10}\,M} = \mathbf{1.2 \times 10^{-5}\,\textit{M}} > 1 \times 10^{-7}\,M$;　acidic

(c)　　$[OH^-] = 100[H^+]$;　$K_w = [H^+] \times 100[H^+] = 100[H^+]^2$;

　　　$[H^+] = (K_w/100)^{1/2} = \mathbf{1.0 \times 10^{-8}\,\textit{M}} < 1 \times 10^{-7}\,M$;　basic

16.6　In pure water at 25°C, $[H^+] = [OH^-] = 1 \times 10^{-7}\,M$.

If $[OH^-] > 1 \times 10^{-7}\,M$, the solution is basic; if $[OH^-] < 1 \times 10^{-7}\,M$, the solution is acidic.

(a)　　$[OH^-] = \dfrac{K_w}{[H^+]} = \dfrac{1.0 \times 10^{-14}}{6 \times 10^{-6}\,M} = 1.67 \times 10^{-9} = \mathbf{2 \times 10^{-9}\,\textit{M}} < 1 \times 10^{-7}\,M$;　acidic

(b)　　$[OH^-] = \dfrac{K_w}{[H^+]} = \dfrac{1.0 \times 10^{-14}}{2.5 \times 10^{-7}\,M} = \mathbf{4.0 \times 10^{-8}\,\textit{M}} < 1 \times 10^{-7}\,M$;　acidic

(c)　　$[H^+] = 5[OH^-]$;　$K_w = 5[OH^-][OH^-] = 5[OH^-]^2$

　　　$[OH^-] = (K_w/5)^{1/2} = \mathbf{4.5 \times 10^{-8}\,\textit{M}} < 1 \times 10^{-7}\,M$;　acidic

16.7　At 37°C, $K_w = 2.4 \times 10^{-14} = [H^+][OH^-]$.

In pure water, $[H^+] = [OH^-]$;　$2.4 \times 10^{-14} = [H^+]^2$;　$[H^+] = (2.4 \times 10^{-14})^{1/2}$

$[H^+] = [OH^-] = 1.5 \times 10^{-7}\,M$

16.8　$K_w = [D^+][OD^-]$;　for pure D_2O, $[D^+] = [OD^-]$; $8.9 \times 10^{-16} = [D^+]^2$;　$[D^+] = [OD^-] = 3.0 \times 10^{-8}\,M$

The pH Scale

16.9　A change of one pH unit (in either direction) is:

$\Delta pH = pH_2 - pH_1 = -(\log[H^+]_2 - \log[H^+]_1) = -\log \dfrac{[H^+]_2}{[H^+]_1} = \pm 1$.　The antilog of +1 is 10; the antilog of -1 is 1×10^{-1}. Thus, a ΔpH of one unit represents an increase or decrease in $[H^+]$ by a factor of 10.

(a)　　$\Delta pH = \pm 2.00$ is a change of $10^{2.00}$;　$[H^+]$ changes by a factor of 100.

(b)　　$\Delta pH = \pm 0.5$ is a change of $10^{0.50}$;　$[H^+]$ changes by a factor of 3.2.

16.10　$[H^+]_A = 500\,[H^+]_B$　From Exercise 16.9, $\Delta pH = -\log \dfrac{[H^+]_B}{[H^+]_A}$

$\Delta pH = -\log \dfrac{[H^+]_B}{500\,[H^+]_B} = -\log \left(\dfrac{1}{500}\right) = 2.70$

The pH of solution A is 2.70 pH units lower than the pH of solution B, because $[H^+]_A$ is 500 times greater than $[H^+]_B$. The greater $[H^+]$, the lower the pH of the solution.

16.11　(a)　　$K_w = [H^+][OH^-]$. If NaOH is added to water, it dissociates into $Na^+(aq)$ and $OH^-(aq)$. This increases $[OH^-]$ and necessarily decreases $[H^+]$. When $[H^+]$ decreases, pH increases.

(b)   pH = -log [H$^+$] = -log (0.005) = 2.3   If pH < 7, the solution is acidic.

(c)   pH = 6.3   pOH = 14.0 - 6.3 = 7.7
        [H$^+$] = 10^{-pH} = 10$^{-6.3}$ = 5 × 10^{-7} M [OH$^-$] = 10^{-pOH} = 10$^{-7.7}$ = 2 × 10^{-8} M

16.12   (a)   K$_w$ = [H$^+$][OH$^-$]. If HNO$_3$ is added to water, it dissociates into H$^+$(aq) and NO$_3^-$ (aq). This increases [H$^+$] and necessarily decreases [OH$^-$].   When [H$^+$] increases, pH decreases.

        (b)   [H$^+$] = K$_w$ / [OH$^-$] = 1.0 × 10^{-14} / 3.7 × 10^{-4} M = 2.7 × 10^{-11} M
                pH = -log (2.7 × 10^{-11}) = 10.57. If pH > 7, the solution is basic.

        (c)   pH = 8.6   pOH = 14.0 - pH = 5.4
                [H$^+$] = 10^{-pH} = 10$^{-8.6}$ = 2.51 × 10^{-9} = 3 × 10^{-9} M [OH$^-$] = 10^{-pOH} = 10$^{-5.4}$ = 4 × 10^{-6} M

16.13

[H$^+$]	[OH$^-$]	pH	pOH	acidic or basic
2.5 × 10^{-4} M	4.0 × 10^{-11} M	3.60	10.40	acidic
1.4 × 10^{-7} M	6.9 × 10^{-8} M	6.84	7.16	acidic
6 × 10^{-4} M	2 × 10^{-11} M	3.2	10.8	acidic
5 × 10^{-9} M	2 × 10^{-6} M	8.3	5.7	basic

16.14

[H$^+$]	[OH$^-$]	pH	pOH	acidic or basic
6.4 × 10^{-6} M	1.6 × 10^{-9} M	5.19	8.81	acidic
1.1 × 10^{-10} M	8.8 × 10^{-5} M	9.94	4.06	basic
3 × 10^{-8} M	3 × 10^{-7} M	7.5	6.5	basic
8 × 10^{-2} M	1 × 10^{-13} M	1.1	12.9	acidic

16.15   (a)   According to the Arrhenius definition, an acid when dissolved in water increases [H$^+$]. According to the Brønsted-Lowry definition, an acid is capable of donating H$^+$, regardless of physical state. The Arrhenius definition of an acid is confined to an aqueous solution; the Brønsted-Lowry definition applies to any physical state.

        (b)   HCl(g) + NH$_3$(g) → NH$_4^+$Cl$^-$(s)   HCl is the B-L (Brønsted-Lowry) acid; it donates an H$^+$ to NH$_3$ to form NH$_4^+$. NH$_3$ is the B-L base; it accepts the H$^+$ from HCl.

16.16   (a)   According to the Arrhenius definition, a base when dissolved in water increases [OH$^-$]. According to the Brønsted-Lowry theory, a base is an H$^+$ acceptor regardless of physical state. A Brønsted-Lowry base is not limited to aqueous solution and need not contain OH$^-$ or produce it in aqueous solution.

        (b)   NH$_3$(g) + H$_2$O(l) ⇌ NH$_4^+$(aq) + OH$^-$(aq)   When NH$_3$ dissolves in water, it accepts H$^+$ from H$_2$O (B-L definition). In doing so, OH$^-$ is produced (Arrhenius definition). Note that the OH$^-$ produced was originally part of the H$_2$O molecule, not part of the NH$_3$ molecule.

16.17   A conjugate base has one less H^+ than its conjugate acid.

(a) ClO_2^- (b) HS^- (c) SO_4^{2-} (d) NH_3

16.18   A conjugate acid has one more H^+ than its conjugate base.

(a) NH_4^+ (b) HIO (c) $HC_2H_3O_2$ (d) $H_2AsO_4^-$

16.19

B-L acid	+	**B-L base**	$\rightleftharpoons$	**Conjugate acid**	+	**Conjugate base**
(a) $NH_4^+(aq)$		$CN^-(aq)$		$HCN(aq)$		$NH_3(aq)$
(b) $H_2O(l)$		$(CH_3)_3N(aq)$		$(CH_3)_3NH^+(aq)$		$OH^-(aq)$
(c) $HCHO_2(aq)$		$PO_4^{3-}(aq)$		$HPO_4^{2-}(aq)$		$CHO_2^-(aq)$

16.20

B-L acid	+	**B-L base**	$\rightleftharpoons$	**Conjugate acid**	+	**Conjugate base**
(a) $HSO_4^-(aq)$		$CO_3^{2-}(aq)$		$HCO_3^-(aq)$		$SO_4^{2-}(aq)$
(b) $H_3O^+(aq)$		$HPO_4^{2-}(aq)$		$H_2PO_4^-(aq)$		$H_2O(l)$
(c) $HCO_3^-(aq)$		$OH^-(aq)$		$H_2O(l)$		$CO_3^{2-}(aq)$

16.21   Acid: $HC_2O_4^-(aq) + H_2O(l) \rightleftharpoons C_2O_4^{2-}(aq) + H_3O^+(aq)$
          B-L acid   B-L base   conj. base   conj. acid

Base: $HC_2O_4^-(aq) + H_2O(l) \rightleftharpoons H_2C_2O_4(aq) + OH^-(aq)$
          B-L base   B-L acid   conj. acid   conj. base

16.22   (a) $H_2PO_4^-(aq) + H_2O(l) \rightleftharpoons H_3PO_4(aq) + OH^-(aq)$

(b) $H_2PO_4^-(aq) + H_2O(l) \rightleftharpoons HPO_4^{2-}(aq) + H_3O^+(aq)$

(c) H_3PO_4 is the conjugate acid of $H_2PO_4^-$

HPO_4^{2-} is the conjugate base of $H_2PO_4^-$

16.23   (a)   weak, NO_2^- (b)   strong, HSO_4^- (c)   weak, PO_4^{3-}
(d)   negligible, CH_3^- (e)   weak, CH_3NH_2

16.24   (a)   weak, $HC_2H_3O_2$ (b)   weak, H_2CO_3 (c)   strong, OH^-
(d)   negligible, HCl (e)   weak, NH_4^+

16.25   (a) HNO_3.   It is one of the seven strong acids (Section 16.5);   it has the more electronegative central atom.

(b) H_2O. When NH_3 and H_2O are combined, as in $NH_3(aq)$, H_2O acts as the B-L acid. It has the greater tendency to donate H^+. For binary hydrides in general, acid strength increases going to the right across a row on the periodic chart; the more polar the H-X bond, the stronger the acid (Section 16.10).

16.26   (a) F^-. H_2SO_4 is a stronger acid than HF, so F^- is the stronger conjugate base.

16 Acid-Base Equilibria

Solutions to Exercises

(b) NH_3. When NH_3 and H_2O are combined, as in $NH_3(aq)$, NH_3 acts as the B-L base, accepting H^+ from H_2O. NH_3 has the greater tendency to accept H^+. For binary hydrides, base strength increases going to the left across a row of the periodic chart (Section 16.10).

16.27   Acid-base equilibria favor formation of the weaker acid and base. Compare the substances acting as acids on opposite sides of the equation. (Bases can also be compared; the conclusion should be the same.)

Base + **Acid** $\rightleftharpoons$ **Conjugate acid** + **Conjugate base**

(a) $NH_2^-(aq)$ + $H_2O(l)$ $\rightleftharpoons$ $NH_3(aq)$ + $OH^-(aq)$

H_2O is a stronger acid than NH_3 (Exercise 16.25), so the equilibrium lies to the right.

(b) $H_2O(l)$ + $HClO_2(aq)$ $\rightleftharpoons$ $H_3O^+(aq)$ + $ClO_2^-(aq)$

H_3O^+ is a stronger acid than $HClO_2$, so the equilibrium lies to the left.

(c) $F^-(aq)$ + $H_3O^+(aq)$ $\rightleftharpoons$ $HF(aq)$ + $H_2O(l)$

H_3O^+ is a stronger acid than HF, so the equilibrium lies to the right.

16.28   Acid-base equilibria favor formation of the weaker acid and base.

Base + **Acid** $\rightleftharpoons$ **Conjugate acid** + **Conjugate base**

(a) $O^{2-}(aq)$ + $H_2O(l)$ $OH^-(aq)$ + $OH^-(aq)$

H_2O is a stronger acid than OH^-, so the equilibrium lies to the right. (The intended base in this problem is oxide, O^{2-} not superoxide, O_2^-. The reaction of superoxide with water is complex and involves more than H^+ transfer.)

(b) $HS^-(aq)$ + $HSO_4^-(aq)$ $\rightleftharpoons$ $H_2S(aq)$ + $SO_4^{2-}(aq)$

According to Figure 16.7, HSO_4^- is a stronger acid than HS^-, so the products are as shown. HSO_4^- is also a stronger acid than H_2S, so the equilibrium lies to the right.

(c) $F^-(aq)$ + $HCO_3^-(aq)$ $\rightleftharpoons$ $HF(aq)$ + $CO_3^{2-}(aq)$

HF is a stronger acid than HCO_3^-, so the equilibrium lies to the left.

Strong Acids and Bases

16.29   (a)   A *strong* acid is completely ionized in aqueous solution; a strong acid is a strong electrolyte.

(b)   For a strong acid such as HCl, $[H^+]$ = initial acid concentration. $[H^+]$ = 0.500 M

(c)   HCl, HBr, HI

315

16.30 **(a)**   A *strong* base is completely ionized (hydrolyzed) in aqueous solution; a strong base is a strong electrolyte.

(b) $Sr(OH)_2$ is a soluble strong base.

$$Sr(OH)_2(aq) \rightarrow Sr^{2+}(aq) + 2OH^-(aq)$$

0.125 M $Sr(OH)_2(aq)$ = 0.250 M OH^-

(c)   Base strength should not be confused with solubility. Base strength describes the tendency of a dissolved molecule (formula unit for ionic compounds such as $Mg(OH)_2$) to ionize into cations and hydroxide ions. $Mg(OH)_2$ is a strong base because each $Mg(OH)_2$ unit that dissolves is ionized into $Mg^{2+}(aq)$ and $OH^-(aq)$. $Mg(OH)_2$ is not very soluble, so relatively few $Mg(OH)_2$ units dissolve when the solid compound is added to water.

16.31   For a strong acid, $[H^+]$ = initial acid concentration.

(a) 1.8×10^{-4} M HBr = 1.8×10^{-4} M H^+; pH = -log (1.8×10^{-4}) = 3.74

(b) $$\frac{1.02\,g\,HNO_3}{0.250\,L\,soln} \times \frac{1\,mol\,HNO_3}{63.02\,g\,HNO_3} = 0.06474 = 0.0647\ M\ HNO_3$$

$[H^+]$ = 0.0647 M; pH = -log (0.06474) = 1.189

(c) $M_c \times L_c = M_d \times L_d$; 0.500 M × 0.00200 L = ? M × 0.500 L

$$M_d = \frac{0.500\ M \times 0.00200}{0.0500} = 0.0200\ M\ HCl$$

$[H^+]$ = 0.0200 M; pH = -log (0.0200) = 1.699

(d) $$[H^+]_{total} = \frac{mol\,H^+\,from\,HBr + mol\,H^+\,from\,HCl}{total\,L\,solution}$$

$$[H^+]_{total} = \frac{(0.0100\ M\ HBr \times 0.0100\ L) + (2.50 \times 10^{-3}\ M \times 0.0200\ L)}{0.0300\ L}$$

$$[H^+]_{total} = \frac{1.00 \times 10^{-4}\,mol\,H^+ + 0.500 \times 10^{-4}\,mol\,H^+}{0.0300\ L} = 5.00 \times 10^{-3}\ M$$

pH = -log $(5.00 \times 10^{-3}\ M)$ = 2.301

16.32   For a strong acid, which is completely ionized, $[H^+]$ = the initial acid concentration.

(a)   0.025 M HNO_3 = 0.025 M H^+; pH = -log (0.025) = 1.60

(b) $$\frac{0.824\,g\,HClO_4}{0.500\,L\,soln} \times \frac{1\,mol\,HClO_4}{100.5\,g\,HClO_4} = 0.01640 = 0.0164\ M\ HClO_4$$

$[H^+]$ = 0.0164 M; pH = -log (0.0164) = 1.785

(c) $M_c \times V_c = M_d \times V_d$

1.5 M HCl $\times$ 5.00 mL HCl = M_d HCl $\times$ 100 mL HCl

M_d HCl = $\dfrac{1.5\, M \times 5.00\ mL}{100\ mL}$ = 0.075 M HCl = 0.075 M H^+

pH = -log (0.075) = 1.1249 = 1.12

(d) $[H^+]_{total} = \dfrac{mol\ H^+\ from\ HCl + mol\ H^+\ from\ HI}{total\ L\ solution}$;   mol = $M \times L$

$[H^+]_{total} = \dfrac{(0.020\, M\ HCl \times 0.010\ L) + (0.010\, M\ HI \times 0.030\ L)}{0.040\ L}$

$[H^+]_{total} = \dfrac{2.0 \times 10^{-4}\ mol\ H^+ + 3.0 \times 10^{-4}\ mol\ H^+}{0.040\ L}$ = 0.0125 = 0.013 M

pH = -log (0.0125) = 1.90

16.33   (a) $[OH^-]$ = 2$[Sr(OH)_2]$ = 2$(3.5 \times 10^{-4}\, M)$ = $7.0 \times 10^{-4}\ M$ OH^- (see Exercise 16.30(b))

pOH = -log (7.0×10^{-4}) = 3.15;   pH = 14 - pOH = 10.85

(b) $\dfrac{1.50\ g\ LiOH}{0.250\ L\ soln} \times \dfrac{1\ mol\ LiOH}{23.95\ g\ LiOH}$ = 0.2505 = 0.251 M LiOH = $[OH^-]$

pOH = -log (0.2505) = 0.601;   pH = 14 - pOH = 13.399

(c) $M_c \times L_c = M_d \times L_d$;   0.095 M $\times$ 0.00100 L = ? M $\times$ 2.00 L

$M_d = \dfrac{0.095\, M \times 0.00100\ L}{2.00\ L}$ = 4.75×10^{-5} = $4.8 \times 10^{-5}\ M$ NaOH = $[OH^-]$

pOH = -log (4.75×10^{-5}) = 4.32;   pH = 14 - pOH = 9.68

(d) $[OH^-]_{total} = \dfrac{mol\ OH^-\ from\ KOH + mol\ OH^-\ from\ Ca(OH)_2}{total\ L\ soln}$

$[OH^-]_{total} = \dfrac{(0.0105\, M \times 0.00500\ L) + 2(3.5 \times 10^{-3} \times 0.0150\ L)}{0.0200\ L}$

$[OH^-]_{total} = \dfrac{5.250 \times 10^{-5}\ mol\ OH^- + 10.5 \times 10^{-5}\ mol\ OH^-}{0.0200\ L}$ = 7.875×10^{-3}

$= 7.9 \times 10^{-3}\ M$

pOH = -log (7.875×10^{-3}) = 2.104;   pH = 14 - pOH = 11.90

$(3.5 \times 10^{-3}\ M$ has 2 sig figs, so the $[OH^-]$ has 2 sig figs and pH and pOH have 2 decimal places.)

16.34   For a strong base, which is completely ionized, $[OH^-]$ = the initial base concentration. Then, $pOH = -log\ [OH^-]$ and $pH = 14 - pOH$.

    (a)   0.050 M KOH = 0.050 M OH^-; $pOH = -log\ (0.050) = 1.30$; $pH = 14 - 1.30 = 12.70$

    (b) $\dfrac{2.33\ g\ NaOH}{0.500\ L} \times \dfrac{1\ mol\ NaOH}{40.00g\ NaOH} = 0.1165 = 0.117\ M\ NaOH = [OH^-]$

          $pOH = -log\ (0.1165) = 0.934$; $pH = 14 - pOH = 13.066$

    c) $M_c \times V_c = M_d \times L_d$

          0.150 M $Ca(OH)_2 \times 10.0$ mL = M_d $Ca(OH)_2 \times 500$ mL

          $M_d\ Ca(OH)_2 = \dfrac{0.150\ M\ Ca(OH)_2 \times 10.0\ mL}{500\ mL} = 0.00300\ M\ Ca(OH)_2$

          $Ca(OH)_2(aq) \rightarrow Ca^{2+}(aq) + 2OH^-(aq)$

          $[OH^-] = 2[Ca(OH)_2] = 2(3.00 \times 10^{-3}\ M) = 6.00 \times 10^{-3}\ M$

          $pOH = -log\ (6.00 \times 10^{-3}) = 2.222$; $pH = 14 - 2.222 = 11.778$

    (d) $[OH^-]_{total} = \dfrac{mol\ OH^-\ from\ Ba(OH)_2 + mol\ OH^-\ from\ NaOH}{total\ L\ solution}$

          $\dfrac{(6.8 \times 10^{-3}\ M \times 0.0300\ L) + 2(0.015\ M \times 0.0100\ L)}{0.0400\ L}$

          $[OH^-]_{total} = \dfrac{3.0 \times 10^{-4}\ mol\ OH^- + 2.0 \times 10^{-4}\ mol\ OH^-}{0.0400\ L} = 0.0125 = 0.013\ M\ OH^-$

          $pOH = -log\ (0.0125) = 1.90$; $pH = 14 - pOH = 12.10$

16.35   Upon dissolving, Li_2O dissociates to form Li^+ and O^{2-}. According to Equation 16.18, O^{2-} is completely protonated in aqueous solution.

    Thus, initial $[Li_2O] = [O_2^-]$; $[OH^-] = 2[O^{2-}] = 2[Li_2O]$

    $[Li_2O] = \dfrac{mol\ Li_2O}{L\ solution} = 2.00\ g\ Li_2O \times \dfrac{1\ mol\ Li_2O}{29.88\ g\ Li_2O} \times \dfrac{1}{0.600\ L} = 0.1116 = 0.112\ M$

    $[OH^-] = 0.2232 = 0.223\ M$; $pOH = 0.651$ $pH = 14.00 - pOH = 13.349$

16.36 $NaH(aq) \rightarrow Na^+(aq) + H^-(aq)$

    $H^-(aq) + H_2O(l) \rightarrow H_2(g) + OH^-(aq)$

    Thus, initial $[NaH] = [OH^-]$

$$[NaH] = \frac{mol\ NaH}{L\ solution} = 5.00\ g\ NaH \times \frac{1\ mol\ NaH}{24.00\ g\ NaH} \times \frac{1}{0.900\ L} = 0.23148 = 0.231\ M$$

$[OH^-] = 0.231\ M$; pOH = -log (0.23148) = 0.635, pH = 14 - pOH = 13.365

Weak Acids

16.37   (a) $HBrO_2(aq) \rightleftharpoons H^+(aq) + BrO_2^-(aq)$; $K_a = \dfrac{[H^+][BrO_2^-]}{[HBrO_2]}$

                $HBrO_2(aq) + H_2O(l) \rightleftharpoons H_3O^+(aq) + BrO_2^-(aq)$; $K_a = \dfrac{[H_3O^+][BrO_2^-]}{[HBrO_2]}$

      (b) $HC_3H_5O_2(aq) \rightleftharpoons H^+(aq) + C_3H_5O_2^-(aq)$; $K_a = \dfrac{[H^+][C_3H_5O_2^-]}{[HC_3H_5O_2]}$

                $HC_3H_5O_2(aq) + H_2O(l) \rightleftharpoons H_3O^+(aq) + C_3H_5O_2^-(aq)$; $K_a = \dfrac{[H_3O^+][C_3H_5O_2^-]}{[HC_3H_5O_2]}$

16.38   (a) $HCHO_2(aq) \rightleftharpoons H^+(aq) + CHO_2^-(aq)$; $K_a = \dfrac{[H^+][CHO_2^-]}{[HCHO_2]}$

                $HCHO_2(aq) + H_2O(l) \rightleftharpoons H_3O^+(aq) + CHO_2^-(aq)$; $K_a = \dfrac{[H_3O^+][CHO_2^-]}{[HCHO_2]}$

      (b) $HSO_4^-(aq) \rightleftharpoons H^+(aq) + SO_4^{2-}(aq)$; $K_a = \dfrac{[H^+][SO_4^{2-}]}{[HSO_4^-]}$

                $HSO_4^-(aq) + H_2O(l) \rightleftharpoons H_3O^+(l) + SO_4^{2-}(aq)$; $K_a = \dfrac{[H_3O^+][SO_4^{2-}]}{[HSO_4^-]}$

16.39 $HC_3H_5O_3(aq) \rightleftharpoons H^+(aq) + C_3H_5O_3^-(aq)$; $K_a = \dfrac{[H^+][C_3H_5O_3^-]}{[HC_3H_5O_3]}$

$[H^+] = [C_3H_5O_3^-] = 10^{-2.44} = 3.63 \times 10^{-3} = 3.6 \times 10^{-3}\ M$

$[HC_3H_5O_3] = 0.10 - 3.63 \times 10^{-3} = 0.0964 = 0.096\ M$

$$K_a = \frac{(3.63 \times 10^{-3})^2}{(0.0964)} = 1.4 \times 10^{-4}$$

16.40 $HC_8H_7O_2(aq) \rightleftharpoons H^+(aq) + C_8H_7O_2^-(aq)$; $K_a = \dfrac{[H^+][C_8H_7O_2^-]}{[HC_8H_7O_2]}$

$[H^+] = [C_8H_7O_2^-] = 10^{-2.68} = 2.09 \times 10^{-3} = 2.1 \times 10^{-3}\ M$

$[HC_8H_7O_2] = 0.085 - 2.09 \times 10^{-3} = 0.0829 = 0.083\ M$

$$K_a = \frac{(2.09 \times 10^{-3})^2}{0.0829} = 5.3 \times 10^{-5}$$

16.41
$$HC_7H_5O_2(aq) \rightleftharpoons H^+(aq) + C_7H_5O_2^-(aq)$$

initial	0.050 M	0	0
equil.	(0.050 - x) M	x M	x M

$$K_a = \frac{[H^+][C_7H_5O_2^-]}{[HC_7H_5O_2]} = \frac{x^2}{(0.050 - x)} \approx \frac{x^2}{0.050} = 6.5 \times 10^{-5}$$

$$x^2 = 0.050 \, (6.5 \times 10^{-5}); \quad x = 1.8 \times 10^{-3} \, M = [H^+] = [H_3O^+] = [C_7H_5O_2^-]$$

$$[HC_7H_5O_2] = 0.050 - 0.0018 = 0.048 \, M$$

$$\frac{1.8 \times 10^{-3} \, M \, H^+}{0.050 \, M \, HC_7H_5O_2} \times 100 = 3.6\% \text{ ionization; the assumption is valid}$$

16.42
$$HClO(aq) \rightleftharpoons H^+(aq) + ClO^-(aq)$$

initial	0.075 M	0	0
equil.	(0.075 - x) M	x M	x M

$$K_a = \frac{[H^+][ClO^-]}{[HClO]} = \frac{x^2}{(0.075 - x)} \approx \frac{x^2}{0.075} = 3.0 \times 10^{-8}$$

$$x^2 = 0.075 \, (3.0 \times 10^{-8}); \quad x = 4.74 \times 10^{-5} = 4.7 \times 10^{-5} \, M = [H^+] = [H_3O^+] = [ClO^-]$$

$$[HClO] = 0.075 - 4.7 \times 10^{-5} = 0.07495 = 0.075 \, M$$

$$\frac{4.7 \times 10^{-5} \, M \, H^+}{0.075 \, M \, HClO} \times 100 = 0.063\% \text{ ionization; the assumption is valid}$$

16.43 (a)
$$HN_3(aq) \rightleftharpoons H^+(aq) + N_3^-(aq)$$

initial	0.175 M	0	0
equil.	(0.175 - x) M	x M	x M

$$K_a = \frac{[H^+][N_3^-]}{[HN_3]} = \frac{(x)(x)}{(0.175 - x)} \approx \frac{x^2}{0.175} = 1.9 \times 10^{-5}$$

$$x^2 = 0.175 \, (1.9 \times 10^{-5}); \quad x = [H^+] = 1.8 \times 10^{-3} \, M, \quad pH = 2.74$$

$$\frac{1.8 \times 10^{-3} \, M \, H^+}{0.175 \, M \, HN_3} \times 100 = 1.0\% \text{ ionization; the assumption is valid}$$

(b)
$$K_a = \frac{[H^+][C_3H_5O_2^-]}{[HC_3H_5O_2]} = \frac{(x)(x)}{(0.040 - x)} \approx \frac{x^2}{0.040} = 1.3 \times 10^{-5}$$

$$x^2 = 0.040 \, (1.3 \times 10^{-5}); \quad x = [H^+] = 7.2 \times 10^{-4} \, M; \quad pH = 3.14$$

$$\frac{7.2 \times 10^{-4} \, M \, H^+}{0.040 \, M \, HC_3H_5O_2} \times 100 = 1.8\% \text{ ionization; the assumption is valid}$$

16.44   (a) $HOCl(aq)$ ⇌ $H^+(aq)$ + $OCl^-(aq)$

   initial   0.045 M 0   0

   equil   (0.045 - x) M x M x M

$$K_a = \frac{[H^+][OCl^-]}{[HOCl]} = \frac{(x)(x)}{(0.045 - x)} \approx \frac{x^2}{0.045} = 3.0 \times 10^{-8}$$

$x^2 = 0.045\,(3.0 \times 10^{-8})$; x = $[H^+]$ = 3.7×10^{-5} M, pH = 4.43

$$\frac{3.7 \times 10^{-5}\,M\,H^+}{0.045\,M\,HOCl} \times 100 = 0.083\%\text{ ionization; the assumption is valid}$$

(b) $$K_a = \frac{[H^+][C_6H_5O^-]}{[C_6H_5OH]} = \frac{(x)(x)}{(0.0068 - x)} \approx \frac{x^2}{0.0068} = 3.0 \times 10^{-10}$$

$x^2 = 0.0068\,(1.3 \times 10^{-10})$; x = $[H^+]$ = 9.4×10^{-7} M, pH = 6.03

Because $[H^+]$ from the ionization of phenol, 9.7×10^{-7} M, is on the order of magnitude of the $[H^+]$ expected from the dissociation of water, $[H^+]$ from the solvent should be considered.

$1.0 \times 10^{-14} = [H^+][OH^-]$; $[OH^-]$ = y, $[H^+]$ = 9.4×10^{-7} + y; (y + 9.4×10^{-7})y = 1.0×10^{-14}

From the quadratic formula, y = 1.1×10^{-8} M. Since this is less than 5% of the $[H^+]$ due to phenol, it can be ignored.

(c) $HONH_2(aq)$ + $H_2O(l)$ ⇌ $HONH_3^+(aq)$ + $OH^-(aq)$

   initial   0.080 M 0   0

   equil   (0.080 - x) M x M x M

$$K_b = \frac{[HONH_3^+][OH^-]}{[HONH_2]} = \frac{(x)(x)}{(0.080 - x)} \approx \frac{x^2}{0.080} = 1.1 \times 10^{-8}$$

$x^2 = 0.080\,(1.1 \times 10^{-8})$; x = $[OH^-]$ = 2.97×10^{-5} = 3.0×10^{-5} M, pH = 9.47

$$\frac{3.0 \times 10^{-5}\,M\,OH^-}{0.080\,M\,HONH_2} \times 100 = 0.038\%\text{ ionization; the assumption is valid}$$

16.45   (a) $HN_3(aq)$ ⇌ $H^+(aq)$ + $N_3^-(aq)$

   initial   0.400 M 0   0

   equil   (0.400 - x) M x M x M

$$K_a = \frac{[H^+][N_3^-]}{[HN_3]} = 1.9 \times 10^{-5}; \quad \frac{x^2}{(0.400 - x)} \approx \frac{x^2}{0.400} = 1.9 \times 10^{-5}$$

x = 0.00276 = 2.8×10^{-3} M = $[H^+]$; % ionization = $\dfrac{2.76 \times 10^{-3}}{0.400} \times 100 = 0.69\%$

(b) $1.9 \times 10^{-5} \approx \dfrac{x^2}{0.100}$; $x = 0.00138 = 1.4 \times 10^{-3}$ M H^+

% ionization $= \dfrac{1.38 \times 10^{-3}\, M\, H^+}{0.100\, M\, HN_3} \times 100 = 1.4\%$

(c) $1.9 \times 10^{-5} \approx \dfrac{x^2}{0.0400}$; $x = 8.72 \times 10^{-4} = 8.7 \times 10^{-4}\, M\, H^+$

% ionization $= \dfrac{8.72 \times 10^{-4}\, M\, H^+}{0.0400\, M\, HN_3} \times 100 = 2.2\%$

Notice that a tenfold dilution [part (a) versus part (c)] leads to a slightly more than threefold increase in percent ionization.

16.46   (a) $HCrO_4^-(aq) \rightleftharpoons H^+(aq) + CrO_4^{2-}(aq)$

$K_a = 3.0 \times 10^{-7} = \dfrac{[H^+][CrO_4^{2-}]}{[HCrO_4^-]} = \dfrac{x^2}{0.250 - x}$

$x^2 = 0.250\,(3.0 \times 10^{-7})$; $x = 2.74 \times 10^{-4} = 2.7 \times 10^{-4}\, M\, H^+$

% ionization $= \dfrac{2.74 \times 10^{-4}\, M\, H^+}{0.250\, M\, HCrO_4^-} \times 100 = 0.11\%$

(b) $\dfrac{x^2}{0.0800} \approx 3.0 \times 10^{-7}$; $x = 1.549 \times 10^{-4} = 1.5 \times 10^{-4}\, M\, H^+$

% ionization $= \dfrac{1.55 \times 10^{-4}\, M\, H^+}{0.0800\, M\, HCrO_4^-} \times 100 = 0.19\%$

(c) $\dfrac{x^2}{0.0200} \approx 3.0 \times 10^{-7}$; $x = 7.746 = 7.7 \times 10^{-5}\, M\, H^+$

% ionization $= \dfrac{7.75 \times 10^{-5}\, M\, H^+}{0.0200\, M\, HCrO_4^-} \times 100 = 0.39\%$

16.47 $[H^+] = 0.094 \times [HX]_{initial} = 0.0188 = 0.019\, M$

(a) $HX(aq) \rightleftharpoons H^+(aq) + X^-(aq)$

initial   0.200 M 0   0

equil.   (0.200 - 0.019) M 0.019 M 0.019 M

$K_a = \dfrac{[H^+][X^-]}{[HX]} = \dfrac{(0.0188)^2}{0.181} = 2.0 \times 10^{-3}$

16.48 $[H^+] = 0.132 \times [CH_2BrCOOH]_{initial} = 0.0132\, M$

                $CH_2BrCOOH(aq) \rightleftharpoons H^+(aq) + CH_2BrCOO^-(aq)$

initial   0.100 M 0   0

equil.   0.087M 0.0132 M 0.0132 M

$$K_a = \frac{[H^+][CH_2BrCOO^-]}{[CH_2BrCOOH]} = \frac{(0.0132)^2}{0.0868} = 2.0 \times 10^{-3}$$

16.49 Let the weak acid be HX. $HX(aq) \rightleftharpoons H^+(aq) + X^-(aq)$

$$K_a = \frac{[H^+][X^-]}{[HX]}; \quad [H^+] = [X^-] = y; \quad K_a = \frac{y^2}{[HX] - y}; \quad \text{assume that \% ionization is small}$$

$$K_a = \frac{y^2}{[HX]}; \quad y = K_a^{1/2}[HX]^{1/2}$$

$$\% \text{ ionization} = \frac{y}{[HX]} \times 100 = \frac{K_a^{1/2}[HX]^{1/2}}{[HX]} \times 100 = \frac{K_a^{1/2}}{[HX]^{1/2}} \times 100$$

That is, percent ionization varies inversely as the square root of concentration HX.

16.50 $HX(aq) \rightleftharpoons H^+(aq) + X^-(aq); \quad K_a = \frac{[H^+][X^-]}{[HX]}$

$[H^+] = [X^-]$; assume the % ionization is small; $K_a = \frac{[H^+]^2}{[HX]}$; $[H^+] = K_a^{1/2}[HX]^{1/2}$

$pH = -\log K_a^{1/2}[HX]^{1/2} = -\log K_a^{1/2} - \log [HX]^{1/2}$; $pH = -1/2 \log K_a - 1/2 \log [HX]$

This is the equation of a straight line; where the intercept is $-1/2 \log K_a$, the slope is $-1/2$, and the independent variable is $\log [HX]$.

16.51

$H_3C_6H_5O_7(aq) \rightleftharpoons H^+(aq) + H_2C_6H_5O_7^-(aq)$		$K_{a1} = 7.4 \times 10^{-4}$
$H_2C_6H_5O_7^-(aq) \rightleftharpoons H^+(aq) + HC_6H_5O_7^{2-}(aq)$		$K_{a2} = 1.7 \times 10^{-5}$
$HC_6H_5O_7^{2-}(aq) \rightleftharpoons H^+(aq) + C_6H_5O_7^{3-}(aq)$		$K_{a3} = 4.0 \times 10^{-7}$

To calculate the pH of a 0.050 M solution, assume initially that only the first ionization is important:

	$H_3C_6H_5O_7(aq)$	$\rightleftharpoons$	$H^+(aq)$	+	$H_2C_6H_5O_7^-(aq)$
initial	0.050 M		0		0
equil.	(0.050 - x) M		x M		x M

$$K_{a1} = \frac{[H^+][H_2C_6H_5O_7^-]}{[H_3C_6H_5O_7]} = \frac{x^2}{(0.050 - x)} = 7.4 \times 10^{-4}$$

$x^2 = (0.050 - x)(7.4 \times 10^{-4}); \quad x^2 \approx (0.050)(7.4 \times 10^{-4}); \quad x = 0.00608 = 6.1 \times 10^{-3} \, M$

Since this value for x is rather large in relation to 0.050, a better approximation for x can be obtained by substituting this first estimate into the expression for x^2, then solving again for x:

$$x^2 = (0.050 - x)(7.4 \times 10^{-4}) = (0.050 - 6.08 \times 10^{-3})(7.4 \times 10^{-4})$$
$$x^2 = 3.2 \times 10^{-5}; \quad x = 5.7 \times 10^{-3} \, M$$

The correction to the value of x, though not large, is significant. (This is the same result obtained from the quadratic formula.) Does the second ionization produce a significant additional concentration of H^+?

$$H_2C_6H_5O_7^- (aq) \rightleftharpoons H^+(aq) + HC_6H_5O_7^{2-} (aq)$$

initial	5.7×10^{-5} M	5.7×10^{-3} M	0
equil.	$(5.7 \times 10^{-3} - y)$	$(5.7 \times 10^{-3} + y)$	y

$$K_{a2} = \frac{[H^+][HC_6H_5O_7^{2-}]}{[H_2C_6H_5O_7^-]} = 1.7 \times 10^{-5}; \quad \frac{(5.7 \times 10^{-3} + y)(y)}{(5.7 \times 10^{-3} - y)} = 1.7 \times 10^{-5}$$

Assume that y is small relative to 5.7×10^{-3}; that is, that additional ionization of $H_2C_6H_5O_7^-$ is small, then

$$\frac{(5.7 \times 10^{-3})y}{(5.7 \times 10^{-3})} = 1.7 \times 10^{-5} \text{ M}; \quad y = 1.7 \times 10^{-5} \text{ M}$$

This value is indeed small compared to 5.7×10^{-3} M. This indicates that the second ionization can be neglected. pH is therefore $-\log [5.7 \times 10^{-3}] = 2.24$.

16.52 $H_6C_4O_6(aq) \rightleftharpoons H^+(aq) + H_5C_4O_6^-(aq) \quad K_{a1} = 1.0 \times 10^{-3}$

$H_5C_4O_6^-(aq) \rightleftharpoons H^+(aq) + H_4C_4O_6^{2-}(aq) \quad K_{a2} = 4.6 \times 10^{-5}$

Begin by calculating the $[H^+]$ from the first ionization. The equilibrium concentrations are $[H^+] = [H_5C_4O_6^-] = x$, $[H_6C_4O_6] = 0.025 - x$.

$$K_{a1} = \frac{[H^+][H_5C_4O_6^-]}{[H_6C_4O_6]} = \frac{x^2}{0.025 - x}; \quad x^2 + 1.0 \times 10^{-3}x - 2.5 \times 10^{-5} = 0$$

Using the quadratic formula, $x = 4.5 \times 10^{-3}$ M H^+ from the first ionization. Next calculate the H^+ contribution from the second ionization.

$$H_5C_4O_6^-(aq) \rightleftharpoons H^+(aq) + H_4C_4O_6^{2-}(aq)$$

initial	4.5×10^{-3}	4.5×10^{-3}	0
equil.	$(4.5 \times 10^{-3} - y)$	$(4.5 \times 10^{-3} + y)$	y

$$K_{a2} = \frac{(4.5 \times 10^{-3} + y)(y)}{(4.5 \times 10^{-3} - y)} = 4.6 \times 10^{-5}; \quad y = 4.6 \times 10^{-5} \text{ M } H^+$$

Since y, 4.6×10^{-5} M H^+, is only 1% of the $[H^+]$ from the first ionization, y can be safely ignored when calculating total $[H^+]$. Thus, pH = $-\log (4.5 \times 10^{-3}) = 2.35$.

Weak Bases; the K_a-K_b Relationship; Acid-Base Properties of Salts

16.53 All Brønsted-Lowry bases contain at least one lone (nonbonded) pair of electrons to attract H^+.

16.54 Organic amines (neutral molecules with lone pairs on N atoms) and anions that are the conjugate bases of weak acids function as weak bases.

16 Acid-Base Equilibria Solutions to Exercises

16.55 (a) $(CH_3)_2NH(aq) + H_2O(l) \rightleftharpoons (CH_3)_2NH_2^+(aq) + OH^-(aq);$ $K_b = \dfrac{[(CH_3)_2NH_2^+][OH^-]}{[(CH_3)_2NH]}$

(b) $CO_3^{2-}(aq) + H_2O(l) \rightleftharpoons HCO_3^-(aq) + OH^-(aq);$ $K_b = \dfrac{[HCO_3^-][OH^-]}{[CO_3^{2-}]}$

(c) $CHO_2^-(aq) + H_2O(l) \rightleftharpoons HCHO_2(aq) + OH^-(aq);$ $K_b = \dfrac{[HCHO_2][OH^-]}{[CHO_2^-]}$

16.56 (a) $C_3H_7NH_2(aq) + H_2O(l) \rightleftharpoons C_3H_7NH_3^+(aq) + OH^-(aq);$ $K_b = \dfrac{[C_3H_7NH_3^+][OH^-]}{[C_3H_7NH_2]}$

(b) $HPO_4^{2-}(aq) + H_2O(l) \rightleftharpoons H_2PO_4^-(aq) + OH^-(aq);$ $K_b = \dfrac{[H_2PO_4^-][OH^-]}{[HPO_4^{2-}]}$

(c) $C_6H_5CO_2^-(aq) + H_2O(l) \rightleftharpoons C_6H_5CO_2H(aq) + OH^-(aq);$ $K_b = \dfrac{[C_6H_5CO_2H][OH^-]}{[C_6H_5CO_2^-]}$

16.57
	$C_2H_5NH_2(aq) + H_2O(l) \rightleftharpoons$	$C_2H_5NH_3^+(aq) +$	$OH^-(aq)$
initial	0.050 M	0	0
equil.	(0.050 - x) M	x M	x M

$K_b = \dfrac{[C_2H_5NH_3^+][OH^-]}{[C_2H_5NH_2]} = \dfrac{(x)(x)}{(0.050-x)} \approx \dfrac{x^2}{0.050} = 6.4 \times 10^{-4}$

$x^2 = 0.050(6.4 \times 10^{-4});$ $x = [OH^-] = 5.7 \times 10^{-3}$ M; pH = 11.76

$\dfrac{5.7 \times 10^{-3} \, M \, OH^-}{0.050 \, M \, C_2H_5NH_2} \times 100 = 11.3\%$ ionization; the assumption is **not** valid

To obtain a more precise result, the K_b expression is rewritten in standard quadratic form and solved via the quadratic formula.

$\dfrac{x^2}{0.050-x} = 6.4 \times 10^{-4};$ $x^2 + 6.4 \times 10^{-4} x - 3.2 \times 10^{-5} = 0$

$x = \dfrac{b \pm \sqrt{b^2 - 4ac}}{2a} = \dfrac{-6.4 \times 10^{-4} \pm \sqrt{(6.4 \times 10^{-4})^2 - 4(1)(-3.2 \times 10^{-5})}}{2}$

$x = 5.346 \times 10^{-3} = 5.3 \times 10^{-3}$ M OH^-; pOH = 2.27, pH = 14.00 - pOH = 11.73

Note that the pH values obtained using the two algebraic techniques are very similar.

16.58
	$C_3H_5O_2^-(aq) + H_2O(l) \rightleftharpoons$	$HC_3H_5O_2(aq) +$	$OH^-(aq)$
initial	0.85 M	0	0
equil.	(0.85 - x) M	x M	x M

$K_b = \dfrac{[HC_3H_5O_2][OH^-]}{[C_3H_5O_2^-]} = \dfrac{x^2}{0.85-x} \approx \dfrac{x^2}{0.85} = 7.7 \times 10^{-10}$

$x^2 = 0.85 \,(7.7 \times 10^{-10})$; $x = [OH^-] = 2.56 \times 10^{-5} = 2.6 \times 10^{-5}\,M$; pH = 9.41

$$\dfrac{2.56 \times 10^{-5}\,M\,OH^-}{0.85\,M\,C_3H_5O_2^-} \times 100 = 0.0030\%\ \text{hydrolysis; the assumption is valid}$$

16.59   (a)   For a conjugate acid/conjugate base pair such as $C_6H_5OH/C_6H_5O^-$, K_b for the conjugate base is always K_w/K_a for the conjugate acid. K_b for the conjugate base can always be calculated from K_a for the conjugate acid, so a separate list of K_b values is not necessary.

        (b) $K_b = K_w/K_a = 1.0 \times 10^{-14}/1.3 \times 10^{-10} = 7.7 \times 10^{-5}$

        (c) K_b for phenolate (7.7×10^{-5}) > K_b for ammonia (1.8×10^{-5}).
             Phenolate is a stronger base than NH_3.

16.60   (a)   We need K_a for the conjugate acid of CO_3^{2-}, K_a for HCO_3^-. K_a for HCO_3^- is K_{a2}.

        (b) $K_b = K_w/K_a = 1.0 \times 10^{-14}/5.6 \times 10^{-11} = 1.8 \times 10^{-4}$

        (c) K_b for CO_3^{2-} (1.8×10^{-4}) > K_b for NH_3 (1.8×10^{-5}).
             CO_3^{2-} is a stronger base than NH_3.

16.61   When the solute in an aqueous solution is a salt, evaluate the acid/base properties of the component ions.

        (a)   NaCN is a soluble salt and thus a strong electrolyte. When it is dissolved in H_2O, it dissociates completely into Na^+ and CN^-. [NaCN] = $[Na^+]$ = $[CN^-]$ = 0.10 M. Na^+ is the conjugate acid of the strong base NaOH and thus does not influence the pH of the solution. CN^-, on the other hand, is the conjugate base of the weak acid HCN and **does** influence the pH of the solution. Like any other weak base, it hydrolyzes water to produce OH^-(aq). Solve the equilibrium problem to determine $[OH^-]$.

$$CN^-(aq) + H_2O(l) \rightleftharpoons HCN(aq) + OH^-(aq)$$

initial	0.10 M	0	0
equil.	(0.10 - x) M	x M	x M

$$K_b \text{ for } CN^- = \frac{[HCN][OH^-]}{[CN^-]} = \frac{K_w}{K_a \text{ for HCN}} = \frac{1 \times 10^{-14}}{4.9 \times 10^{-10}} = 2.04 \times 10^{-5} = 2.0 \times 10^{-5}$$

$2.04 \times 10^{-5} = \dfrac{(x)\,(x)}{(0.10 - x)}$; assume the percent of CN^- that hydrolyzes is small

$x^2 = 0.10\,(2.04 \times 10^{-5})$; $x = [OH^-] = 0.00143 = 1.4 \times 10^{-3}\,M$

pOH = 2.85; pH = 14 - 2.85 = 11.15

(b) $Na_2CO_3(aq) \rightarrow 2Na^+(aq) + CO_3^{2-}(aq)$

CO_3^{2-} is the conjugate base of HCO_3^- and its hydrolysis reaction will determine the $[OH^-]$ and pH of the solution (see similar explanation for NaCN in part (a)). We will assume the process $HCO_3^-(aq) + H_2O(l) \rightleftharpoons H_2CO_3(aq) + OH^-$ will not add significantly to the $[OH^-]$ in solution because $[HCO_3^-(aq)]$ is so small. Solve the equilibrium problem for $[OH^-]$.

	$CO_3^{2-}(aq) + H_2O(l) \rightleftharpoons$	$HCO_3^-(aq) +$	$OH^-(aq)$
initial	0.080 M	0	0
equil.	(0.080 - x) M	x	x

$$K_b = \frac{[HCO_3^-][OH^-]}{[CO_3^{2-}]} = \frac{K_w}{K_a \text{ for } HCO_3^-} = \frac{1.0 \times 10^{-14}}{5.6 \times 10^{-11}} = 1.79 \times 10^{-4} = 1.8 \times 10^{-4}$$

$$1.8 \times 10^{-4} = \frac{x^2}{(0.080 - x)}; \quad x^2 = 0.080 \, (1.79 \times 10^{-4}); \quad x = 0.00378 = 3.8 \times 10^{-3} \, M \, OH^-$$

(Assume x is small compared to 0.080); pOH = 2.42; pH = 14 - 2.42 = 11.58

$$\frac{3.8 \times 10^{-3} \, M \, OH^-}{0.080 \, M \, CO_3^{2-}} \times 100 = 4.75\% \text{ hydrolysis; the assumption is valid.}$$

(c)   For the two salts present, Na^+ and Ca^{2+} are negligible acids. NO_2^- is the conjugate base of HNO_2 and will determine the pH of the solution.

Calculate total $[NO_2^-]$ present initially.

$[NO_2^-]_{total}$ = $[NO_2^-]$ from $NaNO_2$ + $[NO_2^-]$ from $Ca(NO_2)_2$

$[NO_2^-]_{total}$ = 0.10 M + 2(0.20 M) = 0.50 M

The hydrolysis equilibrium is:

	$NO_2^-(aq) + H_2O(l) \rightleftharpoons$	$HNO_2 +$	$OH^-(aq)$
initial	0.50 M	0	0
equil.	(0.50 - x) M	x M	x M

$$K_b = \frac{[HNO_2][OH^-]}{[NO_2^-]} = \frac{K_w}{K_a \text{ for } HNO_2} = \frac{1.0 \times 10^{-14}}{4.5 \times 10^{-4}} = 2.22 \times 10^{-11} = 2.2 \times 10^{-11}$$

$$2.2 \times 10^{-11} = \frac{x^2}{(0.50 - x)} \approx \frac{x^2}{0.50}; \quad x^2 = 0.50 \, (2.22 \times 10^{-11})$$

$x = 3.33 \times 10^{-6} = 3.3 \times 10^{-6} \, M \, OH^-$; pOH = 5.48; pH = 14 - 5.48 = 8.52

16.62 (a)   Proceeding as in 16.61(a):

	$F^-(aq) + H_2O(l) \rightleftharpoons$	$HF(aq) +$	$OH^-(aq)$
initial	0.10 M	0 M	0 M
equil	(0.10 - x) M	x M	x M

$$K_b \text{ for } F^- = \frac{[HF][OH^-]}{[F^-]} = \frac{K_w}{K_a \text{ for } HF} = \frac{1.0 \times 10^{-14}}{6.8 \times 10^{-4}} = 1.47 \times 10^{-11} = 1.5 \times 10^{-11}$$

$$1.5 \times 10^{-11} = \frac{(x)(x)}{(0.10 - x)}; \text{ assume the amount of } F^- \text{ that hydrolyzes is small}$$

$$x^2 = 0.10 (1.47 \times 10^{-11}); \quad x = [OH^-] = 1.213 \times 10^{-6} = 1.2 \times 10^{-6} \, M$$

$$pOH = 5.92; \quad pH = 14 - 5.92 = 8.08$$

(b) $Na_2S(aq) \rightarrow S^{2-}(aq) + 2Na^+(aq)$

         $S^{2-}(aq) + H_2O(l) \rightleftharpoons HS^-(aq) + OH^-(aq)$

Assume the further hydrolysis of HS^- does not significantly contribute to $[OH^-]$. As in part (a) above, $[OH^-] = [HS^-] = x$; $[S^{2-}] = 0.10 \, M$

$$K_b = \frac{[HS^-][OH^-]}{[S^{2-}]} = \frac{K_w}{K_a \text{ for } HS^-} = \frac{1.0 \times 10^{-14}}{1 \times 10^{-19}} = 1 \times 10^5$$

Since $K_b \gg 1$, the equilibrium above lies far to the right and $[OH^-] = [S^{2-}]$.

$[OH^-] = 0.10 \, M$; $pOH = 1.00$, $pH = 13.00$

(c)   As in exercise 16.61(c), calculate total $[C_2H_3O_2^-]$.

         $[C_2H_3O_2^-]_t = [C_2H_3O_2^-]$ from $NaC_2H_3O_2 + [C_2H_3O_2^-]$ from $Ba(C_2H_3O_2)_2$

         $[C_2H_3O_2^-]_t = 0.085 \, M + 2(0.055 \, M) = 0.195 \, M$

The hydrolysis equilibrium is $C_2H_3O_2^-(aq) + H_2O(l) \rightleftharpoons HC_2H_3O_2(aq) + OH^-(aq)$

$$K_b = \frac{[HC_2H_3O_2][OH^-]}{[C_2H_3O_2^-]} = \frac{K_w}{K_a \text{ for } HC_2H_3O_2} = \frac{1.0 \times 10^{-14}}{1.8 \times 10^{-5}} = 5.56 \times 10^{-10}$$

$$= 5.6 \times 10^{-10}$$

$[OH^-] = [HC_2H_3O_2] = x$, $[C_2H_3O_2^-] = 0.195 - x$

$$K_b = 5.56 \times 10^{-10} = \frac{x^2}{(0.195 - x)}; \text{ assume } x \text{ is small compared to } 0.195 \, M$$

$$x^2 = 0.195 (5.56 \times 10^{-10}); \quad x = [OH^-] = 1.04 \times 10^{-5} = 1.0 \times 10^{-5}$$

$$pH = 14 + \log(1.04 \times 10^{-5}) = 9.02$$

16.63 (a)   acidic; NH_4^+ is a weak acid, Br^- is negligible.

        (b)   acidic; Fe^{3+} is a highly charged metal cation and a Lewis acid; Cl^- is negligible.

        (c)   basic; CO_3^{2-} is the conjugate base of HCO_3^-; Na^+ is negligible.

        (d)   neutral; both K^+ and ClO_4^- are negligible.

        (e)   acidic; $HC_2O_4^-$ is amphoteric, but K_a for the acid dissociation (6.4×10^{-5}) is much greater than K_b for the base hydrolysis ($1.0 \times 10^{-14} / 5.9 \times 10^{-2} = 1.7 \times 10^{-13}$).

16.64 (a) neutral; both Cs^+ and Br^- are negligible.

(b) acidic; Al^{3+} is a highly charged metal cation and a Lewis acid; NO_3^- is negligible.

(c) basic; CN^- is the conjugate base of HCN; K^+ is negligible.

(d) acidic; $CH_3NH_3^+$ is the conjugate acid of CH_3NH_2; Cl^- is negligible.

(e) acidic; HSO_4^- is a negligible base, but a fairly strong acid ($K_a = 1.2 \times 10^{-2}$).

16.65 The solution will be basic because of the hydrolysis of the sorbate anion, $C_6H_7O_2^-$. Calculate the initial molarity of $C_6H_7O_2^-$.

$$\frac{4.93\,g\,KC_6H_7O_2}{0.500\,L} \times \frac{1\,mol\,KC_6H_7O_2}{150.2\,g\,KC_6H_7O_2} = 0.065646 = 0.0656\,M\,KC_6H_7O_2$$

$[C_6H_7O_2^-] = [KC_6H_7O_2] = 0.0656\,M$

$$C_6H_7O_2^-(aq) + H_2O(l) \rightleftharpoons HC_6H_7O_2(aq) + OH^-(aq)$$

initial 0.0656 M · 0 · 0

equil. (0.0656 - x) M · x M · x M

$$K_b = \frac{[HC_6H_7O_2][OH^-]}{[C_6H_7O_2^-]} = \frac{K_w}{K_a\,for\,HC_6H_7O_2} = \frac{1.0 \times 10^{-14}}{1.7 \times 10^{-5}} = 5.88 \times 10^{-10} = 5.9 \times 10^{-10}$$

$$5.88 \times 10^{-10} = \frac{x^2}{0.0656 - x} \approx \frac{x^2}{0.0656};\ x^2 = 0.0656(5.88 \times 10^{-10})$$

$x = [OH^-] = 6.21 \times 10^{-6} = 6.2 \times 10^{-6}\,M$; pOH = 5.21; pH = 14 - pOH = 8.79

16.66 The solution is basic because of the hydrolysis of PO_4^{3-}. The molarity of PO_4^{3-} is

$$\frac{50.0\,g\,Na_3PO_4}{1.00\,L\,soln} \times \frac{1\,mol\,Na_3PO_4}{163.9\,g\,Na_3PO_4} = 0.3051 = 0.305\,M\,PO_4^{3-}$$

$$PO_4^{3-}(aq) + H_2O(l) \rightleftharpoons HPO_4^{2-}(aq) + OH^-(aq)$$

$$K_b = \frac{[HPO_4^{2-}][OH^-]}{[PO_4^{3-}]} = \frac{K_w}{K_a\,for\,HPO_4^{2-}} = \frac{1.0 \times 10^{-14}}{4.2 \times 10^{-13}} = 0.0238 = 2.4 \times 10^{-2}$$

Ignoring the further hydrolysis of HPO_4^{2-}.

$[OH^-] = [HPO_4^{2-}] = x$, $[PO_4^{3-}] = 0.305 - x$

$$2.4 \times 10^{-2} = \frac{x^2}{(0.305 - x)};\ x^2 + 2.4 \times 10^{-2}x - 0.0073 = 0$$

Since K_b is relatively large, we will not assume x is small compared to 0.305.

$$x = \frac{-0.024 \pm \sqrt{(0.024)^2 - 4(1)(-0.0073)}}{2(1)} = \frac{-0.024 \pm \sqrt{0.0299}}{2(1)}$$

$x = 0.074\,M\,OH^-$; pH = 14 + log(0.074) = 12.87

Acid-Base Character and Chemical Structure

16.67   (a)   As the electronegativity of the central atom (X) increases, more electron density is withdrawn from the X-O and O-H bonds, respectively. In water, the O-H bond is ionized to a greater extent and the strength of the oxyacid increases.

      (b)   As the number of nonprotonated oxygen atoms in the molecule increases, they withdraw electron density from the other bonds in the molecule and the strength of the oxyacid increases.

16.68   (a)   Acid strength increases as the polarity of the X-H bond increases and decreases as the strength of the X-H bond increases.

      (b)   Assuming the element, X, is more electronegative than H, as the electronegativity of X increases, the X-H bond becomes more polar and the strength of the acid increases. This trend holds true as electronegativiy increases across a row of the periodic chart. However, as electronegativity decreases going down a family, acid strength increases because the strength of the H-X bond decreases, even though the H-X bond becomes less polar.

16.69   (a) HNO_3 is a stronger acid than HNO_2 because it has one more nonprotonated oxygen atom, and thus a higher oxidation number on N.

      (b)   For binary hydrides, acid strength increases going down a family, so H_2S is a stronger acid than H_2O.

      (c) H_2SO_4 is a stronger acid because H^+ is much more tightly held by the anion HSO_4^-.

      (d)   For oxyacids, the greater the electronegativity of the central atom, the stronger the acid, so H_2SO_4 is a stronger acid than H_2SeO_4.

      (e) CCl_3COOH is stronger because the electronegative Cl atoms withdraw electron density from other parts of the molecule, which weakens the O-H bond and makes H^+ easier to remove.

16.70   (a)   For binary hydrides, acid strength increases going across a row, so HCl is a stronger acid than H_2S.

      (b)   For oxyacids, the more electronegative the central atom, the stronger the acid, so H_3PO_4 is a stronger acid than H_3AsO_4.

      (c) $HBrO_3$ has one more nonprotonated oxygen and a higher oxidation number on Br, so it is a stronger acid than $HBrO_2$.

      (d)   The first dissociation of a polyprotic acid is always stronger because H^+ is more tightly held by an anion, so $H_2C_2O_4$ is a stronger acid than $HC_2O_4^-$.

(e)   The conjugate base of benzoic acid, $C_7H_5O_2^-$, is stabilized by resonance, while the conjugate base of phenol, $C_6H_5O^-$, is not. $HC_7H_5O_2$ has greater tendency to form its conjugate base and is the stronger acid.

16.71   (a) BrO^- (HClO is the stronger acid due to a more electronegative central atom, so BrO^- is the stronger base.)

         (b) BrO^- ($HBrO_2$ has more nonprotonated O atoms and is the stronger acid, so BrO^- is the stronger base.)

         (c) HPO_4^{2-} (larger negative charge, greater attraction for H^+)

16.72   (a) NO_2^- (HNO_3 is the stronger acid because it has more nonprotonated O atoms, so NO_2^- is the stronger base.)

         (b) AsO_4^{3-} (Because P is more electronegative, H_3PO_4 is the stronger acid, so AsO_4^{3-} is the stronger base.)

         (c) CO_3^{2-} (The more negative the anion, the stronger the attraction for H^+.)

16.73   (a)   True

         (b)   False. In a series of acids that have the same central atom, acid strength increases with the number of nonprotonated oxygen atoms bonded to the central atom.

         (c)   False. H_2Te is a stronger acid than H_2S because the H-Te bond is longer, weaker and more easily dissociated than the H-S bond.

16.74   (a)   True.

         (b)   False. For oxyacids with the same structure but different central atom, the acid strength **increases** as the electronegativity of the central atom increases.

         (c)   False. HF is a weak acid, weaker than the other hydrogen halides, primarily because the H-F bond energy is exceptionally high.

Lewis Acids and Bases

16.75
Theory	**Acid**	**Base**
Arrhenius	forms H^+ ions in water	produces OH^- in water
Brønsted-Lowry	proton (H^+) donor	proton acceptor
Lewis	electron pair acceptor	electron pair donor

The Brønsted-Lowry theory is more general than Arrhenius's definition, because it is based on a unified model for the processes responsible for acidic or basic character, and it shows the relationships between these processes. The Lewis theory is more general still because it does not restrict the acidic species to compounds having ionizable hydrogen. Any substance that can be viewed as an electron-pair acceptor is defined as a Lewis acid.

16.76 (a)　HBr acts as a proton donor, a Brønsted-Lowry acid, toward the water molecule, which acts as a proton acceptor, a Brønsted-Lowry base. The conjugate acid H_3O^+ (H^+(aq)) and conjugate base Br^- are formed in the reaction.

(b)　Hydride ion, H^- acts like a Brønsted-Lowry base, accepting a proton from a water molecule to form H_2(g) and the conjugate base OH^-. The balanced equation is

$$NaH(s) + H_2O(l) \rightarrow H_2(g) + Na^+(aq) + OH^-(aq)$$

(c)　Pyridine acts like a Lewis base, donating an electron pair to SO_2, which acts as a Lewis acid. The SO_2 is able to act as an acceptor by shifting a pair of electrons from S to O:

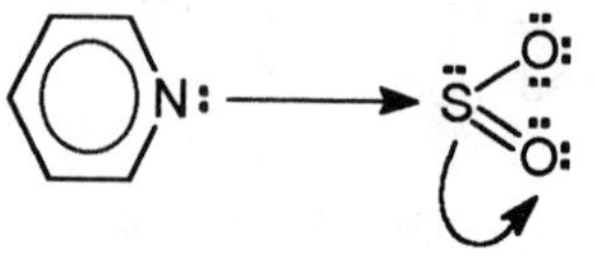

(d)　In terms of Lewis acid-base theory, the reaction of SO_2 with water is very similar to the reaction of SO_2 with pyridine. SO_2 acts as an electron pair acceptor (Lewis acid), and water acts as the electron pair donor (Lewis base).

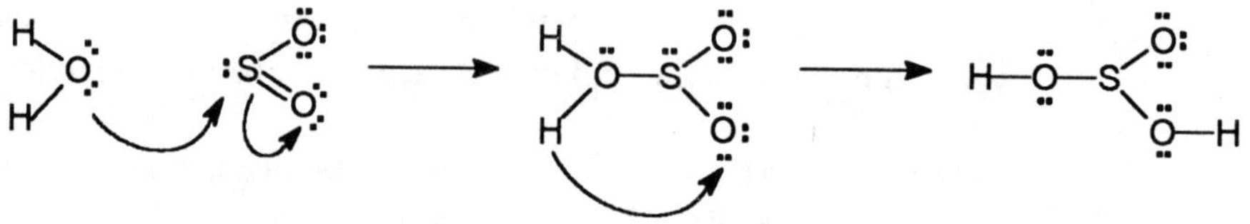

16.77

	Lewis Acid	Lewis Base
(a)	$Fe(ClO_4)_3$ or Fe^{3+}	H_2O
(b)	H_2O	CN^-
(c)	BF_3	$(CH_3)_3N$
(d)	HIO	NH_2^-

16.78

	Lewis Acid	Lewis Base
(a)	HNO_2 (or H^+)	OH^-
(b)	$FeBr_3$ (Fe^{3+})	Br^-
(c)	Zn^{2+}	NH_3
(d)	SO_2	H_2O

16.79 (a)　Cu^{2+}, higher cation charge

(b)　Fe^{3+}, higher cation charge

(c)　Al^{3+}, smaller cation radius, same charge

16.80   (a) $ZnBr_2$, smaller cation radius, same charge

        (b) $Cu(NO_3)_2$, higher cation charge

        (c) $NiBr_2$, smaller cation radius, same charge

Additional Exercises

16.81   (a) $H_2O(l) \rightarrow H^+(aq) + OH^-(aq)$; $H^+(aq) + OH^-(aq) \rightarrow H_2O(l)$

        (b)   As OH^- is removed from solution, more H_2O must dissociate to replace it, increasing $[H^+]$ as well.

16.82 $pK_w = 13.76$, $K_w = 10^{-13.76} = 1.738 \times 10^{-14} = 1.7 \times 10^{-14}$

       $[H^+] = [OH^-]$; $K_w = [H^+]^2$; $1.738 \times 10^{-14} = [H^+]^2$; $[H^+] = 1.3 \times 10^{-7} \, M$

16.83 $pH_1 = -\log [H^+]_1$; $[H^+]_1 = K_w / [OH^-]_1$; $[OH^-]_2 = 10[OH^-]_1$

       $[H^+]_2 = K_w / 10[OH^-]_1$; $10 [H^+]_2 = K_w / [OH^-]_1 = [H^+]_1$

       $10 [H^+]_2 = [H^+]_1$; $[H^+]_2 = [H^+]_1 / 10$; $pH_2 = -\log[H^+]_1 /10$

       $pH_2 = -(\log [H^+]_1 - \log 10) = -\log [H^+]_1 + 1 = pH_1 + 1$

       The pH increases by 1 pH unit

16.84   The equilibrium shifts in the forward direction with reaction between OH^- and H^+. The solution will be blue, corresponding to an excess of the conjugate base form, Bb^-, in solution.

16.85   (a)   A high O_2 concentration displaces protons from Hb, and thus produces a more acidic solution, with lower pH.

        (b) $[H^+]$ = antilog $(-7.4) = 4.0 \times 10^{-8} \, M$. The blood is very slightly basic.

        (c)   The equilibrium indicates that a high $[H^+]$ shifts the equilibrium toward the proton-bound form HbH^+, which means a **lower** concentration of HbO_2 in the blood. Thus the ability of hemoglobin to transport oxygen is impeded.

16.86   The solution with the higher pH has the lower $[H^+]$.

        (a)   For solutions with equal concentrations, the weaker acid will have a lower $[H^+]$ and higher pH.

        (b)   The acid with $K_a = 8 \times 10^{-6}$ is the weaker acid, so it has the higher pH.

        (c)   The base with $pK_b = 4.5$ is the stronger base, has greater $[OH^-]$ and smaller $[H^+]$, so higher pH.

16.87   A pH of 3.25 corresponds to $[H^+] = 5.6 \times 10^{-4} \, M$.

        (a)   Because HCl is a strong acid, $[HCl] = 5.62 \times 10^{-4} \, M = 5.6 \times 10^{-4} \, M$

            $(0.200 \, L)(5.62 \times 10^{-4} \, mol/L) = 1.1 \times 10^{-4} \, mol \, HCl$

(b) K_a for $HC_7H_5O_2$ is 6.5×10^{-5} (benzoic acid, Appendix D)

$$\frac{(5.62 \times 10^{-4})^2}{x} = 6.5 \times 10^{-5}; \; x = 4.86 \times 10^{-3} = 4.9 \times 10^{-3} \; M$$

$$0.200 \; L \times \frac{4.86 \times 10^{-3} \; mol \; HC_7H_5O_2}{1 \; L} = 9.7 \times 10^{-4} \; mol \; HC_7H_5O_2$$

(c)   HF has $K_a = 6.8 \times 10^{-4}$

$$\frac{(5.62 \times 10^{-4})^2}{x} = 6.8 \times 10^{-4}; \; x = 4.64 \times 10^{-4} = 4.6 \times 10^{-4} \; M$$

This is the amount that remains in solution **after** ionization to form H^+ and F^-. The total HF is the sum of the ionized and nonionized portions:

$$4.64 \times 10^{-4} \; M + 5.62 \times 10^{-4} \; M = 1.026 \times 10^{-3} = 1.03 \times 10^{-3} \; M$$

$$0.200 \; L \times \frac{1.03 \times 10^{-3} \; mol \; HF}{1 \; L} = 2.06 \times 10^{-4} \; mol \; HF$$

16.88 $K_a = \dfrac{[H^+][C_8H_5O_4^{2-}]}{[HC_8H_5O_4^-]}$; $[H^+] = [C_8H_5O_4^{2-}] = 10^{-pH} = 10^{-4.24} = 5.754 \times 10^{-5} = 5.8 \times 10^{-5} \; M$

$$[HC_8H_5O_4^-] = 525 \; mg \; HC_8H_5O_4^- \times \frac{1 \; g}{1000 \; mg} \times \frac{1 \; mol \; HC_8H_5O_4^-}{166.1 \; g \; HC_8H_5O_4^-} \times \frac{1}{0.250 \; L}$$

$$= 0.01264 = 0.0126 \; M$$

$$K_a = \frac{[H^+][C_8H_5O_4^{2-}]}{[HC_8H_5O_4^-]} = \frac{(5.754 \times 10^{-5})^2}{0.01264} = 2.6 \times 10^{-7}$$

16.89   Let $[H^+] = [NC_7H_4SO_3^-] = z$. $K_a = $ -antilog $(11.68) = 2.09 \times 10^{-12} = 2.1 \times 10^{-12}$

$$\frac{z^2}{0.10 - z} \approx \frac{z^2}{0.10} = 2.09 \times 10^{-12} \quad z = [H^+] = 4.6 \times 10^{-7} \; M; \quad pH = 6.34$$

Notice that the pH is not much lower than it is in pure water. A correction for the contribution from ionization of water itself to the concentration of $[H^+]$ can be made, but it turns out that this contribution can be neglected even in this case where the source of acidity is not very great.

16.90   Calculate the initial concentration of $HC_9H_7O_4$.

$$\frac{325 \; mg}{1 \; tablet} \times 2 \; tablets \times \frac{1 \; g}{1000 \; mg} \times \frac{1 \; mol \; HC_9H_7O_4}{180.2 \; g \; HC_9H_7O_4} = 0.003607 = 0.00361 \; mol \; HC_9H_7O_4$$

$$\frac{0.003607 \; mol \; HC_9H_7O_4}{0.250 \; L} = 0.01443 = 0.0144 \; M \; HC_9H_7O_4$$

	$HC_9H_7O_4(aq)$	$\rightleftharpoons$	$C_9H_7O_4^-(aq)$	$+$	$H^+(aq)$
initial	0.0144 M		0 M		0 M
equil	(0.0144 - x)		x M		x M

$$K_a = 3.3 \times 10^{-4} = \frac{[H^+][C_7H_9O_4^-]}{[HC_7H_9O_4]} = \frac{x^2}{(0.0144 - x)}$$

Assuming x is small compared to 0.0144,

$$x^2 = 0.01443\ (3.3 \times 10^{-4}); \quad x = [H^+] = 2.2 \times 10^{-3}\ M$$

$$\frac{2.2 \times 10^{-3}\ M\ H^+}{0.01443\ M\ HC_9H_7O_4} \times 100 = 15\%\ \text{ionization}; \quad \text{the assumption is not valid}$$

Using the quadratic formula, $x^2 + 3.3 \times 10^{-4}\ x - 4.76 \times 10^{-6} = 0$

$$x = \frac{-3.3 \times 10^{-4} \pm \sqrt{(3.3 \times 10^{-4})^2 - 4(1)(-4.76 \times 10^{-6})}}{2(1)} = \frac{-3.3 \times 10^{-4} \pm \sqrt{1.915 \times 10^{-5}}}{2}$$

$$x = 2.02 \times 10^{-3} = 2.0 \times 10^{-3}\ M\ H^+; \quad pH = -\log(2.02 \times 10^{-3}) = 2.69$$

16.91 (a) In $1 \times 10^{-8}\ M\ HNO_3(aq)$, $[H^+]$ due to HNO_3 is $1 \times 10^{-8}\ M$, but there is also $[H^+]$ due to the autoionization of H_2O. Total $[H^+] > 1 \times 10^{-8}$, so pH < 8.

(b) Let $x = [OH^-] = [H^+]$ due to autoionization.
Then $K_w = 1.0 \times 10^{-14} = [H^+][OH^-] = (1 \times 10^{-8} + x)(x)$; $x^2 + 1 \times 10^{-8}\ x - 1 \times 10^{-14} = 0$
From the quadratic formula, $x = 9.5 \times 10^{-8}\ M\ H^+$ due to autoionization.

Total $[H^+] = 9.5 \times 10^{-8}\ M$ (from autoionization) + $1 \times 10^{-8}\ M$ (from HNO_3)
$= 1.05 \times 10^{-7}\ M\ H^+$; pH = 6.98 (to 1 decimal place, pH = 7.0)

16.92 (a) $H_2X \rightarrow H^+ + HX^-$
Assuming HX^- does not ionize, $[H^+] = 0.050\ M$, pH = 1.30

(b) $H_2X \rightarrow 2H^+ + X^-$; $0.050\ M\ H_2X = 0.10\ M\ H^+$; pH = 1.00

(c) The observed pH of a 0.050 M solution of H_2X is only slightly less than 1.30, the pH assuming no ionization of HX^-. HX^- is not completely ionized; H_2X, which is completely ionized, is a stronger acid than HX^-.

(d) Since H_2X is a strong acid, HX^- has no tendency to act like a base. HX^- does act like a weak acid, so a solution of $NaHX$ would be acidic.

16.93 Considering the stepwise dissociation of H_3PO_4:

$$H_3PO_4(aq) \rightleftharpoons H^+(aq) + H_2PO_4^-(aq)$$

initial	0.100	0	0
equil.	(0.100 - x) M	x	x

$$K_{a1} = \frac{[H^+][H_2PO_4^-]}{[H_3PO_4]} = \frac{x^2}{(0.100 - x)} = 7.5 \times 10^{-3}; \quad x^2 + 7.5 \times 10^{-3}\ x - 7.5 \times 10^{-4} = 0$$

$$x = \frac{-7.5 \times 10^{-3} \pm \sqrt{(7.5 \times 10^{-3})^2 - 4(1)(-7.5 \times 10^{-4})}}{2(1)} = \frac{-7.5 \times 10^{-3} \pm \sqrt{0.00306}}{2}$$

$x = 0.024 \ M \ H^+$, $0.024 \ M \ H_2PO_4^-$ available for further ionization

$$H_2PO_4^-(aq) \rightleftharpoons H^+(aq) + HPO_4^{2-}(aq)$$

initial	0.024 M	0.024 M	0 M
equil.	(0.024 - y) M	(0.024 + y) M	y M

$$K_{a2} = \frac{[H^+][HPO_4^{2-}]}{[H_2PO_4^-]} = \frac{(y)(0.024 + y)}{(0.024 - y)} = 6.2 \times 10^{-8}$$

Since K_{a2} is very small, assume x is small compared to 0.024 M.

$$6.2 \times 10^{-8} = \frac{0.024 \ y}{0.024}; \ y = [HPO_4^{2-}] = 6.2 \times 10^{-8} \ M$$

$$[H_2PO_4^-] = (2.4 \times 10^{-2} \ M + 6.2 \times 10^{-8} \ M) \approx 2.4 \times 10^{-2} \ M$$

$$[H^+] = (2.4 \times 10^{-2} \ M + 6.2 \times 10^{-8} \ M) \approx 2.4 \times 10^{-2} \ M$$

$[HPO_4^{2-}]$ available for further ionization = $6.2 \times 10^{-8} \ M$

$$HPO_4^{2-} \rightleftharpoons H^+(aq) + PO_4^{3-}(aq)$$

initial	6.2×10^{-8} M	2.4×10^{-2} M	0 M
equil.	$(6.2 \times 10^{-8} - z)$ M	$(2.4 \times 10^{-2} + z)$ M	z M

$$K_{a3} = \frac{[H^+][PO_4^{3-}]}{[HPO_4^{2-}]} = \frac{2.4 \times 10^{-2} + z)(z)}{(6.2 \times 10^{-8} - z)} = 4.2 \times 10^{-13}$$

Assuming z is small compared to 6.2×10^{-8} (and 2.4×10^{-2}),

$$\frac{(2.4 \times 10^{-2})(z)}{(6.2 \times 10^{-8})} = 4.2 \times 10^{-13}; \ z = 1.1 \times 10^{-18} \ M \ PO_4^{3-}$$

The contribution of z to $[HPO_4^{2-}]$ and $[H^+]$ is negligible. In summary, after all dissociation steps have reached quilibrium:

$$[H^+] = 2.4 \times 10^{-2} \ M, \ [H_2PO_4^-] = 2.4 \times 10^{-2} \ M, \ [HPO_4^{2-}] = 6.2 \times 10^{-8} \ M,$$

$$[PO_4^{3-}] = 1.1 \times 10^{-18} \ M$$

Note that the first ionization step is the major source of H^+ and the others are important as sources of HPO_4^{2-} and PO_4^{3-}. The $[PO_4^{3-}]$ is **very** small at equilibrium.

16.94

$$C_{10}H_{15}ON(aq) + H_2O(l) \rightleftharpoons C_{10}H_{15}ONH^+(aq) + OH^-(aq)$$

initial	0.035 M	0	0
equil.	(0.035- y) M	y M	y M

$$K_b = \frac{[C_{10}H_{15}ONH^+][OH^-]}{[C_{10}H_{15}ON]} = \frac{y^2}{(0.035 - y)} = 1.4 \times 10^{-4}$$

$\dfrac{y^2}{0.035} \approx 1.4 \times 10^{-4}$; $y = 2.2 \times 10^{-3}\ M = [OH^-]$; pOH = 2.66, pH = 11.34

This $[OH^-]$ represents approximately 6% ionization, but use of the quadratic formula gives $[OH^-] = 2.1 \times 10^{-3}$ and pH = 11.33, values not significantly different from those obtained using the approximation.

K_a for $C_{10}H_{15}ONH^+$ is $\dfrac{1.0 \times 10^{-14}}{1.4 \times 10^{-4}} = 7.1 \times 10^{-11}$; $pK_a = 10.15$

16.95 $C_{18}H_{21}NO_3(aq)\ +\ H_2O(l) \rightleftharpoons C_{18}H_{21}NO_3H^+(aq)\ +\ OH^-(aq)$

initial   0.0050 *M*   0   0

equil.   (0.0050 - x) *M*   x *M*   x *M*

$K_b = \dfrac{C_{18}H_{21}NO_3H^+][OH^-]}{[C_{18}H_{21}NO_3]} = \dfrac{x^2}{0.0050 - x}$

pOH = 14.00 - 9.95 = 4.05; $[OH^-] = 10^{-4.05} = 8.91 \times 10^{-5} = 8.9 \times 10^{-5}\ M$

$K_b = \dfrac{(8.91 \times 10^{-5})^2}{(0.0050 - 8.91 \times 10^{-5})} = 1.6 \times 10^{-6}$

16.96   Call each compound in the neutral form Q.

Then, $Q(aq) + H_2O(l) \rightleftharpoons QH^+(aq) + OH^-$. $K_b = [QH^+]\,[OH^-]\,/\,[Q]$

The ratio in question is $[QH^+]\,/\,[Q]$, which equals $K_b\,/\,[OH^-]$ for each compound. At pH = 2.5, pOH = 11.5, $[OH^-]$ = antilog (-11.5) = $3.16 \times 10^{-12} = 3 \times 10^{-12}\ M$. Now calculate $K_b\,/[OH^-]$ for each compound:

Nicotine $\dfrac{[QH^+]}{[Q]} = 7 \times 10^{-7}\,/\,3.16 \times 10^{-12} = 2 \times 10^5$

Caffeine $\dfrac{[QH^+]}{[Q]} = 4 \times 10^{-14}\,/\,3.16 \times 10^{-12} = 1 \times 10^{-2}$

Strychnine $\dfrac{[QH^+]}{[Q]} = 1 \times 10^{-6}\,/\,3.16 \times 10^{-12} = 3 \times 10^5$

Quinine $\dfrac{[QH^+]}{[Q]} = 1 \times 10^{-6}\,/\,3.16 \times 10^{-12} = 3.5 \times 10^5$

For all the compounds except caffeine the protonated form is much higher concentration than the neutral form. However, for caffeine, a very weak base, the neutral form dominates.

16.97   (a)   Consider the formation of the zwitterion as a series of steps (Hess's Law).

$$NH_2\text{-}CH_2\text{-}COOH + H_2O \rightleftharpoons NH_2\text{-}CH_2\text{-}COO^- + H_3O^+ \qquad K_a$$

$$NH_2\text{-}CH_2\text{-}COOH + H_2O \rightleftharpoons {}^+NH_3\text{-}CH_2\text{-}COOH + OH^- \qquad K_b$$

$$H_3O^+ + OH^- \rightleftharpoons 2H_2O \qquad 1/K_w$$

$$NH_2\text{-}CH_2\text{-}COOH \rightleftharpoons {}^+NH_3\text{-}CH_2\text{-}COO^- \qquad \frac{K_a \times K_b}{K_w}$$

$$K = \frac{K_a \times K_b}{K_w} = \frac{(4.3 \times 10^{-3})(6.0 \times 10^{-5})}{1.0 \times 10^{-14}} = 2.6 \times 10^7$$

The large value of K indicates that formation of the zwitterion is favorable. The assumption is that the same $NH_2\text{-}CH_2\text{-}COOH$ molecule is acting like an acid ($\text{-}COOH \rightarrow H^+ + \text{-}COO^-$) and a base ($\text{-}NH_2 + H^+ \rightarrow NH_3^+$), simultaneously. Glycine is both a stronger acid and a stronger base than water, so the H^+ transfer should be intramolecular, as long as there are no other acids or bases in the solution.

(b)   Since glycine exists as the zwitterion in aqueous solution, the pH is determined by the equilibrium below.

$${}^+NH_3\text{-}CH_2\text{-}COO^- + H_2O \rightleftharpoons NH_2\text{-}CH_2\text{-}COO^- + H_3O^+$$

$$K_a = \frac{[NH_2\text{-}CH_2\text{-}COO^-][H_3O^+]}{[{}^+NH_3\text{-}CH_2\text{-}COO^-]} = \frac{K_w}{K_b} = \frac{1.0 \times 10^{-14}}{6.0 \times 10^{-5}} = 1.67 \times 10^{-10} = 1.7 \times 10^{-10}$$

$$x = [H_3O^+] = [NH_2\text{-}CH_2\text{-}COO^-]; \quad K_a = 1.67 \times 10^{-10} = \frac{(x)(x)}{(0.050 - x)} \approx \frac{x^2}{0.050}$$

$$x = [H_3O^+] = 2.89 \times 10^{-6} = 2.9 \times 10^{-6} \ M, \quad pH = 5.54$$

(c)   In a strongly acidic solution the CO_2^- function would be protonated, so glycine would exist as ${}^+H_3NCH_2COOH$. In strongly basic solution the $\text{-}NH_3^+$ group would be deprotonated, so glycine would be in the form $H_2NCH_2CO_2^-$.

16.98   The general Lewis structure for these acids is

where X = H or C

Replacement of H by the more electronegative chlorine atoms causes the central carbon to become more positively charged, thus withdrawing more electrons from the attached COOH group, in turn causing the O-H bond to be more polar, so that H^+ is more readily ionized.

To calculate pH proceed as usual, except that the full quadratic form must be used for all but acetic acid.

Acid	pH
acetic	2.87
chloroacetic	1.95
dichloroacetic	1.36
trichloroacetic	1.1

Integrative Exercises

16.99 At 25°C, $[H^+] = [OH^-] = 1.0 \times 10^{-7} M$

$$\frac{1.0 \times 10^{-7} \text{ mol H}^+}{1 \text{ L H}_2\text{O}} \times 0.0010 \text{ L} \times \frac{6.022 \times 10^{23} \text{ H}^+ \text{ ions}}{\text{mol H}^+} = 6.0 \times 10^{13} \text{ H}^+ \text{ ions}$$

16.100 $[H^+] = 10^{-pH} = 10^{-2} = 1 \times 10^{-2} M H^+$; $1 \times 10^{-2} M \times 0.450 \text{ L} = 4.5 \times 10^{-3} = 5 \times 10^{-3}$ mol H^+

$$HCl(aq) + HCO_3^-(aq) \rightarrow Cl^-(aq) + H_2O(l) + CO_2(g)$$

$$4.5 \times 10^{-3} \text{ mol H}^+ = 4.5 \times 10^{-3} \text{ mol HCO}_3^- \times \frac{84.01 \text{ g NaHCO}_3}{1 \text{ mol HCO}_3^-} = 0.378 = 0.4 \text{ g NaHCO}_3$$

16.101 Strategy: Use PV = nRT to calculate mol SO_2, and thus mol H_2SO_3 and M H_2SO_3. Solve the equilibrium problem to find $[H^+]$ and pH.

$$n = \frac{PV}{RT} = \frac{1.0 \text{ atm} \times 3.9 \text{ L SO}_2}{293 \text{ K}} \times \frac{K \cdot \text{mol}}{0.08206 \text{ L} \cdot \text{atm}} = 0.162 = 0.16 \text{ mol SO}_2$$

From the given reaction, mol SO_2 = mol H_2SO_3. 0.16 mol $H_2SO_3/1.0$ L = 0.16 M H_2SO_3

$$H_2SO_3(aq) \rightleftharpoons H^+(aq) + HSO_3^-(aq) \quad K_{a1} = 1.7 \times 10^{-2}$$
$$HSO_3^-(aq) \rightleftharpoons H^+(aq) + SO_3^{2-}(aq) \quad K_{a2} = 6.4 \times 10^{-8}$$

$$K_{a1} = 1.7 \times 10^{-2} = \frac{[H^+][HSO_3^-]}{[H_2SO_3]} = \frac{x^2}{0.162 - x}; \text{ since } K_{a1} \text{ is relatively large, use the quadratic.}$$

$$x^2 + 1.7 \times 10^{-2} x - 2.75 \times 10^{-3} = 0; \quad x = \frac{-1.7 \times 10^{-2} \pm \sqrt{(1.7 \times 10^{-2})^2 - 4(1)(-2.75 \times 10^{-3})}}{2}$$

$x = 0.0447 = 0.045 M H^+$; pH = 1.35

Assumptions: 1) SO_2 is an ideal gas

2) The volume of SO_2(aq) is 1.0 L. That is, there is no change in volume when 3.9 L of SO_2(g) dissolve in 1.0 L of water.

3) Because K_{a2} is small, the second dissociation of H_2SO_3 does not contribute significantly to $[H^+]$ and pH.

By using the quadratic formula, we avoided assuming that the % ionization of H_2SO_3(aq) was small relative to the initial concentration of H_2SO_3(aq).

16.102 (a)   24 valence e^-, 12 e^- pairs

$$:\!\overset{\cdot\cdot}{\underset{}{Cl}}\!-\!Al\!-\!\overset{\cdot\cdot}{\underset{}{Cl}}\!:$$
$$|$$
$$:\!\overset{\cdot\cdot}{\underset{\cdot\cdot}{Cl}}\!:$$

The formal charges on all atoms are zero. Structures with multiple bonds lead to nonzero formal charges. There are 3 VSEPR e^- pairs about Al. The electron pair geometry and molecular structure are trigonal planar.

(b)   The Al atom in $AlCl_3$ has an incomplete octet and is electron deficient. It "needs" to accept another electron pair, to act like a Lewis acid.

(c)

Both the Al and N atoms in the product have tetrahedral geometry.

(d)   The Lewis theory is most appropriate. H^+ and $AlCl_3$ are both electron pair acceptors, Lewis acids.

16.103 Strategy: Use acid ionization equilibrium to calculate the total moles of particles in solution. Use density to calculate kg solvent. From the molality (m) of the solution, calculate ΔT_b and T_b.

	$HSO_4^-(aq)$	$\rightleftharpoons$	$H^+(aq)$	+	$SO_4^{2-}(aq)$
initial	0.10 M		0		0
equil.	0.10 - x M		x M		x M
	0.071 M		0.029 M		0.029 M

$$K_a = 1.2 \times 10^{-2} = \frac{[H^+][SO_4^{2-}]}{[HSO_4^-]} = \frac{x^2}{0.10 - x}; \quad K_a \text{ is relatively large, so use the quadratic.}$$

$$x^2 + 0.012\,x - 0.0012 = 0; \quad x = \frac{-0.012 \pm \sqrt{(0.012)^2 - 4(1)(-0.0012)}}{2}; \quad x = 0.029\ M\ H^+,\ SO_4^{2-}$$

Total ion concentration = 0.10 M Na^+ + 0.071 M HSO_4^- + 0.029 M H^+ + 0.029 M SO_4^{2-}

$$= 0.229 = 0.23\ M.$$

Assume 100.0 mL of solution. 1.002 g/mL × 100.0 mL = 100.2 g solution.

$$0.10\ M\ NaHSO_4 \times 0.1000\ L = 0.010\ \text{mol NaHSO}_4 \times \frac{120.1\ \text{g NaHSO}_4}{\text{mol NaHSO}_4}$$

$$= 1.201 = 12\ \text{g NaHSO}_4$$

100.2 g soln - 1.201 g $NaHSO_4$ = 99.0 g = 0.099 kg H_2O

$$m = \frac{\text{mol ions}}{\text{kg H}_2\text{O}} = \frac{0.229\,M \times 0.1000\,\text{L}}{0.0990\,\text{kg}} = 0.231 = 0.23\,m \text{ ions}$$

$$\Delta T_b = K_b(m) = 0.52°C/m\,(0.23\,m) = +0.12°C; \quad T_b = 100.0 + 0.12 = 100.1°C$$

16.104 The **apparent** molality of the solution is given by

$$1.90°C \times \frac{1\,m}{1.86°C} = 1.02\,m$$

The ionization equilibrium, and molalities of the various species are:

$$HF(aq) \rightleftharpoons H^+(aq) + F^-(aq)$$

equil. $1.00 - x\,m$ \qquad $x\,m$ \qquad $x\,m$

The total molality is thus $(1.00 - x + x + x) = 1.00 + x = 1.02$. Thus, $x = 0.02$. For such a dilute aqueous solution, molality and molarity are essentially identical (Section 13.2). Thus, we can write

$$\% \text{ ionization} = \frac{[H^+]}{[HF]} \times 100 = \frac{0.02}{1.00} \times 100 = 2\%$$

(b) \qquad Assuming $M \approx m$, $K_a = \frac{[H^+][F^-]}{[HF]} = \frac{(0.02)^2}{0.98} = 4 \times 10^{-4}$.

The precision of the K_a is not very good; to improve upon it, we would need to know the magnitude of the freezing point depression more precisely.

16.105 Calculate M of the solution from osmotic pressure, and K_b using the equilibrium expression for the hydrolysis of cocaine. Let Coc = cocaine and CocH$^+$ be the conjugate acid of cocaine.

$$\pi = MRT; \quad M = \pi/RT = \frac{52.7\,\text{torr}}{288\,\text{K}} \times \frac{1\,\text{atm}}{760\,\text{torr}} \times \frac{\text{mol} \cdot \text{K}}{0.08206\,\text{L} \cdot \text{atm}}$$

$$= 0.002934 = 2.93 \times 10^{-3}\,M\,\text{Coc}$$

pH = 8.53; pOH = 14 - pH = 5.47; $[OH^-] = 10^{-5.47} = 3.39 \times 10^{-6} = 3.4 \times 10^{-6}\,M$

$$Coc(aq) + H_2O(l) \rightleftharpoons CocH^+(aq) + OH^-(aq)$$

initial $2.93 \times 10^{-3}\,M$ \qquad\qquad 0 \qquad\qquad 0

equil. $(2.93 \times 10^{-3} - 3.4 \times 10^{-6})\,M$ \qquad $3.4 \times 10^{-6}\,M$ \qquad $3.4 \times 10^{-6}\,M$

$$K_b = \frac{[CocH^+][OH^-]}{[Coc]} = \frac{(3.39 \times 10^{-6})^2}{(2.934 \times 10^{-3} - 3.39 \times 10^{-6})} = 3.9 \times 10^{-9}$$

Note that % hydrolysis is small in this solution, so "x", $3.4 \times 10^{-6}\,M$, is small compared to $2.93 \times 10^{-3}\,M$ and could be ignored in the denominator of the calculation.

16.106 (a)   rate = $k[IO_3^-][SO_3^{2-}][H^+]$

     (b) $\Delta pH = pH_2 - pH_1 = 3.50 - 5.00 = -1.50$

              $\Delta pH = -\log [H^+]_2 - (-\log [H^+]_1); \ -\Delta pH = \log [H^+]_2 - \log [H^+]_1$

              $-\Delta pH = \log [H^+]_2 / [H^+]_1; \ [H^+]_2 / [H^+]_1 = 10^{-\Delta pH}$

              $[H^+]_2/[H^+]_1 = 10^{1.50} = 31.6 = 32$. The rate will increase by a factor of 32 if $[H^+]$

              increases by a factor of 32. The reaction goes faster at lower pH.

     (c)   Since H^+ does not appear in the overall reaction, it is either a catalyst or an intermediate. An intermediate is produced and then consumed during a reaction, so its contribution to the rate law can usually be written in terms of concentrations of other reactants (Sample Exercise 14.11). A catalyst is present at the beginning and end of a reaction and can appear in the rate law if it participates in the rate-determining step (Exercise 14.62). This reaction is pH dependent because H^+ is a homogeneous catalyst that participates in the rate-determining step.

16.107 (a)   (i) $HCO_3^-(aq) \rightleftharpoons H^+(aq) + CO_3^{2-}(aq)$ $K_1 = K_{a2}$ for $H_2CO_3 = 5.6 \times 10^{-11}$

                   $H^+(aq) + OH^-(aq) \rightleftharpoons H_2O(l)$ $K_2 = 1/K_w = 1 \times 10^{14}$

                _____

                   $HCO_3^-(aq) + OH^-(aq) \rightleftharpoons CO_3^{2-}(aq) + H_2O(l)$ $K = K_1 \times K_2 = 5.6 \times 10^3$

              (ii) $NH_4^+(aq) \rightleftharpoons H^+(aq) + NH_3(aq)$ $K_1 = K_a$ for $NH_4^+ = 5.6 \times 10^{-10}$

                   $CO_3^{2-}(aq) + H^+(aq) \rightleftharpoons HCO_3^-(aq)$ $K_2 = 1/K_{a2}$ for $H_2CO_3 = 1.8 \times 10^{10}$

                _____

                   $NH_4^+(aq) + CO_3^{2-}(aq) \rightleftharpoons HCO_3^-(aq) + NH_3(aq)$ $K = K_1 \times K_2 = 10$

     (b)   Both (i) and (ii) have K > 1, although K =10 is not **much** greater than 1. Both could be written with a single arrow. (This is true in general when a strong acid or strong base, $H^+(aq)$ or $OH^-(aq)$, is a reactant.)

17 Additional Aspects of Equilibria

Common-Ion Effect

17.1 (a) The extent of dissociation of a weak electrolyte is decreased when a strong electrolyte containing an ion in common with the weak electrolyte is added to it.

(b) NaOCl

17.2 (a) For a generic weak base B, $K_b = \dfrac{[BH^+][OH^-]}{[B]}$. If an external source of BH^+ such as BH^+Cl^- is added to a solution of B(aq), $[BH^+]$ increases, decreasing $[OH^-]$ and increasing [B], effectively suppressing the ionization (hydrolysis) of B.

(b) NH_4Cl

17.3 In general, when an acid is added to a solution, pH decreases; when a base is added to a solution, pH increases.

(a) pH increases; NO_2^- decreases the ionization of HNO_2 and decreases $[H^+]$.

(b) pH decreases; $CH_3NH_3^+$ decreases the ionization (hydrolysis) of CH_3NH_2 and decreases $[OH^-]$

(c) pH increases; CHO_2^- decreases the ionization of $HCHO_2$ and decreases $[H^+]$

(d) no change; Br^- is a negligible base and does not affect the 100% ionization of the strong acid HBr.

(e) pH decreases; the pertinent equilibrium is
$C_2H_3O_2^-$ (aq) + H_2O(l) $\rightleftharpoons$ $HC_2H_3O_2$ + OH^-(aq).
HCl reacts with OH^-(aq), decreasing $[OH^-]$ and pH.

17.4 In general, when an acid is added to a solution pH decreases; when a base is added to a solution, pH increases.

(a) pH increases; $C_7H_5O_2^-$ decreases ionization of $HC_7H_5O_2$ and decreases pH.

(b) pH decreases; $C_5H_5NH^+$ decreases ionization (hydrolysis) of C_5H_5N and decreases $[OH^-]$.

(c) pH increases; NH_3 reacts with HCl, decreasing $[H^+]$.

(d) pH increases; HCO_3^- decreases ionization of H_2CO_3 and decreases $[H^+]$.

(e) no change; ClO_4^- is a negligible base and Na^+ is a negligible acid.

17.5   (a) $HC_3H_5O_2(aq) \rightleftharpoons H^+(aq) + C_3H_5O_2^-(aq)$

     i   0.16 M 0.080 M

     c   -x   +x   +x

     e   (0.16 - x) M +x M (0.080 + x) M

$$K_a = 1.3 \times 10^{-5} = \frac{[H^+][C_3H_5O_2^-]}{[HC_3H_5O_2]} = \frac{(x)(0.080 + x)}{(0.16 - x)}$$

Assume x is small compared to 0.080 and 0.16.

$$1.3 \times 10^{-5} = \frac{0.080\,x}{0.16}; \quad x = 2.6 \times 10^{-5} = [H^+], \text{ pH} = 4.59$$

   (b) $(CH_3)_3N(aq) + H_2O(l) \rightleftharpoons (CH_3)_3NH^+(aq) + OH^-(aq)$

     i   0.15 M 0.12 M

     c   -x   +x   +x

     e   (0.15 - x) M (0.12 + x) M +x M

$$K_b = 6.4 \times 10^{-5} = \frac{[OH^-][(CH_3)_3NH^+]}{[(CH_3)_3N]} = \frac{(x)(0.12 + x)}{(0.15 - x)} \approx \frac{0.12\,x}{0.15}$$

$$x = 8.0 \times 10^{-5} = [OH^-], \text{ pOH} = 4.10, \text{ pH} = 14.00 - 4.10 = 9.90$$

17.6   (a) $HCHO_2(aq) \rightleftharpoons H^+(aq) + CHO_2^-(aq)$

     i   0.180 M 0.100 M

     c   -x   +x   +x

     e   (0.180 - x) M +x M (0.100 + x) M

$$K_a = 1.8 \times 10^{-4} = \frac{[H^+][CHO_2^-]}{[HCHO_2]} = \frac{(x)(0.100 + x)}{(0.180 - x)} \approx \frac{0.100\,x}{0.180}$$

$$x = 3.2 \times 10^{-4}\,M = [H^+], \text{ pH} = 3.49$$

Since the extent of ionization of a weak acid or base is suppressed by the presence of a conjugate salt, the 5% rule usually holds true in buffer solutions.

   (b) $C_5H_5N(aq) + H_2O(l) \rightleftharpoons C_5H_5NH^+(aq) + OH^-(aq)$

     i   0.0750 M 0.0500 M

     c   -x   +x   +x

     e   (0.0750 - x) M (0.0500 + x) M +x M

$$K_b = 1.7 \times 10^{-9} = \frac{[C_5H_5NH^+][OH^-]}{[C_5H_5N]} = \frac{(0.0500 + x)(x)}{(0.0750 - x)} \approx \frac{0.0500\,x}{0.0750}$$

$$x = 2.6 \times 10^{-9}\,M = [OH^-], \text{ pOH} = 8.59, \text{ pH} = 14.00 - 8.59 = 5.41$$

17.7 $HBu(aq) \rightleftharpoons H^+(aq) + Bu^-(aq)$ $K_a = \dfrac{[H^+][Bu^-]}{[HBu]} = 1.5 \times 10^{-5}$

     equil (a) $0.10 - x\,M$ $x\,M$ $x\,M$

     equil (b) $0.10 - x\,M$ $x\,M$ $0.050 + x\,M$

(a) $K_a = 1.5 \times 10^{-5} = \dfrac{x^2}{0.10 - x} \approx \dfrac{x^2}{0.10}$; $x = [H^+] = 1.225 \times 10^{-3} = 1.2 \times 10^{-3}\,M$

        % ionization $= \dfrac{1.2 \times 10^{-3}\,M\,H^+}{0.10\,M\,HBu} \times 100 = 1.2\%$

(b) $K_a = 1.5 \times 10^{-5} = \dfrac{(x)(0.050 + x)}{0.10 - x} \approx \dfrac{0.050\,x}{0.10}$; $x = 3.0 \times 10^{-5}\,M\,H^+$

        % ionization $= \dfrac{3.0 \times 10^{-5}\,M\,H^+}{0.10\,M\,HBu} \times 100 = 0.030\%$ ionization

17.8 $HLac(aq) \rightleftharpoons H^+(aq) + Lac^-(aq)$ $K_a = \dfrac{[H^+][Lac^-]}{[HLac]} = 1.4 \times 10^{-4}$

     equil (a) $0.10 - x\,M$ $x\,M$ $x\,M$

     equil (b) $0.10 - x\,M$ $x\,M$ $0.010 + x\,M$

(a) $K_a = 1.4 \times 10^{-4} = \dfrac{x^2}{0.10 - x} \approx \dfrac{x^2}{0.10}$; $x = [H^+] = 3.74 \times 10^{-3}\,M = 3.7 \times 10^{-3}\,M$

        % ionization $= \dfrac{3.7 \times 10^{-3}\,M\,H^+}{0.10\,M\,Lac} \times 100 = 3.7\%$

(b) $K_a = 1.4 \times 10^{-4} = \dfrac{(x)(0.050 + x)}{0.10 - x} \approx \dfrac{0.010\,x}{0.10}$; $x = 1.4 \times 10^{-3}\,M\,H^+$

        % ionization $= \dfrac{1.4 \times 10^{-3}\,M\,H^+}{0.10\,M\,Lac} \times 100 = 1.4\%$ ionization

Buffers

17.9 $HC_2H_3O_2$ and $NaC_2H_3O_2$ are a weak conjugate acid/conjugate base pair which act as a buffer because unionized $HC_2H_3O_2$ reacts with added base, while $C_2H_3O_2^-$ combines with added acid, leaving $[H^+]$ relatively unchanged. Although HCl and KCl are a conjugate acid/conjugate base pair, Cl^- is a negligible base. That is, it has no tendency to combine with added acid to form unionized HCl. Any added acid simply increases $[H^+]$ in an HCl/KCl mixture. In general, the conjugate bases of strong acids are negligible and mixtures of strong acids and their conjugate salts do not act as buffers.

17.10   (a)   The pH of a buffer is determined by K_a for the conjugate acid present and the **ratio** of conjugate base concentration to conjugate acid concentration.

          (b)   The buffering capacity of a buffer is determined by the absolute concentrations of the conjugate acid and conjugate base present. For example, a buffer that is 1.0 M in both $HC_2H_3O_2$ and $C_2H_3O_2^-$ has a higher capacity than one that is 0.10 M in the two components.

17.11   Assume that % ionization is small in these buffers (Exercises 17.7 and 17.8).

(a) $K_a = \dfrac{[H^+][CHO_2^-]}{[HCHO_2]}$; $[H^+] = \dfrac{K_a[HCHO_2]}{[CHO_2^-]} \approx \dfrac{1.8 \times 10^{-4} (0.20)}{(0.15)}$

   $[H^+] = 2.40 \times 10^{-4} = 2.4 \times 10^{-4}\,M$; pH = -log $(2.40 \times 10^{-4}) = 3.62$

(b)   mol = $M \times$ L; total volume = 85 mL + 95 mL = 180 mL = 0.180 L

   $[H^+] = \dfrac{K_a\,(0.16\,M \times 0.085\,L)/0.180\,L}{(0.15\,M \times 0.095\,L)/0.180\,L} = \dfrac{1.8 \times 10^{-4}(0.0136)}{0.01425}$

   (The total volume of the solution cancels in this calculation.)
   $[H^+] = 1.718 \times 10^{-4} = 1.7 \times 10^{-4}\,M$; pH = -log $(1.718 \times 10^{-4}) = 3.77$

17.12   Assume that % ionization is small in these buffers (Exercises 17.7 and 17.8).

(a)   The conjugate acid in this buffer is HCO_3^-, so use K_{a2} for H_2CO_3, 5.6×10^{-11}

   $K_a = \dfrac{[H^+][CO_3^{2-}]}{[HCO_3^-]}$; $[H^+] = \dfrac{K_a[HCO_3^-]}{[CO_3^{2-}]} = \dfrac{5.6 \times 10^{-11}(0.15)}{(0.10)}$

   $[H^+] = 8.40 \times 10^{-11} = 8.4 \times 10^{-11}\,M$; pH = -log $(8.40 \times 10^{-11}\,M) = 10.08$

(b)   mol = $M \times$ L; total volume = 120 mL = 0.120 L

   $[H^+] = \dfrac{K_a\,(0.16\,M \times 0.055\,L)/0.120\,L}{(0.11\,M \times 0.065\,L)/0.120\,L} = \dfrac{5.6 \times 10^{-11}(0.0088)}{0.00715}$

   $[H^+] = 6.892 \times 10^{-11} = 6.9 \times 10^{-11}\,M$; pH = -log$(6.892 \times 10^{-11}) = 10.16$

17.13   (a) $HC_2H_3O_2(aq) \rightleftharpoons H^+(aq) + C_2H_3O_2^-(aq)$; $K_a = 1.8 \times 10^{-5} = \dfrac{[H^+][C_2H_3O_2^-]}{[HC_2H_3O_2]}$

   $[HC_2H_3O_2] = \dfrac{20.0\,g\,HC_2H_3O_2}{2.00\,L\,soln} \times \dfrac{1\,mol\,HC_2H_3O_2}{60.05\,g\,HC_2H_3O_2} = 0.167\,M$

   $[C_2H_3O_2^-] = \dfrac{20.0\,g\,NaC_2H_3O_2}{2.00\,L\,soln} \times \dfrac{1\,mol\,NaC_2H_3O_2}{82.04\,g\,NaC_2H_3O_2} = 0.122\,M$

   $[H^+] = \dfrac{K_a[HC_2H_3O_2]}{[C_2H_3O_2^-]} = \dfrac{1.8 \times 10^{-5}(0.167 - x)}{(0.122 + x)} \approx \dfrac{1.8 \times 10^{-5}(0.167)}{(0.122)}$

   $[H^+] = 2.4843 \times 10^{-5} = 2.5 \times 10^{-5}\,M$, pH = 4.60

(b) $Na^+(aq) + C_2H_3O_2^-(aq) + H^+(aq) + Cl^-(aq) \rightarrow HC_2H_3O_2(aq) + Na^+(aq) + Cl^-(aq)$

(c) $HC_2H_3O_2(aq) + Na^+(aq) + OH^-(aq) \rightarrow C_2H_3O_2^-(aq) + H_2O(l) + Na^+(aq)$

17.14 NH_4^+/NH_3 is a basic buffer. Either the hydrolysis of NH_3 or the dissociation of NH_4^+ can be used to determine the pH of the buffer. Using the dissociation of NH_4^+ leads directly to $[H^+]$ and facilitates use of the Henderson-Hasselbach relationship.

(a) $NH_4^+(aq) \rightleftharpoons H^+(aq) + NH_3(aq)$

$$K_a = \frac{K_w}{K_b} = \frac{1.0 \times 10^{-14}}{1.8 \times 10^{-5}} = 5.56 \times 10^{-10} = 5.6 \times 10^{-10}$$

$$[NH_3] = \frac{5.0 \text{ g } NH_3}{2.50 \text{ L soln}} \times \frac{1 \text{ mol } NH_3}{17.0 \text{ g } NH_3} = 0.118 = 0.12 \; M \; NH_3$$

$$[NH_4^+] = \frac{20.0 \text{ g } NH_4Cl}{2.50 \text{ L}} \times \frac{1 \text{ mol } NH_4Cl}{53.50 \text{ g } NH_4Cl} = 0.1495 = 0.50 \; M \; NH_4^+$$

$$K_a = \frac{[H^+][NH_3]}{[NH_4^+]}; \; [H^+] = \frac{K_a[NH_4^+]}{[NH_3]} = \frac{5.56 \times 10^{-10}(0.1495 - x)}{(0.118 + x)} \approx \frac{5.56 \times 10^{-10}(0.1495)}{(0.118)}$$

$[H^+] = 7.044 \times 10^{-10} = 7.0 \times 10^{-10} \; M, \; pH = 9.15$

(b) $NH_3(aq) + H^+(aq) + NO_3^-(aq) \rightarrow NH_4^+(aq) + NO_3^-(aq)$

(c) $NH_4^+(aq) + Cl^-(aq) + K^+(aq) + OH^-(aq) \rightarrow NH_3(aq) + H_2O(l) + Cl^-(aq) + K^+(aq)$

17.15   In this problem, $[BrO^-]$ is the unknown.

pH = 8.80, $[H^+] = 10^{-8.80} = 1.585 \times 10^{-9} = 1.6 \times 10^{-9} \; M$

$[HBrO] = 0.200 - 1.6 \times 10^{-9} \approx 0.200 \; M$

$$K_a = 2.5 \times 10^{-9} = \frac{1.585 \times 10^{-9}[BrO^-]}{0.200}; \; [BrO^-] = 0.3155 = 0.32 \; M$$

For 1.00 L, 0.32 mol NaBrO are needed.

17.16 $HC_3H_5O_3(aq) \rightleftharpoons H^+(aq) + C_3H_5O_3^-(aq)$

$$[H^+] = \frac{K_a[HC_3H_5O_3]}{[C_3H_5O_3^-]}; \; [H^+] = 10^{-3.90} = 1.26 \times 10^{-4} \; M$$

$[HC_3H_5O_3] = 0.0150 \; M$; calculate $[C_3H_5O_3^-]$

$$[C_3H_5O_3^-] = \frac{K_a[HC_3H_5O_3]}{[H^+]} = \frac{1.4 \times 10^{-4}(0.0150)}{1.26 \times 10^{-4}} = 0.01667 = 1.7 \times 10^{-2} \; M$$

$$\frac{0.01667 \text{ mol } NaC_3H_5O_3}{1.00 \text{ L}} \times \frac{112.1 \text{ g } NaC_3H_5O_3}{1 \text{ mol } NaC_3H_5O_3} = 1.9 \text{ g } NaC_3H_5O_3$$

17.17   (a) $K_a = \frac{[H^+][C_2H_3O_2^-]}{[HC_2H_3O_2]}; \; [H^+] = \frac{K_a[HC_2H_3O_2]}{[C_2H_3O_2^-]}$

$$[H^+] \approx \frac{1.8 \times 10^{-5}(0.11)}{(0.15)} = 1.320 \times 10^{-5} = 1.3 \times 10^{-5} \; M; \; pH = 4.88$$

(b) $HC_2H_3O_2(aq) + KOH(aq) \rightarrow C_2H_3O_2^-(aq) + H_2O(l) + K^+(aq)$

0.11 mol	0.02 mol	0.15 mol
-0.02 mol	-0.02 mol	+0.02 mol
0.09 mol	0 mol	0.17 mol

$$[H^+] = \frac{1.8 \times 10^{-5}\,(0.09\text{ mol/0.100 L})}{(0.17\text{ mol/0.100 L})} = 9.53 \times 10^{-6} = 1 \times 10^{-5}\ M;\ pH = 5.02 = 5.0$$

(c) $C_2H_3O_2^-(aq) + HCl(aq) \rightarrow HC_2H_3O_2(aq) + Cl^-(aq)$

0.15 mol	0.02 mol	0.11 mol
-0.02 mol	-0.02 mol	+0.02 mol
0.13 mol	0 mol	0.13 mol

$$[H^+] = \frac{1.8 \times 10^{-5}\,(0.13\text{ mol/0.100 L})}{(0.13\text{ mol/0.100 L})} = 1.8 \times 10^{-5}\ M;\ pH = 4.74$$

17.18 (a) $K_a = \dfrac{[H^+][C_3H_5O_2^-]}{[HC_3H_5O_2]}$; $[H^+] = \dfrac{K_a[HC_3H_5O_2]}{[C_3H_5O_2^-]}$

Since this expression contains a ratio of concentrations, we can ignore total volume and work directly with moles.

$$[H^+] = \frac{1.3 \times 10^{-5}\,(0.14 - x)}{(0.12 + x)} \approx \frac{1.3 \times 10^{-5}\,(0.14)}{0.12} = 1.517 \times 10^{-5}$$
$$= 1.5 \times 10^{-5}\ M,\ pH = 4.82$$

(b) $HC_3H_5O_2(aq) + OH^-(aq) \rightarrow C_2H_3O_2^-(aq) + H_2O(l)$

0.14 mol	0.01 mol	0.12 mol
-0.01 mol	-0.01 mol	+0.01 mol
0.13 mol	0 mol	0.13 mol

$$[H^+] \approx \frac{1.3 \times 10^{-5}\,(0.13)}{(0.13)} = 1.3 \times 10^{-5}\ M;\ pH = 4.89$$

(c) $C_3H_5O_2^-(aq) + HNO_3(aq) \rightarrow HC_3H_5O_2(aq) + Cl^-(aq)$

0.12 mol	0.01 mol	0.14 mol
-0.01 mol	-0.01 mol	+0.01 mol
0.11 mol	0 mol	0.15 mol

$$[H^+] \approx \frac{1.3 \times 10^{-5}\,(0.15)}{(0.11)} = 1.77 \times 10^{-3} = 1.8 \times 10^{-3}\ M;\ pH = 4.75$$

17.19 $H_2CO_3(aq) \rightleftharpoons H^+(aq) + HCO_3^-(aq)$ $K_a = \dfrac{[H^+][HCO_3^-]}{[H_2CO_3]}$; $\dfrac{[HCO_3^-]}{[H_2CO_3]} = \dfrac{K_a}{[H^+]}$

(a)   at pH = 7.4, $[H^+] = 10^{-7.4} = 4.0 \times 10^{-8}\ M$; $\dfrac{[HCO_3^-]}{[H_2CO_3]} = \dfrac{4.3 \times 10^{-7}}{4.0 \times 10^{-8}} = 11$

(b)   at pH = 7.1, $[H^+] = 7.9 \times 10^{-8}\ M$; $\dfrac{[HCO_3^-]}{[H_2CO_3]} = 5.4$

17.20 $\dfrac{6.5 \text{ g NaH}_2\text{PO}_4}{0.355 \text{ L soln}} \times \dfrac{1 \text{ mol NaH}_2\text{PO}_4}{120 \text{ g NaH}_2\text{PO}_4} = 0.153 = 0.15 \text{ } M$

$\dfrac{8.0 \text{ g Na}_2\text{HPO}_4}{0.355 \text{ L soln}} \times \dfrac{1 \text{ mol NaH}_2\text{PO}_4}{142 \text{ g Na}_2\text{HPO}_4} = 0.159 = 0.16 \text{ } M$

Using Equation [17.9] to find the pH of the buffer:

$$\text{pH} = -\log (6.2 \times 10^{-8}) + \log \dfrac{0.159}{0.153} = 7.2076 + 0.0167 = 7.22$$

Acid-Base Titrations

17.21   (a) $40.0 \text{ mL HNO}_3 \times \dfrac{0.0350 \text{ mol HNO}_3}{1000 \text{ mL soln}} \times \dfrac{1 \text{ mol NaOH}}{1 \text{ mol HNO}_3} \times \dfrac{1000 \text{ mL soln}}{0.0350 \text{ mol NaOH}}$

$= 40.0 \text{ mL NaOH soln}$

(b) $65.0 \text{ mL HBr} \times \dfrac{0.0620 \text{ } M \text{ HBr}}{1000 \text{ mL soln}} \times \dfrac{1 \text{ mol NaOH}}{1 \text{ mol HBr}} \times \dfrac{1000 \text{ mL soln}}{0.0350 \text{ mol NaOH}}$

$= 115 \text{ mL NaOH soln}$

(c) $\dfrac{1.65 \text{ g HCl}}{1 \text{ L soln}} \times \dfrac{1 \text{ mol HCl}}{36.46 \text{ g HCl}} = 0.04526 = 0.0453 \text{ } M \text{ HCl}$

$80.0 \text{ mL HCl} \times \dfrac{0.04526 \text{ mol HCl}}{1000 \text{ mL}} \times \dfrac{1 \text{ mol NaOH}}{1 \text{ mol HCL}} \times \dfrac{1000 \text{ mL soln}}{0.0350 \text{ mol NaOH}}$

$= 103.45 = 103 \text{ mL NaOH soln}$

17.22   (a) $40.0 \text{ mL NaOH} \times \dfrac{0.0750 \text{ mol NaOH}}{1000 \text{ mL soln}} \times \dfrac{1 \text{ mol HCl}}{1 \text{ mol NaOH}} \times \dfrac{1000 \text{ mL soln}}{0.0750 \text{ mol HCl}}$

$= 40.0 \text{ mL HCl soln}$

(b) $38.2 \text{ mL KOH} \times \dfrac{0.105 \text{ mol KOH}}{1000 \text{ mL soln}} \times \dfrac{1 \text{ mol HCl}}{1 \text{ mol KOH}} \times \dfrac{1000 \text{ mL soln}}{0.0750 \text{ mol HCl}}$

$= 53.5 \text{ mL HCl soln}$

(c) $50.0 \text{ mL} \times \dfrac{1.65 \text{ g NaOH}}{1000 \text{ mL}} \times \dfrac{1 \text{ mol NaOH}}{40.00 \text{ g NaOH}} \times \dfrac{1 \text{ mol HCl}}{1 \text{ mol NaOH}} \times \dfrac{1000 \text{ mL soln}}{0.0750 \text{ mol HCl}}$

$= 27.5 \text{ mL HCl soln}$

17.23   Construct a table similar to Table 17.1.

moles $H^+ = M_{\text{HBr}} \times L_{\text{HBr}} = 0.200 \text{ } M \times 0.0200 \text{ L} = 4.00 \times 10^{-3} \text{ mol}$

moles $OH^- = M_{\text{NaOH}} \times L_{\text{NaOH}} = 0.200 \text{ } M \times L_{\text{NaOH}}$

	mL_{HBr}	mL_{NaOH}	Total Volume	Moles H^+	Moles OH^-	Molarity Excess Ion	pH
(a)	20.0	15.0	35.0	4.00×10^{-3}	3.00×10^{-3}	$0.0286(H^+)$	1.544
(b)	20.0	19.9	39.9	4.00×10^{-3}	3.98×10^{-3}	$5 \times 10^{-4}(H^+)$	3.3
(c)	20.0	20.0	40.0	4.00×10^{-3}	4.00×10^{-3}	$1 \times 10^{-7}(H^+)$	7.0
(d)	20.0	20.1	40.1	4.00×10^{-3}	4.02×10^{-3}	$5 \times 10^{-4}(H^+)$	10.7
(e)	20.0	35.0	55.0	4.00×10^{-3}	7.00×10^{-3}	$0.0545(OH^-)$	12.737

molarity of excess ion = moles ion / total vol in L

(a) $$\frac{4.00 \times 10^{-3} \text{ mol H}^+ - 3.00 \times 10^{-3} \text{ mol OH}^-}{0.0350 \text{ L}} = 0.0286 \ M \ H^+$$

(b) $$\frac{4.00 \times 10^{-3} \text{ mol H}^+ - 3.98 \times 10^{-3} \text{ mol OH}^-}{0.0339 \text{ L}} = 5.01 \times 10^{-4} = 5 \times 10^{-4} \ M \ H^+$$

(c) equivalence point, mol H^+ = mol OH^-

NaBr does not hydrolyze, so $[H^+] = [OH^-] = 1 \times 10^{-7} \ M$

(d) $$\frac{4.02 \times 10^{-3} \text{ mol H}^+ - 4.00 \times 10^{-3} \text{ mol OH}^-}{0.041 \text{ L}} = 4.88 \times 10^{-4} = 5 \times 10^{-4} \ M \ OH^-$$

(e) $$\frac{7.00 \times 10^{-3} \text{ mol H}^+ - 4.00 \times 10^{-3} \text{ mol OH}^-}{0.0550 \text{ L}} = 0.054545 = 0.0545 \ M \ OH^-$$

17.24 Construct a table similar to Table 17.1

moles $OH^- = M_{KOH} \times L_{KOH} = 0.200 \ M \times 0.0300 \ L = 6.00 \times 10^{-3}$ mol

moles $H^+ = M_{HClO_4} \times L_{HClO_4} = 0.0150M \times L_{HClO_4}$

	mL_{NaOH}	mL_{HClO_4}	Total Volume	Moles OH^-	Moles H^+	Molarity Excess Ion	pH
(a)	30.0	30.0	60.0	6.00×10^{-3}	4.50×10^{-3}	$0.0250(OH^-)$	12.398
(b)	30.0	39.5	69.5	6.00×10^{-3}	5.93×10^{-3}	$1 \times 10^{-3}(OH^-)$	11.0
(c)	30.0	39.9	69.9	6.00×10^{-3}	5.99×10^{-3}	$1 \times 10^{-4}(OH^-)$	10.2
(d)	30.0	40.0	70.0	6.00×10^{-3}	6.00×10^{-3}	$1 \times 10^{-7}(H^+)$	7.0
(e)	30.0	40.1	70.1	6.00×10^{-3}	6.02×10^{-3}	$3 \times 10^{-4}(H^+)$	3.5

molarity of excess ion = $\dfrac{\text{moles ion}}{\text{total vol in L}}$

(a) $$\frac{6.00 \times 10^{-3} \text{ mol OH}^- - 4.50 \times 10^{-3} \text{ mol H}^+}{0.0600 \text{ L}} = 0.0250 \ M \ OH^-$$

(b) $$\frac{6.00 \times 10^{-3} \text{ mol OH}^- - 5.93 \times 10^{-3} \text{ mol H}^+}{0.0695 \text{ L}} = 1.01 \times 10^{-3} = 1 \times 10^{-3} \ M \ OH^-$$

(c) $\dfrac{6.00 \times 10^{-3}\,\text{mol OH}^- - 5.99 \times 10^{-3}\,\text{mol H}^+}{0.0699\,\text{L}} = 1.43 \times 10^{-4} = 1 \times 10^{-4}\ M\ \text{OH}^-$

(d) equivalence point, mol OH^- = mol H^+

$KClO_4$ does not hydrolyze, so $[\text{H}^+] = [\text{OH}^-] = 1 \times 10^{-7}\ M$

(e) $\dfrac{6.00 \times 10^{-3}\,\text{mol OH}^- - 6.02 \times 10^{-3}\,\text{mol H}^+}{0.0701\,\text{L}} = 2.85 \times 10^{-4} = 3 \times 10^{-4}\ M\ \text{H}^+$

17.25 (a) The quantity of base required to reach the equivalence point is the same in the two titrations.

(b) The pH is higher initially in the titration of a weak acid.

(c) The pH is higher at the equivalence point in the titration of a weak acid.

(d) The pH in excess base is essentially the same for the two cases.

(e) In titrating a weak acid, one needs an indicator that changes at a higher pH than for the strong acid titration. The choice is more critical because the **change** in pH close to the equivalence point is smaller for the weak acid titration.

17.26 (a) At the equivalence point, moles HX added = moles B initially present =
$0.1\ M \times 0.03\ \text{L} = 0.003$ moles HX added.

(b) BH^+

(c) Both K_a for BH^+ and concentration BH^+ determine pH at the equivalence point.

(d) Because the pH at the equivalence point will be less than 7, methyl red would be more appropriate.

17.27 (a) At 0 mL, only weak acid, $HC_2H_3O_2$, is present in solution. Using the acid dissociation equilibrium

$$HC_2H_3O_2(aq) \rightleftharpoons H^+(aq) + C_2H_3O_2^-(aq)$$

initial	0.150 M	0	0
equil	0.150 - x M	x M	x M

$K_a = \dfrac{[\text{H}^+][C_2H_3O_2^-]}{[HC_2H_3O_2]} = 1.8 \times 10^{-5}$ (Appendix D)

$1.8 \times 10^{-5} = \dfrac{x^2}{(0.150 - x)} \approx \dfrac{x^2}{0.150}$; $x^2 = 2.7 \times 10^{-6}$; $x = [\text{H}^+] = 0.001643$
$= 1.6 \times 10^{-3}$; pH = 2.78

(b)-(f) Calculate the moles of each component after the acid-base reaction takes place. Moles $HC_2H_3O_2$ originally present = $M \times L = 0.150\ M \times 0.0500\ \text{L} = 7.50 \times 10^{-3}$ mol. Moles NaOH added = $M \times L = 0.0150\ M \times y$ mL.

$$NaOH(aq)\quad +\quad HC_2H_3O_2(aq)\quad \rightarrow\ Na^+C_2H_3O_2^-(aq)\ +\ H_2O(l)$$

$(0.150\ M \times 0.0250\ \text{L}) =$

(b)	before rx	3.75×10^{-3} mol	7.50×10^{-3} mol	
	after rx	**0**	**3.75×10^{-3} mol**	**3.75×10^{-3} mol**

(c)   before rx $\dfrac{(0.150\ M \times 0.0490\ L) =}{7.35 \times 10^{-3}\ mol}$ $7.50 \times 10^{-3}\ mol$

         after rx   **0**   **0.15×10^{-3} mol**   **7.35×10^{-3} mol**

(d)   before rx $\dfrac{(0.150\ M \times 0.0500\ L) =}{7.50 \times 10^{-3}\ mol}$ $7.50 \times 10^{-3}\ mol$

         after rx   **0**   **0**   **7.50×10^{-3} mol**

(e)   before rx $\dfrac{(0.150\ M \times 0.0510\ L) =}{7.65 \times 10^{-3}\ mol}$ $7.50 \times 10^{-3}\ mol$

         after rx   **0.15×10^{-3} mol**   **0**   **7.50×10^{-3} mol**

(f)   before rx $\dfrac{(0.150\ M \times 0.0750\ L) =}{11.25 \times 10^{-3}\ mol}$ $7.50 \times 10^{-3}\ mol$

         after rx   **3.75×10^{-3} mol**   **0**   **7.50×10^{-3} mol**

Calculate the molarity of each species (M = mol/L) and solve the appropriate equilibrium problem in each part.

(b) $V_T = 50.0\ mL\ HC_2H_3O_2 + 25.0\ mL\ NaOH = 75.0\ mL = 0.0750\ L$

$$[HC_2H_3O_2] = \frac{3.75 \times 10^{-3}\ mol}{0.0750} = 0.0500\ M$$

$$[C_2H_3O_2^-] = \frac{3.75 \times 10^{-3}\ mol}{0.0750} = 0.0500\ M$$

$$HC_2H_3O_2(aq) \ \rightleftharpoons\ H^+(aq)\ +\ C_2H_3O_2^-(aq)$$

equil $0.0500 - x\ M$ $x\ M$ $0.0500 + x\ M$

$$K_a = \frac{[H^+][C_2H_3O_2^-]}{[HC_2H_3O_2]};\quad [H^+] = \frac{K_a\,[HC_2H_3O_2]}{[C_2H_3O_2^-]}$$

$$[H^+] = \frac{1.8 \times 10^{-5}\,(0.0500 - x)}{(0.0500 + x)} = 1.8 \times 10^{-5}\ M\,H^+;\ pH = 4.74$$

(c) $[HC_2H_3O_2] = \dfrac{0.15 \times 10^{-3}\ mol}{0.0990\ L} = 0.001515 = 1.5\ \times 10^{-3}\ M$

$$[C_2H_3O_2^-] = \frac{7.35 \times 10^{-3}\ mol}{0.0990\ L} = 0.07424 = 0.074\ M$$

$$[H^+] = \frac{1.8 \times 10^{-5}\,(1.515 \times 10^{-3} - x)}{(0.07424 + x)} \approx 3.7 \times 10^{-7}\ M\,H^+;\ pH = 6.43$$

(d)   At the equivalence point, only $C_3H_5O_2^-$ is present.

$$[C_2H_3O_2^-] = \frac{7.50 \times 10^{-3}\ mol}{0.100\ L} = 0.0750\ M$$

The pertinent equilibrium is the base hydrolysis of $C_2H_3O_2^-$.

$$C_2H_3O_2^- \text{ (aq)} + H_2O(l) \rightleftharpoons HC_2H_3O_2 \text{ (aq)} + OH^- \text{(aq)}$$

initial	0.0750 M	0	0
equil	0.0750 - x M	x	x

$$K_b = \frac{K_w}{K_a \text{ for } HC_2H_3O_2} = \frac{1.0 \times 10^{-14}}{1.8 \times 10^{-5}} = 5.56 \times 10^{-10} = 5.6 \times 10^{-10} = \frac{[HC_2H_3O_2][OH^-]}{[C_2H_3O_2^-]}$$

$$5.56 \times 10^{-10} = \frac{x^2}{0.0750 - x}; \quad x^2 \approx 5.56 \times 10^{-10} \, (0.0750); \quad x = 6.458 \times 10^{-6}$$

$$= 6.5 \times 10^{-10} \; M \, OH^-$$

$$pOH = -\log(6.458 \times 10^{-6}) = 5.19; \quad pH = 14.00 - pOH = 8.81$$

(e) After the equivalence point, the excess strong base determines the pOH and pH. The $[OH^-]$ from the hydrolysis of $C_2H_3O_2^-$ is small and can be ignored.

$$[OH^-] = \frac{0.15 \times 10^{-3} \text{ mol}}{0.101 \text{ L}} = 1.485 \times 10^{-3} = 1.5 \times 10^{-3} \; M; \quad pOH = 2.83$$

$$pH = 14.00 - 2.83 = 11.17$$

(f) $$[OH^-] = \frac{3.75 \times 10^{-3} \text{ mol}}{0.125 \text{ L}} = 0.0300 \; M \, OH^-; \quad pOH = 1.52; \quad pH = 14.00 - 1.52 = 12.48$$

17.28 (a) Weak base problem: $$K_b = 1.8 \times 10^{-5} = \frac{[NH_4^+][OH^-]}{[NH_3]}$$

At equilibrium, $[OH^-] = x$, $[NH_3] = (0.100 - x)$; $[NH_4^+] = x$

$$1.8 \times 10^{-5} = \frac{x^2}{(0.100 - x)} \approx \frac{x^2}{0.100}; \quad x = [OH^-] = 1.342 \times 10^{-3} = 1.3 \times 10^{-3} \; M$$

$$pH = 14.00 - 2.87 = 11.13$$

(b-f) Calculate mol NH_3 and mol NH_4^+ after the acid-base reaction takes place.
0.100 $M \, NH_3 \times 0.0600$ L = 6.00×10^{-3} mol NH_3 present initially.

		NH_3(aq)	+	HCl(aq)	$\rightarrow$	NH_4^+(aq) + Cl$^-$(aq)
				(0.150 $M \times$ 0.0200 L) =		
(b)	before rx	6.00×10^{-3} mol		3.00×10^{-3} mol		0 mol
	after rx	**3.00×10^{-3} mol**		**0 mol**		**3.00×10^{-3} mol**
				(0.150 $M \times$ 0.0395 L) =		
(c)	before rx	6.00×10^{-3} mol		5.93×10^{-3} mol		0 mol
	after rx	**0.07×10^{-3} mol**		**0 mol**		**5.93×10^{-3} mol**
				(0.150 $M \times$ 0.0400 L) =		
(d)	before rx	6.00×10^{-3} mol		6.00×10^{-3} mol		0 mol
	after rx	**0 mol**		**0 mol**		**6.00×10^{-3} mol**

$$(0.150\ M \times 0.0405\ L) =$$

(e)   before rx 6.00×10^{-3} mol 6.08×10^{-3} mol   0 mol

        after rx   **0 mol**   **0.08×10^{-3} mol**   **6.00×10^{-3} mol**

$$(0.150\ M \times 0.0600\ L) =$$

(f)   before rx 6.00×10^{-3} mol 9.00×10^{-3} mol   0 mol

        after rx   **0 mol**   **3.00×10^{-3} mol**   **6.00×10^{-3} mol**

(b)   Using the acid dissociation equilibrium for NH_4^+ (so that we calculate $[H^+]$ directly), $NH_4^+(aq) \rightleftharpoons H^+(aq) + NH_3(aq)$

$$K_a = \frac{[H^+][NH_3]}{[NH_4^+]} = \frac{K_w}{K_b\ \text{for}\ NH_3} = \frac{1.0 \times 10^{-14}}{1.8 \times 10^{-5}} = 5.56 \times 10^{-10} = 5.6 \times 10^{-10}$$

$$[NH_3] = \frac{3.00 \times 10^{-3}\ \text{mol}}{0.0800\ L} = 0.0375\ M; \quad [NH_4^+] = \frac{3.00 \times 10^{-3}\ \text{mol}}{0.0800\ L} = 0.0375\ M$$

$$[H^+] = \frac{5.56 \times 10^{-10}\ [NH_4^+]}{[NH_3]} \approx \frac{5.56 \times 10^{-10}\ (0.0375)}{(0.0375)} = 5.56 \times 10^{-10}; \quad pH = 9.25$$

(We will assume $[H^+]$ is small compared to $[NH_3]$ and $[NH_4^+]$.)

(c) $[NH_3] = \dfrac{7.0 \times 10^{-5}\ \text{mol}}{0.0995\ L} = 7.04 \times 10^{-4} = 7 \times 10^{-4}\ M; \quad [NH_4^+] = \dfrac{5.93 \times 10^{-3}\ \text{mol}}{0.0995\ L}$

$$= 5.96 \times 10^{-2}\ M$$

$$[H^+] = \frac{5.56 \times 10^{-10}\ (5.96 \times 10^{-2})}{(7.04 \times 10^{-4})} = 4.71 \times 10^{-8} = 5 \times 10^{-8}\ M; \quad pH = 7.3$$

(d)   At the equivalence point, $[H^+] = [NH_3] = x$

$$[NH_4^+] = \frac{6.00 \times 10^{-3}\ M}{0.100\ L} = 0.0600\ M$$

$$5.56 \times 10^{-10} = \frac{x^2}{0.060}; \quad x = [H^+] = 5.776 \times 10^{-6} = 5.8 \times 10^{-6}; \quad pH = 5.24$$

(e)   Past the equivalence point, $[H^+]$ from the excess HCl determines the pH.

$$[H^+] = \frac{8 \times 10^{-5}\ \text{mol}}{0.101\ L} = 7.92 \times 10^{-4} = 8 \times 10^{-4}\ M; \quad pH = 3.1$$

(f) $[H^+] = \dfrac{3.00 \times 10^{-3}\ \text{mol}}{0.120\ L} = 0.0250\ M; \quad pH = 1.602$

17.29   The volume of 0.200 M HBr required in all cases equals the volume of base and the final volume = $2V_{base}$. The concentration of the salt produced at the equivalence point =

$$\frac{0.200\ M \times V_{base}}{2V_{base}} = 0.100\ M.$$

(a)   0.100 M NaBr, pH = 7.00

(b) $0.100\ M\ HONH_3^+Br^-$; $HONH_3^+(aq) \rightleftharpoons H^+(aq) + HONH_2$

                                     [equil] $0.100 - x$ x x

$$K_a = \frac{[H^+][HONH_2]}{[HONH_3^+]} = \frac{K_w}{K_b} = \frac{1.0 \times 10^{-14}}{1.1 \times 10^{-8}} = 9.09 \times 10^{-7} = 9.1 \times 10^{-7}$$

Assume x is small with respect to [salt].

$K_a = x^2 / 0.100$; $x = [H^+] = 3.02 \times 10^{-4} = 3.0 \times 10^{-4}\ M$, pH = 3.52

(c) $0.100\ M\ C_6H_5NH_3^+Br^-$.   Proceeding as in (b):

$$K_a = \frac{[H^+][C_6H_5NH_2]}{[C_6H_5NH_3^+]} = \frac{K_w}{K_b} = 2.33 \times 10^{-5} = 2.3 \times 10^{-5}$$

$[H^+]^2 = 0.100(2.33 \times 10^{-5})$; $[H^+] = 1.52 \times 10^{-3} = 1.5 \times 10^{-3}\ M$,   pH = 2.82

17.30   The volume of NaOH solution required in all cases is

$$V_{base} = \frac{V_{acid} \times M_{acid}}{M_{base}} = \frac{(0.100)\ V_{acid}}{(0.080)} = 1.25\ V_{acid}$$

The total volume at the equivalence point is $V_{base} + V_{acid} = 2.25\ V_{acid}$.
The concentration of the salt at the equivalence point is

$$\frac{M_{acid}\ V_{acid}}{2.25\ V_{acid}} = \frac{0.100}{2.25} = 0.0444\ M$$

(a) $0.0444\ M$ NaBr, pH = 7.00

(b) $0.0444\ M\ Na^+C_3H_5O_3^-$; $C_3H_5O_3^-(aq) + H_2O(l) \rightleftharpoons HC_3H_5O_3(aq) + OH^-(aq)$

$$K_b = \frac{[HC_3H_5O_3][OH^-]}{[C_3H_5O_3^-]} = \frac{K_w}{K_a} = \frac{1.0 \times 10^{-14}}{1.4 \times 10^{-4}} = 7.14 \times 10^{-11} = 7.1 \times 10^{-11}$$

$[HC_3H_5O_3] = [OH^-]$; $[C_3H_5O_3^-] \approx 0.0444$

$[OH^-]^2 \approx 0.0444(7.14 \times 10^{-11})$; $[OH^-] = 1.78 \times 10^{-6} = 1.8 \times 10^{-6}\ M$,

                                            pOH = 5.75; pH = 8.25

(c) $CrO_4^{-2}(aq) + H_2O(l) \rightleftharpoons HCrO_4^-(aq) + OH^-(aq)$

$$K_b = \frac{[HCrO_4^-][OH^-]}{[CrO_4^{2-}]} = \frac{K_w}{K_a} = 3.33 \times 10^{-8} = 3.3 \times 10^{-8}$$

$[OH^-]^2 \approx 0.0444(3.33 \times 10^{-8})$; $[OH^-] = 3.845 \times 10^{-5} = 3.8 \times 10^{-5}$, pH = 9.58

Solubility Equilibria

17.31 (a) The concentration of undissolved solid does not appear in the solubility produce expression because it is constant as long as there is solid present. Concentration is a ratio of moles solid to volume of the solid; solids occupy a specific volume not dependent on the solution volume. As the amount (moles) of solid changes, the volume changes proportionally, so that the ratio of moles solid to volume solid is constant.

(b) $K_{sp} = [Ag^+][I^-]$; $K_{sp} = [Ba^{2+}][CO_3^{2-}]$; $K_{sp} = [Cu^+]^2[S^{2-}]$
$K_{sp} = [Ce^{3+}][F^-]^3$; $K_{sp} = [Ca^{2+}]^3[PO_4^{3-}]^2$

17.32 (a) *Solubility* is the amount (grams, moles) of solute that will dissolve in a certain volume of solution. *Solubility-product constant* is an equilibrium constant, the product of the molar concentrations of all the dissolved ions in solution.

(b) $K_{sp} = [Cu^{2+}][S^{2-}]$; $K_{sp} = [Ni^{2+}][C_2O_4^{2-}]$; $K_{sp} = [Ag^+]^2[SO_4^{2-}]$
$K_{sp} = [Co^{3+}][OH^-]^3$; $K_{sp} = [Fe^{2+}]^3[AsO_4^{3-}]^2$

17.33 (a) $CaF_2(s) \rightleftharpoons Ca^{2+}(aq) + 2F^-(aq)$; $K_{sp} = [Ca^{2+}][F^-]^2$

The molar solubility is the moles of CaF_2 that dissolve per liter of solution. Each mole of CaF_2 produces **1** mol $Ca^{2+}(aq)$ and **2** mol $F^-(aq)$.
$[Ca^{2+}] = 1.24 \times 10^{-3}$ M; $[F^-] = 2 \times 1.24 \times 10^{-3}$ $M = 2.48 \times 10^{-3}$ M
$K_{sp} = (1.24 \times 10^{-3})(2.48 \times 10^{-3})^2 = 7.63 \times 10^{-9}$

(b) $SrF_2(s) \rightleftharpoons Sr^{2+}(aq) + 2F^-(aq)$; $K_{sp} = [Sr^{2+}][F^-]^2$

Transform the gram solubility to molar solubility.

$$\frac{1.1 \times 10^{-2}\,g\,SrF_2}{0.100\,L} \times \frac{1\,mol\,SrF_2}{125.6\,g\,SrF_2} = 8.76 \times 10^{-4} = 8.8 \times 10^{-4}\,mol\,SrF_2/L$$

$[Sr^{2+}] = 8.76 \times 10^{-4}$ M; $[F^-] = 2(8.76 \times 10^{-4}$ $M)$

$K_{sp} = (8.76 \times 10^{-4})(2(8.76 \times 10^{-4}))^2 = 2.7 \times 10^{-9}$

(c) $Ba(IO_3)_2(s) \rightleftharpoons Ba^{2+}(aq) + 2IO_3^-(aq)$; $K_{sp} = [Ba^{2+}][IO_3^-]^2$

Since 1 mole of dissolved $Ba(IO_3)_2$ produces 1 mole of Ba^{2+}, the molar solubility of

$Ba(IO_3)_2 = [Ba^{2+}]$. Let $x = [Ba^{2+}]$; $[IO_3^-] = 2x$

$K_{sp} = 6.0 \times 10^{-10} = (x)(2x)^2$; $4x^3 = 6.0 \times 10^{-10}$; $x^3 = 1.5 \times 10^{-10}$; $x = 5.3 \times 10^{-4}$ M

The molar solubility of $Ba(IO_3)_2$ is 5.3×10^{-4} mol/L.

17.34   (a) $PbBr_2(s) \rightleftharpoons Pb^{2+}(aq) + 2Br^-(aq)$

$K_{sp} = [Pb^{2+}][Br^-]^2$; $[Pb^{2+}] = 1.0 \times 10^{-2} M$, $[Br^-] = 2.0 \times 10^{-2} M$

$K_{sp} = (1.0 \times 10^{-2} M)(2.0 \times 10^{-2} M)^2 = 4.0 \times 10^{-6}$

(b) $AgIO_3(s) \rightleftharpoons Ag^+(aq) + IO_3^-(aq)$; $K_{sp} = [Ag^+][IO_3^-]$

$$[Ag^+] = [IO_3^-] = \frac{0.0490 \text{ g } AgIO_3}{1.00 \text{ L soln}} \times \frac{1 \text{ mol } AgIO_3}{282.8 \text{ g } AgIO_3} = 1.733 \times 10^{-4} = 1.73 \times 10^{-4} M$$

$K_{sp} = (1.733 \times 10^{-4} M)(1.733 \times 10^{-4} M) = 3.00 \times 10^{-8}$

(c) $Cu(OH)_2(s) \rightleftharpoons Cu^{2+}(aq) + 2OH^-(aq)$; $K_{sp} = [Cu^{2+}][OH^-]^2$

$[Cu^{2+}] = x$, $[OH^-] = 2x$; $K_{sp} = 2.2 \times 10^{-20} = (x)(2x)^2$

$2.2 \times 10^{-20} = 4x^3$; $x = [Cu^{2+}] = 1.765 \times 10^{-7} = 1.8 \times 10^{-7} M$

$$\frac{1.765 \times 10^{-7} \text{ mol } Cu(OH)_2}{1 \text{ L}} \times \frac{97.56 \text{ g } Cu(OH)_2}{1 \text{ mol } Cu(OH)_2} = 1.7 \times 10^{-5} \text{ g } Cu(OH)_2$$

However, $[OH^-]$ from $Cu(OH)_2 = 3.53 \times 10^{-7}$ M; this is similar to $[OH^-]$ from the autoionization of water.

$K_w = [H^+][OH^-]$; $[H^+] = y$, $[OH^-] = (3.53 \times 10^{-7} + y)$

$1.0 \times 10^{-14} = y(3.53 \times 10^{-7} + y)$; $y^2 + 3.53 \times 10^{-7} \, y - 1.0 \times 10^{-14}$

$$y = \frac{-3.53 \times 10^{-7} \pm \sqrt{(3.53 \times 10^{-7})^2 - 4(1)(-1.0 \times 10^{-14})}}{2}; \quad y = 2.64 \times 10^{-8}$$

$[OH^-]_{total} = 3.53 \times 10^{-7} M + 0.264 \times 10^{-7} M = 3.79 \times 10^{-7} M$

Recalculating $[Cu^{2+}]$ and thus molar solubility of $Cu(OH)_2(s)$:

$2.2 \times 10^{-20} = x(3.79 \times 10^{-7})^2$; $x = 1.53 \times 10^{-7} M \, Cu^{2+}$

$$\frac{1.53 \times 10^{-7} \text{ mol } Cu(OH)_2(s)}{1 \text{ L}} \times \frac{97.56 \text{ g } Cu(OH)_2}{1 \text{ mol } Cu(OH)_2} = 1.5 \times 10^{-5} \text{ g } Cu(OH)_2$$

Note that the presence of OH^- as a common ion decreases the water solubility of $Cu(OH)_2$.

17.35 $CaC_2O_4(s) \rightleftharpoons Ca^{2+}(aq) + C_2O_4^{2-}(aq)$; $K_{sp} = [Ca^{2+}][C_2O_4^{2-}]$

$$[Ca^{2+}] = [C_2O_4^{2-}] = \frac{0.0061 \text{ g } CaC_2O_4}{1.00 \text{ L soln}} \times \frac{1 \text{ mol } CaC_2O_4}{128.1 \text{ g } CaC_2O_4} = 4.76 \times 10^{-5} = 4.8 \times 10^{-5} M$$

$K_{sp} = (4.76 \times 10^{-5} M)(4.76 \times 10^{-5} M) = 2.3 \times 10^{-9}$

17.36 $PbI_2(s) \rightleftharpoons Pb^{2+}(aq) + 2I^-(aq)$; $K_{sp} = [Pb^{2+}][I^-]^2$

$$[Pb^{2+}] = \frac{0.54 \text{ g } PbI_2}{1.00 \text{ L soln}} \times \frac{1 \text{ mol } PbI_2}{461.0 \text{ g } PbI_2} = 1.17 \times 10^{-3} = 1.2 \times 10^{-3} M$$

$[I^-] = 2[Pb^{2+}]$; $K_{sp} = [Pb^{2+}](2[Pb^{2+}])^2 = 4[Pb^{2+}]^3 = 4(1.17 \times 10^{-3})^3 = 6.4 \times 10^{-9}$

17.37 (a) $AgBr(s) \rightleftharpoons Ag^+(aq) + Br^-(aq); \ K_{sp} = [Ag^+][Br^-] = 5.0 \times 10^{-13}$

molar solubility = $x = [Ag^+] = [Br^-]; \ K_{sp} = x^2$

$x = (5.0 \times 10^{-13})^{1/2}; \ x = 7.1 \times 10^{-7} \ mol \ AgBr/L$

(b)   Molar solubility = $x = [Br^-]; \ [Ag^+] = 0.030 \ M + x$

$K_{sp} = (0.030 + x)(x) \approx 0.030(x)$

$5.0 \times 10^{-13} = 0.030(x); \ x = 1.7 \times 10^{-11} \ mol \ AgBr/L$

(c)   Molar solubility = $x = [Ag^+]$

There are two sources of Br^-: $NaBr(0.50 \ M)$ and $AgBr(x \ M)$

$K_{sp} = (x)(0.50 + x);$ Assuming x is small compared to 0.50 M

$5.0 \times 10^{-13} = 0.50 \ (x); \ x \approx 1.0 \times 10^{-12} \ mol \ AgBr/L$

17.38 (a) $CaF_2(s) \rightleftharpoons Ca^{2+}(aq) + 2F^-(aq); \ K_{sp} = [Ca^{2+}][F^-]^2 = 3.9 \times 10^{-11}$

molar solubility = $x = [Ca^{2+}]; \ [F^-] = 2[Ca^{2+}] = 2x$

$K_{sp} = (x)(2x)^2; \ 3.9 \times 10^{-11} = 4x^3, \ x = (9.75 \times 10^{-12})^{1/3}, \ x = 2.14 \times 10^{-4}$

$= 2.1 \times 10^{-4} \ M \ Ca^{2+}$

$$\frac{2.14 \times 10^{-4} \ mol \ CaF_2}{1 \ L} \times \frac{78.1 \ g \ CaF_2}{1 \ mol} = 0.017 \ g \ CaF_2/L$$

(b)   Molar solubility = $x = [Ca^{2+}]$

There are two sources of F^-: $KF(0.15 \ M)$ and $CaF_2 \ (2x \ M)$.

$K_{sp} = (x)(0.15 + 2x)^2;$ Assuming x is small compared to 0.15 M,

$3.9 \times 10^{-11} = 0.0225 \ x, \ x = 1.73 \times 10^{-9} = 1.7 \times 10^{-9} \ M \ Ca^{2+}$

$$\frac{1.73 \times 10^{-9} \ mol \ CaF_2}{1 \ L} \times \frac{78.1 \ g \ CaF_2}{1 \ mol} = \frac{1.4 \times 10^{-7} \ g \ CaF_2}{1 \ L} = 0.14 \ \mu g/L$$

$= 0.14 \ ppb \ CaF_2$

(c)   Molar solubility = $x, \ [F^-] = 2x, \ [Ca^{2+}] = 0.080 \ M + x$

$K_{sp} = (0.080 + x)(2x)^2 = 0.080 \ (2x)^2; \ 3.9 \times 10^{-11} = 0.32 \ x^2$

$x = (1.22 \times 10^{-10})^{1/2} = 1.10 \times 10^{-5} = 1.1 \times 10^{-5} \ M \ CaF_2$

$$\frac{1.10 \times 10^{-5} \ mol \ CaF_2}{L} \times \frac{78.1 \ g \ CaF_2}{1 \ mol} = \frac{8.6 \times 10^{-4} \ g \ CaF_2}{L} = \frac{0.86 \ mg \ CaF_2}{L}$$

17.39 $Cu(OH)_2(s) \rightleftharpoons Cu^{2+}(aq) + 2OH^-(aq); \ K_{sp} = 2.2 \times 10^{-20}$

Since the $[OH^-]$ is set by the pH of the solution, the solubility of $Cu(OH)_2$ is just $[Cu^{2+}]$.

(a) $pH = 7.0, \ pOH = 14 - pH = 7.0, \ [OH^-] = 10^{-pOH} = 1.0 \times 10^{-7} \ M$

$K_{sp} = 2.2 \times 10^{-20} = [Cu^{2+}](1.0 \times 10^{-7})^2; \ [Cu^{2+}] = \dfrac{2.2 \times 10^{-20}}{1.0 \times 10^{-14}} = 2.2 \times 10^{-6} \ M$

(In pure water, [OH$^-$] from Cu(OH)$_2$ is similar to (OH$^-$) from the autoionization of water, resulting in a cubic equation for [Cu^{2+}]. The solubility of Cu(OH)$_2$ at pH = 7.0 is actually greater than the solubility in pure water.)

(b)   pH = 9.0, pOH = 5.0, [OH$^-$] = 1.0 × 10^{-5}

$$K_{sp} = 2.2 \times 10^{-20} = [Cu^{2+}][1.0 \times 10^{-5}]^2; \quad [Cu^{2+}] = \frac{2.2 \times 10^{-20}}{1.0 \times 10^{-10}} = 2.2 \times 10^{-10} \ M$$

(c)   pH = 11.0, pOH = 3.0, [OH$^-$] = 1.0 × 10^{-3}

$$K_{sp} = 2.2 \times 10^{-20} = [Cu^{2+}][1.0 \times 10^{-3}]^2; \quad [Cu^{2+}] = \frac{2.2 \times 10^{-20}}{1.0 \times 10^{-6}} = 2.2 \times 10^{-14} \ M$$

17.40 Mn(OH)$_2$(s) $\rightleftharpoons$ Mn^{2+}(aq) + 2OH$^-$(aq); K$_{sp}$ = 1.9 × 10^{-13}

Since [OH$^-$] is set by the pH of the solution, the solubility of Mn(OH)$_2$ is just [Mn^{2+}].

(a)   pH = 7.0, pOH = 14 - pH = 7.0, [OH$^-$] = 10^{-pOH} = 1.0 × 10^{-7} M

$$K_{sp} = 1.9 \times 10^{-13} = [Mn^{2+}](1.0 \times 10^{-7})^2; \quad [Mn^{2+}] = \frac{1.9 \times 10^{-13}}{1.0 \times 10^{-14}} = 19 \ M$$

Note that the solubility of Mn(OH)$_2$ in pure water is 3.6 × 10^{-5} M, and the pH of the resulting solution is 9.0. The relatively low pH of a solution buffered to pH 7.0 actually increases the solubility of Mn(OH)$_2$.

(b)   pH = 9.5, pOH = 4.5, [OH$^-$] = 3.16 × 10^{-5} = 3.2 × 10^{-5} M

$$K_{sp} = 1.9 \times 10^{-13} = [Mn^{2+}](3.16 \times 10^{-5})^2; \quad [Mn^{2+}] = \frac{1.9 \times 10^{-13}}{1.0 \times 10^{-9}} = 1.9 \times 10^{-4} \ M$$

(c)   pH = 11.8, pOH = 2.2, [OH$^-$] = 6.31 × 10^{-3} = 6.3 × 10^{-3} M

$$K_{sp} = 1.9 \times 10^{-13} = [Mn^{2+}](6.31 \times 10^{-3})^2; \quad [Mn^{2+}] = \frac{1.9 \times 10^{-13}}{3.98 \times 10^{-5}} = 4.8 \times 10^{-9} \ M$$

17.41 Let the molar solubility of CaF$_2$ = x. K$_{sp}$ = 3.9 × 10^{-11}

$$BaF_2 = y \quad K_{sp} = 1.0 \times 10^{-6}$$

[F$^-$] = 0.010 M + x + y; assume x and y are small compared to 0.010.

CaF$_2$; K$_{sp}$ = [Ca^{2+}][F$^-$]^2; 3.9 × 10^{-11} ≈ (x) (0.010)^2; x = 3.9 × 10^{-7} mol CaF$_2$/L

BaF$_2$; K$_{sp}$ = [Ca^{2+}][F$^-$]^2; 1.0 × 10^{-6} ≈ (y) (0.010)^2; y = 1.0 × 10^{-2} mol BaF$_2$/L

$$\frac{[Ca^{2+}]}{[Ba^{2+}]} = \frac{3.9 \times 10^{-7} \ M}{1.0 \times 10^{-2} \ M} = 3.9 \times 10^{-5} \ \text{mol } Ca^{2+} / \ 1.0 \text{ mol } Ba^{2+}$$

17.42 [Ca^{2+}][CO$_3^{2-}$] = 2.8 × 10^{-9} ; [Fe^{2+}][CO$_3^{2-}$] = 3.2 × 10^{-11}

Since [CO$_3^{2-}$] is the same for both equilibria:

$$[CO_3^{2-}] = \frac{2.8 \times 10^{-9}}{[Ca^{2+}]} = \frac{3.2 \times 10^{-11}}{[Fe^{2+}]}; \ \text{rearranging} \ \frac{[Ca^{2+}]}{[Fe^{2+}]} = \frac{2.8 \times 10^{-9}}{3.2 \times 10^{-11}} = 88$$

17.43 $K_{sp} = [Ba^{2+}][MnO_4^-]^2 = 2.5 \times 10^{-10}$

$[MnO_4^-]^2 = 2.5 \times 10^{-10} / 2.0 \times 10^{-8} = 0.0125$; $[MnO_4^-] = \sqrt{0.0125} = 0.11\ M$

17.44   (a) $Ce(IO_3)_3(s) \rightleftharpoons Ce^{3+}(aq) + 3IO_3^-(aq)$; $K_{sp} = [Ce^{3+}][IO_3^-]^3 = 3.2 \times 10^{-10}$

Let $[Ce^{3+}] = x$; then $[IO_3^-] = 3x$; $(x)(3x)^3 = 3.2 \times 10^{-10}$; $27x^4 = 3.2 \times 10^{-10}$

$x = 1.855 \times 10^{-3}\ M$ = molar solubility of $Ce(IO_3)_3$ in water

(b)   If $[Ce^{3+}] = 1.855 \times 10^{-4}\ M$, let $[IO_3^-] = y$; $(1.855 \times 10^{-4})(y)^3 = 3.2 \times 10^{-10}$

$y = 0.01199 = 1.2 \times 10^{-2}\ M$

This is the **total** concentration of IO_3^- in solution. Of this, a part comes from $Ce(IO_3)_3$. The contribution from this source is $3(1.855 \times 10^{-4}\ M) = 5.6 \times 10^4\ M$. Thus, the $NaIO_3$ concentration would need to be $(1.2 \times 10^{-2} - 5.6 \times 10^{-4})\ M = 0.0114 = 1.1 \times 10^{-2}\ M$.

17.45   If the anion in the slightly soluble salt is the conjugate base of a strong acid, there will be no reaction.

(a) $MnS(s) + 2H^+(aq) \rightarrow H_2S(aq) + Mn^{2+}(aq)$

(b) $PbF_2(s) + 2H^+(aq) \rightarrow 2HF(aq) + Pb^{2+}(aq)$

(c) $AuCl_3(s) + H^+(aq) \rightarrow$ no reaction

(d) $Hg_2C_2O_4(s) + 2H^+(aq) \rightarrow H_2C_2O_4(aq) + Hg_2^{2+}(aq)$

(e) $CuBr(s) + H^+(aq) \rightarrow$ no reaction

17.46   If the anion of the salt is the conjugate base of a weak acid, it will combine with H^+, reducing the concentration of the free anion in solution, thereby causing more salt to dissolve. More soluble in acid: (a) $ZnCO_3$ (b) LaF_3 (d) $AgCN$ (e) $Ba_3(PO_4)_2$

17.47   The formation equilibrium is

$$Cu^{2+}(aq) + 4NH_3(aq) \rightleftharpoons Cu(NH_3)_4^{2+}(aq) \quad K_f = \frac{[Cu(NH_3)_4^{2+}]}{[Cu^{2+}][NH_3]^4} = 5 \times 10^{12}$$

Assuming that nearly all the Cu^{2+} is in the form $Cu(NH_3)_4^{2+}$

$[Cu(NH_3)_4^{2+}] = 1 \times 10^{-3}\ M$; $[Cu^{2+}] = x$; $[NH_3] = 0.10\ M$

$5 \times 10^{12} = \dfrac{(1 \times 10^{-3})}{x(0.10)^4}$; $x = 2 \times 10^{-12}\ M = [Cu^{2+}]$

17.48 $NiC_2O_4(s) \rightleftharpoons Ni^{2+}(aq) + C_2O_4^{2-}(aq)$; $K_{sp} = [Ni^{2+}][C_2O_4^{2-}] = 4 \times 10^{-10}$

When the salt has just dissolved, $[C_2O_4^{2-}]$ will be 0.020 M. Thus $[Ni^{2+}]$ must be less than $4 \times 10^{-10} / 0.020 = 2 \times 10^{-8}\ M$. To achieve this low $[Ni^{2+}]$ we must complex the Ni^{2+} ion with NH_3: $Ni^{2+}(aq) + 6NH_3(aq) \rightleftharpoons Ni(NH_3)_6^{2+}(aq)$. Essentially all $Ni(II)$ is in the form of the complex, so $[Ni(NH_3)_6^{2+}] = 0.020$. Find K_f for $Ni(NH_3)_6^{2+}$ in Table 17.2.

$$K_f = \frac{[Ni(NH_3)_6^{2+}]}{[Ni^{2+}][NH_3]^6} = \frac{(0.020)}{(2 \times 10^{-8})[NH_3]^6} = 5.5 \times 10^8; \quad [NH_3]^6 = 1.82 \times 10^{-3};$$

$$[NH_3] = 0.349 = 0.3 \ M$$

17.49 $Ag\,I(s) \rightleftharpoons Ag^+(aq) + I^-(aq)$

$Ag^+(aq) + 2CN^-(aq \rightleftharpoons Ag(CN)_2^-(aq)$

$\overline{Ag\,I(s) + 2CN^-(aq) \rightleftharpoons Ag(CN)_2^-(aq) + I^-(aq)}$

$$K = K_{sp} \times K_f = [Ag^+][I^-] \times \frac{[Ag(CN)_2^-]}{[Ag^+][CN^-]^2} = (8.3 \times 10^{-17})(1 \times 10^{-21}) = 8 \times 10^4$$

17.50 $Ag_2S(s) \rightleftharpoons 2Ag^+(aq) + S^{2-}(aq)$ K_{sp}

$S^{2-}(aq) + 2H^+(aq) \rightleftharpoons H_2S(aq)$ $1/(K_{a1} \times K_{a2})$

$2[Ag^+(aq) + 2Cl^-(aq) \rightleftharpoons AgCl_2^-(aq)]$ K_f^2

Add: $\overline{Ag_2S(s) + 2H^+(aq) + 4Cl^-(aq) \rightarrow 2AgCl_2^-(aq) + H_2S(aq)}$

$$K = \frac{K_{sp} \times K_f^2}{K_{a1} \times K_{a2}} = \frac{(6 \times 10^{-51})(1.1 \times 10^5)^2}{(9.5 \times 10^{-8})(1 \times 10^{-19})} = 7.64 \times 10^{-15} = 8 \times 10^{-15}$$

Precipitation; Qualitative Analysis

17.51   Precipitation conditions: will Q (see Chapter 15) exceed K_{sp} for the compound?

     (a)   In base, Mn^{2+} can form $Mn(OH)_2(s)$.

           $Mn(OH)_2(s) \rightleftharpoons Mn^{2+}(aq) + 2OH^-(aq); \quad K_{sp} = [Mn^{2+}][OH^-]^2$

           $Q = [Mn^{2+}][OH^-]^2; \quad [Mn^{2+}] = 0.050 \ M; \quad pOH = 6; \quad [OH^-] = 10^{-6} = 1 \times 10^{-6} \ M$

           $Q = (0.050)(1 \times 10^{-6})^2 = 5 \times 10^{-14}; \quad K_{sp} = 1.9 \times 10^{-13}$ (Appendix D)

           $Q < K_{sp}$, no $Mn(OH)_2$ precipitates.

     (b) $Ag_2SO_4(s) \rightleftharpoons 2Ag^+(aq) + SO_4^{2-}(aq); \quad K_{sp} = [Ag^+]^2[SO_4^{2-}]$

           $[Ag^+] = \dfrac{0.010 \ M \times 100 \ mL}{120 \ mL} = 8.33 \times 10^{-3} = 8.3 \times 10^{-3} \ M$

           $[SO_4^{2-}] = \dfrac{0.050 \ M \times 20 \ mL}{120 \ mL} = 8.33 \times 10^{-3} = 8.3 \times 10^{-3} \ M$

           $Q = (8.33 \times 10^{-3})^2(8.33 \times 10^{-3}) = 5.8 \times 10^{-7}; \quad K_{sp} = 1.4 \times 10^{-5}$

           $Q < K_{sp}$, no Ag_2SO_4 precipitates.

17.52 (a) $Co(OH)_2(s) \rightleftharpoons Co^{2+}(aq) + 2OH^-(aq)$; $K_{sp} = [Co^{2+}][OH^-]^2 = 1.6 \times 10^{-15}$

pH = 8.5; pOH = 14 - 8.5 = 5.5; $[OH^-] = 10^{-5.5} = 3.16 \times 10^{-6} = 3 \times 10^{-6}$ M

$(Q = (0.020)(3.16 \times 10^{-6})^2 = 2 \times 10^{-13}$; $Q > K_{sp}$, $Co(OH)_2$ will precipitate

(b) $AgIO_3(s) \rightleftharpoons Ag^+(aq) + IO_3^-(aq)$; $K_{sp} = [Ag^+][IO_3^-] = 3.0 \times 10^{-8}$

$$[Ag^+] = \frac{0.010\ M\ Ag^+ \times 0.100\ L}{0.110\ L} = 9.09 \times 10^{-3} = 9.1 \times 10^{-3}\ M$$

$$[IO_3^-] = \frac{0.015\ M\ IO_3^- \times 0.100\ L}{0.110\ L} = 1.36 \times 10^{-3} = 1.4 \times 10^{-3}\ M$$

$Q = (9.09 \times 10^{-3})(1.36 \times 10^{-3}) = 1.2 \times 10^{-5}$; $Q > K_{sp}$, $AgIO_3$ will precipitate

17.53 $Ni(OH)_2(s) \rightleftharpoons Ni^{2+}(aq) + 2OH^-(aq)$; $K_{sp} = [Ni^{2+}][OH^-]^2 = 1.6 \times 10^{-14}$

At equilibrium, $[Ni^{2+}][OH^-]^2 = 1.6 \times 10^{-14}$. Change $[Ni^{2+}]$ to mol/L and solve for $[OH^-]$.

$$\frac{1\ \mu g\ Ni^{2+}}{1\ L} \times \frac{1 \times 10^{-6}\ g}{1\ \mu g} \times \frac{1\ mol\ Ni^{2+}}{58.7\ g\ Ni^{2+}} = 1.70 \times 10^{-8} = 2 \times 10^{-8}\ M\ Ni^{2+}$$

$1.6 \times 10^{-14} = (1.70 \times 10^{-8})[OH^-]^2$; $[OH^-]^2 = 9.39 \times 10^{-7}$; $[OH^-] = 9.69 \times 10^{-4}$

$$= 1 \times 10^{-3}$$

pOH = 3.01; pH = 14.0 - 3.01 = 10.99 = 11.0

17.54 $AgCl(s) \rightleftharpoons Ag^+(aq) + Cl^-(aq)$; $K_{sp} = [Ag^+][Cl^-] = 1.8 \times 10^{-10}$

$$[Ag^+] = \frac{0.10\ M \times 0.2\ mL}{50\ mL} = 4 \times 10^{-4}\ M; \quad [Cl^-] = \frac{1.8 \times 10^{-10}}{4 \times 10^{-4}\ M} = 4.5 \times 10^{-7} = 5 \times 10^{-7}\ M$$

$$\frac{4.5 \times 10^{-7}\ mol\ Cl^-}{1\ L} \times \frac{35.45\ g\ Cl^-}{1\ mol\ Cl^-} \times 0.050\ L = 8 \times 10^{-7}\ g\ Cl^-$$

17.55 Calculate $[I^-]$ needed to initiate precipitation of each ion. The cation that requires lower $[I^-]$ will precipitate first.

Ag^+: $K_{sp} = [Ag^+][I^-]$; $8.3 \times 10^{-17} = (2.0 \times 10^{-4})[I^-]$; $[I^-] = \dfrac{8.3 \times 10^{17}}{2.0 \times 10^{-4}} = 4.2 \times 10^{-13}\ M\ I^-$

Pb^{2+}: $K_{sp} = [Pb^{2+}][I^-]^2$; $1.4 \times 10^{-8} = (1.5 \times 10^{-3})[I^-]^2$; $[I^-] = \left(\dfrac{1.4 \times 10^{-8}}{1.5 \times 10^{-3}}\right)^{1/2}$

$$= 3.1 \times 10^{-3}\ M\ I^-$$

Ag I will precipitate first, at $[I^-] = 4.2 \times 10^{-13}\ M$.

17.56 (a) $BaSO_4$: $K_{sp} = [Ba^{2+}][SO_4^{2-}] = 1.1 \times 10^{-10}$ Precipitation will begin when $Q = K_{sp}$.

$1.1 \times 10^{-10} = (0.010)[SO_4^{2-}]$; $[SO_4^{2-}] = 1.1 \times 10^{-8}$

$SrSO_4$: $K_{sp} = [Sr^{2+}][SO_4^{2-}] = 2.8 \times 10^{-7}$

$2.8 \times 10^{-7} = (0.010)[SO_4^{2-}]$; $[SO_4^{2-}] = 2.8 \times 10^{-5}$

The $[SO_4^{2-}]$ necessary to begin precipitation is the smaller of the two values, $1.1 \times 10^{-8}\ M\ SO_4^{2-}$.

(b) Ba^{2+} precipitates first, because it requires the smaller $[SO_4^{2-}]$.

(c) Sr^{2+} will begin to precipitate when $[SO_4^{2-}]$ reaches $2.8 \times 10^{-5}\ M$.

17.57   The first two experiments eliminate Group 1 and 2 ions (Figure 17.22). The fact that no insoluble carbonates form in the filtrate from the third experiment rules out Group 4 ions. The ions which might be in the sample are those of Group 3, that is, Al^{3+}, Fe^{2+}, Zn^{2+}, Cr^{3+}, Ni^{2+}, Co^{2+}, or Mn^{2+}, and those of Group 5, NH_4^+, Na^+ or K^+.

17.58   Initial solubility in water rules out CdS and HgO. Formation of a precipitate on addition of HCl indicates the presence of $Pb(NO_3)_2$ (formation of $PbCl_2$). Formation of a precipitate on addition of H_2S at pH 1 probably indicates $Cd(NO_3)_2$ (formation of CdS). (This test can be misleading because enough Pb^{2+} can remain in solution after filtering $PbCl_2$ to lead to visible precipitation of PbS.) Absence of a precipitate on addition of H_2S at pH 8 indicates that $ZnSO_4$ is not present. The yellow flame test indicates presence of Na^+. In summary, $Pb(NO_3)_2$ and Na_2SO_4 are definitely present, $Cd(NO_3)_2$ is probably present, and CdS, HgO and $ZnSO_4$ are definitely absent.

17.59   (a)   Make the solution acidic using $0.5\ M$ HCl; saturate with H_2S. CdS will precipitate, ZnS will not.

(b)   Add excess base; $Fe(OH)_3(s)$ precipitates, but Cr^{3+} forms the soluble complex $Cr(OH)_4^-$

(c)   Add $(NH_4)_2HPO_4$; Mg^{2+} precipitates as $MgNH_4PO_4$, K^+ remains in solution.

(d)   Add $6\ M$ HCl, precipitate Ag^+ as AgCl(s).

17.60   (a)   Make the solution slightly basic and saturate with H_2S; CdS will precipitate, Na^+ remains in solution.

(b)   Make the solution acidic, saturate with H_2S; CuS will precipitate, Mg^{2+} remains in solution.

(c)   Add HCl, $PbCl_2$ precipitates. (It is best to carry out the reaction in an ice-water bath to reduce the solubility of $PbCl_2$.)

(d)   Add dilute HCl; AgCl precipitates, Hg^{2+} remains in solution.

17.61   (a)   Because phosphoric acid is a weak acid, the concentration of free PO_4^{3-}(aq) in an aqueous phosphate solution is low except in strongly basic media. In less basic media the solubility product of the phosphates that one wishes to precipitate is not exceeded.

(b) K_{sp} for those cations in Group 3 is much larger. Thus, to exceed K_{sp} a higher $[S^{2-}]$ is required. This is achieved by making the solution more basic.

(c)   They should all redissolve in strongly acidic solution. e.g., in $12\ M$ HCl (all the chlorides of Group 3 metals are soluble).

17.62 The addition of $(NH_4)_2HPO_4$ could result in precipitation of salts from metal ions of the other groups. The $(NH_4)_2HPO_4$ will render the solution basic, so metal hydroxides could form as well as insoluble phosphates. It is essential to separate the metal ions of a group from other metal ions before carrying out the specific tests for that group.

Additional Exercises

17.63 The equilibrium of interest is

$$HC_5H_3O_3(aq) \rightleftharpoons H^+(aq) + C_5H_3O_3^-(aq); \quad K_a = 6.76 \times 10^{-4} = \frac{[H^+][C_5H_3O_3^-]}{[HC_5H_3O_3]}$$

Begin by calculating $[HC_5H_3O_3]$ and $[C_5H_3O_3^-]$ for each case.

(a)
$$\frac{35.0 \text{ g } HC_5H_3O_3}{0.250 \text{ L soln}} \times \frac{1 \text{ mol } HC_5H_3O_3}{112.1 \text{ g } HC_5H_3O_3} = 1.249 = 1.25 \text{ M } HC_5H_3O_3$$

$$\frac{30.0 \text{ g } NaC_5H_3O_3}{0.250 \text{ L soln}} \times \frac{1 \text{ mol } NaC_5H_3O_3}{134.1 \text{ g } NaC_5H_3O_3} = 0.8949 = 0.895 \text{ M } C_5H_3O_3^-$$

$$[H^+] = \frac{K_a[HC_5H_3O_3]}{[C_5H_3O_3^-]} = \frac{6.76 \times 10^{-4}(1.249 - x)}{(0.8949 + x)} \approx \frac{6.76 \times 10^{-4}(1.249)}{(0.8949)}$$

$$[H^+] = 9.43 \times 10^{-4} \text{ M}, \quad pH = 3.025$$

(b) For dilution, $M_1V_1 = M_2V_2$

$$[HC_5H_3O_3] = \frac{0.250 \text{ M} \times 30.0 \text{ mL}}{125 \text{ mL}} = 0.0600 \text{ M}$$

$$[C_5H_3O_3^-] = \frac{0.220 \text{ M} \times 20.0 \text{ mL}}{125 \text{ mL}} = 0.0352 \text{ M}$$

$$[H^+] \approx \frac{6.76 \times 10^{-4}(0.0600)}{0.0352} = 1.15 \times 10^{-3} \text{ M}, \quad pH = 2.938$$

(yes, $[H^+]$ is < 5% of 0.0352 M)

(c) $0.0850 \text{ M} \times 0.500 \text{ L} = 0.0425 \text{ mol } HC_5H_3O_3$

$1.65 \text{ M} \times 0.0500 \text{ L} = 0.0825 \text{ mol NaOH}$

	$HC_5H_3O_3(aq)$ +	NaOH(aq)	→	$NaC_5H_3O_3(aq)$ +	$H_2O(l)$
initial	0.0425 mol	0.0825 mol			
reaction	-0.0425 mol	-0.0425mol		+0.0425 mol	
after	0 mol	0.0400 mol		0.0425 mol	

The strong base NaOH dominates the pH; the contribution of $C_5H_3O_3^-$ is negligible. This combination would be "after the equivalence point" of a titration. The total volume is 0.550 L.

$$[OH^-] = \frac{0.0400 \text{ mol}}{0.550 \text{ L}} = 0.0727 \text{ } M; \quad pOH = 1.138, \quad pH = 12.862$$

17.64 From Equation [17.9] we see that when the acid and base forms are present in equal concentrations the pH = pK_a. Thus $pK_a = 7.80$.

17.65 $K_a = \dfrac{[H^+][In^-]}{[HIn]}$; at pH = 4.68, [HIn] = [In^-]; [H^+] = K_a; pH = pK_a = 4.68

17.66 (a) $HA(aq) + B(aq) \rightleftharpoons HB^+(aq) + A^-(aq)$ $\quad K_c = \dfrac{[HB^+][A^-]}{[HA][B]}$

(b) Note that the solution is slightly basic because B is a stronger base than HA is an acid. (Or, equivalently, that A^- is a stronger base than HB^+ is an acid.) Thus, a little of the A^- is used up in reaction: $A^-(aq) + H_2O(l) \rightleftharpoons HA(aq) + OH^-(aq)$. Since pH is not very far from neutral, it is reasonable to assume that the reaction in part (a) has gone far to the right, and that $[A^-] \approx [HB^+]$ and $[HA] \approx [B]$. Then

$$K_a = \frac{[A^-][H^+]}{[HA]} = 5.0 \times 10^{-6}; \text{ when pH = 8.8, } [H^+] = 1.58 \times 10^{-9} = 1.6 \times 10^{-9} \text{ } M$$

$$\frac{[A^-]}{[HA]} = 5.0 \times 10^{-6} / 1.58 \times 10^{-9} = 3.155 \times 10^3 = 3.2 \times 10^3$$

From the assumptions above, $\dfrac{[A^-]}{[HA]} = \dfrac{[HB^+]}{[B]}$, so $K_c \approx \dfrac{[A^-]^2}{[HA]^2} = 9.953 \times 10^6 = 1.0 \times 10^7$

(c) K_b for the reaction $B(aq) + H_2O(l) \rightleftharpoons BH^+(aq) + OH^-(aq)$ can be calculated by noting that the equilibrium constant for the reaction in part (a) can be written as $K_c = K_a (HA) \times K_b (B) / K_w$. (You should prove this to yourself.) Then,

$$K_b (B) = \frac{K_c \times K_w}{K_a (HA)} = \frac{(9.953 \times 10^6)(1.0 \times 10^{-14})}{5.0 \times 10^{-6}} = 2.0 \times 10^{-2}$$

K_b (B) is larger than K_a (HA), as it must be if the solution is basic.

17.67 (a) $K_a = \dfrac{[H^+][CHO_2^-]}{[HCHO_2]}$; $[H^+] = \dfrac{K_a[HCHO_2]}{[CHO_2^-]}$

Buffer A: $[HCHO_2] = [CHO_2^-] = \dfrac{1.00 \text{ mol}}{1.00 \text{ L}} = 1.00 \text{ } M$

$$[H^+] = \frac{1.8 \times 10^{-4} (1.00 \text{ } M)}{(1.00 \text{ } M)} = 1.8 \times 10^{-4} \text{ } M, \quad pH = 3.74$$

Buffer B: $[HCHO_2] = [CHO_2^-] = \dfrac{0.010 \text{ mol}}{1.00 \text{ L}} = 0.010 \text{ } M$

$$[H^+] = \frac{1.8 \times 10^{-4} (0.010 \text{ } M)}{(0.010 \text{ } M)} = 1.8 \times 10^{-4} \text{ } M, \quad pH = 3.74$$

The pH of a buffer is determined by the identity of the conjugate acid/conjugate base pair (that is, the relevant K_a value) and the **ratio of concentrations** of the conjugate acid and conjugate base. The absolute concentrations of the components is not relevant. The pH values of the two buffers are equal because they both contain $HCHO_2$ and $NaCHO_2$ and the $[HCHO_2] / [CHO_2^-]$ **ratio** is the same in both solutions.

(b)   Buffer capacity is determined by the absolute amount of conjugate acid and conjugate base available to absorb strong acid (H^+) or strong base (OH^-) that is added to the buffer. Buffer A has the greater capacity because it contains the greater absolute concentrations of $HCHO_2$ and CHO_2^-.

(c)   Buffer A: CHO_2^- +   HCl $\rightarrow$ $HCHO_2$ + Cl^-

                     1.00 mol   0.001 mol   1.00 mol

                     0.999 mol   0   1.001 mol

$$[H^+] = \frac{1.8 \times 10^{-4}\,(1.001)}{(0.999)} = 1.8 \times 10^{-4}\ M, \ pH = 3.74$$

(In a buffer calculation, volumes cancel and we can substitute moles directly into the K_a expression.)

     Buffer B: CHO_2^- +   HCl $\rightarrow$ $HCHO_2$ + Cl^-

                     0.010 mol   0.001 mol   0.010 mol

                     0.009 mol   0   0.011 mol

$$[H^+] = \frac{1.8 \times 10^{-4}\,(0.011)}{(0.009)} = 2.2 \times 10^{-4}\ M, \ pH = 3.66$$

(d)   Buffer A: $1.00\ M$ HCl $\times$ 0.010 L = 0.010 mol H^+ added

             mol $HCHO_2$ = 1.00 + 0.010 = 1.01 mol

             mol CHO_2^- = 1.00 - 0.010   = 0.99 mol

$$[H^+] = \frac{1.8 \times 10^{-4}\,(1.01)}{(0.99)} = 1.8 \times 10^{-4}\ M, \ pH = 3.74$$

     Buffer B: mol $HCHO_2$ = 0.010 + 0.010 = 0.020 mol = 0.020 M

                   mol CHO_2^- = 0.010 - 0.010 = 0.000 mol

The solution is no longer a buffer; the only source of CHO_2^- is the dissociation of $HCHO_2$.

$$K_a = \frac{[H^+][CHO_2^-]}{[HCHO_2]} = \frac{x^2}{(0.020-x)\ M}$$

The extent of ionization is greater than 5%; from the quadratic formula, x = $[H^+]$ = 1.8×10^{-3}, pH = 2.74.

(e)   Adding 10 mL of 1.00 M HCl to buffer B exceeded its capacity, while the pH of buffer A was unaffected. This is quantitative confirmation that buffer A has a significantly greater capacity than buffer B. In fact, 1.0 L of 1.0 M HCl would be required to exceed the capacity of buffer A. Buffer A, with 100 times more $HCHO_2$ and CHO_2^- has 100 times the capacity of buffer B.

17.68   The pH of a buffer is centered around pK_a for its conjugate acid. For the bases in Table D.2, pK_a for the conjugate acids = 14 - pK_b. 14 - pK_b = 10.6; pK_b = 3.4, K_b = $10^{-3.4}$ = 4 × 10^{-4}. Select two bases with K_b values near 4 × 10^{-4}.

Methylamine, dimethylamine and ethylamine have K_b values closest to 4 × 10^{-4}, and ammonia and trimethylamine would probably also work. We will select methylamine and dimethylamine. (We could also select very weak acids with pK_a = 10.6, K_a = $10^{-10.6}$ ≈ 2.5 × 10^{-11}. Either HIO or HCO_3^- would be appropriate.)

In general, $BH^+(aq) \rightleftharpoons B(aq) + H^+(aq)$

$$K_a = \frac{[B][H^+]}{[BH^+]}; \quad [H^+] = \frac{K_a[BH^+]}{[B]}; \quad [H^+] = 10^{-10.6} = 2.51 \times 10^{-11} \, M$$

For methylamine, $K_a = \dfrac{1.0 \times 10^{-14}}{4.4 \times 10^{-4}} = 2.272 \times 10^{-11} = 2.3 \times 10^{-11}$;

$$\frac{[BH^+]}{[B]} = \frac{[H^+]}{K_a} = \frac{2.51 \times 10^{-11}}{2.27 \times 10^{-11}} = 1.1$$

The ratio of $[CH_3NH_3^+]$ to $[CH_3NH_2]$ is 1.1 to 1.

For dimethylamine, $K_a = \dfrac{1.0 \times 10^{-14}}{5.4 \times 10^{-4}} = 1.852 \times 10^{-11} = 1.9 \times 10^{-11}$;

$$\frac{[BH^+]}{[B]} = \frac{[H^+]}{K_a} = \frac{2.51 \times 10^{-11}}{1.85 \times 10^{-11}} = 1.4$$

The ratio of $[(CH_3)_2NH_2^+]$ to $[(CH_3)_2NH]$ is 1.4 to 1. (The stronger base requires more of its conjugate acid to achieve a buffer of the same pH.)

17.69 $\dfrac{0.30 \text{ mol } HC_2H_3O_2}{1 \text{ L soln}} \times 0.750 \text{ L} = 0.225 = 0.23 \text{ mol } HC_2H_3O_2$

$0.225 \text{ mol } HC_2H_3O_2 \times \dfrac{60.05 \text{ g } HC_2H_3O_2}{1 \text{ mol } HC_2H_3O_2} \times \dfrac{1 \text{ g gl acetic acid}}{0.99 \text{ g } HC_2H_3O_2} \times \dfrac{1.00 \text{ mL gl acetic acid}}{1.05 \text{ g gl acetic acid}}$

$$= 13 \text{ mL glacial acetic acid}$$

At pH 4.50, $[H^+] = 10^{-4.50} = 3.16 \times 10^{-5} = 3.2 \times 10^{-5} \, M$; this is small compared to 0.30 M $HC_2H_3O_2$.

$$K_a = \frac{(3.16 \times 10^{-5})[C_2H_3O_2^-]}{0.30} = 1.8 \times 10^{-5}; \quad [C_2H_3O_2^-] = 0.171 = 0.17 \, M$$

$\dfrac{0.171 \text{ mol } NaC_2H_3O_2}{1 \text{ L soln}} \times 0.750 \text{ L} \times \dfrac{82.03 \text{ g } NaC_2H_3O_2}{1 \text{ mol } NaC_2H_3O_2} = 10.52 = 11 \text{ g } NaC_2H_3O_2$

17.70   **(a)**   For a monoprotic acid (one H^+ per mole of acid), at the equivalence point
moles OH^- added = moles H^+ originally present

$$M_B \times L_B = \text{g acid/molar mass}$$

$$MM = \frac{\text{g acid}}{M_B \times L_B} = \frac{0.1355 \text{ g}}{0.0950 \, M \times 0.0193 \text{ L}} = 73.90 = 73.9 \text{ g/mol}$$

    (b)   initial mol HA = $\dfrac{0.1355 \text{ g}}{73.9 \text{ g/mol}} = 1.834 \times 10^{-3} = 1.83 \times 10^{-3}$ mol HA

mol OH^- added to pH 5.10 = $0.0950 \, M \times 0.0120$ L = 1.14×10^{-3} mol OH^-

	HA(aq)	+	NaOH(aq)	$\rightarrow$	NaA(aq) + H_2O
before rx	1.834×10^{-3} mol		1.14×10^{-3} mol		0
change	-1.14×10^{-3} mol		-1.14×10^{-3} mol		$+1.14 \times 10^{-3}$ mol
after rx	0.694×10^{-3} mol		0		1.14×10^{-3} mol

$$[HA] = \frac{0.694 \times 10^{-3} \text{ mol}}{0.0370 \text{ L}} = 0.01874 = 0.0187 \, M$$

$$[A^-] = \frac{1.14 \times 10^{-3} \text{ mol}}{0.0370 \text{ L}} = 0.03081 = 0.0308 \, M; \quad [H^+] = 10^{-5.10} = 7.94 \times 10^{-6} \, M$$

The mixture after reaction (a buffer) can be described by the acid dissociation
equilibrium

	HA(aq)	$\rightleftharpoons$	H^+(aq)	+	A^-(aq)
initial	0.0187 M		0		0.0308 M
equil	$(0.0187 - 7.94 \times 10^{-6} \, M)$		$7.94 \times 10^{-6} \, M$		$(0.0308 + 7.94 \times 10^{-6}) \, M$

$$K_a = \frac{[H^+][A^-]}{[HA]} \approx \frac{(7.94 \times 10^{-6})(0.0308)}{(0.0187)} = 1.3 \times 10^{-5}$$

(Although we have carried 3 figures through the calculation to avoid rounding errors,
the data indicate an answer with 2 significant figures.)

17.71   **(a)** $\dfrac{0.4885 \text{ g KHP}}{0.100 \text{ L}} \times \dfrac{1 \text{ mol KHP}}{204.2 \text{ g KHP}} = 0.02392 = 0.0239 \, M \, P^{2-}$ at the equivalence point

The pH at the equivalence point is determined by the hydrolysis of P^{2-}.

$$P^{2-}(aq) + H_2O(l) \rightleftharpoons HP^-(aq) + OH^-(aq)$$

$$K_b = \frac{[HP^-][OH^-]}{[P^{2-}]} = \frac{K_w}{K_a \text{ for } HP^-} = \frac{1.0 \times 10^{-14}}{3.1 \times 10^{-6}} = 3.23 \times 10^{-9} = 3.2 \times 10^{-9}$$

$$3.23 \times 10^{-9} = \frac{x^2}{(0.02392 - x)} \approx \frac{x^2}{0.2392}; \quad X = [OH^-] = 8.8 \times 10^{-6} \, M$$

pH = 14 - 5.06 = 8.94. From Figure 16.4, either phenolphthalein (pH 8.2 - 10.0) or
thymol blue (pH 8.0 - 9.6) could be used to detect the equivalence point.

Phenolphthalein is usually the indicator of choice because the colorless to pink change is easier to see.

(b) $0.4885 \text{ g KHP} \times \dfrac{1 \text{ mol KHP}}{204.2 \text{ g KHP}} \times \dfrac{1 \text{ mol NaOH}}{1 \text{ mol KHP}} \times \dfrac{1}{0.03855 \text{ L NaOH}}$

$$= 0.06206 \; M \text{ NaOH}$$

17.72   (a)   Initially, the solution is 0.100 M in SO_3^{2-}.

$$SO_3^{2-}(aq) + H_2O(l) \rightleftharpoons HSO_3^-(aq) + OH^-(aq)$$

$$K_b = \frac{[HSO_3^-][OH^-]}{[SO_3^{2-}]} = \frac{K_w}{K_a[HSO_3^-]} = 1.56 \times 10^{-7} = 1.6 \times 10^{-7}$$

Proceeding in the usual way for a weak base, calculate $[OH^-] = 1.25 \times 10^{-4}$

$$= 1.3 \times 10^{-4} \; M, \text{ pH} = 10.10.$$

(b)   It will require 50.00 mL of 0.100 M HCl to reach the first equivalence point, at which point HSO_3^- is the predominant species.

(c)   An additional 50.00 mL are required to react with HSO_3^- to form H_2SO_3. At the second equivalence point there is a 0.0500 M solution of H_2SO_3. By the usual procedure for a weak acid:

$$H_2SO_3(aq) \rightleftharpoons H^+(aq) + HSO_3^-(aq)$$

$$K_a = \frac{[H^+][HSO_3^-]}{[H_2SO_3]} = 1.7 \times 10^{-2}; \quad \frac{(x)^2}{(0.050 - x)} = 1.7 \times 10^{-2}$$

Use the full quadratic equation form to solve for x = $[H^+] = 2.2 \times 10^{-2}$; pH = 1.66.

17.73   The reaction involved is HA(aq) + OH^-(aq) $\rightleftharpoons$ A^-(aq) + H_2O(l). We thus have 0.080 mol A^- and 0.12 mol HA in a total volume of 1.0 L, so the "initial" molarities of A^- and HA are 0.080 M and 0.12 M, respectively. The weak acid equilibrium of interest is

$$HA(aq) \rightleftharpoons H^+(aq) + A^-(aq)$$

(a) $K_a = \dfrac{[H^+][A^-]}{[HA]}$; $[H^+] = 10^{-4.80} = 1.58 \times 10^{-5} = 1.6 \times 10^{-5} \; M$

Assuming $[H^+]$ is small compared to [HA] and $[A^-]$,

$$K_a \approx \frac{(1.58 \times 10^{-5})(0.080)}{(0.12)} = 1.06 \times 10^{-5} = 1.1 \times 10^{-5}, \quad pK_a = 4.98$$

(b)   At pH = 5.00, $[H^+] = 1.0 \times 10^{-5} \; M$. Let b = extra moles NaOH.

[HA] = 0.12 - b, $[A^-]$ = 0.080 + b

$$1.06 \times 10^{-5} \approx \frac{(1.0 \times 10^{-5})(0.080 + b)}{(0.12 - b)}; \quad 2.06 \times 10^{-5} b = 4.72 \times 10^{-7};$$

b = 0.023 mol NaOH

17.74 Assume that H_3PO_4 will react with NaOH in a stepwise fashion. (This is not unreasonable, since the three K_a values for H_3PO_4 are significantly different.)

	$H_3PO_4(aq)$	+	$NaOH(aq)$	→	$H_2PO_4^-(aq)$	+	$Na^+(aq)$	+	$H_2O(l)$
before	0.20 mol		0.30 mol		0 mol				
after	0 mol		0.10 mol		0.20 mol				

	$H_2PO_4^-(aq)$	+	$NaOH(aq)$	→	$HPO_4^-(aq)$	+	$Na^+(aq)$	+	$H_2O(l)$
before	0.20 mol		0.10 mol		0.25 mol				
after	0.10 mol		0		0.35 mol				

Thus, after all NaOH has reacted, the resulting 1.00 L solution is a buffer containing 0.10 mol $H_2PO_4^-$ and 0.35 mol HPO_4^{2-}. $H_2PO_4^-(aq) \rightleftharpoons H^+(aq) + HPO_4^{2-}(aq)$

$$K_a = 6.2 \times 10^{-8} = \frac{[HPO_4^{2-}][H^+]}{[H_2PO_4^-]}; \quad [H^+] = \frac{6.2 \times 10^{-8}\,(0.10\,M)}{0.35\,M} = 1.77 \times 10^{-8} = 1.8 \times 10^{-8}\,M;$$

$$pH = 7.75$$

17.75 The simplest way to obtain the desired buffer mixture would be to titrate the H_3PO_4 solution with NaOH, using a pH meter, until the pH had risen to 7.20. From the values of K_a for phosphoric acid, we can guess that the equilibrium of importance will be the second acid dissociation.

$$H_2PO_4^-(aq) \rightleftharpoons HPO_4^{2-}(aq) + H^+(aq); \quad K_a = 6.2 \times 10^{-8}, \quad pK_a = 7.21$$

Using Equation [17.19]

$$7.20 = 7.21 + \log\frac{[HPO_4^{2-}]}{[H_2PO_4^-]}; \quad \frac{[HPO_4^{2-}]}{[H_2PO_4^-]} = 0.98$$

That is, at pH 7.20 we have added enough 1.0 M NaOH to the solution to have produced approximately equal concentrations of the two anions $H_2PO_4^-$ and HPO_4^{2-}. If we had begun with 100 mL of the 1.0 M H_3PO_4 solution we would have added about 150 mL of 1.0 M NaOH, so the total volume of solution would be 250 mL. The total moles phosphate present = 1.0 M × 0.100 L = 0.10 mol, so mol $H_2PO_4^-$ = mol HPO_4^{2-} = 0.050 mol. The concentration of each component is 0.050 mol/0.25 L = 0.20 M.

17.76 The pH of a buffer system is centered around pK_a for the conjugate acid component. For a diprotic acid, two conjugate acid/conjugate base pairs are possible.

$$H_2X(aq) \rightleftharpoons H^+(aq) + HX^-(aq); \quad K_{a1} = 2.0 \times 10^{-2}; \quad pK_{a1} = 1.70$$

$$HX^-(aq) \rightleftharpoons H^+(aq) + X^{2-}(aq); \quad K_{a2} = 5.0 \times 10^{-7}; \quad pK_{a2} = 6.30$$

Clearly HX^- / X^{2-} is the more appropriate combination for preparing a buffer with pH = 6.50. The $[H^+]$ in this buffer = $10^{-6.50}$ = 3.16×10^{-7} = 3.2×10^{-7} M. Using the K_{a2} expression to calculate the $[X^{2-}] / [HX]$ ratio:

$$K_{a2} = \frac{[H^+][X^{2-}]}{[HX^-]}; \quad \frac{K_{a2}}{[H^+]} = \frac{[X^{2-}]}{[HX]} = \frac{5.0 \times 10^{-7}}{3.16 \times 10^{-7}} = 1.58 = 1.6$$

Since X^{2-} and HX^- are present in the same solution, the ratio of concentrations is also a ratio of moles.

$$\frac{[X^{2-}]}{[HX]} = \left(\frac{mol\ X^{2-}/L\ soln}{mol\ HX^-/L\ soln}\right) = \frac{mol\ X^{2-}}{mol\ HX^-} = 1.58;\ mol\ X^{2-} = (1.58)\ mol\ HX^-$$

In the 1.0 L of 1.0 M H_2X, there is 1.0 mol of X^{2-} containing material.

Thus, mol HX^- + 1.58 (mol HX^-) = 1.0 mol. 2.58 (mol HX^-) = 1.0;
mol HX^- = 1.0 / 2.58 = 0.39 mol HX^- ; mol X^{2-} = 1.0 - 0.39 = 0.61 mol X^{2-} .

Thus enough 1.0 M NaOH must be added to produce 0.39 mol HX^- and 0.61 mol X^{2-}.

Considering the neutralization in a step-wise fashion (see discussion of titrations of polyprotic acids in Section 17.3),

	$H_2X(aq)$	+	$NaOH(aq)$	→	$HX^-(aq)$	+	$H_2O(l)$
before	1.0 mol		1 mol		0		
after	0		0		1.0 mol		

	$HX^-(aq)$	+	$NaOH(aq)$	→	$X^{2-}(aq)$	+	$H_2O(l)$
before	1.0		0.61				
change	-0.61		-0.61		+0.61		
after	0.39		0		0.61		

Starting with 1.0 mol of H_2X, 1.0 mol of NaOH is added to completely convert it to 1.0 mol of HX^-. Of that 1.0 mol of HX^-, 0.61 mol must be converted to 0.61 mol X^{2-}. The total moles of NaOH added is (1.00 + 0.61) = 1.61 mol NaOH.

$$L\ NaOH = \frac{mol\ NaOH}{M\ NaOH} = \frac{1.61\ mol}{1.0\ M} = 1.6\ L\ of\ 1.0\ M\ NaOH$$

17.77 $C_3H_5O_3^-$ will be formed by reaction of $HC_3H_5O_3$ with NaOH.
0.1000 M × 0.05000 L = 5.000 × 10^{-3} mol $HC_3H_5O_3$; b = mol NaOH needed

	$HC_3H_5O_3$	+	$NaOH$	→	$C_3H_5O_3^-$	+	H_2O	+	Na^+
initial	5.000 × 10^{-3}		b mol						
rx	-b mol		-b mol		+b mol				
after rx	5.000 × 10^{-3} - b mol		0		b mol				

$$K_a = \frac{[H^+][C_3H_5O_3^-]}{[HC_3H_5O_3]};\ K_a = 1.4 \times 10^{-4};\ [H^+] = 10^{-pH} = 10^{-3.50} = 3.16 \times 10^{-4} = 3.2 \times 10^{-4}\ M$$

Since solution volume is the same for $HC_3H_5O_3$ and $C_3H_5O_3^-$, we can use moles in the equation for $[H^+]$.

$$K_a = 1.4 \times 10^{-4} = \frac{3.16 \times 10^{-4}\ (b)}{(5.000 \times 10^{-3} - b)}$$

0.4427 $(5.000 \times 10^{-3} - b) = b$, $2.214 \times 10^{-3} = 1.4427 b$, $b = 1.53 \times 10^{-3}$

$$= 1.5 \times 10^{-3} \text{ mol OH}^-$$

(The precision of K_a dictates that the result has 2 sig figs.)

Substituting this result into the K_a expression gives $[H^+] = 3.27 \times 10^{-4}$.

(Using 1.53×10^{-3} mol OH$^-$ (3 sig figs) gives, $[H^+] = 3.16 \times 10^{-4}$, a more reassuring cross check.)

Calculate volume NaOH required from M = mol/L.

1.53×10^{-3} mol OH$^-$ $\times \dfrac{1 \text{ L}}{1.000 \text{ mol}} \times \dfrac{1 \text{ μL}}{1 \times 10^{-6} \text{ L}} = 1.5 \times 10^3$ μL (1.5 mL)

17.78 (a)   CdS: 8.0×10^{-28}; CuS: 6×10^{-37} CdS has greater molar solubility.

(b)   PbCO$_3$: 7.4×10^{-14}; BaCrO$_4$: 1.2×10^{-10} BaCrO$_4$ has greater molar solubility.

(c)   Hg(OH)$_2$: 3.0×10^{-26} Fe(OH)$_3$: 4×10^{-38}

Since the stoichiometry of the two complexes is not the same, K_{sp} values can't be compared directly; molar solubilities must be calculated from K_{sp} values. For both compounds, [OH$^-$] from the autoionization of water is large compared to [OH$^-$] from dissociation of the salt.

Hg(OH)$_2$: $K_{sp} = 3.0 \times 10^{-26} = [Hg^{2+}][OH^-]^2$, $[Hg^{2+}] = x$, $[OH^-] = 1.0 \times 10^{-7}$
       $3.0 \times 10^{-26} = (x)(1.0 \times 10^{-7})^2$, $x = 3.0 \times 10^{-12}$ M

Fe(OH)$_3$: $K_{sp} = 4 \times 10^{-38} = [Fe^{3+}][OH^-]^3$, $[Fe^{3+}] = x$, $[OH^-] = 1.0 \times 10^{-7}$
       $4 \times 10^{-38} = (x)(1.0 \times 10^{-7})^3$, $x = 4 \times 10^{-17}$ M

Hg(OH)$_2$ has greater molar solubility than Fe(OH)$_3$.

(d)   Again, molar solubilities must be calculated for comparison.
Ba$_3$(PO$_4$)$_2$: $K_{sp} = 3.4 \times 10^{-23} = [Ba^{2+}]^3[PO_4^{3-}]^2$
       x = molar solubility, $[Ba^{2+}] = 3x$, $[PO_4^{3-}] = 2x$
       $3.4 \times 10^{-23} = (3x)^3(2x)^2 = 108x^5$, $x = 1.3 \times 10^{-5}$ M

NiC$_2$O$_4$: $K_{sp} = 4 \times 10^{-10} = [Ni^{2+}][C_2O_4^{2-}]$, $x = [Ni^{2+}] = [C_2O_4^{2-}]$
       $4 \times 10^{-10} = x^2$, $x = 2 \times 10^{-5}$ M

NiC$_2$O$_4$ has slightly greater molar solubility than Ba$_3$(PO$_4$)$_2$, but the solubilities are much closer than one might expect from a cursory inspection of the K_{sp} values.

17.79 Ce(OH)$_3$(s) $\rightleftharpoons$ Ce^{3+}(aq) + 3OH$^-$(aq); $K_{sp} = 2.0 \times 10^{-20} = [Ce^{3+}][OH^-]^3$

$x = [Ce^{3+}]$, $3x = [OH^-]$; $2.0 \times 10^{-20} = (x)(3x)^3 = 27x^4$; $x = 5.22 \times 10^{-6} = 5.2 \times 10^{-6}$ M

$[OH^-] = 3x = 1.565 \times 10^{-5} = 1.6 \times 10^{-5}$ M, pH = 9.19

([OH$^-$] from Ce(OH)$_3$ is large compared to [OH$^-$] from the autoionization of H$_2$O.)

17.80   (a) $K_{sp} = [Fe^{3+}][OH^-]^3 = 4 \times 10^{-38}$, $[Fe^{3+}] = x$, pH =4.0, $[OH^-] = 1 \times 10^{-10}$

            $(x)(1 \times 10^{-10})^3 = 4 \times 10^{-38}$, $x = 4 \times 10^{-8}$ M

$$\frac{4 \times 10^{-8} \text{ mol Fe(OH)}_3}{1 \text{ L}} \times \frac{107 \text{ g Fe(OH)}_3}{1 \text{ mol Fe(OH)}_3} = 4 \times 10^{-6} \text{ g Fe(OH)}_3 / \text{L}$$

    (b) $[Fe^{3+}] = x$, pH =10.0, $[OH^-] = 1 \times 10^{-4}$

            $(x)(1 \times 10^{-4})^3 = 4 \times 10^{-38}$; $x = 4 \times 10^{-26}$ M

$$\frac{4 \times 10^{-26} \text{ mol Fe(OH)}_3}{1 \text{ L}} \times \frac{107 \text{ g Fe(OH)}_3}{1 \text{ mol Fe(OH)}_3} = 4 \times 10^{-24} \text{ g Fe(OH)}_3 / \text{L}$$

          $Fe(OH)_3$ is much more soluble at the lower pH value.

17.81 $PbSO_4(s) \rightleftharpoons Pb^{2+}(aq) + SO_4^{2-}(aq)$; $K_{sp} = 1.6 \times 10^{-8} = [Pb^{2+}][SO_4^{2-}]$

       $SrSO_4(s) \rightleftharpoons Sr^{2+}(aq) + SO_4^{2-}(aq)$; $K_{sp} = 2.8 \times 10^{-7} = [Sr^{2+}][SO_4^{2-}]$

       Let $x = [Pb^{2+}]$, $y = [Sr^{2+}]$, $x + y = [SO_4^{2-}]$

$$\frac{x(x+y)}{y(x+y)} = \frac{1.6 \times 10^{-8}}{2.8 \times 10^{-7}}; \frac{x}{y} = 0.0571 = 0.057; x = 0.57 y$$

       $y(0.057 y+y) = 2.8 \times 10^{-7}$; $1.057 y^2 = 2.8 \times 10^{-7}$; $y = 5.146 \times 10^{-4} = 5.1 \times 10^{-4}$

       $x = 0.057 y$; $x = 0.057(5.1 \times 10^{-4}) = 2.9 \times 10^{-5}$

       $[Pb^{2+}] = 2.9 \times 10^{-5}$ M, $[Sr^{2+}] = 5.1 \times 10^{-4} M$, $[SO_4^{2-}] = 5.4 \times 10^{-4} M$

17.82   To determine precipitation conditions, we must know K_{sp} for $CaF_2(s)$ and calculate Q under the specified conditions. $K_{sp} = 3.9 \times 10^{-11} = [Ca^{2+}][F^-]^2$

       $[Ca^{2+}]$ and $[F^-]$: The term 1 ppb means 1 part per billion or 1 g solute per billion g solution. Assuming that the density of this very dilute solution is the density of water:

$$1 \text{ ppb} = \frac{1 \text{ g solute}}{1 \times 10^9 \text{ g solution}} \times \frac{1 \text{ g solution}}{1 \text{ mL solution}} \times \frac{1 \times 10^3 \text{ mL}}{1 \text{ L}} = \frac{1 \times 10^{-6} \text{ g solute}}{1 \text{ L solution}}$$

$$\frac{1 \times 10^{-6} \text{ g solute}}{1 \text{ L solution}} \times \frac{1 \text{ μg}}{1 \times 10^{-6} \text{ g}} = 1 \text{ μg} / 1 \text{ L}$$

$$8 \text{ ppb Ca}^{2+} \times \frac{1 \text{ μg}}{1 \text{ L}} = \frac{8 \text{ μg Ca}^{2+}}{1 \text{ L}} = \frac{8 \times 10^{-6} \text{ g Ca}^{2+}}{1 \text{ L}} \times \frac{1 \text{ mol Ca}^{2+}}{40 \text{ g}} = 2 \times 10^{-7} \text{ M Ca}^{2+}$$

$$1 \text{ ppb F}^- \times \frac{1 \text{ μg}}{1 \text{ L}} = \frac{1 \text{ μg F}^-}{1 \text{ L}} = \frac{1 \times 10^{-6} \text{ g F}^-}{1 \text{ L}} \times \frac{1 \text{ mol F}^-}{19.0 \text{ g}} = 5 \times 10^{-8} \text{ M F}^-$$

       $Q = [Ca^{2+}][F^-]^2 = (2 \times 10^{-7})(5 \times 10^{-8})^2 = 5 \times 10^{-22}$

       $5 \times 10^{-22} < 3.9 \times 10^{-22}$, $Q < K_{sp}$, no CaF_2 will precipitate

17.83 $MgC_2O_4(s) \rightleftharpoons Mg^{2+}(aq) + C_2O_4^{2-}(aq)$

$K_{sp} = [Mg^{2+}][C_2O_4^{2-}] = 8.6 \times 10^{-5}$

If $[Mg^{2+}]$ is to be 3.0×10^{-2} M, $[C_2O_4^{2-}] = 8.6 \times 10^{-5} / 3.0 \times 10^{-2} = 2.87 \times 10^{-3} = 2.9 \times 10^{-3}$ M

The oxalate ion undergoes hydrolysis:

$C_2O_4^{2-}(aq) + H_2O(l) \rightleftharpoons HC_2O_4^-(aq) + OH^-(aq)$

$K_b = \dfrac{[HC_2O_4^-][OH^-]}{[C_2O_4^{2-}]} = 1.0 \times 10^{-14} / 6.4 \times 10^{-5} = 1.56 \times 10^{-10} = 1.6 \times 10^{-10}$

$[Mg^{2+}] = 3.0 \times 10^{-2}$ M, $[C_2O_4^{2-}] = 2.87 \times 10^{-3} = 2.9 \times 10^{-3}$ M

$[HC_2O_4^-] = (3.0 \times 10^{-2} - 2.87 \times 10^{-3})$ M $= 2.71 \times 10^{-2} = 2.7 \times 10^{-2}$ M

$[OH^-] = 1.56 \times 10^{-10} \times \dfrac{[C_2O_4^{2-}]}{[HC_2O_4^-]} = 1.56 \times 10^{-10} \times \dfrac{(2.87 \times 10^{-3})}{(2.71 \times 10^{-2})} = 1.65 \times 10^{-11}$

$= 1.7 \times 10^{-11}$ M; pH $= 3.22$

17.84   The pertinent equilibria are

$CuCO_3(s) \rightleftharpoons Cu^{2+}(aq) + CO_3^{2-}(aq)$ $K_{sp} = 1.4 \times 10^{-10}$

$CO_3^{2-}(aq) + H_2O(l) \rightleftharpoons HCO_3^-(aq) + OH^-(aq)$ $K_b = 1.79 \times 10^{-4} = 1.8 \times 10^{-4}$

$CuCO_3(s) + H_2O(l) \rightleftharpoons Cu^{2+}(aq) + HCO_3^-(aq) + OH^-(aq)$ $K = K_{sp} \times K_b = 2.5 \times 10^{-14}$

At pH $= 7.5$, pOH $= 6.5$, $[OH^-] = 10^{-6.5} = 3.16 \times 10^{-7} = 3.2 \times 10^{-7}$ M.

What is the relationship among $[Cu^{2+}]$, $[CO_3^{2-}]$ and $[HCO_3^-]$? For each $Cu^{2+}(aq)$, there is one $CO_3^{2-}(aq)$, which then hydrolyzes.

$K_b = \dfrac{[HCO_3^-][OH^-]}{[CO_3^{2-}]}$; $\dfrac{[HCO_3^-]}{[CO_3^{2-}]} = \dfrac{K_b}{[OH^-]} = \dfrac{1.79 \times 10^{-4}}{3.16 \times 10^{-7}} = 5.6 \times 10^2$

Essentially all CO_3^{2-} becomes HCO_3^-. Assume $[Cu^{2+}] = [HCO_3^-] = x$

$K = 2.5 \times 10^{-14} = [Cu^{2+}][HCO_3^-][OH^-] = x^2 (3.16 \times 10^{-7})$; $x = 2.8 \times 10^{-4}$ M

Solubility $= [Cu^{2+}] = 2.8 \times 10^{-4}$ M.

17.85 $PbCl_2(s) \rightleftharpoons Pb^{2+}(aq) + 2Cl^-(aq)$ $K_{sp} = 1.6 \times 10^{-5}$

$K_{sp} = 1.6 \times 10^{-5} = [Pb^{2+}][Cl^-]^2 = [Pb^{2+}][0.1]^2$; $[Pb^{2+}] = 1.6 \times 10^{-3}$ M $= 2 \times 10^{-3}$ M

17.86   The student failed to account for the hydrolysis of the AsO_4^{3-} ion. If there were no hydrolysis, $[Mg^{2+}]$ would indeed be 1.5 times that of $[AsO_4^{3-}]$. However, as the reaction $AsO_4^{3-}(aq) + H_2O(l) \rightleftharpoons HAsO_4^{2-}(aq) + OH^-(aq)$ proceeds, the ion product $[Mg^{2+}]^3[AsO_4^{3-}]^2$ falls below the value for K_{sp}. More $Mg_3(AsO_4)_2$ dissolves, more hydrolysis occurs, and so on, until an equilibrium is reached. At this point $[Mg^{2+}]$ is much larger than $[AsO_4^{3-}]$. However, it is exactly 1.5 times the **total** concentration of all arsenic-containing species. That is,

$$[Mg^{2+}] = 1.5\ ([AsO_4^{3-}] + [HAsO_4^{2-}] + [H_2AsO_4^-] + [H_3AsO_4])$$

17.87 (a) $K_{sp} = [Cr^{3+}][OH^-]^3 = 6.3 \times 10^{-31}$, $[Cr^{3+}] = x$, $[OH^-] = 1.0 \times 10^{-7}$

$(x)(1.0 \times 10^{-7})^3 = 6.3 \times 10^{-31}$, $x = 6.3 \times 10^{-10}$ M

(b) In pure water, $[Cr^{3+}] = x$, $[OH^-] = 3x$

$x(3x)^3 = 6.3 \times 10^{-31}$, $27x^4 = 6.3 \times 10^{-31}$, $x = [Cr^{3+}] = 1.2 \times 10^{-8}$ M

$[OH^-] = 3.6 \times 10^{-8}$ M

But this is on the order of magnitude of the contribution from H_2O.

$[OH^-] = 1.0 \times 10^{-7} + 3x$; $(x)(1.0 \times 10^{-7} + 3x)^3 = 6.3 \times 10^{-31}$

Since this is a high order equation, try successive approximation.

Begin with $x = 1.0 \times 10^{-8}$

$(1.0 \times 10^{-8})(1.0 \times 10^{-7} + 0.3 \times 10^{-7})^3 = (1.0 \times 10^{-8})(1.3 \times 10^{-7})^3 = 2.2 \times 10^{-29}$

Try 1.0×10^{-9} $(1.0 \times 10^{-9})(1.03 \times 10^{-7})^3$ $= 1.1 \times 10^{-30}$

Try 6.0×10^{-10} $(6.0 \times 10^{-10})(1.018 \times 10^{-7})^3 = 6.3 \times 10^{-31}$

The $[OH^-]$ from the autoionization of water, 1.0×10^{-7}, decreases the solubility of $Cr(OH)_3$ enough so the $[OH^-]$ from the dissociation of the solid is small compared to 1.0×10^{-7}. The solubility in pure water is essentially the same as the solubility in a solution buffered to pH 7.

17.88 $Zn(OH)_2(s) \rightleftharpoons Zn^{2+}(aq) + 2OH^-(aq)$ $K_{sp} = 1.2 \times 10^{-17}$

$Zn^{2+}(aq) + 4OH^-(aq) \rightleftharpoons Zn(OH)_4^{2-}(aq)$ $K_f = 4.6 \times 10^{17}$

$Zn(OH)_2(s) + 2OH^-(aq) \rightleftharpoons Zn(OH)_4^{2-}(aq)$ $K = K_{sp} \times K_f = 5.5$

$$K = 5.52 = 5.5 = \frac{[Zn(OH)_4^{2-}]}{[OH^-]^2}$$

If 0.010 mol $Zn(OH)_2$ dissolves, 0.010 mol $Zn(OH)_4^{2-}$ should be present at equilibrium.

$$[OH^-]^2 = \frac{(0.010)}{5.52}; \quad [OH^-] = 0.043\ M \quad [OH^-] \geq 0.043\ M \text{ or pH} \geq 12.63$$

17.89 The two competing equilibria are

$Au^+(aq) + 2CN^-(aq) \rightleftharpoons Au(CN)_2^-(aq)$ $K_f = 2.0 \times 10^{38}$

$AuCl(s) \rightleftharpoons Au^+(aq) + Cl^-(aq)$ $K_{sp} = 2.0 \times 10^{-13}$

$AuCl(s) + 2CN^-(aq) \rightleftharpoons Au(CN)_2^-(aq) + Cl^-(aq)$ $K = K_f \times K_{sp} = 4.0 \times 10^{25}$

K is large and the reaction essentially goes to completion.

$$0.080\ M \times 1.0\ L = 0.080\ mol\ CN^- \times \frac{1\ mol\ AuCl}{2\ mol\ CN^-} = 0.040\ mol\ AuCl(s)$$

Integrative Exercises

17.90 (a)   Complete ionic:

$H^+(aq) + Cl^-(aq) + Na^+(aq) + NO_2^-(aq) \rightarrow HNO_2(aq) + Na^+(aq) + Cl^-(aq)$

Na^+ and Cl^- are spectator ions.

Net ionic: $H^+(aq) + NO_2^-(aq) \rightleftharpoons HNO_2(aq)$

(b)   The net ionic equation in part (a) is the reverse of the dissociation of HNO_2.

$$K = \frac{1}{K_a} = \frac{1}{4.5 \times 10^{-4}} = 2.22 \times 10^3 = 2.2 \times 10^3$$

(c)   For Na^+ and Cl^-, this is just a dilution problem.

$M_1V_1 = M_2V_2$; V_2 is 50.0 mL + 50.0 mL = 100.0 mL

Cl^-: $\dfrac{0.15\,M \times 50.0\,\text{mL}}{100.0\,\text{mL}} = 0.075\,M$; Na^+: $\dfrac{0.15\,M \times 50.0\,\text{mL}}{100.0\,\text{mL}} = 0.075\,M$

H^+ and NO_2^- react to form HNO_2. Since K >> 1, the reaction essentially goes to completion.

$0.15\,M \times 0.0500\,\text{mL} = 7.5 \times 10^{-3}$ mol H^+

$0.15\,M \times 0.0500\,\text{mL} = 7.5 \times 10^{-3}$ mol NO_2^-
___

$= 7.5 \times 10^{-3}$ mol HNO_2

Solve the weak acid problem to determine $[H^+]$, $[NO_2^-]$ and $[HNO_2]$ at equilibrium.

$K_a = \dfrac{[H^+][NO_2^-]}{[HNO_2]}$; $[H^+] = [NO_2^-] = x\,M$; $[HNO_2] = \dfrac{(7.5 \times 10^{-3} - x)\,\text{mol}}{0.100\,\text{L}} = (0.075 - x)\,M$

$4.5 \times 10^{-4} = \dfrac{x^2}{(0.075-x)} \approx \dfrac{x^2}{0.075}$; $x = 5.8 \times 10^{-3}\,M$

$\dfrac{[H^+]}{[HNO_2]} \times 100 = \dfrac{5.8 \times 10^{-3}}{0.075} \times 100 = 7.8\%$ dissociation

Using the quadratic formula to determine x: $x^2 + 4.5 \times 10^{-4}x - 3.38 \times 10^{-5} = 0$

$$x = \frac{-4.5 \times 10^{-4} + \sqrt{(4.5 \times 10^{-4})^2 - 4(1)(-3.38 \times 10^{-5})}}{2}$$

$x = 5.58 \times 10^{-3} = 5.6 \times 10^{-3}\,M\,H^+$ and NO^{-2} (Note that $[H^+]$ is not very different from the value obtained using the assumption.) $[HNO_2] = 0.075 - 0.0056 = 0.069\,M$

In summary:

$[Na^+] = [Cl^-] = 0.075\,M$, $[HNO_2] = 0.069\,M$, $[H^+] = [NO_2^-] = 0.0056\,M$

17.91   (a)   For a monoprotic acid (one H^+ per mole of acid), at the equivalence point

moles OH^- added = moles H^+ originally present

$$M_B \times L_B = \text{g acid/molar mass}$$

$$MM = \frac{\text{g acid}}{M_B \times L_B} = \frac{0.1044\ g}{0.0500\ M \times 0.0220\ L} = 94.48 = 94.5\ g/mol$$

(b)   11.05 mL is exactly half-way to the equivalence point (22.10 mL). When half of the unknown acid is neutralized, $[HA] = [A^-]$, $[H^+] = K_a$ and $pH = pK_a$.

$$K_a = 10^{-4.89} = 1.3 \times 10^{-5}$$

(c)   From Appendix D, Table D.1, acids with K_a values close to 1.3×10^{-5} are

name	K_a	formula	molar mass
propionic	1.3×10^{-5}	$HC_3H_5O_2$	74.1
butanoic	1.5×10^{-5}	$HC_4H_7O_2$	88.1
acetic	1.8×10^{-5}	$HC_2H_3O_2$	60.1
hydrazoic	1.9×10^{-5}	HN_3	43.0

Of these, butanoic has the closest match for K_a and molar mass, but the match is not good.

17.92 $n = \dfrac{PV}{RT} = 735\ torr \times \dfrac{1\ atm}{760\ torr} \times \dfrac{7.5\ L}{295\ K} \times \dfrac{K \cdot mol}{0.08206\ L \cdot atm} = 0.300 = 0.30\ mol\ NH_3$

$0.40\ M \times 0.50\ L = 0.20\ mol\ HCl$

	$HCl(aq)$ +	$NH_3(g)$ →	$NH_4^+(aq)$ +	$Cl^-(aq)$
before	0.20 mol	0.30 mol		
after	0	0.10 mol	0.20 mol	0.20 mol

The solution will be a buffer because of the substantial concentrations of NH_3 and NH_4^+ present. Use K_a for NH_4^+ to describe the equilibrium.

	$NH_4^+(aq)$ ⇌	$NH_3(aq)$ +	$H^+(aq)$
equil.	0.20 - x	0.10 + x	x

$$K_a = \frac{1.0 \times 10^{-14}}{1.8 \times 10^{-5}} = 5.56 \times 10^{-10} = 5.6 \times 10^{-10}\ ; \quad K_a = \frac{[NH_3][H^+]}{[NH_4^+]}; \quad [H^+] = \frac{K_a[NH_4^+]}{[NH_3]}$$

Since this expression contains a ratio of concentrations, volume will cancel and we can substitute moles directly. Assume x is small compared to 0.10 and 0.20.

$$[H^+] = \frac{5.56 \times 10^{-10}\ (0.20)}{(0.10)} = 1.111 \times 10^{-9} = 1.1 \times 10^{-9}\ M,\ pH = 8.95$$

17.93   Calculate the initial M of aspirin in the stomach and solve the equilibrium problem to find equilibrium concentrations of $C_8H_7O_2COOH$ and $C_8H_7O_2COO^-$. At pH = 2, $[H^+] = 1 \times 10^{-2}$.

$$\frac{325 \text{ mg}}{\text{tablet}} \times 2 \text{ tablets} \times \frac{1 \text{ g}}{1000 \text{ mg}} \times \frac{1 \text{ mol } C_8H_7O_2COOH}{180.2 \text{ g } C_8H_7O_2COOH} \times \frac{1}{1 \text{ L}} = 3.61 \times 10^{-3} = 4 \times 10^{-3} \; M$$

$$C_8H_7O_2COOH(aq) \rightleftharpoons C_8H_7O_2COO^-(aq) + H^+(aq)$$

initial	$3.61 \times 10^{-3} \; M$	0	$1 \times 10^{-2} \; M$
equil	$(3.61 \times 10^{-3} - x) \; M$	$x \; M$	$(1 \times 10^{-2} + x) \; M$

$$K_a = 3 \times 10^{-5} = \frac{[H^+][C_8H_7O_2COO^-]}{[C_8H_7O_2COOH]} = \frac{(0.01 + x)(x)}{(3.61 \times 10^{-3} - x)} \approx \frac{0.01x}{3.61 \times 10^{-3}}$$

$$x = [C_8H_7O_2COO^-] = 1.08 \times 10^{-5} = 1 \times 10^{-5} \; M$$

$$\% \text{ ionization} = \frac{1.08 \times 10^{-5} \; M \; C_8H_7O_2COO^-}{3.61 \times 10^{-3} \; M \; C_8H_7O_2COOH} \times 100 = 0.3\%$$

(% ionization is small, so the assumption was valid.)

% aspirin molecules = 100.0% - 0.3% = 99.7% molecules

17.94 $\pi = MRT$, $M = \dfrac{\pi}{RT} = \dfrac{21 \text{ torr}}{298 \text{ K}} \times \dfrac{1 \text{ atm}}{760 \text{ torr}} \times \dfrac{K \cdot mol}{0.08206 \; L \cdot atm} = 1.13 \times 10^{-3} = 1.1 \; M$

$$SrSO_4(s) \rightleftharpoons Sr^{2+}(aq) + SO_4^{2-}(aq); \quad K_{sp} = [Sr^{2+}][SO_4^{2-}]$$

The total particle concentration is $1.13 \times 10^{-3} \; M$. Each mole of $SrSO_4$ that dissolves produces 2 mol of ions, so $[Sr^{2+}] = [SO_4^{2-}] = 1.13 \times 10^{-3} \; M / 2 = 5.65 \times 10^{-4} = 5.7 \times 10^{-4} \; M$.

$$K_{sp} = (5.65 \times 10^{-4})^2 = 3.2 \times 10^{-7}$$

17.95   For very dilute aqueous solutions, assume the solution density is 1 g/mL.

$$ppb = \frac{g \text{ solute}}{10^9 \text{ g solution}} = \frac{1 \times 10^{-6} \text{ g solute}}{1 \times 10^3 \text{ g solution}} = \frac{\mu g \text{ solute}}{L \text{ solution}}$$

(a) $K_{sp} = [Ag^+][Cl^-] = 1.8 \times 10^{-10}$; $[Ag^+] = (1.8 \times 10^{-10})^{1/2} = 1.34 \times 10^{-5} = 1.3 \times 10^{-5} \; M$

$$\frac{1.34 \times 10^{-5} \text{ mol Ag}^+}{L} \times \frac{107.9 \text{ g Ag}^+}{1 \text{ mol Ag}^+} \times \frac{1 \mu g}{1 \times 10^{-6} g} = \frac{1.4 \times 10^3 \, \mu g \text{ Ag}^+}{L}$$
$$= 1.4 \times 10^3 \text{ ppb} = 1.4 \text{ ppm}$$

(b) $K_{sp} = [Ag^+][Br^-] = 5.0 \times 10^{-13}$; $[Ag^+] = (5.0 \times 10^{-13})^{1/2} = 7.07 \times 10^{-7} = 7.1 \times 10^{-7} \; M$

$$\frac{7.07 \times 10^{-7} \text{ mol Ag}^+}{L} \times \frac{107.9 \text{ g Ag}^+}{1 \text{ mol Ag}^+} \times \frac{1 \mu g}{1 \times 10^{-6} g} = 76 \text{ ppb}$$

(c) $K_{sp} = [Ag^+][I^-] = 8.3 \times 10^{-17}$; $[Ag^+] = (8.3 \times 10^{-17})^{1/2} = 9.11 \times 10^{-9} = 9.1 \times 10^{-9} \; M$

$$\frac{9.11 \times 10^{-9} \text{ mol Ag}^+}{L} \times \frac{107.9 \text{ g Ag}^+}{1 \text{ mol Ag}^+} \times \frac{1 \mu g}{1 \times 10^{-6} g} = 0.98 \text{ ppb}$$

AgBr(s) would maintain $[Ag^+]$ in the correct range.

18 Chemistry of the Environment

Earth's Atmosphere

18.1 (a) The temperature profile of the atmosphere (Figure 18.1) is the basis of its division into regions. The center of each peak or trough in the temperature profile corresponds to a new region.

(b) Troposphere, 0-12 km; stratosphere, 12-50 km; mesosphere, 50-85 km; thermosphere, 85-110 km.

18.2 (a) The troposphere is the 12 km layer of the atmosphere immediately above the earth's surface. The tropopause is the boundary between the troposphere and the stratosphere. In the troposphere, temperature decreases with increasing altitude while in the stratosphere, it increases with increasing altitude. The tropopause is the dividing point, the first minimum in the temperature profile of the atmosphere.

(b) Atmospheric pressure in the troposphere ranges from 1.0 atm to 0.4 atm, while pressure in the stratosphere ranges from 0.4 atm to 0.001 atm. Gas density (g/L) is directly proportional to pressure. The much lower density of the stratosphere means it has the smaller mass, despite having a larger volume than the troposphere.

18.3 $P_{Ar} = \chi_{Ar} \cdot P_{atm}$; $P_{Ar} = 0.00934 \ (765 \ torr) = 7.15 \ torr$

$P_{Ne} = \chi_{Ne} \cdot P_{atm}$; $P_{Ne} = 0.00001818 \ (765 \ torr) = 0.0139 \ torr$

18.4 $0.37 \ ppm \ O_3 = \dfrac{0.37 \ mol \ O_3}{1 \times 10^6 \ mol \ air} = 3.7 \times 10^{-7} = \chi_{O_3}$

$P_{O_3} = \chi_{O_3} \cdot P_{atm} = 3.7 \times 10^{-7} \ (650 \ torr) = 2.4 \times 10^{-4} \ torr$

18.5 $P_{Xe} = \chi_{Xe} \cdot P_{atm}$; $P_{Xe} = 8.7 \times 10^{-8} \ (0.94 \ atm) = 8.178 \times 10^{-8} = 8.2 \times 10^{-8} \ atm$

$n_{Xe} \dfrac{P_{Xe}V}{RT} = \dfrac{8.178 \times 10^{-8} \ atm \times 1.0 \ L}{300 \ K} \times \dfrac{K \cdot mol}{0.08206 \ L \cdot atm} = 3.322 \times 10^{-9} = 3.3 \times 10^{-9} \ mol \ Xe$

$3.322 \times 10^{-9} \ mol \ Xe \times \dfrac{6.022 \times 10^{23} \ atoms}{1 \ mol} = 2.0 \times 10^{15} \ Xe \ atoms$

18.6   6.0 ppm CO = $\dfrac{6.0 \text{ mol CO}}{1 \times 10^6 \text{ mol air}}$ = 6.0×10^{-6} = χ_{CO}

$P_{CO} = \chi_{CO} \cdot P_{atm} = 6.0 \times 10^{-6} \times 735 \text{ torr} \times \dfrac{1 \text{ atm}}{760 \text{ torr}} = 5.80 \times 10^{-6} = 5.8 \times 10^{-6}$ atm

$n = \dfrac{PV}{RT} = \dfrac{5.80 \times 10^{-6} \text{ atm} \times 1.0 \text{ L}}{293 \text{ K}} \times \dfrac{K \cdot mol}{0.08206 \text{ L} \cdot atm} = 2.412 \times 10^{-7} = 2.4 \times 10^{-7}$ mol CO

2.412×10^{-7} mol CO $\times \dfrac{6.022 \times 10^{23} \text{ molecules}}{mol} = 1.453 \times 10^{17} = 1.5 \times 10^{17}$ molecules CO

The Upper Atmosphere; Ozone

18.7 $\dfrac{210 \times 10^3 \text{ J}}{1 \text{ mol}} \times \dfrac{1 \text{ mol}}{6.022 \times 10^{23} \text{ molecules}} = 3.487 \times 10^{-19} = 3.49 \times 10^{-19}$ J/molecule

$\lambda = c/\nu$ We also have that $E = h\nu$, so $\nu = E/h$. Thus,

$\lambda = \dfrac{hc}{E} = \dfrac{(6.626 \times 10^{-34} \text{ J} \cdot sec)(3.00 \times 10^8 \text{ m/sec})}{3.487 \times 10^{-19} \text{ J}} = 5.70 \times 10^{-7}$ m = 570 nm

18.8 $\dfrac{339 \times 10^3 \text{ J}}{1 \text{ mol}} \times \dfrac{1 \text{ mol}}{6.022 \times 10^{23} \text{ molecules}} = 5.6294 \times 10^{-19} = 5.63 \times 10^{-19}$ J/molecule

$\lambda = \dfrac{hc}{E} = \dfrac{(6.626 \times 10^{-34} \text{ J} \cdot sec)(3.00 \times 10^8 \text{ m/sec})}{5.6294 \times 10^{-19} \text{ J}} = 353$ nm

$\dfrac{293 \times 10^3 \text{ J}}{1 \text{ mol}} \times \dfrac{1 \text{ mol}}{6.022 \times 10^{23} \text{ molecules}} = 4.8655 \times 10^{-19} = 4.87 \times 10^{-19}$ J/molecule

$\lambda = \dfrac{(6.626 \times 10^{-34} \text{ J} \cdot sec)(3.00 \times 10^8 \text{ m/sec})}{4.8655 \times 10^{-19} \text{ J}} = 409$ nm

Photons of wavelengths longer than 409 nm cannot cause rupture of the C-Cl bond in either CF_3Cl or CCl_4. Photons with wavelengths between 409 and 353 nm can cause C-Cl bond rupture in CCl_4, but not in CF_3Cl.

18.9   The bond dissociation energy of N_2, 947 kJ/mol, is much higher than that of O_2, 495 kJ/mol. Photons with a wavelength short enough to photodissociate N_2 are not as abundant as the ultraviolet photons which lead to photodissociation of O_2. Also, N_2 does not absorb these photons as readily as O_2 so even if a short-wavelength photon is available, it may not be absorbed by an N_2 molecule.

18.10   Photoionization of O_2 requires 1205 kJ/mol. Photodissociation requires only 495 kJ/mol . At lower elevations, solar radiation with wavelengths corresponding to 1205 kJ/mol or shorter has already been absorbed, while the longer wavelength radiation has passed through relatively well. Below 90 km, the increased concentration of O_2 and the availability of longer wavelength radiation cause the photodissociation process to dominate.

18.11 (a) Oxygen atoms exist longer at 120 km because there are fewer particles (atoms and molecules) at this altitude and thus fewer collisions and subsequent reactions that consume O atoms.

(b) Ozone is the primary absorber of high energy ultraviolet radiation in the 200-310 nm range. If this radiation were not absorbed in the stratosphere, plants and animals at the earth's surface would be seriously and adversely affected.

18.12 (a) The highest rate of ozone, O_3, formation occurs at about 50 km, near the stratopause. The formation of ozone is an exothermic process as M* carries excess energy away from the O_3 molecule. The heat energy from the formation of O_3 causes the temperature to be higher near the stratopause than the lower altitude tropopause.

(b) The first step in the formation of O_3 is the photodissociation of O_2 to form two O atoms. Then, an O atom and an O_2 molecule collide to form O_3^*, a species with excess energy. If no other collisions occur, O_3^* spontaneously decomposes. If a carrier molecule such as N_2 or O_2 collides with O_3^* and removes the excess energy, O_3 is formed. It is the energy carried by M* that contributes to the temperature maximum at 50 km altitude.

18.13 *CFC* stands for chlorofluorocarbon, a class of compounds that contain chlorine, fluorine and carbon. A common CFC is Freon-12, CF_2Cl_2.

18.14 A *hydrofluorocarbon* is a compound that contains hydrogen, fluorine and carbon; it contains hydrogen in place of chlorine. HFCs are potentially less harmful than CFCs because photodissociation does not produce Cl atoms, which catalyze the destruction of ozone, O_3.

18.15 (a) HCl(g), $ClONO_2$(g)

(b) Neither HCl nor $ClONO_2$ react directly with ozone. The chlorine that is present in the "chlorine reservoir" does not participate in the destruction of ozone. Thus, the larger the "chlorine reservoir," the slower the rate of ozone depletion.

18.16 In the presence of polar stratospheric clouds during the Antarctic winter, HCl and $ClONO_2$ react to form Cl_2 (Equation 18.13):

$$HCl(g) + ClONO_2(g) \rightarrow Cl_2(g) + HNO_3(g)$$

Cl_2 remains in the stratosphere until the season changes and the appropriate wavelength of radiation is available to photodissociate Cl_2:

$$Cl_2(g) \xrightarrow{h\nu} 2Cl(g)$$

Free chlorine atoms then react with O_3 to form O_2 and ClO, which goes on to regenerate Cl atoms in a cycle that may destroy thousands of O_3 molecules before the Cl atoms become part of the chlorine reservoir again. This theory agrees with the seasonal variations in ozone concentration.

Chemistry of the Troposphere

18.17   (a)   CO binds with hemoglobin in the blood to block O_2 transport to the cells; people with CO poisoning suffocate from lack of O_2.

       (b) SO_2 is corrosive to lung tissue and contributes to higher levels of respiratory disease and shorter life expectancy, especially for people with other respiratory problems such as asthma. It also is a major source of acid rain, which damages forests and wildlife in natural waters.

       (c) O_3 is extremely reactive and toxic because of its ability to form free radicals upon reaction with organic molecules in the body. It is particularly dangerous for asthma suffers, exercisers and the elderly. O_3 can also react with organic compounds in polluted air to form peroxyacylnitrates, which cause eye irritation and breathing difficulties.

18.18 $2CO + O_2 \rightarrow 2CO_2$ $2SO_2 + O_2 \rightarrow 2SO_3$

         $CO_2 + H_2O \rightarrow H_2CO_3$ $SO_3 + H_2O \rightarrow H_2SO_4$

Oxidation of CO to CO_2 and SO_2 to SO_3 produces gases which readily dissolve in atmospheric moisture to form acid rain. CO_2 and SO_3 are actually more dangerous to the environment because of their potential to form acid rain.

18.19   CO in unpolluted air is typically 0.05 ppm, whereas in urban air CO is about 10 ppm. A major source is automobile exhaust. SO_2 is less than 0.01 ppm in unpolluted air and on the order of 0.08 ppm in urban air. A major source is coal and oil-burning power plants, but there is also some SO_2 in auto exhausts. NO is about 0.01 ppm in unpolluted air and about 0.05 ppm in urban air. It comes mainly from auto exhausts.

18.20   (a)   Methane, CH_4, arises from decomposition of organic matter by certain microorganisms; it also escapes from underground gas deposits.

       (b) SO_2 is released in volcanic gases, and also is produced by bacterial action on decomposing vegetable and animal matter.

       (c)   Nitric oxide, NO, results from oxidation of decomposing organic matter, and is formed in lightning flashes.

       (d)   CO is a possible product of some vegetable matter decay.

18.21   All oxides of nonmetals produce acid solutions when dissolved in water. Sulfur oxides are produced naturally during volcanic eruptions and carbon oxides are products of combustion and metabolism. These dissolved gases cause rainwater to be naturally acidic.

18.22   (a) $H_2SO_4(aq) + Fe(s) \rightarrow FeSO_4(aq) + H_2(g)$
               (The actual product may be red rust, Fe_2O_3, formed by the further oxidation of Fe^{2+} by $O_2(g)$.)

(b) $H_2SO_4(aq) + CaCO_3(s) \rightarrow CaSO_4(s) + H_2O(l) + CO_2(g)$

18.23 Among the components of coal are sulfur-containing organic compounds. Combustion (oxidation) of these molecules produces $SO_2(g)$. Formation of $SO_3(g)$ requires further oxidation of SO_2 according to the reaction

$$2SO_2(g) + O_2(g) \rightleftharpoons 2 SO_3(g)$$

Oxidation of $SO_2(g)$ to $SO_3(g)$ is significant but not complete, perhaps because C consumes most of the available $O_2(g)$.

18.24 $SO_2(g) + H_2O(l) \rightarrow H_2SO_3(aq)$
$SO_3(g) + H_2O(l) \rightarrow H_2SO_4(aq)$
$3NO_2(g) + H_2O(l) \rightarrow 2HNO_3(aq) + NO(g)$
(This is an oxidation-reduction as well as Lewis acid-base reaction.)

18.25 (a)   Visible (Figure 6.4)

(b) $E_{photon} = hc/\lambda = \dfrac{6.626 \times 10^{-34}\,J\bullet s \times 3.00 \times 10^8\,m/s}{420 \times 10^{-9}\,m} = 4.733 \times 10^{-19}$

$= 4.73 \times 10^{-19}$ J/photon

$\dfrac{4.733 \times 10^{-19}\,J}{1\,photon} \times \dfrac{6.022 \times 10^{23}\,photons}{1\,mol} \times \dfrac{1\,kJ}{1000\,J} = 285$ kJ/mol

18.26 (a)   Ultraviolet (Figure 6.4)

(b) $E_{photon} = hc/\lambda = \dfrac{6.626 \times 10^{-34}\,J\bullet s \times 3.00 \times 10^8\,m/s}{335 \times 10^{-9}\,m} = 5.934 \times 10^{-19}$

$= 5.93 \times 10^{-19}$ J/photon

$\dfrac{5.934 \times 10^{-19}\,J}{1\,photon} \times \dfrac{6.022 \times 10^{23}\,photons}{1\,mol} \times \dfrac{1\,kJ}{1000\,J} = 357$ kJ/mol

(c)   The average C-H bond energy from Table 8.4 is 413 kJ/mol. The energy calculated in part (b), 357 kJ/mol, is the energy required to break 1 mol of C-H bonds in formaldehyde, CH_2O. The C-H bond energy in CH_2O must be less than the "average" C-H bond energy.

18.27 Most of the energy entering the atmosphere from the sun is in the form of visible radiation, while most of the energy leaving the earth is in the form of infrared radiation. CO_2 is transparent to the incoming visible radiation, but absorbs the outgoing infrared radiation.

18.28 A *greenhouse gas* functions like the glass in a greenhouse. In the atmosphere, greenhouse gases absorb infrared radiation given off by the earth and send it back to the earth's surface, where it is detected as heat. The main greenhouse gases are $H_2O(g)$ and $CO_2(g)$ but $CH_4(g)$ and other trace gases act similarly.

The World Ocean

18.29 A salinity of 5 denotes that there are 5 g of dry salt per kg of water.

$$\frac{5 \text{ g NaCl}}{1 \text{ kg soln}} \times \frac{1 \text{ kg soln}}{1 \text{ L soln}} \times \frac{1 \text{ mol NaCl}}{58 \text{ g NaCl}} \times \frac{1 \text{ mol Na}^+}{1 \text{ mol NaCl}} = 0.09 \ M \text{ Na}^+$$

18.30 If the phosphorous is present as phosphate, there is a 1:1 ratio between the molarity of phosphorus and molarity of phosphate. Thus, we can calculate the molarity based on the given mass of P.

$$\frac{0.07 \text{ g P}}{1 \times 10^6 \text{ g H}_2\text{O}} \times \frac{1 \text{ mol P}}{31 \text{ g P}} \times \frac{1 \text{ mol PO}_4^{3-}}{1 \text{ mol P}} \times \frac{1 \times 10^3 \text{ g H}_2\text{O}}{1 \text{ L H}_2\text{O}} = 2.26 \times 10^{-6}$$

$$= 2.3 \times 10^{-6} \ M \text{ PO}_4^{3-}$$

18.31 $$1 \times 10^{11} \text{ g Br} \times \frac{1 \times 10^3 \text{ g H}_2\text{O}}{0.067 \text{ g Br}} \times \frac{1 \text{ L H}_2\text{O}}{1 \times 10^3 \text{ g H}_2\text{O}} = 1.5 \times 10^{12} \text{ L H}_2\text{O}$$

Because the process is only 10% efficient, ten times this much, or 1.5×10^{13} L H_2O, must be processed.

18.32 $$4.0 \times 10^7 \text{ g Mg(OH)}_2 \times \frac{1 \text{ mol Mg(OH)}_2}{58.3 \text{ g Mg(OH)}_2} \times \frac{1 \text{ mol CaO}}{1 \text{ mol Mg(OH)}_2} \times \frac{56.1 \text{ g CaO}}{1 \text{ mol CaO}} = 3.8 \times 10^7 \text{ g CaO}$$

Fresh Water

18.33 (a) Decomposition of organic matter by aerobic bacteria depletes dissolved O_2. A low dissolved oxygen concentration indicates the presence of organic pollutants.

(b) According to Section 13.4, the solubility of $O_2(g)$ (or any gas) in water decreases with increasing temperature.

18.34 (a) Aerobic conditions lead to oxidized products: CO_2, HCO_3^-, H_2O, SO_4^{2-}, NO_3^-, HPO_4^{2-}, $H_2PO_4^-$.

(b) Anaerobic conditions lead to less oxidized products: $CH_4(g)$, $H_2S(g)$, $NH_3(g)$, $PH_3(g)$

18.35 $$1.0 \text{ g C}_{18}\text{H}_{29}\text{O}_3\text{S}^- \times \frac{1 \text{ mol C}_{18}\text{H}_{29}\text{O}_3\text{S}^-}{325 \text{ g C}_{18}\text{H}_{29}\text{O}_3\text{S}^-} \times \frac{51 \text{ mol O}_2}{2 \text{ mol C}_{18}\text{H}_{29}\text{O}_3\text{S}^-} \times \frac{32.0 \text{ g O}_2}{1 \text{ mol O}_2} = 2.5 \text{ g O}_2$$

Notice that the mass of O_2 required is 2.5 times greater than the mass of biodegradable material.

18.36 $$50,000 \text{ persons} \times \frac{59 \text{ g O}_2}{1 \text{ person}} \times \frac{1 \times 10^6 \text{ g H}_2\text{O}}{9 \text{ g O}_2} \times \frac{1 \text{ L H}_2\text{O}}{1 \times 10^3 \text{ g H}_2\text{O}} = 3.3 \times 10^8 \text{ L H}_2\text{O}$$

18.37 $$\text{Ca}^{2+}(aq) + 2\text{HCO}_3^-(aq) \rightarrow \text{CaCO}_3(s) + \text{CO}_2(g) + \text{H}_2\text{O}(l)$$

18.38 $Mg^{2+}(aq) + Ca(OH)_2(s) \rightarrow Mg(OH)_2(s) + Ca^{2+}(aq)$

The excess $Ca^{2+}(aq)$ is removed as $CaCO_3$ (See solution 18.37) by naturally occurring bicarbonate or added Na_2CO_3.

18.39 $Ca(OH)_2$ is added to remove Ca^{2+} as $CaCO_3(s)$, and Na_2CO_3 removes the remaining Ca^{2+}. $Ca^{2+}(aq) + 2HCO_3^-(aq) + [Ca^{2+}(aq) + 2OH^-(aq)] \rightarrow 2CaCO_3(s) + 2H_2O(l)$. One mole $Ca(OH)_2$ is needed for each 2 moles of $HCO_3^-(aq)$ present. If there are 7.0×10^{-4} mol $HCO_3^-(aq)$ per liter, we must add 3.5×10^{-4} mol $Ca(OH)_2$ per liter, or a total of 0.35 mol $Ca(OH)_2$ for 10^3 L. This reaction removes 3.5×10^{-4} mol of the original Ca^{2+} from each liter of solution, leaving 1.5×10^{-4} M $Ca^{2+}(aq)$. To remove this $Ca^{2+}(aq)$, we add 1.5×10^{-4} mol Na_2CO_3 per liter, or a total of 0.15 mol Na_2CO_3, forming $CaCO_3(s)$.

18.40 $Ca(OH)_2$ is added to remove Ca^{2+} as $CaCO_3(s)$, and Na_2CO_3 removes the remaining Ca^{2+}. $Ca^{2+}(aq) + 2HCO_3^-(aq) + [Ca^{2+}(aq) + 2OH^-(aq)] \rightarrow 2CaCO_3(s) + 2H_2O(l)$. One mole $Ca(OH)_2$ is needed for each 2 moles of $HCO_3^-(aq)$ present.

$$5.0 \times 10^7 \text{ L } H_2O \times \frac{1.7 \times 10^{-3} \text{ mol } HCO_3^-}{1 \text{ L } H_2O} \times \frac{1 \text{ mol } Ca(OH)_2}{2 \text{ mol } HCO_3^-} \times \frac{74 \text{ g } Ca(OH)_2}{1 \text{ mol } Ca(OH)_2}$$
$$= 3.1 \times 10^6 \text{ g } Ca(OH)_2$$

This operation reduces the Ca^{2+} concentration from 5.7×10^{-3} M to $(5.7 \times 10^{-3} - 1.7 \times 10^{-3})$ $M = 4.0 \times 10^{-3}$ M. Next we must add sufficient Na_2CO_3 to further reduce $[Ca^{2+}]$ to 1.1×10^{-3} M. We thus need to reduce $[Ca^{2+}]$ by $(4.0 \times 10^{-3} - 1.1 \times 10^{-3})$ $M = 2.9 \times 10^{-3}$ M. $Ca^{2+}(aq) + CO_3^{-2}(aq) \rightarrow CaCO_3(s)$.

$$5.0 \times 10^7 \text{ L } H_2O \times \frac{2.9 \times 10^{-3} \text{ mol } Ca^{2+}}{1 \text{ L } H_2O} \times \frac{1 \text{ mol } Na_2CO_3}{1 \text{ mol } Ca^{2+}} \times \frac{106 \text{ g } Na_2CO_3}{1 \text{ mol } Na_2CO_3}$$
$$= 1.5 \times 10^7 \text{ g } Na_2CO_3$$

Additional Exercises

18.41   (a)   *Acid rain* is rain with a larger $[H^+]$ and thus a lower pH than expected. The additional H^+ is produced by the dissolution of sulfur and nitrogen oxides such as $SO_3(g)$ and $NO_2(g)$ in rain droplets to form sulfuric and nitric acid, $H_2SO_4(aq)$ and $HNO_3(aq)$.

       (b)   The *greenhouse effect* is warming of the atmosphere caused by heat trapping gases such as $CO_2(g)$ and $H_2O(g)$. That is, these gases absorb infrared or "heat" radiation emitted from the earth's surface and serve to maintain a relatively constant temperature on the surface. A significant increase in the amount of atmospheric CO_2 (from burning fossil fuels and other sources) could cause a corresponding increase in the average surface temperature and drastically change the global climate.

       (c)   *Photochemical smog* is an unpleasant collection of atmospheric pollutants initiated by photochemical dissociation of NO_2 to form NO and O atoms. The major components are $NO(g)$, $NO_2(g)$, $CO(g)$ and unburned hydrocarbons, all produced by automobile engines, and $O_3(g)$, ozone.

(d)   The *ozone hole* is a region of depleted O_3 in the stratosphere over Antarctica. It is caused by reactions between O_3 and Cl atoms originating from chlorofluorocarbons (CFC's), CF_xCl_{4-x}. Depletion of the ozone layer would allow damaging ultraviolet radiation disruptive to the plant and animal life in our ecosystem to reach earth.

18.42   Avg. mol wt. at the surface = 40.0(0.20) + 16.0(0.35) + 32.0(0.45) = 28.0 = 28 g/mol.

Next, calculate the percentage composition at 200 km. The fractions can be "normalized" by saying that the 0.45 fraction of O_2 is converted into <u>two</u> 0.45 fractions of O atoms, then dividing by the total fractions, 0.20 + 0.35 + 0.45 + 0.45 = 1.45:

$$\text{Avg. mol. wt.} = \frac{40.0(0.20) + 16.0(0.35) + 16.0(0.90)}{1.45} = 19 \text{ g/mol}$$

18.43   The ionization energies of metal atoms are much lower than those of any of the atomic or molecular ions present at 120 km; that is, O_2^+, O^+, NO^+. Thus, any of these ions react with metal atoms as illustrated for NO^+: $M + NO^+ \rightarrow M^+ + NO$.

18.44   (a)   The production of Cl atoms in the stratosphere is the result of the photodissociation of a C-Cl bond in the chlorofluorocarbon molecule.

$$CF_2Cl_2(g) \xrightarrow{h\nu} CF_2Cl(g) + Cl(g)$$

According to Table 8.4, the bond dissociation energy of a C-Br bond is 276 kJ/mol, while the value for a C-Cl bond is 328 kJ/mol. Photodissociation of $CBrF_3$ to form Br atoms requires less energy than the production of Cl atoms and should occur readily in the stratosphere.

(b) $CBrF_3(g) \xrightarrow{h\nu} CF_3(g) + Br(g)$ Also, under certain conditions

                                                $BrO(g) + BrO(g) \rightarrow Br_2O_2(g)$

         $Br(g) + O_3(g) \rightarrow BrO(g) + O_2(g)$ $Br_2O_2(g) + h\nu \rightarrow O_2(g) + 2Br(g)$

18.45   (a) HNO_3 is a major component in acid rain.

(b)   While it removes CO, the reaction produces NO_2. The photodissociation of NO_2 to form O atoms is the first step in the formation of tropospheric ozone and photochemical smog.

(c)   Again, NO_2 is the initiator of photochemical smog. Also, methoxyl radical, OCH_3, is a reactive species capable of initiating other undesirable reactions.

18.46   (a)   Anthracite coal would be more expensive than bituminous coal because the supply is lower and the demand is higher. Also, since anthracite coal requires less treatment of the combustion products, it might be cost effective even at a higher purchase price.

(b)   Burning bituminous coal releases $SO_2(g)$ into the atmosphere where it can be oxidized to $SO_3(g)$. In the troposphere, these gases can be carried from their point of origin to other parts of the country or the world. The gases dissolve in atmospheric water droplets to produce H_2SO_3 and H_2SO_4. When the water droplets collect, acid rain falls at a point far from the power plant that released the combustion products.

18.47 From section 18.4:

$$N_2(g) + O_2(g) \rightleftharpoons 2NO(g) \qquad \Delta H = +180.8 \text{ kJ} \quad (1)$$

$$2NO(g) + O_2(g) \rightleftharpoons 2NO_2(g) \qquad \Delta H = -113.1 \text{ kJ} \quad (2)$$

In an endothermic reaction, heat is a reactant. As the temperature of the reaction increases, the addition of heat favors formation of products and the value of K increases. The reverse is true for exothermic reactions; as temperature increases, the value of K decreases. Thus, K for reaction (1), which is endothermic, increases with increasing temperature and K for reaction (2), which is exothermic, decreases with increasing temperature.

18.48 Oxygen is present in the atmosphere to the extent of 209,000 parts per million. If CO binds 210 times more effectively than O_2, then the **effective** concentration of CO is 210×86 ppm $= 18,000$ ppm. The fraction of carboxyhemoglobin in the blood leaving the lungs is thus $\dfrac{18,000}{18,000 + 209,000} = 0.079$. Thus, 7.9 percent of the blood is in the form of carboxyhemoglobin, 92.1 percent as the O_2-bound oxyhemoglobin.

18.49 (a) $CH_4(g) + 2O_2(g) \rightarrow CO_2(g) + 2H_2O(g)$

(b) $2CH_4(g) + 3O_2(g) \rightarrow 2CO(g) + 4H_2O(g)$

(c) vol $CH_4 \rightarrow$ vol $O_2 \rightarrow$ volume air ($\chi_{O_2} = 0.20948$)

Equal volumes of gases at the same temperature and pressure contain equal numbers of moles (Avogadro's law). If 2 moles of O_2 are required for 1 mole of CH_4, 2.0 L of pure O_2 are needed to burn 1.0 L of CH_4.

$$\text{vol } O_2 = \chi_{O_2} \times \text{vol}_{\text{air}} = \frac{\text{vol } O_2}{\chi_{O_2}} = \frac{2.0 \text{ L}}{0.20948} = 9.5 \text{ L air}$$

(d) Since incomplete combustion requires less O_2 per mole of CH_4, running a home furnace with too little O_2 or too much CH_4 decreases the $O_2 : CH_4$ ratio and encourages the production of toxic CO(g) rather than $CO_2(g)$.

18.50 (a) According to Section 13.4, the solubility of gases in water decreases with increasing temperature. Thus, the solubility of $CO_2(g)$ in the ocean would decrease if the temperature of the ocean increased.

(b) If the solubility of $CO_2(g)$ in the ocean decreased because of global warming, more $CO_2(g)$ would be released into the atmosphere, perpetuating a cycle of increasing temperature and concomitant release of $CO_2(g)$ from the ocean.

18.51 (a) $NO(g) + h\nu \rightarrow N(g) + O(g)$

(b) $NO(g) + h\nu \rightarrow NO^+(g) + e^-$

(c) $NO(g) + O_3(g) \rightarrow NO_2(g) + O_2(g)$

(d) $3NO_2(g) + H_2O(l) \rightarrow 2HNO_3(aq) + NO(g)$

18.52   (a) CO_3^{2-} is a relatively strong Brønsted base and produces OH^- in aqueous solution according to the hydrolysis reaction:

$$CO_3^{2-}(aq) + H_2O(l) \rightleftharpoons HCO_3^-(aq) + OH^-(aq), \quad K_b = 1.8 \times 10^{-4}$$

If $[OH^-(aq)]$ is sufficient to exceed K_{sp} for $Mg(OH)_2$, the solid will precipitate.

(b) $\dfrac{150 \text{ mg Mg}^{2+}}{1 \text{ kg soln}} \times \dfrac{1 \text{ g Mg}^{2+}}{1000 \text{ mg Mg}^{2+}} \times \dfrac{1 \text{ kg soln}}{1 \text{ L soln}} \times \dfrac{1 \text{ mol Mg}^{2+}}{24.305 \text{ g Mg}^{2+}} = 6.172 \times 10^{-3}$

$$= 6.17 \times 10^{-3} \; M \text{ Mg}^{2+}$$

$\dfrac{5.0 \text{ g Na}_2CO_3}{1.0 \text{ L soln}} \times \dfrac{1 \text{ mol CO}_3^{2-}}{106.0 \text{ g Na}_2CO_3} = 0.0472 = 0.047 \; M \text{ CO}_3^{2-}$

$K_b = 1.8 \times 10^{-4} = \dfrac{[HCO_3^-][OH^-]}{[CO_3^{2-}]} \approx \dfrac{x^2}{0.0472}; \quad x = [OH^-] = 2.915 \times 10^{-3}$

$$= 2.9 \times 10^{-3} \; M$$

(This represents 6.2% hydrolysis, but the result will not be significantly different using the quadratic formula.)

$Q = [Mg^{2+}][OH^-]^2 = (6.172 \times 10^{-3})(2.915 \times 10^{-3})^2 = 5.2 \times 10^{-8}$

K_{sp} for $Mg(OH)_2 = 1.8 \times 10^{-11}$; $Q > K_{sp}$, so $Mg(OH)_2$ will precipitate.

18.53   (a)   In Figure 18.18, the aeration step speeds oxidation of any remaining dissolved organic matter by aerobic bacteria and ensures an adequate concentration of dissolved O_2 (to prevent build-up of anaerobic bacteria) in the processed water supplied to users.

(b) Cl_2 is used as an antibacterial agent in water treatment because it is more convenient than O_3. Cl_2 can be stored and transported as a liquid, whereas O_3 must be generated on-site as a gas.

(c)   Boiling the water (100°C) would kill any *cryptosporidium* bacteria present. This measure should have been possible for all water destined for human consumption or hygiene.

18.54   (a) $Cl_2(g) + H_2O(l) \rightarrow HOCl(aq) + H^+(aq) + Cl^-(aq)$

(b) $Ca^{2+}(aq) + 2HCO_3^-(aq) \xrightarrow{\Delta} CaCO_3(s) + CO_2(g) + H_2O(l)$

18.55   Because NO has an odd electron, like Cl(g), it could act as a catalyst for decomposition of ozone in the stratosphere. The increased destruction of ozone by NO would result in less absorption of short wavelength UV radiation now being screened out primarily by the ozone. Radiation of this wavelength is known to be harmful to humans; it causes skin cancer. There is evidence that many plants don't tolerate it very well either, though more research is needed to test this idea.

In Chapter 22 the oxidation of NO to NO_2 by oxygen is described. On dissolving in water, NO_2 disproportionates into NO_3^-(aq) and NO(g). Thus, over time the NO in the troposphere will be converted into NO_3^- which is in turn incorporated into soils.

Integrative Exercises

18.56   (a) 0.021 ppm $NO_2 = \dfrac{0.021 \text{ mol } NO_2}{1 \times 10^6 \text{ mol air}} = 2.1 \times 10^{-8} = \chi_{NO_2}$

                $P_{NO_2} = \chi_{NO_2} \bullet P_{atm} = 2.1 \times 10^{-8} (745 \text{ torr}) = 1.565 \times 10^{-5} = 1.6 \times 10^{-5}$ torr

    (b) $n = \dfrac{PV}{RT}$; molecules $= n \times \dfrac{6.022 \times 10^{23} \text{ molecules}}{\text{mol}} = \dfrac{PV}{RT} \times \dfrac{6.022 \times 10^{23} \text{ molecules}}{\text{mol}}$

           $V = 15 \text{ ft} \times 14 \text{ ft} \times 8 \text{ ft} \times \dfrac{12^3 \text{ in}^3}{\text{ft}^3} \times \dfrac{2.54^3 \text{ cm}^3}{\text{in}^3} \times \dfrac{1 \text{ L}}{1000 \text{ cm}^3} = 4.757 \times 10^4 = 5 \times 10^4$ L

       $1.565 \times 10^{-5} \text{ torr} \times \dfrac{1 \text{ atm}}{760 \text{ torr}} \times \dfrac{4.757 \times 10^4 \text{ L}}{293 \text{ K}} \times \dfrac{K \bullet mol}{0.08206 \text{ L} \bullet atm}$

                    $\times \dfrac{6.022 \times 10^{23} \text{ molecules}}{\text{mol}} = 2.453 \times 10^{19} = 2 \times 10^{19}$ molecules

18.57   mi/gal $\rightarrow$ gal C_8H_{18} $\rightarrow$ g C_8H_{18} $\rightarrow$ mol C_8H_{18} $\xrightarrow[\text{ratio}]{\text{mol}}$ mol CO_2 $\rightarrow$ g CO_2 $\rightarrow$ kg CO_2

    $1.2 \times 10^4 \text{ mi} \times \dfrac{1 \text{ gal}}{21 \text{ mi}} \times \dfrac{4 \text{ qt}}{1 \text{ gal}} \times \dfrac{1 \text{ L}}{1.057 \text{ qt}} \times \dfrac{1000 \text{ mL}}{1 \text{ L}} \times \dfrac{0.682 \text{ g}}{1 \text{ mL}} = 1.475 \times 10^6$

                                     $= 1.5 \times 10^6$ g C_8H_{18}

    $2C_8H_{18}(l) + 25O_2(g) \rightarrow 16CO_2(g) + 18H_2O(l)$

    1.475×10^6 g $C_8H_{18} \times \dfrac{1 \text{ mol } C_8H_{18}}{114.2 \text{ g } C_8H_{18}} \times \dfrac{16 \text{ mol } CO_2}{2 \text{ mol } C_8H_{18}} \times \dfrac{44.01 \text{ g } CO_2}{1 \text{ mol } CO_2} \times \dfrac{1 \text{ kg}}{1000 \text{ g}}$

                                       $= 4.5 \times 10^3$ kg CO_2

18.58   (a) $8{,}376{,}726$ tons coal $\times \dfrac{83 \text{ ton C}}{100 \text{ ton coal}} \times \dfrac{44.01 \text{ ton } CO_2}{12.01 \text{ ton C}} = 2.5 \times 10^7$ ton CO_2

           $8{,}376{,}726$ tons coal $\times \dfrac{2.5 \text{ ton S}}{100 \text{ ton coal}} \times \dfrac{64.07 \text{ ton } SO_2}{32.07 \text{ ton S}} = 4.18 \times 10^5$

                                     $= 4.2 \times 10^5$ ton SO_2

    (b) $CaO(s) + SO_2(g) \rightarrow CaSO_3(s)$

         4.18×10^5 ton $SO_2 \times \dfrac{55 \text{ ton } SO_2 \text{ removed}}{100 \text{ ton } SO_2 \text{ produced}} \times \dfrac{120.15 \text{ ton } CaSO_3}{64.07 \text{ ton } SO_2}$

                                    $= 4.3 \times 10^5$ ton $CaSO_3$

18.59   For reaction 1: $\Delta H = 102 - 101 - (142.3) = -141$ kJ/mol
        For reaction 2: $\Delta H = 101 - 102 - (247.5) = -249$ kJ/mol

Because both reactions are distinctly exothermic it is possible that the $ClO-ClO_2$ pair could be a catalyst for the decomposition of ozone.

18.60   (a)   Assume the density of water at 20°C is the same as at 25°C.

$$1.00 \text{ gal} \times \frac{4 \text{ qt}}{1 \text{ gal}} \times \frac{1 \text{ L}}{1.057 \text{ qt}} \times \frac{1000 \text{ mL}}{1 \text{ L}} \times \frac{0.99707 \text{ g } H_2O}{1 \text{ mL}} = 3773$$

$$= 3.77 \times 10^3 \text{ g } H_2O$$

The $H_2O(l)$ must be heated from 20°C to 100°C and then vaporized at 100°C.

$$3.773 \times 10^3 \text{ g } H_2O \times \frac{4.184 \text{ J}}{\text{g °C}} \times 80 \text{ °C} \times \frac{1 \text{ kJ}}{1000 \text{ J}} = 1263 = 1.3 \times 10^3 \text{ kJ}$$

$$3.773 \times 10^3 \text{ g } H_2O \times \frac{1 \text{ mol } H_2O}{18.02 \text{ g } H_2O} \times \frac{40.67 \text{ kJ}}{\text{mol } H_2O} = 8516 = 8.52 \times 10^3 \text{ kJ}$$

energy = 1263 kJ + 8516 kJ = 9779 = 9.8×10^3 kJ/gal H_2O

        (b)   According to Exercise 5.8, 1 kwh = 3.6×10^6 J.

$$\frac{9779 \text{ kJ}}{\text{gal } H_2O} \times \frac{1000 \text{ J}}{\text{kJ}} \times \frac{1 \text{ kwh}}{3.6 \times 10^6 \text{ J}} \times \frac{\$0.075}{\text{kwh}} = \$0.20/\text{gal}$$

        (c) $\dfrac{\$0.20}{\$1.05} \times 100 = 19\%$ of the total cost is energy

        (d)   This analysis underestimates the cost of energy for vaporization, because it assumes 100% efficiency, that there is perfect transfer of the 9.8×10^3 kJ from the surroundings (heating device, etc.) to the 1.00 gal of water. Energy transfer is never 100% efficient.

                Also, vaporization is not the only step in the production of distilled water. Energy is required to pump the water, produce containers, fill containers, transport the bottled water, etc.

18.61   Osmotic pressure is determined by total concentration of dissolved particles. From Table 18.6, $M_T = 1.13$. The minimum pressure needed to initiate reverse osmosis is $\pi = MRT$, Equation [13.13]. Assuming 298 K,

$$\pi = MRT = \frac{1.13 \text{ mol}}{1 \text{ L}} \times 298 \text{ K} \times \frac{0.08206 \text{ L} \cdot \text{atm}}{\text{K} \cdot \text{mol}} = 27.6 \text{ atm}$$

18.62   (a)   A rate constant of $M^{-1}s^{-1}$ is indicative of a reaction that is second order overall. For the reaction given, the rate law is probably rate = $k[O][O_3]$. (Although rate = $k[O]^2$ or $k[O_3]^2$ are possibilities, it is difficult to envision a mechanism consistent with either one that would result in two molecules of O_2 being produced.)

(b)   Yes. Most atmospheric processes are initiated by collision. One could imagine an activated complex of four O atoms collapsing to form two O_2 molecules. Also, the rate constant is large, which is less likely for a multistep process. The reaction is analogous to the destruction of O_3 by Cl atoms (Equation 18.7), which is also second order with a large rate constant.

(c)   According to the Arrhenius equation, $k = Ae^{-E_a/RT}$. Thus, the larger the value of k, the smaller the activation energy, E_a. The value of the rate constant for this reaction is large, so the activation energy is small.

(d) $\Delta H_f^\circ = 2\Delta H_f^\circ\, O_2(g) - \Delta H_f^\circ\, O(g) - \Delta H_f^\circ\, O_3(g)$

$\Delta H_f^\circ = 0 - 247.5\ kJ - 142.3\ kJ = -389.8\ kJ$

The reaction is exothermic, so energy is released; the reaction would raise the temperature of the stratosphere.

18.63   From the composition of air at sea-level (Table 18.1) calculate the partial pressures of $N_2(g)$ and $O_2(g)$ in the original sample.

$P_x = \chi_x \cdot P_T$; $P_{N_2} = 0.78084\ (1.0\ atm) = 0.78\ atm$; $P_{O_2} = 0.20948\ (1.0\ atm) = 0.21\ atm$

	$N_2(g)$	+	$O_2(g)$	⇌	2NO(g)
initial	0.78 atm		0.21 atm		0
charge	-x		-x		+2x
equil	(0.78-x) atm		(0.21-x) atm		2x atm

$$K_p = \frac{P_{NO}^2}{P_{N_2} \times P_{O_2}} = \frac{(2x)^2}{(0.78-x)(0.21-x)} = \frac{4x^2}{0.164 - 0.99x + x^2} = \frac{4x^2}{0.164 - 0.99x + x^2} = 0.05$$

$0.05\,(0.164 - 0.99x + x^2) = 4x^2$; $0 = 3.95x^2 + 0.05x - 0.0082$

Using the quadratic formula, $x = \dfrac{-b \pm \sqrt{b^2 - 4ac}}{2a} = \dfrac{-0.05 \pm \sqrt{(0.05)^2 - 4(3.95)(-0.0082)}}{2(3.95)}$

$x = \dfrac{-0.05 \pm \sqrt{0.0025 + 0.1296}}{7.90} = \dfrac{-0.05 \pm 0.363}{7.90}$

The negative result is meaningless; x = 0.04 atm; $P_{NO} = 2x = 0.08$ atm

Assuming that the total pressure of the gaseous mixture at equilibrium is still 1.0 atm,

$\chi_{NO} = P_{NO}/P_T = 0.08\ atm/1.0\ atm = 0.08$

ppm for gases = $\chi \times 10^6$ (see Section 18.1)

$ppm_{CO} = 0.08 \times 10^6 = 8 \times 10^4$ ppm

18.64   Initial pressures: $P_{N_2} = 0.78 (1.5 \text{ atm}) = 1.17 = 1.2 \text{ atm}$; $P_{O_2} = 0.21 (1.5 \text{ atm}) = 0.315$
$$= 0.32 \text{ atm}$$

$$P_{NO} = \frac{3200 \text{ atm}}{1 \times 10^6 \text{ atm}} = 1.5 \text{ atm} = 4.8 \times 10^{-3} \text{ atm}$$

Calculate Q to determine which direction the reaction will proceed.

$$Q = \frac{P_{NO}^2}{P_{N_2} \times P_{O_2}} = \frac{(4.8 \times 10^{-3})^2}{(1.17)(0.315)} = 6.3 \times 10^{-5}, \quad K_p = 1.0 \times 10^{-5}$$

$Q > K_p$, the reaction proceeds to the left.

	$N_2(g)$	$+$	$O_2(g)$	$\rightleftharpoons$	$2NO(g)$
initial	1.17 atm		0.315 atm		4.8×10^{-3} atm
change	+x		+x		-2x
equil	(1.17 + x)atm		(0.315 + x)atm		$(4.8 \times 10^{-3} - 2x)$ atm

$$K_p = 1.0 \times 10^{-5} = \frac{(4.8 \times 10^{-3} - 2x)^2}{(1.17 + x)(0.315 + x)}$$

Assume x is small compared to 1.17 and 0.315 (but not 4.8×10^{-3}).

P_{NO} at equilibrium $= (4.8 \times 10^{-3} - 2x) = y$

$$K_p = 1.0 \times 10^{-5} = \frac{y^2}{(1.17)(0.315)}; \quad y^2 = 3.686 \times 10^{-6}; \quad y = 1.92 \times 10^{-3} = 1.9 \times 10^{-3} \text{ atm}$$

Concentration, $M = \text{mol/L} = n/V = p/RT$;

$$[N_2] = \frac{1.17 \text{ atm}}{1173 \text{ K}} \times \frac{K \cdot \text{mol}}{0.08206 \text{ L} \cdot \text{atm}} = 1.2 \times 10^{-2} \text{ } M$$

$$[O_2] = \frac{0.315 \text{ atm}}{1173 \text{ K}} \times \frac{K \cdot \text{mol}}{0.08206 \text{ L} \cdot \text{atm}} = 3.3 \times 10^{-3} \text{ } M$$

$$[NO] = \frac{1.92 \times 10^{-3} \text{ atm}}{1173 \text{ K}} \times \frac{K \cdot \text{mol}}{0.0821 \text{ L} \cdot \text{atm}} = 2.0 \times 10^{-5} \text{ } M \text{ (or 1300 ppm)}$$

Note that the effect of attaining equilibrium is to reduce the concentration of NO, without significantly changing the concentrations of N_2 and O_2.

18.65   Calculate $[H_2SO_4]$ required to produce a solution with pH = 2.5. From the volume of rainfall, calculate the amount of H_2SO_4 present.

$$[H^+] = 10^{-2.5} = 3.16 \times 10^{-3} = 3 \times 10^{-3} \text{ } M$$

$$H_2SO_4(aq) \rightarrow H^+(aq) + HSO_4^-(aq)$$

equil $x\ M$ $x\ M$

$$HSO_4^- \rightleftharpoons H^+(aq) + SO_4^{2-} \qquad K_{a2} = 1.2 \times 10^{-2}$$

equil $(x-y)\ M$ $(x+y)\ M$ $y\ M$

The first ionization of H_2SO_4 is complete; the second is not.

$[H^+] = x + y = 3.16 \times 10^{-3}\ M$; $x = 3.16 \times 10^{-3} - y$

$$K_a = \frac{[H^+][SO_4^{2-}]}{[HSO_4^-]};\quad 0.012 = \frac{(3.16 \times 10^{-3})(y)}{(3.16 \times 10^{-3} - 2y)}$$

$3.795 \times 10^{-5} - 0.024\,y = 3.16 \times 10^{-3}\,y$; $3.795 \times 10^{-5} = 0.0272\,y$; $y = 1.4 \times 10^{-3}\ M$

$x = [H_2SO_4] = 3.16 \times 10^{-3}\ M - 1.4 \times 10^{-3}\ M = 1.76 \times 10^{-3}\ M = 2 \times 10^{-3}\ M\ H_2SO_4$

$$V = 1.0\ \text{in} \times 150\ \text{mi}^2 \times \frac{5280^2\ \text{ft}^2}{\text{mi}^2} \times \frac{12\ \text{in}^2}{\text{ft}^2} \times \frac{2.54^3\ \text{cm}^3}{\text{in}^3} \times \frac{1\ L}{1000\ \text{cm}^3} = 9.868 \times 10^9$$
$$= 9.9 \times 10^9\ L$$

$$\frac{1.76 \times 10^{-3}\ \text{mol}\ H_2SO_4}{1\ L\ \text{rainfall}} \times 9.868 \times 10^9\ L \times \frac{98.1\ g\ H_2SO_4}{1\ \text{mol}\ H_2SO_4} \times \frac{1\ kg}{1000\ g} = 1.70 \times 10^6$$
$$= 2 \times 10^6\ kg\ H_2SO_4$$

18.66   (a)   According to Table 18.1, the mole fraction of CO_2 in air is 0.000355.

$P_{CO_2} = \chi_{CO_2} \cdot P_{atm} = 0.000355\,(1.00\ \text{atm}) = 3.55 \times 10^{-4}\ \text{atm}$

$C_{CO_2} = kP_{CO_2} = 3.1 \times 10^{-2}\ M/\text{atm} \times 3.55 \times 10^{-4}\ \text{atm} = 1.10 \times 10^{-5} = 1.1 \times 10^{-5}\ M$

(b) H_2CO_3 is a weak acid, so the $[H^+]$ is regulated by the equilibria:

$$H_2CO_3(aq) \rightleftharpoons H^+(aq) + HCO_3^-(aq) \qquad K_{a1} = 4.3 \times 10^{-7}$$

$$HCO_3^-(aq) \rightleftharpoons H^+(aq) + CO_3^{2-}(aq) \qquad K_{a2} = 5.6 \times 10^{-11}$$

Since the value of K_{a2} is small compared to K_{a1}, we will assume that most of the $H^+(aq)$ is produced by the first dissociation.

$K_{a1} = 4.3 \times 10^{-7} = \dfrac{[H^+][HCO_3^-]}{[H_2CO_3]}$; $[H^+] = [HCO_3^-] = x$, $[H_2CO_3] = 1.1 \times 10^{-5} - x$

Since K_{a1} and $[H_2CO_3]$ have similar values, we cannot assume x is small compared to 1.1×10^{-5}.

$$4.3 \times 10^{-7} = \frac{x^2}{(1.1 \times 10^{-5} - x)};\quad 4.7 \times 10^{-12} - 4.3 \times 10^{-7}\,x = x^2$$

$$0 = x^2 + 4.3 \times 10^{-7} - 4.7 \times 10^{-12}$$

$$x = \frac{-4.3 \times 10^{-7} \pm \sqrt{(4.3 \times 10^{-7})^2 - 4(1)(-4.73 \times 10^{-12})}}{2(1)}$$

$$x = \frac{-4.3 \times 10^{-7} \pm \sqrt{1.85 \times 10^{-13} + 1.89 \times 10^{-11}}}{2} = \frac{-4.3 \times 10^{-7} \pm 4.37 \times 10^{-6}}{2}$$

The negative result is meaningless; $x = 1.97 \times 10^{-6} = 2.0 \times 10^{-6}$ $M\,H^+$; pH = 5.71

Since this $[H^+]$ is quite small, the $[H^+]$ from the autoionization of water might be significant. Calculation shows that for $[H^+] = 2.0 \times 10^{-6}$ M, from H_2CO_3, $[H^+]$ from $H_2O = 5.2 \times 10^{-9}$ M, which we can ignore.

18.67   (a) $Al(OH)_3(s) \rightleftharpoons Al^{3+}(aq) + 3OH^-(aq)$ $K_{sp} = 3.7 \times 10^{-15} = [Al^{3+}][OH^-]^3$

This is a precipitation conditions problem. At what $[OH^-]$ (we can get pH from $[OH^-]$) will $Q = 3.7 \times 10^{-15}$, the requirement for the onset of precipitation?

$Q = 3.7 \times 10^{-15} = [Al^{3+}][OH^-]^3$. Find the molar concentration of $Al_2(SO_4)_3$ and thus $[Al^{3+}]$.

$$\frac{2.0\text{ lb }Al_2(SO_4)_3}{1000\text{ gal }H_2O} \times \frac{453.6\text{ g}}{1\text{ lb}} \times \frac{1\text{ mol }Al_2(SO_4)_3}{342.2\text{ g }Al_2(SO_4)_3} \times \frac{1\text{ gal}}{4\text{ qt}} \times \frac{1\text{ qt}}{0.946\text{ L}}$$

$$= 7.0 \times 10^{-4}\ M\,Al_2(SO_4)_3 = 1.4 \times 10^{-3}\ M\,Al^{3+}$$

$Q = 3.7 \times 10^{-15} = (1.4 \times 10^{-3})[OH^-]^3$; $[OH^-]^3 = 2.64 \times 10^{-12}$

$[OH^-] = 1.38 \times 10^{-4} = 1.4 \times 10^{-4}\ M$;   pOH = 3.86;   pH = 14 − 3.86 = 10.14

(b) $CaO(s) + H_2O(l) \rightarrow Ca^{2+}(aq) + 2OH^-(aq)$; $[OH^-] = 1.38 \times 10^{-4}$ mol/L

$$\text{mol }OH^- = \frac{1.38 \times 10^{-4}\text{ mol}}{1\text{ L}} \times 1000\text{ gal} \times \frac{4\text{ qt}}{1\text{ gal}} \times \frac{0.946\text{ L}}{1\text{ qt}} = 0.523 = 0.52\text{ mol }OH^-$$

$$0.523\text{ mol }OH^- \times \frac{1\text{ mol }CaO}{2\text{ mol }OH^-} \times \frac{56.1\text{ g }CaO}{1\text{ mol }CaO} \times \frac{1\text{ lb}}{453.6\text{ g}} = 0.032\text{ lb }CaO$$

19 Chemical Thermodynamics

Spontaneity and Entropy

19.1   Spontaneous: a, d, e; nonspontaneous: b, c

19.2   Spontaneous: b, c, d; nonspontaneous: a, e

19.3   (a) $NH_4NO_3(s)$ dissolves in water, as in a chemical cold pack. Naphthalene (moth balls) sublimes at room temperature.

       (b)   Melting of a solid is spontaneous above its melting point but nonspontaneous below its melting point.

19.4   Berthelot's suggestion is incorrect. Some spontaneous processes that are nonexothermic are expansion of certain pressurized gases, dissolving of one liquid in another, and dissolving of many salts in water.

19.5   (a)   At or below $0°C$ (b) above $0°C$

       (c)   An increase in the entropy of the universe always accompanies a spontaneous process. When water freezes, ΔS_{sys} decreases, but because the reaction is exothermic, ΔS_{surr} increases. Below $0°C$, the increase in ΔS_{surr} predominates, and the process is spontaneous. Above $0°C$, the increase in ΔS_{surr} does not compensate for the decrease in ΔS_{sys}, and the process is not spontaneous.

19.6   (a)   Vaporization of any substance is endothermic.

       (b)   The vaporization of ethanol is spontaneous above $78.5°C$.

       (c)   The process is nonspontaneous below $78.5°C$.

19.7   (a)   E is a state function.

       (b)   The quantities q and w depend on path.

       (c)   If the process is *reversible* the system can move back and forth from one state to the other by the same path.

19.8   (a) $\Delta E (1 \rightarrow 2) = -\Delta E(2 \rightarrow 1)$

       (b)   We can say nothing about the values of q and w because we have no information about the paths.

(c)   If the changes of state are reversible, the two paths are the same and $w (1 \rightarrow 2) =$ $-w (2 \rightarrow 1)$.

Entropy and the Second Law

19.9   (a)   If the entropy of the system increases, the final state is more disordered than the initial state. However, we cannot predict whether the process is spontaneous. A positive ΔS_{system} does not guarantee spontaneity. For an irreversible spontaneous process, $\Delta S_{univ} > 0$; ΔS_{system} may be positive or negative.

(b) ΔS is positive for Exercise 19.1 (a).

19.10   (a) $\Delta S < 0$

(b) ΔS is positive for Exercise 19.2 (a), (b) and (e). At room temperature and 1 atm pressure, H_2O is a liquid, so there are more moles of gas in the products and $\Delta S > 0$ for the reaction $CO_2(g) + 2H_2O(l) \rightarrow CH_4(g) + 2O_2(g)$.

19.11   S increases in (a), (b) and (c); S decreases in (d).

19.12   More disorder is associated with forming a gas because of the large volume expansion and increased motional freedom associated with a gas.

19.13   (a) $Hg(l) \rightarrow Hg(s)$, ΔS is negative

(b)   Fusion is melting, the opposite of freezing.

$\Delta H = -\Delta H_{fus} = -2.33 \text{ kJ/mol} = -2.33 \times 10^3 \text{ J/mol}$

$\Delta G = \Delta H - T\Delta S = 0; \quad \Delta S = \Delta H/T$

$\Delta S = -2.33 \times 10^3 \text{ J/mol} / (273.15 - 38.9) \text{ K} = -9.95 \text{ J/mol} \cdot \text{K}$

19.14   (a) $Br_2(l) \rightarrow Br_2(g)$, ΔS is positive

(b) $\Delta H = 50.0 \text{ g } Br_2 \times \dfrac{1 \text{ mol } Br_2}{159.8 \text{ g } Br_2} \times \dfrac{29.6 \text{ kJ}}{\text{mol } Br_2} = +9.26 \text{ kJ} = 9.26 \times 10^3 \text{ J}$

$\Delta S = \Delta H/T = 9.26 \times 10^3 \text{ J}/(273.15 + 58.8) \text{ K} = 27.9 \text{ J/K}$

19.15   (a) $\Delta S = nR \ln (V_2/V_1) = 1.50 \text{ mol} \times 8.314 \text{ J/mol} (\ln 90.0 \text{ L} / 20.0 \text{ L}) = +18.8 \text{ J}$

(b)   Yes. Expansion provides more possible positions and greater motional freedom for the gas molecules, so disorder or entropy of the system increases.

19.16 $\Delta S = nR \ln (V_2/V_1) = 2.40 \text{ mol} \times 8.314 \text{ J/mol} (\ln 1/10) = -45.9 \text{ J}$

The sign of ΔS is negative because compression results in fewer possible positions for the gas molecules and a more ordered state.

19.17 (a) For a spontaneous process, the entropy of the universe increases; for a reversible process, the entropy of the universe does not change.

(b) In a reversible process, $\Delta S_{system} + \Delta S_{surroundings} = 0$. If ΔS_{system} is positive, $\Delta S_{surroundings}$ must be negative.

(c) Since $\Delta S_{universe}$ must be positive for a spontaneous process, $\Delta S_{surroundings}$ must be positive and greater than 75 J/K.

19.18 (a) For a spontaneous process, $\Delta S_{universe} > 0$. For a reversible process, $\Delta S_{universe} = 0$.

(b) $\Delta S_{surroundings}$ is positive and greater than the magnitude of the decrease in ΔS_{system}.

(c) $\Delta S_{system} = -34$ J/K.

19.19 (a) The entropy of a pure crystalline substance at absolute zero is zero.

(b) In *translational* motion, the entire molecule moves in a single direction; in *rotational* motion, the molecule rotates or spins around a fixed axis. *Vibrational* motion is reciprocating motion. The bonds within a molecule stretch and bend, but the average position of the atoms does not change.

(c)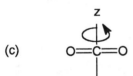

19.20 (a) A value of 0 for entropy requires that the substance is a perfectly crystalline (not disordered or amorphous, see Chapter 11) solid at a temperature of 0 K.

(b) If the molecule has two or more atoms, the thermal energy can be distributed as translational, vibrational or rotational motion. If the gas is monatomic, only translational motion is possible.

(c)

19.21 (a) $O_2(g)$ at 0.5 atm (larger volume and more motional freedom)

(b) $Br_2(g)$ (gases have higher entropy due primarily to much larger volume)

(c) 1 mol of $N_2(g)$ in 22.4 L (larger volume provides more motional freedom)

(d) $CO_2(g)$ (more motional freedom)

19.22 (a) 1 mol of $Cl_2(g)$ at 373 K and 1 atm (higher temperature, greater molecular speeds, more overall motion)

(b) 1 mol $H_2O(g)$ at 100°C, 1 atm (larger volume occupied by $H_2O(g)$)

(c) 2 mol of $O(g)$ at 300 K, 60.0 L (more total particles, larger volume in which to move)

(d) 1 mol of $KNO_3(aq)$ at 40°C (more motional freedom in aqueous solution)

19.23 (a) ΔS positive (moles of gas increase)

(b) ΔS negative (increased order)

(c) ΔS positive (gas produced, increased disorder)

(d) ΔS negative (fewer moles of gas, increased order)

19.24 (a) ΔS negative (increased order)   (b) ΔS positive (increased disorder)

(c) ΔS negative (increased order)   (d) ΔS negative (increased order)

19.25 (a) $I_2(s)$, 116.73 J/mol·K; $I_2(g)$, 260.57 J/mol·K. In general, the gas phase of a substance has a larger $S°$ than the solid phase because of the greater volume and motional freedom of the molecules.

(b) NaBr(s), 86.82 J/mol·K; NaBr(aq), 141 J/mol·K. Ions in solution have much greater motional freedom than ions in a solid lattice.

(c) C (diamond), 2.43 J/mol·K; C (graphite) 5.69 J/mol·K. Diamond is a network covalent solid with each C atom tetrahedrally bound to four other C atoms. Graphite consists of sheets of fused planar 6-membered rings with each C atom bound in a trigonal planar arrangement to three other C atoms. The internal entropy in graphite is greater because there is translational freedom among the planar sheets of C atoms while there is very little vibrational freedom within the network covalent diamond lattice.

(d) 1 mol of $H_2(g)$, 130.58 J/K; 2 mol of H(g), 2(114.60) = 229.20 J/K. More particles have a greater number of degrees of freedom.

19.26 (a) $S°$ for $POCl_3(l)$ is 222 J/mol·K and for $POCl_3(g)$ is 325 J/mol·K. Molecules in the gas phase occupy a larger volume and have more motional freedom than molecules in the liquid state.

(b) $S°$ for HCl(aq) is 56.5 J/mol·K and for HCl(g) is 186.69 J/mol·K. A dissolved gas has much less motional freedom than in the gas phase. This decrease in available volume outweighs the increase in the number of particles when HCl ionizes in aqueous solution.

(c) 1 mol $CaCO_3(s)$, 92.88 J/K; 1 mol CaO(s) + 1 mol $CO_2(g)$, (39.75 + 213.6) = 253.4 J/K. The second member of the pair has more total particles and half of them are in the gas phase for greater total motional freedom. Note that 1 mol of $CaCO_3(s)$ has greater entropy than 1 mol of CaO(s), because of the more ways to store energy in the more complex CO_3^{2-} anion.

(d) 1 mol $C_6H_6(g)$, 269.2 J/K; 3 mol $C_2H_2(g)$, 3(200.8) = 602.4 J/K. The second member of the pair has more total particles and more moles of gas. More particles in a larger volume have more possible arrangements (states) and greater entropy.

19.27 (a) $\Delta S° = S° \, C_2H_6(g) - S° \, C_2H_4(g) - S° \, H_2(g)$
              $= 229.5 - 219.4 - 130.58 = -120.5$ J/K

$\Delta S°$ is negative because there are fewer moles of gas in the products.

(b) $\Delta S° = 2S° \, NO_2(g) - \Delta S° \, N_2O_4(g) = 2(240.45) - 304.3 = +176.6$ J/K

$\Delta S°$ is positive because there are more moles of gas in the products.

(c) $\Delta S° = \Delta S° \, BeO(s) + \Delta S° \, H_2O(g) - \Delta S° \, Be(OH)_2(s)$
              $= 13.77 + 188.83 - 50.21 = +152.39$ J/K

$\Delta S°$ is positive because the product contains more total particles and more moles of gas.

(d) $\Delta S° = 2S° \, CO_2(g) + 4S° \, H_2O(g) - 2S° \, CH_3OH(g) - 3S° \, O_2(g)$
              $= 2(213.6) + 4(188.83) - 2(237.6) - 3(205.0) = +92.3$ J/K

$\Delta S°$ is positive because the product contains more total particles and more moles of gas.

19.28 (a) $\Delta S° = S° \, NH_4Cl(s) - S° \, NH_3(g) - S° \, HCl(g)$
              $= 94.6 - 192.5 - 186.69 = -284.6$ J/K

$\Delta S°$ is negative because the products contain fewer (no) moles of gas.

(b) $\Delta S° = 2S° \, Al_2O_3(s) - 4S°Al(s) - 3S° \, O_2(g)$
              $= 2(51.00) - 4(28.32) - 3(205.0) = -626.3$ J/K

$\Delta S°$ is negative because the products contain fewer (no) moles of gas.

(c) $\Delta S° = S° \, MgCl_2(s) + 2S°H_2O(l) - S° \, Mg(OH)_2(s) - 2S° \, HCl(g)$
              $= 89.6 + 2(69.91) - 63.24 - 2(186.69) = -207.20$ J/K

$\Delta S°$ is negative because the products contain fewer (no) moles of gas.

(d) $\Delta S° = S° \, C_2H_6(g) + S° \, H_2(g) - 2S° \, CH_4(g)$
              $= 229.5 + 130.58 - 2(186.3) = -12.5$ J/K

$\Delta S°$ is very small because there are the same number of moles of gas in the products and reactants. The slight decrease is related to the relatively small $S°$ value for $H_2(g)$, which has fewer degrees of freedom than molecules with more than two atoms.

Gibbs Free Energy

19.29 (a) $\Delta G = \Delta H - T\Delta S$

(b)   If ΔG is positive, the process is nonspontaneous, but the reverse process is spontaneous.

(c)   There is no relationship between ΔG and rate of reaction. A spontaneous reaction, one with a $-\Delta G$, may occur at a very slow rate. For example: $2H_2(g) + O_2(g) \rightarrow 2H_2O(g)$, $\Delta G = -457$ kJ is very slow if not initiated by a spark.

19.30   (a)   The *standard* free energy change, $\Delta G°$, represents the free energy change for the process when all reactants and products are in their standard states. When any or all reactants or products are not in their standard states, the free energy is represented simply as ΔG. The value for ΔG thus depends on the specific states of all reactants and products.

      (b)   When $\Delta G = 0$, the system is at equilibrium.

      (c)   The sign and magnitude of ΔG give no information about rate; we cannot predict whether the reaction will occur rapidly.

19.31 $\Delta G° = \Delta H° - T\Delta S° = -1.06 \times 10^5$ J $- 298$ K$(58$ J/K$) = -1.23 \times 10^5$ J $= -123$ kJ
         The process is highly spontaneous. We do not expect $H_2O_2(g)$ to be stable at 298 K.

19.32   (a) $\Delta H° = -1216.3$ kJ $- [-553.5$ kJ $- 393.5$ kJ$] = -269.3$ kJ

                $\Delta G° = -1137.6$ kJ $- [-525.1$kJ $- 394.4$ KJ$] = -218.1$ kJ

                $\Delta S° = 112.1 - [70.42 + 213.6] = -171.9$ J/K

                $\Delta G° = -269.3$ kJ $- 298(-0.1719)$ kJ $= -218.1$ kJ

      (b) $\Delta H° = 2(-435.9) + 3(0) - 2(-391.2) = -89.4$ kJ

                $\Delta G° = 2(-408.3) + 3(0) - 2(-289.9) = -236.8$ kJ

                $\Delta S° = 2(82.7) + 3(205.0) - 2(143.0) = 494.4$ J/K

                $\Delta G° = -89.4$ kJ $- 298(0.4944)$ kJ $= -236.7$ kJ

      (c) $\Delta H° = 4(-241.82) + 2(-393.5) - [3(0) + 2(-238.6)] = -1277.1$ kJ

                $\Delta G° = 4(-228.57) + 2(-394.4) - [3(0) + 2(-166.20) = -1370.7$ kJ

                $\Delta S° = 4(188.83) + 2(213.6) - [3(205.0) + 2(126.8)] = 313.9$ J/K

                $\Delta G° = -1277.1$ kJ $- 298(0.3139)$ kJ $= -1370.6$ kJ

      (d) $\Delta H° = 0 + 90.37 - [127.7 + 52.6] = -89.9$ kJ

                $\Delta G° = 0 + 86.71 - [105.7 + 66.3] = -85.3$ kJ

                $\Delta S° = 223.0 + 210.6 - [165.2 + 264] = +4.4$ J/K

                $\Delta G° = -89.9$ kJ $- 298(0.0044)$ kJ $= -91.2$ kJ

                (The discrepancy in $\Delta G°$ values is due to experimental uncertainties in the tabulated thermodynamic properties.)

19.33   (a) $\Delta G° = 2\Delta G°$ HCl(g) $- [\Delta G°$ $H_2(g) + \Delta G°$ $Cl_2(g)]$
                  $= 2(-95.27$ kJ$) - 0 - 0 = -190.5$ kJ, spontaneous

      (b) $\Delta G° = \Delta G°$ MgO(s) $+ 2\Delta G°$ HCl(g) $- [\Delta G°$ $MgCl_2(s) + \Delta G°$ $H_2O(l)]$
                  $= -569.6 + 2(-95.27) - [-592.1 + (-237.13)] = +69.1$ kJ, nonspontaneous

(c) $\Delta G° = \Delta G° \ N_2H_4(g) + \Delta G° \ H_2(g) - 2 \ \Delta G° \ NH_3(g)$
        $= 159.4 + 0 - 2(-16.66) = +192.7$ kJ, nonspontaneous

(d) $\Delta G° = 2\Delta G° \ NO(g) + \Delta G° \ Cl_2(g) - 2\Delta G° \ NOCl(g)$
        $= 2(86.71) + 0 - 2(66.3) = +40.8$ kJ, nonspontaneous

19.34   (a) $\Delta G° = 2\Delta G° \ SO_3(g) - 2\Delta G° \ SO_2(g) - \Delta G° \ O_2(g)$
                $= 2(-370.4) - 2(-300.4) - 0 = -140.0$ kJ, spontaneous

        (b) $\Delta G° = 3\Delta G° \ NO(g) - \Delta G° \ NO_2(g) - \Delta G° \ N_2O(g)$
                $= 3(86.71) - (51.84) - (103.59) = +104.70$ kJ, nonspontaneous

        (c) $\Delta G° = 4\Delta G° \ FeCl_3(s) + 3\Delta G° \ O_2(g) - 6\Delta G° \ Cl_2(g) - 2\Delta G° \ Fe_2O_3(s)$
                $= 4(-334) + 3(0) - 6(0) - 2(-740.98) = +146$ kJ, nonspontaneous

        (d) $\Delta G° = \Delta G° \ S(s) + 2\Delta G° \ H_2O(g) - \Delta G° \ SO_2(g) - 2\Delta G° \ H_2(g)$
                $= 0 + 2(-228.57) - (-300.4) - 2(0) = -156.7$ kJ, spontaneous

19.35   (a) ΔG is negative at low temperatures, positive at high temperature. That is, the reaction proceeds in the forward direction spontaneously at lower temperatures but spontaneously reverses at higher temperatures.

        (b) ΔG is positive at all temperatures. The reaction is nonspontaneous in the forward direction at all temperatures.

        (c) ΔG is positive at low temperatures, negative at high temperatures. That is, the reaction will proceed spontaneously in the forward direction at high temperature.

19.36 $\Delta G° = \Delta H° - T\Delta S°$

        (a) $\Delta G° = -844$ kJ $- 298$ K$(-0.165$ kJ/K$) = -795$ kJ, spontaneous

        (b) $\Delta G° = +572$ kJ $- 298$ K$(0.179$ kJ/K$) = +519$ kJ, nonspontaneous

                To be spontaneous, ΔG must be negative $(\Delta G < 0)$.

                Thus, $\Delta H° - T\Delta S° < 0$; $\Delta H° < T\Delta S°$; $T > \Delta H°/\Delta S°$; $T > \dfrac{572 \text{ kJ}}{0.179 \text{ kJ/K}} = 3,200$ K

19.37   At 330 K, $\Delta G < 0$; $\Delta G = \Delta H - T\Delta S < 0$

        23 kJ $- 330$ K $(\Delta S) < 0$; 23 kJ < 330 K (ΔS); $\Delta S > 23$ kJ/300 K

        $\Delta S > 0.070$ kJ/K or $\Delta S > +70$ J/K

19.38   At $-10°$C or 263 K, $\Delta G > 0$. $\Delta G = \Delta H - T\Delta S > 0$

        $\Delta H - 263$ K $(87$ J/K$) > 0$; $\Delta H > +2.3 \times 10^4$ J; $\Delta H > +23$ kJ

19.39   (a) $\Delta G = \Delta H - T\Delta S; 0 = 45$ kJ $- T(125$ J/K$)$; 45×10^3 J $= T(125$ J/K$)$

                $T = 45 \times 10^3$ J/125 J/K $= 360$ K

        (b)   spontaneous

19.40 ΔG is negative when $T\Delta S > \Delta H$ or $T > \Delta H/\Delta S$.

$\Delta H° = \Delta H° \ CH_4(g) + \Delta H° \ CO_2(g) - \Delta H° \ CH_3COOH(l)$
$= -74.8 + (-393.5) - (-487.0) = +18.7 \ kJ$

$\Delta S° = S° \ CH_4(g) + S° \ CO_2(g) - S° \ CH_3COOH(l) = +186.3 + 213.6 - 159.8 = +240.1 \ J/K$

$T > \dfrac{18.7 \ kJ}{0.2401 \ kJ/K} = 77.9 \ K$

The reaction is spontaneous above 77.9 K (-195°C).

19.41   (a) $\Delta H° = \Delta H° \ CH_3OH(g) - \Delta H° \ CO(g) - 2\Delta H° \ H_2(g)$
$= -201.2 - (-110.5) - 2(0) = -90.7 \ kJ$

$\Delta S° = S° \ CH_3OH(g) - S° \ CO(g) - 2S° \ H_2(g)$
$= 237.6 - 197.9 - 2(130.58) = -221.46 = -221.5 \ J/K$

(b)   Since $\Delta S°$ is negative, $-T\Delta S$ will become more positive and $\Delta G°$ will become more positive with increasing temperature.

(c)   The data in Appendix C are tabulated at 298 K.
$\Delta G° = \Delta G° \ CH_3OH(g) - \Delta G° \ CO(g) - 2\Delta G° \ H_2(g)$
$= -161.9 - (-137.2) - 2(0) = -24.7 \ kJ$

For comparison, $\Delta G° = \Delta H° - T\Delta S° = -90.7 \ kJ - (298 \ K)(-0.22146 \ kJ/K)$
$\Delta G° = -90.7 \ kJ + 66.0 \ kJ = -24.7 \ kJ$

The reaction is spontaneous under standard conditions at 298 K.

(d) $\Delta G° = \Delta H° - T\Delta S° = -90.7 \ kJ - 500 \ K(-0.22146 \ kJ/K) = +20.0 \ kJ$
The reaction is nonspontaneous under standard conditions at 500 K.

19.42   (a)   Calculate $\Delta H°$ and $\Delta S°$ to determine the sign of $T\Delta S°$.

$\Delta H° = 3\Delta H_f° \ NO(g) - \Delta H_f° \ NO_2(g) + \Delta H_f° \ N_2O(g)$
$= 3(90.37) - 33.84 - 81.6 = 155.7 \ kJ$

$\Delta S° = 3S° \ NO(g) - S° \ NO_2(g) - S° \ N_2O(g)$
$= 3(210.62) - 240.45 - 220.0 = 171.4 \ J/K$

$\Delta G° = \Delta H° - T\Delta S°$. Since $\Delta S°$ is positive, $-T\Delta S°$ becomes more negative as T increases and $\Delta G°$ becomes more negative.

(b) $\Delta G° = \Delta H° - T\Delta S° = 155.7 kJ - (800 \ K)(171.4 \ J/K)$
$\Delta G° = 155.7 \ kJ - 137 \ kJ = 19 \ kJ$

Since $\Delta G°$ is positive at 800 K, the reaction is not spontaneous at this temperature.

(c) ΔG° = 155.7 kJ - (1000 K)(171.4 J/K) = 155.7 kJ - 171.4 kJ = -15.7 kJ

ΔG° is negative at 1000 K and the reaction is spontaneous at this temperature.

19.43 (a) $C_2H_2(g) + 5/2\ O_2(g) \rightarrow 2CO_2(g) + H_2O(l)$

$\Delta H^\circ = 2\Delta H^\circ\ CO_2(g) + \Delta H^\circ\ H_2O(l) - \Delta H^\circ\ C_2H_2(g) - 5/2\Delta H^\circ\ O_2(g)$

= 2(-393.5) - 285.83 - 226.7 = -1299.5 kJ produced/mol C_2H_2 burned

(b) $w_{max} = \Delta G^\circ = 2\Delta G^\circ\ CO_2(g) + \Delta G^\circ\ H_2O(l) - \Delta G^\circ\ C_2H_2(g) - \Delta G^\circ\ O_2(g)$

= 2(-394.4) - 237.13 - 209.2 = -1235.1 kJ

The negative sign indicates that the system does work on the surroundings; the system can accomplish a maximum of 1235.1 kJ of work on its surroundings.

19.44 (a) $CH_4(g) + 2O_2(g) \rightarrow CO_2(g) + 2H_2O(l)$

$\Delta H^\circ = 2\Delta H^\circ\ H_2O(l) + \Delta H^\circ\ CO_2(g) - \Delta H^\circ\ CH_4(g) - 2\Delta H^\circ\ O_2(g)$

= 2(-285.83) - 393.5 - (-74.8 + 0) = -890.4 kJ produced/mol CH_4 burned

(b) $w_{max} = \Delta G^\circ = 2\Delta G^\circ\ H_2O(l) + \Delta G^\circ\ CO_2(g) - \Delta G^\circ\ CH_4(g) - 2\ \Delta G^\circ\ O_2(g)$

= 2(-237.13) - 394.4 + 50.8 = -817.9 kJ

The negative sign indicates the the system does work on the surroundings; the system can accomplish a maximum of 817.9 kJ of work on its surroundings.

Free Energy and Equilibrium

19.45 Use $\Delta G = \Delta G^\circ + RT\ lnQ$, where Q is the reaction quotient. When the numerator in Q increases, ln Q becomes more positive, in turn ΔG is more positive. When the denominator in Q becomes larger, Q decreases, ln Q becomes smaller, or more negative, and ΔG becomes smaller or more negative. (a) ΔG becomes more positive; (b) ΔG becomes more negative; (c) ΔG becomes more positive

19.46 Consider the relationship $\Delta G = \Delta G^\circ + RT\ lnQ$ where Q is the reaction quotient.

(a) $H_2(g)$ appears in the denominator of Q for this reaction. An increase in pressure of H_2 decreases Q and ΔG becomes smaller or more negative. Increasing the concentration of a reactant increases the tendency for a reaction to occur.

(b) $H_2(g)$ appears in the numerator of Q for this reaction. Increasing the pressure of H_2 increases Q and ΔG becomes more positive. Increasing the concentration of a product decreases the tendency for the reaction to occur.

(c) $H_2(g)$ appears in the denominator of Q for this reaction. An increase in pressure of H_2 decreases Q and ΔG becomes smaller or more negative.

19.47 (a) $\Delta G^\circ = \Delta G^\circ\ N_2O_4(g) - 2\Delta G^\circ\ NO_2(g) = 98.28 - 2(51.84) = -5.40$ kJ

(b) $\Delta G = \Delta G° + RT \ln P_{N_2O_4} / P_{NO_2}^2$

$$= -5.40 \text{ kJ} + \frac{8.314 \text{ J}}{1 \text{ K} \cdot \text{mol}} \times 298 \ln 0.10/(2.00)^2 = -14.54 \text{ kJ}$$

19.48 (a) $\Delta G° = 2\Delta G° \, CO_2(g) - 2\Delta G° \, CO(g) - \Delta G° \, O_2(g)$

$\Delta G° = 2(-394.4) - 2(-137.2) - 0 = -514.4 \text{ kJ}$

(b) $\Delta G = \Delta G° + RT \ln \dfrac{P_{CO_2}^2}{P_{O_2} \times P_{CO}^2}$

$$= -514.4 \text{ kJ} + \frac{8.314 \text{ J}}{1 \text{ K} \cdot \text{mol}} \times 298 \text{ K} \times \ln \frac{(0.10)^2}{300 \, (6.0)^2}$$

$$= -514.4 \text{ kJ} - 34.4 \text{ kJ} = -548.8 \text{ kJ}$$

19.49 $\Delta G° = -RT \ln K_p$, Equation 19.20; $\ln K_p = -\Delta G°/RT$

(a) $\Delta G° = 2\Delta G° \, HI(g) - \Delta G° \, H_2(g) - \Delta G° \, I_2(g)$
      $= 2(1.30) - 19.37 = -16.77 \text{ kJ}$

$$\ln K_p = \frac{-(-16.77 \text{ kJ}) \times 10^3 \text{ J/kJ}}{8.314 \text{ J/K} \times 298 \text{ K}} = 6.76876 = 6.769; \quad K_p = 870$$

(b) $\Delta G° = \Delta G° \, C_2H_4(g) + \Delta G° \, H_2O(g) - \Delta G° \, C_2H_5OH(g)$

      $= 68.11 - 228.57 - (-168.5) = 8.04 = 8.0 \text{ kJ}$

$$\ln K_p = \frac{-8.04 \text{ kJ} \times 10^3 \text{ J/kJ}}{8.314 \text{ J/K} \times 298 \text{ K}} = -3.24511 = -3.2; \quad K_p = 0.04$$

(c) $\Delta G° = \Delta G° \, C_6H_6(g) - 3\Delta G° \, C_2H_2(g) = 129.7 - 3(209.2) = -497.9 \text{ kJ}$

$$\ln K_p = \frac{-\Delta G°}{RT} = \frac{-(-497.9 \text{ kJ}) \times 10^3 \text{ J/kJ}}{8.314 \text{ J/K} \times 298 \text{ K}} = 200.963 = 201.0; \quad K_p = 2 \times 10^{87}$$

19.50 $\Delta G° = -RT \ln K_p$; $\ln K_p = -\Delta G° / RT$; at 298 K, RT = 2.4776 = 2.478 kJ

(a) $\Delta G° = \Delta G° \, NaOH(s) + \Delta G° \, CO_2(g) - \Delta G° \, NaHCO_3(s)$
      $= -379.5 + (-394.4) - (-851.8) = +77.9 \text{ kJ}$

$$\ln K_p = \frac{-\Delta G°}{RT} = \frac{-77.9 \text{ kJ}}{2.478 \text{ kJ}} = -31.442 = -31.4; \quad K_p = 2 \times 10^{-14}$$

$K_p = P_{CO_2} = 2 \times 10^{-14}$

(b) $\Delta G° = 2\Delta G° \, HCl(g) + \Delta G° \, Br_2(g) - 2\Delta G° \, HBr(g) - \Delta G° \, Cl_2(g)$
      $= 2(-95.27) + 3.14 - 2(-53.22) - 0 = -80.96 \text{ kJ}$

$$\ln K_p = \frac{-(-80.96)}{2.4776} = +32.68; \quad K_p = 1.6 \times 10^{14}$$

$$K_p = \frac{P_{HCl}^2 \times P_{Br_2}}{P_{HBr}^2 \times P_{Cl_2}} = 1.6 \times 10^{-14}$$

(c)   From Exercise 19.34(a), $\Delta G°$ at 298 K = -140.0 kJ.

$$\ln K_p = \frac{-\Delta G°}{RT} = \frac{-(-140.0)}{2.4776} = 56.51; \quad K_p = 3.5 \times 10^{24}$$

$$K_p = \frac{P_{SO_3}^2}{P_{SO_2}^2 \times P_{O_2}} = 3.5 \times 10^{24}$$

19.51 $K_p = P_{CO_2}$. Calculate $\Delta G°$ at the two temperatures using $\Delta G° = \Delta H° - T\Delta S°$ and then calculate K_p and P_{CO_2}.

$$\Delta H° = \Delta H°_f \; CaO(s) + \Delta H°_f \; CO_2(g) - \Delta H°_f \; CaCO_3(s)$$
$$= -635.5 + -393.5 - (-1207.1) = +178.1 \; kJ$$

$$\Delta S° = S° \; CaO(s) + \Delta S° \; CO_2(g) - S° \; CaCO_3(s)$$
$$= 39.75 + 213.6 - 92.88 = +160.47 \; J/K = 0.1605 \; kJ/K$$

(a) ΔG at 298 K = 178.1 kJ - 298 K (0.16047 kJ/K) = 130.28 = 130.3 kJ

$$\ln K_p = \frac{-\Delta G°}{RT} = \frac{-130.28 \times 10^3 \; J}{8.314 \; J/K \times 298 \; K} = -52.5837 = -52.58$$

$$K_p = 1.5 \times 10^{-23}; \quad P_{CO_2} = 1.5 \times 10^{-23} \; atm$$

(b) ΔG at 800 K = 178.1 kJ - 800 K (0.16047 kJ) = 49.724 = 49.7 kJ

$$\ln K_p = \frac{-\Delta G°}{RT} = \frac{-49.724 \times 10^3 \; J}{8.314 \; J/K \times 800 \; K} = -7.4759 = -7.48$$

$$K_p = 5.7 \times 10^{-4}; \quad P_{CO_2} = 5.7 \times 10^{-4} \; atm$$

19.52 (a) $\Delta G° = \Delta G° \; PbO(s) + \Delta G° \; CO_2(g) - \Delta G° \; PbCO_3(s)$

$$= -187.9 - 394.4 - (-625.5) = 43.2 \; kJ$$

$$\ln K_p = \frac{-\Delta G°}{RT} = \frac{-43.2 \times 10^3 \; J}{8.314 \; J/K \times 298 \; K} = -17.436 = -17.4$$

$$K_p = P_{CO_2} = 3 \times 10^{-8} \; atm$$

(b) $\Delta H° = \Delta H° \; PbO(s) + \Delta H° \; CO_2(g) - \Delta H° \; PbCO_3(s)$

$$= -217.3 - 393.5 + 699.1 = 88.3 \; kJ$$

$$\Delta S° = S° \; PbO(s) + S° \; CO_2(g) - S° \; PbCO_3(s)$$

$$= 68.70 + 213.6 - 131.0 = 151.3 \; J/K \; or \; 0.1513 \; kJ/K$$

$$\Delta G° = \Delta H° - T\Delta S° = 88.3 \; kJ - 773 \; K (0.1513 \; kJ) = -28.655 = -28.7 \; kJ$$

$$\ln K_p = \frac{-(-28.655 \times 10^3 \; J)}{8.314 \; J/K \times 773 \; K} = 4.4587 = 4.46$$

$$K_p = P_{CO_2} = 86 \; atm$$

19.53   (a) $\Delta G° = -RT \ln K_a = -(8.314)(298) \ln (6.5 \times 10^{-5}) = 23.89 = +23.9$ kJ

       (b)   By definition $\Delta G = 0$, when the system is at equilibrium.

       (c) $\Delta G = \Delta G° + RT \ln Q$

$$= +23.87 \text{ kJ} + (8.314 \times 10^{-3})(298) \times \ln \frac{(3.0 \times 10^{-3})(2.0 \times 10^{-5})}{(0.10)} = -11.6 \text{ kJ}$$

The fact that ΔG is negative tells us that this particular mixture will move spontaneously in the direction of further ionization.

19.54   (a) $\Delta G° = -RT \ln K_b = -(8.314)(298) \ln (1.8 \times 10^{-5}) = 2.707 \times 10^4 \text{ J} = 27.1$ kJ

       (b)   By definition $\Delta G = 0$ at equilibrium

       (c) $\Delta G = \Delta G° + RT \ln ([NH_4^+][OH^-]/[NH_3])$

$$\Delta G = 27.07 \text{ kJ} + (8.314 \times 10^{-3})(298) \times \ln \frac{(0.10)(0.05)}{0.10} = 19.6 \text{ kJ}$$

Additional Exercises

19.55   (a)   False. The essential question is whether the reaction proceeds far to the right before arriving at equilibrium. The position of equilibrium, which is the essential aspect, is not only dependent on ΔH but the entropy change as well.

       (b)   True

       (c)   False. Spontaneity relates to the position of equilibrium in a process, not to the rate at which that equilibrium is approached.

       (d)   True

       (e)   False. Such a process **might** be spontaneous, but would not necessarily be so. Spontaneous processes are those that are exothermic and/or that lead to increased disorder in the system.

19.56

Process	ΔH	ΔS
(a)	+	+
(b)	−	−
(c)	+	+
(d)	+	+
(e)	−	+

19.57   There is no inconsistency. The second law states that in any spontaneous process there is an increase in the entropy of the universe. While there may be a decrease in entropy of the system, as in the present case, this decrease is more than offset by an increase in entropy of the surroundings.

19.58   Melting = -183°C = 90 K; boiling = -89°C = 184 K; S° = 0 at 0 K; S° = 229.5 J at 298 K

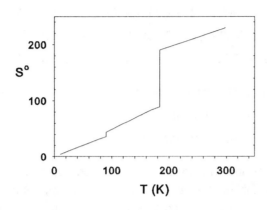

19.59   (a)   Formation reactions are the synthesis of 1 mole of compound from elements in their standard states.

$$1/2 \ N_2(g) + 3/2 \ H_2(g) \ \rightarrow NH_3(g)$$

$$C(s) + 2Cl_2(g) \ \rightarrow CCl_4(l)$$

$$K(s) + 1/2 \ N_2(g) + 3/2 \ O_2(g) \ \rightarrow KNO_3(s)$$

In each of these formation reactions, there are fewer moles of gas in the products than the reactants, so we expect $\Delta S°$ to be negative. If $\Delta G_f° = \Delta H_f° - T\Delta S_f°$ and $\Delta S_f°$ are negative, $-T\Delta S_f°$ is positive and $\Delta G_f°$ is more positive than $\Delta H_f°$.

(b) $$C(s) + 1/2 \ O_2(g) \ \rightarrow CO(g)$$

In this reaction, there are more moles of gas in products, $\Delta S_f°$ is positive, $-T\Delta S_f°$ is negative and $\Delta G_f°$ is more negative than $\Delta H_f°$.

19.60   (a)   (i) $$2RbCl(s) + 3O_2(g) \ \rightarrow \ 2RbClO_3(s)$$

$$\Delta H° = 2\Delta H_f° \ RbClO_3(s) - 3\Delta H_f° \ O_2(g) - 2\Delta H_f° \ RbCl(s)$$

$$= 2(-392.4) - 3(0) - 2(-430.5) = +76.2 \ kJ$$

$$\Delta S° = 2(152) - 3(205.0) - 2(92) = -495 \ J/K = -0.495 \ kJ/K$$

$$\Delta G° = 2(-292.0) - 3(0) - 2(-412.0) = +240.0 \ kJ$$

(ii) $$C_2H_2(g) + 4Cl_2(g) \ \rightarrow 2CCl_4(l) + H_2(g)$$

$$\Delta H° = 2\Delta H_f° \ CCl_4(l) + \Delta H_f° \ H_2(g) - \Delta H_f° \ C_2H_2(g) - 4\Delta H_f° \ Cl_2(g)$$

$$= 2(-139.3) + 0 - (226.7) - 4(0) = -505.3 \ kJ$$

$$\Delta S° = 2(214.4) + 130.58 - (200.8) - 4(222.96) = -533.3 \ J/K = -0.5333 \ kJ/K$$

$$\Delta G° = 2(-68.6) + 0 - (209.2) - 4(0) = -346.4 \ kJ$$

      (iii) $TiCl_4(l) + 2H_2O(l) \rightarrow TiO_2(s) + 4HCl(aq)$

            $\Delta H° = \Delta H_f° \, TiO_2(s) + 4\Delta H° \, HCl(g) - \Delta H° \, TiCl_4(l) - 2\Delta H° \, H_2O(l)$

                $= -944.7 + 4(-92.30) - (-804.2) - 2(-285.83) = +62.0 \text{ kJ}$

            $\Delta S° = 50.29 + 4(186.69) - 221.9 - 2(69.91) = 435.3 \text{ J/K} = +0.4353 \text{ kJ/K}$

            $\Delta G° = -889.4 + 4(-95.27) - (-728.1) - 2(-237.13) = -68.1 \text{ kJ}$

(b)   (i) $\Delta G°$ is (+), nonspontaneous

      (ii) $\Delta G°$ is (-), spontaneous

      (iii) $\Delta G°$ is (-), spontaneous

(c)   In each case the manner in which free energy change varies with temperature depends mainly on ΔS: $\Delta G = \Delta H - T\Delta S$. When ΔS is substantially positive, ΔG becomes more negative as temperature increases. When ΔS is substantially negative, ΔG becomes more positive as temperature increases.

      (i) $\Delta S°$ is negative, $\Delta G°$ becomes more positive with increasing temperature.

      (iii) $\Delta S°$ is negative, $\Delta G°$ becomes more positive with increasing temperature. (The reaction will become nonspontaneous at some temperature.)

      (iii) $\Delta S°$ is positive, $\Delta G°$ becomes more negative with increasing temperature.

19.61 $\Delta G = \Delta G° + RT \ln Q$

(a) $Q = \dfrac{P_{NH_3}^2}{P_{N_2} \times P_{H_2}^3} = \dfrac{(4.7)^2}{(8.5)(2.4)^3} = 0.188 = 0.19$

      $\Delta G° = 2\Delta G° \, NH_3(g) - \Delta G° \, N_2(g) - 3\Delta G° \, H_2(g)$

          $= 2(-16.66) - 0 - 3(0) = -33.32 \text{ kJ}$

      $\Delta G = -33.32 \text{ kJ} + \dfrac{8.314 \times 10^{-3} \text{ kJ}}{\text{K} \cdot \text{mol}} \times 298 \text{ K} \times \ln(0.188)$

      $\Delta G = -33.32 - 4.14 = -37.46 \text{ kJ}$

(b) $Q = \dfrac{P_{N_2}^3 \times P_{H_2O}^4}{P_{N_2H_4}^2 \times P_{NO_2}^2} = \dfrac{(2.5)^3 (1.2)^4}{(1.0 \times 10^{-3})^2 (1.0 \times 10^{-3})^2} = 3.24 \times 10^{13} = 3.2 \times 10^{13}$

      $\Delta G° = 3\Delta G_f° \, N_2(g) + 4\Delta G_f° \, H_2O(g) - 2\Delta G_f° \, N_2H_4(g) - 2\Delta G_f° \, NO_2(g)$

          $= 0 + 4(-228.61) - 2(159.4) - 2(51.84) = -1336.9 \text{ kJ}$

      $\Delta G = -1336.9 \text{ kJ} + 2.478 \ln 3.24 \times 10^{13} = -1259.8 \text{ kJ}$

(c) $Q = \dfrac{P_{N_2} \times P_{H_2}^2}{P_{N_2H_4}} = \dfrac{(1 \times 10^{-3})(2 \times 10^{-4})^2}{6.0} = 6.67 \times 10^{-12} = 7 \times 10^{-12}$

$$\Delta G° = \Delta G°_f\ N_2(g) + 2\Delta G°_f\ H_2(g) - \Delta G°_f\ N_2H_4$$

$$= 0 + 2(0) - 159.4 = -159.4\ kJ$$

$$\Delta G = -159.4\ kJ + 2.478\ \ln 6.67 \times 10^{-12} = -223.2\ kJ$$

19.62 $\Delta G° = -RT\ \ln K_p$

     (a) $\Delta G° = -8.314\ J/K \times 718\ K \times \ln 50.2 = -23.4\ kJ$

     (b) $\Delta G° = -8.314\ J/K \times 1073\ K \times \ln(9.1 \times 10^2) = -60.8\ kJ$

19.63

Reaction	(a) Sign of $\Delta H°$	(a) Sign of $\Delta S°$	(b) K > 1?	(c) Variation in K as Temp. Increases
(i)	-	-	yes	decrease
(ii)	+	+	no	increase
(iii)	+	+	no	increase
(iv)	+	+	no	increase

     (a)   Note that positive $\Delta H°$ leads to a smaller value of K, while positive $\Delta S°$ increases the value of K, particularly at high temperatures.

19.64   (a) $Ag_2O(s) \rightarrow 2Ag(s) + 1/2\ O_2(g)$

     (b) $\Delta G° = 2\Delta G°_f\ Ag(s) + 1/2\ \Delta G°_f\ O_2(g) - \Delta G°_f\ Ag_2O(s)$

            $= 2(0) + 1/2(0) - (-11.20) = +11.20\ kJ$

     (c) $K_p = \sqrt{P_{O_2}}$; $\Delta G° = -RT\ \ln K_p$; $\ln K_p = -\Delta G°/RT$

            $\ln K_p = \dfrac{-11.20 \times 10^3\ J}{8.314\ J/K \times 298\ K} = -4.52055 = -4.521$; $K_p = 1.0883 \times 10^{-2} = 1.09 \times 10^{-2}$

            $P_{O_2} = K_p^2 = 1.18 \times 10^{-4}\ atm = 9.00 \times 10^{-2}\ torr$

19.65   (a) $K = \dfrac{\chi_{CH_3COOH}}{\chi_{CH_3OH}\ P_{CO}}$

            $\Delta G° = -RT\ \ln K$; $\ln K_p = -\Delta G/RT$

            $\Delta G° = \Delta G°_f\ CH_3COOH(l) - \Delta G°_f\ CH_3OH(l) - \Delta G°_f\ CO(g)$

               $= -392.4 - (-166.23) - (-137.2) = -89.0\ kJ$

            $\ln K_p = \dfrac{-(-89.0\ kJ)}{(8.314 \times 10^{-3}\ kJ/K)\ (298\ K)} = 35.922 = 35.9$; $K_p = 4 \times 10^{15}$

     (b) $\Delta H° = \Delta H°_f\ CH_3COOH(l) - \Delta H°_f\ CH_3OH(l) - \Delta H°_f\ CO(g)$

               $= -487.0 - (-238.6) - (-100.5) = -137.9\ kJ$

The reaction is exothermic, so the value of K will decrease with increasing temperature, and the mole fraction of CH_3COOH will also decrease. Elevated temperatures must be used to increase the speed of the reaction. Thermodynamics cannot predict the rate at which a reaction reaches equilibrium.

(c) $\Delta G° = -RT \ln K$; $K = 1$, $\ln K = 0$, $\Delta G° = 0$

$\Delta G° = \Delta H - T\Delta S$; when $\Delta G° = 0$, $\Delta H° = T\Delta S°$

$\Delta S° = $ S° $CH_3COOH(l) - $ S° $CH_3OH(l) - $ S° $CO(g)$

$= 159.8 - 126.8 - 197.9 = -164.9$ J/K $= -0.1649$ kJ/K

-137.9 kJ $= T(-0.1649$ kJ/K$)$, $T = 836.3$ K

The equilibrium favors products up to 836 K or 563 °C, so the elevated temperatures to increase the rate of reaction can be safely employed.

19.66   (a)   First calculate $\Delta G°$ for each reaction:

For $C_6H_{12}O_6(s) + 6O_2(g) \rightleftharpoons 6CO_2(g) + 6H_2O(l)$ (A)

$\Delta G° = 6(-237.13) + 6(-394.4) - (-910.4) = -2879$ kJ

For $C_6H_{12}O_6(s) \rightleftharpoons 2C_2H_5OH(l) + 2CO_2(g)$ (B)

$\Delta G° = 2(-394.4) + 2(-174.8) - (-910.4) = -228$ kJ

For (A), $\ln K = 2879 \times 10^3/(8.314)(298) = 1162$; $K = 5 \times 10^{504}$

For (B), $\ln K = 228 \times 10^3/(8.314)(298) = 92.026 = 92.0$; $K = 9 \times 10^{39}$

(b)   Both these values for K are unimaginably large. However, K for reaction (A) is larger, because $\Delta G°$ is more negative. The magnitude of the work that can be accomplished by coupling a reaction to its surroundings is measured by ΔG. According to the calculations above, considerably more work can in principle be obtained from reaction (A), because $\Delta G°$ is more negative.

19.67   (a) $\Delta G° = -RT \ln K_p$ (Equation 19.20); $\ln K_p = -\Delta G°/RT$

Use $\Delta G° = \Delta H° - T\Delta S°$ to get $\Delta G°$ at the two temperatures. Calculate $\Delta H°$ and $\Delta S°$ using data in Appendix C.

$2CH_4(g) \rightarrow C_2H_6(g) + H_2(g)$

$\Delta H° = \Delta H_f° \, C_2H_6(g) + \Delta H_f° \, H_2(g) - 2\Delta H_f° \, CH_4(g) = -84.68 + 0 - 2(-74.8) = 64.92$
$= 64.9$ kJ

$\Delta S° = $ S° $C_2H_6(g) + $ S° $H_2(g) - 2$S° $CH_4(g) = 229.5 + 130.58 - 2(186.3) = -12.52$
$= -12.5$ J/K

at 298 K, $\Delta G = 64.92$ kJ $- 298$ K$(-12.52 \times 10^{-3}$ kJ/K$) = 68.65 = 68.7$ kJ

$\ln K_p = \dfrac{-68.65 \text{ kJ}}{(8.314 \times 10^{-3} \text{ kJ/K})(298 \text{ K})} = -27.709 = -27.7$, $K_p = 9.25 \times 10^{-13} = 9 \times 10^{-13}$

at 773 K, $\Delta G = 64.9$ kJ $- 773$ K$(-12.52 \times 10^{-3}$ J/K$) = 74.598 = 74.6$ kJ

$$\ln K_p = \frac{-74.598 \text{ kJ}}{(8.314 \times 10^{-3} \text{ kJ/K})(773 \text{ K})} = -11.607 = -11.6, \quad K_p = 9.1 \times 10^{-6} = 9 \times 10^{-6}$$

Because the reaction is endothermic, the value of K_p increases with an increase in temperature.

$2CH_4(g) + 1/2\, O_2(g) \rightarrow C_2H_6(g) + H_2O(g)$

$\Delta H° = \Delta H_f° \; C_2H_6(g) + \Delta H_f° \; H_2O(g) - 2\Delta H_f° \; CH_4(g) - 1/2\, \Delta H_f° \; O_2(g)$

    $= -84.68 + (-241.82) - 2(-74.8) - 1/2\,(0) = -176.9$ kJ

$\Delta S° = S° \; C_2H_6(g) + \; S° \; H_2O(g) - 2S° \; CH_4(g) - 1/2\; S° \; O_2(g)$

    $= 229.5 + 188.83 - 2(186.3) - 1/2\,(205.0) = -56.77 = -56.8$ J/K

at 298 K, $\Delta G = -176.9$ kJ $- 298$ K$(-56.77 \times 10^{-3}$ kJ/K$) = -159.98 = -160.0$ kJ

$$\ln K_p = \frac{-(-159.98 \text{ kJ})}{(8.314 \times 10^{-3} \text{ kJ/K})(298 \text{ K})} = 64.571 = 64.6; \quad K_p = 1 \times 10^{28}$$

at 773 K, $\Delta G = -176.9$ kJ $- 773$ K $(-56.77 \times 10^{-3}$ kJ/K$) = -133.02 = -133.0$ kJ

$$\ln K_p = \frac{-(-133.02 \text{ kJ})}{(8.314 \times 10^{-3} \text{ kJ/K})(773 \text{ K})} = 20.698 = 20.7; \quad K_p = 1 \times 10^{9}$$

Because this reaction is exothermic, the value of K_p decreases with increasing temperature.

(b)   The difference in $\Delta G°$ for the two reactions is primarily enthalpic; the first reaction is endothermic and the second exothermic. Both reactions have $-\Delta S°$, which inhibits spontaneity.

(c)   This is an example of coupling a useful but nonspontaneous reaction with a spontaneous one to spontaneously produce a desired product.

      $2CH_4(g) \rightarrow C_2H_6(g) + H_2(g)$ $\Delta G_{298}° = +68.7$ kJ, nonspontaneous

      $H_2(g) + 1/2\, O_2(g) \rightarrow H_2O(g)$ $\Delta G_{298}° = -228.57$ kJ, spontaneous

      _____

      $2CH_4(g) + 1/2\, O_2(g) \rightarrow C_2H_6(g) + H_2O(g)$ $\Delta G_{298}° = -160.0$ kJ, spontaneous

(d) $CH_4(g) + 2O_2(g) \rightarrow CO_2(g) + 2H_2O(g)$

19.68   (a)   The equilibrium of interest here can be written as:
      K^+ (plasma) $\rightleftharpoons$ K^+ (muscle)

      Since an aqueous solution is involved in both cases, assume that the equilibrium constant for the above process is exactly 1, that is, $\Delta G° = 0$. However, ΔG is not zero because the concentrations are not the same on both sides of the membrane. Use Equation [19.20] to calculate ΔG:

$$\Delta G = \Delta G^\circ + RT \ln \frac{[K^+(\text{muscle})]}{[K^+(\text{plasma})]}$$

$$= 0 + (8.314)(310) \ln \frac{(0.15)}{(5.0 \times 10^{-3})} = 8.77 \text{ kJ}$$

(b) Note that ΔG is positive. This means that work must be done on the system (blood plasma plus muscle cells) to move the K^+ ions "uphill," as it were. The minimum amount of work possible is given by the value for ΔG. This value represents the minimum amount of work required to transfer one mole of K^+ ions from the blood plasma at 5×10^{-3} *M* to muscle cell fluids at 0.15 *M*, assuming constancy of concentrations. In practice, a larger than minimum amount of work is required.

19.69 ΔG° for the metabolism of glucose is:

$6\Delta G^\circ$ $CO_2(g)$ + $6\Delta G^\circ$ $H_2O(l)$ - $\Delta G^\circ C_6H_{12}O_6(s)$ - $6\Delta G^\circ$ $O_2(g)$

$\Delta G^\circ = 6(-394.4) + 6(-237.13) - (-910.4 \text{ kJ} + 0) = -2878.8 \text{ kJ}$

moles ATP = -2878.8 kJ × 1 mol ATP / $(-30.5$ kJ$)$ = 94.4 mol ATP / mol glucose

19.70 (a) To obtain ΔH° from the equilibrium constant data, graph ln K at various temperatures vs 1/T, being sure to employ absolute temperature. The slope of the linear relationship that should result is $-\Delta H^\circ/R$; thus, ΔH° is easily calculated.

(b) Use $\Delta G^\circ = \Delta H^\circ - T\Delta S^\circ$ and $\Delta G^\circ = -RT \ln K$. Substituting the second expression into the first, we obtain

$$-RT \ln K = \Delta H^\circ - T\Delta S^\circ; \quad \ln K = \frac{-\Delta H^\circ}{RT} - \frac{-\Delta S^\circ}{R} = \frac{-\Delta H^\circ}{RT} + \frac{\Delta S^\circ}{R}$$

Thus, the constant in the equation given in the exercise is $\Delta S^\circ/R$.

19.71 $S = k \ln W$ (Equation 19.8), $k = R/N$, $W \propto V^m$

$\Delta S = S_2 - S_1$; $S_1 = k \ln W_1$, $S_2 = k \ln W_2$

$\Delta S = k \ln W_2 - k \ln W_1$; $W_2 = cV_2{}^m$; $W_1 = cV_1{}^m$

(The number of particles, m, is the same in both states.)

$\Delta S = k \ln cV_2{}^m - cV_1{}^m$; $\ln a^b = b \ln a$

$\Delta S = k m \ln cV_2 - k m \ln cV_1$; $\ln a - \ln b = \ln (a/b)$

$$\Delta S = k m \ln \left(\frac{cV_2}{cV_1} \right) = k m \ln \left(\frac{V_2}{V_1} \right) = \frac{R}{N} m \ln \left(\frac{V_2}{V_1} \right)$$

$$\frac{m}{N} = \frac{\text{particles}}{6.022 \times 10^{23}} = n(\text{mol}); \quad \Delta S = nR \ln \left(\frac{V_2}{V_1} \right)$$

Integrative Exercises

19.72 (a) According to Exercise 19.13, $\Delta S = \Delta H/T$ for a phase change.

acetone: $\Delta S^\circ_{vap} = \Delta H^\circ_{vap}/T = 29.1 \text{ kJ/mol}/329.25 \text{ K} = 88.4 \text{ J/mol} \cdot \text{K}$

dimethyl ether: $\Delta S^\circ_{vap} = 21.5 \text{ kJ/mol}/248.35 \text{ K} = 86.6 \text{ J/mol} \cdot \text{K}$

ethanol: $\Delta S^\circ_{vap} = 38.6 \text{ kJ/mol}/351.6 \text{ K} = 110 \text{ J/mol} \cdot \text{K}$

octane: $\Delta S^\circ_{vap} = 34.4 \text{ kJ/mol}/398.75 \text{ K} = 86.3 \text{ J/mol} \cdot \text{K}$

pyridine: $\Delta S^\circ_{vap} = 35.1 \text{ kJ/mol}/388.45 \text{ K} = 90.4 \text{ J/mol} \cdot \text{K}$

(b) Ethanol is the only liquid listed that doesn't follow *Trouton's rule* and it is also the only substance that exhibits hydrogen bonding in the pure liquid. Hydrogen bonding leads to more ordering in the liquid state and a greater than usual increase in entropy upon vaporization. The rule appears to hold for liquids with London dispersion forces (octane) and ordinary dipole-dipole forces (acetone, dimethyl ether, pyridine), but not for those with hydrogen bonding.

(c) Owing to strong hydrogen bonding interactions, water probably does not obey Trouton's rule.

From Appendix B, ΔH°_{vap} at $100°C = 40.67 \text{ kJ/mol}$.

$\Delta S^\circ_{vap} = 40.67 \text{ kJ/mol}/373.15 \text{ K} = 109.0 \text{ J/mol} \cdot \text{K}$

(d) Use $\Delta S^\circ_{vap} = 88 \text{ J/mol} \cdot \text{K}$, the middle of the range for Trouton's rule, to estimate ΔH°_{vap} for chlorobenzene.

$\Delta H^\circ_{vap} = \Delta S^\circ_{vap} \times T = 88 \text{ J/mol} \cdot \text{K} \times 404.95 \text{ K} = 36 \text{ kJ/mol}$

19.73 (a) $O_2 (g) \xrightarrow{h\nu} 2O(g)$; ΔS increases because there are more moles of gas in the products.

(b) $O_2(g) + O(g) \rightarrow O_3(g)$, ΔS decreases because there are fewer moles of gas in the products.

(c) ΔS increases as the gas molecules diffuse into the larger volume of the stratosphere; there are more possible positions and therefore more motional freedom.

(d) $SO_3(g) + H_2O(l) \rightarrow H_2SO_4(aq)$; ΔS decreases because the motion of SO_3 molecules in aqueous solution is much more restricted than in the gas phase.

(e) $NaCl(aq) \rightarrow NaCl(s) + H_2O(l)$; ΔS decreases as the mixture (seawater, greater disorder) is separated into pure substances (fewer possible arrangements, more order).

19.74   (a)   Yes, mixing of gases is a spontaneous process. When gases mix, many new positions become available to the gas particles. The number of possible states (W) for a random mixture of gases is much larger than the number of states for the separated gases and ΔS for mixing is positive. Assuming no significant change in enthalpy, the increase in entropy results in a spontaneous process.

(b)   Consider the expansion of the two gases separately. Then, $P_T = P_{He} + P_{Ar}$.

He: $P_1V_1 = P_2V_2$; 1.30 atm × 1.75 L = P_2 × 5.75 L; P_2 = 0.3957 = 0.396 atm

Ar: $P_1V_1 = P_2V_2$; 735 torr × 4.00 L = P_2 × 5.75 L; P_2 = 511.3 = 511 torr

$$P_T = P_{He} + P_{Ar} = 0.3957 \text{ atm} + 511.3 \text{ torr} \times \frac{1 \text{ atm}}{760 \text{ torr}} = 1.0684 = 1.068 \text{ atm}$$

(c) P_{He} = 0.396 atm

(d)   According to Equation [19.2], for isothermal expansion,

$$\Delta S = nR \ln(V_2/V_1) \quad \Delta S_T = \Delta S_{He} + \Delta S_{Ar}; \quad n = PV/RT$$

$$n_{He} = 1.30 \text{ atm} \times 1.75 \text{ L} \times \frac{\text{mol} \cdot \text{K}}{0.08206 \text{ L} \cdot \text{atm}} \times \frac{1}{328 \text{ K}} = 0.08452 = 0.0845 \text{ mol He}$$

$$\Delta S_{He} = 0.08452 \text{ mol He} \times \frac{8.314 \text{ J}}{\text{mol} \cdot \text{K}} \times \ln(5.75 \text{ L} / 1.75 \text{ L}) = 0.8360 = 0.836 \text{ J/K}$$

$$n_{Ar} = 735 \text{ torr} \times \frac{1 \text{ atm}}{760 \text{ torr}} \times 4.00 \text{ L} \times \frac{\text{mol} \times \text{K}}{0.08206 \text{ L} \cdot \text{atm}} \times \frac{1}{328 \text{ K}} = 0.1437$$
$$= 0.144 \text{ mol Ar}$$

$$\Delta S_{Ar} = 0.1437 \text{ mol Ar} \times \frac{8.314 \text{ J}}{\text{mol} \cdot \text{K}} \times \ln(5.75 \text{ L} / 4.00 \text{ L}) = 0.4336 = 0.434 \text{ J/K}$$

$$\Delta S_T = \Delta S_{He} + \Delta S_{Ar} = 0.8360 \text{ J/K} + 0.4336 \text{ J/K} = 1.2696 = 1.270 \text{ J/K}$$

(e)   When the two gases are separated, the entropy of the system decreases but the entropy of surroundings increases as heat is removed from the system to effect condensation. The second law requires only that $\Delta S_{universe}$ is ≥ 0.

19.75   (a)   He(g) atoms are spheres that can only increase kinetic energy through translational motion.

(b) H_2(g) molecules (H-H) can increase kinetic energy through translational, rotational and vibrational motion.

(c)   According to the kinetic molecular theory outlined in Section 10.7, an ideal-gas "molecule" is a volumeless point. An ideal-gas molecule can only increase kinetic energy by translational motion, so average kinetic energy (ϵ) and root mean square speed (u) are related by Equation [10.18], $\epsilon = 1/2mu^2$. This is essentially true for monatomic gases like He(g), but not for molecular gases like H_2(g) that can increase kinetic energy by rotation and vibration as well as translation. For H_2(g), not all kinetic energy is expressed as molecular speed, so it behaves less ideally than He(g).

19.76   (a) ΔH for the process is zero if N_2 behaves as an ideal gas. The process can be written as N_2 (1 atm) $\rightarrow$ N_2 (0.5 atm).

        (b)   The gas can be returned to its original condition by exerting a force on the piston to push the gas back into flask A. Again ΔH is zero.

        (c)   The overall changes in ΔG and ΔH for the system are zero, as must be the case if the system is returned to its original condition, because both enthalpy and free energy are state functions.

        (d)   In the expansion, the system does no work on its surroundings. In the compression phase, the surroundings must do work on the system. This work corresponds to the force exerted on the piston as it moves through chamber B, compressing the gas.

        (e)   The work done by the surroundings on the system in the compression of the gas results in an increase in entropy of the surroundings. Thus the overall entropy of the universe increases.

19.77   (a) $Ag(s) + 1/2\,N_2(g) + 3/2\,O_2(g) \rightarrow AgNO_3(s)$; ΔS decreases because there are fewer moles of gas in the product.

        (b) $\Delta G_f^{\circ} = \Delta H_f^{\circ} - T\Delta S_f^{\circ}$; $\Delta S_f^{\circ} = (\Delta G_f^{\circ} - \Delta H_f^{\circ}) / (-T) = (\Delta H_f^{\circ} - \Delta G_f^{\circ}) / T$

              $\Delta S_f^{\circ} = -124.4$ kJ $-$ (-33.4 kJ) / 298 K $= -0.305$ kJ/K $= -305$ J/K

              ΔS_f° is relatively large and negative, as anticipated from part (a).

        (c)   Dissolving of $AgNO_3$ can be expressed as

              $AgNO_3(s) \rightarrow AgNO_3$ (aq, 1 m)

              $\Delta H^{\circ} = \Delta H_f^{\circ}\ AgNO_3(aq) - \Delta H_f^{\circ}\ AgNO_3(s) = -101.7 - (-124.4) = +22.7$ kJ

              $\Delta H^{\circ} = \Delta H_f^{\circ}\ MgSO_4(aq) - \Delta H_f^{\circ}\ MgSO_4(s) = -1374.8 - (-1283.7) = -91.1$ kJ

              Dissolving $AgNO_3(s)$ is endothermic ($+\Delta H^{\circ}$), but dissolving $MgSO_4(s)$ is exothermic ($-\Delta H^{\circ}$).

        (d) $AgNO_3$: $\Delta G^{\circ} = \Delta G_f^{\circ}\ AgNO_3(aq) - \Delta G_f^{\circ}\ AgNO_3(s) = -34.2 - (-33.4) = -0.8$ kJ

              $\Delta S^{\circ} = (\Delta H^{\circ} - \Delta G^{\circ}) / T = [22.7$ kJ $- (-0.8$ kJ$)] / 298$ K $= 0.0789$ kJ/K $= 78.9$ J/K

              $MgSO_4$: $\Delta G^{\circ} = \Delta G_f^{\circ}\ MgSO_4(aq) - \Delta G_f^{\circ}\ MgSO_4(s) = -1198.4 - (-1169.6) = -28.8$ kJ

              $\Delta S^{\circ} = (\Delta H^{\circ} - \Delta G^{\circ}) / T = [-91.1$ kJ $- (-28.8$ kJ$)] / 298$ K $= -0.209$ kJ/K $= -209$ J/K

        (e)   In general, we expect dissolving a crystalline solid to be accompanied by an increase in positional disorder and an increase in entropy; this is the case for $AgNO_3$ (ΔS° = + 78.9 J/K). However, for dissolving $MgSO_4(s)$, there is a substantial decrease in entropy (ΔS = -209 J/K). According to Section 13.5, ion-pairing is a significant phenomenon in electrolyte solutions, particularly in concentrated solutions where the charges of the ions are greater than 1. According to Table 13.6, a 0.1 m $MgSO_4$ solution has a van't Hoff factor of 1.21. That is, for each mole of $MgSO_4$ that dissolves, there are only 1.21 moles of "particles" in solution instead of 2 moles of particles. For a 1 m solution, the factor is even smaller. Also, the exothermic enthalpy of mixing indicates substantial interactions between solute and solvent. Substantial ion-pairing coupled with ion-dipole interactions with H_2O molecules lead to a decrease in entropy for $MgSO_4(aq)$ relative to $MgSO_4(s)$.

19.78 (a) $K_p = P_{NO_2}^2 / P_{N_2O_4}$

Assume equal amounts means equal number of moles. For gases, P = n(RT/V). In an equilibrium mixture, RT/V is a constant, so moles of gas is directly proportional to partial pressure. Gases with equal partial pressures will have equal moles of gas present. The condition $P_{NO_2} = P_{N_2O_4}$ leads to the expression $K_p = P_{NO_2}$. Thus, for any value of (or P_T), the condition of equal moles of the two gases can be achieved at some temperature. Assume $P_{NO_2} = P_{N_2O_4} = 1.0$ atm.

$K_p = \dfrac{(1.0)^2}{1.0} = 1.0$; $\ln K_p = 0$; $\Delta G° = 0 = \Delta H° - T\Delta S°$; $T = \Delta H°/\Delta S°$

$\Delta H° = 2\Delta H_f° NO_2(g) - \Delta H_f° N_2O_4(g) = 2(33.84) - 9.66 = +58.02$ kJ

$\Delta S° = 2S° NO_2(g) - S° N_2O_4(g) = 2(240.45) - 304.3 = 0.1766$ kJ/K

$T = \dfrac{58.02 \text{ kJ}}{0.1766 \text{ kJ/K}} = 328.5$ K or 55.5°C

(b) $P_T = 1.00$ atm; $P_{N_2O_4} = x$, $P_{NO_2} = 2x$; $x + 2x = 1.00$ atm

$x = P_{N_2O_4} = 0.3333 = 0.333$ atm; $P_{NO_2} = 0.6667 = 0.667$ atm

$K_p = \dfrac{(0.6667)^2}{0.3333} = 1.334 = 1.33$; $\Delta G° = -RT \ln K = \Delta H° - T\Delta S°$

$-(8.314 \times 10^{-3}$ kJ/K)(ln 1.334) T = 58.02 kJ - (0.1766 kJ/K) T

$(-0.00239$ kJ/K) T + (0.1766 kJ/K) T = 58.02 kJ

$(0.1742$ kJ/K) T = 58.02 kJ; T = 333.0 K

(c) $P_T = 10.00$ atm; $x + 2x = 10.00$ atm

$x = P_{N_2O_4} = 3.3333 = 3.333$ atm; $P_{NO_2} = 6.6667 = 6.667$ atm

$K_p = \dfrac{(6.6667)^2}{3.3333} = 13.334 = 13.33$; $-RT \ln K = \Delta H° - T\Delta S°$

$-(8.314 \times 10^{-3}$ kJ/K)(ln 13.334) T = 58.02 kJ - (0.1766 kJ/K) T

$(-0.02154$ kJ/K) T + (0.1766 kJ/K) T = 58.02 kJ

$(0.15506$ kJ/K) T = 58.02 kJ; T - 374.2 K

(d)   The reaction is endothermic, so an increase in the value of K_p as calculated in parts (b) and (c) should be accompanied by an increase in T.

19.79   (a) $\Delta G° = 3\Delta G_f° \, S(s) + 2\Delta G_f° \, H_2O(g) - \Delta G_f° \, SO_2(g) - 2\Delta G_f° \, H_2S(g)$

                $= 3(0) + 2(-228.57) - (-300.4) - 2(-33.01) = -90.72 = -90.7 \text{ kJ}$

           $\ln K = \dfrac{-\Delta G°}{RT} = \dfrac{-(-90.72 \text{ kJ})}{(8.314 \times 10^{-3} \text{ kJ/K})(298 \text{ K})} = 36.6165 = 36.6; \; K = 7.99 \times 10^{15}$

                                                                      $= 8 \times 10^{15}$

   (b)   The reaction is highly spontaneous at 298 K and feasible in principle. However, use of $H_2S(g)$ produces a severe safety hazard for workers and the surrounding community.

   (c) $P_{H_2O} = \dfrac{25 \text{ torr}}{760 \text{ torr/atm}} = 0.033 \text{ atm}$

          $K = \dfrac{P_{H_2O}^2}{P_{SO_2} \times P_{H_2S}^2}; \; P_{SO_2} = P_{H_2S} = x \text{ atm}$

          $K = 7.99 \times 10^{15} = \dfrac{(0.033)^2}{x(x)^2}; \; x^3 = \dfrac{(0.033)^2}{7.99 \times 10^{15}}$

          $x = 5 \times 10^{-7} \text{ atm}$

   (d) $\Delta H° = 3\Delta H_f° \, S(s) + 2\Delta H_f° \, H_2O(g) - \Delta H_f° \, SO_2(g) - 2\Delta H_f° \, H_2S(g)$

             $= 3(0) + 2(-241.82) - (-296.9) - 2(-20.17) = -146.4 \text{ kJ}$

          $\Delta S° = 3S° \, S(s) + 2S° \, H_2O(g) - S° \, SO_2(g) - 2S° \, H_2S(g)$

             $= 3(31.88) + 2(188.83) - 248.5 - 2(205.6) = -186.4 \text{ J/K}$

       The reaction is exothermic ($-\Delta H$) and has a negative $\Delta S°$, so the value of K will decrease and the reaction will become nonspontaneous at some higher temperature. The process will be less effective at elevated temperatures.

19.80   The activated complex in Figure 14.12 is a single "particle" or entity that contains 4 atoms. It is formed from two molecules, A_2 and B_2, that must collide with exactly the correct energy and orientation to form the single entity. There are many fewer degrees of freedom for the activated complex than the separate reactant molecules, so the *entropy of activation* is negative.

19.81   (a) $\Delta S_{sys} = n R \ln (V_2 / V_1)$

      (b)   For any reversible process, $\Delta S_{univ} = 0$.

      (c) $\Delta S = q_{rev} / T = n R \ln (V_2 / V_1); \; q_{rev} = n RT \ln (V_2 / V_1)$

          For expansion, $V_2 > V_1$, $\ln (V_2 / V_1) > 0$, $q_{rev} > 0$, so heat is transferred into the system.

      (d) $\Delta E = 0$ for this expansion.

      (e) $\Delta E = q + w; \; 0 = q + w; \; w = -q = -nRT \ln (V_2 / V_1)$

          For expansion, $q > 0$, so $w < 0$; the system does work on the surroundings.

      (f)   For a compression, the surroundings do work on the system, so $w > 0$ and $q < 0$. Since q_{sys} decreases, q_{surr} increases and the cylinder of the pump gets warm.

20 Electrochemistry

Oxidation-Reduction Reactions

20.1 (a) *Oxidation* is the loss of electrons.

(b) The electrons appear on the products side (right side) of an oxidation half-reaction.

(c) The *oxidant* is the reactant that is reduced; it gains the electrons that are lost by the substance being oxidized.

20.2 (a) *Reduction* is the gain of electrons.

(b) The electrons appear on the reactants side (left side) of a reduction half-reaction.

(c) The *reductant* is the reactant that is oxidized; it provides the electrons that are gained by the substance being reduced.

20.3 (a) I is reduced from +5 to 0; C is oxidized from +2 to +4.

(b) Hg is reduced from +2 to 0; N is oxidized from -2 to 0.

(c) N is reduced from +5 to +2; S is oxidized from -2 to 0.

(d) Cl is reduced from +4 to +3; O is oxidized from -1 to 0.

20.4 (a) No oxidation-reduction

(b) N is both oxidized and reduced; it is reduced from +4 to +2 and oxidized from +4 to +5.

(c) No oxidation-reduction

(d) S is reduced from +6 to +4; Br is oxidized from -1 to 0.

20.5 (a) $2PbS(s) + 3O_2(g) \xrightarrow{\Delta} 2PbO(s) + 2SO_2(g)$

(b) O_2 is the oxidant, it undergoes reduction from the 0 to the -2 state. S is the reductant; it is oxidized from the -2 to the +4 state. The oxidation state of Pb is unchanged in the reaction.

20.6 (a) $2N_2H_4(g) + N_2O_4(g) \rightarrow 3N_2(g) + 4H_2O(l)$

(b) N_2O_4 serves as the oxidizing agent; it is itself reduced. N_2H_4 serves as the reducing agent; it is itself oxidized.

20.7 (a) $Co^{2+}(aq) \rightarrow Co^{3+}(aq) + 1e^-$ (oxidation)

(b) $H_2O_2(aq) \rightarrow O_2(g) + 2H^+(aq) + 2e^-$ (oxidation)

(c) $ClO_3^-(aq) + 6H^+(aq) + 6e^- \rightarrow Cl^-(aq) + 3H_2O(l)$ (reduction)

(d) $4OH^-(aq) \rightarrow O_2(g) + 2H_2O(l) + 4e^-$ (oxidation)

(e) $SO_3^{2-}(aq) + 2OH^-(aq) \rightarrow SO_4^{2-}(aq) + H_2O(l) + 2e^-$ (oxidation)

20.8 (a) $Sn^{4+}(aq) + 2e^- \rightarrow Sn^{2+}(aq)$ (reduction)

(b) $Mn^{2+}(aq) + 2H_2O(l) \rightarrow MnO_2(s) + 4H^+(aq) + 2e^-$ (oxidation)

(c) $NO_3^-(aq) + 4H^+(aq) + 3e^- \rightarrow NO(g) + 2H_2O(l)$ (reduction)

(d) $ClO^-(aq) + H_2O(l) + 2e^- \rightarrow Cl^-(aq) + 2OH^-(aq)$ (reduction)

(e) $Cr(OH)_3(s) + 5OH^-(aq) \rightarrow CrO_4^{2-}(aq) + 4H_2O(l) + 3e^-$ (oxidation)

20.9 (a) $Pb(OH)_4^{2-}(aq) + ClO^-(aq) \rightarrow PbO_2(s) + Cl^-(aq) + 2OH^-(aq) + H_2O(l)$

(b) The half reactions are:

$$3H_2O(l) + Tl_2O_3(s) + 4e^- \rightarrow 2TlOH(s) + 4OH^-(aq)$$
$$2[2NH_2OH(aq) + 2OH^-(aq) \rightarrow N_2(g) + 4H_2O(l) + 2e^-]$$

Net: $\overline{Tl_2O_3(s) + 4NH_2OH(aq) \rightarrow 2TlOH(s) + 2N_2(g) + 5H_2O(l)}$

(c)

$$2[Cr_2O_7^{2-}(aq) + 14H^+(aq) + 6e^- \rightarrow 2Cr^{3+}(aq) + 7H_2O(l)]$$
$$3[CH_3OH(aq) + H_2O(l) \rightarrow HCO_2H(aq) + 4H^+(aq) + 4e^-]$$

Net: $\overline{2Cr_2O_7^{2-}(aq) + 3CH_3OH(aq) + 16H^+(aq) \rightarrow 4Cr^{3+}(aq) + 3HCO_2H(aq) + 11H_2O(l)}$

(d)

$$2[MnO_4^-(aq) + 8H^+(aq) + 5e^- \rightarrow Mn^{2+}(aq) + 4H_2O(l)]$$
$$5[2Cl^-(aq) \rightarrow Cl_2(g) + 2e^-]$$

Net: $\overline{2MnO_4^-(aq) + 10Cl^-(aq) + 16H^+(aq) \rightarrow 2Mn^{2+}(aq) + 5Cl_2(g) + 8H_2O(l)}$

(e) $H_2O_2(aq) + 2e^- \rightarrow O_2(g) + 2H^+(aq)$

Since the reaction is in base, the H^+ can be "neutralized" by adding $2OH^-$ to each side of the equation to give $H_2O_2(aq) + 2OH^-(aq) + 2e^- \rightarrow O_2(g) + 2H_2O(l)$. The other half reaction is $2[ClO_2(aq) + e^- \rightarrow ClO_2^-(aq)]$.

Net: $H_2O_2(aq) + 2ClO_2(aq) + 2OH^-(aq) \rightarrow O_2(g) + 2ClO_2^-(aq) + 2H_2O(l)$

(f)

$$3[NO_2^-(aq) + H_2O(l) \rightarrow NO_3^-(aq) + 2H^+(aq) + 2e^-]$$
$$Cr_2O_7^{2-}(aq) + 14H^+(aq) + 6e^- \rightarrow 2Cr^{3+}(aq) + 7H_2O(l)$$

Net: $\overline{3NO_2^-(aq) + Cr_2O_7^{2-}(aq) + 8H^+(aq) \rightarrow 3NO_3^-(aq) + 2Cr^{3+}(aq) + 4H_2O(l)}$

20.10 (a) $Cr_2O_7^{2-}(aq) + I^-(aq) + 8H^+ \rightarrow 2Cr^{3+}(aq) + IO_3^-(aq) + 4H_2O(l)$

(b) $4MnO_4^-(aq) + 5CH_3OH(aq) + 12H^+(aq) \rightarrow 4Mn^{2+}(aq) + 5HCO_2H(aq) + 11H_2O(l)$

(c) $4As(s) + 3ClO_3^-(aq) + 3H^+(aq) + 6H_2O(l) \rightarrow 4H_3AsO_3(aq) + 3HClO(aq)$

(d) $As_2O_3(s) + 2NO_3^-(aq) + 2H_2O(l) + 2H^+(aq) \rightarrow 2H_3AsO_4(aq) + N_2O_3(aq)$

(e) $2MnO_4^-(aq) + Br^-(aq) + H_2O(l) \rightarrow 2MnO_2(s) + BrO_3^-(aq) + 2OH^-(aq)$

(f) $4H_2O_2(aq) + Cl_2O_7(aq) + 2OH^-(aq) \rightarrow 2ClO_2^-(aq) + 4O_2(g) + 5H_2O(l)$

Voltaic Cells; Cell Potential

20.11 (a)   The reaction $Cu^{2+}(aq) + Zn(s) \rightarrow Cu(s) + Zn^{2+}(aq)$ is occurring in both Figures. In Figure 20.3, the reactants are in contact, and the concentrations of the ions in solution aren't specified. In Figure 20.4, the oxidation half-reaction and reduction half-reaction are occurring in separate compartments, joined by a porous connector. The concentrations of the two solutions are initially 1.0 *M*. In Figure 20.4, electrical current is isolated and flows through the voltmeter. In Figure 20.3, the flow of electrons cannot be isolated or utilized.

       (b)   In the cathode compartment of the voltaic cell in Figure 20.4, Cu^{2+} cations are reduced to Cu atoms, decreasing the number of positively charged particles in the compartment. Na^+ cations are drawn into the compartment to maintain charge balance as Cu^{2+} ions are removed.

20.12 (a)   The salt bridge provides a mechanism by which ions not directly involved in the redox reaction can migrate into the anode and cathode compartments to maintain charge neutrality of the solutions. Ionic conduction within the cell, through the salt bridge, completes the cell circuit.

       (b)   In the anode compartment of Figure 20.4, $Zn°$ atoms are oxidized to Zn^{2+} cations, increasing the number of positively charged particles in the compartment. NO_3^- anions migrate into the compartment to maintain charge balance as Zn^{2+} ions are produced.

20.13 (a) $Zn(s) \rightarrow Zn^{2+}(aq) + 2e^-$; $Ni^{2+}(aq) + 2e^- \rightarrow Ni(s)$

       (b)   Zn(s) is the anode; Ni(s) is the cathode.

       (c)   Zn(s) is negative (-); Ni(s) is positive (+).

       (d)   Electrons flow from the Zn(-) electrode toward the Ni(+) electrode.

       (e)   Cations migrate toward the Ni(s) cathode; anions migrate toward the Zn(s) anode.

20.14 (a) $Ag^+(aq) + 1e^- \rightarrow Ag(s)$; $Ni(s) \rightarrow Ni^{2+}(aq) + 2e^-$

       (b)   Ni(s) is the anode, Ag(s) is the cathode.

       (c)   Ni(s) is negative; Ag(s) is positive.

       (d)   Electrons flow from the Ni(-) electrode to the Ag(+) electrode.

       (e)   Cations migrate toward the Ag(s) cathode, anions migrate toward the Ni(s) anode.

20.15 (a)   *Electromotive force*, emf, is the driving force that causes electrons to flow through the external circuit of a voltaic cell. It is the potential energy difference between an electron at the anode and an electron at the cathode.

        (b)   One *volt* is the potential energy difference required to impart 1 J of energy to a charge of 1 coulomb. 1 V = 1 J/C.

        (c)   The *standard cell potential* is the cell potential (emf) measured at standard conditions, 1 *M* aqueous solutions and gases at 1 atm pressure.

20.16 (a)   Electrons generated at the Zn anode flow through the external circuit to the Cu cathode because their potential energy at the cathode is lower then their potential energy at the anode. This potential energy difference is the driving force of a voltaic cell.

        (b)   *Electrical potential* is a potential energy difference measured in volts, the energy difference per unit charge. 1 V = 1 J/C

        (c)   A standard hydrogen electrode (SHE) is an electrode designed to produce the reference half-reaction
$$2H^+(aq, 1\ M) + 2e^- \rightarrow H_2\ (g, 1\ atm);\quad E° = 0\ V$$
It consists of a platinum electrode immersed in 1 *M* H^+(aq) and encased in an open glass tube with H_2(g) at 1 atm bubbling over the electrode.

20.17 (a)   A *standard reduction potential* is the relative potential of a reduction half-reaction measured at standard conditions, 1 *M* aqueous solutions and 1 atm gas pressure.

        (b)   The reference half-reaction for determining standard reduction potentials is $2H^+$ (aq, 1 *M*) + $2e^- \rightarrow H_2$ (g, 1 atm). By definition, $E°_{red}$ for this half-reaction is 0.

        (c)   The reduction of Ag^+(aq) to Ag(s) is much more energetically favorable, because it has a substantially more positive $E°_{red}$ (0.799 V) than the reduction of Sn^{2+}(aq) to Sn(s) (-0.136 V).

20.18 (a)   It is not possible to measure the standard reduction potential of a single half-reaction because each voltaic cell consists of two half-reactions and only the potential of a complete cell can be measured.

        (b)   The standard reduction potential of a half-reaction is measured by combining it with a reference half-reaction of known potential and measuring the cell potential. Assuming the half-reaction of interest is the reduction half-reaction: $E°_{cell}$ = $E°_{red}$(cathode) - $E°_{red}$(anode) = $E°_{red}$(unknown) - $E°_{red}$(reference); $E°_{red}$(unknown) = $E°_{cell}$ + $E°_{red}$ (reference).

        (c) Na^+(aq) + $e^- \rightarrow$ Na(s)   E° = -2.71 V

               Ca^{2+}(aq) + $2e^- \rightarrow$ Ca(s)   E° = -2.87 V

The reduction of $Ca^{2+}(aq)$ to $Ca(s)$ is the more difficult reduction because it has a more negative $E°$ value.

20.19 (a)   The two half-reactions are:

$Tl^{3+}(aq) + 2e^- \rightarrow Tl^+(aq)$ cathode $E°_{red}$ = ?

$2(Cr^{2+}(aq) \rightarrow Cr^{3+}(aq) + e^-)$ anode $E°_{red}$ = -0.41

(b) $E°_{cell} = E°_{red}$ (cathode) - $E°_{red}$ (anode); 1.19 V = $E°_{red}$ - (-0.41 V);

$E°_{red}$ = 1.19 V - 0.41 V = 0.78 V

(c)

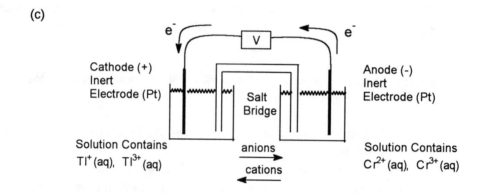

Note that because $Cr^{2+}(aq)$ is readily oxidized, it would be necessary to keep oxygen out of the right-hand cell compartment.

20.20 (a) $PdCl_4^{2-}(aq) + 2e^- \rightarrow Pd(s) + 4Cl^-$ cathode $E°_{red}$ = ?

$Cd(s) \rightarrow Cd^{2+}(aq) + 2e^-$ anode $E°_{red}$ = -0.403 V

(b) $E°_{cell} = E°_{red}$ (cathode) - $E°_{red}$ (anode); 1.03 V = $E°_{red}$ - (-0.403 V);

$E°_{red}$ = 1.03 V - 0.403 = 0.63 V

(c)

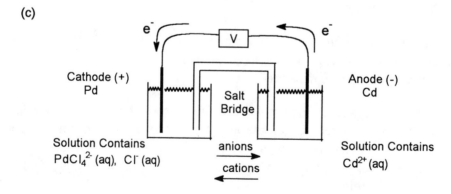

20.21 (a) $Cl_2(g) + 2e^- \rightarrow 2Cl^-(aq)$ $E°_{red}$ = 1.359 V

$2I^-(aq) \rightarrow I_2(s) + 2e^-$ $E°_{red}$ = 0.536 V

$E°$ = 1.359 V - 0.536 V = 0.823 V

(b) $2[Fe^{3+}(aq) + 1e^- \rightarrow Fe^{2+}(aq)]$ $E^\circ_{red} = 0.771$ V

$Hg(l) \rightarrow Hg^{2+}(aq) + 2e^-$ $E^\circ_{red} = 0.854$ V

$E^\circ = 0.771V - 0.854$ V $= -0.083$ V

(c) $Cu^+(aq) + 1e^- \rightarrow Cu^\circ(s)$ $E^\circ_{red} = 0.521$ V

$Cu^+(aq) \rightarrow Cu^{2+}(aq) + 1e^-$ $E^\circ_{red} = 0.153$ V

$E^\circ = 0.521$ V $- 0.153$ V $= 0.368$ V

20.22 (a) $Mn^{2+}(aq) \rightarrow Mn^\circ(s)$ $E^\circ_{red} = -1.18$ V

$Ca(s) \rightarrow Ca^{2+}(aq) + 2e^-$ $E^\circ_{red} = -2.87$ V

$E^\circ = -1.18$ V $- (-2.87$ V$) = 1.69$ V

(b) $2[Al^{3+}(aq) + 3e^- \rightarrow Al^\circ(s)]$ $E^\circ_{red} = -1.66$ V

$3[Zn(s) \rightarrow Zn^{2+} + 2e^-]$ $E^\circ_{red} = -0.763$ V

$E^\circ = -1.66$ V $- (-0.763$ V$) = -0.90$ V

(c) $Co^{2+}(aq) + 2e^- \rightarrow Co(s)$ $E^\circ_{red} = -0.277$ V

$2[Co^{2+}(aq) \rightarrow Co^{3+}(aq) + 1e^-]$ $E^\circ_{red} = 1.842$ V

$E^\circ = -0.277$ V $- 1.842$ V $= -2.119$ V

20.23 (a) $Hg^{2+}(aq) + 2e^- \rightarrow Hg(l)$ $E^\circ_{red} = 0.854$ V

$Mn(s) \rightarrow Mn^{2+}(aq) + 2e^-$ $E^\circ_{red} = -1.18$ V

$Hg^{2+}(aq) + Mn(s) \rightarrow Hg(l) + Mn^{2+}(aq)$ $E^\circ = 0.854 - (-1.18) = 2.03$ V

(b)   The reverse of the two reactions above would produce the smallest (most negative) cell emf, -2.03 V. The combination with the smallest positive cell emf is

$Hg^{2+}(aq) + 2e^- \rightarrow Hg(l)$ $E^\circ_{red} = 0.854$ V

$2Cu(s) \rightarrow 2Cu^+(aq) + 2e^-$ $E^\circ_{red} = 0.521$ V

$Hg^{2+}(aq) + 2Cu(s) \rightarrow Hg(l) + 2Cu^+(aq)$ $E^\circ = 0.854 - 0.521 = 0.333$ V

20.24 (a) $2[Au(s) + 4Br^-(aq) \rightarrow AuBr_4^-(aq) + 3e^-]$ $E^\circ_{red} = -0.858$ V

$3[2e^- + IO^-(aq) + H_2O(l) \rightarrow I^-(aq) + 2OH^-(aq)]$ $E^\circ_{red} = 0.49$ V

$2Au(s) + 8Br^-(aq) + 3IO^-(aq) + 3H_2O(l) \rightarrow 2AuBr_4^-(aq) + 3I^-(aq) + 6OH^-(aq)$

$E^\circ = 0.49 - (-0.858) = 1.35$ V

(b)   The reverse of the two reactions above would produce the smallest (most negative) cell emf, -1.35 V. The combination with the smallest positive cell emf is

$$2Eu^{2+}(aq) + Sn^{2+}(aq) \rightarrow 2Eu^{3+}(aq) + Sn(s)$$

$$E° = -0.14 - (-0.43) = 0.29 \text{ V}$$

20.25   The reduction half-reactions are:

$$Fe^{3+}(aq) + 1e^- \rightarrow Fe^{2+}(aq) \qquad E°_{red} = 0.771 \text{ V}$$

$$H_2O_2(aq) + 2H^+(aq) + 2e^- \rightarrow 2H_2O(l) \qquad E°_{red} = 1.776 \text{ V}$$

(a)   Because it has the larger $E°_{red}$, the second half-reaction occurs more readily.

(b)   Since the second half-reaction occurs more readily, it is the reduction half-reaction and occurs at the cathode of the complete cell.

(c) $Fe^{2+}(aq) \rightarrow Fe^{3+}(aq) + 1e^-$

(d) $E°_{cell} = 1.776 \text{ V} - 0.771 \text{ V} = 1.005 \text{ V}$

20.26   (a)   The half-reactions are:

$$2H^+(aq) + 2e^- \rightarrow H_2(g) \qquad E°_{red} = 0.00 \text{ V}$$

$$Al(s) \rightarrow Al^{3+}(aq) + 3e^- \qquad E°_{red} = -1.66 \text{ V}$$

Because it has the larger $E°_{red}$ (0.00 V vs -1.66 V), the first half-reaction is the reduction half-reaction in a voltaic cell. The standard hydrogen electrode (SHE) is the cathode and the Al strip is the anode.

(b)   The Al strip will lose mass as the reaction proceeds, because Al(s) is transformed to $Al^{3+}(aq)$.

(c) $E°_{cell} = 0.00 \text{ V} - (-1.66 \text{ V}) = 1.66 \text{ V}$

20.27   The reduction half-reactions are:

$$Cu^{2+}(aq) + 2e^- \rightarrow Cu(s) \qquad E° = 0.337 \text{ V}$$

$$Sn^{2+}(aq) + 2e^- \rightarrow Sn(s) \qquad E° = -0.136 \text{ V}$$

(a)   It is evident that Cu^{2+} is more readily reduced. Therefore, Cu serves as the cathode, Sn as the anode.

(b)   The copper electrode gains mass as Cu is plated out, the Sn electrode loses mass as Sn is oxidized.

(c)   The overall cell reaction is $Cu^{2+}(aq) + Sn(s) \rightarrow Cu(s) + Sn^{2+}(aq)$

(d) $E° = 0.337 - (-0.136) = 0.473 \text{ V}$

20.28   (a)   The two half-reactions are:

$$Pb^{2+}(aq) + 2e^- \rightarrow Pb°(s) \qquad E° = -0.126 \text{ V}$$

$$Cl_2(g) + 2e^- \rightarrow 2Cl^-(aq) \qquad E° = 1.359 \text{ V}$$

Because $E°$ for the reduction of Cl_2 is higher, the reduction of Cl_2 occurs at the Pt cathode. The Pb electrode is the anode.

(b)   Neither electrode gains mass, because the Cl^- ions produced at the cathode are not deposited on the Pt, and the Pb anode loses mass as Pb^{2+} (aq) is produced.

(c) $Cl_2(g) + Pb(s) \rightarrow Pb^{2+}(aq) + 2Cl^-(aq)$

(d) $E° = 1.359\ V - (-0.126\ V) = 1.485\ V$

Oxidizing and Reducing Agents; Spontaneity

20.29 (a)   Top. The reduction half-reactions near the top of Table 20.1 are most likely to occur; a strong oxidant is most likely to be reduced.

(b)   Left. An oxidant is reduced, so it is a reactant in a reduction half-reaction.

20.30 (a)   Negative. A strong reductant is likely to be oxidized, thus having a negative reduction potential.

(b)   Right. Reducing agents are likely to be oxidized, and thus to be in a low oxidation state; the products of reduction half-reactions are in lower oxidation states than reactants.

20.31 In each case, choose the half-reaction with the more positive reduction potential and with the given substance on the left.

(a) $Br_2(l)$ (1.065 V vs 0.536 V)   (b) $Ag^+(aq)$ (0.799 V vs 0.222 V)
(c) $Ce^{4+}(aq)$ (1.61 V vs 1.359 V)   (d) MnO_4^- (aq, acidic) (1.51 V vs 1.33 V)

20.32 In each case choose the half-reaction with the more negative reduction potential and the given substance on the right.

(a)   Al(s) (-1.66 V vs -0.28 V)

(b)   K(s) (-2.925 V vs -2.37 V)

(c) $Fe(CN)_6^{4-}$ (aq) (0.36 V vs 0.68 v)

(d) H_2(g, basic) (-0.83 V vs 0.000 V)

20.33 (a)   The strongest oxidizing agent is the species most readily reduced, as evidenced by a large, positive reduction potential. That species is H_2O_2. The weakest oxidizing agent is the species that least readily accepts an electron. We expect that it will be very difficult to reduce Zn(s); indeed, Zn(s) acts as a comparatively strong **reducing** agent. No potential is listed for reduction of Zn(s), but we can safely assume that it is less readily reduced than any of the other species present.

(b)   The strongest reducing agent is the species most easily oxidized (the largest negative reduction potential). Zn, E_{red}° = -0.76 V, is the strongest reducing agent and F^-, E_{red}° = 2.87 V, is the weakest.

20.34 (a)   Arranged in order of increasing strength as oxidizing agents (and increasing reduction potential):

$$Cu^{2+}(aq) \; < \; O_2(g) \; < \; Cr_2O_7^{2-}(aq) \; < \; Cl_2(g) \; < \; H_2O_2(aq)$$

(b)   Arranged in order of increasing strength as reducing agents (and decreasing reduction potential):

$$H_2O_2(aq) \; < \; I^-(aq) < Sn^{2+}(aq) < Zn(s) < Al(s)$$

20.35 Any of the **reduced** species in Table 20.1 from a half-reaction with a reduction potential more negative than -0.43 V will reduce Eu^{3+} to Eu^{2+}. These include Zn(s), $H_2(g)$, etc. Fe(s) is questionable.

20.36 Any oxidized species with a reduction potential greater than 0.59 V will oxidize RuO_4^{2-} to RuO_4^-. MnO_4^- is on the borderline, but any species above it on Table 20.1 would be strong enough.

20.37 (a)   The more positive the emf of a reaction the more spontaneous the reaction.

(b)   Reactions (a) and (c) in Exercise 20.21 are spontaneous.

20.38 (a) ΔG° = -nFE°. The more positive the emf of a reaction, the more negative the value of ΔG.

(b)   In Exercise 20.22, only the reaction in part (a) has a positive E° value and is spontaneous.

20.39 ΔG° = -nFE°

(a) ΔG° = -2 mol e^- × $\dfrac{96,500 \text{ J}}{\text{V} \cdot \text{mol } e^-}$ × $\dfrac{1 \text{ kJ}}{1000 \text{ J}}$ × 0.823 V = -159 kJ

(b) ΔG° = -2(96.5)(-0.083) = 16 kJ

(c) ΔG° = -1(96.5)(0.368) = -35.5 kJ

20.40 ΔG° = -nFE°

(a) ΔG° = -2 mol e^- × $\dfrac{96,500 \text{ J}}{\text{V} \cdot \text{mol } e^-}$ × $\dfrac{1 \text{ kJ}}{1000 \text{ J}}$ × 1.69 V = -326 kJ

(b) ΔG° = -6(96.5)(-0.90) = 5.2 × 10^2 kJ

(c) ΔG° = -2(96.5)(-2.119) = 409.0 kJ

20.41   (a) $2Fe^{2+}(aq) + S_2O_6^{2-}(aq) + 4H^+(aq) \rightarrow 2Fe^{3+}(aq) + 2H_2SO_3(aq)$

            $E° = 0.60\ V - 0.77\ V = -0.17\ V$

            $2Fe^{2+}(aq) + N_2O(aq) + 2H^+(aq) \rightarrow 2Fe^{3+}(aq) + N_2(g) + H_2O(l)$

            $E° = -1.77\ V - 0.77\ V = -2.54\ V$

            $Fe^{2+}(aq) + VO_2^+(aq) + 2H^+(aq) \rightarrow Fe^{3+}(aq) + VO^{2+}(aq) + H_2O(l)$

            $E° = 1.00\ V - 0.77\ V = +0.23\ V$

   (b) $\Delta G° = -nFE°$ For the first reaction,

            $\Delta G° = -2\ mol \times \dfrac{96,500\ J}{1\ V \cdot mol} \times (-0.17\ V) = 3.3 \times 10^5\ J$ or $33\ kJ$

            For the second reaction, $\Delta G° = -2(96,500)(-2.54) = 4.90 \times 10^2\ kJ$

            For the third reaction, $\Delta G° = -1(96,500)(0.23) = -22\ kJ$

20.42   (a) $2I^-(aq) \rightarrow I_2(s) + 2e^-$ $E°_{red} = 0.536\ V$

               $Hg_2^{2+}(aq) + 2e^- \rightarrow 2Hg(l)$ $E°_{red} = 0.789\ V$

           ———

             $2I^-(aq) + Hg_2^{2+}(aq) \rightarrow I_2(s) + 2Hg(l)$ $E° = 0.789 - 0.536 = 0.253\ V$

             $\Delta G° = -nFE° = -2\ mol\ e^- \times \dfrac{96.5\ kJ}{V \cdot mol\ e^-} \times 0.253\ V = -48.8\ kJ$

   (b) $3[Cu^+(aq) \rightarrow Cu^{2+}(aq) + 1e^-]$ $E°_{red} = 0.153\ V$

               $NO_3^-(aq) + 4H^+(aq) + 3e^- \rightarrow NO(g) + H_2O(l)$ $E°_{red} = 0.96\ V$

           ———

             $3Cu^+(aq) + NO_3^-(aq) + 4H^+(aq) \rightarrow 3Cu^{2+}(aq) + NO(g) + 2H_2O(l)$

             $E° = 0.96 - 0.153 = 0.81\ V$; $\Delta G° = -3(96.5)(0.81) = -2.3 \times 10^2\ kJ$

   (c) $2[Cr(OH)_3(s) + 5OH^-(aq) \rightarrow CrO_4^{2-}(aq) + 4H_2O(l) + 3e^-]$ $E°_{red} = -0.13\ V$

               $3[ClO^-(aq) + H_2O(l) + 2e^- \rightarrow Cl^-(aq) + 2OH^-(aq)]$ $E°_{red} = 0.89\ V$

           ———

             $2Cr(OH)_3(s) + 3ClO^-(aq) + 4OH^-(aq) \rightarrow 2CrO_4^{2-}(aq) + 3Cl^-(aq) + 5H_2O(l)$

             $E° = 0.89 - (-0.13) = 1.02\ V$; $\Delta G° = -6(96.5)(1.02) = -591\ kJ$

The Nernst Equation

20.43 $Zn(s) + 2H^+(aq) \rightarrow Zn^{2+}(aq) + H_2(g)$; $E = E° - \dfrac{0.0592}{n} \log Q$; $Q = \dfrac{[Zn^{2+}]P_{H_2}}{[H^+]^2}$

     (a) P_{H_2} decreases, Q decreases, E increases

     (b)   No effect (Zn(s) does not appear in Q expression.)

     (c) $[H^+]$ increases, Q decreases, E increases

     (d)   No effect ($NaNO_3$ does not participate in the reaction.)

20.44 $Al(s) + 3Ag^+(aq) \rightarrow Al^{3+}(aq) + 3Ag(s)$; $E = E° - \dfrac{0.0592}{n} \log Q$; $Q = \dfrac{[Al^{3+}]}{[Ag^+]^3}$

     Any change that causes the reaction to be less spontaneous (that causes Q to increase and ultimately shifts the equilibrium to the left) will result in a less positive value for $E°$.

(a) Decreases E by increasing $[Al^{3+}]$ on the right side of the equation.

(b) No effect; the "concentrations" of pure solids and liquids do not influence the value of K for a heterogeneous equilibrium.

(c) Decreases E; adding water decreases the concentration of Ag^+, which increases Q.

(d) No effect; the concentration of Ag^+ and the value of Q are unchanged.

20.45 (a)

$$3[Mn^{2+}(aq) + 2e^- \rightarrow Mn(s)] \qquad E^\circ_{red} = -1.18 \text{ V}$$
$$2[Al(s) \rightarrow Al^{3+}(aq) + 3e^-] \qquad E^\circ_{red} = -1.66 \text{ V}$$

$$3Mn^{2+}(aq) + 2Al(s) \rightarrow 3Mn(s) + 2Al^{3+}(aq) \qquad E^\circ = -1.18 - (-1.66) = 0.48 \text{ V}$$

(b) $$E = E^\circ - \frac{0.0592}{n} \log \frac{[Al^{3+}]^2}{[Mn^{2+}]^3}; \quad n = 6$$

$$E = 0.48 - \frac{0.0592}{6} \log \frac{(1.5)^2}{(0.10)^3} = 0.48 - \frac{0.0592}{6} \log (2.25 \times 10^3)$$

$$E = 0.48 - \frac{0.0592}{6} (3.352) = 0.48 - 0.033 = 0.45 \text{ V}$$

20.46 (a)

$$3[Ce^{4+}(aq) + 1e^- \rightarrow Ce^{3+}(aq)] \qquad E^\circ_{red} = 1.61 \text{ V}$$
$$Cr(s) \rightarrow Cr^{3+}(aq) + 3e- \qquad E^\circ_{red} = -0.74 \text{ V}$$

$$3Ce^{4+}(aq) + Cr(s) \rightarrow 3Ce^{3+}(aq) + Cr^{3+}(aq) \qquad E^\circ = 1.61 - (-0.74) = 2.35 \text{ V}$$

(b) $$E = E^\circ - \frac{0.0592}{n} \log \frac{[Ce^{3+}]^3 [Cr^{3+}]}{[Ce^{4+}]^3}; \quad n = 3$$

$$E = 2.35 - \frac{0.0592}{3} \log \frac{(0.010)^3 (0.010)}{(1.5)^3} = 2.35 - \frac{0.0592}{3} \log (2.963 \times 10^{-9})$$

$$E = 2.35 - \frac{0.0592 (-8.53)}{3} = 2.35 + 0.168 = 2.52 \text{ V}$$

20.47 (a)

$$4[Fe^{2+}(aq) \rightarrow Fe^{3+}(aq) + 1e^-] \qquad E^\circ_{red} = 0.771 \text{ V}$$
$$O_2(g) + 4H^+(aq) + 4e^- \rightarrow 2H_2O(l) \qquad E^\circ_{red} = 1.23 \text{ V}$$

$$4Fe^{2+}(aq) + O_2(g) + 4H^+(aq) \rightarrow 4Fe^{3+}(aq) + 2H_2O(l) \qquad E^\circ = 1.23 - 0.771 = 0.46 \text{ V}$$

(b) $$E = E^\circ - \frac{0.0592}{n} \log \frac{[Fe^{3+}]^4}{[Fe^{2+}]^4 [H^+]^4 P_{O_2}}; \quad n = 4, [H^+] = 1.00 \times 10^{-3} M$$

$$E = 0.46 \text{ V} - \frac{0.0592}{4} \log \frac{(1.0 \times 10^{-3})^4}{(2.0)^4 (1.0 \times 10^{-3})^4 (0.50)} = 0.46 - \frac{0.0592}{4} \log (0.125)$$

$$E = 0.46 - \frac{0.0592}{4} (-0.903) = 0.46 + 0.0134 = 0.47 \text{ V}$$

20.48 (a)

$$2[Fe^{3+}(aq) + 1\,e^- \rightarrow Fe^{2+}(aq)] \qquad\qquad E^{\circ}_{red} = 0.771\ V$$

$$H_2(g) \rightarrow 2H^+(aq) + 2e^- \qquad\qquad E^{\circ}_{red} = 0.000\ V$$

$$2Fe^{3+}(aq) + H_2(g) \rightarrow 2Fe^{2+}(aq) + 2H^+(aq) \qquad E^{\circ} = 0.771 - 0.000 = 0.771\ V$$

(b)

$$E = E^{\circ} - \frac{0.0592}{n} \log \frac{[Fe^{2+}]^2 [H^+]^2}{[Fe^{3+}]^2\, P_{H_2}}; \quad [H^+] = 10^{-pH} = 1.0 \times 10^{-4},\ n = 2$$

$$E = 0.771 - \frac{0.0592}{2} \log \frac{(0.010)^2\,(1.0 \times 10^{-4})^2}{(0.50)^2\,(0.25)} = 0.771 - \frac{0.0592}{2} \log(1.6 \times 10^{-11})$$

$$E = 0.771 - \frac{0.0592\,(-10.80)}{2} = 0.771 + 0.320 = 1.091\ V$$

20.49 $E = E^{\circ} - \dfrac{0.0592}{2} \log \dfrac{[H_2][Zn^{2+}]}{[H^+]^2}; \quad E^{\circ} = 0.0\ V - (-0.763\ V) = 0.763\ V$

$$0.720 = 0.763 - \frac{0.0592}{2} \times (\log[H_2][Zn^{2+}] - 2\log[H^+]) = 0.763 - \frac{0.0592}{2} \times (1.00 - 2\log[H^+])$$

$$0.720 = 0.763 + 0.0296 + 0.0592 \log[H^+]; \quad \log[H^+] = \frac{0.720 - 0.763 - 0.0296}{0.0592}$$

$$\log[H^+] = -1.226 = -1.2; \quad [H^+] = 0.0594 = 0.06\ M;\quad pH = 1.2$$

20.50 $E^{\circ} = -0.136\ V - (-0.126\ V) = -0.010\ V;\ n = 2$

$$0.22 = -0.010 - \frac{0.0592}{2} \log \frac{[Pb^{2+}]}{[Sn^{2+}]} = -0.010 - \frac{0.0592}{2} \log \frac{[Pb^{2+}]}{1.00}$$

$$\log[Pb^{2+}] = \frac{-0.23\,(2)}{0.0592} = -7.770 = -7.8; \quad [Pb^{2+}] = 1.7 \times 10^{-8} = 2 \times 10^{-8}\ M$$

For $PbSO_4(s)$, $K_{sp} = [Pb^{2+}][SO_4^{2-}] = (1.0)(1.7 \times 10^{-8}) = 1.7 \times 10^{-8}$

20.51 $E^{\circ} = \dfrac{0.0592\ V}{n} \log K;\quad \log K = \dfrac{nE^{\circ}}{0.0592\ V}$

(a) $E^{\circ} = 0.763 - 0.136 = 0.627\ V;\ n = 2\ (Sn^{+2} + 2e^- \rightarrow Sn)$

$$\log K = \frac{2(0.627)}{0.0592} = 21.182 = 21.2; \quad K = 1.52 \times 10^{21} = 2 \times 10^{21}$$

(b) $E^{\circ} = 0.277 - (0) = 0.277\ V;\ n = 2\ (2H^+ + 2e^- \rightarrow H_2)$

$$\log K = \frac{2(0.277)}{0.0592} = 9.358 = 9.36; \quad K = 2.3 \times 10^9$$

(c) $E^{\circ} = -1.065 - (-1.51) = 0.44\ V;\ n = 10\ (2MnO_4^- + 10\,e^- \rightarrow 2Mn^{+2})$

$$\log K = \frac{10(0.44)}{0.0592} = 74.324 \approx 74; \quad K = 2.1 \times 10^{74} = 10^{74}$$

20.52 $E° = \dfrac{0.0592\ V}{n} \log K;\ \ \log K = \dfrac{nE°}{0.0592\ V}$

(a) $E° = 1.00\ V - (-0.28\ V) = 1.28\ V;\ n = 2\ (Ni^{+2} + 2\ e^- \rightarrow Ni)$

$\log K = \dfrac{2(1.28\ V)}{0.0592\ V} = 43.243 = 43.2;\ K = 1.75 \times 10^{43} = 2 \times 10^{43}$

(b) $E° = 1.61\ V - 0.32\ V = 1.29\ V;\ n = 3\ (3Ce^{4+} + 3\ e^- \rightarrow 3Ce^{3+})$

$\log K = \dfrac{3(1.29)}{0.0592} = 65.372 = 65.4;\ K = 2.35 \times 10^{65} = 2 \times 10^{65}$

(c) $E° = 0.36\ V - (-0.23\ V) = 0.59\ V;\ n = 4\ (4Fe(CN)_6{}^{3-} + 4\ e^- \rightarrow 4Fe(CN)_6{}^{4-})$

$\log K = \dfrac{4(0.59)}{0.0592} = 39.865 = 40;\ K = 7 \times 10^{39} = 10^{40}$

20.53 $E° = \dfrac{0.0592}{n} \log K;\ \ \log K = \dfrac{nE°}{0.0592\ V}$

(a) $\log K = \dfrac{1(0.35\ V)}{0.0592\ V} = 5.912 = 5.9;\ K = 8.2 \times 10^5 = 8 \times 10^5\ (= 10^6)$

(b) $\log K = \dfrac{2(0.35\ V)}{0.0592\ V} = 11.824 = 12;\ K = 6.7 \times 10^{11} = 10^{12}$

(c) $\log K = \dfrac{3(0.35\ V)}{0.0592\ V} = 17.736 = 18;\ K = 5.5 \times 10^{17} = 10^{18}$

20.54 $E° = \dfrac{0.0592\ V}{n} \log K;\ \ n = \dfrac{0.0592\ V}{E°} \log K$

$n = \dfrac{0.0592\ V}{0.21\ V} \log 1.31 \times 10^7;\ n = 2$

Commercial Voltaic Cells

20.55   The overall cell reaction is:

$Zn(s) + 2NH_4{}^+(aq) + 2MnO_2(s) \rightarrow Mn_2O_3(s) + 2NH_3(aq) + H_2O(l)$

$120\ g\ Zn \times \dfrac{1\ mol\ Zn}{65.39\ g\ Zn} \times \dfrac{2\ mol\ MnO_2}{1\ mol\ Zn} \times \dfrac{86.94\ g\ MnO_2}{1\ mol\ MnO_2} = 319\ g\ MnO_2$

20.56   The overall cell reaction (Equation [20.19]) is:

$Pb(s) + PbO_2(s) + 4H^+(aq) + 2SO_4{}^{2-}(aq) \rightarrow 2PbSO_4(s) + 2H_2O(l)$

$600\ g\ Pb \times \dfrac{1\ mol\ Pb}{207.2\ g\ Pb} \times \dfrac{1\ mol\ PbO_2}{1\ mol\ Pb} \times \dfrac{239.2\ g\ PbO_2}{1\ mol\ PbO_2} = 693\ g\ PbO_2$

20.57   (a)   In discharge: $Cd(s) + NiO_2(s) + 2H_2O(l) \rightleftharpoons Cd(OH)_2(s) + Ni(OH)_2(s)$

             In charging, the reverse reaction occurs.

      (b) $E° = 0.49 \text{ V} - (-0.76 \text{ V}) = 1.25 \text{ V}$

20.58   (a) $HgO(s) + Zn(s) \rightarrow Hg(l) + ZnO(s)$

      (b) $E°_{cell} = E°_{red}$ (cathode) $- E°_{red}$ (anode); $E°_{red}$ (anode) $= E°_{red} - E°_{cell} = 0.098 - 1.35$

                                                            $= -1.25 \text{ V}$

      (c) $E°_{red}$ is different from $Zn^{2+}(aq) + 2e^- \rightarrow Zn(s)$ (-0.76 V) because in the battery the process happens in the presence of base and Zn^{2+} is stabilized as $ZnO(s)$. Stabilization of the product of a reaction increases the driving force for the reaction, so $E°$ is more positive.

20.59 $E°_{red}$ for Cd (-0.40 V) is less negative than $E°_{red}$ for Zn (-0.76 V), so E_{cell} will have a smaller (less positive) value.

20.60   The alkali metals Li and Na have much more metallic character than Zn, Cd, Pb or Ni. The reduction potentials for Li and Na are thus more negative, leading to greater overall cell potentials for the battery. Also, Li and Na have lower density, so greater total energy for a battery could be achieved for a given total mass of material. One disadvantage is that Na and Li are very reactive and the cell reactions are difficult to control.

Electrolysis; Electrical Work

20.61   (a)   The products are different because in aqueous electrolysis water is reduced in preference to Mg^{2+}.

      (b) $MgCl_2(l) \rightarrow Mg(l) + Cl_2(g)$
             $2Cl^-(aq) + 2H_2O(l) \rightarrow Cl_2(g) + H_2(g) + 2OH^-(aq)$

             The aqueous solution electrolysis is entirely analogous to that for NaCl(aq), Section 20.8.

      (c) $Mg^{2+}(aq) + 2e^- \rightarrow Mg(s)$ $E°_{red} = -2.37 \text{ V}$
             $2Cl^-(aq) \rightarrow Cl_2(g) + 2e^-$ $E°_{red} = 1.359 \text{ V}$

             ————————————————————————————————————

             $MgCl_2(aq) \rightarrow Mg(s) + Cl_2(g)$ $E° = -2.37 - 1.359 = -3.73 \text{ V}$

             $H_2O(l) + 2e^- \rightarrow H_2(g) + 2OH^-(aq)$ $E°_{red} = -0.83 \text{ V}$
             $2Cl^-(aq) \rightarrow Cl_2(g) + 2e^-$ $E°_{red} = 1.359 \text{ V}$

             ————————————————————————————————————

             $2Cl^-(aq) + 2H_2O(l) \rightarrow Cl_2(g) + H_2(g) + 2OH^-(aq)$ $E° = -0.83 - 1.359 = -2.19 \text{ V}$

             The minus signs mean that voltage must be applied in order for the reaction to occur.

20.62   (a)   anode: $2Br^-(l) \rightarrow Br_2(l) + 2e^-$

                 cathode: $Al^{3+}(l) \rightarrow Al(l) + 3e^-$ (Al(s) melts at a lower temperature than AlBr(s).)

      (b)   anode: $2Br^-(l) \rightarrow Br_2(l) + 2e^-$

                 cathode: $2H_2O(l) + 2e^- \rightarrow H_2(g) + 2OH^-(aq)$

                 The reduction potential for water is less negative than that for $Al^{3+}(aq)$, so water is reduced in preference to Al^{3+}.

      (c)   The minimum emf values for the two possibilities in aqueous solution are:
                 Al^{3+} is reduced: -1.66 V - 1.065 V = -2.73 V
                 H_2O is reduced: -0.83 V - 1.065 V = -1.90 V

                 The minus signs indicate that voltage must be applied.

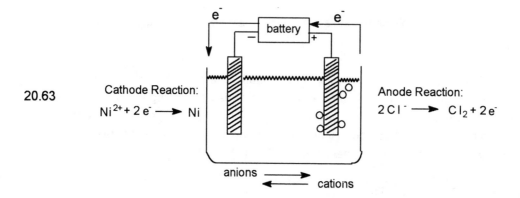

20.63   Cathode Reaction:

$Ni^{2+} + 2e^- \longrightarrow Ni$

                 Anode Reaction:

$2Cl^- \longrightarrow Cl_2 + 2e^-$

anions $\longrightarrow$
$\longleftarrow$ cations

Chlorine is produced in preference to oxidation of water because of a large overvoltage for O_2 formation.

20.64

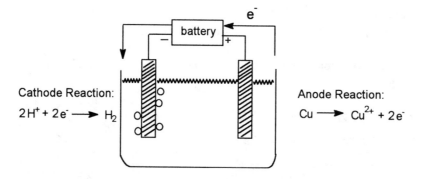

              Cathode Reaction:

$2H^+ + 2e^- \longrightarrow H_2$

                 Anode Reaction:

$Cu \longrightarrow Cu^{2+} + 2e^-$

Overall cell reaction: $Cu(s) + 2H^+(aq) \rightarrow Cu^{2+}(aq) + H_2(g)$ $E° = -0.34$ V
E_{min}, the **minimum** voltage required to produce the cell reaction under standard conditions, is 0.34 V. Note, however, that this value is at standard conditions $[H^+] = 1$ M, and $[Cu^{2+}] = 1$ M. We also ignore the overvoltage, which is likely to be several tenths of a volt.

20.65   Coulombs = amps•s, since this is a $3e^-$ reduction, each mole of Cr(s) requires 3 Faradays.

(a) $13.5 \text{ A} \times 3.00 \text{ d} \times \dfrac{24 \text{ hr}}{1 \text{ d}} \times \dfrac{60 \text{ min}}{1 \text{ hr}} \times \dfrac{60 \text{ s}}{1 \text{ min}} \times \dfrac{1 \text{ C}}{1 \text{ amp•s}} \times \dfrac{1 \text{ F}}{96,500 \text{ C}}$

         $\times \dfrac{1 \text{ mol Cr}}{3 \text{ F}} \times \dfrac{52.00 \text{ g Cr}}{1 \text{ mol Cr}} = 629 \text{ g Cr(s)}$

(b) $1.00 \text{ mol Cr} \times \dfrac{3 \text{ F}}{1 \text{ mol Cr}} \times \dfrac{96,500 \text{ C}}{\text{F}} \times \dfrac{1 \text{ amp•s}}{1 \text{ C}} \times \dfrac{1}{12.0 \text{ hr}} \times \dfrac{1 \text{ hr}}{60 \text{ min}} \times \dfrac{1 \text{ min}}{60 \text{ s}}$

                                                                     $= 6.70 \text{ A}$

20.66   Coulombs = amps•s; since this is a $2e^-$ reduction, each mole of Cr(s) requires 2 Faradays.

(a) $1.80 \text{ A} \times 16.7 \text{ hr} \times \dfrac{60 \text{ min}}{1 \text{ hr}} \times \dfrac{60 \text{ s}}{1 \text{ min}} \times \dfrac{1 \text{ C}}{1 \text{ amp•s}} \times \dfrac{1 \text{ F}}{96,500 \text{ C}} \times \dfrac{1 \text{ mol Mg}}{2 \text{ F}} \times \dfrac{24.31 \text{ g Mg}}{1 \text{ mol Mg}}$

                                                                     $= 13.6 \text{ g Mg}$

(b) $5.00 \text{ g Mg} \times \dfrac{1 \text{ mol Mg}}{24.31 \text{ g Mg}} \times \dfrac{2 \text{ F}}{1 \text{ mol Mg}} \times \dfrac{96,500 \text{ C}}{\text{F}} \times \dfrac{1 \text{ amp•s}}{\text{C}} \times \dfrac{1 \text{ min}}{60 \text{ s}} \times \dfrac{1}{2.00 \text{ A}}$

                                                                     $= 331 \text{ min}$

20.67   (a) $7.50 \text{ amp} \times 100 \text{ min} \times \dfrac{60 \text{ s}}{1 \text{ min}} \times \dfrac{1 \text{ C}}{1 \text{ amp•s}} \times \dfrac{1 \text{ Faraday}}{96,500 \text{ C}} \times \dfrac{1 \text{ mol Cl}_2}{2 \text{ Faraday}}$

           $\times \dfrac{22.400 \text{ L Cl}_2}{1 \text{ mol Cl}_2} = 5.22 \text{ L Cl}_2$

(b)   From the balanced equation (section 20.8), we see that 2 mol NaOH are formed per mol Cl_2. Proceeding as in (a), but replacing the last factor by (2 mol NaOH/1 mol Cl_2), we obtain 0.466 mol NaOH.

20.68   The cell reaction is presumed to be:

$$Sn^{2+}(aq) + H_2O(l) \rightarrow Sn(s) + 1/2\, O_2(g) + 2H^+(aq)$$

Calculate mol Sn^{2+} reduced:

$$4.60 \text{ amp} \times 30.0 \text{ min} \times \dfrac{60 \text{ s}}{1 \text{ min}} \times \dfrac{1 \text{ C}}{1 \text{ amp•s}} \times \dfrac{1 \text{ Faraday}}{96,500 \text{ C}} \times \dfrac{1 \text{ mol Sn}^{2+}}{2 \text{ Faraday}} = 0.0429 \text{ mol Sn}^{2+}$$

Initially there were $\dfrac{0.600 \text{ mol}}{\text{L}} \times (0.500 \text{ L}) = 0.300 \text{ mol}$. Thus, following electrolysis, there are $0.300 - 0.043 = 0.257 \text{ mol Sn}^{2+}$.

$[Sn^{2+}] = \dfrac{0.257 \text{ mol}}{0.500 \text{ L}} = 0.514 \; M$

Electrolysis also produces $2(0.0429) = 0.0858 \text{ mol } H^+(aq)$.

Thus $[H^+] = \dfrac{0.0858 \text{ mol}}{0.500 \text{ L}} = 0.172 \; M$. The concentration of SO_4^{2-} remains unchanged.

20.69 For a voltaic cell at standard conditions, $w_{max} = \Delta G° = -nFE°$.

$$I_2(s) + 2e^- \rightarrow 2I^-(aq) \qquad\qquad E°_{red} = 0.536 \text{ V}$$

$$Sn(s) \rightarrow Sn^{2+}(aq) + 2e^- \qquad\qquad E°_{red} = -0.136 \text{ V}$$

$$I_2(s) + Sn(s) \rightarrow 2I^-(aq) + Sn^{2+}(aq) \qquad E° = 0.536 - (-0.136) = 0.672 \text{ V}$$

$$w_{max} = -2(96.5)(0.672) = -129.7 = -130 \text{ kJ/mol Sn}$$

$$\frac{-129.7 \text{ kJ}}{\text{mol Sn(s)}} \times 0.460 \text{ mol Sn} = -59.7 \text{ kJ}$$

20.70 $w_{max} = \Delta G° = -nFE°$. Since the cell is at standard conditions, calculate $E°$.

$$2H^+(aq) + 2e^- \rightarrow H_2(g) \qquad\qquad E°_{red} = 0.000 \text{ V}$$

$$Cd(s) \rightarrow Cd^{2+}(aq) + 2e^- \qquad\qquad E°_{red} = -0.403 \text{ V}$$

$$2H^+(aq) + Cd(s) \rightarrow H_2(g) + Cd^{2+}(aq) \qquad E°_{cell} = 0.000 - (-0.403) = 0.403 \text{ V}$$

$$w_{max} = -2(96.5)(0.403) = -77.78 = -77.8 \text{ kJ/mol Cd(s)}$$

$$\frac{-77.78 \text{ kJ}}{\text{mol Cd(s)}} \times 0.780 \text{ mol Cd(s)} = -60.7 \text{ kJ} \quad \text{(The (-) sign means work is done } \textbf{by} \text{ the cell.)}$$

20.71 (a) $90{,}000 \text{ amp} \times 16 \text{ hr} \times \dfrac{3600 \text{ s}}{1 \text{ hr}} \times \dfrac{1 \text{ C}}{1 \text{ amp} \cdot \text{s}} \times \dfrac{1 \text{ Faraday}}{96{,}500 \text{ C}} \times \dfrac{1 \text{ mol Mg}}{2 \text{ Faraday}}$

$\times \dfrac{24.3 \text{ g Mg}}{1 \text{ mol Mg}} \times 0.50 = 3.3 \times 10^5 \text{ g Mg}$

(b) $\text{Coulombs} = 9 \times 10^4 \text{ amp} \times 16 \text{ hr} \times \dfrac{3600 \text{ s}}{1 \text{ hr}} \times \dfrac{1 \text{ C}}{1 \text{ amp} \cdot \text{s}} = 5.184 \times 10^9 = 5.2 \times 10^9 \text{ C}$

$\text{kilowatt-hours} = 5.184 \times 10^9 \text{ C} \times 4.20 \text{ V} \times \dfrac{1 \text{ J}}{1 \text{ C} \cdot \text{V}} \times \dfrac{1 \text{ kWh}}{3.6 \times 10^6 \text{ J}} = 6.0 \times 10^3 \text{ kWh}$

20.72 (a) $66{,}000 \text{ amp} \times 12 \text{ hr} \times \dfrac{3600 \text{ s}}{1 \text{ hr}} \times \dfrac{1 \text{ C}}{1 \text{ amp} \cdot \text{s}} \times \dfrac{1 \text{ Faraday}}{96{,}500 \text{ C}} \times \dfrac{1 \text{ mol Li}}{2 \text{ Faraday}}$

$\times \dfrac{6.94 \text{ g Li}}{1 \text{ mol Li}} \times 0.85 = 1.7 \times 10^5 \text{ g Li}$

(b) If the cell is 85% efficient, $\dfrac{96{,}500 \text{ C}}{\text{Faraday}} \times \dfrac{1 \text{ Faraday}}{0.85 \text{ mol}} = 1.135 \times 10^5$

$$= 1.1 \times 10^5 \text{ C/mol Li required}$$

$\text{Energy} = 5.50 \text{ V} \times \dfrac{1.135 \times 10^5 \text{ C}}{\text{mol Li}} \times \dfrac{1 \text{ J}}{1 \text{ C} \cdot \text{V}} \times \dfrac{1 \text{ kWh}}{3.6 \times 10^6 \text{ J}} = 0.17 \text{ kWh}$

Corrosion

20.73 (a)   anode: $Fe(s) \rightarrow Fe^{2+}(aq) + 2e^-$

cathode: $O_2(g) + 4H^+(aq) + 4e^- \rightarrow 2H_2O(l)$

(b) $2Fe^{2+}(aq) + 3H_2O(l) + 3H_2O(l) \rightarrow Fe_2O_3 \cdot 3H_2O(s) + 6H^+(aq) + 2e^-$

$O_2(g) + 4H^+(aq) + 4e^- \rightarrow 2H_2O(l)$

(Multiply the oxidation half-reaction by two to balance electrons and obtain the overall balanced reaction.)

20.74 (a) $CO_2(g)$ is a Lewis acid and produces an acidic aqueous solution. Atmospheric $CO_2(g)$ could maintain acidic conditions at the surface of the iron, encouraging corrosion.

(b) $O_2(g) + 4H^+(aq) + 4e^- \rightarrow 2H_2O(l)$ $E^{\circ}_{red} = 1.23$ V

$4[Ag(s) \rightarrow Ag^+(aq) + 1e^-]$ $E^{\circ}_{red} = 0.799$ V

$\overline{\qquad\qquad\qquad\qquad\qquad\qquad\qquad\qquad\qquad\qquad\qquad\qquad\qquad}$

$4Ag(s) + O_2(g) + 4H^+(aq) \rightarrow 4Ag^+(aq) + 2H_2O(l)$ $E^{\circ} = 1.23 - 0.799 = 0.43$ V

In acidic solution, the air oxidation of $Ag(s)$ has a postive E° value and is therefore spontaneous.

20.75 No. To act as a sacrificial anode, a metal must have a more negative reduction potential (be easier to oxidize) than Fe^{2+}. E°_{red} $Sn^{2+} = -0.14$ V, E°_{red} $Fe^{2+} = -0.44$ V. Tin gives protection by providing a complete cover for iron.

20.76 No. To afford cathodic protection, a metal must be more difficult to reduce (have a more negative reduction potential) than Fe^{2+}. E°_{red} $Co^{2+} = -0.28$ V, E°_{red} $Fe^{2+} = -0.44$ V.

20.77 The pH of an aqueous medium is an important factor in corrosion. Note from Section 20.10 that an increased $[H^+]$ shifts the equilibrium to the right. When $[H^+]$ is depressed by the presence of a base, the reduction of O_2 is less favorable, and corrosion slows down. The added amines serve the function of keeping $[H^+]$ low.

20.78 It is well established that corrosion occurs most readily when the metal surface is in contact with water. Thus, moisture is a requirement for corrosion. Corrosion also occurs more readily in acid solution, because O_2 has a more positive reduction potential in the presence of $H^+(aq)$. SO_2 and its oxidation products dissolve in water to produce acidic solutions, which encourage corrosion. The anodic and cathodic reactions for the corrosion of Ni are:

$Ni(s) \rightarrow Ni^{2+}(aq) + 2e^-$ $E^{\circ}_{red} = -0.28$ V

$O_2(g) + 4H^+(aq) + 4e^- \rightarrow 2H_2O(l)$ $E^{\circ}_{red} = 1.23$ V

Nickel(II) oxide, $NiO(s)$, can form by the dry air oxidation of Ni. This NiO coating serves to protect against further corrosion. However, NiO dissolves in acidic solutions such as those produced by SO_2 or SO_3, according to the reaction:

$NiO(s) + 2H^+(aq) \rightarrow Ni^{2+}(aq) + H_2O(l)$

This exposes $Ni(s)$ to further wet corrosion.

Additional Exercises

20.79 (a)

$$2HBrO_3(aq) + 10H^+(aq) + 10e^- \rightarrow Br_2(l) + 6H_2O(l)$$
$$10[Fe^{2+}(aq) \qquad\qquad \rightarrow Fe^{3+}(aq) + 1e^-]$$
$$\overline{}$$
$$10Fe^{2+}(aq) + 2HBrO_3(aq) + 10H^+(aq) \rightarrow 10Fe^{3+}(aq) + Br_2(l) + 6H_2O(l)$$

(b)

$$2[NO_3^-(aq) + 4H^+(aq) + 3e^- \rightarrow NO(g) + 2H_2O(l)]$$
$$3[Cu(s) \qquad\qquad \rightarrow Cu^{2+}(aq) + 2e^-]$$
$$\overline{}$$
$$3Cu(s) + 2NO_3^-(aq) + 8H^+(aq) \rightarrow 3Cu^{2+}(aq) + 2NO(g) + 4H_2O(l)$$

(c)

$$3IO_3^-(aq) + 18H^+(aq) + 16e^- \rightarrow I_3^-(aq) + 9H_2O(l)$$
$$\qquad\qquad + 18OH^-(aq) \qquad\rightarrow \qquad\qquad + 18OH^-(aq)$$
$$\overline{}$$
$$3IO_3^-(aq) + 9H_2O(l) + 16e^- \rightarrow I_3^-(aq) + 18OH^-(aq)$$
$$8[3I^-(aq) \qquad\qquad \rightarrow I_3^-(aq) + 2e^-]$$
$$\overline{}$$
$$1/3\,[3IO_3^-(aq) + 24I^-(aq) + 9H_2O(l) \rightarrow 9I_3^-(aq) + 18OH^-(aq)]$$
$$IO_3^-(aq) + 8I^-(aq) + 3H_2O(l) \rightarrow 3I_3^-(aq) + 6OH^-(aq)$$

20.80 (a)

$$MnO_4^{2-}(aq) + 4H^+(aq) + 2e^- \rightarrow MnO_2(s) + 2H_2O(l)$$
$$2\,[MnO_4^{2-}(aq) \rightarrow MnO_4^-(aq) + 1e^-]$$
$$\overline{}$$
$$3\,MnO_4^{2-}(aq) + 4H^+(aq) \rightarrow 2\,MnO_4^-(aq) + MnO_2(s) + 2H_2O(l)$$

(b)

$$H_2SO_3(aq) + 4H^+(aq) + 4e^- \rightarrow S(s) + 3H_2O(l)$$
$$2[H_2SO_3(aq) + H_2O \rightarrow HSO_4^-(aq) + 3H^+(aq) + 2e^-]$$
$$\overline{}$$
$$3H_2SO_3(aq) \rightarrow S(s) + 2HSO_4^-(aq) + 2H^+(aq) + H_2O(l)$$

(c)

$$Cl_2(aq) + 2H_2O(l) \rightarrow 2ClO^-(aq) + 4H^+(aq) + 2e^-$$
$$4OH^-(aq) \qquad\qquad + 4OH^-(aq)$$
$$\overline{}$$
$$Cl_2(aq) + 4OH^-(aq) \rightarrow 2ClO^-(aq) + 2H_2O(l) + 2e^-$$
$$Cl_2(aq) + 2e^- \rightarrow 2Cl^-(aq)$$
$$\overline{}$$
$$1/2[2Cl_2(aq) + 4OH^-(aq) \rightarrow 2Cl^-(aq) + 2ClO^-(aq) + 2H_2O(l)]$$
$$Cl_2(aq) + 2OH^-(aq) \rightarrow Cl^-(aq) + ClO^-(aq) + H_2O(l)$$

20.81 (a) Anode reaction: $\qquad\qquad\qquad\qquad 5Fe^{2+}(aq) \rightarrow 5Fe^{3+}(aq) + 5e^-$

Cathode reaction: $8H^+(aq) + MnO_4^-(aq) + 5e^- \rightarrow Mn^{2+}(aq) + 4H_2O(l)$

Electrons move from the Pt electrode in the $Fe^{2+}(aq)$ solution to the Pt electrode in the $MnO_4^-(aq)$ solution. Anions migrate through the salt bridge from the cathode beaker to the anode beaker; cations migrate in the opposite direction. The electrode in the iron-containing beaker has a negative sign, that in the MnO_4^- beaker has a positive sign.

(b)   Using Table 20.1, $E° = 1.51 \text{ V} - 0.771 \text{ V} = 0.74 \text{ V}$.

20.82 (a)

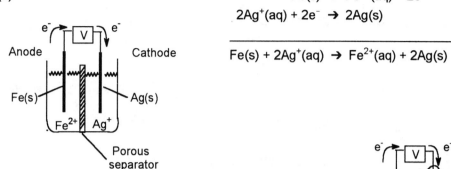

$$Fe(s) \rightarrow Fe^{2+}(aq) + 2e^-$$
$$2Ag^+(aq) + 2e^- \rightarrow 2Ag(s)$$

___

$$Fe(s) + 2Ag^+(aq) \rightarrow Fe^{2+}(aq) + 2Ag(s)$$

(b)

$$Zn(s) \rightarrow Zn^{2+}(s) + 2e^-$$
$$2H^+(aq) + 2e^- \rightarrow H_2(g)$$

___

$$Zn(s) + 2H^+(aq) \rightarrow Zn^{2+}(aq) + H_2(g)$$

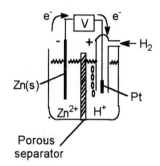

(c) $Cu|Cu^{2+}||ClO_3^-, Cl^-|Pt$ Here, both the oxidized and reduced forms of the cathode solution are in the same phase, so we separate them by a comma, and then indicate an inert electrode.

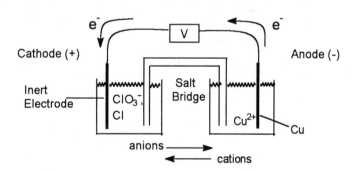

20.83 (a)   Clearly, Zn^{2+} is more difficult to reduce than Ag^+. Therefore, zinc is oxidized to Zn^{2+}, silver oxide is reduced to metallic silver. (The silver ion actually undergoes the reduction.)

(b)   Electrons move away from the anode ($Zn \rightarrow Zn^{2+} + 2e^-$) during cell operation. Thus, the anode is negatively charged. The cathode is positively charged; it is here that electrons are drawn to reduce the Ag_2O during cell operation.

(c) $Zn(s) \rightarrow Zn^{2+}(aq) + 2e^-$ $E^°_{red} = -0.763 \text{ V}$

               $Ag_2O(s) + H_2O(l) \rightarrow 2Ag(s) + 2OH^-(aq)$ $E^°_{red} = 0.344 \text{ V}$

___

$$Zn(s) + Ag_2O(s) + H_2O(l) \rightarrow Zn^{2+}(aq) + 2Ag(s) + 2OH^-(aq)$$

$E° = 0.344 - (-0.763) = 1.107 \text{ V}$

20.84 $2[Rh^{3+}(aq) + 3e^- \rightarrow Rh(s)]$ $E^\circ_{red} = ?$

        $3[Cd(s) \rightarrow Cd^{2+}(aq) + 2e^-]$ $E^\circ_{red} = -0.403$ V

     $2Rh^{3+}(aq) + 3Cd(s) \rightarrow 2Rh(s) + 3Cd^{2+}(aq)$ $E^\circ = 1.20$ V

(b)   Cd(s) is the anode, and Rh(s) is the cathode.

(c)   The cell is at standard conditions. $E^\circ_{cell} = E^\circ_{red}$ (cathode) - E°_{red} (anode)

     $E^\circ_{red} = E^\circ_{cell} + E^\circ_{red}$ (anode) $= 1.20$ V $- 0.403$ V $= 0.80$ V

(d) $\Delta G^\circ = -nFE^\circ = 6(96.5)(1.20) = -695$ kJ

20.85   We need in each case to determine whether E° is positive (spontaneous) or negative (nonspontaneous).

(a) $E^\circ = 0.672$ V, spontaneous   (b) $E^\circ = -0.82$ V, nonspontaneous

(c) $E^\circ = 0.55$ V, spontaneous   (d) $E^\circ = 0.12$ V, nonspontaneous

20.86   (a)   The reduction potential for $O_2(g)$ in the presence of acid is 1.23 V. $O_2(g)$ cannot oxidize Au(s) to $Au^+(aq)$ or $Au^{3+}(aq)$, even in the presence of acid.

(b)   The substances need a reduction potential greater than 1.50 V. These include $Co^{3+}(aq)$, $F_2(g)$, $H_2O(aq)$ and $O_3(g)$. Marginal oxidizing agents (those with reduction potential near 1.50 V) from Appendix E are $BrO_3^-(aq)$, $Ce^{4+}(aq)$, HClO(aq), $MnO_4^-(aq)$ and $PbO_2(s)$.

(c) $3[Au^+(aq) + 1e^- \rightarrow Au(s)]$ $E^\circ_{red} = 1.69$ V

       $Au(s) \rightarrow Au^{3+}(aq) + 3e^-$ $E^\circ_{red} = 1.50$ V

     $3Au^+(aq) \rightarrow 2Au(s) + Au^{3+}(aq)$ $E^\circ = 1.69 - 1.50 = +0.19$ V

     If E° is positive, ΔG° is negative and the disproportionation of $Au^+(aq)$ is spontaneous.

(d)   Since the Au^+ spontaneously disproportionates, the product of the reaction is AuF_3.

     $3[F_2(g) + 2e^- \rightarrow 2F^-(aq)]$ $E^\circ_{red} = 2.87$ V

       $2[Au(s) \rightarrow Au^{3+}(aq) + 3e^-]$ $E^\circ_{red} = 1.50$ V

     $2Au(s) + 3F_2(g) \rightarrow 2AuF_3(aq)$ $E^\circ = 2.87 - 1.50 = 1.37$ V

20.87 (a) $2[Ag^+(aq) + 1e^- \rightarrow Ag(s)]$ $E^°_{red} = 0.80$ V

$Ni(s) \rightarrow Ni^{2+}(aq) + 2e^-$ $E^°_{red} = -0.28$

$2Ag^+(aq) + Ni(s) \rightarrow 2Ag(s) + Ni^{2+}(aq)$ $E^° = 0.80 - (-0.28) = 1.08$ V

(b)   As the reaction proceeds, $Ni^{2+}(aq)$ is produced, so $[Ni^{2+}]$ increases as the cell operates.

(c) $E = E^° - \dfrac{0.0592}{n} \log K;$ $1.14 = 1.08 - \dfrac{0.0592}{2} \log \dfrac{[Ni^{2+}]}{[Ag^+]^2}$

$-\dfrac{0.06(2)}{0.0592} = \log(0.0050) - \log[Ag^+]^2;$ $\log[Ag^+]^2 = \log(0.0050) + \dfrac{0.06(2)}{0.0592}$

$\log[Ag^+]^2 = -2.301 + 2.027 = -0.274;$ $[Ag^+]^2 = 0.532$ $M;$ $[Ag^+] = 0.729 = 0.73$ M

(Strictly speaking, 2.027 has only 1 sig fig, yielding $\log[Ag^+] \approx 0$, $[Ag^+] \approx 1$. This result contains no useful information.)

20.88 The cell half-reactions are: $Cr(s) \rightarrow Cr^{3+}(aq, Conc\ 1) + 3e^-$

$Cr^{3+}(aq, Conc\ 2) + 3e^- \rightarrow Cr(s)$

$Cr^{3+}(aq, Conc\ 2) \rightarrow Cr^{3+}(aq, Conc\ 1)$

$E = 0 - \dfrac{0.0592}{3} \log \dfrac{[Cr^{3+}]_1}{[Cr^{3+}]_2};$ $(E^° = 0)$ Let $[Cr^{3+}]_1 = 0.040$ M, $[Cr^{3+}]_2 = 2.0$ M

$E = -\dfrac{0.0592}{3} \log \dfrac{0.040}{2.0} = +0.034$ V

The fact that E is positive means that the cell reaction is spontaneous as written: $Cr^{3+} (0.040\ M) \rightarrow Cr^{3+} (2.0\ M)$. This means that the cathode is in the compartment containing $Cr^{3+}(2.0\ M)$; the anode is in the compartment containing the dilute $(0.040\ M)$ solution. For concentration cells in general, oxidation occurs in the compartment with the dilute solution and reduction in the compartment with the concentrated solution, so that the concentrations of cations in the two compartments become more similar.

20.89 (a) $I_2(s) + 2e^- \rightarrow 2\ I^-(aq)$ $E^°_{red} = 0.536$ V

$2[Cu(s) \rightarrow Cu^+(aq) + 1e^-]$ $E^°_{red} = 0.521$ V

$I_2(s) + 2Cu(s) \rightarrow 2Cu^+(aq) + 2\ I^-(aq)$ $E^° = 0.536 - 0.521 = 0.015$ V

$E = E^° - \dfrac{0.0592}{n} \log Q = 0.015 - \dfrac{0.0592}{2} \log [Cu^+]^2\ [I^-]^2$

$E = +0.015 - \dfrac{0.0592}{2} \log (2.1)^2(3.2)^2 = +0.015 - 0.049 = -0.034$ V

(b)   Since the cell potential is negative at these concentration conditions, the cell would be spontaneous in the opposite direction and the inert electrode in the I_2/I^- compartment would be the anode; $Cu(s)$ would be the cathode.

(c)   No. At standard conditions the cell reaction is as written in part (a) and $Cu(s)$ is the anode.

(d) $E = 0$, $E° = \dfrac{0.0592}{2} \log (1.4)^2 [I^-]^2$; $\dfrac{2(0.015)}{0.0592} = \log (1.4)^2 + 2 \log [I^-]$;

$\log[I^-] = 0.107 = 0.11$; $[I^-] = 10^{0.107} = 1.28 = 1.3 \, M \, I^-$

20.90 The reaction can be written as a sum of the steps:

$$Pb^{2+}(aq) + 2e^- \rightarrow Pb(s) \qquad\qquad E°_{red} = -0.126 \, V$$

$$PbS(s) \rightarrow Pb^{2+}(aq) + S^{2-}(aq) \qquad E° = \, ?$$

$$PbS(s) + 2e^- \rightarrow Pb(s) + S^{2-}(aq) \qquad\qquad E°_{red} = \, ?$$

"$E°$" for the second step can be calculated from K_{sp}.

$$E° = \dfrac{0.0592}{n} \log K_{sp} = \dfrac{0.0592}{2} \log (8.0 \times 10^{-28}) = \dfrac{0.0592}{2} (-27.10) = -0.802 \, V$$

$E°$ for the half-reaction = $-0.126 \, V + (-0.802 \, V) = -0.928 \, V$

Calculating an imaginary $E°$ for a nonredox process like step 2 may be a disturbing idea. Alternatively, one could calculate K for step 1 (5.4×10^{-5}), K for the reaction in question ($K = K_1 \times K_{sp} = 4.4 \times 10^{-32}$) and then $E°$ for the half-reaction. The result is the same.

20.91 $Cu^+(aq) + 1e^- \rightarrow Cu^{2+}(aq) \qquad\qquad E°_{red} = -0.153 \, V$

        $Cu^+(aq) \rightarrow Cu°(s) + 1e^- \qquad\qquad E°_{red} = -0.521 \, V$

    $2Cu^+(aq) \rightarrow Cu°(s) + Cu^{2+}(aq) \qquad E° = -0.153 - (-0.521) = 0.368 \, V$

$$E° = \dfrac{0.0592}{n} \log K; \quad \log K = \dfrac{nE°}{0.0592} = \dfrac{1(0.368)}{0.0592} = 6.216 = 6.22$$

$K = 10^{6.216} = 1.6 \times 10^6$

20.92 (a)   Total volume of $Cr = 2.3 \times 10^{-4} \, m \times 0.32 \, m^2 = 7.36 \times 10^{-5} = 7.4 \times 10^{-5} \, m^3$

        mol $Cr = 7.36 \times 10^{-5} \, m^3 \, Cr \times \dfrac{100^3 \, cm^3}{1 \, m^3} \times \dfrac{7.20 \, g \, Cr}{1 \, cm^3} \times \dfrac{1 \, mol \, Cr}{52.0 \, g \, Cr} = 10.2$

                                            = 10 mol Cr

    The electrode reaction is:

    $CrO_4^{2-}(aq) + 4H_2O(l) + 6e^- \rightarrow Cr(s) + 8OH^-(aq)$

    Coulombs requires = 10.2 mol $Cr \times \dfrac{6 \, Faraday}{1 \, mol \, Cr} \times \dfrac{96,500 \, C}{1 \, Faraday} = 5.90 \times 10^6 = 5.9 \times 10^6 \, C$

(b) $5.90 \times 10^6 \text{ C} \times \dfrac{1 \text{ amp} \cdot \text{s}}{1 \text{ C}} \times \dfrac{1}{6.0 \text{ s}} = 9.8 \times 10^5 \text{ amp}$

20.93 $3.20 \text{ amp} \times 40 \text{ min} \times \dfrac{60 \text{ s}}{1 \text{ min}} \times \dfrac{1 \text{ C}}{1 \text{ amp} \cdot \text{s}} \times \dfrac{1 \text{ Faraday}}{96{,}500 \text{ C}} \times \dfrac{1 \text{ mol e}^-}{1 \text{ Faraday}} = 0.0796 = 0.080 \text{ mol e}^-$

$\dfrac{4.57 \text{ g In}}{0.0796 \text{ mol e}^-} \times \dfrac{1 \text{ mol In}}{114.8 \text{ g In}} = 0.50 \text{ mol In}/1 \text{ mol e}^-$

This result tells us that In must be in the +2 oxidation state in the molten halide.

20.94   (a) $1.0 \text{ lb Al} \times \dfrac{453.6 \text{ g}}{1 \text{ lb}} \times \dfrac{1 \text{ mol Al}}{26.98 \text{ g Al}} \times \dfrac{3 \text{ F}}{1 \text{ mol Al}} \times \dfrac{96{,}500 \text{ C}}{\text{F}} \times \dfrac{1 \text{ amp} \cdot \text{s}}{\text{C}} \times \dfrac{1}{11.2 \text{ A}}$

$\times \dfrac{1 \text{ min}}{60 \text{ s}} \times \dfrac{1 \text{ hr}}{60 \text{ min}} = 1.2 \times 10^2 \text{ hr}$

(b) $\text{Coulombs} = 453.6 \text{ g Al} \times \dfrac{1 \text{ mol Al}}{26.98 \text{ g Al}} \times \dfrac{3 \text{ F}}{1 \text{ mol Al}} \times \dfrac{96{,}500 \text{ C}}{\text{F}} = 4.867 \times 10^6$

$= 4.9 \times 10^6 \text{ C}$

$\text{kWh} = 4.867 \times 10^6 \text{ C} \times 6.0 \text{ V} \times \dfrac{1 \text{ J}}{\text{C} \cdot \text{V}} \times \dfrac{1 \text{ kWh}}{3.6 \times 10^6 \text{ J}} = 8.1 \text{ kWh}$

This is the amount of electrical power expended if the cell is 100% efficient.
8.1 kWh/0.40 = **20 kWh** required to produce 1.0 lb of Al.

20.95   (a)   anode: $2\text{Cl}^-(l) \rightarrow \text{Cl}_2(g) + 2\text{e}^-$
            cathode: $\text{K}^+(l) + 1\text{e}^- \rightarrow \text{K}(l)$

(b)   anode: $2\text{Cl}^-(l) \rightarrow \text{Cl}_2(g) + 2\text{e}^-$
            cathode: $2\text{H}_2\text{O}(l) + 2\text{e}^- \rightarrow \text{H}_2(g) + 2\text{OH}^-(aq)$

(c)   anode: $2\text{H}_2\text{O}(l) \rightarrow \text{O}_2(g) + 4\text{H}^+(aq) + 4\text{e}^-$
            cathode: $2\text{H}_2\text{O}(l) + 2\text{e}^- \rightarrow \text{H}_2(g) + 2\text{OH}^-(aq)$

The oxidation and reduction of $\text{H}_2\text{O}(l)$ occur preferentially because they require less energy than the oxidation of $\text{F}^-(aq)$ or reduction of $\text{K}^+(aq)$.

20.96 $24.0 \text{ hr} \times 20.0 \text{ amp} \times \dfrac{3600 \text{ s}}{1 \text{ hr}} \times \dfrac{1 \text{ mol e}^-}{96{,}500 \text{ amp} \cdot \text{s}} \times \dfrac{4 \text{ mol NaBO}_3}{8 \text{ mol e}^-} \times \dfrac{81.80 \text{ g NaBO}_3}{1 \text{ mol NaBO}_3}$

$= 732 \text{ g NaBO}_3$

20.97   (a) $\text{Ni}^{2+}(aq) + 2\text{e}^- \rightarrow \text{Ni}(s) \qquad E^\circ_{red} = -0.28 \text{ V}$

$2\text{Br}^-(aq) \rightarrow \text{Br}_2(l) + 2\text{e}^- \qquad E^\circ_{red} = 1.065 \text{ V}$

$E^\circ = -0.28 - 1.065 = -1.35 \text{ V}; \quad E_{min} = 1.35 \text{ V}$

(b) A larger voltage than the calculated E_{min} is required to overcome cell resistance, if the reaction is to proceed at a reasonable rate. There may also be an overvoltage for reduction of $Br^-(aq)$ at Pt, although this should not be large.

20.98 (a) The work obtainable is given by the product of the voltage, which has units of J/C, times the number of Coulombs of electricity produced:

$$w_{max} = 300 \text{ amp} \cdot \text{hr} \times \frac{3600 \text{ s}}{1 \text{ hr}} \times \frac{1 \text{ C}}{1 \text{ amp} \cdot \text{s}} \times \frac{6 \text{ J}}{1 \text{ C}} \times \frac{1 \text{ kWh}}{3.6 \times 10^6 \text{ J}} = 1.8 \text{ kWh} \approx 2 \text{ kWh}$$

(b) This maximum amount of work is never realized because some of the electrical energy is dissipated in overcoming the internal resistance of the battery; because the cell voltage does not remain constant as the reaction proceeds; because the systems to which the electrical energy is delivered are not capable of completely converting electrical energy into work.

20.99 (a) The major reason "deep discharge" causes battery failure is loss of materials from the electrode surfaces, during either discharge or charging. When PbO_2 is almost completely removed from the cathode during discharge, as it is converted to $PbSO_4$ there may be some flaking off of $PbSO_4$. This solid settles in the bottom of the cell, and is no longer able to take part in the cell reaction. It may eventually cause shorting between the plates.

(b) In an auto system, a running engine powers a generator and voltage regulator which together keep the battery fully charged, or nearly so. Whatever charge is used to start the engine is replaced while the auto is running. Except when lights are left on accidentally, or the battery is run down in cold weather starting, it carries nearly a full charge.

20.100 The ship's hull should be made negative. By keeping an excess of electrons in the metal of the ship, the tendency for iron to undergo oxidation, with release of electrons, is reduced. The ship, as a negatively charged "electrode," becomes the site of reduction, rather than oxidation, in an electrolytic process.

20.101 (a) $7 \times 10^8 \text{ mol } H_2 \times \dfrac{2 \text{ F}}{1 \text{ mol } H_2} \times \dfrac{96,500 \text{ C}}{1 \text{ F}} = 1.35 \times 10^{14} = 1 \times 10^{14} \text{ C}$

(b)
$$2H_2O(l) \rightarrow O_2(g) + 4H^+(aq) + 4e^- \quad E^\circ_{red} = 1.23 \text{ V}$$
$$2[H^+(aq) + 2e^- \rightarrow H_2(g)] \quad E^\circ_{red} = 0 \text{ V}$$

$$2H_2O(l) \rightarrow O_2(g) + 2H_2(g) \quad E^\circ = 0.00 - 1.23 = -1.23 \text{ V}$$

$$E = E^\circ - \frac{0.0592}{4} \log [O_2][H_2]^2 = -1.23 \text{ V} - \frac{0.0592}{4} \log (300)^3$$

$$E = -1.23 \text{ V} - 0.11 \text{ V} = -1.34 \text{ V}; \quad E_{min} = 1.34 \text{ V}$$

(c) $\text{Energy} = nFE = 2(7 \times 10^8 \text{ mol})(1.34 \text{ V}) \dfrac{96,500 \text{ J}}{\text{V} \cdot \text{mol}} = 1.81 \times 10^{14} = 2 \times 10^{14} \text{ J}$

(d) 1.81×10^{14} J $\times \dfrac{1 \text{ kWh}}{3.6 \times 10^6 \text{ J}} \times \dfrac{\$0.23}{\text{kWh}} = \$1.16 \times 10^7 = \$1 \times 10^7$

It will cost more than ten million dollars for the electricity alone.

Integrative Exercises

20.102 $2[NO_3^-(aq) + 4H^+(aq) + 3e^- \rightarrow NO(g) + 2H_2O(l)]$ $E^{\circ}_{red} = 0.96$ V

                          $3[Cu(s) \rightarrow Cu^{2+}(aq) + 2e^-]$ $E^{\circ}_{red} = 0.34$ V

$3Cu(s) + 2NO_3^-(aq) + 8H^+(aq) \rightarrow 3Cu^{2+}(aq) + 2NO(g) + 4H_2O(l)$

$E^{\circ} = 0.96 - 0.34 = 0.62$ V

$2H^+(aq) + 2e^- \rightarrow H_2(g)$ $E^{\circ}_{red} = 0$ V

            $Cu(s) \rightarrow Cu^{2+}(aq) + 2e^-$ $E^{\circ}_{red} = 0.34$ V

$Cu(s) + 2H^+(aq) \rightarrow Cu^{2+}(aq) + H_2(g)$ $E^{\circ} = 0 - 0.34 = -0.34$ V

The overall cell potential for the oxidation of Cu(s) by HNO_3 is positive and the reaction is spontaneous. The cell potential for the oxidation of Cu(s) by HCl is negative and the reaction is nonspontaneous. Note that in the reaction with HNO_3, it is NO_3^- that is reduced, not H^+.

20.103 In an electrode process where a gas is involved, the electrode is usually inert; it does not participate directly in the reaction. The electrode does act as an electron carrier, a heterogeneous catalyst for the reaction. The reactant is adsorbed onto the surface of the electrode at an active site. For all heterogeneous catalysts, the preparation of the catalyst determines the number of active sites and the effectiveness of the catalyst. In this case, it determines the rate of the redox process at the electrode.

20.104 (a)   The oxidation potential of A is equal in magnitude but opposite in sign to the reduction potential of A^+.

        (b)   Li(s) has the highest oxidation potential, Au(s) the lowest.

        (c)   The relationship is resonable because both oxidation potential and ionization energy describe removing electrons from a substance. Ionization energy is a property of gas phase atoms or ions, while oxidation potential is a property of the bulk material.

20.105 (a) $NO_3^-(aq) + 4H^+(aq) + 3e^- \rightarrow NO(g) + 2H_2O(l)$ $E^{\circ}_{red} = 0.96$ V

                              $Au(s) \rightarrow Au^{3+}(aq) + 3e^-$ $E^{\circ}_{red} = 1.498$ V

        $Au(s) + NO_3^-(aq) + 4H^+(aq) \rightarrow Au^{3+}(aq) + NO(g) + 2H_2O(l)$

        $E^{\circ} = 0.96 - 1.498 = -0.54$ V; E° is negative, the reaction is not spontaneous.

(b)

$$3[2H^+(aq) + 2e^- \rightarrow H_2(g)] \qquad\qquad E^{\circ}_{red} = 0.000 \text{ V}$$
$$2[Au(s) + 4Cl^-(aq) \rightarrow AuCl_4^-(aq) + 3e^-] \qquad E^{\circ}_{red} = 1.002 \text{ V}$$

$$2Au(s) + 6H^+(aq) + 8Cl^-(aq) \rightarrow 2AuCl_4^-(aq) + 3H_2(g)$$

$E^{\circ} = 0.000 - 1.002 = -1.002$ V; E° is negative, the reaction is not spontaneous.

(c)

$$NO_3^-(aq) + 4H^+(aq) + 3e^- \rightarrow NO(g) + 2H_2O(l) \qquad E^{\circ}_{red} = 0.96 \text{ V}$$
$$Au(s) + 4Cl^-(aq) \rightarrow AuCl_4^-(aq) + 3e^- \qquad E^{\circ}_{red} = 1.002 \text{ V}$$

$$Au(s) + NO_3^-(aq) + 4Cl^-(aq) + 4H^+(aq) \rightarrow AuCl_4^-(aq) + NO(g) + 2H_2O(l)$$

$E^{\circ} = 0.96 - 1.002 = -0.04$; E° is small but negative, the process is not spontaneous.

(d)
$$E = E^{\circ} - \frac{0.0592}{3} \log \frac{[AuCl_4^-] P_{NO}}{[NO_3^-][Cl^-]^4[H^+]^4}$$

If $[H^+]$, $[Cl^-]$ and $[NO_3^-]$ are much greater than 1.0 M, the log term is negative and the correction to E° is positive. If the correction term is greater than 0.042 V, the value of E is positive and the reaction at nonstandard conditions is spontaneous.

20.106 The two half-reactions in the electrolysis of $H_2O(l)$ are:

$$2[2H_2O(l) + 2e^- \rightarrow H_2(g) + 2OH^-]$$
$$2H_2O(l) \rightarrow O_2(g) + 4H^+ + 4e^-$$

$$2H_2O(l) \rightarrow 2H_2(g) + O_2(g)$$

4 mol e^-/2 mol $H_2(g)$ or 2 mol e^-/mol $H_2(g)$

Using partial pressures and the ideal-gas law, calculate the mol $H_2(g)$ produced, and the current required to do so.

$P_T = P_{H_2} + P_{H_2O}$. From Appendix B, P_{H_2O} at 25.5°C is approximately 24.5 torr.

$P_{H_2} = 768$ torr - 24.5 torr = 743.5 = 744 torr

$$n = PV/RT = \frac{(743.5/760) \text{ atm} \times 0.0123 \text{ L}}{298.5 \text{ K} \times 0.08206 \text{ L·atm/mol·K}} = 4.912 \times 10^{-4} = 4.91 \times 10^{-4} \text{ mol } H_2$$

$$4.912 \times 10^{-4} \text{ mol } H_2 \times \frac{2 \text{ mol } e^-}{\text{mol } H_2} \times \frac{96,500 \text{ C}}{1 \text{ mol } e^-} \times \frac{1 \text{ amp·s}}{1 \text{ C}} \times \frac{1 \text{ min}}{60 \text{ s}} \times \frac{1}{2.00 \text{ min}}$$

$$= 0.790 \text{ amp}$$

20.107 First balance the equation:

$$4CyFe^{2+}(aq) + O_2(g) + 4H^+(aq) \rightarrow 4CyFe^{3+}(aq) + 2H_2O(l); \ E = +0.60 \ V; \ n = 4$$

(a)   From Equation [20.11] we can calculate ΔG for the process under the conditions specified for the measured potential E:

$$\Delta G = -nFE = -(4 \text{ mol } e^-) \times \frac{96.5 \text{ kJ}}{1 \text{ V} \cdot \text{mol } e^-}(0.60 \text{ V}) = -231.6 = -232 \text{ kJ}$$

(b)   The moles of ATP synthesized per mole of O_2 is given by:

$$\frac{231.6 \text{ kJ}}{O_2 \text{ molecule}} \times \frac{1 \text{ mol ATP formed}}{37.7 \text{ kJ}} = \text{approximately 6 mol ATP/mol } O_2$$

20.108
$$AgSCN(s) + e^- \rightarrow Ag(s) + SCN^-(aq) \qquad E^\circ_{red} = 0.0895 \ V$$
$$Ag(s) \rightarrow Ag^+(aq) + e^- \qquad E^\circ_{red} = 0.799 \ V$$

$$AgSCN(s) \rightarrow Ag^+(aq) + SCN^-(aq) \qquad E^\circ = 0.0895 - 0.799 = -0.710 \ V$$

$$E^\circ = \frac{0.0592}{n} \log K_{sp}; \ \log K_{sp} = \frac{(-0.710)(1)}{0.0592} = -11.993 = -12.0$$

$$K_{sp} = 10^{-11.993} = 1.02 \times 10^{-12} = 1 \times 10^{-12}$$

20.109 (a)

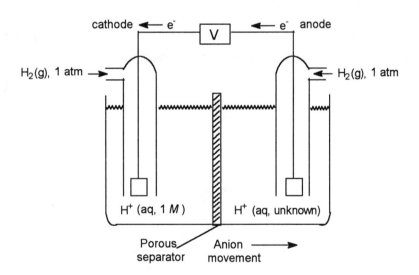

(b)
$$2H^+(aq, 1 \ M) + 2e^- \rightarrow H_2(g) \qquad E^\circ_{red} = 0$$
$$H_2(g) \rightarrow 2H^+(aq, 1 \ M) + 2e^- \qquad E^\circ_{red} = 0$$

$$2H^+(aq, 1 \ M) + H_2(g) \rightarrow 2H^+(aq, 1 \ M) + H_2(g) \qquad E^\circ = 0$$

(c)   At standard conditions, $[H^+] = 1 \ M$, pH = 0

(d)　　$E = E° - \dfrac{0.0592}{2} \log = \dfrac{[H^+(\text{unknown})]^2 P_{H_2}}{[H^+(1\ M)]^2 P_{H_2}}$

$E = 0 - \dfrac{0.0592}{2} \log [H^+]^2 = \dfrac{0.0592}{2} \times 2(-\log [H^+])$

$E = 0.0592(5.0) = 0.30\ V = 0.0592\ \text{pH}$

(e)　　E_{cell} charges $0.0592(0.01) = 0.000592 = 0.0006\ V$ for each 0.01 pH unit. The voltmeter would have to be precise to at least 0.001 V to detect a charge of 0.01 pH units.

20.110　(a)

$\text{Ag}^+(aq) + e^- \rightarrow \text{Ag}(s)$　　　　　　$E^°_{red} = 0.799\ V$

$\text{Fe}^{2+}(aq) \rightarrow \text{Fe}^{3+}(aq) + 1e^-$　　　$E^°_{red} = 0.771\ V$

$\text{Ag}^+(aq) + \text{Fe}^{2+}(aq) \rightarrow \text{Ag}(s) + \text{Fe}^{3+}(aq)$

$E° = 0.799\ V - 0.771\ V = 0.028\ V$

(b)　　$\text{Ag}^+(aq)$ is reduced at the cathode and $\text{Fe}^{2+}(aq)$ is oxidized at the anode.

(c)　　$\Delta G° = -nFE° = -(1)(96.5)(0.028) = -2.7\ kJ$

$\Delta S° = S°\ \text{Ag}(s) + S°\ \text{Fe}^{3+}(aq) - S°\text{Ag}^+(aq) - S°\ \text{Fe}^{2+}(aq)$

$= 42.55\ J + 293.3\ J - 73.93\ J - 113.4\ J = 148.5\ J$

$\Delta G° = \Delta H° - T\Delta S°$　Since $\Delta S°$ is positive, $\Delta G°$ will become more negative and $E°$ will become more positive as temperature is increased.

21 Nuclear Chemistry

Radioactivity

21.1 p = protons, n = neutrons, e = electrons; number of protons = atomic number; number of neutrons = mass number - atomic number

(a) $^{59}_{28}$Ni: 28p, 31n   (b) ^{94}Zr: 40p, 54n   (c) ^{18}O: 8p, 10n

21.2 p = protons, n = neutrons, e = electrons; number of protons = atomic number; number of neutrons = mass number - atomic number

(a) $^{107}_{47}$Ag: 47p, 60n   (b) ^{31}P: 15p, 16n   (c) ^{115}In: 49p, 66n

21.3 (a) $^{1}_{1}$p or $^{1}_{1}$H   (b) $^{0}_{1}$e   (c) $^{0}_{-1}\beta$ or $^{0}_{-1}$e

21.4 (a) $^{1}_{0}$n   (b) $^{0}_{-1}$e   (c) $^{4}_{2}$He

21.5 (a) $^{214}_{83}$Bi $\rightarrow$ $^{214}_{84}$Po + $^{0}_{-1}$e   (b) $^{195}_{79}$Au + $^{0}_{-1}$e (orbital electron) $\rightarrow$ $^{195}_{78}$Pt

(c) $^{38}_{19}$K $\rightarrow$ $^{38}_{18}$Ar + $^{0}_{1}$e   (d) $^{242}_{94}$Pu $\rightarrow$ $^{238}_{92}$U + $^{4}_{2}$He

21.6 (a) $^{131}_{53}$I $\rightarrow$ $^{131}_{54}$Xe + $^{0}_{-1}$e   or $^{0}_{-1}\beta$

(b) $^{230}_{90}$Th $\rightarrow$ $^{226}_{88}$Ra + $^{4}_{2}$He; α emission is the only decay process that reduces the atomic number by two.

(c) $^{181}_{74}$W + $^{0}_{-1}$e (orbital electron) $\rightarrow$ $^{181}_{73}$Ta   (d) $^{13}_{7}$N $\rightarrow$ $^{13}_{6}$C + $^{0}_{1}$e

21.7 The total mass number change is (235-207) = 28. Since each α particle accompanies a change of -4 in mass number, whereas emission of a β particle does not correspond to a mass change, there are 7 α particle emissions. The change in atomic number in the series is 10. Each α particle results in an atomic number lower by two. The 7 α particle emissions alone would cause a decrease of 14 in atomic number. Each β particle emissions raises the atomic number by one. To obtain the observed lowering of 10 in the series, there must be 4 β emissions.

21.8 This decay series represents a change of (237-209 =) 28 mass units. Since only alpha emissions change the nuclear mass, and each changes the mass by four, there must be a total of 7 α emissions. Each alpha emission causes a decrease of two in atomic number. Therefore, the 7 alpha emissions, by themselves, should have caused a decrease in atomic number of 14. The series as a whole involves a decrease of 10 in atomic number. Thus, there must be a total of 4 β emissions, each of which increase atomic number by one.

21.9   (a) $^{24}_{11}Na \rightarrow {}^{24}_{12}Mg + {}^{0}_{-1}e$;   a β particle is produced

(b) $^{188}_{80}Hg \rightarrow {}^{188}_{79}Au + {}^{0}_{1}e$;   a positron is produced

(c) $^{122}_{53}I \rightarrow {}^{122}_{54}Xe + {}^{0}_{-1}e$;   a β particle is produced

(d) $^{242}_{94}Pu \rightarrow {}^{238}_{92}U + {}^{4}_{2}He$;   an α particle is produced

2.10   (a) $^{211}_{82}Pb \rightarrow {}^{211}_{83}Bi + {}^{0}_{-1}\beta$

(b) $^{50}_{25}Mn \rightarrow {}^{50}_{24}Cr + {}^{0}_{1}e$

(c) $^{179}_{74}W + {}^{0}_{-1}e \rightarrow {}^{179}_{73}Ta$

(d) $^{230}_{90}Th \rightarrow {}^{226}_{88}Ra + {}^{4}_{2}He$

Nuclear Stability

21.11   (a) $^{8}_{5}B$ - low neutron/proton ratio, positron emission

(b) $^{68}_{29}Cu$ - high neutron/proton ratio, beta emission

(c) $^{241}_{93}Np$ - high neutron/proton ratio, beta emission

(Even though ^{241}Np has an atomic number ≥ 84, the most common decay pathway for nuclides with neutron/proton ratios higher than the isotope listed on the periodic chart is beta decay.)

(d) $^{39}_{17}Cl$ - high neutron/proton ratio, beta emission

21.12   (a) $^{66}_{32}Ge$ - lower neutron/proton ratio, positron emission or electron capture

(b) $^{105}_{45}Rh$ - high neutron/proton ratio, beta emission

(c) $^{137}_{53}I$ - high neutron/proton ratio, beta emission

(d) $^{133}_{58}Ce$ - low neutron/proton ratio, electron capture or positron emission

21.13   (a)   It is on the edge - slightly low neutron/proton ratio; could be a positron emitter or undergo orbital electron capture.

(b)   No - somewhat high neutron/proton ratio; beta emitter.

(c)   Yes.

(d)   No - high atomic number; alpha emitter.

21.14   (a)   It is on the edge - slightly low neutron/proton ratio; could be a positron emitter or undergo orbital electron capture.

        (b)   No - low neutron/proton ratio; should be a positron emitter, or possibly undergo orbital electron capture.

        (c)   No - high neutron/proton ratio; should be a beta emitter.

        (d)   No - high atomic number; it should be a alpha emitter.

21.15   Use the criteria listed in Table 21.3.

        (a)   Stable: $^{39}_{19}K$ odd proton, even neutron more abundant than odd proton, odd neutron; 20 neutrons is a magic number.

        (b)   Stable: $^{209}_{83}Bi$ odd proton, even neutron more abundant than odd proton, odd neutron; 126 neutrons is a magic number.

        (c)   Stable: $^{25}_{12}Mg$ even though $^{24}_{10}Ne$ is an even proton, even neutron nuclide, it has a very high neutron/proton ratio and lies outside the band of stability.

21.16   Use the criteria listed in Table 21.3.

        (a) $^{19}_{9}F$ odd proton, even neutron more abundant

        (b) $^{80}_{34}Se$ even-even more abundant

        (c) $^{56}_{26}Fe$ even-even more abundant

        (d) $^{118}_{50}Sn$ even-even more abundant than odd-odd

21.17   (a) $^{4}_{2}He$ (c) $^{40}_{20}Ca$ (e) $^{208}_{82}Pb$

        (d) $^{58}_{28}Ni$ has a magic number of protons, but not neutrons.

21.18 $^{112}_{50}Sn$ has a magic number of protons and even numbers of protons and neutrons, good indications of nuclear stability. $^{112}_{49}In$ has no magic numbers and odd numbers of protons and neutrons, indicators of nuclear instability or radioactivity.

21.19   Radioactive:
       $^{14}_{8}O$, $^{115}_{52}Te$ - low neutron/proton ratio; $^{208}_{84}Po$ - atomic number $\geq$ 84

       $^{84}_{34}Se$ - even though this is an even proton, even neutron nuclide with a magic number of neutrons, the neutron/proton ratio is so high that the nuclide is radioactive.

       Stable: $^{32}_{16}S$ - even proton, even neutron, stable neutron/proton ratio

21.20   The criterion employed in judging whether the nucleus is likely to be radioactive is the position of the nucleus on the plot shown in Figure 21.2. If the neutron/proton ratio is too high or low, or if the atomic number exceeds 83, the nucleus will be radioactive. Radioactive: (b) low neutron/proton ratio; (c) low neutron/proton ratio; (e) high atomic number; stable: (a) and (d).

Nuclear Transmutations

21.21 Protons and alpha particles are positively charged and must be moving very fast to overcome electrostatic forces which would repel them from the target nucleus. Neutrons are electrically neutral and not repelled by the nucleus.

21.22 A major difference is that the charge on the nitrogen nucleus, +7, is much smaller than on the gold nucleus, +79. Thus, the alpha particle could more easily penetrate the coulomb barrier (that is, the repulsive energy barrier due to like charges) to make contact with the nitrogen nucleus than the gold nucleus. Rutherford used alpha particles that were being emitted from some radioactive source. He did not have access to machines that can accelerate particles to very high energy. It would be necessary to do just that to observe reaction of an alpha particle with a gold nucleus.

21.23 (a) $^{252}_{98}Cf + ^{10}_{5}B \rightarrow 3\,^{1}_{0}n + ^{259}_{103}Lw$ (b) $^{2}_{1}H + ^{3}_{2}He \rightarrow ^{4}_{2}He + ^{1}_{1}H$

       (c) $^{1}_{1}H + ^{11}_{5}B \rightarrow 3\,^{4}_{2}He$ (d) $^{122}_{53}I \rightarrow ^{122}_{54}Xe + ^{0}_{-1}e$ (e) $^{59}_{26}Fe \rightarrow ^{0}_{-1}e + ^{59}_{27}Co$

21.24 (a) $^{32}_{16}S + ^{1}_{0}n \rightarrow ^{1}_{1}p + ^{32}_{15}P$ (b) $^{7}_{4}Be + ^{0}_{-1}e$ (orbital electron) $\rightarrow ^{7}_{3}Li$

       (c) $^{187}_{75}Re \rightarrow ^{187}_{76}Os + ^{0}_{-1}e$ (d) $^{98}_{42}Mo + ^{2}_{1}H \rightarrow ^{1}_{0}n + ^{99}_{43}Tc$

       (e) $^{235}_{92}U + ^{1}_{0}n \rightarrow ^{135}_{54}Xe + ^{99}_{38}Sr + 2\,^{1}_{0}n$

21.25 (a) $^{14}_{7}N + ^{1}_{1}H \rightarrow ^{11}_{6}C + ^{4}_{2}He$ (b) $^{14}_{7}N + ^{4}_{2}He \rightarrow ^{17}_{8}O + ^{1}_{1}H$

       (c) $^{59}_{26}Fe + ^{4}_{2}He \rightarrow ^{63}_{29}Cu + ^{0}_{-1}e$

21.26 (a) $^{238}_{92}U + ^{1}_{0}n \rightarrow ^{239}_{92}U + ^{0}_{0}Y$ (b) $^{14}_{7}N + ^{1}_{1}H \rightarrow ^{11}_{6}C + ^{4}_{2}He$

       (c) $^{18}_{8}O + ^{1}_{0}n \rightarrow ^{19}_{9}F + ^{0}_{-1}e$

Rates of Radioactive Decay

21.27 Chemical reactions do not affect the character of atomic nuclei. The energy changes involved in chemical reactions are much too small to allow us to alter nuclear properties via chemical processes. Therefore, the nuclei that are formed in a nuclear reaction will continue to emit radioactivity regardless of any chemical changes we bring to bear. However, we can hope to use chemical means to separate radioactive substances, or remove them from foods or a portion of the environment.

21.28 The suggestion is not reasonable. The energies of nuclear states are very large relative to ordinary temperatures. Thus, merely changing the temperature by less than 100 K would not be expected to significantly affect the behavior of nuclei with regard to nuclear decay rates.

21.29 After 12.3 yr, one half-life, there are $(1/2)48.0 = 24.0$ mg. 49.2 yr is exactly four half-lives. There are then $(48.0)(1/2)^4 = 3.0$ mg tritium remaining.

21.30 Calculate the decay constant, k, and then $t_{1/2}$.

$$k = -\frac{1}{t}\ln\frac{N_t}{N_o} = -\frac{1}{165\text{ min}}\ln\frac{0.125\text{ g}}{1.000\text{ g}} = 0.01260 = 0.0126\text{ min}^{-1}$$

Using Equation [21.20],

$t_{1/2} = 0.693/k = 0.693/0.01260\text{ min}^{-1} = 55.0\text{ min}$

21.31 Using Equation [21.19],

$$k = \frac{-1}{t}\ln\frac{N_t}{N_o} = \frac{-1}{1.00\text{ yr}}\times\ln\frac{2921}{3012} = 0.03068 = 0.0307\text{ yr}^{-1}$$

Using Equation [21.20],

$t_{1/2} = 0.693/k = 0.693/(0.03068\text{ yr}^{-1}) = 22.6\text{ yr}$

21.32 $\ln\dfrac{N_t}{N_o} = -kt$; solve for k

$$k = -\ln\frac{N_t}{N_o}\times\frac{1}{t} = -\ln\frac{3130}{4600}\times\frac{1}{30.0} = 0.01283 = 0.0128\text{ d}^{-1};\ t_{1/2} = 0.693/0.01283\text{ d}^{-1} = 54.0\text{ d}$$

21.33 $k = 0.693/t_{1/2} = 0.693/27.8\text{ d} = 0.02493 = 0.0249\text{ d}^{-1}$

$$t = \frac{-1}{k}\ln\frac{N_t}{N_o} = \frac{-1}{0.02493\text{ d}^{-1}}\ln\frac{0.75}{1.85} = 36.2\text{ d}$$

21.34 $K = 0.693/t_{1/2} = 0.693/5.26\text{ yr} = 0.1317 = 0.132\text{ yr}^{-1};\ N_t/N_o = 0.75$

$$t = \frac{-1}{k}\ln\frac{N_t}{N_o} = -(1/0.1317\text{ yr}^{-1})\ln(0.75) = 2.18\text{ yr}$$

2.18 yr = 26.2 mo = 797 d. The source will have to be replaced sometime in January of 1999.

21.35 $^{226}_{88}\text{Ra} \rightarrow\ ^{222}_{86}\text{Rn} +\ ^{4}_{2}\text{He}$

1 α particle is produced for each ^{226}Ra that decays. Calculate the mass of ^{226}Ra remaining after 1.0 min, calculate by subtraction the mass that has decayed, and use Avogadro's number to get the number of $^{4}_{2}\text{He}$ particles.

Calculate k in min^{-1}. $1622\text{ yr}\times\dfrac{365\text{ d}}{1\text{ yr}}\times\dfrac{24\text{ hr}}{1\text{ d}}\times\dfrac{60\text{ min}}{1\text{ hr}} = 8.525\times10^8\text{ min}$

$$k = \frac{0.693}{t_{1/2}} = \frac{0.693}{8.525\times10^8\text{ min}} = 8.129\times10^{-10}\text{ min}^{-1}$$

$$\ln\frac{N_t}{N_o} = -kt = (-8.129\times10^{-10}\text{ min}^{-1})(1.0\text{ min}) = -8.129\times10^{-10}$$

$$\frac{N_t}{N_o} = (1.000 - 8.129\times10^{-10});\ \text{(don't round here!)}$$

$N_t = 5.0 \times 10^{-3}$ g $(1.00 - 8.129 \times 10^{-10})$ The amount that decays is $N_o - N_t$:

$$5.0 \times 10^{-3} \text{ g} - [5.0 \times 10^{-3} (1.00 - 8.129 \times 10^{-10})] = 5.0 \times 10^{-3} \text{ g} (8.129 \times 10^{-10})$$

$$= 4.065 \times 10^{-12} = 4.1 \times 10^{-12} \text{ g Ra}$$

$$[N_o - N_t] = 4.065 \times 10^{-12} \text{ g Ra} \times \frac{1 \text{ mol Ra}}{226.0 \text{ g Ra}} \times \frac{6.022 \times 10^{23} \text{ Ra atoms}}{1 \text{ mol Ra}} \times \frac{1 \; {}^{4}_{2}\text{He}}{1 \text{ Ra atom}}$$

$$= 1.1 \times 10^{10} \; \alpha \text{ particles emitted in 1 min}$$

21.36 Proceeding as in Exercise 21.35, calculate k in s^{-1}.

$$5.26 \text{ yr} \times \frac{365 \text{ d}}{1 \text{ yr}} \times \frac{24 \text{ hr}}{1 \text{ d}} \times \frac{3600 \text{ sec}}{1 \text{ hr}} = 1.659 \times 10^8 = 1.66 \times 10^8 \text{ s}$$

$$k = \frac{0.693}{t_{1/2}} = \frac{0.693}{1.659 \times 10^8} = 4.178 \times 10^{-9} = 4.18 \times 10^{-9} \text{ s}^{-1}$$

$$\ln \frac{N_t}{N_o} = -kt = -(4.178 \times 10^{-9} \text{ s}^{-1})(10.0 \text{ s}) = -4.18 \times 10^{-8}$$

$$\frac{N_t}{N_o} = e^{-4.18 \times 10^{-8}} = (1.000 - 4.18 \times 10^{-8}); \quad N_t = 650 \times 10^{-6} \text{ g} (1.000 - 4.18 \times 10^{-8})$$

The amount that decays is $N_o - N_t$:

$$650 \times 10^{-6} \text{ g} - [650 \times 10^{-6} \text{ g} (1.000 - 4.18 \times 10^{-8})] = 650 \times 10^{-6} \text{ g} (4.18 \times 10^{-8})$$

$$= 2.717 \times 10^{-11} = 2.72 \times 10^{-11} \text{ g Co}$$

$$N_o - N_t = 2.717 \times 10^{-11} \text{ g Co} \times \frac{1 \text{ mol Co}}{60 \text{ g Co}} \times \frac{6.022 \times 10^{23} \text{ Co atoms}}{1 \text{ mol Co}} \times \frac{1 \; \beta}{1 \text{ Ra atom}}$$

$$= 2.73 \times 10^{11} \; \beta \text{ particles emitted in 10.0 seconds}$$

21.37 Calculate k in yr^{-1} and solve Equation 21.19 for t. $N_o = 15.2$/min/g, $N_t = 8.9$/min/g

$$k = 0.693/t_{1/2} = 0.693/5.73 \times 10^3 \text{ yr} = 1.209 \times 10^{-4} = 1.21 \times 10^{-4} \text{ yr}^{-1}$$

$$t = \frac{-1}{k} \ln \frac{N_t}{N_o} = \frac{-1}{1.209 \times 10^{-4}} \ln \frac{8.9}{15.2} = 4.43 \times 10^3 \text{ yr}$$

21.38 $t = \dfrac{-1}{k} \ln \dfrac{N_t}{N_o}$; $k = 0.693/5.73 \times 10^3 \text{ yr} = 1.209 \times 10^{-4} = 1.21 \times 10^{-4} \text{ yr}^{-1}$

$$t = \frac{-1}{1.209 \times 10^{-4} \text{ yr}^{-1}} \ln \frac{25.8}{31.7} = 1.70 \times 10^3 \text{ yr}$$

21.39 Follow the procedure outlined in Sample Exercise 21.7. The original quantity of ^{238}U is 50.0 mg plus the amount that gave rise to 14.0 mg of ^{206}Pb. This amount is 14.0(238/206) = 16.2 mg.

$k = 0.693/4.5 \times 10^9 \text{ yr} = 1.54 \times 10^{-10} = 1.5 \times 10^{-10} \text{ yr}^{-1}$

$$t = \frac{-1}{k} \ln \frac{N_t}{N_o} = \frac{-1}{1.54 \times 10^{-10} \text{ yr}^{-1}} \ln \frac{50.0}{66.2} = 1.8 \times 10^9 \text{ yr}$$

21.40 $k = 0.693/1.27 \times 10^9 \text{ yr} = 5.457 \times 10^{-10} = 5.46 \times 10^{-10} \text{ yr}^{-1}$

If the mass of ^{40}Ar is 3.6 times that of ^{40}K, then the original mass of ^{40}K must have been 3.6 + 1 = 4.6 times that now present.

$$t = \frac{-1}{5.457 \times 10^{-10} \text{ yr}^{-1}} \times \ln \frac{1}{(4.6)} = 2.8 \times 10^9 \text{ yr}$$

Energy Changes

21.41 $\Delta E = c^2 \Delta m; \quad \Delta m = \Delta E / c^2; \quad 1 \text{ J} = \text{kg} \cdot \text{m}^2/\text{s}^2$

$$\Delta m = \frac{393.5 \times 10^3 \text{ kg} \cdot \text{m}^2/\text{s}^2}{(2.9979 \times 10^8 \text{ m/s})^2} \times \frac{1000 \text{ g}}{1 \text{ kg}} = 4.378 \times 10^{-9} \text{ g}$$

21.42 $\Delta E = c^2 \Delta m = (3.0 \times 10^8 \text{ m/s})^2 \times 0.1 \text{ mg} \times \dfrac{1 \text{ g}}{1000 \text{ mg}} \times \dfrac{1 \text{ kg}}{1000 \text{ g}} \times \dfrac{1 \text{ kJ}}{1000 \text{ J}} = 9 \times 10^6 \text{ kJ}$

21.43 $\Delta m = 8(1.0072765 \text{ amu}) + 8(1.0086655 \text{ amu}) - 15.99052 \text{ amu} = 0.137016$
                                                              $= 0.13702 \text{ amu}$

$$\Delta E = (2.99795 \times 10^8 \text{ m/s})^2 \times 0.137016 \text{ amu} \times \frac{1 \text{ g}}{6.02214 \times 10^{23} \text{ amu}} \times \frac{1 \text{ kg}}{1 \times 10^3 \text{ g}}$$

        $= 2.04489 \times 10^{-11} = 2.0449 \times 10^{-11} \text{ J} / {}^{18}\text{O nucleus required}$

$$2.04489 \times 10^{-11} \frac{\text{J}}{\text{nucleus}} \times \frac{6.02214 \times 10^{23} \text{ atoms}}{\text{mol}} = 1.2315 \times 10^{13} \text{ J/mol } {}^{18}\text{O}$$

21.44 $\Delta m = 16(1.0072765 \text{ amu}) + 16(1.0086655 \text{ amu}) - 31.95452 \text{ amu} = 0.300552$
                                                               $= 0.30055 \text{ amu}$

$$\Delta E = (2.99795 \times 10^8 \text{ m/s})^2 \times 0.300552 \text{ amu} \times \frac{1 \text{ g}}{6.02214 \times 10^{23} \text{ amu}} \times \frac{1 \text{ kg}}{1000 \text{ g}}$$

        $= 4.48557 \times 10^{-11} = 4.4856 \times 10^{-11} \text{ J} / {}^{32}\text{S nucleus required}$

$$4.48557 \times 10^{-11} \frac{\text{J}}{\text{nucleus}} \times \frac{6.02214 \times 10^{23} \text{ nuclei}}{\text{mol}} = 2.7013 \times 10^{13} \text{ J/mol } {}^{32}\text{S binding energy}$$

21.45   In each case, calculate the mass defect, total nuclear binding energy and then binding energy per nucleon.

(a) $3(1.0072765) + 4(1.0086655) = 7.0564915$ amu

mass defect $= 7.0564915 - 7.01600 = 0.04049$ amu

$$\Delta E = 0.04049 \text{ amu} \times \frac{1 \text{ g}}{6.02214 \times 10^{23} \text{ amu}} \times \frac{1 \text{ kg}}{1000 \text{ g}} \times \frac{8.98768 \times 10^{16} \text{ m}^2}{\text{s}^2}$$

$= 6.0429 \times 10^{-12} = 6.043 \times 10^{-12}$ J

binding energy/nucleon $= 6.0429 \times 10^{-12}$ J $/ 7 = 8.633 \times 10^{-13}$ J/nucleon

(b) $30(1.0072765) + 34(1.0086655) = 64.512922$ amu

mass defect $= 64.512922 - 63.92914 = 0.58378$ amu

$$\Delta E = 0.58378 \text{ amu} \times \frac{1 \text{ g}}{6.02214 \times 10^{23} \text{ amu}} \times \frac{1 \text{ kg}}{1000 \text{ g}} \times \frac{8.98768 \times 10^{16} \text{ m}^2}{\text{s}^2}$$

$= 8.71256 \times 10^{-11} = 8.7126 \times 10^{-11}$ J

binding energy/ nucleon $= 8.71256 \times 10^{-11}$ J $/ 64 = 1.3613 \times 10^{-12}$ J/nucleon

(c) $90(1.0072765) + 142(1.0086655) = 233.885386$ amu

mass defect $= 233.885386 - 232.0382 = 1.8472$ amu

$$\Delta E = 1.8472 \text{ amu} \times \frac{1 \text{ g}}{6.02214 \times 10^{23} \text{ amu}} \times \frac{1 \text{ kg}}{1000 \text{ g}} \times \frac{8.98768 \times 10^{16} \text{ m}^2}{\text{s}^2}$$

$= 2.75683 \times 10^{-10} = 2.7568 \times 10^{-10}$ J

binding energy/nucleon $= 2.75683 \times 10^{-10}$ J $/ 232 = 1.1883 \times 10^{-12}$ J/nucleon

21.46   In each case, calculate the mass defect, total nuclear binding energy and then binding energy per nucleon.

(a) $6(1.0072765) + 6(1.0086655) = 12.095652$ amu

mass defect $= 12.095652 - 12.00000 = 0.09565$ amu

$$\Delta E = 0.09565 \text{ amu} \times \frac{1 \text{ g}}{6.02214 \times 10^{23} \text{ amu}} \times \frac{1 \text{ kg}}{1000 \text{ g}} \times \frac{8.98768 \times 10^{16} \text{ m}^2}{\text{s}^2}$$

$= 1.4275 \times 10^{-11} = 1.428 \times 10^{-11}$ J

binding energy/nucleon $= 1.4275 \times 10^{-11}$ J $/ 12 = 1.190 \times 10^{-12}$ J/nucleon

(b) $27(1.0072765) + 32(1.0086655) = 59.473762$ amu

mass defect $= 59.473762 - 58.9332 = 0.5406$ amu

$$\Delta E = 0.5406 \text{ amu} \times \frac{1 \text{ g}}{6.02214 \times 10^{23} \text{ amu}} \times \frac{1 \text{ kg}}{1000 \text{ g}} \times \frac{8.98768 \times 10^{16} \text{ m}^2}{\text{s}^2}$$

$$= 8.0681 \times 10^{-11} = 8.068 \times 10^{-11} \text{ J}$$

binding energy/nucleon = 8.0681×10^{-11} J / 59 = 1.367×10^{-12} J/nucleon

(c) $82(1.0072765) + 124(1.0086655) = 207.671195$

mass defect = $207.671195 - 205.97447 = 1.69673$

$$\Delta E = 1.69673 \text{ amu} \times \frac{1 \text{ g}}{6.02214 \times 10^{23} \text{ amu}} \times \frac{1 \text{ kg}}{1000 \text{ g}} \times \frac{8.98768 \times 10^{16} \text{ m}^2}{\text{s}^2}$$

$$= 2.532266 \times 10^{-10} = 2.53227 \times 10^{-10}$$

binding energy/nucleon = 2.532266×10^{-10} J / 206 = 1.22926×10^{-12} J/nucleon

21.47 $\dfrac{1.07 \times 10^{16} \text{ kJ}}{1 \text{ min}} \times \dfrac{60 \text{ min}}{1 \text{ hr}} \times \dfrac{24 \text{ hr}}{1.0 \text{ day}} = 1.541 \times 10^{19} \dfrac{\text{kJ}}{\text{day}} = 1.5 \times 10^{22}$ J/day

$$\Delta m = \frac{1.541 \times 10^{22} \text{ kg} \cdot \text{m}^2/\text{s}^2/\text{d}}{(2.998 \times 10^8 \text{ m/s})^2} = 1.714 \times 10^5 = 1.7 \times 10^5 \text{ kg/d}$$

Now calculate the mass change in the given nuclear reaction:

$\Delta m = 140.9140 + 91.9218 + 2(1.00867) - 235.0439 = -0.19076$ amu.

Converting from atoms to moles and amu to grams, it requires 235 g ^{235}U to produce energy equivalent to a change in mass of 0.191 g.

Now 0.10% of 1.714×10^5 kg is 1.714×10^2 kg = 1.714×10^5 g. Then,

$$1.714 \times 10^5 \text{ g} \times \frac{235 \text{ g } ^{235}\text{U}}{0.191 \text{ g}} = 2.11 \times 10^8 \text{ g } ^{235}\text{U}$$

This is about 230 tons of ^{235}U **per day**.

21.48   (a) $\Delta m = 4.00260 + 1.00867 - 3.01605 - 2.01410 = -0.01888$ amu

$$\Delta E = 0.01888 \text{ amu} \times \frac{1 \text{ g}}{1 \text{ amu}} \times \frac{1 \text{ kg}}{10^3 \text{ g}} \times (2.9979 \times 10^8 \text{ m/sec})^2 = 1.697 \times 10^{12} \text{ J/mol}$$

(b) $\Delta m = 3.01603 + 1.00867 - 2(2.01410) = -3.50 \times 10^{-3}$ amu

$\Delta E = 3.15 \times 10^{11}$ J/mol

(c) $\Delta m = 4.00260 + 1.00782 - 3.01603 - 2.01410 = -1.97 \times 10^{-2}$ amu

$\Delta E = 1.77 \times 10^{12}$ J/mol

21.49   We can use Figure 21.13 to see that the binding energy per nucleon (which gives rise to the mass defect) is greatest for nuclei of mass numbers around 50. Thus (a) $^{59}_{27}$Co should possess the greatest mass defect per nucleon.

21.50   In a fission reactor absorption of neutrons causes a single heavy nucleus to undergo fission, producing two medium mass nuclei. These are seen in Figure 21.13 to have a larger total mass defect than the starting single heavier nucleus, so energy is released.

Effects and Uses of Radioisotopes

21.51   The ^{59}Fe would be incorporated into the diet component, which in turn is fed to the rabbits. After a time blood samples could be removed from the animals, the red blood cells separated, and the radioactivity of the sample measured. If the iron in the dietary compound has been incorporated into blood hemoglobin, the blood cell sample should show beta emission. Samples could be taken at various times to determine the rate of iron uptake, rate of loss of the iron from the blood, and so forth.

21.52   (a)   Add ^{36}Cl to water as a chloride salt. Then dissolve ordinary CCl_3COOH. After a time, distill the volatile materials away from the salt; CCl_3COOH is volatile, and will distill with water. Count radioactivity in the volatile material. If chlorine exchange has occurred, there will be radioactivity.

(b)   Prepare a saturated solution of $BaCl_2$ containing a small amount of solid $BaCl_2$. Add to this solution solid $BaCl_2$ containing ^{36}Cl. If the solid-solution equilibrium is dynamic, some of the ^{36}Cl in the solid will find itself in solution as chloride ion. After allowing some time for equilibrium to become established, filter the solution, count radioactivity in the solution that is separated from the solid. If there were no dynamic equilibrium, the $^{36}Cl^-$ would remain in the added solid, since the solution is already saturated before the addition of more solid.

(c)   Utilize ^{36}Cl in soils of various pH values; grow plants for a given period of time. Remove plants, and directly measure radioactivity in samples from stems, leaves and so forth, or reduce the volume of plant sample by some form of digestion and evaporation of solution to give a dry residue that can be counted.

21.53   (a)   *Control rods* control neutron flux so that there are enough neutrons to sustain the chain reaction but not so many that the core overheats.

(b)   A *moderator* slows neutrons so that they are more easily captured by fissioning nuclei.

21.54   (a)   In a *chain reaction*, one neutron initiates a nuclear transformation that produces more than one neutron. The product neutrons initiate more transformations, so that the reaction is self-sustaining.

(b)   *Critical mass* is the mass of fissionable material required to sustain a chain reaction so that only one product neutron is effective at initiating a new transformation.

21.55   (a) $^{235}_{92}U + ^{1}_{0}n \rightarrow ^{160}_{62}Sm + ^{72}_{30}Zn + 4\,^{1}_{0}n$ (b) $^{239}_{94}Pu + ^{1}_{0}n \rightarrow ^{144}_{58}Ce + ^{94}_{36}Kr + 2\,^{1}_{0}n$

21.56   (a) $^{2}_{1}H + ^{2}_{1}H \rightarrow ^{3}_{2}He + ^{1}_{0}n$ (b) $^{233}_{92}U + ^{1}_{0}n \rightarrow ^{133}_{51}Sb + ^{98}_{41}Nb + 3\,^{1}_{0}n$

21.57   The extremely high temperature is required to overcome the electrostatic charge repulsions between the nuclei so that they come together to react.

21.58   Fusion reactors are attractive because, unlike fission reactors, they produce little or no radioactive waste. The problem is, no known material can withstand the extremely high temperatures required for fusion; the fusion reaction cannot be contained with current technology.

21.59   •OH is a free radical; it contains an unpaired (free) electron, which makes it an extremely reactive species. (As an odd electron molecule, it violates the octet rule.) It can react with almost any particle (atom, molecule, ion) to acquire an electron and become OH⁻. This often starts a disruptive chain of reactions, each producing a different free radical.

Hydroxide ion, OH⁻, on the other hand, will be attracted to cations or the positive end of a polar molecule. It's most common reaction is ubiquitous and innocuous: $H^+ + OH^- \rightarrow H_2O$. The acid-base reactions of OH⁻ are usually much less disruptive to the organism than the chain of redox reactions initiated by •OH radical.

21.60

H_2O^+ is a free radical because it contains seven valence electrons. Around the central O atom there are two bonding and one nonbonding electron pairs and a single unpaired electron (3(2) + 1 = 7 valence electrons).

The reaction above involves homolytic cleavage of an O-H bond. That is, one of the bonding electrons goes with H to form H_3O^+, and the other bonding electron stays with the radical •OH.

21.61   (a)   1 Ci = 3.7×10^{10} disintegrations(dis)/s; 1 Bq = 1 dis/s

$$8.7 \text{ mCi} \times \frac{1 \text{ Ci}}{1000 \text{ mCi}} \times \frac{3.7 \times 10^{10} \text{ dis/s}}{\text{Ci}} = 3.22 \times 10^8 = 3.2 \times 10^8 \text{ dis/s} = 3.2 \times 10^8 \text{ Bq}$$

(b)   1 rad = 1×10^{-2} J/kg; 1 Gy = 1 J/kg = 100 rad. From part (a), the activity of the source is 3.2×10^8 dis/s.

$$3.22 \times 10^8 \text{ dis/s} \times 2.0 \text{ s} \times 0.65 \times \frac{9.12 \times 10^{-13} \text{ J}}{\text{dis}} \times \frac{1}{0.250 \text{ kg}} = 1.53 \times 10^{-3}$$

$$= 1.5 \times 10^{-3} \text{ J/kg}$$

$$1.5 \times 10^{-3} \text{ J/kg} \times \frac{1 \text{ rad}}{1 \times 10^{-2} \text{ J/kg}} \times \frac{1000 \text{ mrad}}{\text{rad}} = 1.5 \times 10^2 \text{ mrad}$$

$$1.5 \times 10^{-3} \text{ J/kg} \times \frac{1 \text{ Gy}}{1 \text{ J/kg}} = 1.5 \times 10^{-3} \text{ Gy}$$

(c)   rem = rad (RBE); Sv = Gy (RBE) = 100 rem

mrem = 1.53×10^2 mrad (9.5) = 1.45×10^3 = 1.5×10^3 mrem (or 1.5 rem)

Sv = 1.53×10^{-3} Gy (9.5) = 1.45×10^{-2} = 1.5×10^{-2} Sv

21.62   (a)   1 Ci = 3.7×10^{10} dis/s; 1 Bq = 1 dis/s

$28 \text{ mCi} \times \dfrac{1 \text{ Ci}}{1000 \text{ mCi}} \times 3.7 \times 10^{10} \text{ dis/s} = 1.04 \times 10^9 = 1.0 \times 10^9 \text{ dis/s} = 1.0 \times 10^9 \text{ Bq}$

(b)   1 Gy = 1 J/kg; 1 Gy = 100 rad

$1.04 \times 10^9 \text{ dis/s} \times 95 \text{ s} \times 0.050 \times \dfrac{8.75 \times 10^{-14} \text{ J}}{\text{dis}} \times \dfrac{1}{65 \text{ kg}} = 6.62 \times 10^{-6}$

$= 6.6 \times 10^{-6}$ J/kg

$6.6 \times 10^{-6} \text{ J/kg} \times \dfrac{1 \text{ Gy}}{1 \text{ J/kg}} = 6.6 \times 10^{-6} \text{ Gy}; 6.6 \times 10^{-6} \text{ Gy} \times \dfrac{100 \text{ rad}}{1 \text{ Gy}} = 6.6 \times 10^{-4} \text{ rad}$

(c)   rem = rad (RBE); Sv = Gy (RBE)

$6.6 \times 10^{-4} \text{ rad} (1.0) = 6.6 \times 10^{-4} \text{ rem} \times \dfrac{1000 \text{ mrem}}{1 \text{ rem}} = 0.66 \text{ mrem}$

$6.6 \times 10^{-6} \text{ Gy} (1.0) = 6.6 \times 10^{-6} \text{ Sv}$

(d)   From Figure 21.23, the average annual background radiation is 360 mrem, or about 1 mrem/day. This 0.66 mrem exposure is less than the average background radiation for a day.

Additional Exercises

21.63 $^{222}_{86}\text{Rn} \rightarrow X + 3\,^4_2\text{He} + 2\,^0_{-1}\beta$

This corresponds to a reduction in mass number of $(3 \times 4 =)$ 12 and a reduction in atomic number of $(3 \times 2 - 2) = 4$. The stable nucleus is $^{210}_{82}\text{Pb}$. [This is part of the sequence in Figure 21.4.]

21.64 $^1_0\text{n} \rightarrow\ ^1_1\text{p} +\ ^0_{-1}\text{e}$

The other product of neutron decay is an electron.

21.65   The most massive radionuclides will have the highest neutron/proton ratios. Thus, they are most likely to decay by a process that lowers this ratio, beta emission. The least massive nuclides, on the other hand, will decay by a process that increases the neutron/proton ratio, positron emission or orbital electron capture.

21.66   (a) $^{36}_{17}\text{Cl} \rightarrow\ ^{36}_{18}\text{Ar} +\ ^0_{-1}\text{e}$

(b)   According to Table 21.3, nuclei with even numbers of both protons and neutrons, or an even number of one kind of nucleon, are more stable. ^{35}Cl and ^{37}Cl both have an odd number of protons **but** an even number of neutrons. ^{36}Cl has an odd number of protons and neutrons (17 p, 19 n), so it is less stable than the other two isotopes. Also, ^{37}Cl has 20 neutrons, a nuclear closed shell.

21.67   (a) $^{6}_{3}Li + ^{56}_{28}Ni \rightarrow ^{62}_{31}Ga$

       (b) $^{40}_{20}Ca + ^{248}_{96}Cm \rightarrow ^{288}_{116}X$

       (c) $^{88}_{38}Sr + ^{84}_{36}Kr \rightarrow ^{116}_{46}Pd + ^{56}_{26}Ni$

       (d) $^{40}_{20}Ca + ^{238}_{92}U \rightarrow ^{70}_{30}Zn + 4\,^{1}_{0}n + 2\,^{102}_{41}Nb$

21.68   This is similar to Exercises 21.35 and 21.36.

$^{209}_{84}Po \rightarrow ^{205}_{82}Pb + ^{4}_{2}He$

Each ^{209}Po nucleus that decays is 1 disintegration. Calculate the mass of ^{209}Po remaining after 1.0 s, calculate by subtraction the mass that has decayed, and use Avogadro's number to get the number of nuclei that have decayed.

Calculate k in s^{-1}. $105\ yr \times \dfrac{365\ d}{1\ yr} \times \dfrac{24\ hr}{1\ d} \times \dfrac{60\ min}{1\ hr} \times \dfrac{60\ s}{1\ min} = 3.311 \times 10^{9} = 3.31 \times 10^{9}\ s$

$k = 0.693 / t_{1/2} = 0.693/3.311 \times 10^{9}\ s = 2.093 \times 10^{-10} = 2.09 \times 10^{-10}\ s^{-1}$

$\ln(N_t / N_o) - kt = -(2.093 \times 10^{-10}\ s^{-1})(1.0\ s) = -2.093 \times 10^{-10}$

$N_t / N_o = e^{-2.093 \times 10^{-10}} = (1.00 - 2.093 \times 10^{-10}); \quad N_t = 1.0 \times 10^{-12}\ g\ (1.000 - 2.093 \times 10^{-10})$

The amount that decays is $N_o - N_t$:

$1.0 \times 10^{-12}\ g - [1.0 \times 10^{-12}\ g\ (1.000 - 2.093 \times 10^{-10})] = 1.0 \times 10^{-12}\ g\ (2.093 \times 10^{-10})$

$$= 2.093 \times 10^{-22} = 2.1 \times 10^{-22}\ g\ Po$$

$N_o - N_t = 2.093 \times 10^{-22}\ g\ Po \times \dfrac{1\ mol\ Po}{209\ g\ Po} \times \dfrac{6.022 \times 10^{23}\ Po\ atoms}{1\ mol\ Po} = 0.603 = 0.60\ dis$

This is 0.60 disintegrations in 1.0 s, or approximately 1 dis/s.

$0.603\ dis/s \times \dfrac{1\ Ci}{3.7 \times 10^{10}\ dis/s} = 1.6 \times 10^{-10}\ Ci\ (2.7 \times 10^{-11}\ Ci\ based\ on\ 1\ dis/s.)$

21.69

Time (hr)	N_t (dis/min)	ln N_t
0	180	5.193
2.5	130	4.868
5.0	104	4.644
7.5	77	4.34
10.0	59	4.08
12.5	46	3.83
17.5	24	3.18

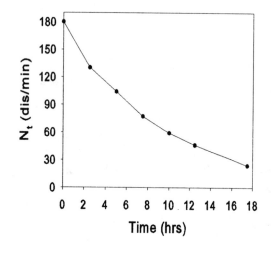

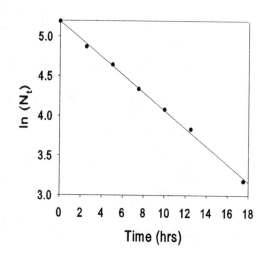

The plot on the left is a graph of activity (disintegrations per minute) vs. time. Choose $t_{1/2}$ at the time where $N_t = 1/2 \ N_o = 90$ dis/min. $t_{1/2} \approx 6.0$ hr.

Rearrange Equation 21.19 to obtain the linear relationship shown on the right.

$\ln(N_t / N_o) = -kt$; $\ln N_t - \ln N_o = -kt$; $\ln N_t = -kt + \ln N_o$

The slope of this line = $-k$ = 0.11; $t_{1/2}$ = 0.693/0.11 = 6.3 hr.

21.70 $1 \times 10^{-6} \text{ curie} \times \dfrac{3.7 \times 10^{10} \text{ dis/s}}{\text{curie}} = 3.7 \times 10^4 \text{ dis/s}$

rate = 3.7×10^4 nuclei/s = kN

$k = \dfrac{0.693}{t_{1/2}} = \dfrac{0.693}{28.8 \text{ yr}} \times \dfrac{1 \text{ yr}}{365 \times 24 \times 3600 \text{ sec}} = 7.630 \times 10^{-10} = 7.63 \times 10^{-10} \text{ s}^{-1}$

3.7×10^4 nuclei/s = $(7.63 \times 10^{-10}/\text{s})$ N; N = 4.849×10^{13} = 4.8×10^{13} nuclei

mass ^{90}Sr = 4.849×10^{13} nuclei $\times \dfrac{90 \text{ g Sr}}{6.022 \times 10^{23} \text{ nuclei}} = 7.2 \times 10^{-9}$ g Sr

21.71 First calculate k in s^{-1}

$$k = \frac{0.693}{2.4 \times 10^4 \, yr} \times \frac{1 \, yr}{365 \times 24 \times 3600 \, s} = 9.16 \times 10^{-13} = 9.2 \times 10^{-13} \, s^{-1}$$

Now calculate N:

$$N = 0.500 \, g \, Pu \times \frac{1 \, mol \, Pu}{239 \, g \, Pu} \times \frac{6.022 \times 10^{23} \, Pu \, atoms}{1 \, mol \, Pu} = 1.26 \times 10^{21} = 1.3 \times 10^{21} \, Pu \, atoms$$

$$rate = (9.16 \times 10^{-13} \, s^{-1})(1.26 \times 10^{21} \, Pu \, atoms) = 1.15 \times 10^9 = 1.2 \times 10^9 \, dis/s$$

21.72 Assume that no depletion of iodide from the water due to plant uptake has occurred. Then the activity after 32 days would be:

$$k = 0.693/t_{1/2} = 0.693/8.1 \, d = 0.0856 = 0.086 \, d^{-1}$$

$$\ln \frac{N_t}{N_o} = -(0.0856 \, d^{-1})(32 \, d) = -2.739 = -2.7; \quad \frac{N_t}{N_o} = 0.0646 = 0.06$$

We thus expect $N_t = 0.0646(89) = 5.7$ counts/min. This is just the **observed** activity; we can assume that the plants did not absorb iodide, because that would have resulted in a lower level of remaining activity.

21.73 First, calculate k in s^{-1}

$$k = \frac{0.693}{12.26 \, yr} \times \frac{1 \, yr}{365 \, d} \times \frac{1 \, d}{24 \, hr} \times \frac{1 \, hr}{3600 \, sec} = 1.7924 \times 10^{-9} = 1.792 \times 10^{-9} \, s^{-1}$$

From Equation [21.18], $1.50 \times 10^3 \, s^{-1} = (1.7924 \times 10^{-9} \, s^{-1})(N)$;

$N = 8.369 \times 10^{11} = 8.37 \times 10^{11}$ In 26.0 g of water, there are

$$26.0 \, g \, H_2O \times \frac{1 \, mol \, H_2O}{18.02 \, g \, H_2O} \times \frac{6.022 \times 10^{23} \, H_2O}{1 \, mol \, H_2O} \times \frac{2 \, H}{1 \, H_2O} = 1.738 \times 10^{24}$$

$$= 1.74 \times 10^{24} \, H \, atoms$$

The mole fraction of $_1^3H$ atoms in the sample is thus

$$8.369 \times 10^{11}/1.738 \times 10^{24} = 4.816 \times 10^{-13} = 4.82 \times 10^{-13}$$

21.74 Because of the relationship $\Delta E = \Delta mc^2$, the mass defect (Δm) is directly related to the binding energy (ΔE) of the nucleus.

^7Be - 4p, 3n; $4(1.0072765) + 3(1.0086655) = 7.05510 \, amu$

Total mass defect = $7.0551 - 7.0147 = 0.0404 \, amu$

$0.0404 \, amu/7 \, nucleons = 5.77 \times 10^{-3} \, amu/nucleon$

$$\Delta E = \Delta m \times c^2 = \frac{5.77 \times 10^{-3} \, amu}{nucleon} \times \frac{1 \, g}{6.022 \times 10^{23} \, amu} \times \frac{1 \, kg}{1 \times 10^3 \, g} \times \frac{8.988 \times 10^{16} \, m^2}{sec^2}$$

$$= \frac{5.77 \times 10^{-3} \, amu}{nucleon} \times \frac{1.4925 \times 10^{-10} \, J}{1 \, amu} = 8.61 \times 10^{-13} \, J/nucleon$$

^{9}Be - 4p, 5n; 4(1.0072765) + 5(1.0086655) = 9.07243 amu

Total mass defect = 9.0724 - 9.0100 = 0.06243 = 0.0624 amu

0.0624 amu/9 nucleons = 6.937 × 10^{-3} = 6.94 × 10^{-3} amu/nucleon

6.937 × 10^{-3} amu/nucleon × 1.4925 × 10^{-10} J/amu = 1.035 × 10^{-12} = 1.04 × 10^{-12} J/nucleon

^{10}Be - 4p, 6n; 4(1.0072765) + 6(1.0086655) = 10.0810 amu

Total mass defect = 10.0811 - 10.0113 = 0.0698 amu

0.0698 amu/10 nucleons = 6.98 × 10^{-3} amu/nucleon

6.98 × 10^{-3} amu/nucleon × 1.4925 × 10^{-10} J/amu = 1.042 × 10^{-12} = 1.04 × 10^{-12} J/nucleon

The binding energies/nucleon for ^{9}Be and ^{10}Be are very similar; that for ^{10}Be is slightly higher.

21.75 $\Delta m = \Delta E/c^2$; $\Delta m = \dfrac{3.9 \times 10^{26} \text{ kg} \cdot \text{m/s}^2}{(3.00 \times 10^8 \text{ m/s})^2} = 4.3 \times 10^9$ kg

This is the mass lost in one second.

21.76 1000 Mwatts × $\dfrac{1 \times 10^6 \text{ watts}}{1 \text{ Mwatt}}$ × $\dfrac{1 \text{ J}}{1 \text{ watt} \cdot \text{s}}$ × $\dfrac{1\ ^{235}\text{U atom}}{3 \times 10^{-11} \text{ J}}$ × $\dfrac{1 \text{ mol U}}{6.02 \times 10^{23} \text{ atoms}}$

× $\dfrac{235 \text{ g U}}{1 \text{ mol}}$ × $\dfrac{3600 \text{ s}}{1 \text{ hr}}$ × $\dfrac{24 \text{ hr}}{1 \text{ d}}$ × $\dfrac{365 \text{ d}}{1 \text{ yr}}$ × $\dfrac{40}{100}$ (efficiency) = 1.64 × 10^5

= 2 × 10^5 g U/ yr = 200 kg U/yr

21.77 2 × 10^{-12} curies × $\dfrac{3.7 \times 10^{10} \text{ dis/s}}{1 \text{ curie}}$ = 7.4 × 10^{-2} = 7 × 10^{-2} dis/s

$\dfrac{7.4 \times 10^{-2} \text{ dis/s}}{75 \text{ kg}}$ × $\dfrac{8 \times 10^{-13} \text{ J}}{\text{dis}}$ × $\dfrac{1 \text{ rad}}{1 \times 10^{-2} \text{ J/g}}$ × $\dfrac{3600 \text{ s}}{\text{hr}}$ × $\dfrac{24 \text{ hr}}{1 \text{ d}}$

× $\dfrac{365 \text{ d}}{1 \text{ yr}}$ = 2.49 × 10^{-6} = 2 × 10^{-6} rad/yr

Recall that there are 10 rem/rad for alpha particles.

$\dfrac{2.49 \times 10^{-6} \text{ rad}}{1 \text{ yr}}$ × $\dfrac{10 \text{ rem}}{1 \text{ rad}}$ = 2.49 × 10^{-5} = 2 × 10^{-5} rem/yr

Integrative Exercises

21.78 Calculate N, the number of ^{36}Cl nuclei, the value of k in s^{-1}, and the activity in dis/s.

49.5 mg NaClO$_4$ × $\dfrac{1 \text{ g}}{1000 \text{ mg}}$ × $\dfrac{1 \text{ mol NaClO}_4}{122.44 \text{ g NaClO}_4}$ × $\dfrac{1 \text{ mol Cl}}{1 \text{ mol NaClO}_4}$ × $\dfrac{6.022 \times 10^{23} \text{ Cl atoms}}{\text{mol Cl}}$

× $\dfrac{31\ ^{36}\text{Cl atoms}}{100 \text{ Cl atoms}}$ = 7.547 × 10^{19} = 7.55 × 10^{19} ^{36}Cl atoms

$$k = 0.693/t_{1/2} = \frac{0.693}{3.1 \times 10^5 \text{ yr}} \times \frac{1}{365 \times 24 \times 3600 \text{ s}} = 7.09 \times 10^{-14} = 7.1 \times 10^{-14} \text{ s}^{-1}$$

$$\text{rate} = kN = (7.09 \times 10^{-14} \text{ s}^{-1})(7.547 \times 10^{19} \text{ nuclei}) = 5.35 \times 10^6 = 5.4 \times 10^6 \text{ dis/s}$$

21.79 Calculate the amount of energy produced by the nuclear fusion reaction, the enthalpy of combustion, $\Delta H°$, of C_3H_8, and then the mass of C_3H_8 required.

Δm for the reaction $4\ {}_1^1H \rightarrow {}_2^4He + 2\ {}_1^0e$ is $4(1.00782) - 4.00260$ amu $- 2(5.4858 \times 10^{-4}$ amu$)$

$$= 0.027583 = 0.02758 \text{ amu}$$

$$\Delta E = \Delta mc^2 = 0.027583 \text{ amu} \times \frac{1 \text{ g}}{6.02214 \times 10^{23} \text{ amu}} \times \frac{1 \text{ kg}}{1000 \text{ g}} \times (2.997946 \times 10^8 \text{ m/s})^2$$

$$= 4.1164 \times 10^{-12} = 4.116 \times 10^{-12} \text{ J} / 4\ {}^1H \text{ nuclei}$$

$$1.0 \text{ g } {}^1H \times \frac{1\ {}^1H \text{ nucleus}}{1.00782 \text{ amu}} \times \frac{6.02214 \times 10^{23} \text{ amu}}{\text{g}} \times \frac{4.1164 \times 10^{-12} \text{ J}}{4\ {}^1H \text{ nuclei}}$$

$$= 6.149 \times 10^{11} \text{ J} = 6.1 \times 10^8 \text{ kJ produced by the fusion of 1.0 g } {}^1H.$$

$$C_3H_8(g) + 5O_2(g) \rightarrow 3CO_2(g) + 4H_2O(g)$$

$$\Delta H° = 3(-393.5 \text{ kJ}) + 4(-241.82 \text{ kJ}) - (-103.85) - (0) = -2043.9 \text{ kJ}$$

$$6.149 \times 10^8 \text{ kJ} \times \frac{1 \text{ mol } C_3H_8(g)}{2043.9 \text{ kJ}} \times \frac{44.094 \text{ g } C_3H_8}{\text{mol } C_3H_8} = 1.33 \times 10^7 \text{ g} = 1.3 \times 10^4 \text{ kg } C_3H_8$$

13,000 kg $C_3H_8(g)$ would have to be burned to produce the same amount of energy as fusion of 1.0 g ^1H.

21.80 (a) $0.18 \text{ Ci} \times \frac{3.7 \times 10^{10} \text{ dis/s}}{\text{Ci}} \times \frac{3600 \text{ s}}{Li} \times \frac{24 \text{ hr}}{d} \times 235 \text{ d} = 1.35 \times 10^{17}$

$$= 1.4 \times 10^{17} \ \alpha \text{ particles}$$

(b) $P = nRT/V = 1.35 \times 10^{17} \text{ He atoms} \times \frac{1 \text{ mol He}}{6.022 \times 10^{23} \text{ atoms}} \times \frac{295 \text{ K}}{0.0150 \text{ L}} \times \frac{0.08206 \text{ L} \cdot \text{atm}}{\text{K} \cdot \text{mol}}$

$$= 3.62 \times 10^{-4} = 3.6 \times 10^{-4} \text{ atm} = 0.28 \text{ torr}$$

21.81 Calculate N_t in dis/min/g C from 1.5×10^{-2} dis/0.788 g $CaCO_3$. $N_o = 15.2$ dis/min/g C. Calculate k from $t_{1/2}$, calculate t from $\ln N_t / N_o = -kt$.

$$C(s) + O_2(g) \rightarrow CO_2(g) + Ca(OH_2)(aq) \rightarrow CaCO_3(s) + H_2O(l)$$

1 C atom $\rightarrow$ 1 $CaCO_3$ molecule

$$\frac{1.5 \times 10^{-2} \text{ Bq}}{0.788 \text{ g } CaCO_3} \times \frac{1 \text{ dis/s}}{1 \text{ Bq}} \times \frac{60 \text{ s}}{1 \text{ min}} \times \frac{100.1 \text{ g } CaCO_3}{12.01 \text{ g C}} = 9.52 = 9.5 \text{ dis/min/g C}$$

$$k = 0.693/t_{1/2} = 0.693/5.73 \times 10^3 \text{ yr} = 1.209 \times 10^{-4} = 1.21 \times 10^{-4} \text{ yr}^{-1}$$

$$t = -\frac{1}{k} \ln \frac{N_t}{N_o} = \frac{-1}{1.209 \times 10^{-4} \text{ yr}^{-1}} \ln \frac{9.52 \text{ dis/min/g C}}{15.2 \text{ dis/min/g C}} = 3.87 \times 10^3 \text{ yr}$$

21.82   Calculate the energy required to ionize one H_2O molecule, and then the wavelength associated with this energy.

$$\frac{1216 \text{ kJ}}{\text{mol } H_2O} \times \frac{1000 \text{ J}}{1 \text{ kJ}} \times \frac{1 \text{ mol } H_2O}{6.02214 \times 10^{23} \text{ molecules}} = 2.0192 \times 10^{-18}$$

$$= 2.019 \times 10^{-18} \text{ J/molecule}$$

$$\lambda = hc/E = \frac{6.626 \times 10^{-34} \text{ J} \cdot \text{s} \times 2.998 \times 10^8 \text{ m/s}}{2.0192 \times 10^{-18} \text{ J}} = 9.838 \times 10^{-8} \text{ m} = 98.38 \text{ nm}$$

21.83   Determine the wavelengths of the photons by first calculating the energy equivalent of the mass of an electron or positron. (Since **two** photons are formed by annihilation of **two** particles of equal mass, we need to calculate the energy equivalent of just one particle.) The mass of an electron is 9.109×10^{-31} kg.

$$\Delta E = (9.109 \times 10^{-31} \text{ kg}) \times (2.998 \times 10^8 \text{ m/s})^2 = 8.187 \times 10^{-14} \text{ J}$$

Also, $\Delta E = h\nu$; $\Delta E = hc/\lambda$; $\lambda = hc/\Delta E$

$$\lambda = \frac{(6.626 \times 10^{-34} \text{ J} \cdot \text{s})(2.998 \times 10^8 \text{ m/s})}{8.187 \times 10^{-14} \text{ J}} = 2.426 \times 10^{-12} \text{ m} = 2.426 \times 10^{-3} \text{ nm}$$

This is a very short wavelength indeed; it lies at the short wavelength end of the range of observed gamma ray wavelengths (see Figure 6.4).

21.84   (a) $Ba(NO_3)_2(aq) + Na_2SO_4(aq) \rightarrow BaSO_4(s) + 2NaNO_3(aq)$

         (b) $1.25 \text{ mmol } Ba^{2+} + 1.25 \text{ mmol } SO_4^{2-} \rightarrow 1.25 \text{ mmol } BaSO_4$

                Neither reactant is in excess, so the activity of the filtrate is due entirely to $[SO_4^{2-}]$ from ionization of $BaSO_4(s)$. Calculate $[SO_4^{2-}]$ in the filtrate by comparing the activity of the filtrate to the activity of the reactant.

$$\frac{0.050 \, M \, SO_4^{2-}}{1.22 \times 10^6 \text{ Bq/mL}} = \frac{x \, M \text{ filtrate}}{250 \text{ Bq/mL}}$$

$$[SO_4^{2-}] \text{ in the filtrate} = 1.0246 \times 10^{-5} = 1.0 \times 10^{-5} \, M$$

$$K_{sp} = [Ba^{2+}][SO_4^{2-}]; \quad [SO_4^{2-}] = [Ba^{2+}]$$

$$K_{sp} = (1.0246 \times 10^{-5})^2 = 1.0498 \times 10^{-10} = 1.0 \times 10^{-10}$$

22 Chemistry of the Nonmetals

Periodic Trends and Chemical Reactions

22.1   Metals: Sr, Ce, Rh;   nonmetals: Se, Kr;   metalloid: Sb

22.2   Metals: Zr, In, Cs;   nonmetals: Xe, Br;   metalloid: Ge

22.3   Bismuth should be most metallic. It has a metallic luster and a relatively low ionization energy. Bi_2O_3 is soluble in acid but not in base, characteristic of the basic properties of the oxide of a metal rather than the oxide of a nonmetal.

22.4   B is most nonmetallic in Group 3A. Some properties which demonstrate this are its low electrical and thermal conductivity, lack of metallic luster (it is a yellow powder), and the acidity of its oxide. Al, immediately below B, has the properties of a metal, good electrical and thermal conductivity, luster, and a basic oxide.

22.5   (a)   Cl   (b) K

       (c)   K in the gas phase (lowest ionization energy), Li in aqueous solution (most positive $E°$ value)

       (d)   Ne; Ne and Ar are difficult to compare because they do not form compounds and their radii are not measured in the same way as other elements. However, Ne is several rows to the right of C and surely has a smaller atomic radius. The next smallest is C.

       (e)   C

22.6   (a)   O (b) Br (c) Ba (d) O (e) Co

22.7   (a)   Nitrogen is too small to accommodate five fluorine atoms about it. The P and As atoms are larger. Furthermore, P and As have available 3d and 4d orbitals, respectively, to form hybrid orbitals that can accommodate more than an octet of electrons about the central atom.

       (b)   Si does not readily form π bonds, which would be necessary to satisfy the octet rule for both atoms in SiO.

       (c)   A reducing agent is a substance that readily loses electrons. As has a lower electronegativity than N; that is, it more readily gives up electrons to an acceptor and is more easily oxidized.

22.8   (a)   Nitrogen is a highly electronegative element. In HNO_3 it is in its highest oxidation state, +5, and thus is more readily reduced than phosphorus, which forms stable P-O bonds.

        (b)   The difference between the third row element and the second lies in the smaller size of C as compared with Si, and the fact that Si has 3d orbitals available to form an sp^3d^2 hybrid set that can accommodate more than an octet of electrons.

        (c)   Two of the carbon compounds, C_2H_4 and C_2H_2, contain C-C π bonds. Si does not readily form π bonds (to itself or other atoms), so Si_2H_4 and Si_2H_2 are not known as stable compounds.

22.9   (a) $NaNH_2(s) + H_2O(l) \rightarrow NH_3(aq) + Na^+(aq) + OH^-(aq)$

        (b) $2C_3H_7OH(l) + 9O_2(g) \rightarrow 6CO_2(g) + 8H_2O(l)$

        (c) $NiO(s) + C(s) \rightarrow CO(g) + Ni(s)$ or $2NiO(s) + C(s) \rightarrow CO_2(g) + 2Ni(s)$

        (d) $AlP(s) + 3H_2O(l) \rightarrow PH_3(g) + Al(OH)_3(s)$

        (e) $Na_2S(s) + 2HCl(aq) \rightarrow H_2S(g) + 2Na^+(aq) + 2Cl^-(aq)$

22.10   (a) $NaOCH_3(s) + H_2O(l) \rightarrow NaOH(aq) + CH_3OH(aq)$

        (b) $Na_2O(s) + 2HC_2H_3O_2(aq) \rightarrow 2NaC_2H_3O_2(aq) + H_2O(l)$

        (c) $WO_3(s) + H_2(g) \rightarrow W(s) + 3H_2O(g)$

        (d) $4NH_2OH(l) + O_2(g) \rightarrow 6H_2O(l) + 2N_2(g)$

        (e) $Al_4C_3(s) + 12H_2O(l) \rightarrow 4Al(OH)_3(s) + 3CH_4(g)$

Hydrogen, the Noble Gases, and the Halogens

22.11 ^1_1H - protium; ^2_1H - deuterium; ^3_1H - tritium

22.12   Tritium is radioactive. $^3_1H \rightarrow \, ^3_2He + \, ^0_{-1}e$

22.13   Like other elements in group 1A, hydrogen has only one valence electron. Like other elements in group 7A, hydrogen needs only one electron to complete its valence shell. The most common oxidation number of H is +1, like the group 1A elements; H can also exist in the -1 oxidation state, a state common to the group 7A elements.

22.14   Hydrogen does not have a strictly comparable electronic arrangement. There are no closed shells of electrons underlying the valence shell. When hydrogen completes its valence shell it has only two electrons therein, a characteristic it shares only with He.

22.15   (a) $Mg(s) + 2H^+(aq) \rightarrow Mg^{2+}(aq) + H_2(g)$

        (b) $C(s) + H_2O(g) \xrightarrow{1000\,^\circ C} CO(g) + H_2(g)$

(c) $CH_4(g) + H_2O(g) \xrightarrow{1100\,^{\circ}C} CO(g) + 3H_2(g)$

22.16 (a) Electrolysis of brine; reaction of carbon with steam; reaction of methane with steam; byproduct in petroleum refining

(b) Synthesis of ammonia; synthesis of methanol; reducing agent; hydrogenation of unsaturated vegetable oils

22.17 (a) $NaH(s) + H_2O(l) \rightarrow NaOH(aq) + H_2(g)$

(b) $Fe(s) + H_2SO_4(aq) \rightarrow Fe^{2+}(aq) + H_2(g) + SO_4^{2-}(aq)$

(c) $H_2(g) + Br_2(g) \rightarrow 2HBr(g)$

(d) $Na(l) + H_2(g) \rightarrow 2NaH(s)$

(e) $PbO(s) + H_2(g) \xrightarrow{\Delta} Pb(s) + H_2O(g)$

22.18 (a) $2Al(s) + 6H^+(aq) \rightarrow 2Al^{3+}(aq) + 3H_2(g)$

(b) $Mg(s) + H_2O(g) \rightarrow MgO(s) + H_2(g)$

(c) $MnO_2(s) + H_2(g) \rightarrow MnO(s) + H_2O(g)$

(d) $CaH_2(s) + H_2O(l) \rightarrow Ca(OH)_2(aq) + 2H_2(g)$

22.19 (a) Ionic (metal hydride)   (b) molecular (nonmetal hydride)

(c) metallic (nonstoichiometric transition metal hydride)

22.20 (a) Molecular   (b) ionic   (c) metallic

22.21 Xenon is larger, and can more readily accommodate an expanded octet. More important is the lower ionization energy of xenon; because the valence electrons are a greater average distance from the nucleus, they are more readily promoted to a state in which the Xe atom can form bonds with fluorine.

22.22 The noble gases are colorless, odorless, diamagnetic, inert gases. They make up a very small fraction of the atmosphere and do not exist in naturally occurring compounds. Because they are rare and unreactive, they were difficult to detect.

22.23 (a) IO_3^-, +5   (b) $H\underline{Br}O_3$, +5   (c) $\underline{BrF}_3$; Br, +3; F, -1

(d) $NaO\underline{Cl}$, +1   (e) $H\underline{I}O_2$, +3   (f) $\underline{Xe}O_3$, +6

22.24 (a) $Ca\underline{Br}_2$, -1   (b) $HC\underline{l}O_4$, +7   (c) $\underline{XeOF}_4$; Xe, +6; F, -1

(d) $\underline{Cl}O_2^-$, +3   (e) $H\underline{Br}O$, +1   (f) $\underline{IF}_5$; I, +5; F, -1

22.25 (a) potassium chlorate   (b) calcium iodate   (c) aluminum chloride

(d) bromic acid   (e) paraperiodic acid   (f) xenon tetrafluoride

22.26 (a)   iron(II) perchlorate   (b) chlorous acid   (c) xenon difluoride

        (d)   iodine pentafluoride   (e) xenon trioxide   (f) hydrobromic acid

22.27 (a)   Van der Waals intermolecular attractive forces increase with increasing numbers of electrons in the atoms.

        (b) F_2 reacts with water: $F_2(g) + H_2O(l) \rightarrow 2HF(aq) + O_2(g)$. That is, fluorine is too strong an oxidizing agent to exist in water.

        (c)   HF has extensive hydrogen bonding.

        (d)   Oxidizing power is related to electronegativity. Electronegativity decreases in the order given.

22.28 (a)   The more electronegative the central atom, the greater the extent to which it withdraws charge from oxygen, in turn making the O-H bond more polar, and enhancing ionization of H^+.

        (b)   HF reacts with the silica which is a major component of glass:
$6HF(aq) + SiO_2(s) \rightarrow SiF_6^{2-}(aq) + 2H_2O(l) + 2H^+(aq)$

        (c)   Iodide is oxidized by sulfuric acid, as described in Equation 23.10.

        (d)   The major factor is size; there is not room about Br for the three chlorides plus the two unshared electron pairs that would occupy the bromine valence shell orbitals.

        (e) I_2 reacts with I^- to form the triiodide ion, I_3^-.

22.29 (a) $Br_2(l) + 2OH^-(aq) \rightarrow BrO^-(aq) + Br^-(aq) + H_2O(l)$

        (b) $Cl_2(g) + 2Br^-(aq) \rightarrow Br_2(l) + 2Cl^-(aq)$

        (c) $Br_2(l) + H_2O_2(aq) \rightarrow 2Br^-(aq) + O_2(g) + 2H^+(aq)$

22.30 (a) $3CaBr_2(s) + 2H_3PO_4(l) \rightarrow Ca_3(PO_4)_2(s) + 6HBr(g)$

        (b) $AlBr_3(s) + 3H_2O(l) \rightarrow Al(OH)_3(s) + 3HBr(g)$

        (c) $2HF(aq) + CaCO_3(s) \rightarrow CaF_2(s) + H_2O(l) + CO_2(g)$

22.31 (a) $PBr_5(l) + 4H_2O(l) \rightarrow H_3PO_4(aq) + 5H^+(aq) + 5Br^-(aq)$

        (b) $IF_5(l) + 3H_2O(l) \rightarrow H^+(aq) + IO_3^-(aq) + 5HF(aq)$

        (c) $SiBr_4(l) + 4H_2O(l) \rightarrow Si(OH)_4(s) + 4H^+(aq) + 4Br^-(aq)$

        (d) $2F_2(g) + 2H_2O(l) \rightarrow 4HF(aq) + O_2(g)$

        (e) $2ClO_2(g) + H_2O(l) \rightarrow H^+(aq) + ClO_3^-(aq) + HClO_2(aq)$

        (f) $HI(g) \rightarrow H^+(aq) + I^-(aq)$

22.32 (a) $CaF_2(s) + H_2SO_4(l) \rightarrow 2HF(g) + CaSO_4(s)$

        (b) $2I^-(aq) + Cl_2(g) \rightarrow I_2(s) + 2Cl^-(aq)$

        (c) $Xe(g) + 2F_2(g) \rightarrow XeF_4(s)$

(d) $Ca^{2+}(aq) + 2OH^-(aq) + Cl_2(aq) \rightarrow ClO^-(aq) + Cl^-(aq) + H_2O(l) + Ca^{2+}(aq)$

(e) $S_8(s) + 24F_2(g) \rightarrow 8SF_6(g)$ or $S_8(s) + 24BrF_5(l) \rightarrow 8SF_6(g) + 24BrF_3(l)$

(f) $Cl_2(g) + 2OH^-(aq) + 2Na^+(aq) \rightarrow ClO^-(aq) + Cl^-(aq) + H_2O(l)$

       NaClO (very unstable) can be isolated by cooling a concentrated solution.

22.33   (a)   linear   (b) square-planar   (c) trigonal pyramidal

        (d)   octahedral about the central iodine   (e) square-planar

22.34

(For clarity, the three unshared electron pairs on each F in the anion are omitted.) The VSEPR model predicts a bent structure for the cation, and an octahedral geometry about Sb for the anion.

Oxygen and the Group 6A Elements

22.35   (a)   As an oxidizing agent in steel-making; to bleach pulp and paper; in oxyacetylene torches; in medicine to assist in breathing

        (b)   Synthesis of pharmaceuticals, lubricants and other organic compounds where C=C bonds are cleaved; in water treatment

22.36

Ozone has two resonance forms (Section 8.7); the molecular structure is bent, with an O-O-O bond angle of approximately 120°. The π bond in ozone is delocalized over the entire molecule; neither individual O-O bond is a full double bond, so the observed O-O distance of 1.28 Å is greater than the 1.21 Å distance in O_2, which has a full O-O double bond.

22.37   (a) $CaO(s) + H_2O(l) \rightarrow Ca^{2+}(aq) + 2OH^-(aq)$

        (b) $Al_2O_3(s) + 6H^+(aq) \rightarrow 2Al^{3+}(aq) + 3H_2O(l)$

        (c) $Na_2O_2(s) + 2H_2O(l) \rightarrow 2Na^+(aq) + 2OH^-(aq) + H_2O_2(aq)$

        (d) $N_2O_3(g) + H_2O(l) \rightarrow 2HNO_2(aq)$

        (e) $2KO_2(s) + 2H_2O(l) \rightarrow 2K^+(aq) + 2OH^-(aq) + O_2(g) + H_2O_2(aq)$

        (f) $NO(g) + O_3(g) \rightarrow NO_2(g) + O_2(g)$

22.38   (a) $2HgO(s) \xrightarrow{\Delta} Hg(l) + O_2(g)$

        (b) $2Cu(NO_3)_2(s) \xrightarrow{\Delta} 2CuO(s) + 4NO_2(g) + O_2(g)$

(c) $PbS(s) + 4O_3(g) \rightarrow PbSO_4(s) + 4O_2(g)$

(d) $2ZnS(s) + 3O_2(g) \rightarrow 2ZnO(s) + 2SO_2(g)$

(e) $2K_2O_2(s) + 2CO_2(g) \rightarrow 2K_2CO_3(s) + O_2(g)$

(f) $2Ag(s) + O_3(g) \rightarrow Ag_2O(s) + O_2(g)$

22.39 (a)   Neutral   (b) acidic (oxide of a nonmetal)

(c)   basic (oxide of a metal)   (d) amphoteric

22.40 (a) Mn_2O_7 (higher oxidation state of Mn)

(b) SnO_2 (higher oxidation state of Sn)

(c) SO_3 (higher oxidation state of S)

(d) SO_2 (more nonmetallic character of S)

(e) Ga_2O_3 (more nonmetallic character of Ga)

(f) SO_2 (more nonmetallic character of S)

22.41 (a) $\underline{Se}O_3$, +6   (b) $H_6\underline{Te}O_6$, +6   (c) $Zn\underline{Se}O_4$, +6   (d) $\underline{S}F_4$, +4

(e) $H_2\underline{S}$, -2   (f) $H_2\underline{S}O_3$, +4

22.42 (a) $H_2\underline{Se}O_3$, +4   (b) $KH\underline{S}O_3$, +4   (c) $H_2\underline{Te}$, -2   (d) $C\underline{S}_2$, -2

(e) $Ca\underline{S}O_4$, +6   (f) $Na_2\underline{S}_2O_3$, +2

22.43 (a)   Potassium thiosulfate   (b) aluminum sulfide   (c) sodium hydrogen selenite

(d)   selenium hexafluoride

22.44 (a)   Hydrogen selenide   (b) iron pyrite (iron(II) persulfide)

(c)   sodium hydrogen sulfate (or sodium bisulfate)   (d) sodium selenate

22.45 The half reaction for oxidation in all these cases is:

$H_2S(aq) \rightarrow S(s) + 2H^+ + 2e^-$ (The product could be written as $S_8(s)$, but this is not necessary. In fact it is not necessarily the case that S_8 would be formed, rather than some other allotropic form of the element.)

(a) $2Fe^{3+}(aq) + H_2S(aq) \rightarrow 2Fe^{2+}(aq) + S(s) + 2H^+(aq)$

(b) $Br_2(l) + H_2S(aq) \rightarrow 2Br^-(aq) + S(s) + 2H^+(aq)$

(c) $2MnO_4^-(aq) + 6H^+(aq) + 5H_2S(aq) \rightarrow 2Mn^{2+}(aq) + 5S(s) + 8H_2O(l)$

(d) $2NO_3^-(aq) + H_2S(aq) + 2H^+(aq) \rightarrow 2NO_2(aq) + S(s) + 2H_2O(l)$

22.46 (a) $2[MnO_4^-(aq) + 8H^+(aq) + 5e^- \rightarrow Mn^{2+}(aq) + 4H_2O(l)]$

$5[H_2SO_3(aq) + H_2O(l) \rightarrow SO_4^{2-}(aq) + 4H^+(aq) + 2e^-]$

$2MnO_4^-(aq) + 5H_2SO_3(aq) \rightarrow 2MnSO_4(s) + 5SO_4^{2-}(aq) + 3H_2O(l) + 4H^+(aq)$

(b)　　　　　$Cr_2O_7^{2-}(aq) + 14H^+(aq) + 6e^- \rightarrow 2Cr^{3+}(aq) + 7H_2O(l)$

　　　　　　　$3[H_2SO_3(aq) + H_2O(l) \rightarrow SO_4^{2-}(aq) + 4H^+(aq) + 2e^-]$

　$\overline{Cr_2O_7^{2-}(aq) + 3H_2SO_3(aq) + 2H^+(aq) \rightarrow 2Cr^{3+}(aq) + 3SO_4^{2-}(aq) + 4H_2O(l)}$

(c)　　　　　　　　$Hg_2^{2+}(aq) + 2e^- \rightarrow 2Hg(l)$

　　　　　　$H_2SO_3(aq) + H_2O(l) \rightarrow SO_4^{2-}(aq) + 4H^+(aq) + 2e^-$

　$\overline{Hg_2^{2+}(aq) + H_2SO_3(aq) + H_2O(l) \rightarrow 2Hg(l) + SO_4^{2-}(aq) + 4H^+(aq)}$

22.47　(a)

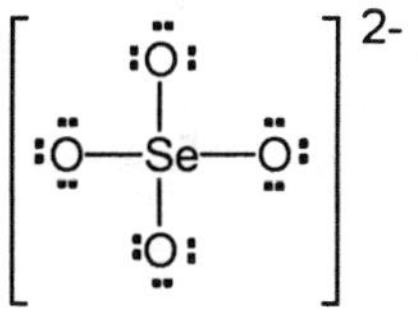

tetrahedral

(b)

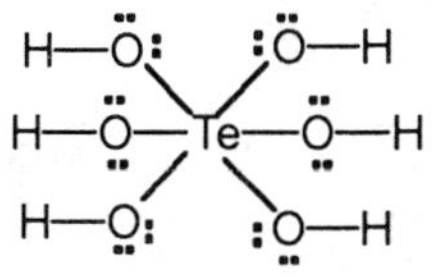

octahedral

(c)

bent

(d)

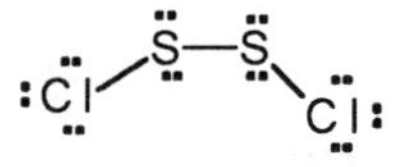

bent (free rotation around S-S bond)

(e)

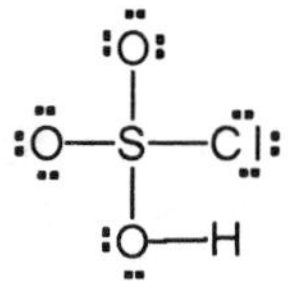

tetrahedral

22.48　SF_4, 34 e^-　　　　　　　　　　SF_5^-,　42 e^-

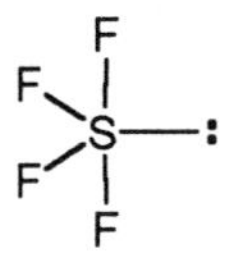

(lone pairs on F atoms omitted for clarity)

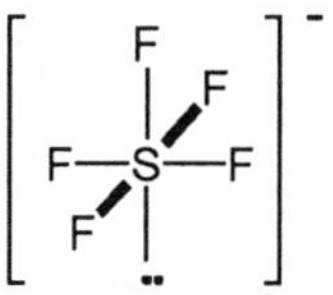

trigonal bipyramidal electron pair geometry

see saw molecular geometry

octahedral electron pair geometry

square pyramidal molecular geometry

22.49　(a)　$SeO_2(s) + H_2O(l) \rightarrow H_2SeO_3(aq) \rightleftharpoons H^+(aq) + HSeO_3^-(aq)$

　　　(b)　$ZnS(s) + 2H^+(aq) \rightarrow Zn^{2+}(aq) + H_2S(g)$

　　　(c)　$8SO_3^{2-}(aq) + S_8(s) \rightarrow 8S_2O_3^{2-}(aq)$

　　　(d)　$Se(s) + 2H_2SO_4(l) \xrightarrow{\Delta} SeO_2(g) + 2SO_2(g) + 2H_2O(g)$

　　　(e)　$SO_3(aq) + H_2SO_4(l) \rightarrow H_2S_2O_7(l)$

　　　(f)　$H_2SeO_3(aq) + H_2O_2(aq) \rightarrow 2H^+(aq) + SeO_4^{2-}(aq) + H_2O(l)$

22.50   (a) $H_2SeO_3(aq) + N_2H_4(aq) \rightarrow Se(s) + N_2(g) + 3H_2O(l)$

       (b) $H_6TeO_6(s) \xrightarrow{\Delta} TeO_3(s) + 3H_2O(l)$

       (c) $Al_2Se_3(s) + 6H^+(aq) \rightarrow 2Al^{3+}(aq) + 3H_2Se(g)$

       (d) $Cl_2(aq) + S_2O_3^{2-}(aq) + H_2O(l) \rightarrow 2Cl^-(aq) + S(s) + SO_4^{2-}(aq) + 2H^+(aq)$

Nitrogen and the Group 5A Elements

22.51   (a) $H\underline{N}O_2$, +3   (b) $\underline{N}_2H_4$, -2   (c) $KC\underline{N}$, -3   (d) $Na\underline{N}O_3$, +5

       (e) $\underline{N}H_4Cl$, -3   (f) $Li_3\underline{N}$, -3

22.52   (a) $Na\underline{N}O_2$, +3   (b) $\underline{N}H_3$, -3   (c) $\underline{N}_2O$, +1   (d) $NaC\underline{N}$, -3

       (e) $H\underline{N}O_3$, +5   (f) $\underline{N}O_2$, +4

22.53   (a)   :Ö=N̈—Ö—H

          The molecule is bent around the central oxygen and nitrogen atoms; the four atoms need not lie in a plane.

       (b)   [:N̈=N=N̈:]⁻ ⟷ [:N≡N—N̈:]⁻ ⟷ [:N̈—N≡N:]⁻

          The molecule is linear.

       (c)   [H—N—N: with H,H on left N top/bottom and H,H on right N top/bottom]⁺

       (d)   [:Ö—N=Ö with :Ö: on top]⁻

          The geometry is tetrahedral around the left nitrogen, trigonal pyramidal around the right.

          (three equivalent resonance forms) The ion is trigonal planar.

22.54   (a)

$$\left[\begin{array}{c} H \\ | \\ H-N-H \\ | \\ H \end{array} \right]^{+}$$

tetrahedral

(b) $\ddot{:}\underset{\cdot\cdot}{O}-\overset{\overset{\displaystyle :\ddot{O}:}{\|}}{N}-\overset{\cdot\cdot}{\underset{\cdot\cdot}{O}}-H \longleftrightarrow \overset{\displaystyle :\ddot{O}:}{O}=N-\overset{\cdot\cdot}{\underset{\cdot\cdot}{O}}-H \longleftrightarrow :\overset{\cdot\cdot}{\underset{\cdot\cdot}{O}}-N\overset{\displaystyle :\ddot{O}:}{=}\overset{\cdot\cdot}{O}-H$

The geometry around nitrogen is trigonal planar, but the hydrogen atom is not required to lie in this plane. The third resonance form makes a much smaller contribution to the structure than the first two.

(c) $:\ddot{N}=N=\ddot{O}: \longleftrightarrow :N\equiv N-\ddot{\underset{\cdot\cdot}{O}}: \longleftrightarrow :\ddot{N}-N\equiv O:$

The molecule is linear. Again, the third resonance form makes less contribution to the structure because of the high formal charges involved.

(d) $\ddot{O}=N-\ddot{\underset{\cdot\cdot}{O}}: \longleftrightarrow :\ddot{\underset{\cdot\cdot}{O}}-N=\ddot{O}:$

The molecule is bent (nonlinear).

22.55   (a) $Mg_3N_2(s) + 6H_2O(l) \rightarrow 3Mg(OH)_2(s) + 2NH_3(aq)$

(b) $2NO(g) + O_2(g) \rightarrow 2NO_2(g)$

(c) $4NH_3(g) + 3O_2(g) \xrightarrow{\Delta} 2N_2(g) + 6H_2O(g)$

(d) $NaNH_2(s) + H_2O(l) \rightarrow Na^+(aq) + OH^-(aq) + NH_3(aq)$

22.56   (a) $N_2O_5(g) + H_2O(l) \rightarrow 2H^+(aq) + 2NO_3^-(aq)$

(b) $Li_3N(s) + 3H_2O(l) \rightarrow NH_3(aq) + 3LiOH(aq)$

(c) $NH_3(aq) + H^+(aq) \rightarrow NH_4^+(aq)$

(d) $N_2H_4(l) + O_2(g) \rightarrow N_2(g) + 2H_2O(g)$

22.57   (a) $4Zn(s) + 2NO_3^-(aq) + 10H^+(aq) \rightarrow 4Zn^{2+}(aq) + N_2O(g) + 5H_2O(l)$

(b) $4NO_3^-(aq) + S(s) + 4H^+(aq) \rightarrow 4NO_2(g) + SO_2(g) + 2H_2O(l)$

(or $6NO_3^-(aq) + S(s) + 4H^+(aq) \rightarrow 6NO_2(g) + SO_4^{2-}(aq) + 2H_2O(l)$

(c) $2NO_3^-(aq) + 3SO_2(g) + 2H_2O(l) \rightarrow 2NO(g) + 3SO_4^{2-}(aq) + 4H^+(aq)$

22.58   (a) $N_2H_4(g) + 5F_2(g) \rightarrow 2NF_3(g) + 4HF(g)$

(b) $4CrO_4^{2-}(aq) + 3N_2H_4(aq) + 4H_2O(l) \rightarrow 4Cr(OH)_4^-(aq) + 4OH^-(aq) + 3N_2(g)$

473

(c) $Cu^{2+}(aq) + 2e^- \rightarrow Cu(s)$

$2NH_2OH(aq) \rightarrow N_2(g) + 2H_2O(l) + 2H^+(aq) + 2e^-$

$Cu^{2+}(aq) + 2NH_2OH(aq) \rightarrow Cu(s) + N_2(g) + 2H_2O(l) + 2H^+(aq)$

22.59 (a) $2NO_3^-(aq) + 12H^+(aq) + 10e^- \rightarrow N_2(g) + 6H_2O(l)$ $E^\circ_{red} = +1.25$ V

(b) $2NH_4^+(aq) \rightarrow N_2(g) + 8H^+(aq) + 6e^-$ $E^\circ_{red} = 0.27$ V

22.60 (a) $NO_3^-(aq) + 4H^+(aq) + 3e^- \rightarrow NO(g) + 2H_2O(l)$ $E^\circ_{red} = +0.96$ V

(b) $HNO_2(aq) \rightarrow NO_2(g) + H^+(aq) + 1e^-$ $E^\circ_{red} = 1.12$ V

22.61 (a) sodium phosphide   (b) arsenic acid   (c) tetraphosphorus decaoxide

(d) arsenic pentafluoride

22.62 (a) potassium arsenide   (b) phosphorous tribromide

(c) diantimony trioxide or antimony(III) oxide   (d) sodium dihydrogen arsenate

22.63 (a) $H_3\underline{P}O_4$, +5 (b) $H_3\underline{As}O_3$, +3 (c) $\underline{Sb}_2S_3$, +3 (d) $Ca(H_2\underline{P}O_4)_2$, +5 (e) $K_3\underline{P}$, -3

22.64 (a) $H_3\underline{P}O_3$, +3 (b) $H_4\underline{P}_2O_7$, +5 (c) $\underline{Sb}Cl_3$, +3 (d) $Mg_3(\underline{As}O_4)_2$, +5 (e) $\underline{P}_2O_5$, +5

22.65 (a) Phosphorus is a larger atom and can more easily accommodate five surrounding atoms and an expanded octet of electrons than nitrogen can. Also, P has energetically "available" 3d orbitals which participate in the bonding, but nitrogen does not.

(b) Only one of the three hydrogens in H_3PO_2 is bonded to oxygen. The other two are bonded directly to phosphorus and are not easily ionized because the P-H bond is not very polar.

(c) PH_3 is a weaker base than H_2O (PH_4^+ is a stronger acid than H_3O^+). Any attempt to add H^+ to PH_3 in the presence of H_2O merely causes protonation of H_2O.

(d) Antimony is more metallic in character than phosphorus. Because it is larger than P it has less attraction for additional electrons (it is a weaker Lewis acid). Thus, the reaction stops before complete hydrolysis has occurred.

(e) White phosphorus consists of P_4 molecules, with P-P-P bond angles of 60°. Each P atom has four VSEPR pairs of electrons, so the predicted electron pair geometry is tetrahedral and the preferred bond angle is 109°. Because of the severely strained bond angles in P_4 molecules, white phosphorus is highly reactive.

22.66 (a) Only two of the hydrogens in H_3PO_3 are bound to oxygen. The third is attached directly to phosphorus, and not readily ionized, because the H-P bond is not very polar.

(b)   The smaller, more electronegative nitrogen withdraws more electron density from the O-H bond, making it more polar and more likely to ionize.

(c)   Phosphate rock consists of $Ca_3(PO_4)_2$, which is only slightly soluble in water. The phosphorus is unavailable for plant use.

(d) N_2 can form stable π bonds to complete the octet of both N atoms. Because phosphorus atoms are larger than nitrogen atoms, they do not form stable π bonds with themselves and must form σ bonds with several other phosphorus atoms (producing P_4 tetrahedral or sheet structures) to complete their octets.

(e)   In solution Na_3PO_4 is completely dissociated into Na^+ and PO_4^{3-}. PO_4^{3-}, the conjugate base of the very weak acid HPO_4^{2-}, has a K_b of 2.4×10^{-2} and produces a considerable amount of OH^- by hydrolysis of H_2O.

22.67 (a) $2Ca_3(PO_4)_2(s) + 6SiO_2(s) + 10C(s) \rightarrow P_4(g) + 6CaSiO_3(l) + 10CO_2(g)$

(b) $3H_2O(l) + PCl_3(l) \rightarrow H_3PO_3(aq) + 3H^+(aq) + 3Cl^-(aq)$

(c) $6Cl_2(g) + P_4(s) \rightarrow 4PCl_3(l)$

(d) $P_4O_{10}(s) + 6H_2O(l) \rightarrow 4H_3PO_4(aq)$

22.68 (a) $PCl_5(l) + AsF_3(l) \rightarrow PF_3(g) + AsCl_5(s)$

(b) $As_2O_3(s) + 3H_2O(l) \rightarrow 2H_3AsO_3(aq)$

(c) $2H_3PO_4(l) \xrightarrow{\Delta} H_4P_2O_7(l) + H_2O(g)$

(d) $As(s) + HNO_3(aq) + H_2O(l) \rightarrow H_3AsO_3(aq) + NO(g)$

Carbon, the Other Group 4A Elements, and Boron

22.69 (a)   HCN   (b) SiC   (c) $CaCO_3$ (d) CaC_2

22.70 (a) H_2CO_3 (b) NaCN   (c) $KHCO_3$ (d) C_2H_2

22.71 (a) $[:C\equiv N:]^-$ (b) $:C\equiv O:$ (c) $[:C\equiv C:]^{2-}$

(d) $\ddot{\underset{..}{S}}=C=\ddot{\underset{..}{S}}$ (e) $\ddot{\underset{..}{O}}=C=\ddot{\underset{..}{O}}$ (f)

one of three equivalent
resonance structures

22.72 (a) The CH_3 carbon is tetrahedral; it employs an sp^3 hybrid orbital set. The other two carbons have a linear geometry about them; they employ an sp hybrid orbital set.

(b) The carbon in CN^- uses an sp hybrid orbital set. The Lewis structure of CN^- is $[:C \equiv N:]^-$. The lone pair and the C-N σ bond electrons occupy the sp hybrid orbitals. The other two p orbitals are employed in π bonding to N.

(c) The carbon in CS_2 has an sp hybrid orbital set, consistent with the linear geometry.

(d) In C_2H_6 each carbon is in a tetrahedral environment of one C-C and three C-H bonds. It employs sp^3 hybrid orbitals.

22.73 (a) $ZnCO_3(s) \xrightarrow{\Delta} ZnO(s) + CO_2(g)$

(b) $BaC_2(s) + 2H_2O(l) \rightarrow Ba^{2+}(aq) + 2OH^-(aq) + C_2H_2(g)$

(c) $C_2H_4(g) + 3O_2(g) \rightarrow 2CO_2(g) + 2H_2O(g)$

(d) $2CH_3OH(l) + 3O_2(g) \rightarrow 2CO_2(g) + 4H_2O(g)$

(e) $NaCN(s) + H^+(aq) \rightarrow Na^+(aq) + HCN(g)$

22.74 (a) $CO_2(g) + OH^-(aq) \rightarrow HCO_3^-(aq)$

(b) $NaHCO_3(s) + H^+(aq) \rightarrow Na^+(aq) + H_2O(l) + CO_2(g)$

(c) $2CaO(s) + 5C(s) \xrightarrow{\Delta} 2CaC_2(s) + CO_2(g)$

(d) $C(s) + H_2O(g) \xrightarrow{\Delta} H_2(g) + CO(g)$

(e) $CuO(s) + CO(g) \rightarrow Cu(s) + CO_2(g)$

22.75 (a) $2CH_4(g) + 2NH_3(g) + 3O_2(g) \xrightarrow[\text{cat}]{800°C} 2HCN(g) + 6H_2O(g)$

(b) $NaHCO_3(s) + H^+(aq) \rightarrow CO_2(g) + H_2O(l) + Na^+(aq)$

(c) $2BaCO_3(s) + O_2(g) + 2SO_2(g) \rightarrow 2BaSO_4(s) + 2CO_2(g)$

22.76 (a) $2Mg(s) + CO_2(g) \rightarrow 2MgO(s) + C(s)$

(b) $6CO_2(g) + 6H_2O(l) \xrightarrow{h\nu} C_6H_{12}O_6(aq) + 6O_2(g)$

(c) $CO_3^{2-}(aq) + H_2O(l) \rightarrow HCO_3^-(aq) + OH^-(aq)$

22.77 (a) $\underline{Si}O_2$, +4 (b) $\underline{Ge}Cl_4$, +4 (c) $Na\underline{B}H_4$, +3 (d) $\underline{Sn}Cl_2$, +2 (e) $\underline{B}_2H_6$, -3

22.78 (a) $H_3\underline{B}O_3$, +3 (b) $\underline{Si}Br_4$, +4 (c) $\underline{Pb}Cl_2$, +2 or $\underline{Pb}Cl_2$, + 4
(d) $Na_2\underline{B}_4O_7 \cdot 10H_2O$, +3 (e) $\underline{B}_2O_3$, +3

22.79   (a)   Carbon   (b) lead   (c) silicon

22.80   (a)   Carbon   (b) lead   (c) germanium

22.81 $GeCl_4(l) + 2H_2O(g) \rightarrow GeO_2(s) + 4HCl(g)$

        $SiCl_4(g) + 2H_2O(g) \rightarrow SiO_2(s) + 4HCl(g)$

22.82   (a) $GeCl_4(l) + Ge(s) \rightarrow 2GeCl_2(l)$

       (b)   In $GeCl_2$ there are six electrons about Ge. According to the VSEPR model the three electron pairs should be in a trigonal plane. One of the electron pairs is unshared, the other two are in the Ge-Cl bonds. The molecule should therefore appear bent, with a Cl-Ge-Cl angle of somewhat less than 120°.

22.83   (a) SiO_4^{4-} (b) SiO_3^{2-} (c) SiO_3^{2-}

22.84   (a)

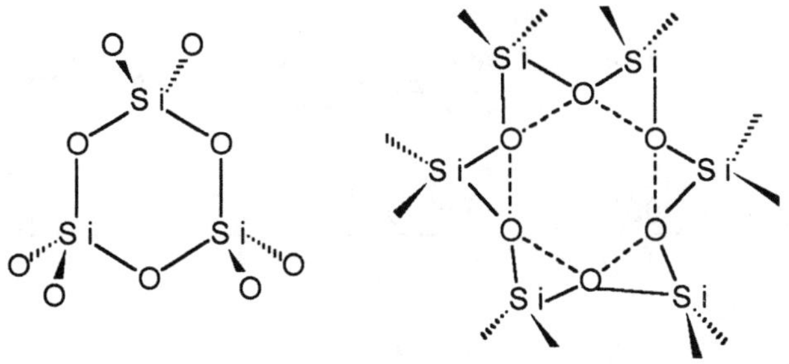

       (b) $Si_3O_9^{6-}$ $Si_6O_{18}^{12-}$

22.85   (a)   Diborane (Figure 22.55 and below) has bridging H atoms linking the two B atoms. The structure of ethane shown below has the C atoms bound directly, with no bridging atoms.

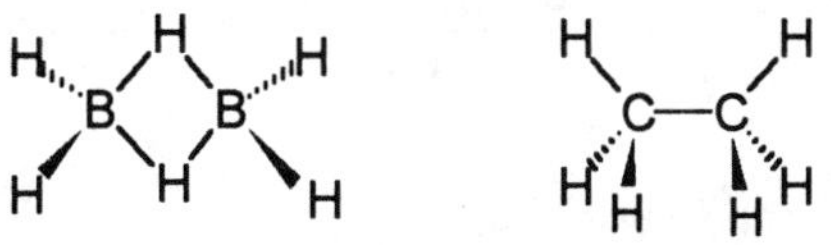

       (b) B_2H_6 is an electron deficient molecule. It has 12 valence electrons, while C_2H_6 has 14 valence electrons. The 6 valence electron pairs in B_2H_6 are all involved in B-H sigma bonding, so the only way to satisfy the octet rule at B is to have the bridging H atoms shown in Figure 22.55.

       (c)   A hydride ion, H^-, has two electrons while an H atom has one. The term *hydridic* indicates that the H atoms in B_2H_6 have more than the usual amount of electron density for a covalently bound H atom.

22.86 (a) $B_2H_6(g) + 6H_2O(l) \rightarrow 2H_3BO_3(aq) + 6H_2(g)$

       (b) $4H_3BO_3(s) \xrightarrow{\Delta} H_2B_4O_7(s) + 5H_2O(g)$

       (c) $B_2O_3(s) + 3H_2O(l) \rightarrow 2H_3BO_3(aq)$

Additional Exercises

22.87 First find how many moles of gas occupy 1 ft^3 at STP.

$$1\text{ ft}^3 \times \left(\frac{12\text{ in}}{1\text{ ft}}\right)^3 \times \left(\frac{2.54\text{ cm}}{1\text{ in}}\right)^3 \times \frac{1\text{ L}}{10^3\text{ cm}^3} = 28.317 = 28.3\text{ L}$$

$$\text{at STP } n = \frac{(1\text{ atm})(28.317\text{ L})}{(0.08206\text{ L}\cdot\text{atm/mol}\cdot\text{K})(273\text{ K})} = 1.264 = 1.26\text{ mol/ft}^3$$

       (a) $2.2 \times 10^8\text{ kg H}_2 \times \dfrac{1000\text{ g}}{1\text{ kg}} \times \dfrac{1\text{ mol H}_2}{2.016\text{ g H}_2} \times \dfrac{1\text{ ft}^3}{1.264\text{ mol}} = 8.6 \times 10^{10}\text{ ft}^3\text{ H}_2$

       (b) $2.4 \times 10^{10}\text{ kg N}_2 \times \dfrac{1000\text{ g}}{1\text{ kg}} \times \dfrac{1\text{ mol N}_2}{28.01\text{ g N}_2} \times \dfrac{1\text{ ft}^3}{1.264\text{ mol}} = 6.8 \times 10^{11}\text{ ft}^3\text{ N}_2$

       (c) $1.7 \times 10^{10}\text{ kg O}_2 \times \dfrac{1000\text{ g}}{1\text{ kg}} \times \dfrac{1\text{ mol O}_2}{32.00\text{ g O}_2} \times \dfrac{1\text{ ft}^3}{1.264\text{ mol}} = 4.2 \times 10^{11}\text{ ft}^3\text{ O}_2$

22.88 (a) $1.00\text{ kg FeTi} \times \dfrac{1\text{ mol FeTi}}{103.7\text{ g FeTi}} \times \dfrac{1000\text{ g}}{1\text{ kg}} \times \dfrac{1\text{ mol H}_2}{1\text{ mol FeTi}} \times \dfrac{2.016\text{ g H}}{1\text{ mol H}_2} = 19.44$

                                                                                        $= 19.4\text{ g H}$

       (b) $V = \dfrac{19.44\text{ g H}_2}{2.016\text{ g/mol H}_2} \times \dfrac{0.08206\text{ L}\cdot\text{atm}}{\text{mol}\cdot\text{K}} \times \dfrac{273\text{ K}}{1\text{ atm}} = 216\text{ L}$

22.89 (a)   React an ionic nitride with D_2O, e.g.,

                $Mg_3N_2(s) + 6D_2O(l) \rightarrow 2ND_3(aq) + 3Mg(OD)_2(s)$

       (b)   React SO_3 with D_2O: $SO_3(g) + D_2O(l) \rightleftharpoons D_2SO_4(aq)$

       (c)   React Na_2O with D_2O: $Na_2O(s) + D_2O(l) \rightarrow 2NaOD(aq)$

       (d)   Dissolve $N_2O_5(g)$ in D_2O: $N_2O_5(g) + D_2O(l) \rightarrow 2DNO_3(aq)$

       (e)   React CaC_2 with D_2O: $CaC_2(s) + 2D_2O(l) \rightarrow Ca^{2+}(aq) + 2OD^-(aq) + C_2D_2(g)$

       (f)   Add NaCN to the D_2SO_4 solution prepared in (b):

                $NaCN(s) + D^+(aq) \xrightarrow{\Delta} DCN(aq) + Na^+(aq)$

                The DCN can be removed as gas from the reaction.

22.90 $2XeO_3(s) \rightarrow 2Xe(g) + 3O_2(g)$

$0.654 \text{ g } XeO_3 \times \dfrac{1 \text{ mol } XeO_3}{179.1 \text{ g } XeO_3} \times \dfrac{5 \text{ mol gas}}{2 \text{ mol } XeO_3} = 9.129 \times 10^{-3} = 9.13 \times 10^{-3} \text{ mol gas}$

$P = \dfrac{(9.129 \times 10^{-3} \text{ mol})(0.08206 \text{ L} \cdot \text{atm/mol} \cdot \text{K})(321 \text{ K})}{0.452 \text{ L}} = 0.532 \text{ atm}$

22.91 (a) $2Na(l) + 2HCl(g) \rightarrow 2NaCl(s) + H_2(g)$

(b) $H_2SO_3(aq) + Br_2(l) + H_2O(l) \rightarrow HSO_4^-(aq) + 2Br^-(aq) + 3H^+(aq)$
Note that two strong acids are formed, $H_2SO_4(aq)$ and $HBr(aq)$.
H_2SO_4 is not volatile, and remains behind when the HBr is distilled.

(c)   The half -reactions are:

$12OH^-(aq) + Br_2(l) \rightarrow 2BrO_3^-(aq) + 6H_2O(l) + 10e^-$

$5[ClO^-(aq) + H_2O(l) + 2e^- \rightarrow Cl^-(aq) + 2OH^-(aq)]$

$2OH^-(aq) + Br_2(l) + 5ClO^-(aq) \rightarrow 5Cl^-(aq) + 2BrO_3^-(aq) + H_2O(l)$

Now take account of the formation of $KBrO_3(s)$:

$2K^+(aq) + 2BrO_3^-(aq) \rightarrow 2KBrO_3(s)$

$2K^+(aq) + 2OH^-(aq) + Br_2(l) + 5ClO_3^-(aq) \rightarrow 5Cl^-(aq) + 2KBrO_3(s) + H_2O(l)$

(d)   The half-reactions are:

$2BrO_3^-(aq) + 12H^+(aq) + 10e^- \rightarrow Br_2(l) + 6H_2O(l)$

$5[H_2SO_3(aq) + H_2O(l) \rightarrow HSO_4^-(aq) + 3H^+(aq) + 2e^-]$

$2BrO_3^-(aq) + 5H_2SO_3(aq) \rightarrow Br_2(l) + 5HSO_4^-(aq) + 3H^+(aq) + H_2O(l)$

(e) $3UCl_4(s) + 4ClF_3(g) \rightarrow 3UF_4(g) + 8Cl_2(g)$

22.92 $BrO_3^-(aq) + XeF_2(aq) + H_2O(l) \rightarrow Xe(g) + 2HF(aq) + BrO_4^-(aq)$

22.93 $I_2 < F_2 < Br_2 < Cl_2$ The lower-than-expected value for F_2 can be ascribed to repulsions between the unshared electron pairs on the fluorine atoms at the short F-F distance necessary to give good overlap for bonding.

22.94 Substances that will burn in O_2: SiH_4, CO, Mg.

The others, SiO_2, CO_2 and CaO, have Si, C and Ca in maximum oxidation states, so O_2 cannot act as an oxidizing agent.

22.95 (a) $SO_2(g) + H_2O(l) \rightleftharpoons H_2SO_3(aq)$

(b) $Cl_2O(g) + H_2O(l) \rightleftharpoons 2HClO(aq)$

(c) $Na_2O(s) + H_2O(l) \rightarrow 2Na^{2+}(aq) + 2OH^-(aq)$

(d) $BaC_2(s) + 2H_2O(l) \rightarrow Ba^{2+}(aq) + 2OH^-(aq) + C_2H_2(g)$

(e) $2RbO_2(s) + 2H_2O(l) \rightarrow 2Rb^+(aq) + 2OH^-(aq) + O_2(g) + H_2O_2(aq)$

(f) $Mg_3N_2(s) + 6H_2O(l) \rightarrow 3Mg(OH)_2(s) + 2NH_3(g)$

(g) $Na_2O_2(s) + 2H_2O \rightarrow H_2O_2(aq) + 2NaOH(aq)$

(h) $NaH(s) + H_2O \rightarrow NaOH(aq) + H_2(g)$

22.96 (a) $H_2SO_4 - H_2O \rightarrow SO_3$ (b) $2HClO_3 - H_2O \rightarrow Cl_2O_5$

(c) $2HNO_2 - H_2O \rightarrow N_2O_3$ (d) $H_2CO_3 - H_2O \rightarrow CO_2$

(e) $2H_3PO_4 - 3H_2O \rightarrow P_2O_5$

22.97 $8Fe(s) + S_8(s) \rightarrow 8FeS(s)$

$S_8(s) + 16F_2(g) \rightarrow 8SF_4(g)$ or $S_8(s) + 24F_2(g) \rightarrow 8SF_6(g)$

$S_8(s) + 8O_2(g) \rightarrow 8SO_2(g)$

$S_8(s) + 8H_2(g) \rightarrow 8H_2S(g)$

Sulfur acts as an oxidizing agent in reactions with Fe or H_2 and as a reducing agent in reactions with O_2 or F_2. Incidentally, these reactions are often written using the symbol S rather than S_8 for sulfur.

22.98 The valence shell electron arrangement controls the upper and lower limits of oxidation state. For group 6 the configuration is ns^2np^4. Thus addition of two electrons makes a closed shell. "Loss" of all six also leaves a closed shell. The limits are -2 to +6.

22.99 $S(g) + O_2(g) \rightarrow SO_2(g)$ $\Delta H = -296.9$ kJ   (1)

$SO_2(g) + 1/2\ O_2(g) \rightarrow SO_3(g)$ $\Delta H = -98.3$ kJ   (2)

$SO_3(g) + H_2O(l) \rightarrow H_2SO_4(aq)$ $\Delta H = -130$ kJ   (3)

_____

$S(g) + 3/2\ O_2(g) + H_2O(l) \rightarrow H_2SO_4(aq)$ $\Delta H = -525$ kJ

$1\ \text{ton}\ H_2SO_4 \times \dfrac{2000\ \text{lb}}{\text{ton}} \times \dfrac{453.6\ \text{g}}{1\ \text{lb}} \times \dfrac{1\ \text{mol}\ H_2SO_4}{98.09\ \text{g}} \times \dfrac{-525\ \text{kJ}}{\text{mol}\ H_2SO_4}$

$= -4.86 \times 10^6$ kJ of heat/ton H_2SO_4

22.100 (a) $2H_2Se(g) + O_2(g) \rightarrow 2H_2O(g) + 2Se(s)$

(b) $\Delta G^\circ = 2\Delta G^\circ_f\ H_2O(g) - 2\Delta G^\circ_f\ H_2Se(g)$

$= 2(-228.57\ \text{kJ}) - 2(15.9\ \text{kJ}) = -488.9$

$\Delta G^\circ = -RT\ \ln K$

$\ln K = \dfrac{-(-488.9 \times 10^3\ \text{J})}{8.314\ \text{J/K} \cdot \text{mol} \times 298\ \text{K}} = 197.33 = 197.3;\ \ K = 5 \times 10^{85}$

22.101 (a) $PO_4^{3-}, +5;\ \ NO_3^-, +5$

(b)   The Lewis structure for NO_4^{3-} would be:

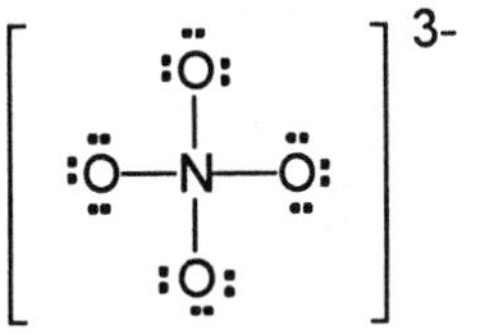

The formal charge on N is +1 and on each O atom is -1. The four electronegative oxygen atoms withdraw electron density, leaving the nitrogen deficient. Since N can form a maximum of four bonds, it cannot form a π bond with one or more of the O atoms to regain electron density, as the P atom in PO_4^{3-} does. Also, the short N-O distance would lead to a tight tetrahedron of O atoms subject to steric repulsion.

22.102 $(CH_3)_2N_2H_2(g) + 2N_2O_4(g) \rightarrow 2CO_2(g) + 3N_2(g) + 4H_2O(g)$

$$4.0 \text{ tons } (CH_3)_2N_2H_2 \times \frac{2000 \text{ lb}}{1 \text{ ton}} \times \frac{453.6 \text{ g}}{1 \text{ lb}} \times \frac{1 \text{ mol } (CH_3)_2N_2H_2}{60.10 \text{ g } (CH_3)_2N_2H_2}$$

$$\times \frac{2 \text{ mol } N_2O_4}{1 \text{ mol } (CH_3)_2N_2H_2} \times \frac{92.02 \text{ g } N_2O_4}{1 \text{ mol } N_2O_4} \times \frac{1 \text{ lb}}{453.6 \text{ g}} \times \frac{1 \text{ ton}}{2000 \text{ lb}} = 12 \text{ tons } N_2O_4$$

22.103 P_4, P_4O_6 and P_4O_{10} all contain a tetrahedron of phosphorus atoms with P-P-P angles of approximately 60°. In the acids containing phosphorus in the +5 oxidation state, H_3PO_4, $H_4P_2O_7$, and $(HPO_3)_n$, P is bound to four O atoms with tetrahedral geometry and approximate 109° bond angles around phosphorus.

22.104 (a)   Use the formula that relates the length of a side of a triangle to the lengths of the other two sides and the opposite angle: $a^2 = b^2 + c^2 - 2bc \cos A$. In this case, a = b = the P-O distance, and A is the P-O-P angle:

For P_4O_6, $a^2 = 2b^2(1 - \cos A) = 2(1.65)^2(1 - \cos 127.5°)$ a = 2.96 Å

For P_4O_{10}, $a^2 = 2b^2(1 - \cos A) = 2(1.60)^2(1 - \cos 124.5°)$ a = 2.83 Å

(b)   The shorter P-P distance in P_4O_{10} is due both to the sharper angle and to the shorter P-O distance. The latter may be due to the fact that phosphorus has a higher effective charge in P_4O_{10}, so the P atom is slightly contracted as compared with P_4O_6 because of its higher oxidation state.

22.105 The equation for this reaction is:

$$CaC_2(s) + 2H_2O(l) \rightarrow Ca^{2+}(aq) + 2OH^-(aq) + C_2H_2(g)$$

$$10.0 \text{ g CaC}_2 \times \frac{1 \text{ mol CaC}_2}{64.10 \text{ g CaC}_2} \times \frac{1 \text{ mol C}_2H_2}{1 \text{ mol CaC}_2} = 0.1562 = 0.156 \text{ mol C}_2H_2(g)$$

$$V = 0.1562 \text{ mol C}_2H_2 \times \frac{0.08206 \text{ L} \cdot \text{atm}}{\text{mol} \cdot \text{K}} \times \frac{300 \text{ K}}{(720/760) \text{ atm}} = 4.06 \text{ L}$$

22.106 $GeO_2(s) + C(s) \xrightarrow{\Delta} Ge(l) + CO_2(g)$

$Ge(l) + 2Cl_2(g) \rightarrow GeCl_4(l)$

$GeCl_4(l) + 2H_2O(l) \rightarrow GeO_2(s) + 4HCl(g)$

$GeO_2(s) + 2H_2(g) \rightarrow Ge(s) + 2H_2O(l)$

22.107 (a) $Li_3N(s) + 3H_2O(l) \rightarrow 3Li^+(aq) + 3OH^-(aq) + NH_3(aq)$

(b) $NH_3(aq) + H_2O(l) \rightleftharpoons NH_4^+(aq) + OH^-(aq)$

(c) $3NO_2(g) + H_2O(l) \rightarrow NO(g) + 2H^+(aq) + 2NO_3^-(aq)$

(d) $2NO_2(g) \rightleftharpoons N_2O_4(g)$

(e) $4NH_3(g) + 5O_2(g) \rightarrow 4NO(g) + 6H_2O(g)$

(f) $2CO(g) + O_2(g) \rightarrow 2CO_2(g)$

(g) $H_2CO_3(aq) \xrightarrow{\Delta} H_2O(g) + CO_2(g)$

(h) $Ni(s) + CO(g) \rightarrow NiO(s) + C(s)$

(i) $CS_2(g) + O_2(g) \rightarrow CO_2(g) + S_2(g)$

(j) $CaO(s) + SO_2(g) \rightarrow CaSO_3(s)$

(k) $2Na(s) + 2H_2O(l) \rightarrow 2NaOH(aq) + H_2(g)$

(l) $CH_4(g) + H_2O(g) \xrightarrow{\Delta} CO(g) + 3H_2(g)$

(m) $LiH(s) + H_2O(l) \rightarrow LiOH(aq) + H_2(g)$

(n) $Fe_2O_3(s) + 3H_2(g) \rightarrow 2Fe(s) + 3H_2O(g)$

22.108 (a) $2[5e^- + MnO_4^-(aq) + 8H^+(aq) \rightarrow Mn^{2+}(aq) + 4H_2O(l)]$

$5[H_2O_2(aq) \rightarrow O_2(g) + 2H^+(aq) + 2e^-]$

$$2MnO_4^-(aq) + 5H_2O_2(aq) + 6H^+(aq) \rightarrow 2Mn^{2+}(aq) + 5O_2(g) + 8H_2O(l)$$

(b) $2[Fe^{2+}(aq) \rightarrow Fe^{3+}(aq) + e^-]$

$H_2O_2(aq) + 2H^+(aq) + 2e^- \rightarrow 2H_2O(l)$

$$2Fe^{2+}(aq) + H_2O_2(aq) + 2H^+(aq) \rightarrow 2Fe^{3+}(aq) + 2H_2O(l)$$

(c) $2I^-(aq) \rightarrow I_2(s) + 2e^-$

$H_2O_2(aq) + 2H^+(aq) + 2e^- \rightarrow 2H_2O(l)$

$$2I^-(aq) + H_2O_2(aq) + 2H^+(aq) \rightarrow I_2(s) + 2H_2O(l)$$

(d) $MnO_2(s) + 4H^+(aq) + 2e^- \rightarrow Mn^{2+}(aq) + 2H_2O(l)$

   $H_2O_2(aq) \rightarrow O_2(g) + 2H^+(aq) + 2e^-$

   ―――――――――――――――――――――――――――――――――――

   $MnO_2(s) + 2H^+(aq) + H_2O_2(aq) \rightarrow Mn^{2+}(aq) + 2H_2O(l) + O_2(g)$

(e) $2I^-(aq) \rightarrow I_2(s) + 2e^-$

   $O_3(g) + H_2O(l) + 2e^- \rightarrow O_2(g) + 2OH^-(aq)$

   ―――――――――――――――――――――――――――――――――――

   $2I^-(aq) + O_3(g) + H_2O(l) \rightarrow O_2(g) + I_2(s) + 2OH^-(aq)$

22.109 Assume that the reactions occur in basic solution. The half-reaction for reduction of H_2O_2 is in all cases $H_2O_2(aq) + 2e^- \rightarrow 2OH^-(aq)$.

(a) $H_2O_2(aq) + S^{2-}(aq) \rightarrow 2OH^-(aq) + S(s)$

(b) $SO_2(g) + 2OH^-(aq) + H_2O_2(aq) \rightarrow SO_4^{2-}(aq) + 2H_2O(l)$

(c) $NO_2^-(aq) + H_2O_2(aq) \rightarrow NO_3^-(aq) + H_2O(l)$

(d) $As_2O_3(s) + 2H_2O_2(aq) + 6OH^-(aq) \rightarrow 2AsO_4^{3-}(aq) + 5H_2O(l)$

(e)   This reaction must be occurring in acidic solution, since $Fe(OH)_3$ would form if the solution were basic. The half-reactions are:

   $2H^+(aq) + H_2O_2(aq) + 2e^- \rightarrow 2H_2O(l)$

   $2[Fe^{2+}(aq) \rightarrow Fe^{3+}(aq) + e^-]$

   ―――――――――――――――――――――――――――――――――――

   $2Fe^{2+}(aq) + H_2O_2(aq) + 2H^+(aq) \rightarrow 2Fe^{3+}(aq) + 2H_2O(l)$

22.110 From Appendix C, we need only ΔH_f° for F(g), so that we can estimate ΔH for the process:

   $F_2(g) \rightarrow F(g) + F(g);$ $\Delta H^\circ = +160$ kJ.

   $XeF_2(g) \rightarrow Xe(g) + F_2(g)$ $-\Delta H_f^\circ = +109$ kJ

   $F_2(g) \rightarrow 2F(g)$ $\Delta H^\circ = +160$ kJ

   ―――――――――――――――――――――――――――――――――――

   $XeF_2(g) \rightarrow Xe(g) + 2F(g)$ $\Delta H^\circ = +269$ kJ

The average Xe-F bond enthalpy is thus 269/2 = 134 kJ. Similarly,

   $XeF_4(g) \rightarrow Xe(g) + 2F_2(g)$ $-\Delta H_f^\circ = +218$ kJ

   $2F_2(g) \rightarrow 4F(g)$ $\Delta H^\circ = 320$ kJ

   ―――――――――――――――――――――――――――――――――――

   $XeF_4(g) \rightarrow Xe(g) + 4F(g)$ $\Delta H^\circ = 538$ kJ

Average Xe-F bond energy = 538/4 = 134 kJ

   $XeF_6(g) \rightarrow Xe(g) + 3F_2(g)$ $-\Delta H_f^\circ = 298$ kJ

   $3F_2(g) \rightarrow 6F(g)$ $\Delta H^\circ = 480$ kJ

   ―――――――――――――――――――――――――――――――――――

   $XeF_6(g) \rightarrow Xe(g) + 6F(g)$ $\Delta H^\circ = 778$ kJ

Average Xe-F bond energy = 778/6 = 130 kJ

The average bond enthalpies are: XeF_2: 134 kJ, XeF_4: 134 kJ, XeF_6: 130 kJ. They are remarkably constant in the series.

22.111

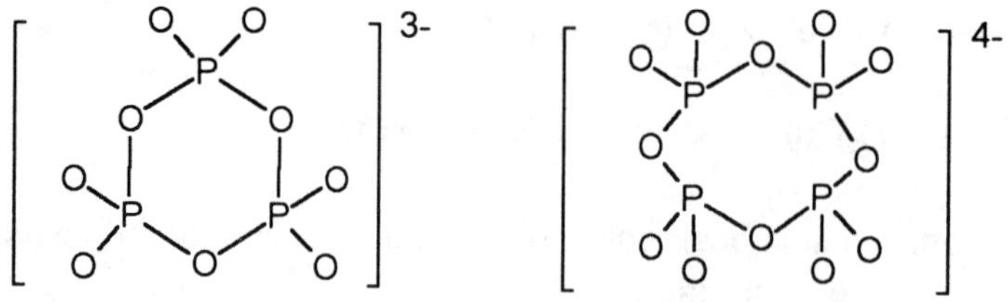

In both structures there are unshared pairs on all oxygens to give octets and the geometry around each P is approximately tetrahedral.

Integrative Exercises

22.112 (a) $H_2(g) + 1/2\,O_2(g) \rightarrow H_2O(l)$; $\Delta H = -285.83$ kJ

$CH_4(g) + 2O_2(g) \rightarrow CO_2(g) + 2H_2O(l)$

$\Delta H = 2(-285.83) - 393.5 - (-74.8) = -890.4$ kJ

(b) for H_2: $\dfrac{-285.83\text{ kJ}}{1\text{ mol }H_2} \times \dfrac{1\text{ mol }H_2}{2.0159\text{ g }H_2} = -141.79$ kJ/g H_2

for CH_4: $\dfrac{-890.4\text{ kJ}}{1\text{ mol }CH_4} \times \dfrac{1\text{ mol }CH_4}{16.043\text{ g }CH_4} = -55.50$ kJ/g CH_4

(c) Find the number of moles of gas that occupy 1 m^3 at STP:

$n = \dfrac{1\text{ atm} \times 1\text{ m}^3}{273\text{ K}} \times \dfrac{1\text{ K}\cdot\text{mol}}{0.08206\text{ L}\cdot\text{atm}} \times \left[\dfrac{100\text{ cm}}{1\text{ m}}\right]^3 \times \dfrac{1\text{ L}}{10^3\text{ cm}^3} = 44.64$ mol

for H_2: $\dfrac{-285.83\text{ kJ}}{1\text{ mol }H_2} \times \dfrac{44.64\text{ mol }H_2}{1\text{ m}^3\,H_2} = 1.276 \times 10^4$ kJ/m^3 H_2

for CH_4: $\dfrac{-890.4\text{ kJ}}{1\text{ mol }CH_4} \times \dfrac{44.64\text{ mol }CH_4}{1\text{ m}^3\,CH_4} = 3.975 \times 10^4$ kJ/m^3 CH_4

22.113 From Equations [22.10] and [22.11] note that four moles of H_2 are formed from one mole of CH_4. Ammonia synthesis involves the reaction: $N_2(g) + 3H_2(g) \rightleftharpoons 2NH_3(g)$

1.7×10^{10} kg $NH_3 \times \dfrac{1000\text{ g}}{1\text{ kg}} \times \dfrac{1\text{ mol }NH_3}{17.03\text{ g }NH_3} \times \dfrac{3\text{ mol }H_2}{2\text{ mol }NH_3}$

$\times \dfrac{1\text{ mol }CH_4}{4\text{ mol }H_2} \times \dfrac{16.04\text{ g }CH_4}{1\text{ mol }CH_4} \times \dfrac{1\text{ kg}}{1000\text{ g}} = 6.0 \times 10^9$ kg CH_4

Twice this, or 1.2×10^{10} kg CH_4 are required to produce the NH_3.

22.114 First calculate the molar solubility of Cl_2 in water.

$$n = \frac{1(0.310\,L)}{\dfrac{0.08206\,L\cdot atm}{1\,mol\cdot K} \times 273\,K} = 0.01384 = 0.0138\,mol\,Cl_2; \quad M = \frac{0.01384\,mol}{0.100\,L} = 0.1384$$
$$= 0.138\,M$$

$$K = \frac{[Cl^-][HOCl][H^+]}{[Cl_2]} = 4.7 \times 10^{-4}$$

$[Cl^-] = [HOCl] = [H^+]$ Let this quantity = x. Then, $\dfrac{x^3}{(0.1384 - x)} = 4.7 \times 10^{-4}$

Assuming that x is small compared with 0.1384:

$x^3 = (0.1384)(4.7 \times 10^{-4}) = 6.504 \times 10^{-5}; \; x = 0.0402 = 0.040\,M$

We can correct the denominator using this value, to get a better estimate of x:

$$\frac{x^3}{0.1384 - 0.0402} = 4.7 \times 10^{-4}; \; x = 0.0359 = 0.036\,M$$

One more round of approximation gives x = 0.0364 = 0.036 *M*. This is the equilibrium concentration of HClO.

22.115 (a) $N_2H_4(g) + O_2(g) \rightarrow N_2(g) + 2H_2O(l)$

     (b) $\Delta H° = \Delta H°_f\,N_2(g) + 2\Delta H°_f\,H_2O(l) - \Delta H°_f\,N_2H_4(aq) - \Delta H°_f\,O_2(g)$

               $= 0 + 2(-285.83) - 95.40 - 0 = -667.06\,kJ$

     (c) $\dfrac{9.1\,g\,O_2}{1 \times 10^6\,g\,H_2O} \times \dfrac{1.0\,g\,H_2O}{1\,mL\,H_2O} \times \dfrac{1000\,mL}{1\,L} \times 3.0 \times 10^4\,L = 273 = 2.7 \times 10^2\,g\,O_2$

         $2.73 \times 10^2\,g\,O_2 \times \dfrac{1\,mol\,O_2}{32.00\,g\,O_2} \times \dfrac{1\,mol\,N_2H_4}{1\,mol\,O_2} \times \dfrac{32.05\,g\,N_2H_4}{1\,mol\,N_2H_4} = 2.7 \times 10^2\,g\,N_2H_4$

22.116 $N_2H_5^+(aq) \rightarrow N_2(g) + 5H^+(aq) + 4e^-$ $E°_{red} = -0.23\,V$

Reduction of the metal should occur when $E°_{red}$ of the metal ion is more positive than about -0.15 V. This is the case for Sn^{2+} (marginal), Cu^{2+} and Ag^+.

22.117 First write the balanced equation to give the number of moles of gaseous products per mole of hydrazine.

     (A) $(CH_3)_2NNH_2 + 2N_2O_4 \rightarrow 3N_2(g) + 4H_2O(g) + 2CO_2(g)$

     (B) $(CH_3)HNNH_2 + 5/4\,N_2O_4 \rightarrow 9/4\,N_2(g) + 3H_2O(g) + CO_2(g)$

In case (A) there are nine moles gas per one mole $(CH_3)_2NNH_2$ plus two moles N_2O_4. The total mass of reactants is 60 + 2(92) = 244 g. Thus, there are

$$\frac{9\,mol\,gas}{244\,g\,reactants} = \frac{0.0369\,mol\,gas}{1\,g\,reactants}$$

In case (B) there are 6.25 moles of gaseous product per one mole $(CH_3)HNNH_2$ plus 1.25 moles N_2O_4. The total mass of this amount of reactants is $46.0 + 1.25(92.0) = 161$ g.

$$\frac{6.25 \text{ mol gas}}{161 \text{ g reactants}} = \frac{0.0388 \text{ mol gas}}{1 \text{ g reactants}}$$

Thus the methylhydrazine (B) has marginally greater thrust.

22.118 (a) $SO_2(g) + 2H_2S(s) \rightarrow 3S(s) + 2H_2O(g)$ or, if we assume S_8 is the product,
$8SO_2(g) + 16H_2S(g) \rightarrow 3S_8(s) + 16H_2O(g)$.

(b) $2000 \text{ lb coal} \times \dfrac{0.035 \text{ lb S}}{1 \text{ lb coal}} \times \dfrac{453.6 \text{ g S}}{1 \text{ lb S}} \times \dfrac{1 \text{ mol S}}{32.07 \text{ g S}} \times \dfrac{1 \text{ mol SO}_2}{1 \text{ mol S}} \times \dfrac{2 \text{ mol H}_2\text{S}}{1 \text{ mol SO}_2}$

$$= 1.98 \times 10^3 = 2.0 \times 10^3 \text{ mol H}_2\text{S}$$

$$V = \frac{1.98 \times 10^3 \text{ mol } (0.08206 \text{ L} \cdot \text{atm/mol} \cdot \text{K})(300 \text{ K})}{(740/760) \text{ atm}} = 5.01 \times 10^4 = 5.0 \times 10^4 \text{ L}$$

(c) $1.98 \times 10^3 \text{ mol H}_2\text{S} \times \dfrac{3 \text{ mol S}}{2 \text{ mol H}_2\text{S}} \times \dfrac{32.07 \text{ g S}}{1 \text{ mol S}} = 9.5 \times 10^4 \text{ g S}$

This is about 210 lb S per ton of coal combusted. (However, two-thirds of this comes from the H_2S, which was presumably also obtained from coal.)

22.119 $FeS(s) + 2HCl(aq) \rightarrow FeCl_2(aq) + H_2S(g)$

The maximum allowable concentration of H_2S is 20 ppm or 20 mol $H_2S/1 \times 10^6$ mol air. The total moles air (29.0 g/mol) in the room is

$$n = \frac{1 \text{ atm} \times 2.7 \times 4.3 \times 4.3 \text{ m}^3 \times \left[\dfrac{100 \text{ cm}}{1 \text{ m}}\right]^3 \times \dfrac{1 \text{ L}}{10^3 \text{ cm}^3}}{\dfrac{0.08206 \text{ L} \cdot \text{atm}}{\text{mol} \cdot \text{K}}} = 2.04 \times 10^3 = 2.0 \times 10^3 \text{ mol air}$$

$$= 2.04 \times 10^3 \text{ mol air} \times \frac{20 \text{ mol H}_2\text{S}}{1 \times 10^6 \text{ mol air}} \times \frac{1 \text{ mol FeS}}{1 \text{ mol H}_2\text{S}} \times \frac{87.9 \text{ g FeS}}{1 \text{ mol FeS}} = 3.6 \text{ g FeS}$$

22.120   The reactions can be written as follows:

$H_2(g) + X(\text{std state}) \rightarrow H_2X(g)$ $\qquad \Delta H_f^\circ$

$\qquad 2H(g) \rightarrow H_2(g)$ $\qquad \Delta H_f^\circ(\text{H-H})$

$\qquad X(g) \rightarrow X(\text{std state})$ $\qquad \Delta H_3$

Add: $2H(g) + X(g) \rightarrow H_2X(g)$ $\qquad \Delta H = \Delta H_f^\circ + \Delta H_f^\circ(\text{H-H}) + \Delta H_3$

These are all the necessary ΔH values. Thus,

Compound	ΔH	D H-X
H_2O	ΔH = -242 kJ - 436 kJ - 248 kJ = -926 kJ	463 kJ
H_2S	ΔH = -20 kJ - 436 kJ - 277 kJ = -733 kJ	367 kJ
H_2Se	ΔH = +30 kJ - 436 kJ - 227 kJ = -633 kJ	316 kJ
H_2Te	ΔH = +100 kJ - 436 kJ - 197 kJ = -533 kJ	266 kJ

The average H-X bond energy in each case is just half of ΔH. The H-X bond energy decreases steadily in the series. The origin of this effect is probably the increasing size of the orbital from X with which the hydrogen 1s orbital must overlap.

22.121 (a)

$$HOOC-CH_2-COOH \xrightarrow{P_2O_5} C_3O_2 + 2H_2O$$

(b) 24 valence e⁻, 12 e⁻ pair $\ddot{O}=C=C=C=\ddot{O}$

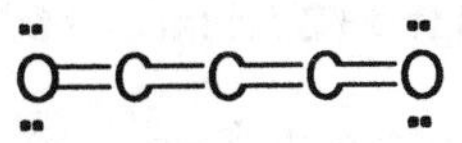

(c) C=O, about 1.23 Å; C=C, 1.34 Å or less. Since consecutive C=C bonds require sp hybrid orbitals on C (as in allene, C_3H_4), we might expect the orbital overlap requirements of this bonding arrangement to require smaller than usual C=C distances.

(d)

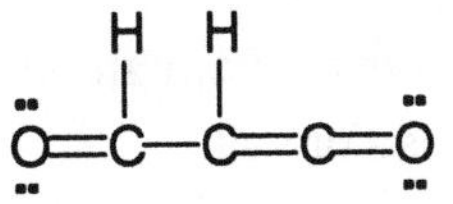

Other possibilities leave either C or O with more than eight electrons, unlikely for these small second row elements.

22.122 BN has the same number of valence electrons per formula unit as carbon. (Three from B, five from N, for an average of four per atom.) To the extent that we can neglect the difference in nuclear charges between B and N, we can think of BN as carbon-like. Indeed, BN takes on the same structural forms as carbon. However, because the B-N bonds are somewhat polar, BN is in fact even harder than diamond.

23 Metals and Metallurgy

Metallurgy

23.1 The important sources of iron are **hematite** (Fe_2O_3) and **magnetite** (Fe_3O_4). The major source of aluminum is **bauxite** ($Al_2O_3 \cdot xH_2O$). In ores, iron is present as the +3 ion, or in both the +2 and +3 states, as in magnetite. Aluminum is always present in the +3 oxidation state.

23.2 Sphalerite is ZnS; Zn is in the +2 oxidation state.

23.3 An ore consists of a little bit of the stuff we want, (chalcopyrite, $CuFeS_2$) and lots of other junk (gangue).

23.4 (a) *Calcination* is heating an ore to decompose the mineral of interest into a simple solid and volatile compound. Calcination usually produces a metal oxide and a gas that is a nonmetal oxide.

(b) *Leaching* is dissolving the mineral of interest to remove it from an ore. The solvent is usually water or an aqueous solution of acid, base or salt.

(c) *Smelting* is heating an ore, often in a reducing atmosphere, to a very high temperature so that two immiscible liquid layers form. The layers are usually the molten metal or metals of interest and slag.

(d) *Slag* is the unwanted layer of the smelting process. It contains molten silicate, aluminate, phosphate or fluoride compounds.

23.5 (a) $2PbS(s) + 3O_2(s) \rightarrow 2PbO(s) + 2SO_2(g)$

(b) $PbCO_3(s) \xrightarrow{\Delta} PbO(s) + CO_2(g)$

(c) $WO_3(s) + 3H_2(g) \rightarrow W(s) + 3H_2O(g)$

(d) $ZnO(s) + CO(g) \rightarrow Zn(l) + CO_2(g)$

23.6 (a) $ZnCO_3(s) \xrightarrow{\Delta} ZnO(s) + CO_2(g)$

(b) $MnO(s) + CO(g) \rightarrow Mn(l) + CO_2(g)$

(c) $Al(OH)_3(s) + OH^-(aq) \rightarrow [Al(OH)_4]^-(aq)$

(d) $TiCl_4(g) + K(l) \rightarrow Ti(s) + 4KCl(s)$

(e) $3CaO(l) + P_2O_5(l) \rightarrow Ca_3(PO_4)_2(l)$

23.7   (a) $SO_3(g)$

        (b) $CO(g)$ provides a reducing environment for the transformation of Pb^{2+} to Pb^0.

        (c) $PbSO_4(s) \rightarrow PbO(s) + SO_3(g)$

                $PbO(s) + CO(g) \rightarrow Pb(s) + CO_2(g)$

23.8   (a) $NiO(s), H_2O(g)$

        (b)   NiO or $Ni(OH)_2$ might be roasted to Ni in a reducing atmosphere such as CO.

        (c) $Ni(OH)_2(s) \rightarrow NiO(s) + H_2O(g)$

                $NiO(s) + CO(g) \rightarrow Ni(s) + CO_2(g)$

                $Ni(OH)_2(s) + CO(g) \rightarrow Ni(s) + CO_2(g) + H_2O(g)$

23.9 $FeO(s) + H_2(g) \rightarrow Fe(s) + H_2O(g)$
        $FeO(s) + CO(g) \rightarrow Fe(s) + CO_2(g)$

        $Fe_2O_3(s) + 3H_2(g) \rightarrow 2Fe(s) + 3H_2O(g)$
        $Fe_2O_3(s) + 3CO(g) \rightarrow 2Fe(s) + 3CO_2(g)$

23.10   The major reducing agent is CO, formed by partial oxidation of the coke (C) with which the furnace is charged.

        $Fe_2O_3(s) + 3CO(g) \rightarrow 2Fe(l) + 3CO_2(g)$
        $Fe_3O_4(s) + 4CO(g) \rightarrow 3Fe(l) + 4CO_2(g)$

23.11   (a)   Air serves primarily to oxidize coke (C) to CO, the main reducing agent in the blast furnace. This exothermic reaction also provides heat for the furnace.

                $2C(s) + O_2(g) \rightarrow 2CO(g)$ $\Delta H = -110$ kJ

        (b)   Limestone, $CaCO_3$, is the source of basic oxide for slag formation.

                $CaCO_3(s) \xrightarrow{\Delta} CaO(s) + CO_2(g)$; $CaO(l) + SiO_2(l) \rightarrow CaSiO_3(l)$

        (c)   Coke is the fuel for the blast furnace, and the source of CO, the major reducing agent in the furnace.

                $2C(s) + O_2(g) \rightarrow 2CO(g)$; $4CO(g) + Fe_3O_4(s) \rightarrow 4CO_2(g) + 3Fe(l)$

        (d)   Water acts as a source of hydrogen, and as a means of controlling temperature. (see Equation [24.9]). $C(s) + H_2O(g) \rightarrow CO(g) + H_2(g)$ $\Delta H = +113$ kJ

23.12   The term *blast* is often associated with an explosion. According to the Chemistry at Work box on Explosives in Chapter 8, an explosive is a liquid or solid that decomposes rapidly and exothermically with the production of large amounts of gas. Several of the reactions in the furnace occur rapidly, exothermically and with the production of gases. The conditions in a blast furnace are similar to those in an explosion, but much more controllable.

23.13   The Bayer process takes advantage of the fact that Al^{3+} is amphoteric, but Fe^{3+} is not. Because it is amphoteric, Al^{3+} reacts with excess OH^- to form the soluble complex ion $Al(OH)_4^-$ while the Fe^{3+} solids cannot. This allows separation of the unwanted iron-containing solids by filtration.

23.14   Gold appears in elemental form in ores. The $O_2(g)$ in air oxidizes $Au(s)$ to $Au^+(aq)$ and the $CN^-(aq)$ stabilizes the Au^+ in the form of the soluble complex ion $Au(CN)_2^-(aq)$. This enables the unwanted solids (gangue) to be filtered away from the gold-containing solution.

23.15   Cobalt could be purified by constructing an electrolysis cell in which the crude metal was the anode and a thin sheet of pure cobalt was the cathode. The electrolysis solution is aqueous with a soluble cobalt salt such as $CoSO_4 \cdot 7H_2O$ serving as the electrolyte. (Other soluble salts with anions that do not participate in the cell reactions could be used.) Anode reaction: $Co(s) \rightarrow Co^{2+}(aq) + 2e^-$; cathode reaction: $Co^{2+}(aq) + 2e^- \rightarrow Co(s)$. Although $E°$ for reduction of $Co^{2+}(aq)$ is slightly negative (-0.277 V), we assume that reduction of water or H^+ does not occur because of a large overvoltage.

23.16 $SnO_2(s) + C(s) \xrightarrow{\Delta} Sn(l) + CO_2(g)$

$Sn(s) \rightarrow Sn^{2+}(aq) + 2e^-$ (anode)

$Sn^{2+}(aq) + 2e^- \rightarrow Sn(s)$ (cathode)

Metals and Alloys

23.17   Sodium is metallic; each atom is bonded to many nearest neighbor atoms by metallic bonding involving just one electron per atom, and delocalized over the entire three-dimensional structure. When sodium metal is distorted, each atom continues to have bonding interactions with many nearest neighbors. In NaCl the ionic forces are strong, and the arrangement of ions in the solid is very regular. When subjected to physical stress, the three-dimensional lattice tends to cleave along the very regular lattice planes, rather than undergo the large distortions characteristic of metals.

23.18   Test electrical and/or thermal conductivity, which should both be high for a metal. Strike a small sample of Si with a hammer; if it flattens into a sheet, it is a metal. If the sample shatters, it is not a metal.

23.19   In the electron-sea model for metallic bonding, the valence electrons of the silver atoms move about the three-dimensional metallic lattice, while the silver atoms maintain regular lattice positions. Under the influence of an applied potential the electrons can move throughout the structure, giving rise to high electrical conductivity. The mobility of the electrons facilitates the transfer of kinetic energy and leads to high thermal conductivity.

23.20   (a) Ductility   (b) In the electron-sea model, electrons are free to move about the lattice and no electron is associated with a particular metal atom. When the copper is shaped into a wire, the positions of the Cu atoms can change without breaking chemical bonds, and the electrons are redistributed to compensate for the change in atomic positions.

23.21   According to the molecular orbital or band theory of metallic bonding, the 4s and 3d atomic orbitals of the fourth row elements are used to form six molecular orbitals, three bonding and three antibonding. The maximum bond order and highest bond strength occurs in metals with six valence electrons (group 6B). Assuming that hardness is directly related to the bond strength between metal atoms (the stronger the bonds, the more difficult it is to distort the solid without disrupting the three-dimensional structure), Cr, with six valence electrons, should have the highest bond strength and greatest hardness in the fourth row.

23.22   The variation in densities reflects shorter metal-metal bond distances. These shorter distances suggest that the extent of metal-metal bonding increases in the series. Thus, it would appear that all the valence electrons in these elements (1, 2, 3 and 4, respectively) are involved in metallic bonding.

23.23   According to band theory, an *insulator* has a completely filled valence band and a large energy gap between the valence band and the nearest empty band; electrons are localized within the lattice. A *conductor* must have a partially filled energy band; a small excitation will promote electrons to previously empty levels within the band and allow them to move freely throughout the lattice, giving rise to the property of conduction. A *semiconductor* has a filled valence band, but the gap between the filled and empty bands is small enough to jump to the empty conduction band. The presence of an impurity may also place an electron in an otherwise empty band (producing an n-type semiconductor), or create a vacancy in an otherwise full band (producing a p-type semiconductor), providing a mechanism for conduction.

23.24   Germanium doped with arsenic (Ge/As) is a better conductor than germanium. Ge, a semiconductor, has a filled valence band but an energetically accessible conduction band. In Ge/As, the additional valence electrons of the doped As atoms cannot fit into the filled valence band and occupy the conduction band. These electrons have access to the vacant orbitals within the conduction band and serve as carriers of electrical current. Ge/As is an *n-type* semiconductor.

23.25   White tin, with a characteristic metallic structure, is expected to be more metallic in character. The electrical conductivity of the white allotropic form is higher because the valence electrons are shared with 12 nearest neighbors rather than being localized in four bonds to nearest neighbors as in gray tin. The Sn-Sn distance should be longer in white tin; there are only four valence electrons from each atom, and 12 nearest neighbors. The **average** tin-tin bond order can, therefore, be only about 1/3, whereas in gray tin the bond order is one. (In gray tin the Sn-Sn distance is 2.81 Å in white tin it is 3.02 Å.)

23.26   Each silver atom has many near neighbors with which it interacts. Because each Ag atom contributes only a few valence level electrons to the bonding, there is a comparatively weak bonding interaction with any one other silver atom. As a result of this relatively weak interaction, the Ag-Ag distance is large. The metal itself has quite high melting and boiling points, because each silver atom has a bonding interaction with **many** neighbors. The total attractive forces experienced by any one Ag atom are very high.

23.27 An *alloy* contains atoms of more than one element and has the properties of a metal. *Solution alloys* are homogeneous mixtures with different kinds of atoms dispersed randomly and uniformly. In *heterogeneous alloys* the components (elements or compounds) are not evenly dispersed and their properties depend not only on composition but methods of preparation. In an *intermetallic compound* the component elements have interacted to form a compound substance, for example, Cu_3As. As with more familiar compounds, these are homogeneous and have definite composition and properties.

23.28 Substitutional and interstitial alloys are both solution alloys. In a *substitutional* alloy, the atoms of the "solute" take positions normally occupied by the "solvent." In an *interstitial* alloy, the atoms of the "solute" occupy the holes or interstitial positions between "solvent" atoms. Substitutional alloys tend to form when solute and solvent atoms are of comparable size and have similar bonding characteristics.

Transition Metals

23.29 Of the properties listed, (b) the first ionization energy and (c) atomic radius are characteristic of isolated atoms. Electrical conductivity (a) and melting point (d) are properties of the bulk metal.

23.30 Heat of fusion, malleability, hardness, and reduction potential (because the reduction process involves formation of M(s)), all relate to bonding in metals and are thus bulk properties.

23.31 The *lanthanide contraction* is the name given to the decrease in atomic size due to the build-up in effective nuclear charge as we move through the lanthanides (elements 58-71) and beyond them. This effect offsets the expected increase in atomic size going from the second to the third transition series. The lanthanide contraction affects size-related properties such as ionization energy, electron affinity and density.

23.32 Zr: $[Kr]5s^24d^2$, Z = 40; Hf: $[Kr]6s^24f^{14}5d^2$, Z = 72

Moving down a family of the periodic chart, atomic size increases because the valence electrons are in a higher principle quantum level (and thus further from the nucleus) and are more effectively shielded from the nuclear charge by a larger core electron cloud. However, the build-up in Z that accompanies the filling of the 4f orbitals causes the valence electrons in Hf to experience a much greater relative nuclear charge than those in La, its neighbor to the left. This increase in Z offsets the usual effect of the increase in *n* value of the valence electrons and the radii of Zr and Hf atoms are similar.

23.33 (a) ScF_3 (b) CoF_3 (c) ZnF_2

23.34 (a) TiO_2 (b) Mn_2O_7 (MnO_4^- is more stable and more common) (c) NiO_2

23.35 Chromium, $[Ar]4s^13d^5$, has six valence-shell electrons, some or all of which can be involved in bonding, leading to multiple stable oxidation states. By contrast, aluminum, $[Ne]3s^23p^1$, has only three valence electrons which are all lost or shared during bonding, producing the +3 state exclusively.

23.36   The electron configurations of the two metals are: Zn: $[Ar]3d^{10}4s^2$; Cu: $[Ar]4s^13d^{10}$ (note exception to the normal filling pattern). The +2 state for Zn and the +1 state for Cu follow directly from their electron configurations, having complete n=3 shells. In Cu, the 4s and 3d orbitals are very similar in energy, as evidenced by the fact that the 3d fills before the 4s. Loss of a second electron is easily achieved, giving the +2 state. Cu and Zn are near the end of the first transition-metal series and have relatively high effective nuclear charges, which prevents them from forming +3 ions.

23.37   (a) Cr^{3+}: $[Ar]3d^3$ (b) Au^{3+}: $[Xe]4f^{14}5d^8$ (c) Ru^{2+}: $[Kr]4d^6$

        (d) Cu^+: $[Ar]3d^{10}$ (e) Mn^{4+}: $[Ar]3d^3$ (f) Ir^{3+}: $[Xe]4f^{14}5d^6$

23.38   (a) Fe^{2+}: $[Ar]3d^6$ (b) Sc^{2+}: $[Ar]3d^1$ (c) Ag^+: $[Kr]4d^{10}$

        (d) Mo^{4+}: $[Kr]4d^2$ (e) Nb^{3+}: $[Kr]4d^2$ (f) Rh^{3+}: $[Kr]4d^6$

23.39   Ease of oxidation decreases from left to right across a period (owing to increasing effective nuclear charge); Ti^{2+} should be more easily oxidized than Ni^{2+}.

23.40   The stronger reducing agent is more easily oxidized; Cr^{2+} is more easily oxidized (see Exercise 23.39), so it is the stronger reducing agent.

23.41   Chromate ion, CrO_4^{2-}, is bright yellow. Dichromate, $Cr_2O_7^{2-}$, is orange and more stable in acid solution than CrO_4^{2-} because of the equilibrium

        $2CrO_4^{2-}(aq) + 2H^+(aq) \rightleftharpoons Cr_2O_7^{2-}(aq) + H_2O(l)$

23.42   (Equation [23.26]) Fe^{2+} is a reducing agent that is readily oxidized to Fe^{3+} in the presence of O_2 from air.

23.43   (a) $Fe(s) + 2HCl(aq) \rightarrow FeCl_2(aq) + H_2(g)$

        (b) $Fe(s) + 4HNO_3(aq) \rightarrow Fe(NO_3)_3(aq) + NO(g) + 2H_2O(l)$
            (See net ionic equation, Equation 23.28) In concentrated nitric acid, the reaction can produce $NO_2(g)$ according to the reaction:
            $Fe(s) + 6HNO_3(aq) \rightarrow Fe(NO_3)_3(aq) + 3NO_2(g) + 3H_2O(l)$

23.44   (a) $Fe^{3+}(aq) + NaOH(aq) \rightarrow Fe(OH)_3(s) + 3Na^+(aq)$

        (b) $FeCO_3(s) + 2HCl(aq) \rightarrow FeCl_2(aq) + CO_2(g) + H_2O(l)$

23.45   The unpaired electrons in a *paramagnetic* material cause it to be weakly attracted into a magnetic field. A *diamagnetic* material, where all electrons are paired, is very weakly repelled by a magnetic field.

23.46   In a *ferromagnetic* substance, the magnetic moments on sites throughout the lattice interact with one another. That is, they are close enough physically, and the overlap of orbitals is such that the individual magnetic sites couple to form a much larger magnetic moment throughout the solid. Because of these interactions, the continued existence of the magnetic moment does not require continued application of an external magnetic field. That is, ferromagnetic materials can form "permanent" magnets, whereas paramagnetic materials cannot.

Additional Exercises

23.47 $PbS(s) + O_2(g) \rightarrow Pb(l) + SO_2(g)$

Regardless of the metal of interest, $SO_2(g)$ is a product of roasting sulfide ores. In an oxygen rich environment, $SO_2(g)$ is oxidized to $SO_3(g)$, which dissolves in $H_2O(l)$ to form sulfuric acid, $H_2SO_4(aq)$. Because of its corrosive nature, $SO_2(g)$ is a dangerous environmental pollutant (Section 18.4) and cannot be freely released into the atmosphere. A sulfuric acid plant near a roasting plant would provide a means for disposing of $SO_2(g)$ that would also generate a profit.

23.48 Al^{3+}, Mg^{2+} and Na^+ all have large negative reduction potentials (Al, Mg and Na are very active metals). A substance with a more negative reduction potential would have to be used to chemically reduce them. All such substances are more expensive and difficult to obtain than Al, Mg and Na. Electrolysis is thus the most cost efficient way to reduce Al^{3+}, Mg^{2+} and Na^+ to their metallic states.

23.49 $HfCl_4(g) + Na(l) \rightarrow 4NaCl(s) + Hf(s)$

23.50   (a) $NiO(s) + 3H^+(aq) \rightarrow Ni^{2+}(aq) + H_2O(l)$

        (b)   The simple answer is that the solid is subjected to acid hydrolysis:
            $CuCo_2S_4(s) + 8H^+(aq) \rightarrow Cu^{2+}(aq) + 2Co^{3+}(aq) + 4H_2S(g)$
            However, in the absence of a strong complexing ligand, Co^{3+} is not stable in water. It oxidizes water according to the following reaction:
            $4Co^{3+}(aq) + 2H_2O(l) \rightarrow 4Co^{2+}(aq) + O_2(g) + 4H^+(aq)$

        (c) $TiO_2(s) + C(s) + 2Cl_2(g) \rightarrow TiCl_4(g) + CO_2(g)$

        (d)   In this reaction O_2 is reduced and sulfide is oxidized. Writing the sulfur product as S_8, the balanced equation is:
            $8ZnS(s) + 4O_2(g) + 16H^+(aq) \rightarrow 8Zn^{2+}(aq) + S_8(s) + 8H_2O(l)$

23.51   (a) $2NH_4VO_3(s) \rightarrow V_2O_5(s) + 2NH_3(g) + H_2O(g)$
        (b) $V_2O_5(s) + SO_2(g) \rightarrow 2VO_2(s) + SO_3(g)$
        (c) $V_2O_5(s) + 2Mg(s) + 10H^+(aq) \rightarrow 2V^{3+}(aq) + 2Mg^{2+}(aq) + 5H_2O(l)$
        (d) $NbCl_5(s) + 3H_2O(l) \rightarrow HNbO_3(s) + 5H^+(aq) + 5Cl^-(aq)$

23.52   Because selenium and tellurium are both nonmetals, we expect them to be difficult to oxidize. Thus, both Se and Te are likely to accumulate as the free elements in the so-called anode slime, along with noble metals that are not oxidized.

23.53 All transition metals have the generic electron configuration $ns^2(n\text{-}1)d^x$. Regardless of the number of d electrons, each transition metal has 2 ns valence electrons that are the first electrons lost when metal ions are formed. Thus, almost every transition metal has a stable +2 oxidation state.

After the 2 ns electrons are lost, a varying number of $(n\text{-}1)d$ electrons can be lost, depending on the identity of the transition metal. The availability of different numbers of d electrons leads to a wide variety of accessible oxidation states for the transition metals.

23.54 The iron accumulates in the solution as $Fe^{2+}(aq)$ from oxidation at the anode. Given that iron can exist in more than one oxidation state and that the presence of air is an important factor, Fe^{2+} is probably being oxidized to $Fe^{3+}(aq)$, Equation [23.26]. Now look at reduction potentials:

$$Ni^{2+}(aq) + 2e^- \rightarrow Ni(s) \qquad E^\circ = -0.28 \text{ V}$$
$$Fe^{3+}(aq) + e^- \rightarrow Fe^{2+}(aq) \qquad E^\circ = 0.77 \text{ V}$$

These data indicate that Fe^{3+} will be reduced to Fe^{2+} at the cathode, instead of Ni^{2+} being reduced to $Ni(s)$. As long as air is present to reoxidize the Fe^{2+} to Fe^{3+}, the iron keeps reappearing as Fe^{3+}, using electrical energy that would otherwise be applied to reduction of Ni^{2+}.

23.55 Assuming that SO_2 and N_2 are the nonmetallic products, two half-reactions can be written:

$$5[MoS_2(s) + 7H_2O(l) \rightarrow MoO_3(s) + 2SO_2(g) + 14H^+(aq) + 14e^-]$$
$$7[12H^+(aq) + 2NO_3^-(aq) + 10e^- \rightarrow N_2(g) + 6H_2O(l)]$$

$$5MoS_2(s) + 14H^+(aq) + 14NO_3^-(aq) \rightarrow 5MoO_3(s) + 10SO_2(g) + 7N_2(g) + 7H_2O(l)$$
$$MoO_3(s) + 2NH_3(aq) + H_2O(l) \rightarrow (NH_4)_2MoO_4(s)$$
$$(NH_4)_2MoO_4(s) \rightarrow 2NH_3(g) + H_2O(g) + MoO_3(s)$$
$$MoO_3(s) + 3H_2(g) \rightarrow Mo(s) + 3H_2O(g)$$

23.56 (a) In a *substitutional alloy* one element replaces another, atom-for-atom, with no disruption in the lattice structure. In an *interstitial alloy* a second component fits into the spaces between the major component.

(b) An *intermetallic compound* is a true chemical substance, with a fixed ratio of numbers of atoms of the different elements present. A *heterogeneous alloy* is a solid mixture; two or more substances (they may be elements or intermetallic compounds) exist together as an intimate mixture of two mutually insoluble solid phases.

(c) A *metal* is a substance in which the bonding produces certain key properties, among them, high electrical and thermal conductivity. Metals have delocalized electrons that are free to move throughout the lattice. *Insulators* possess very low electrical and thermal conductivities. The electrons in insulators are largely confined to localized strong bonds between atoms, or are confined within ions, as in ionic substances that are insulators, such as NaCl.

(d)   In *Na₂, a gaseous molecule*, the two Na atoms are bound together by a localized bond between the two atoms. In *metallic* Na, each Na atom has many near-neighbors. Its one valence electron is delocalized throughout the three-dimensional lattice.

23.57   Silicon has the diamond structure. As with carbon, the four valence electrons of silicon are completely involved in the four localized bonds to its neighbors. There are thus no electrons free to migrate throughout the solid. Titanium exists as a close-packed lattice; each Ti atom has twelve equivalent nearest neighbors. The valence shell electrons cannot be localized between pairs of atoms; rather they are delocalized and mobile throughout the structure. In terms of the band model described in Figure 23.17, Ti has an incompletely occupied allowed energy band. The origin of the different behaviors with regard to structure has to do with the extent of the orbitals in space. Electrons in Si feel a greater attraction to the nucleus than electrons in Ti, so they are more localized.

23.58   The equilibrium of interest is $[ZnL_4] \rightleftharpoons Zn^{2+}(aq) + 4L$ $K = 1/K_f$

Since $Zn(H_2O)_4{}^{2+}$ is $Zn^{2+}(aq)$, its reduction potential is -0.763 V. As the stability (K_f) of the complexes increases, the $[Zn^{2+}(aq)]$ decreases and the reduction potentials become more negative relative to the reduction of $Zn^{2+}(aq)$ to Zn(s).

23.59   In a paramagnetic substance, the unpaired electrons on one atom are not affected by (coupled to) the unpaired electrons on adjacent atoms. The unpaired electrons are in random orientations. The external magnetic field causes rough alignment and therefore weak attraction to the magnetic field. In a ferromagnetic substance, unpaired electrons on adjacent atoms are coupled and aligned in the same direction. This permanent magnetic moment causes the ferromagnetic substance to be strongly attracted into the magnetic field.

23.60 Ni^{2+}: $[Ar]3d^8$ ⥮ ⥮ ⥮ ↑ ↑   paramagnetic
                                    3d

        Cr^{3+}: $[Ar]3d^3$ ↑ ↑ ↑   paramagnetic
                                      3d

Both ions are paramagnetic because they have unpaired electrons in the 3d subshell.

23.61   In a ferromagnetic solid, the magnetic centers are coupled such that the spins of all unpaired electrons are parallel. As the temperature of the solid increases, the average kinetic energy of the atoms increases until the energy of motion overcomes the force aligning the electron spins. The substance becomes paramagnetic; it still has unpaired electrons, but their spins are no longer aligned.

23.62   (a)   Nothing. As noted in Section 20.10, a basic environment (OH⁻) inhibits oxidation of Fe^{2+} to Fe^{3+}, even in the presence of $O_2(g)$. $Cu(OH)_2(s)$ precipitates. Cu^{2+} forms a soluble complex ion with $NH_3(aq)$, but not $OH^-(aq)$.

        (b) $Cu(NO_3)_2(aq) + 2KOH(aq) \rightarrow Cu(OH)_2(s) + 2KNO_3(aq)$

(c) The color of the solution changes from orange ($Cr_2O_7^{2-}$) to yellow (CrO_4^{2-}). The equilibria in question are

$Cr_2O_7^{2-}(aq) + H_2O(l) \rightleftharpoons 2CrO_4^{2-}(aq) + 2H^+(aq)$

As $OH^-(aq)$ is added, it reacts with and removes $H^+(aq)$ from solution, shifting the equilibrium to the right in favor of the yellow CrO_4^{2-}.

23.63 (a) $2NiS(s) + 3O_2(g) \rightarrow 2NiO(s) + 2SO_2(g)$

(b) $2C(s) + O_2(g) \rightarrow 2CO(g)$; $C(s) + H_2O(g) \rightarrow CO(g) + H_2(g)$

$NiO(s) + CO(g) \rightarrow Ni(s) + CO_2(g)$; $NiO(s) + H_2(g) \rightarrow Ni(s) + H_2O(g)$

(c) $Ni(s) + 2HCl(aq) \rightarrow NiCl_2(aq) + H_2(g)$

(d) $NiCl_2(aq) + 2NaOH(aq) \rightarrow Ni(OH)_2(s) + 2NaCl(aq)$

(e) $Ni(OH)_2(s) \rightarrow NiO(s) + H_2O(g)$

23.64 (a) p-type semi-conductor   (b) metallic conductor   (c) insulator

(d) n-type semi-conductor   (e) metallic conductor   (f) insulator

Integrative Exercises

23.65 (a) malleability

(b) $\dfrac{1 \text{ troy oz}}{300 \text{ ft}^2} \times \dfrac{31.1 \text{ g}}{\text{troy oz}} \times \dfrac{1 \text{ ft}^2}{(12 \text{ in})^2} \times \dfrac{1 \text{ in}^2}{(2.54 \text{ cm})^2} \times \dfrac{1 \text{ cm}^3}{19.31 \text{ g}} = 5.779 \times 10^{-6} \text{ cm}$

$5.779 \times 10^{-6} \text{ cm} = \textbf{5.78} \times \textbf{10}^{-5} \textbf{ mm} = 5.78 \times 10^{-8} \text{ m} = 57.8 \text{ nm}$

23.66 The reactions involved are:

$Cu_2(s) + O_2(g) \rightarrow 2Cu(s) + SO_2(g)$

$FeS(s) + O_2(g) \rightarrow Fe(s) + SO_2(g)$

Do this problem by staying with units of kg, and using the ratios of formula weights to arrive at the desired masses.

$2.0 \times 10^4 \text{ kg ore} \times \dfrac{0.32 \text{ kg Cu}_2\text{S}}{1 \text{ kg ore}} \times \dfrac{32.1 \text{ kg S}}{159.2 \text{ kg Cu}_2\text{S}} \times \dfrac{64.1 \text{ kg SO}_2}{32.1 \text{ kg S}} = 2.6 \times 10^3 \text{ kg SO}_2$

$2.0 \times 10^4 \text{ kg ore} \times \dfrac{0.07 \text{ kg FeS}}{1 \text{ kg ore}} \times \dfrac{32.1 \text{ kg S}}{87.9 \text{ kg FeS}} \times \dfrac{64.1 \text{ kg SO}_2}{32.1 \text{ kg S}} = 1.0 \times 10^3 = 1 \times 10^3 \text{ kg SO}_2$

Total SO_2 is about 3.6×10^3 kg SO_2, or, to one significant figure, 4×10^3 kg SO_2.

23.67 $FeO(s) + SiO_2(l) \rightarrow FeSiO_3(l)$

$1.2 \times 10^4 \text{ kg ore} \times \dfrac{0.26 \text{ kg FeO}}{1 \text{ kg ore}} \times \dfrac{1000 \text{ g FeO}}{1 \text{ kg FeO}} \times \dfrac{1 \text{ mol FeO}}{71.9 \text{ g FeO}} \times \dfrac{1 \text{ mol SiO}_2}{1 \text{ mol FeO}}$

$\times \dfrac{60.1 \text{ g SiO}_2}{1 \text{ mol SiO}_2} = 2.6 \times 10^6 \text{ g SiO}_2 = 2.6 \times 10^3 \text{ kg SiO}_2$

23.68 Recall from the discussion in Chapter 13 that like substances tend to be soluble in one another, whereas unlike substances do not. Molten metal consists of atoms that continue to be bound to one another by metallic bonding, even though the substance is liquid. In a slag, on the other hand, the attractive forces are those between ions. The slag phase is a highly polar, ionic medium, whereas the metallic phase is nonpolar, and the attractive interactions are due to metallic bond formation. There is little driving force for materials with such different characteristics to dissolve in one another.

23.69 The first equation indicates that one mole Ni^{2+} is formed from passage of two moles of electrons, and the second equation indicates the same thing. Thus, the simple ratio (1 mol Ni^{2+}/2F).

$$66 \text{ A} \times 8.0 \text{ hr} \times \frac{3600 \text{ s}}{1 \text{ hr}} \times \frac{1 \text{ C}}{1 \text{ A} \cdot \text{s}} \times \frac{1 \text{ F}}{96{,}500 \text{ C}} \times \frac{1 \text{ mol Ni}^{2+}}{2 \text{ F}} \times \frac{58.7 \text{ g Ni}^{2+}}{1 \text{ mol Ni}^{2+}}$$

$$= 5.8 \times 10^2 \text{ g Ni}^{2+}\text{(aq)}$$

23.70 The half-cell reaction is $Cu^{2+}(aq) + 2e^- \rightarrow Cu(s)$

$$0.80 \times 240 \text{ A} \times 10 \text{ hr} \times \frac{3600 \text{ s}}{1 \text{ hr}} \times \frac{1 \text{ C}}{1 \text{ A} \cdot \text{s}} \times \frac{1 \text{ F}}{96{,}500 \text{ C}} \times \frac{1 \text{ mol Cu}}{2 \text{ F}} \times \frac{63.5 \text{ g Cu}}{1 \text{ mol Cu}}$$

$$= 2.3 \times 10^3 \text{ g Cu}$$

23.71 $\Delta G° = \Delta H° - T\Delta S°$ (assume ΔH and ΔS are standard)

(a) $PbO(s) + CO(g) \rightarrow Pb(s) + CO_2(g)$

$\Delta H° = \Delta H_f° \ CO_2(g) + \Delta H_f° \ Pb(s) - \Delta H_f° \ CO(g) - \Delta H_f° \ PbO(s)$

$= -393.5 - (-110.5) - (-217.3) = -65.7 \text{ kJ}$

$\Delta S° = S° \ CO_2(g) + S° \ Pb(s) - S° \ CO(g) - S° \ PbO(s)$

$= 213.6 + 68.85 - 197.9 - 68.70 = 15.8 \text{ J/K}$

$\Delta G° = -65.7 \text{ kJ} - 1473 \text{ K} (0.0158 \text{ kJ/K}) = -89.0 \text{ kJ}$

(b) $Si(s) + 2MnO(s) \rightarrow SiO_2(s) + 2Mn(s)$

$\Delta H° = -910.9 + 0 - 0 - 2(-385.2) = -140.5 \text{ kJ}$

$\Delta S° = 41.84 + 2(32.0) - 18.7 - 2(59.7) = -32.3 \text{ J/K}$

$\Delta G° = -140.5 \text{ kJ} - 1473 \text{ K}(-0.0323 \text{ kJ/K}) = -92.9 \text{ kJ}$

(c) $FeO(s) + H_2(g) \rightarrow Fe(s) + H_2O(g)$

$\Delta H° = -241.82 + 0 - 0 - (-271.9) = +30.1 \text{ kJ}$

$\Delta S° = 27.15 + 188.7 - 60.75 - 130.58 = 24.5 \text{ J/K}$

$\Delta G° = 30.1 \text{ kJ} - 1473 \text{ K} (0.0245 \text{ J/K}) = (30.1 - 36.1) \text{ kJ} = -6.0 \text{ kJ}$

23.72 One important consideration is that molten iron itself not be oxidized too readily. If it were, too much iron oxide would appear in the product. On the other hand, the free energies of formation of the undesirable oxides must be sufficiently large and negative to permit nearly quantitative removal.

23.73 $2[Cu^+(aq) + 1e^- \rightarrow Cu(s)]$ $E^{\circ}_{red} = 0.521 \text{ V}$

          $Cu(s) \rightarrow Cu^{2+}(aq) + 2e^-$ $E^{\circ}_{red} = 0.337 \text{ V}$

$$2Cu^+(aq) \rightarrow Cu^{2+}(aq) + Cu(s) \quad\quad E^{\circ} = (0.521 \text{ V} - 0.337 \text{ V}) = 0.184 \text{ V}$$

According to Equation [20.18], $\log K = \dfrac{nE^{\circ}}{0.0592}$; $n = 2$

$\log K = \dfrac{2(0.184)}{0.0592} = 6.2162 = 6.22$; $K = 1.6 \times 10^6$

23.74 $1 \text{ kg ore} \times \dfrac{700 \text{ g Fe}_3\text{O}_4}{1 \text{ kg ore}} \times \dfrac{1 \text{ mol Fe}_3\text{O}_4}{231.6 \text{ g Fe}_3\text{O}_4} \times \dfrac{3 \text{ mol Fe}}{1 \text{ mol Fe}_3\text{O}_4} \times \dfrac{55.85 \text{ g Fe}}{1 \text{ mol Fe}}$

          $\times \dfrac{100 \text{ g pig iron}}{97.5 \text{ g Fe}} = 519 = 5.2 \times 10^2 \text{ g pig iron}$

Substantial shipping costs could be saved by shipping the pig iron rather than the raw ore. The northern refineries went out of business because the hardwood forests were decimated around the mills, and because reduction with coke became a much better procedure.

23.75 $\Delta G^{\circ} = -RT \ln K$; $\Delta G^{\circ} = \Delta H^{\circ} - T\,\Delta S^{\circ}$

Calculate ΔH° and ΔS° using data from Appendix C, assuming ΔH° and ΔS° remain constant with changing temperature. Then calculate ΔG° and K at the two temperatures.

$\Delta H^{\circ} = 2\Delta H_f CO(g) - \Delta H^{\circ}_f C(s) - \Delta H^{\circ}_f CO_2(g)$
$\Delta H^{\circ} = 2(-110.5) - 0 - (-393.5) = +172.5 \text{ kJ}$

$\Delta S^{\circ} = 2S^{\circ} CO(g) - S^{\circ} C(s) - S^{\circ} CO_2(g)$
      $= 2(197.9) - 5.69 - 213.6 = +176.5 \text{ J/K} = 0.1765 \text{ kJ/K}$

$\Delta G^{\circ}_{298} = 172.5 \text{ kJ} - 298 \text{ K}(0.1765 \text{ kJ/K}) = +119.9 \text{ kJ}$

$\ln K = \dfrac{\Delta G^{\circ}}{-RT} = \dfrac{119.9 \text{ kJ}}{-(8.314 \times 10^{-3} \text{ kJ/K})(298 \text{ K})} = -48.3942 = -48.39$; $K = 9.6 \times 10^{-22}$

$\Delta G^{\circ}_{2000} = 172.5 \text{ kJ} - 2000 \text{ K }(0.1765 \text{ kJ/K}) = -180.5 \text{ kJ}$

$\ln K = \dfrac{-180.5}{-(8.314 \times 10^{-3} \text{ kJ/K})(2000 \text{ K})} = 10.8552 = 10.86$; $K = 5.2 \times 10^4$

23.76 $97{,}000 \text{ A} \times 24 \text{ hr} \times \dfrac{3600 \text{ s}}{1 \text{ hr}} \times \dfrac{1 \text{ C}}{1 \text{ A} \cdot \text{s}} \times \dfrac{1 \text{ F}}{96{,}500 \text{ C}} \times \dfrac{1 \text{ mol Mg}}{2 \text{ F}} \times \dfrac{24.31 \text{ g Mg}}{1 \text{ mol Mg}} \times 0.96$

                                       $= 1.0 \times 10^6 \text{ g Mg} = 1.0 \times 10^3 \text{ kg Mg}$

23.77 Calculate the mass of Zn(s) that will be deposited.

$$1.0 \text{ m} \times 50 \text{ m} \times \frac{(100)^2 \text{ cm}^2}{1 \text{ m}^2} \times 0.30 \text{ mm} \times \frac{1 \text{ cm}}{10 \text{ mm}} \times \frac{7.1 \text{ g}}{\text{cm}^3} \times 2 \text{ sides}$$

$$= 2.13 \times 10^5 = 2.1 \times 10^5 \text{ g Zn}$$

$$2.13 \times 10^5 \text{ g Zn} \times \frac{1 \text{ mol Zn}}{65.38 \text{ g Zn}} \times \frac{2 \text{ F}}{0.90 \text{ mol Zn}} \times \frac{96,500 \text{ C}}{\text{F}} = 6.986 \times 10^8 = 7.0 \times 10^8 \text{ C}$$

(2 F/0.90 mol Zn takes the 90% efficiency into account.)

$$6.986 \times 10^8 \text{ C} \times 3.5 \text{ V} \times \frac{1 \text{ J}}{\text{C} \cdot \text{V}} \times \frac{1 \text{ kWh}}{3.6 \times 10^6 \text{ J}} = 679.2 = 6.8 \times 10^2 \text{ kWh}$$

$$679.2 \text{ kWh} \times \frac{\$0.080}{1 \text{ kWh}} = \$54.34 \rightarrow \$54$$

23.78 (a)   (See Exercise 17.49)

$$Ag_2(s) \rightleftharpoons 2Ag^+(aq) + S^{2-}(aq) \qquad\qquad K_{sp}$$

$$2[Ag^+(aq) + 2CN^-(aq) \rightleftharpoons Ag(CN)_2^-] \qquad\qquad K_f^2$$

$$\overline{Ag_2S(s) + 4CN^-(aq) \rightleftharpoons 2Ag(CN)_2^-(aq) + S^{2-}(aq)}$$

$$K = K_{sp} \times K_f^2 = [Ag^+]^2[S^{2-}] \times \frac{[Ag(CN)_2^-]^2}{[Ag^+]^2[CN^-]^4} = (6 \times 10^{-51})(1 \times 10^{21})^2 = 6 \times 10^{-9}$$

(b)   The equilibrium constant for the cyanidation of Ag_2S, 6×10^{-9}, is much less than one and favors the presence of reactants rather than products. The process is not practical.

(c) $$AgCl(s) \rightleftharpoons Ag^+(aq) + Cl^-(aq) \qquad\qquad K_{sp}$$

$$Ag^+(aq) + 2CN^-(aq) \rightleftharpoons Ag(CN)_2^-(aq) \qquad\qquad K_f$$

$$\overline{AgCl(s) + 2CN^-(aq) \rightleftharpoons Ag(CN)_2^-(aq) + Cl^-(aq)}$$

$$K = K_{sp} \times K_f = [Ag^+][Cl^-] \times \frac{[Ag(CN)_2^-]}{[Ag^+][CN^-]^2} = (1.8 \times 10^{-10})(1 \times 10^{21}) = 2 \times 10^{11}$$

Since K >> 1 for this process, it is potentially useful for recovering silver from horn silver. However the magnitude of K says nothing about the rate of reaction. The reaction could be slow and require heat, a catalyst or both to be practical.

24 Chemistry of Coordination Compounds

Structure and Nomenclature

24.1 (a)   Coordination number = 4, oxidation number = +2

     (b)   5, +4   (c) 6, +3   (d) 5, +2   (e) 6, +3   (f) 4, +2

24.2 (a)   Coordination number = 4, oxidation number = +1

     (b)   6, +3   (c) 4, +2   (d) 6, +3   (e) 6, +3   (f) 6, +3

24.3 (a)

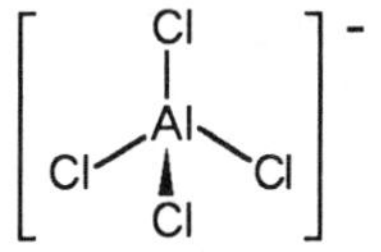

tetrahedral

(b)

linear

(c)

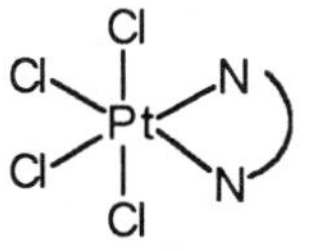

octahedral

(d)

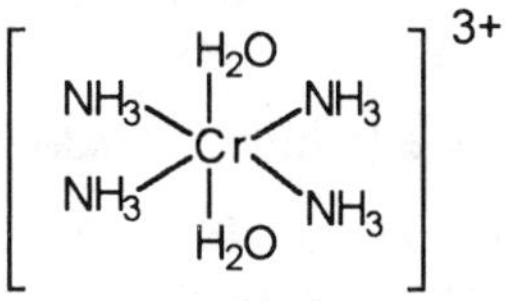

octahedral

24.4 (a)

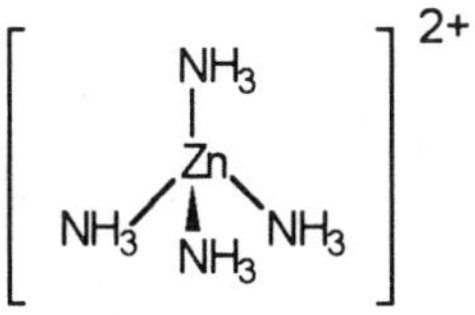

tetrahedral

(b)

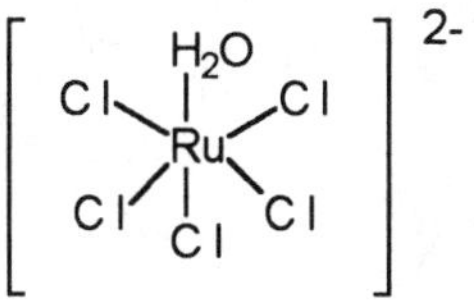

octahedral

(c)

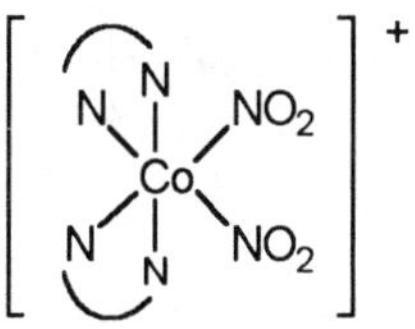

octahedral

(d)

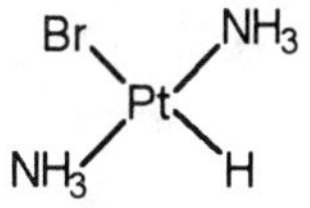

square planar

24.5 [24.3] (a)   tetrachloroaluminate(III)

   (b)   dicyanoargentate(I)

   (c)   tetrachloro(ethylenediamine)platinum(IV)

   (d)   *trans*-tetraamminediaquachromium(III)

[24.4] (a)   tetraamminezinc(II)

   (b)   aquapentachlororuthenate(III)

   (c)   *cis*-bis(ethylenediamine)dinitrocobalt(III)

   (d)   *trans*-diamminebromohydridoplatinum(II)

24.6 (a)   hexaaquanickel(II) bromide

   (b)   potassium dicyanoargentate

   (c)   tetraamminedichlorochromium(III) perchlorate

   (d)   potassium tris(oxalato)ferrate(III)

   (e)   diamminedibromo(ethylenendiamine)cobalt(III) chloride

   (f)   bis(ethylenediamine)palladium(II) diamminetetrabromochromate(III)

24.7 (a) $[Cr(NH_3)_6](NO_3)_3$ (b) $[Co(NH_3)_4CO_3]_2SO_4$ (c) $[Pt(en)_2Cl_2]Br_2$

   (d) $K[V(H_2O)_2Br_4]$ (e) $[Zn(en)_2][HgI_4]$

24.8 (a) $[Mn(H_2O)_5Br]SO_4$ (b) $[Ru(bipy)_3](NO_3)_2$ (c) $[Fe(o\text{-}phen)_2Cl_2]ClO_4$

   (d) $Na[Co(en)Br_4]$ (e) $[Ni(NH_3)_6]_3[Cr(ox)_3]_2$

24.9 (a)   *ortho*-phenanthroline (*o*-phen) is bidentate

   (b)   oxalate (ox), $C_2O_4^{2-}$ is bidentate

   (c)   ethylenediaminetetraacetate, EDTA, is hexadentate

   (d)   ethylenediamine (en) is bidentate

24.10 (a)   4   (b) 4   (c) 6   (d) 6

Isomerism

24.11 (a)

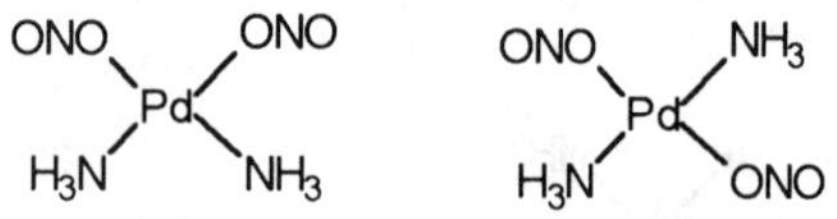

cis   *trans*

(b) $[Pd(NH_3)_2(ONO)_2]$, $[Pd(NH_3)_2(NO_2)_2]$

(c)

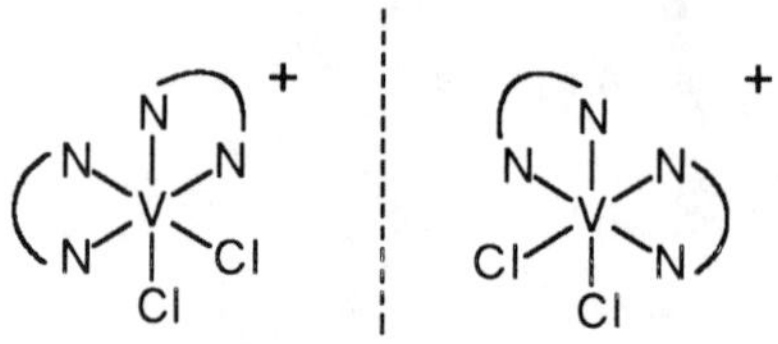

(d) $[Co(NH_3)_4Br_2]Cl$, $[Co(NH_3)_4BrCl]Br$

24.12   (a)

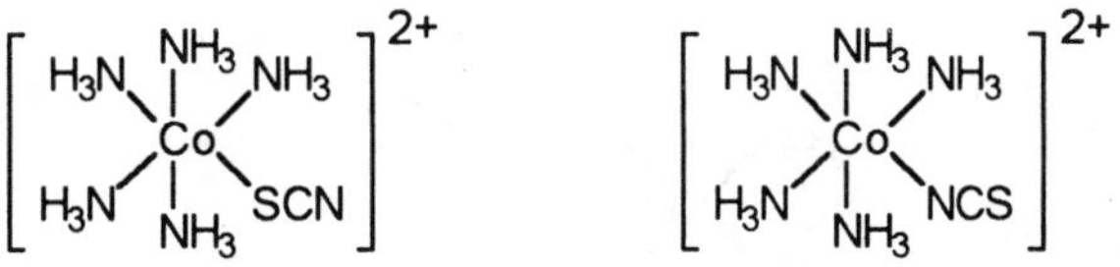

(b)

(c)

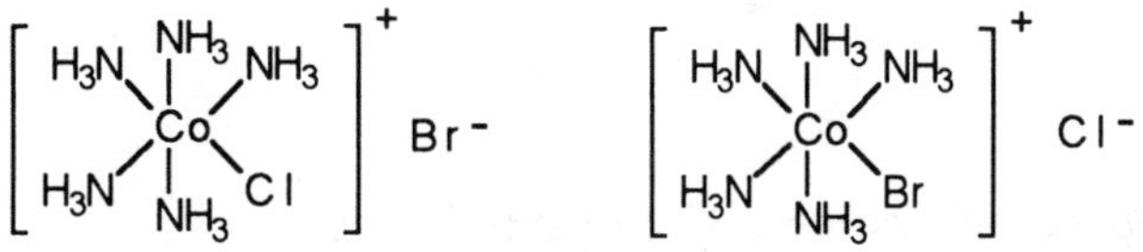

coordination sphere isomerism

24.13

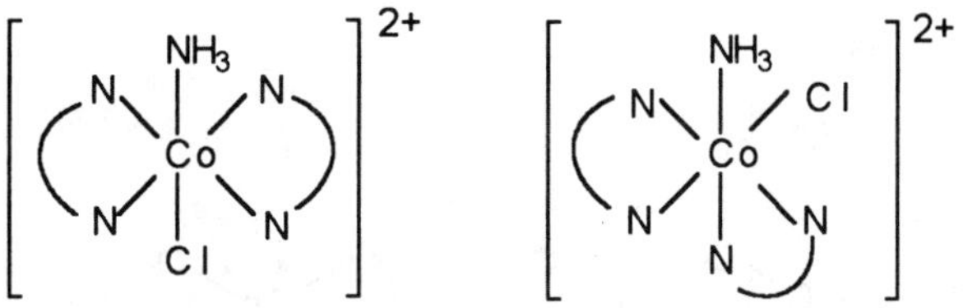

trans   *cis*

The *cis* isomer is chiral.

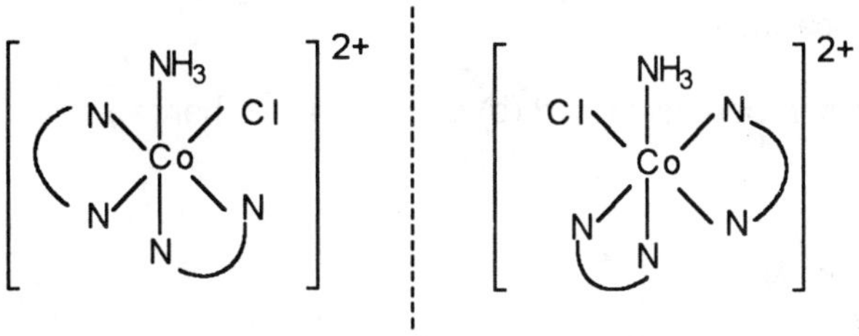

24.14   I

II

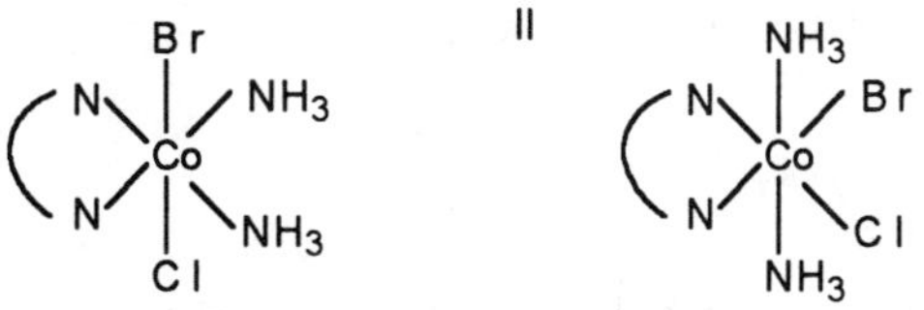

24.14 (continued)

IIIa, IIIb, 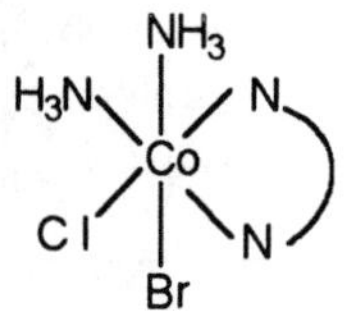 IVa, 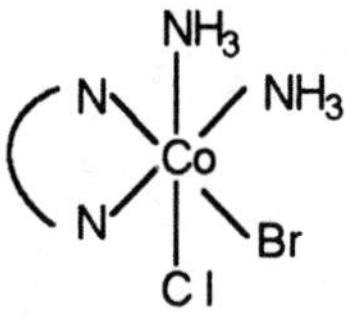 IVb structures

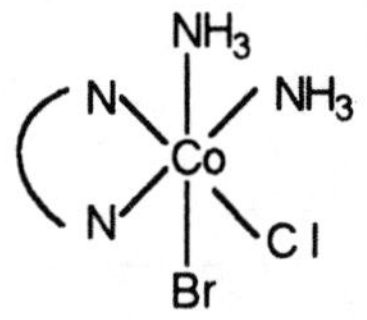

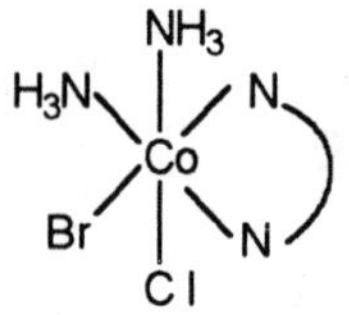

I, II, IIIa and IVa are geometric isomers; IIIa/IIIb and IVa/IVb
are the pairs of optical isomers.

24.15 (a)   only one:

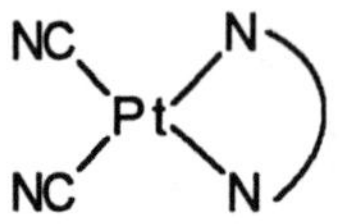

(b)

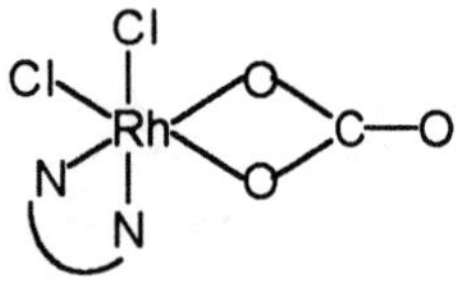

       cis 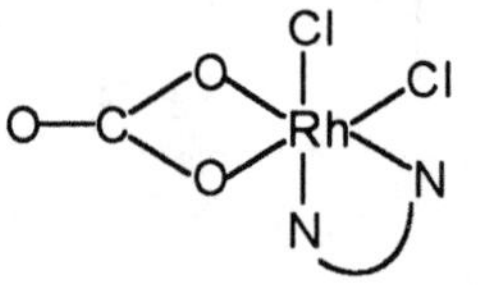   *cis*   *trans*
                     optical isomers

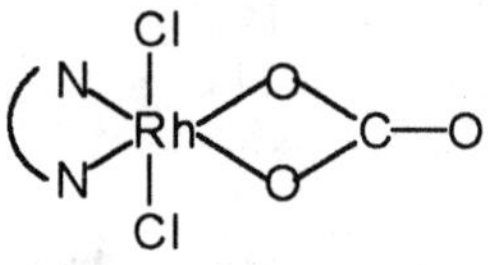

(The three isomeric compounds in part (b) all have a 1− charge.)

(c)   only one:

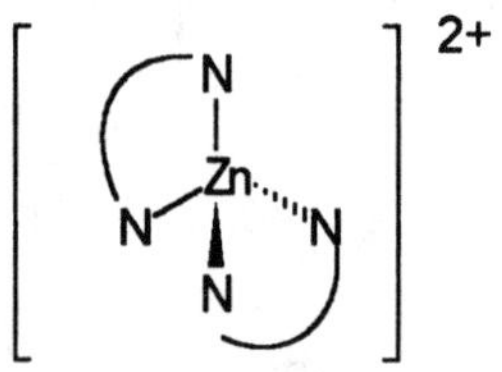

25.16 (a)

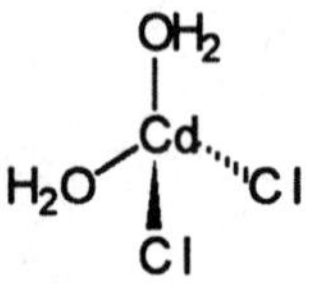

(b)

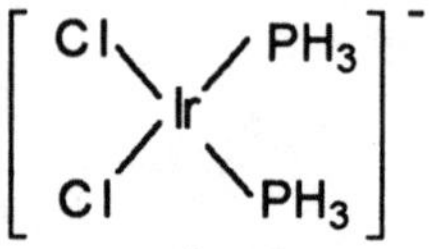

                           cis 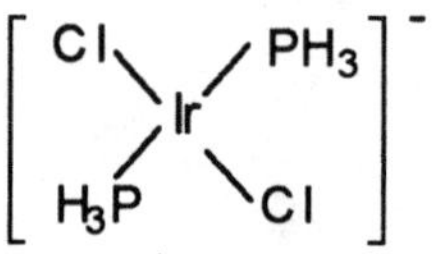   *trans*

(c)

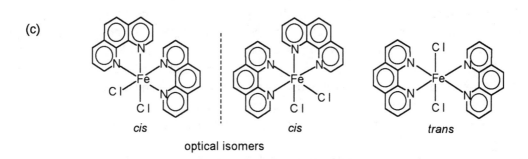

cis | *cis* | *trans*

optical isomers

(The three isomeric complex ions in part (c) each have a 1+ charge.)

Color, Magnetism; Crystal Field Theory

24.17   Color in transition metal compounds arises from electronic transitions between d-orbital energy levels, or d-d transitions. Compounds with d^0 or d^{10} electron configurations are colorless, because d-d transitions are not possible.

    (a) Zn^{2+}, d^{10}, colorless   (b) Cr^{4+}, d^2, colored   (c) Ni^{2+}, d^8, colored

    (d) Al^{3+}, d^0, colorless   (e) Cd^{2+}, d^{10}, colorless   (f) Fe^{2+}, d^6, colored

24.18   Assume that the simple salts dissociate in aqueous solution and that the metal ions exist as aqua complexes. Color arises from d^1 - d^9 metal ions.

    (a) Sc^{3+}, d^0, colorless   (b) Ni^{2+}, d^8, colored   (c) Sr^{2+}, d^0, colorless

    (d) V^{3+}, d^2, colored   (e) Rh^{3+}, d^6, colored   (f) Pb^{2+}, d^{10}, colorless

24.19   (a) Ru^{3+}, d^5 (b) Cu^{2+}, d^9 (c) Co^{3+}, d^6 (d) Mo^{5+}, d^1 (e) Re^{3+}, d^4

24.20   (a) Fe^{4+}, d^4 (b) Co^{2+}, d^7 (c) Ni^{2+}, d^8 (d) Au^{3+}, d^8 (e) Mo^{4+}, d^2

24.21   Blue to blue-violet (Figure 24.23)

24.22   Approximately 460 nm

24.23   Six ligands in an octahedral arrangement are oriented along the x, y and z axes of the metal. These negatively charged ligands (or the negative end of ligand dipoles) have greater electrostatic repulsion with valence electrons in metal orbitals that also lie along these axes, the d_{z^2}, and $d_{x^2-y^2}$. The d_{xy}, d_{xz} and d_{yz} metal orbitals point between the x, y, and z axes, and electrons in these orbitals experience less repulsion with ligand electrons. Thus, in the presence of an octahedral ligand field, the d_{xy}, d_{xz} and d_{xy} metal orbitals are lower in energy than the $d_{x^2-y^2}$ and d_{z^2}.

24.24   The ligands that possess the greatest ability to interact with the central metal atom cause the largest splitting. The properties of ligands that are important are **charge** (usually the more negatively charged ligands produce larger splittings) and **polarizability**, which measures the ability of the ligand to distort its charge distribution as it interacts with the positively charged metal ion.

24.25 Cyanide is a strong field ligand. The d-d electronic transitions occur at relatively high energy, because Δ is large. A yellow color corresponds to absorption of a photon in the violet region of the visible spectrum, between 430 and 400 nm. H_2O is a weaker field ligand than CN^-. The blue or green colors of aqua complexes correspond to absorptions in the region of 620 nm. Clearly, this is a region of lower energy photons than those with characteristic wavelengths in the 430 to 400 nm region. These are very general and imprecise comparisons. Other factors are involved, including whether the complex is high spin or low spin.

24.26 The ions absorb the complement of the color they appear. Green $[Ni(H_2O)_6]^{2+}$ absorbs red light, 650-800 nm. Purple $[Ni(NH_3)_6]^{2+}$ absorbs yellow light, 560-580 nm. Thus, $[Ni(NH_3)_6]^{2+}$ absorbs light with the shorter wavelength. This agrees with the spectrochemical series, which indicates that H_2O will produce a smaller d orbital splitting (Δ) than NH_3. Thus, $[Ni(H_2O)_6]^{2+}$ should absorb light with a smaller energy and longer wavelength.

24.27 (a) Mn - $[Ar]4s^2 3d^5$ (b) Ru - $[Kr]5s^2 4d^6$ (c) Rh - $[Kr]5s^2 4d^7$
          Mn^{3+} - $[Ar]3d^4$ Ru^{3+} - $[Kr]4d^5$ Rh^{3+} - $[Kr]4d^6$

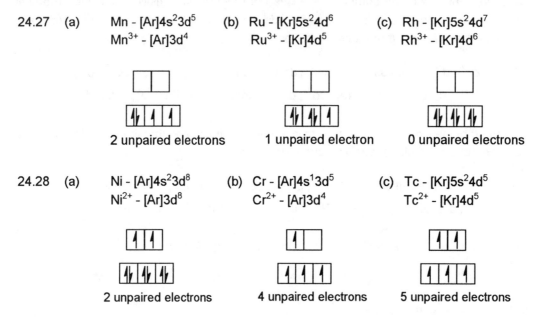

24.28 (a) Ni - $[Ar]4s^2 3d^8$ (b) Cr - $[Ar]4s^1 3d^5$ (c) Tc - $[Kr]5s^2 4d^5$
          Ni^{2+} - $[Ar]3d^8$ Cr^{2+} - $[Ar]3d^4$ Tc^{2+} - $[Kr]4d^5$

24.29 All complexes in this exercise are six-coordinate octahedral.

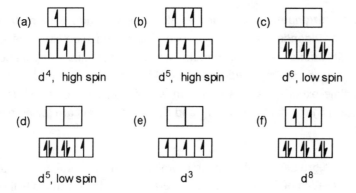

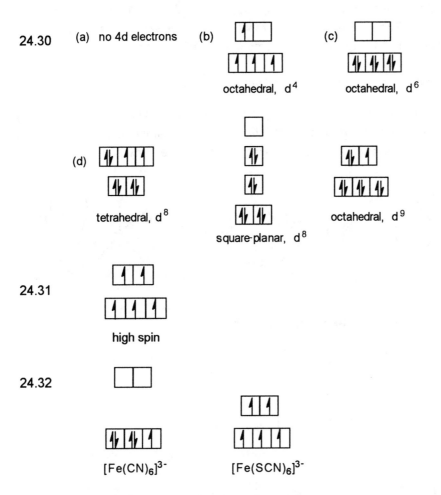

24.30   (a) no 4d electrons   (b)

octahedral, d^4 octahedral, d^6

(d)

tetrahedral, d^8 square-planar, d^8 octahedral, d^9

24.31

high spin

24.32

$[Fe(CN)_6]^{3-}$ $[Fe(SCN)_6]^{3-}$

Both complexes contain Fe^{3+}, a d^5 ion. CN^-, a strong field ligand, produces such a large Δ that the splitting energy is greater than the pairing energy, and the complex is low spin. SCN^- produces a smaller Δ, so it is energetically favorable for d electrons to be unpaired in the higher energy d orbitals.

Additional Exercises

24.33   (a) $K_2[Ni(en)Cl_4]$; $[Ni(en)(H_2O)_2Cl_2]$

(b) $K_2[Ni(CN)_4]$; $[Zn(H_2O)_4](NO_3)_2$; $[Cu(NH_3)_4]SO_4$

(c) $[CoF_6]^{3-}$, high spin; $[Co(NH_3)_6]^{3+}$ or $[Co(CN)_6]^{3-}$, low spin

(d)   thiocyanate, SCN^- or NCS^-; nitrite, NO_2^- or ONO^-

(e) $[Co(en)_2Cl_2]Cl$; see Exercise 24.15(b) and 24.16(c) for other examples.

(f) $[Co(en)_3]Cl_3$, $[Cr(NH_3)_6]Cl_3$

24.34 $[Pt(NH_3)_6]Cl_4$; $[Pt(NH_3)_4Cl_2]Cl_2$; $[Pt(NH_3)_3Cl_3]Cl$; $[Pt(NH_3)_2Cl_4]$; $K[Pt(NH_3)Cl_5]$

24.35   The reaction that occurs produces an ion of higher charge. It would appear from the relative values that the reaction could be

$$[Co(NH_3)_4Br_2]^+(aq) + H_2O(l) \rightarrow [Co(NH_3)_4(H_2O)Br]^{2+}(aq) + Br^-(aq)$$

This reaction would convert the 1:1 electrolyte, $[Co(NH_3)_4Br_2]Br$, to a 1:2 electrolyte, $[Co(NH_3)_3(H_2O)Br]Br_2$. The reaction is not extremely rapid, so we would classify the starting complex as relatively inert, though it does undergo ligand replacement.

24.36   (a)   linkage isomerism   (b) coordination sphere isomerism

24.37   In a square planar complex such as $[Pt(en)_2Cl_2]$, if one pair of ligands is *trans*, the remaining two coordination sites are also *trans* to each other. Ethylenediamine is a relatively short bidentate ligand that cannot occupy *trans* coordination sites, so the *trans* isomer is unknown.

24.38   (a)   (b) 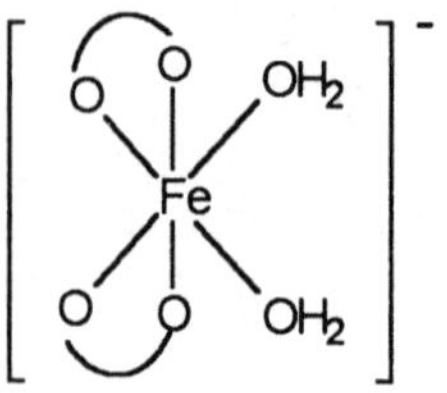   (c)

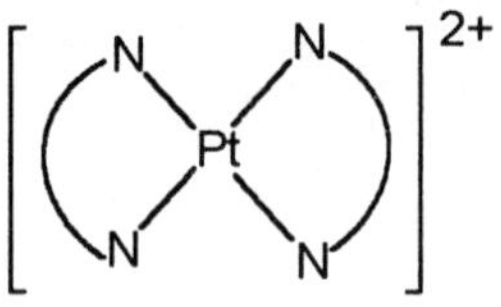

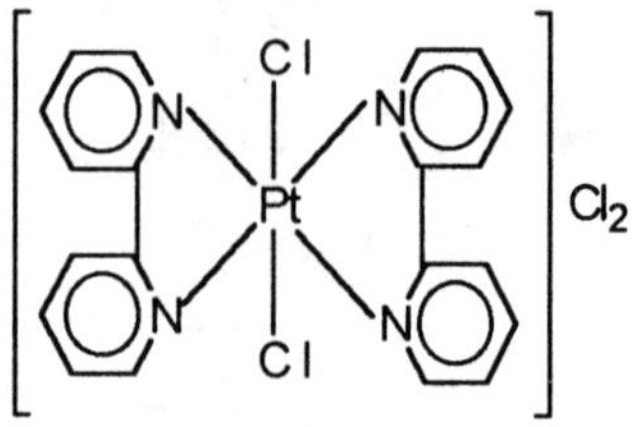

    (d) Hg^{2+} is d^{10}, so the complex is probably tetrahedral.

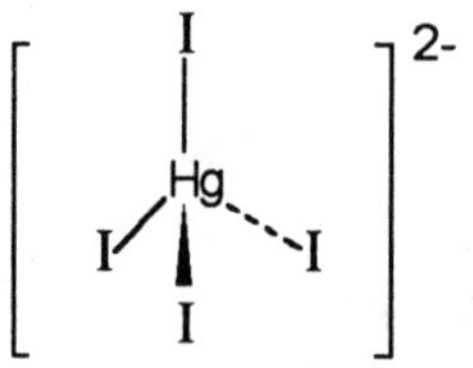

    (e) 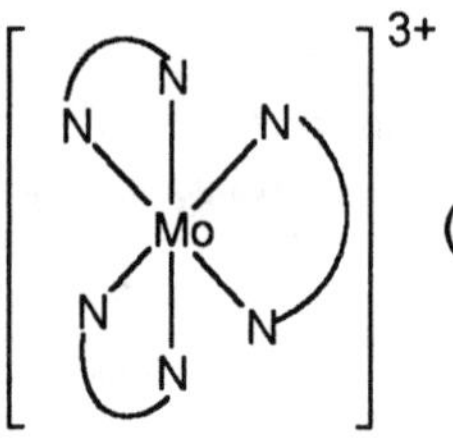   (f) 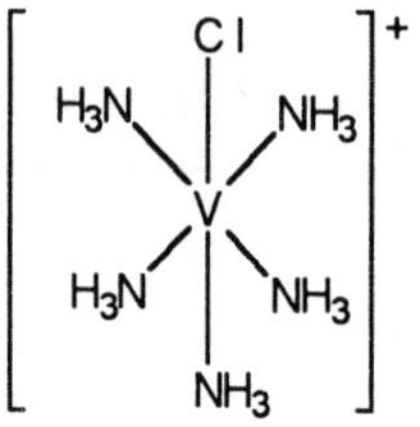   (g)

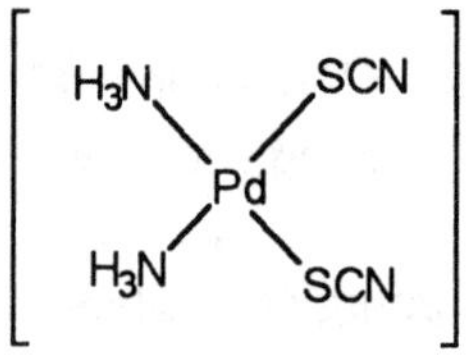

24.39   We will represent the end of the bidentate ligand containing the CF_3 group by a shaded oval, the other end by an open oval:

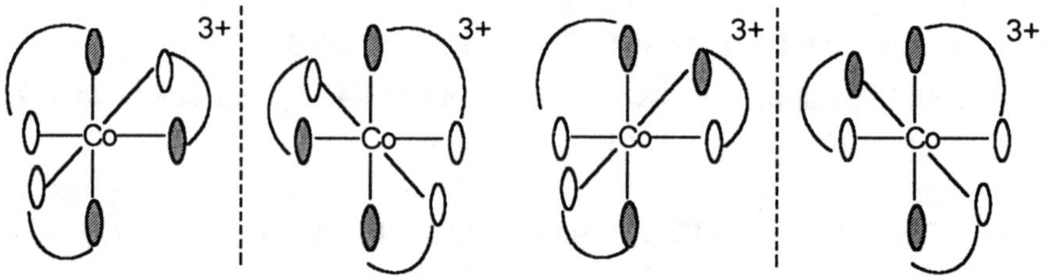

508

24.40 (a) Only one

   (b) Two; (*cis* or *trans* arrangement of N and O ends)

   (c) Four; two are geometrical, the other two are steroisomers of each of these; the figure for solution to Exercise 24.39 applies to this problem as well.

24.41 (a) $AgCl(s) + 2NH_3(aq) \rightarrow [Ag(NH_3)_2]^+(aq) + Cl^-(aq)$

   (b) $[Cr(en)_2Cl_2]Cl(aq) + 2H_2O(l) \rightarrow [Cr(en_2)(H_2O)_2]^{3+} + 3Cl^-(aq)$

      green   brown-orange

     $3Ag^+(aq) + 3Cl^-(aq) \rightarrow 3AgCl(s)$

     $[Cr(en)_2(H_2O)_2]^{3+}$ and $3NO_3^-$ are spectator ions in the second reaction.

   (c) $Zn(OH)_2(s) + 4NH_3(aq) \rightarrow [Zn(NH_3)_4]^{2+}(aq) + 2OH^-(aq)$

   (d) $Co^{2+}(aq) + 4Cl^-(aq) \rightarrow [CoCl_4]^{2-}(aq)$

24.42 The order given is in the direction of increasing Δ in the spectrochemical series. Thus the absorption spectra of $[VF_6]^{3-}$, $[V(NH_3)_6]^{3+}$ and $[V(NO_2)_6]^{3-}$ should exhibit a trend toward absorptions at shorter wavelength (higher energy).

24.43 According to the spectrochemical series, the order of increasing Δ for the ligands is $Cl^- < H_2O < NH_3$. (The tetrahedral Cl^- complex will have an even smaller Δ than an octahedral one.) The smaller the value of Δ, the longer the wavelength of visible light absorbed. The color of light absorbed is the complement of the observed color. A blue complex absorbs orange light (580-650 nm), a pink complex absorbs green light (490-560 nm) and a yellow complex absorbs violet light (400-430 nm). Since $[CoCl_4]^{2-}$ absorbs the longest wavelength, it appears blue. $[Co(H_2O)_6]^{2+}$ absorbs green and appears pink, and $[Co(NH_3)_6]^{3+}$ absorbs violet and appears yellow.

24.44 oxyhemoglobin   deoxyhemoglobin
     $Fe^{2+}: d^6$ $Fe^{2+}: d^6$

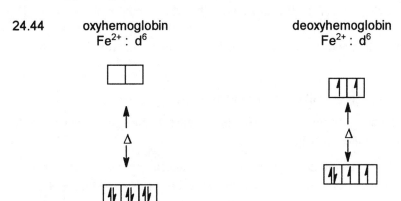

  low spin, no unpaired electrons   high spin, 4 unpaired electrons

  In general, the crystal field splitting, Δ, is greater in low spin than high spin complexes. The energy, Δ, corresponds to the wavelength of light absorbed by the complex and determines its color. Since oxyhemoglobin absorbs higher energy, shorter wavelength light, longer wavelengths remain and the sample appears red. Deoxyhemoglobin absorbs lower energy (orange-red) light, and the sample appears blue.

24.45   (a)   A square-planar complex, d^8 (b)   Octahedral, d^8

(c)   Octahedral, d^4, surely low spin   (d)   Octahedral, d^5; low spin

24.46   (a)   False. The spin pairing energy is **smaller** than Δ in low spin complexes. It is for this reason that electrons pair up in the lower energy orbitals, in spite of the repulsive energy associated with spin pairing, rather than move to the higher energy orbital, which would cost energy in the amount Δ.

(b)   True. Higher metal ion charge causes the ligands to be more strongly attracted, thus producing a larger splitting of the d orbital energies.

(c)   False. Square-planar configurations are associated with a d^8 electron configuration (which Ni^{2+} has) and a ligand that produces a strong field, leading to large separations between the energy levels. Because cyanide, CN^-, is a much stronger field ligand than Cl^-, $[Ni(CN)_4]^{2-}$ is more likely to be square planar than is $[NiCl_4]^{2-}$.

24.47   Longer wavelength absorption corresponds to a smaller splitting of the energies of the d orbitals.

(a) $[CoF_6]^{4-}$, because F^- is a weaker field ligand than CN^-.

(b) $[V(H_2O)_6]^{2+}$, because the metal ion of lower charge attracts the ligands less strongly, thus produces a smaller splitting.

(c) $[MnCl_4]^-$, because in a tetrahedral complex the ligand field is much smaller than in an octahedral complex; furthermore, Cl^- is a much weaker field ligand than CN^-.

24.48   Application of pressure would result in shorter metal ion-oxide distances. This would have the effect of increasing the ligand-electron repulsions, and would result in a larger splitting in the d orbital energies. Thus, application of pressure should result in a shift in the absorption to a higher energy (shorter wavelength).

24.49   The d^3 and d^6 electron configurations in octahedral and tetrahedral fields are shown below:

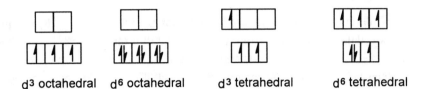

     d^3 octahedral d^6 octahedral d^3 tetrahedral d^6 tetrahedral

In an octahedral environment, d^3 and d^6 strong field configurations have electrons only in the d orbitals which are **lower** in energy than the five degenerate orbitals in the isolated metal atom. In a tetrahedral environment (always weak field, see Section 24.6) d^3 has one electron and d^6 has three electrons in the higher energy d orbitals. Thus the octahedral environment is energetically more favorable for metal ions with d^3 or d^6 configurations.

24.50

$$d_{z^2}, \quad d_{yz}, \quad d_{xz}$$

$$d_{xy}, \quad d_{x^2-y^2}$$

For a d^6 metal ion in a strong ligand field, there would be two unpaired electrons.

Integrative Exercises

24.51   Cobalt(III) complexes are generally inert; that is, they do not rapidly exchange ligands inside the coordination sphere. Therefore, the ions that form precipitates in these two cases are probably outside the coordination sphere. The red complex can be formulated as $[Co(NH_3)_5SO_4]Br$, pentaamminesulfatocobalt(III) bromide, and the violet compound as $[Co(NH_3)_5Br]SO_4$, pentaamminebromocobalt(III) sulfate.

red compound   violet compound

24.52   First determine the empirical formula, assuming that the remaining mass of complex is Pd.

		mol	mole ratios
37.6 g Br × $\dfrac{1\ mol\ Br}{79.904\ g\ Br}$		= 0.4706 mol Br	2
28.3 g C × $\dfrac{1\ mol\ C}{12.01\ g\ C}$		= 2.356 mol C	10
6.60 g N × $\dfrac{1\ mol\ N}{14.01\ g\ N}$		= 0.4711 mol N	2
2.37 g H × $\dfrac{1\ mol\ H}{1.008\ g\ H}$		= 2.351 mol H	10
25.13 g Pd × $\dfrac{1\ mol\ Pd}{106.42\ g\ Pd}$		= 0.2361 mol Pd	1

The chemical formula is $Pd(NC_5H_5)_2Br_2$. This should be a square-planar complex of Pd(II), a nonelectrolyte. Because the dipole moment is zero, we can infer that it must be the *trans*-isomer, *trans*-dibromodipyridinepalladium(II).

24.53 Determine the empirical formula of the complex, assuming the remaining mass is due to oxygen, and a 100 g sample.

$$10.0 \text{ g Mn} \times \frac{1 \text{ mol Mn}}{54.94 \text{ g Mn}} = 0.1820 \text{ mol Mn}; \; 0.182 / 0.182 = 1$$

$$28.6 \text{ g K} \times \frac{1 \text{ mol K}}{39.10 \text{ g K}} = 0.7315 \text{ mol K}; \; 0.732 / 0.182 = 4$$

$$8.8 \text{ g C} \times \frac{1 \text{ mol C}}{12.0 \text{ g C}} = 0.7327 \text{ mol C}; \; 0.733 / 0.182 = 4$$

$$29.2 \text{ g Br} \times \frac{1 \text{ mol Br}}{79.904 \text{ g Br}} = 0.3654 \text{ mol Br}; \; 0.365 / 0.182 = 2$$

$$23.4 \text{ g O} \times \frac{1 \text{ mol O}}{16.00 \text{ g O}} = 1.463 \text{ mol O}; \; 1.46 / 0.182 = 8$$

There are 2 C and 4 O per oxalate ion, for a total of two oxalate ligands in the complex. To match the conductivity of $K_4[Fe(CN)_6]$, the oxalate and bromide ions must be in the coordination sphere of the complex anion. Thus, the compound is $K_4[Mn(ox)_2Br_2]$, potassium dibromobis(oxalato)manganate(II).

24.54 Calculate the concentration of Mg^{2+} alone, and then the concentration of Ca^{2+} by difference. $M \times L = \text{mol}$

$$\frac{0.0104 \text{ mol EDTA}}{1 \text{ L}} \times 0.0187 \text{L} \times \frac{1 \text{ mol Mg}^{2+}}{1 \text{ mol EDTA}} \times \frac{24.31 \text{ g Mg}^{2+}}{1 \text{ mol Mg}^{2+}} \times \frac{1000 \text{ mg}}{\text{g}}$$

$$\times \frac{1}{0.100 \text{ L H}_2\text{O}} = 47.28 = 47.3 \text{ mg Mg}^{2+}/\text{L}$$

$0.0104 \; M \text{ EDTA} \times 0.0315 \text{ L} = \text{mol } (Ca^{2+} + Mg^{2+})$

$0.0104 \; M \text{ EDTA} \times 0.0187 \text{ L} = \text{mol } Mg^{2+}$

$0.0104 \; M \text{ EDTA} \times 0.0128 \text{ L} = \text{mol } Ca^{2+}$

$$0.0104 \; M \text{ EDTA} \times 0.0128 \text{ L} \times \frac{1 \text{ mol Ca}^{2+}}{1 \text{ mol EDTA}} \times \frac{40.08 \text{ g Ca}^{2+}}{1 \text{ mol Ca}^{2+}} \times \frac{1000 \text{ mg}}{\text{g}} \times \frac{1}{0.100 \text{ L H}_2\text{O}}$$

$$= 53.35 = 53.4 \text{ mg Ca}^{2+}/\text{L}$$

24.55 $\dfrac{182 \times 10^3 \text{ J}}{1 \text{ mol}} \times \dfrac{1 \text{ mol}}{6.022 \times 10^{23} \text{ molecules}} = 3.022 \times 10^{-19} = 3.02 \times 10^{-19}$ J/photon

$\Delta E = h\nu = 3.02 \times 10^{-19}$ J; $\nu = \Delta E/h$

$\nu = 3.022 \times 10^{-19} \text{ J}/6.626 \times 10^{-34} \text{ J}\cdot\text{s} = 4.561 \times 10^{14} = 4.56 \times 10^{14} \text{ s}^{-1}$

$\lambda = \dfrac{2.998 \times 10^8 \text{ m/s}}{4.561 \times 10^{14} \text{ s}^{-1}} = 6.57 \times 10^{-7}$ m = 657 nm

We expect that this complex will absorb in the visible, at around 660 nm. It will thus exhibit a blue-green color (Figure 24.23).

24.56 $\Delta E = hc/\lambda = \dfrac{6.626 \times 10^{-34} \text{ J}\cdot\text{s} \times 2.998 \times 10^8 \text{ m/s}}{510 \times 10^{-9} \text{ m}} = 3.895 \times 10^{-19} = 3.90 \times 10^{-19}$ J/photon

$\Delta = 3.895 \times 10^{-19}$ J/photon $\times \dfrac{6.022 \times 10^{23} \text{ photons}}{1 \text{ mol}} \times \dfrac{1 \text{ kJ}}{1000 \text{ J}} = 234.6 = 235$ kJ/mol

24.57   The process can be written:

$$H_2(g) + 2e^- \rightarrow 2H^+(aq) \qquad\qquad E^\circ_{red} = 0.0 \text{ V}$$
$$Cu(s) \rightarrow Cu^{2+} + 2e^- \qquad\qquad E^\circ_{red} = 0.337 \text{ V}$$
$$Cu^{2+}(aq) + 4NH_3(aq) \rightarrow [Cu(NH_3)_4]^{2+}(aq) \qquad\qquad \text{"}E^\circ_f\text{"} = ?$$

$$H_2(g) + Cu(s) + 4NH_3(aq) \rightarrow 2H^+(aq) + [Cu(NH_3)_4]^{2+}(aq) \qquad E = 0.08 \text{ V}$$

$E = E^\circ - RT \ln K;$ $K = \dfrac{[H^+]^2[Cu(NH_3)_4^{2+}]}{P_{H_2}[NH_3]^4}$

$P_{H_2} = 1$ atm, $[H^+] = 1$ M, $[NH_3] = 1$ M, $[Cu(NH_3)_4]^{2+} = 1$ M, $K = 1$

$E = E^\circ - RT \ln(1);$ $E = E^\circ - RT(0);$ $E = E^\circ = 0.08$ V

Since we know E° values for two steps and the overall reaction, we can calculate "E°" for the formation reaction and then K_f, using $E^\circ = \dfrac{0.0592}{n} \log K$ for the step.

$E_{cell} = 0.08$ V $= 0.0$ V $- 0.337$ V $+$ "E°_f"   "E°_f" $= -0.08$ V $+ 0.337$ V $= 0.417$ V $= 0.42$ V

"E°_f" $= \dfrac{0.0592}{n} \log K_f;$ $\log K_f = \dfrac{n(E^\circ_f)}{0.0592} = \dfrac{2(0.417)}{0.0592} = 14.0878 = 14.09$

$K_f = 10^{14.0878} = 1.2 \times 10^{14}$

25 The Chemistry of Life: Organic and Biological Chemistry

Hydrocarbon Structures and Nomenclature

25.1 (a) $CH_3CH_2CH_2CH_2CH_3$, C_5H_{12} (b)

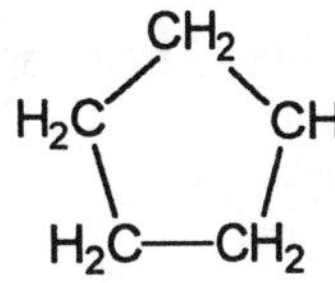

, C_5H_{10}

(c) $CH_2=CHCH_2CH_2CH_3$, C_5H_{10} (d) $HC \equiv C-CH_2CH_2CH_3$, C_5H_8

saturated: (a), (b); unsaturated: (c), (d)

25.2 alkane,

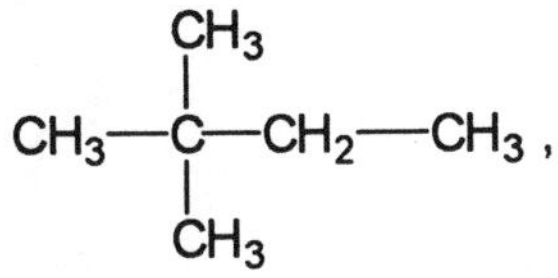

, C_6H_{14}, saturated

cycloalkane,

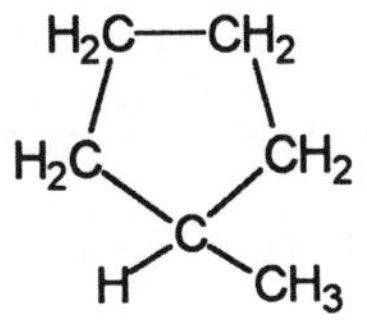

C_6H_{12}, saturated

alkene,

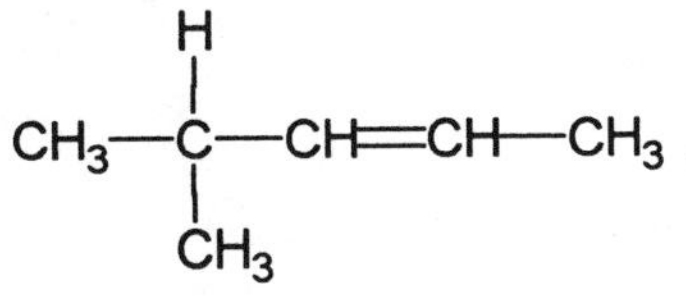

, C_6H_{12}, unsaturated

alkyne, $CH_3-CH_2-C \equiv C-CH_2-CH_3$, C_6H_{10}, unsaturated

aromatic hydrocarbon,

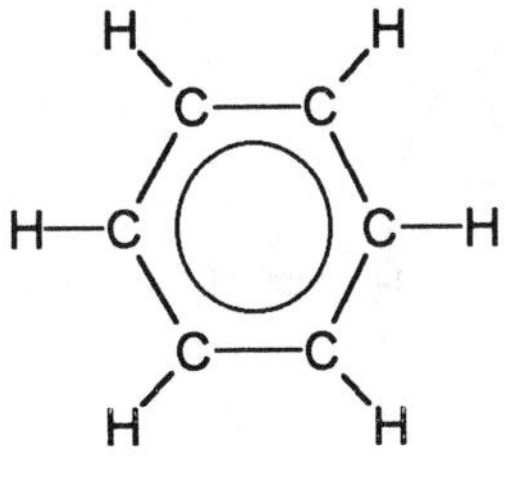

, C_6H_6, unsaturated

25.3 Alkanes have the generic formula C_nH_{2n+2}. C_3H_9 does not have a C : H mole ratio of n : 2n+2; it has too many hydrogen atoms. A molecular formula of C_3H_9 would require at least one carbon atom to form five bonds; carbon never adopts an expanded octet of electrons.

25.4 The structural formula for $CH_3=CHCH_2CH_3$ is

The leftmost C atom has more than four bonds; it violates the octet rule.

25.5 There are five isomers. Their carbon skeletons are as follows:

25.6 (a)

(b)

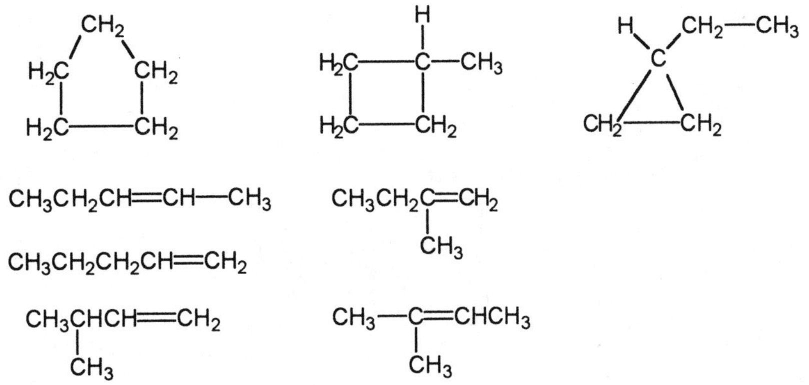

$$CH_3CH_2CH\!=\!CH\!-\!CH_3 \qquad CH_3CH_2C\!=\!CH_2$$
$$\qquad\qquad\qquad\qquad\qquad\qquad\qquad\quad | \atop CH_3$$

$$CH_3CH_2CH_2CH\!=\!CH_2$$

$$CH_3CHCH\!=\!CH_2 \qquad CH_3\!-\!C\!=\!CHCH_3$$
$$\quad\ | \qquad\qquad\qquad\qquad\qquad\ | \atop CH_3 \qquad\qquad\qquad\qquad\qquad CH_3$$

25.7 (a)   2,2,4-trimethylpentane   (b) 3-ethyl-2-methylpentane

(c)   2,3,4-trimethylpentane   (d) 2,3,4-trimethylpentane

(c) and (d) are the same molecule

25.8 (a)   3-ethylpentane   (b) 2,3-dimethylpentane

(c)   3,3-dimethylpentane   (d) 2,3-dimethylpentane

(b) and (d) are the same substance

25.9 (a)   109°   (b) 120°   (c) 180°

25.10 (a) sp^3 (b) sp^2 (c) sp^2 (d) sp

25.11 (a)   2,3-dimethylhexane   (b) 4-ethyl-2,4-dimethylnonane

(c)   3,3,5-trimethylheptane   (d) 3,4,4-trimethylheptane

25.12 (a)

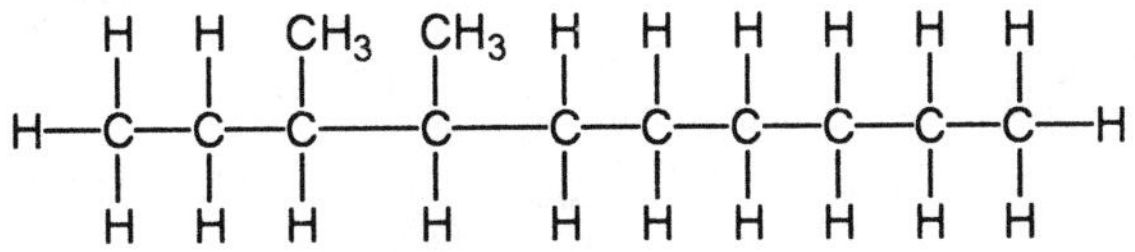

(b)

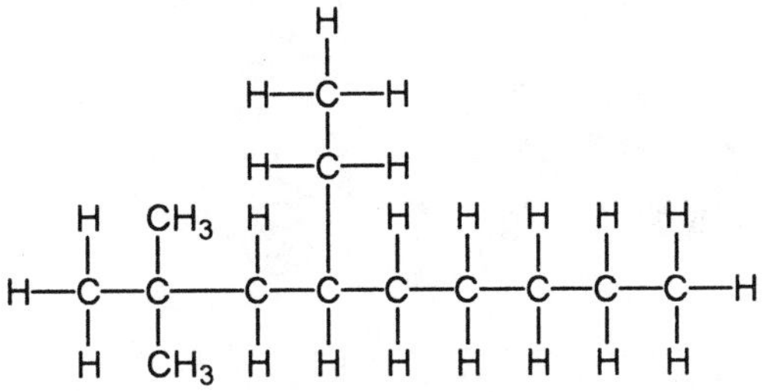

(c)

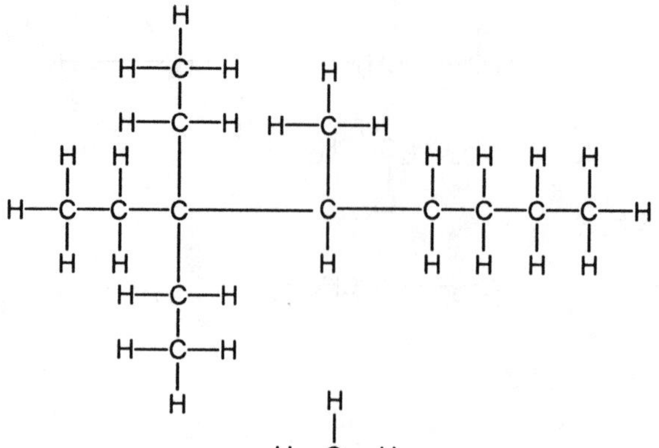

(d)

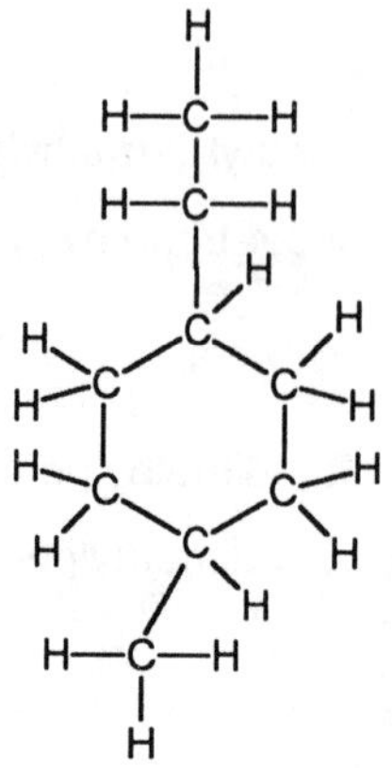

25.13　(a)

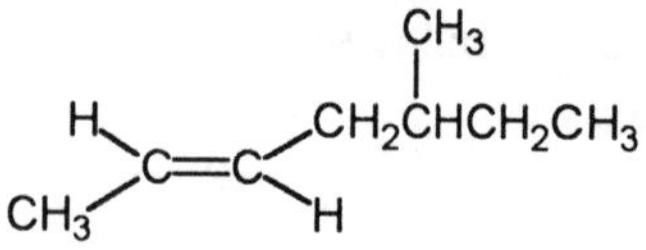

(b) HC≡C—CH₂Cl

(c)

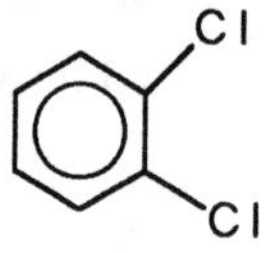

(d) 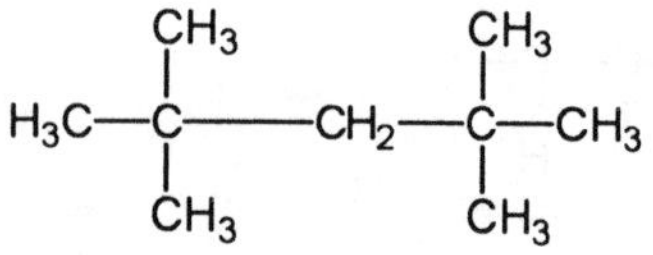 H₃C—C—CH₂—C—CH₃

(e) 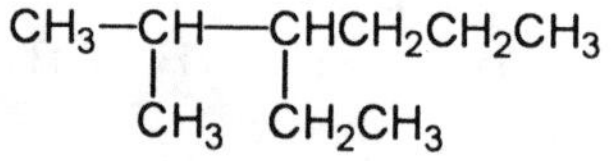 CH₃—CH—CHCH₂CH₂CH₃

(f) 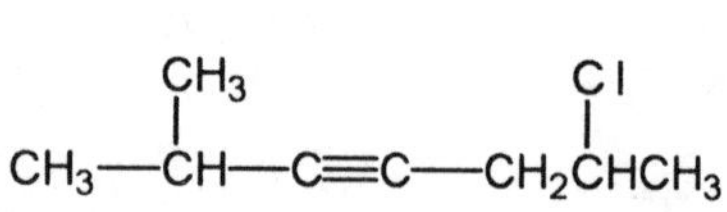 CH₃—CH—C≡C—CH₂CHCH₃

(g)

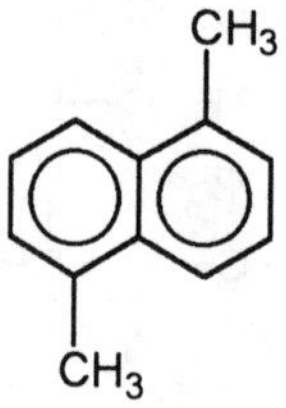

(h) CH₂=CHCH₂CH₂CH₂CH=CH₂

25.14   **(a)** $CH_3\overset{\overset{\displaystyle CH_3}{|}}{\underset{\underset{\displaystyle CH_3}{|}}{C}}CH_2CH_2CH_3$ **(b)** $CH_3\overset{}{C}HCHCH_2CH_2CH_3$ **(c)**

   (d)   **(e)** $CH_3\underset{\underset{\displaystyle Cl}{|}}{C}HCH_2CH_3$ **(f)**

   (g)   **(h)** $CH_3C\equiv C\underset{\underset{\displaystyle CH_3}{|}}{C}HCH_3$

25.15   **(a)**   2,3-dimethylheptane   **(b)** *cis*-6-methyl-3-octene   **(c)** *para*-dibromobenzene

   (d)   4,4-dimethyl-1-hexyne   **(e)** methylcyclobutane

25.16   **(a)**   2,4-dibromopentane   **(b)** 2,5-octadiene   **(c)** 2-phenyl-3,5-dichlorohexane

   (d)   1-chloro-2-cyclopentylpropane   **(e)** *trans*-dichloroethane

25.17   Butene is an alkene, C_4H_8. There are two possible placements for the double bond:

$$CH_2{=}CHCH_2CH_3 \text{ or } CH_3CH{=}CHCH_3$$
$$\text{1-butene} \qquad\qquad \text{2-butene}$$

These two compounds are structural isomers. For 2-butene, there are two different, noninterchangeable ways to construct the carbon skeleton (owing to the absence of free rotation around the double bond). These two compounds are geometric isomers.

     cis-2-butene   *trans*-2-butene

25.18   Each doubly bound carbon atom in an alkene has two unique sites for substitution. These sites cannot be interconverted because rotation about the double bond is restricted; geometric isomerism results. In an alkane, carbon forms only single bonds, so the three remaining sites are interchangeable by rotation about the single bond. Although there is also restricted rotation around the triple bond of an alkyne, there is only one additional bonding site on a triply bound carbon, so no isomerism results.

25.19 $C_3H_4Cl_2$

Cl, H / Cl C=C CH₃	H, Cl / Cl C=C CH₃	Cl, Cl / H C=C CH₃
I	II	III
1,1-dichloropropene	*trans*-1,2-dichloropropene	*cis*-1,2-dichloropropene

H, Cl / H C=C CH₂Cl	H, H / H C=C CHCl₂	(cyclopropane ring)
IV	V	VI
2,3-dichloropropene	3,3-dichloropropene	1,1-dichlorocyclopropane

(cyclopropane ring)	(cyclopropane ring)
VII	VIII
cis-1,2-dichlorocyclopropane	*trans*-1,2-dichlorocyclopropane

Compounds II / III and VII / VIII are geometric isomers; the others are structural isomers (different placement of Cl atoms.)

25.20   (a) no   (b)

cis   *trans*

(c) no   (d) no

25.21   Assuming that each component retains its effective octane number in the mixture (and this isn't always the case), we obtain: octane number = 0.30(0) + 0.70(100) = 70.

25.22   Octane number can be increased by increasing the fraction of branched-chain alkanes or aromatics, since these have high octane numbers. This can be done by cracking. The octane number also can be increased by adding an anti-knock agent such as tetraethyl lead, $Pb(C_2H_5)_4$ (no longer legal), methyl t-butyl ether, MTBE, or an alcohol, methanol or ethanol.

Reactions of Hydrocarbons

25.23   (a)   A combustion reaction is the oxidation-reduction reaction of some substance (fuel) with $O_2(g)$.

$$2C_2H_6(g) + 7O_2(g) \rightarrow 4CO_2(g) + 6H_2O(g)$$

(b)   An addition reaction is the addition of some reagent to the two atoms that form a multiple bond.

(c)   In a substitution reaction, one atom or group of atoms replaces (substitutes for) another atom or group of atoms.

25.24   (a) $CH_2{=}CHCH_2CH_3 + H_2 \longrightarrow CH_3CH_2CH_2CH_3$

(b) $CH_3CH_2CH{=}CHCH_2CH_3 + H_2O \xrightarrow{H_2SO_4} CH_3CH_2CH(OH)CH_2CH_2CH_3$

(c)

25.25   The small 60° C-C-C angles in the cyclopropane ring cause strain that provides a driving force for reactions that result in ring-opening. There is no comparable strain in the five- or six-membered rings.

25.26   The energy gained by forming two σ bonds (or four σ bonds) more than compensates for the loss of one (or two) π bonds when addition occurs to an alkene (or alkyne). However, in benzene the aromatic ring is especially stabilized by the delocalization of π electrons around the ring. It, therefore, requires a substantial activation energy to cause the loss of this aromatic character. The most common reaction in aromatics is substitution rather than addition, since substitution does not result in loss of aromatic character.

25.27   First form an alkyl halide: $C_2H_4(g) + HBr(g) \rightarrow CH_3CH_2Br(l)$; then carry out a Friedel-Crafts reaction:

25.28 Both the *ortho* and *para* isomers are formed; they must be separated by distillation or some other technique.

25.29

		ΔH
$C_{10}H_8(l) + 12O_2(g) \rightarrow 10CO_2(g) + 4H_2O(l)$		-5157 kJ
$-[C_{10}H_{18}(l) + 29/2\ O_2(g) \rightarrow 10CO_2(g) + 9H_2O(l)]$		$-(-6286)$ kJ

$C_{10}H_8(l) + 5H_2O(l) \rightarrow C_{10}H_{18}(l) + 5/2\ O_2(g)$	$+1129$ kJ
$5/2\ O_2(g) + 5H_2(g) \rightarrow 5H_2O(l)$	$5(-285.8)$ kJ

$C_{10}H_8(l) + 5H_2(g) \rightarrow C_{10}H_{18}(l)$	-300 kJ

Compare this with the heat of hydrogenation of ethylene:
$C_2H_4(g) + H_2(g) \rightarrow C_2H_6(g)$; $\Delta H = -84.7 - (52.3) = -137$ kJ. This value applies to just one double bond. For five double bonds, we would expect about -685 kJ. The fact that hydrogenation of napthalene yields only -300 kJ indicates that the overall energy of the napthalene molecule is lower than expected for five isolated double bonds and that there must be some special stability associated with the aromatic system in this molecule.

25.30 Both combustion reactions produce CO_2 and H_2O:

$C_3H_6(g) + 9/2\ O_2(g) \rightarrow 3CO_2(g) + 3H_2O(l)$

$C_5H_{10}(g) + 15/2\ O_2(g) \rightarrow 5CO_2(g) + 5H_2O(l)$

Thus, we can calculate the ΔH_{comb} / CH_2 group for each compound:

$$\frac{\Delta H_{comb}}{CH_2\ \text{group}} = \frac{2089\ \text{kJ/mol}\ C_3H_6}{3CH_2\ \text{groups}} = \frac{693.3\ \text{kJ}}{\text{mol}\ CH_2}; \frac{3317\ \text{kJ/mol}\ C_5H_{10}}{5CH_2\ \text{groups}} = 663.4\ \text{kJ/mol}\ CH_2$$

$\Delta H_{comb}/CH_2$ group for cyclopropane is greater because C_3H_6 contains a strained ring. When combustion occurs, the strain is relieved and the stored energy is released during the reaction.

Functional Groups

25.31 (a) ketone   **(b)** carboxylic acid   **(c)** alcohol   **(d)** ester   **(e)** amide   **(f)** amine

25.32 (a) —NH₂, amine; $\overset{O}{\overset{\|}{-C}}$—OH, carboxylic acid

(b) CH₂=CH—, alkene; —OH, alcohol   (c) $\overset{O}{\overset{\|}{-C}}$—NH—, amide

(d) $\overset{O}{\overset{\|}{-C}}$—, ketone   (e) $\overset{O}{\overset{\|}{-C}}$—H, aldehyde

(f) HC≡C—, alkyne; —OH, alcohol

25.33 propionaldehyde (or propanal): [structure of propanal]

25.34 dimethyl ether: [structure of dimethyl ether]

25.35 About each CH₃ carbon, 109°; about the carbonyl carbon, 120° planar.

25.36 [structure: H—C≡C—C(=O)—O—CH₃ with sp, sp, sp², sp³ labels]

25.37 (a) [structure] 2-propylacetate   (b) CH₃CH₂CH₂O—C(=O)CH₃ 1-propylacetate   (c) CH₃CH₂O—C(=O)—H ethylformate

25.38 (a) CH₃CH₂O—C(=O)—[phenyl] ethylbenzoate   (b) CH₃N(H)—C(=O)CH₂CH₃ N-methylpropionamide   (c) [phenyl]—O—C(=O)CH₃ phenylacetate

25.39 $\text{H}-\overset{\overset{\displaystyle O}{\|}}{\text{C}}-\text{O}-\text{CH}_2\text{CH}_3 + \text{NaOH} \longrightarrow \left[\text{H}-\text{C} \overset{\displaystyle O}{\underset{\displaystyle O}{\diagup}} \right]^- + \text{Na}^+ + \text{CH}_3\text{CH}_2\text{OH}$

25.40 $\text{CH}_3\overset{\overset{\displaystyle O}{\|}}{\text{C}}-\text{O}-\text{CH}_3 + \text{NaOH} \longrightarrow \left[\text{CH}_3\text{C} \overset{\displaystyle O}{\underset{\displaystyle O}{\diagup}} \right]^- + \text{Na}^+ + \text{CH}_3\text{OH}$

25.41 (a) $\text{CH}_3\text{CH}_2\overset{\overset{\displaystyle OH}{|}}{\text{C}}\text{HCH}_3$ (b) $\text{HOCH}_2\text{CH}_2\text{OH}$ (c) $\text{H}-\overset{\overset{\displaystyle O}{\|}}{\text{C}}-\text{OCH}_3$

(d) $\text{CH}_3\text{CH}_2\overset{\overset{\displaystyle O}{\|}}{\text{C}}\text{CH}_2\text{CH}_3$ (e) $\text{CH}_3\text{CH}_2\text{OCH}_2\text{CH}_3$

25.42 (a) $\text{Cl}-\text{CH}_2\text{CH}_2\overset{\overset{\displaystyle O}{\|}}{\text{C}}-\text{H}$ (b) $\text{CH}_3\overset{\overset{\displaystyle O}{\|}}{\text{C}}\text{CHCH}_3$ with CH_3 below   (c) [benzaldehyde with Cl substituent]

(d) $\underset{\text{H}}{\overset{\text{CH}_3\text{OCH}_2}{\diagdown}}\text{C}=\text{C}\underset{\text{H}}{\overset{\text{CH}_3}{\diagup}}$ (e) $\text{CH}_3\text{CH}_2\overset{\overset{\displaystyle O}{\|}}{\text{C}}\text{NC}_2\text{H}_5$ with C_2H_5 below

25.43 (a) methanoic acid   (b) butanoic acid   (c) 3-methylpentanoic acid

25.44 (a) $\text{CH}_3\text{CH}_2\overset{\overset{\displaystyle O}{\|}}{\text{C}}-\text{H}$ (b) $\text{CH}_3\text{CH}_2\text{CH}_2\overset{\overset{\displaystyle O}{\|}}{\text{C}}\text{CH}_3$ (c) $\text{CH}_3\overset{}{\text{CHC}}\overset{\overset{\displaystyle O}{\|}}{\text{C}}\text{CH}_3$ with CH_3 below   (d) $\text{CH}_3\text{CH}_2\text{CHC}\overset{\overset{\displaystyle O}{\|}}{}-\text{H}$ with CH_3 below

Proteins

25.45 (a)   An α-amino acid contains an NH_2 group attached to the carbon that is bound to the carbon of the carboxylic acid function.

(b)   In forming a protein, amino acids undergo a condensation reaction between the amino group and carboxylic acid:

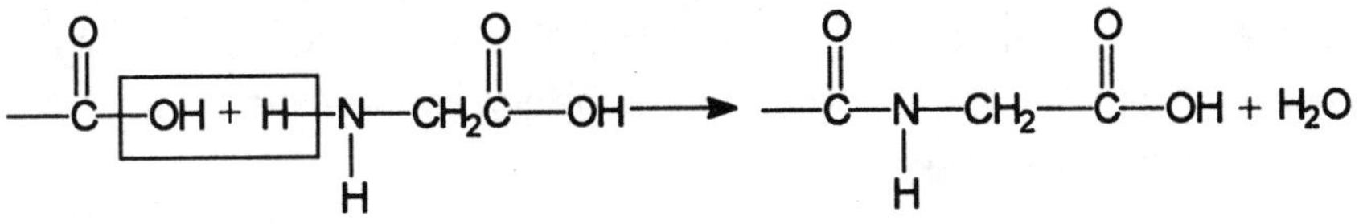

25.46 The side chains possess three characteristics that may be of importance. They may be bulky, and thus impose restraints on where and how the amino acid can undergo reaction. Secondly, the side chain may possess a polar group (e.g., the -OH group in serine), that will in part determine solubility. Finally, the side chain may contain an acidic or basic functional group that will partially determine solubility in acidic or basic medium, and that may become involved in interactions with other amino acids.

25.47 Two dipeptides are possible:

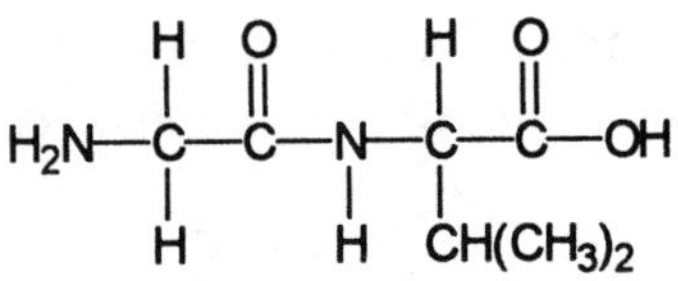

glycylvaline   and   valylglycine

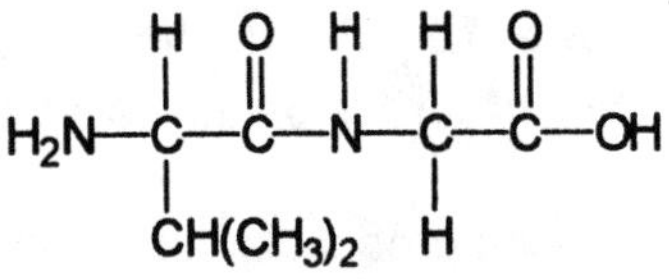

25.48 $H_2NCH(CH_2CO_2H)CO_2H$ + $H_2NCH(CH_2SH)CO_2H$ ⟶

aspartic acid   cysteine

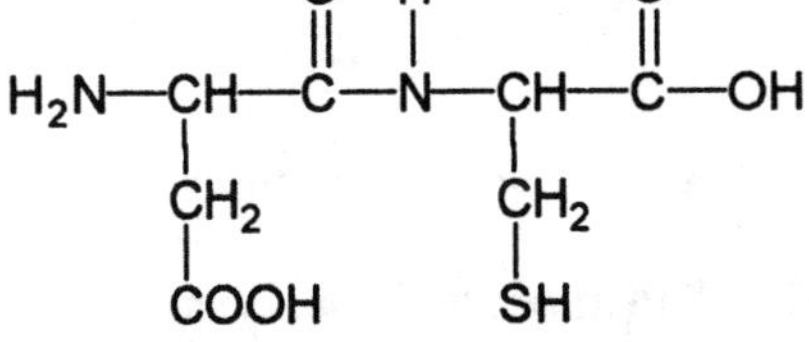

aspartylcysteine

25.49

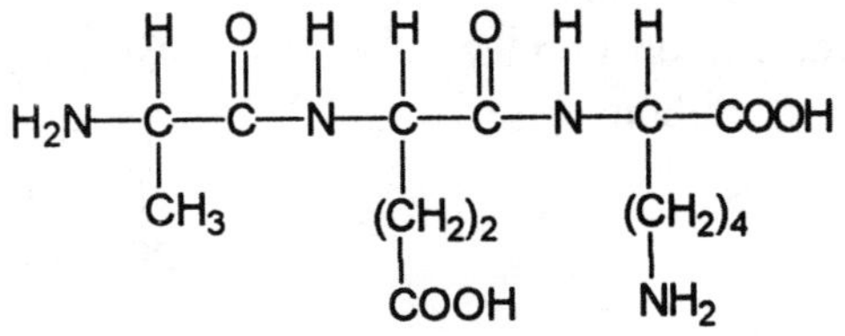

25.50 Alanine, serine

25.51 Eight: ser-ser-ser; ser-ser-phe; ser-phe-ser; phe-ser-ser; ser-phe-phe; phe-ser-phe; phe-phe-ser; phe-phe-phe

25.52 Six: gly-val-ala; gly-ala-val; val-gly-ala; val-ala-gly; ala-gly-val; ala-val-gly

25.53 The *primary structure* of a protein refers to the sequence of amino acids in the chain. Along any particular section of the protein chain the configuration may be helical, or it may be an open chain, or arranged in some other way. This is called the *secondary structure*. The overall shape of the protein molecule is determined by the way the segments of the protein chain fold together, or pack. The interactions which determine the overall shape are referred to as the *tertiary structure*.

25.54 It is quite evident from Figure 25.19 that the hydrogen bonds between an NH group along the chain and the unshared electron pairs of a carbonyl group further along are responsible for maintaining the helix. Indeed, the pitch and general shape of the helix are determined by what specific interactions produce a good hydrogen bonding arrangement.

Carbohydrates

25.55 Glucose exists in solution as a cyclic structure in which the aldehyde function on carbon 1 reacts with the OH group of carbon 5 to form what is called a hemiacetal, Figure 25.22. Carbon atom 1 carries an OH group in the hemiacetal form; in α-glucose this OH group is on the opposite side of the ring as the CH_2OH group on carbon atom 5. In the β (beta) form the OH group on carbon 1 is on the same side of the ring as the CH_2OH group on carbon 5.

The condensation product looks like this:

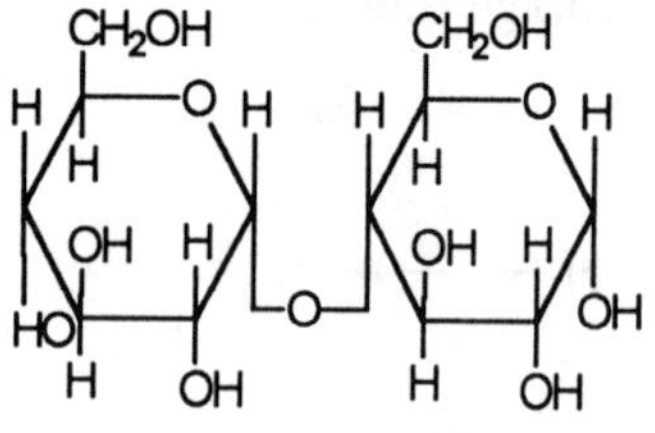

α-linkage β-linkage

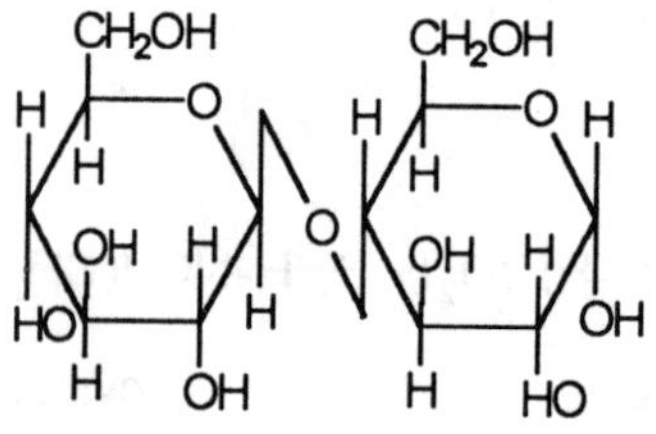

25.56 (a) α-form   (b) β-form   (c) β-form

25.57 The structure is best deduced by comparing galactose with glucose, and inverting the configurations at the appropriate carbon atoms. Recall from Exercise 25.55 that both the β-form (shown here) and the α-form (OH on carbon 1 on the opposite side of ring as the CH_2OH on carbon 5) are possible.

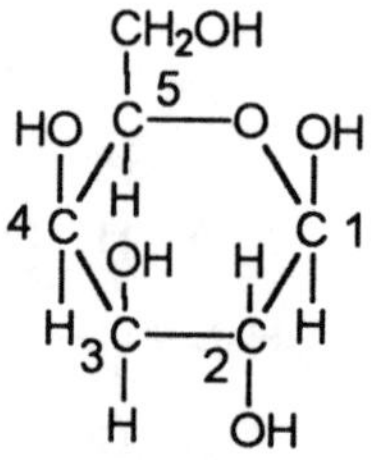

galactose

25.58

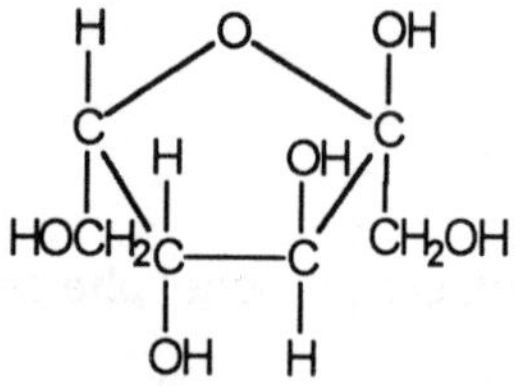

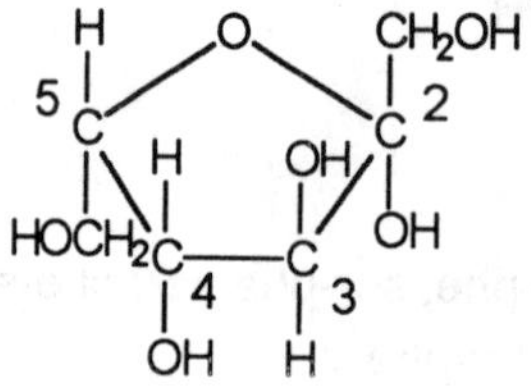

Both the α (left) and β (right) forms are possible.

25.59 In the linear form of galactose shown in Exercise 25.57, carbon atoms 2, 3, 4 and 5 are chiral because they carry four different groups on each. In the ring form (see solution 25.57), carbon atoms 1, 2, 3, 4 and 5 are chiral.

25.60 In the linear form of sorbose shown in Exercise 25.58, numbering from the top down, carbon atoms 3, 4 and 5 are chiral. In the ring form (see Solution 25.58) carbon atoms 2, 3, 4 and 5 are chiral.

Nucleic Acids

25.61 A *nucleotide* consists of a nitrogen-containing aromatic compound, a sugar in the furanose (5-membered) ring form, and a phosphoric acid group. The structure of deoxycytidine monophosphate is shown at right.

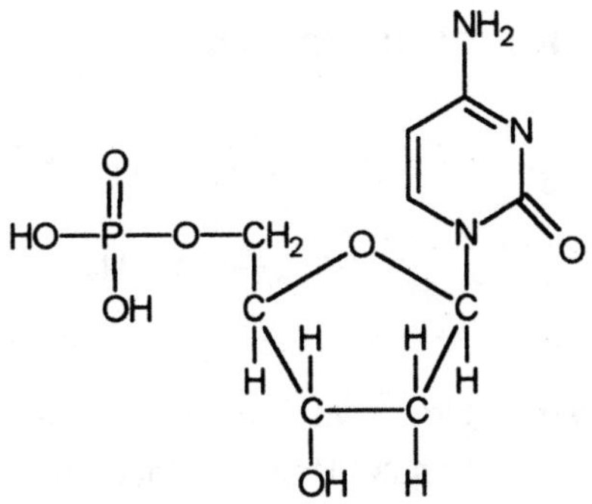

25.62

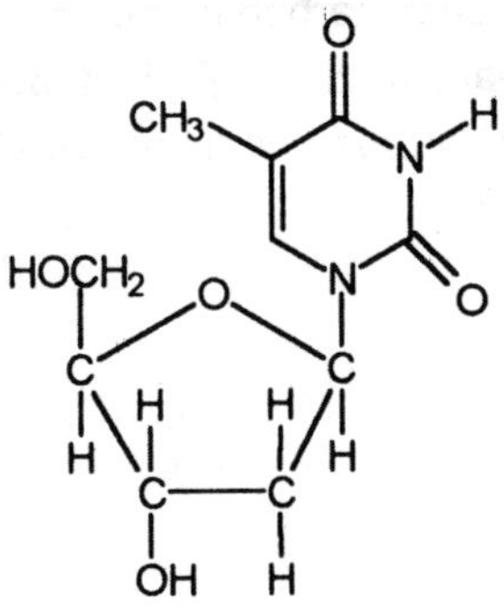

25.63 $C_4H_7O_3CH_2OH + H_3PO_4 \rightarrow C_4H_7O_3CH_2\text{-}O\text{-}PO_3H_2 + H_2O$

25.64

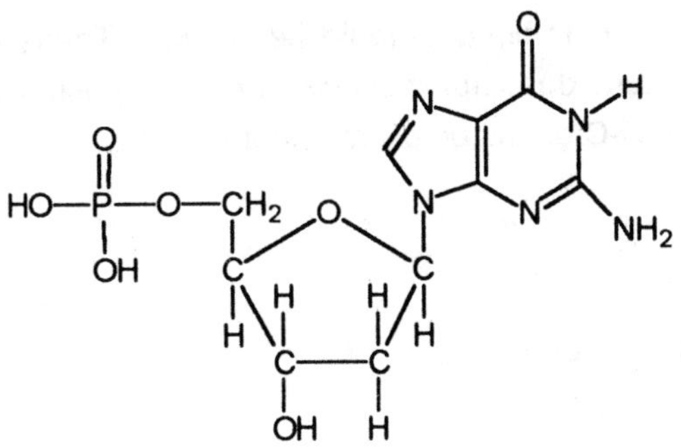

25.65

```
—A—C—T—C—G—A—
  :  :  :  :  :  :
—T—G—A—G—C—T—     ← complementary strand
```

25.66   In the helical structure for DNA, the strands of the polynucleotides are held together by hydrogen-bonding interactions between particular pairs of bases. It happens that adenine and thymine form an especially effective base pair, and that guanine and cytosine are similarly related. Thus, each adenine has a thymine as its opposite number in the other strand, and each guanine has a cytosine as its opposite number. In the overall analysis of the double strand, total adenine must then equal total thymine, and total guanine equals total cytosine.

Additional Exercises

25.67 CH_2=CH—CH=CH_2

HC≡C—CH_2CH_3 CH_3—C≡C—CH_3

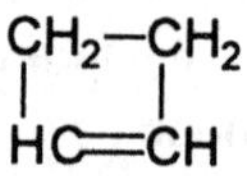

CH_2—CH_2
HC=CH

25.68   No. Hydrocarbons are alkanes in which carbon atoms are involved exclusively in single bonds. There are four bonding electron pairs around each carbon in the chain and the bond angles are 109°. Thus, an unbranched alkane is really a "zig-zag" chain of carbon atoms, with each C atom bound to a maximum of two other C atoms and an appropriate number of H atoms.

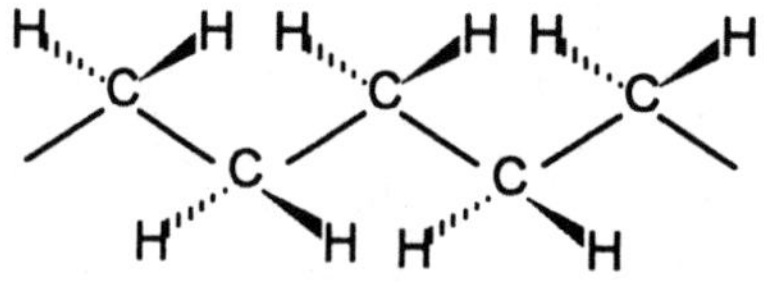

25.69

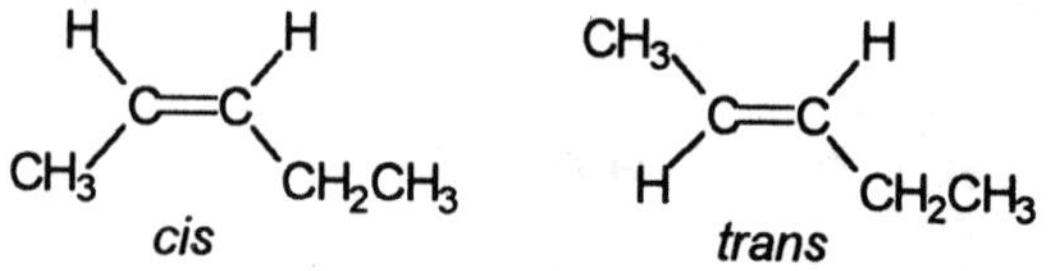

cis   *trans*

To show complete Lewis structures all C-H bonds should be shown. These have been omitted here to save space. Cyclopentene does not show *cis-trans* isomerism because the existence of the ring demands that the C-C bonds be *cis* to one another.

25.70 $CH_3CH_2CH_2$—OH CH_3CHCH_3 with OH CH_3—O—CH_2CH_3

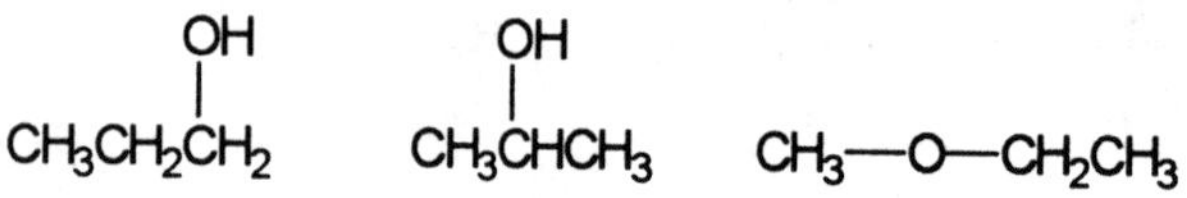

25.71   (a)   2-propanol   (b) dimethyl ether   (c) 3-butene-1-ol
         (d)   3-methyl-3-hexene   (e) 4-methyl-2-pentyne

25.72 The C-Cl bonds in the *trans* compound are pointing in exactly opposite directions. Thus, the C-Cl bond dipoles cancel (Section 9.3). This is not the case in the *cis* compound, as can be seen by writing out the structure:

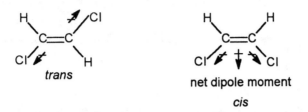

trans

net dipole moment

cis

25.73 Because of the strain in bond angles about the ring, cyclohexyne would not be stable. The alkyne carbons preferentially have a 180° bond angle. However, there are not enough carbons in the ring to make this possible without gross distortions of other bond lengths and angles.

25.74 **(a)** Ether, C—O—C;   alkene,   —CH=CH₂

(b) carboxylic acid, $\overset{\displaystyle O}{\overset{\displaystyle \|}{—C}}$—OH;   ester, $CH_3\overset{\displaystyle O}{\overset{\displaystyle \|}{C}}$—O—

(c) ketone, $\overset{\displaystyle O}{\overset{\displaystyle \|}{—C—}}$;   alkene, —CH=CH— ;   alcohol —C—OH

25.75 **(a)** $CH_3OCH_2CH_3$ **(b)** $CH_3CH_2\overset{\displaystyle O}{\overset{\displaystyle \|}{C}}$—H   **(c)** $CH_3\overset{\displaystyle O}{\overset{\displaystyle \|}{C}}CH_3$

(d) $CH_3CH_2\overset{\displaystyle O}{\overset{\displaystyle \|}{C}}OH$ **(e)** $CH_3\overset{\displaystyle O}{\overset{\displaystyle \|}{C}}$—OCH₃

25.76 **(a)** $CH_3\overset{\displaystyle O}{\overset{\displaystyle \|}{C}}$—OH, C_6H_5OH **(b)** $C_6H_5\overset{\displaystyle O}{\overset{\displaystyle \|}{C}}$—OH, CH_3OH

25.77 **(a)** *cyclobutane*   There is some strain in the four-membered ring because the C-C-C bond angles must be less than the desired 109°.

(b) *cyclohexene*   The C=C bond is capable of addition reactions, for example with Br_2.

(c) *1-hexene*   Whereas the alkene readily undergoes addition reactions, the aromatic hydrocarbon is extremely stable.

(d) *2-hexyne*   Alkynes undergo addition reactions even more readily than alkenes.

25.78 The compound is clearly an alcohol. Its slight solubility in water is consistent with the properties expected of a secondary alcohol with a five-carbon chain. The fact that it is oxidized to a ketone rather than to an aldehyde and then to a carboxylic acid tells us that it is a secondary alcohol:

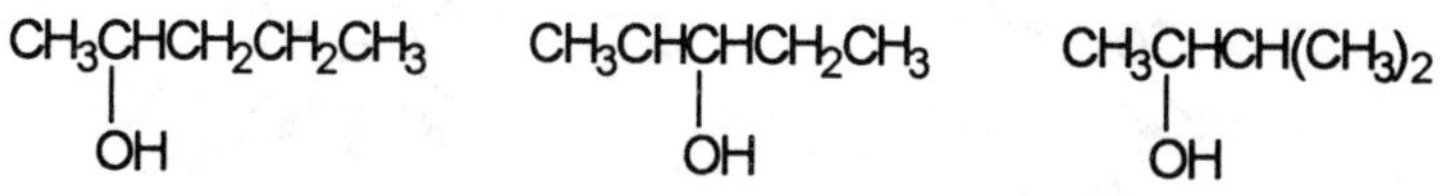

$$CH_3CHCH_2CH_2CH_3 \qquad CH_3CHCHCH_2CH_3 \qquad CH_3CHCH(CH_3)_2$$
$$\quad | \qquad\qquad\qquad\quad | \qquad\qquad\qquad\quad |$$
$$\quad OH \qquad\qquad\qquad\quad OH \qquad\qquad\qquad\quad OH$$

25.79 All growth processes, which require synthesis of new cell materials, require energy. For example, in plants transpiration of water requires energy. In animals, the action of the involuntary muscles, such as the heart, maintenance of a uniform body temperature, synthesis of new cells, motions of limbs, and maintenance of correct body fluid compositions all require energy.

25.80 (a) None

(b) The carbon bearing the secondary -OH has four different groups attached, and is thus chiral.

(c) The carbon bearing the $-NH_2$ group and the carbon bearing the CH_3 group are both chiral.

25.81 (a)   (b)

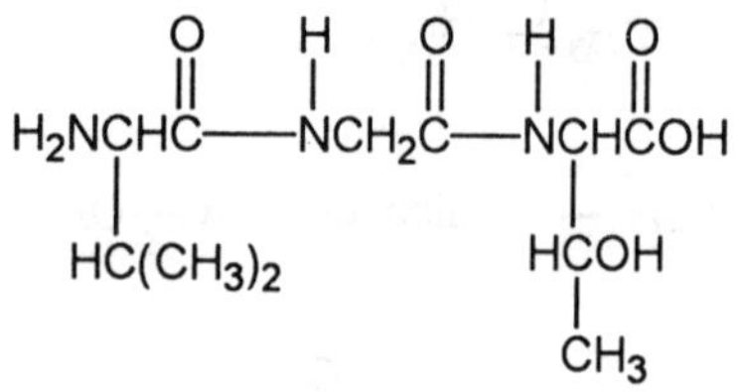

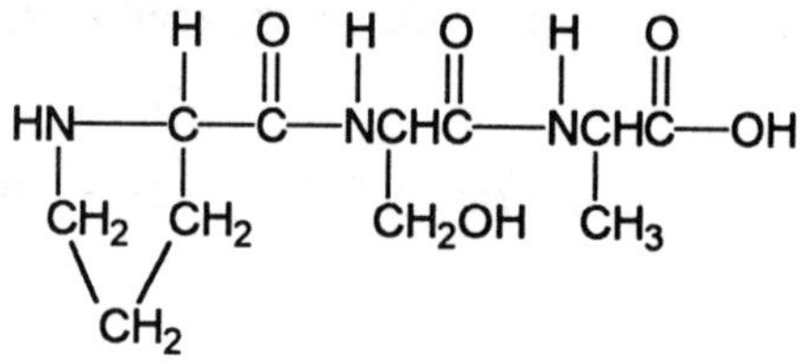

25.82 Glu-cys-gly is the only possible structure.

25.83 (a) Sucrose, maltrose or lactose (see Figure 25.23)

(b) Ribose and deoxyribose (see Section 25.9)

(c) Glucose (Figure 25.22)

(d) Starch and cellulose (Figures 25.24, 25.25 and 25.26)

25.84 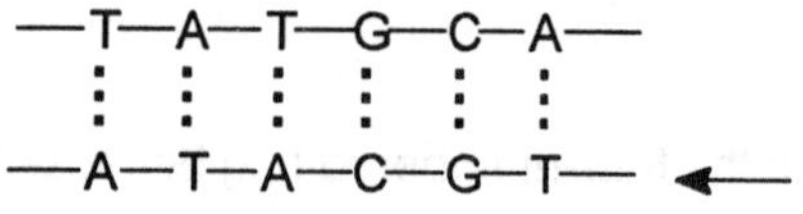
```
—T—A—T—G—C—A—
 :  :  :  :  :  :
—A—T—A—C—G—T—  ←—— complementary strand
```

Integrative Exercises

25.85 CH_3CH_2OH $CH_3\text{-}O\text{-}CH_3$

ethanol   dimethyl ether

Ethanol contains -O-H bonds which form strong intermolecular hydrogen bonds, while

Ethanol contains -O-H bonds which form strong intermolecular hydrogen bonds, while dimethyl ether experiences only weak dipole-dipole and dispersion forces.

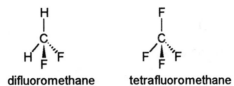

difluoromethane   tetrafluoromethane

CH_2F_2 is a polar molecule, while CF_4 is nonpolar. CH_2F_2 experiences dipole-dipole and dispersion forces, while CF_4 experiences only dispersion forces.

In both cases, stronger intermolecular forces lead to the higher boiling point.

25.86   Determine the empirical formula, molar mass and thus molecular formula of the compound. Confirm with physical data.

$$85.7 \text{ g C} \times \frac{1 \text{ mol C}}{12.01 \text{ g C}} = 7.136 \text{ mol C}; \quad 7.136 / 7.136 = 1$$

$$14.3 \text{ g H} \times \frac{1 \text{ mol H}}{1.008 \text{ g H}} = 14.19 \text{ mol H}; \quad 14.19 / 7.136 \approx 2$$

Empirical formula is CH_2. Using Equation 10.11 (MM = molar mass):

$$MM = \frac{(2.21 \text{ g/L})(0.08206 \text{ L} \cdot \text{atm/mol} \cdot \text{K})(373 \text{ K})}{(735/760) \text{ atm}} = 69.9 \text{ g/mol}$$

The molecular formula is thus C_5H_{10}. The absence of reaction with aqueous Br_2 indicates that the compound is not an alkene, so the compound is probably the cycloalkane, cyclopentane. According to the *Handbook of Chemistry and Physics*, the boiling point of cyclopentane is 49°C at 760 torr. This confirms the identity of the unknown.

25.87   Determine the empirical formula, molar mass and thus molecular formula of the compound. Confirm with physical data.

$$66.7 \text{ g C} \times \frac{1 \text{ mol C}}{12.01 \text{ g C}} = 5.554 \text{ mol C}; \quad 5.554 / 1.388 = 4$$

$$11.2 \text{ g H} \times \frac{1 \text{ mol H}}{1.008 \text{ g H}} = 11.11 \text{ mol H}; \quad 11.11 / 1.388 = 8$$

$$22.2 \text{ g O} \times \frac{1 \text{ mol O}}{16.00 \text{ g O}} = 1.388 \text{ mol O}; \quad 1.388 / 1.388 = 1$$

The empirical formula is C_4H_8O. Using Equation 10.11 (MM = molar mass):

$$MM = \frac{(2.28 \text{ g/L})(0.08206 \text{ L} \cdot \text{atm/mol} \cdot \text{K})(373 \text{ K})}{0.970 \text{ atm}} = 71.9$$

The formula weight of C_4H_8O is 72, so the molecular formula is also C_4H_8O. Since the compound has a carbonyl group and cannot be oxidized to an acid, the only possibility is 2-butanone.

$$\underset{\displaystyle CH_3\overset{\textstyle O}{\overset{\textstyle \|}{C}}CH_2CH_3}{}$$

The boiling point of 2-butanone is 79.6°C, confirming the identification.

25.88   The reaction is: $2NH_2CH_2COOH(aq) \rightarrow NH_2CH_2CONHCH_2COOH(aq) + H_2O(l)$

$\Delta G = (-488) + (-285.83) - 2(-369) = -35.8 = -36$ kJ

25.89   The reaction is approximately:

$\underline{n}C_6H_{12}O_6(aq) \rightarrow [C_6H_{11}O_5]_n(aq) + (n-1)H_2O(l)$

$\Delta G° = n(-662.3 \text{ kJ}) + (n-1)(-285.83 \text{ kJ}) - n(-917.2 \text{ kJ})$

$\quad\quad = n(-285.83 \text{ kJ} - 662.3 + 917.2 \text{ kJ}) + 285.83 \text{ kJ}$

$\Delta G° = n(-30.9) \text{ kJ} + 285.83 \text{ kJ}$

25.90   If a process at standard conditions is spontaneous, $\Delta G°$ is negative. If it is nonspontaneous, $\Delta G°$ is positive. In order for soybeans to grow, the process must be nonspontaneous at ambient temperature, say 30°C. This sets limits for the magnitude of $\Delta S°$.

$\Delta G° = \Delta H° - T\Delta S°$. At 50°C, $\Delta G° < 0$, $\Delta H° - T\Delta S° < 0$.

$-T\Delta S° < -\Delta H°$; $\Delta S° > \Delta H°/T$; $\Delta S° > 238 \times 10^3$ J/323 K = 737 J/K

At 30°C, $\Delta G° > 0$, $\Delta H° - T\Delta S° > 0$, $\Delta S° < \Delta H°/T$

$\Delta S° < 238 \times 10^3$ J/303 K = 785 J/K

737 J/K < $\Delta S°$ < 785 J/K. In round numbers, $\Delta S° \approx 750$ J/K. $\Delta S°$ for this process is large and positive. This indicates that the denatured, uncoiled protein is much less ordered than the viable protein. This is consistent with the greater motional freedom of the uncoiled protein.

25.91   (a)   At low pH, the amine and carboxyl groups are protonated.   At high pH, the amine and carboxyl groups are deprotonated.

$$\underset{\displaystyle \overset{+}{H_3N}-\underset{\underset{\textstyle CH_3}{|}}{\overset{\overset{\textstyle H}{|}}{C}}-\overset{\overset{\textstyle O}{\|}}{C}-OH}{} \qquad\qquad \underset{\displaystyle NH_2-\underset{\underset{\textstyle CH_3}{|}}{\overset{\overset{\textstyle H}{|}}{C}}-\overset{\overset{\textstyle O}{\|}}{C}-O^-}{}$$

(b) $CH_3\overset{\displaystyle O}{\overset{\|}{C}}-OH(aq) \longrightarrow CH_3\overset{\displaystyle O}{\overset{\|}{C}}-O^-(aq) + H^+(aq)$

$K_a = 1.8 \times 10^{-5}$, $pK_a = -\log(1.8 \times 10^{-5}) = 4.74$

The conjugate acid of NH_3 is NH_4^+.

$NH_4^+(aq) \rightarrow NH_3(aq) + H^+(aq)$

$K_a = K_w / K_b = 1 \times 10^{-14} / 1.8 \times 10^{-5} = 5.55 \times 10^{-10} = 5.6 \times 10^{-10}$

$pK_a = -\log(5.55 \times 10^{-10}) = 9.26$

In general, a $-COOH$ group is a stronger acid than a $-[NH_3]^+$ group. The lower pK_a value for amino acids is for the ionization (deprotonation) of the $-COOH$ group and the higher pK_a is for the deprotonation of the $-[NH_3]^+$ group.

25.92 $AMPOH^-(aq) \rightleftharpoons AMPO^{2-}(aq) + H^+(aq)$

$pK_a = 7.21$; $K_a = 10^{-pK_a} = 6.17 \times 10^{-8} = 6.2 \times 10^{-8}$

$K_a = \dfrac{[AMPO^{2-}][H^+]}{[AMPOH^-]} = 6.2 \times 10^{-8}$. When $pH = 7.40$, $[H^+] = 3.98 \times 10^{-8} = 4.0 \times 10^{-8}$.

Then $\dfrac{[AMPOH^-]}{[AMPO^{2-}]} = 3.98 \times 10^{-8} / 6.17 \times 10^{-8} = 0.65$